Communications in Computer and Information Science 2806

Rationale
The CCIS series is devoted to the publication of proceedings of computer science conferences. Its aim is to efficiently disseminate original research results in informatics in printed and electronic form. While the focus is on publication of peer-reviewed full papers presenting mature work, inclusion of reviewed short papers reporting on work in progress is welcome, too. Besides globally relevant meetings with internationally representative program committees guaranteeing a strict peer-reviewing and paper selection process, conferences run by societies or of high regional or national relevance are also considered for publication.

Topics
The topical scope of CCIS spans the entire spectrum of informatics ranging from foundational topics in the theory of computing to information and communications science and technology and a broad variety of interdisciplinary application fields.

Information for Volume Editors and Authors
Publication in CCIS is free of charge. No royalties are paid, however, we offer registered conference participants temporary free access to the online version of the conference proceedings on SpringerLink (http://link.springer.com) by means of an http referrer from the conference website and/or a number of complimentary printed copies, as specified in the official acceptance email of the event.

CCIS proceedings can be published in time for distribution at conferences or as post-proceedings, and delivered in the form of printed books and/or electronically as USBs and/or e-content licenses for accessing proceedings at SpringerLink. Furthermore, CCIS proceedings are included in the CCIS electronic book series hosted in the SpringerLink digital library at http://link.springer.com/bookseries/7899. Conferences publishing in CCIS are allowed to use our online conference service (Meteor) for managing the whole proceedings lifecycle (from submission and reviewing to preparing for publication) free of charge.

Publication process
The language of publication is exclusively English. Authors publishing in CCIS have to sign the Springer CCIS copyright transfer form, however, they are free to use their material published in CCIS for substantially changed, more elaborate subsequent publications elsewhere. For the preparation of the camera-ready papers/files, authors have to strictly adhere to the Springer CCIS Authors' Instructions and are strongly encouraged to use the CCIS LaTeX style files or templates.

Abstracting/Indexing
CCIS is abstracted/indexed in DBLP, Google Scholar, EI-Compendex, Mathematical Reviews, SCImago, Scopus. CCIS volumes are also submitted for the inclusion in ISI Proceedings.

How to start

To start the evaluation of your proposal for inclusion in the CCIS series, please send an e-mail to ccis@springer.com

Emilio Corchado · Héctor Quintián ·
Alicia Troncoso Lora · Esteban Jove ·
Pablo García Bringas · Paolo Fosci ·
Francisco Martínez Álvarez
Editors

Soft Computing Models in Industrial and Environmental Applications

20th International Conference, SOCO 2025
Salamanca, Spain, October 16–17, 2025
Proceedings

Editors
Emilio Corchado
University of Salamanca
Salamanca, Spain

Alicia Troncoso Lora
Pablo de Olavide University
Seville, Spain

Pablo García Bringas
University of Deusto Avda
Bilbao, Spain

Francisco Martínez Álvarez
Pablo de Olavide University
Seville, Spain

Héctor Quintián
University of A Coruña
Ferrol, La Coruña, Spain

Esteban Jove
University of A Coruña
Ferrol, La Coruña, Spain

Paolo Fosci
University of Bergamo
Bergamo, Italy

ISSN 1865-0929 ISSN 1865-0937 (electronic)
Communications in Computer and Information Science
ISBN 978-3-032-19762-7 ISBN 978-3-032-19763-4 (eBook)
https://doi.org/10.1007/978-3-032-19763-4

This Springer imprint is published by the registered company Springer Nature Switzerland AG
The registered company address is: Gewerbestrasse 11, 6330 Cham, Switzerland

Preface

This volume of Lecture Notes in Networks and Systems contains accepted papers presented at the *20th International Conference on Soft Computing Models in Industrial and Environmental Applications* (SOCO 2025), which was held in the beautiful city of Salamanca, Spain, in October 2025.

Soft computing represents a collection or set of computational techniques in machine learning, computer science, and some engineering disciplines which investigate, simulate, and analyse very complex issues and phenomena.

After a single-blind peer-review process in which the 133 submissions each received an average of three reviews, the SOCO 2025 International Program Committee selected 60 papers publication in these conference proceedings, representing an acceptance rate of 45%. In this edition of SOCO, a particular emphasis was put on organising special sessions. Five special sessions were organised related to relevant topics such as Time Series Forecasting in Industrial and Environmental Applications, Efficiency and Explainability in Machine Learning and Soft Computing, Intelligent Techniques Applied to Modelling and Control in Engineering Focused on Marine Energy Systems and Robotics, Quantum Computing, EUropean Transnational Educational, and Machine Learning and Computer Vision in Industry 4.0.

The selection of papers was extremely rigorous to maintain the high quality of the conference. We want to thank the members of the Program Committees for their hard work during the review process. This is a crucial process for creating a high-standard conference; the SOCO conference would not exist without their help.

SOCO enjoyed outstanding keynote speeches by distinguished guest speakers: Ajith Abraham at Sai University (India) and Sung-Bae Cho at Yonsei University (South Korea).

SOCO 2025 has teamed up with *Neurocomputing* (Elsevier) and *Logic Journal of the IGPL* (Oxford University Press) for a suite of special issues, including selected papers from SOCO 2025. Particular thanks go as well to the conference's main sponsors, Startup Olé, the CIBER-OLÉ project (within the National Cybersecurity Industry Promotion Program, framed within the INCIBE Emprende program and financed by INCIBE and the University of Salamanca), the BISITE research group at the University of Salamanca, the CTC research group at the University of A Coruña, and the University of Salamanca. They jointly contributed actively and constructively to the success of this initiative. This activity was carried out in execution of the Strategic Project "Critical infrastructures cybersecure through intelligent modeling of attacks, vulnerabilities and increased security of their IoT devices for the water supply sector" (C061_/23), the result of a collaboration agreement signed between the National Institute of Cybersecurity (INCIBE) and the University of A Coruña. This initiative was carried out within the framework of the funds of the Recovery, Transformation and Resilience Plan, financed by the European Union (Next Generation), the project of the Government of Spain that outlines the roadmap for the modernization of the Spanish economy, the recovery of economic growth and job creation, for solid, inclusive and resilient economic reconstruction after the COVID19 crisis, and to respond to the challenges of the next decade. This activity was also promoted by PID2022-137152NB-I00 funded by MICIU/AEI/10.13039/501100011033 and by ERDF/EU.

We want to thank all the special session organisers, contributing authors, and members of the Program Committees and the Local Organising Committee for their hard and highly valuable work, which contributed to the success of the SOCO 2025 event.

October 2025

Emilio Corchado
Héctor Quintián
Alicia Troncoso Lora
Esteban Jove
Pablo García Bringas
Paolo Fosci
Francisco Martínez Álvarez

Organization

General Chair

Emilio Corchado	University of Salamanca, Spain

International Advisory Committee

Ashraf Saad	Armstrong Atlantic State University, USA
Amy Neustein	Linguistic Technology Systems, USA
Ajith Abraham	Machine Intelligence Research Labs Europe & Sai University, India
Jon G. Hall	Open University, UK
Paulo Novais	Universidade do Minho, Portugal
Amparo Alonso Betanzos	University of A Coruña, Spain
Michael Gabbay	King's College London, UK
Aditya Ghose	University of Wollongong, Australia
Saeid Nahavandi	Deakin University, Australia
Henri Pierreval	LIMOS UMR CNRS 6158 IFMA, France

Program Committee Chairs

Emilio Corchado	University of Salamanca, Spain
Héctor Quintián	University of A Coruña, Spain
Alicia Troncoso Lora	Pablo de Olavide University, Spain
Esteban Jove	University of A Coruña, Spain
Pablo García Bringas	University of Deusto, Spain
Paolo Fosci	University of Bergamo, Italy
Francisco Martínez Álvarez	Pablo de Olavide University, Spain

Program Committee

Agustina Bouchet	University of Oviedo, Spain
Akemi Galvez-Tomida	University of Cantabria, Spain
Álvaro Michelena	University of A Coruña, Spain
Andres Iglesias Prieto	University of Cantabria, Spain

Ángel Arroyo	University of Burgos, Spain
Anton Koval	Luleå University of Technology, Sweden
Antonio Díaz-Longueira	University of A Coruña, Spain
Antonio J. Tallón-Ballesteros	University of Huelva, Spain
Belén Vega Márquez	University of Seville, Spain
Bogdan Okreša Đurić	University of Zagreb, Croatia
Carlos Pereira	ISEC, Portugal
Damian Krenczyk	Silesian University of Technology, Poland
Daniel Honc	University of Pardubice, Czechia
Daniel Urda	University of Burgos, Spain
Daniela Perdukova	Technical Univerzity of Kosice, Slovakia
David Griol	University of Granada, Spain
David Gutiérrez-Avilés	University of Seville, Spain
Dragan Simic	University of Novi Sad, Serbia
Eduardo Solteiro Pires	University of Trás-os-Montes and Alto Douro, Portugal
Eloy Irigoyen	University of the Basque Country, Spain
Enol García González	University of Oviedo, Spain
Enrique De La Cal Marín	University of Oviedo, Spain
Enrique Onieva	University of Deusto, Spain
Esteban Jove	University of A Coruña, Spain
Fernando Martins	University of Oviedo, Spain
Fernando Ribeiro	Polytechnic Institute of Castelo Branco, Portugal
Francisco Martínez-Álvarez	Pablo de Olavide University, Spain
Franjo Jovic	University of Osijek, Croatia
Giuseppe Psaila	University of Bergamo, Italy
Ireneusz Czarnowski	Gdynia Maritime University, Poland
Isaias Garcia	University of León, Spain
Jan Cieslinski	University of Bialystok, Poland
Jaroslav Marek	University of Pardubice, Czechia
Jaume Jordán	Polytechnic University of Valencia, Spain
Javier Díez-González	University of León, Spain
Javier Sanchis Saez	Polytechnic University of Valencia, Spain
Jiri Pospichal	University of Ss. Cyril and Methodius in Trnava, Slovakia
Jorge Barbosa	Instituto Superior de Engenharia de Coimbra, Portugal
Jose Dorronsoro	Autonomous University of Madrid, Spain
José Aveleira Mata	Research Institute of Applied Sciences in Cybersecurity, Spain
José F. Torres	Pablo de Olavide University, Spain
José Valente de Oliveira	Universidade do Algarve, Portugal

Jose Carlos Metrolho	Polytechnic Institute of Castelo Branco, Portugal
José Enrique Sánchez López	University of Seville, Spain
Jose Luis Calvo-Rolle	University of A Coruña, Spain
Jose M. Molina	University of Carlos III of Madrid, Spain
Jose Manuel Lopez-Guede	University of the Basque Country, Spain
José Ramón Villar	University of Oviedo, Spain
José-Luis Casteleiro-Roca	University of A Coruña, Spain
Juan Albino Mendez	University of La Laguna, Spain
Juan M. Alberola	Polytechnic University of Valencia, Spain
Lidia Sánchez-González	University of León, Spain
Manuel Graña	University of the Basque Country, Spain
Manuel Carranza García	University of Seville, Spain
Marcin Paprzycki	IBS PAN and WSM, Poland
Maria Fuente	University of Valladolid, Spain
María Martínez-Ballesteros	University of Seville, Spain
Maria Teresa Godinho	Polytechnic Institute of Beja, Portugal
Mario Villar	University of Oviedo, Spain
Matilde Santos	Complutense University of Madrid, Spain
Mehmet Emin Aydin	University of the West of England, UK
Michal Wozniak	Wrocław University of Technology, Poland
Miguel Carvajal	University of Huelva, Spain
Pablo García Bringas	University of Deusto, Spain
Panagiotis Kyratsis	University of Western Macedonia, Greece
Paulo Moura Oliveira	University of Trás-os-Montes and Alto Douro, Portugal
Petr Dolezel	University of Pardubice, Czechia
Petrica Pop	Technical University of Cluj-Napoca, North University Center at Baia Mare, Romania
Qing Tan	Athabasca University, Canada
Santiago Porras Alfonso	University of Burgos, Spain
Sebastian Saniuk	University of Zielona Góra, Poland
Stefano Pizzuti	Energy New Technologies and Sustainable Economic Development Agency (ENEA), Italy
Valeriu-Manuel Ionescu	University of Pitesti, Romania
Wei-Chiang Hong	Asia Eastern University of Science and Technology, Taiwan
Zita Vale	Instituto Superior de Engenharia do Porto, Portugal

Special Sessions

Time Series Forecasting in Industrial and Environmental Applications

Program Committee

José F. Torres (Organizer)	Pablo de Olavide University, Spain
Federico Divina (Organizer)	Pablo de Olavide University, Spain
Mario Giacobini (Organizer)	University of Turin, Italy
Julio César Mello Román (Organizer)	Universidad Nacional de Asunción, Paraguay
Miguel García Torres (Organizer)	Pablo de Olavide University, Spain
Andrés Manuel Chacón Maldonado (Organizer)	Pablo de Olavide University, Spain
Adrián Gil Gamboa (Organizer)	Pablo de Olavide University, Spain
Ángela R. Troncoso-García	Pablo de Olavide University, Spain
David Gutiérrez-Avilés	University of Seville, Spain
Francesc Rodríguez-Díaz	Pablo de Olavide University, Spain
Francisco Martínez-Álvarez	Pablo de Olavide University, Spain

Efficiency and Explainability in Machine Learning and Soft Computing

Program Committee

Ángela R. Troncoso-García (Organizer)	Pablo de Olavide University, Spain
José Enrique Sánchez López (Organizer)	University of Seville, Spain
Manuel Carranza García (Organizer)	University of Seville, Spain
Manuel Jesús Jiménez Navarro (Organizer)	University of Seville, Spain
María Martínez-Ballesteros (Organizer)	University of Seville, Spain
Pablo Reina-Jiménez (Organizer)	University of Seville, Spain
Ana Rodríguez López	University of Seville, Spain
Belén Vega Márquez	University of Seville, Spain
Daniel Mateos-García	University of Seville, Spain
José María Luna Romera	University of Seville, Spain

Intelligent Techniques Applied to Modelling and Control in Engineering Focused on Marine Energy Systems and Robotics

Program Committee

Bowen Zhou (Organizer)	Northeastern University, China
Fares Mzougui (Organizer)	University of Victoria, Canada
J. Enrique Sierra García (Organizer)	University of Burgos, Spain
Luís Conde Bento (Organizer)	Polytechnic Institute of Leiria, Portugal
Matilde Santos Peñas (Organizer)	Complutense University of Madrid, Spain
Payam Aboutalebi (Organizer)	University of the Basque Country, Spain
Antonio Moreira	Polytechnic Institute of Cávado and Ave, Portugal
Asier Ibeas	Autonomous University of Barcelona, Spain
Cesar Guevara	CUNEF University, Spain
Fernando S. Pacheco	Häme University of Applied Sciences, Finland
Gonzalo Farias	Pontifical Catholic University of Valparaíso, Spain
Guangdi Li	Northeastern University, China
Sahbi Abderrahim	University of Gabes, Tunisia
Xiaowei Zhao	University of Warwick, UK

Quantum Computing

Program Committee

Alicia Troncoso (Organizer)	Pablo de Olavide University, Spain
David Gutiérrez-Avilés (Organizer)	University of Seville, Spain
Francesc Rodríguez-Díaz (Organizer)	Pablo de Olavide University, Spain
Francisco Martínez-Álvarez (Organizer)	Pablo de Olavide University, Spain
Sebastian Grillo (Organizer)	Autonomous University of Asunción, Paraguay
Wojciech Bozejko (Organizer)	Wrocław University of Technology, Poland
Antonio Morales-Esteban	University of Seville, Spain
Khawaja Asim	Pakistan Institute of Engineering and Applied Sciences, Pakistan
Naeem Ullah	University of Engineering and Technology, Taxila, Pakistan

EUropean Transnational Educational

Program Committee

Héctor Quintián (Organizer)	University of A Coruña, Spain
Ana Rosa Pereira Borges	Coimbra Polytechnic Institute, ISEC, Portugal
Antonio Morales-Esteban	University of Seville, Spain
Hugo Sanjurjo-González	University of Deusto, Spain
José Manuel Galán	University of Burgos, Spain
José-Lázaro Amaro-Mellado	University of Seville, Spain
Juan J. Gude	University of Deusto, Spain
Juan Pavón	Complutense University of Madrid, Spain
Julián Estévez	University of the Basque Country, Spain
Maria Jose Marcelino	University of Coimbra, Portugal
Maria Victoria Requena	University of Seville, Spain

Machine Learning and Computer Vision in Industry 4.0

Program Committee

Enrique Dominguez (Organizer)	University of Málaga, Spain
Jose Garcia Rodriguez (Organizer)	University of Alicante, Spain
Ramón Moreno (Organizer)	Grupo Antolin, Spain
Andres Fuster-Guillo	University of Alicante, Spain
Antonio-Maciá Lillo	University of Alicante, Spain
David Mulero	University of Alicante, Spain
David Ortiz-Perez	University of Alicante, Spain
Esteban José Palomo	University of Málaga, Spain
Ezequiel López-Rubio	University of Málaga, Spain
Higinio Mora	University of Alicante, Spain
Javier Rodriguez-Juan	University of Alicante, Spain
Jesus Benito-Picazo	University of Málaga, Spain
Jorge Azorin-Lopez	University of Alicante, Spain
Jorge García-González	University of Málaga, Spain
Jose David Fernandez-Rodriguez	University of Málaga, Spain
Jose Manuel Lopez-Guede	University of the Basque Country, Spain
Juan Miguel Ortiz-De-Lazcano-Lobato	University of Málaga, Spain
Karl Thurnhofer-Hemsi	University of Málaga, Spain
Manuel Benavent-Lledo	University of Alicante, Spain

Marcelo Saval-Calvo	University of Alicante, Spain
Miguel A. Molina-Cabello	University of Málaga, Spain
Nahuel Emiliano García D'Urso	University of Alicante, Spain
Rafael M. Luque-Baena	University of Extremadura, Spain

SOCO 2025 Organizing Committee Chairs

Emilio Corchado	University of Salamanca, Spain
Héctor Quintián	University of A Coruña, Spain

SOCO 2025 Organizing Committee

José Luis Calvo Rolle	University of A Coruña, Spain
Esteban Jove	University of A Coruña, Spain
José Luis Casteleiro Roca	University of A Coruña, Spain
Francisco Zayas Gato	University of A Coruña, Spain
Álvaro Michelena	University of A Coruña, Spain
Míriam Timiraos Díaz	University of A Coruña, Spain
Antonio Javier Díaz Longueira	University of A Coruña, Spain
Paula Patricia Arcano Bea	University of A Coruña, Spain
Manuel Rubiños Trelles	University of A Coruña, Spain
Marta María Álvarez Crespo	University of A Coruña, Spain
Alejandro Vidal Bralo	University of A Coruña, Spain
Agustín García Fisher	University of A Coruña, Spain
Iker Pastor López	University of Deusto, Spain

Contents

Special Session: Intelligent Techniques Applied to Modelling and Control in Engineering Focused on Marine Energy Systems and Robotics

Special Session: Quantum Computing

Optimization, Metaheuristics and Applied Math

An Analysis of Surrogate Models Based on Machine Learning for the Problem of Static Ambulance Allocation

Esther Ródenas-Moreno[1](✉), Victor Sánchez-Anguix[1], Fulgencia Villa[1], and Juan M. Alberola[2]

[1] Instituto Tecnológico de Informática, Grupo de Sistemas de Optimización Aplicada, Ciudad Politécnica de la Innovación, Universitat Politècnica de València, Camino de Vera s/n, 46022 Valencia, Spain
esromo@ade.upv.es, {vsanchez,mfuvilju}@eio.upv.es
[2] Valencian Research Institute for Artificial Intelligence, Universitat Politècnica de València, Camino de Vera, s/n, 46022 Valencia, Spain
jalberola@dsic.upv.es

Abstract. Response time is defined as the interval between receiving an emergency call and arriving at the incident site. It is a crucial metric in ambulance allocation problems, which involve determining optimal initial locations, also known as base stations, for deploying ambulances. These allocations of ambulances to base stations are typically assessed based on the average response time. While simulation is an effective method for calculating such average response time, its computational cost increases significantly with the number of execution runs, especially when used alongside metaheuristics, which usually require several thousands of simulations to explore the solution space. This paper analyses the effectiveness of various machine learning models as a foundation for constructing surrogate models in this field.

Keywords: surrogate models · simulation · metaheuristic algorithms · ambulance allocation

1 Introduction

Emergency Medical Services (EMS) are responsible for managing, deploying, and maintaining the fleet of Emergency Medical Vehicles (EMVs) within a given region. Upon receiving an emergency call, trained professionals assess the situation to determine whether the deployment of an EMV is required. The time between the reception of the emergency call and the arrival of the EMV to the patient's location can have a critical impact on clinical outcomes, particularly in time-sensitive conditions. As a result, response time, the interval from the initial call to the arrival of the EMV, is used as a key performance indicator for evaluating the quality and efficiency of the EMS system.

E. Corchado et al. (Eds.): SOCO 2025, CCIS 2806, pp. 3–12, 2026.
https://doi.org/10.1007/978-3-032-19763-4_1

The static ambulance allocation problem seeks to determine the optimal configuration of the EMVs base locations to achieve specific objectives that vary from maximising territorial coverage to minimising emergency response times [4,6,15]. However, solving this problem is inherently complex from a human's cognitive capabilities, as the scale and computational costs make it expensive to explore the solution space. To overcome this challenge, many experts have turned to heuristic and metaheuristic algorithms as mechanisms to find good decisions in this domain [7,10].

Nevertheless, while these algorithms offer a promising approach to the static ambulance allocation problem, additional consideration must be taken when trying to minimise the average response time for calls, as it is difficult to quantify as an analytical objective function in closed form, given the dynamic aspect of the system. Over the past 50 years, research on ambulance allocation has evolved significantly, moving from simple, static models to dynamic, simulation-based approaches that better reflect real-world complexities [4,6]. Early models have been developed to include simulations [16], such as those utilising Gaussian mixture models for spatial demand [12], genetic algorithms [17], and realistic operational practices aimed at improving EMS systems [18]. Recent advancements feature discrete trace-based simulations that are informed by actual emergency data, as well as the integration of metaheuristics like tabu search and memetic algorithms [8,14].

The primary drawback of all these proposals is the necessity of using a simulator to calculate the value of the objective function, which is computationally expensive. Regardless of the chosen solving method, the simulator significantly increases the cost of exploring the solution space. Scholarly research is increasingly incorporating machine learning techniques to improve travel time predictions [2,11] and to facilitate faster optimisation through surrogate modelling [5]. However, as far as we know, these hybrid approaches do not tackle the issue of the dynamic objective function in metaheuristic algorithms.

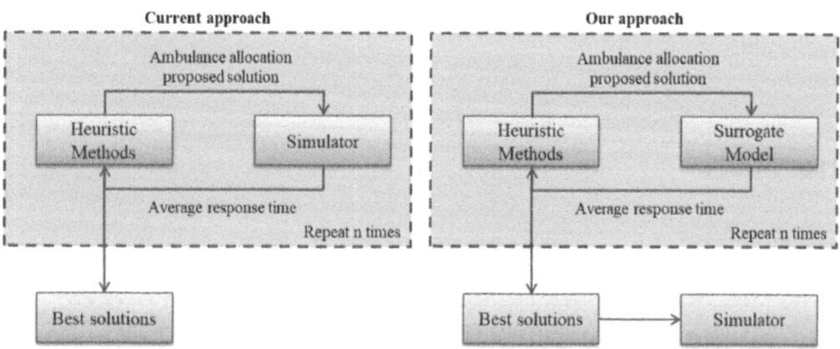

Fig. 1. Structure of the proposed approach vs the current approach.

This study proposes using surrogate models to reduce the computational cost of combining heuristic and metaheuristic algorithms with simulation tools in static ambulance allocation problems. The goal is to determine which surrogate model offers better goodness-of-fit to improve computational efficiency, allowing more candidate solutions to be evaluated within the same time frame. An outline of our proposed solution compared to existing approaches can be seen in Fig. 1. In this approach, a surrogate model is used to assess the quality of solutions proposed by heuristic methods, while a simulator is reserved for evaluating only the best solutions obtained from the selected method. Multiple machine learning models were trained and evaluated using real-world data from the province of Valencia. Finally, using Spearman's correlation coefficient, a comparison of the models is made.

This paper is structured as follows: Sect. 2 provides a formal definition of the problem addressed in this study. Section 3 presents a case study featuring actual call data from the province of Valencia. Section 4 outlines the experimental design, while Sect. 5 analyses the results. Finally, Sect. 6 concludes with the key findings and suggests directions for future research.

2 Problem Description

This paper analyses and identifies which machine learning models can be used as surrogate models for evaluating the efficiency of solutions to the static ambulance allocation problem, specifically focusing on advanced life support (ALS) vehicles. These ambulances are typically deployed for the most time-sensitive emergencies. In this section, we provide a formal definition of the static ambulance allocation problem, as well as defining what we consider a surrogate model for this problem.

Let $B = \{1, 2, \ldots, n\}$ represent the set of prospective base stations where ambulances may be stationed and ready to be deployed to emergency locations. These bases typically consist of health-related public buildings that offer parking, medical supplies, and staff rooms, among other facilities, for ambulances and serve as starting points for their deployment. These locations have been provided by the EMS of the use case. Similarly, let $V = \{1, 2, \ldots, m\}$ represent the set of advanced life support vehicles available for service and that need to be assigned to one of the bases in B. It is assumed that the number of bases, n, significantly exceeds the number of ambulances, m, such that $n \gg m$. A feasible solution X to this issue consists of selecting m unique base stations from B as starting and idle points for ambulances. For the purposes of this paper, it is assumed that only one ambulance can be stationed at each base. That is, it involves selecting $X \subseteq B$ so that $|X| = m$.

The average response time is a critical indicator in emergency medical services and is often used as the objective function to minimise in the static ambulance allocation problem. This average response time needs to be calculated over a set of demand scenarios $\mathcal{D} = \{D_1, \ldots, D_l\}$, where each demand scenario D_k is a set of emergency calls with its respective arrival time and geographical location. For simplicity, let us assume that $RT(c, D_i, X)$ is a function that calculates the simulated response time for call $c \in D_i$, provided its order in the demand scenario D_i, and a solution to the static ambulance allocation problem X. Then, the average response time of the system can be calculated as:

$$f(X) = \frac{1}{\sum_{D_k \in \mathcal{D}} |D_k|} \sum_{D_k \in \mathcal{D}} \sum_{c \in D_k} RT(c, D_k, X) \quad (1)$$

Since response time is influenced by various factors, including the spatial distribution of emergencies and the configuration of the road network, it is difficult to analytically calculate $f(X)$. As a result, simulators are frequently used as evaluation tools, as they offer a more realistic approximation of the solution's response time. However, simulators become significantly more computationally expensive as they attempt to incorporate more realistic features, which, combined with metaheuristic algorithms, significantly increase the computation costs of the problem.

An alternative solution to this issue is the use of surrogate models, which allow for the optimisation of costly objective functions by constructing cheaper approximations of the input-output behaviour of the original objective function. In this paper, we aim to identify the best surrogate model $\hat{f}(X) \approx f(X)$ constructed based on nine different machine learning techniques. For the surrogate models in this work, we assume the same input as that of the original approximated function, since it is the most general approach and it assumes no specific knowledge about the case study.

3 Case Study

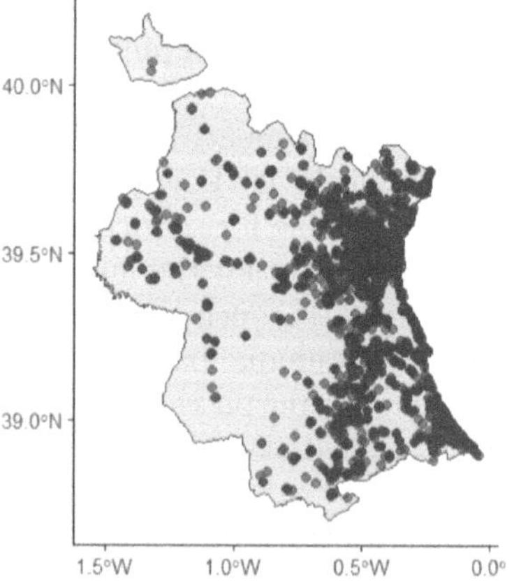

Fig. 2. Emergency calls in the province of Valencia in 2019.

This paper analyses a real-life case study from the Province of Valencia, Spain. In Spain, the public healthcare system assigns EMS responsibilities to each Autonomous Community. As a result, the Valencian regional government is responsible for developing an EMS protocol to effectively serve a population of 2,710,808 people (as of 2024) across an area of 10,813 square kilometres. To accomplish this, they operate a fleet of 20 advanced life support ambulances, which are to be distributed among 678 potential base stations. These base stations include hospitals, health centres, and various other public health facilities that provide the necessary resources for the emergency response teams assigned to each ambulance.

We consider a database of 365 distinct real-life demand scenarios, each representing emergencies that occurred on a specific day from January 1st to December 31st, 2019. This database includes approximately 20,000 emergency calls that required the deployment of an advanced life support ambulance, which can be seen in Fig. 2.

4 Methodology

In this paper, we evaluate various machine learning models to determine their effectiveness as surrogate models for the ambulance allocation problem. Our objective is not to completely eliminate the use of simulators; rather, we aim to limit their application to the best solutions identified through the selected metaheuristic. This approach allows for a more extensive exploration of the solution space within a set timeframe, while still ensuring accurate mean response times for the selected solutions.

Therefore, we aim to evaluate surrogate models based on their accuracy, while also considering their training and prediction times. To do this, the process was divided into five distinct steps for clarity, as illustrated in Fig. 3. It should be highlighted that the methodology employed in this section aims to evaluate surrogate models in a similar setting to the one in which they would be employed in a metaheuristic.

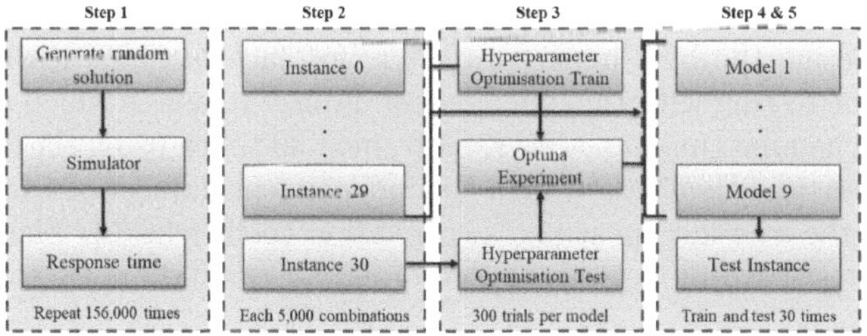

Fig. 3. Experiment design architecture.

The first step involved creating a database of potential solutions and their respective response times to create several datasets that allow us to train and evaluate the surrogate models. To accomplish this, we sampled 156,000 random prospective solutions $X \subset B$, with $|X| = 20$. Each solution consists of a Boolean vector of size $n = 678$, containing $m = 20$ positions marked as 1. This vector represents the 20 selected base stations to which the 20 advanced life support ambulances are assigned. Currently, ambulance assignments are made randomly to simulate an initial sampling made by a metaheuristic method, with only one ambulance allocated per base station.

After generating all the solutions, we employ an EMS simulator, in this case JEMSS [13], to calculate the average response time in minutes of each solution. The simulator processes calls from demand scenarios using the FIFO (first in, first out) method and assigns the nearest available ambulance. Additionally, the road network utilised in the simulation is realistic, consisting of 18,428 nodes and 50,521 arcs.

After completing all simulations, the next step involved dividing sampled solutions into different datasets. Particularly, we created 31 datasets consisting of 5,000 solutions with their corresponding average response time $f(X)$. Additionally, a final independent group of 1,000 instances was formed to test the goodness of the models after all of the model tuning. We created several datasets to analyse the stochastic effect of random sampling on the performance of the different surrogate models.

In Step 3, we focused on determining the optimal hyperparameter configurations for each machine learning model. To achieve this, we conducted experiments using Bayesian optimisation with the Optuna package in Python [3]. The nine machine learning techniques considered in this study include CatBoost, Decision Tree, Elastic Net, LASSO, LightGBM, Neural Networks, Random Forest, Support Vector Machine, and XGBoost. To assess the quality of a hyperparameter configuration, we conducted 30 different trainings by using each time one of the first 30 datasets of 5,000 solutions as training, and the 31st dataset of 5,000 solutions as a test. We employed Spearman's correlation as a quality metric. The reason to choose this metric rather than the root mean squared error is that, in terms of optimisation, most of the time we just need to be able to rank and compare solutions. The average Spearman's correlation of the 30 trainings is employed as a metric to maximise with Bayesian optimisation.

After selecting the best hyperparameters for each machine learning model, we conduct the testing with the independent dataset to assess which model offers a better fit. The nine tuned models were trained 30 times using the 30 groups of 5,000 instances, and then tested with the independent dataset of 1,000 solutions according to the average Spearman's correlation coefficient [9]. The time taken to train and test each model was also recorded for evaluation.

5 Results

This section presents the results of the pipeline and methodology proposed in the previous section by focusing on the real-life call data from the Province of Valencia. It should be reminded that, we evaluate the performance of each model using Spearman's correlation coefficient, as we believe that in this context it is more important to rank the solutions correctly than to have a higher accuracy in the prediction.

Table 1. Summary of model performance: Spearman score and timing metrics.

Model	Avg. Spearman	Avg. Train Time (s)	Avg. Predict Time (s)
CatBoost	0.455249	50.901611	0.194343
Decision Tree	0.175534	0.025072	0.001743
Elastic Net	0.453624	0.151204	0.009596
LASSO	0.444781	0.707018	0.003065
LightGBM	0.503947	3.110737	0.014452
Neural Networks	0.440969	7.884855	0.004196
Random Forest	0.411055	6.197633	0.197852
SVM	0.465711	16.605795	4.180410
XGBoost	0.460506	4.789927	0.018197

Table 1 presents the average Spearman score from 30 trials for each tuned machine learning model, along with the time required for training and prediction. All models, except the Decision-Tree-based model, show a moderate correlation. LightGBM exhibits the highest Spearman's correlation coefficient value, followed by Support Vector Machine and XGBoost. In contrast, the Decision-Tree model shows a low correlation between real and predicted values.

To assess whether the differences in the average Spearman correlation among the models are statistically significant, an analysis of variance (ANOVA) was performed. The ANOVA test produced a p-value of 2.9×10^{-175}, indicating that there are statistically significant differences in the Spearman correlations between the models. However, the ANOVA test does not indicate which specific models differ from the others.

To identify differences between models, we perform a posthoc test using Tukey's honestly significant difference (HSD) intervals, which allow for pairwise comparisons of means across a set of samples [1]. The results of the test are shown in Fig. 4. Please note that the Decision Tree method has been excluded to better analyse the differences between the best-performing models.

Tukey's HSD test evaluates whether the difference between two means is statistically significant, taking into account the multiple comparisons being made. The dots represent the average Spearman correlation coefficient for each model, and the lines indicate the confidence intervals for those averages. If the confidence intervals overlap, it suggests that there are no statistically significant differences between the models. The results show that the LightGBM method has the highest Spearman correlation coefficient, and since its confidence interval does not overlap with any others, it can be concluded that it is significantly better than the rest.

Furthermore, referring to the prediction times collected in Table 1, we can emphasise that, although it is not the fastest model, its performance is comparable to the other models in terms of speed. Therefore, it can be concluded that

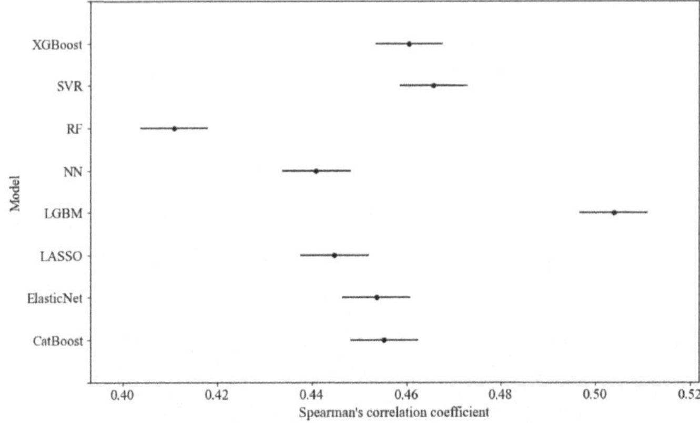

Fig. 4. Tukey's HSD test for Spearman correlation coefficient.

LightGBM is the most effective technique for constructing surrogate models for the ambulance allocation problem in the province of Valencia.

The final configuration of hyperparameters selected for this LightGBM-based model employed the traditional Gradient Boosting Decision Tree as the boosting type. It featured 60 leaves, 800 estimators, a learning rate of approximately 0.023, and a subsample rate of around 0.7. All features were utilised, with a regularisation alpha set to 0 and a regularisation lambda of approximately 9.72. The same approach of experimentation was applied to the other machine learning techniques.

6 Conclusions

This paper analyses various machine learning models as a basis for developing surrogate models in the field of ambulance allocation problems. Using the response time as an objective function requires the use of a simulator to calculate its value, which becomes progressively more complex as additional real-life characteristics are taken into account. Furthermore, combining heuristic algorithms with simulators significantly raises the computational complexity of the problem, as each potential solution requires evaluation through time-consuming simulations. Given the high dimensionality of the solution space, this process may involve assessing thousands or even millions of configurations, especially when considering multiple algorithms. Using a surrogate model to rank solutions allows for the evaluation of a significantly larger number of candidate solutions within the same time frame.

To evaluate the performance of the analysed models as surrogate models, we first created several datasets of randomly sampled prospective solutions (i.e., allocations of ambulances to base stations). We then assessed each solution using an emergency medical service simulator to calculate its average response time to emergency calls in minutes. We evaluated nine machine learning models: CatBoost, Decision Tree, Elastic Net, LASSO, LightGBM, Neural Networks, Random Forest, Support Vector Machine, and XGBoost. To determine their optimal hyperparameter configurations, we used Bayesian optimisation through the Optuna framework. Finally, we trained the fine-tuned models and compared the results against the independent test set. The results suggest that LightGBM may provide a better surrogate solution in terms of Spearman's correlation while also having a competitive training and prediction time.

In the future, we plan to study alternative solution vectors as an input for the surrogate model. Rather than using a Boolean vector directly representing the solution of a metaheuristic, we aim to characterise solutions with aggregated information (e.g., covered population, distance between stations, etc.). We also plan to combine the use of these surrogate models with heuristic algorithms and evaluate its performance with studies that directly employ a simulator as an objective function.

Acknowledgments. This work is partially supported by MINECO/FEDER RTI2018-095390-B-C31 project of the Spanish government, project TED2021-131295B-C32 from the State Research Agency, DIGITAL2022 CLOUDAI02/S8760000 from the European Commission, project INNVA1/2024/91 funded by Agencia Valenciana de la Innovación, project PID2021-123673OB-C31 COSASS and project OPRES PID2021-124975OB-I00 partially funded by the Spanish Ministry of Science and Innovation and FEDER funds. Esther Ródenas is supported by the predoctoral contract (CIACIF/2023/084) from Generalitat Valenciana.

References

1. Abdi, H., Williams, L.J.: Tukey's honestly significant difference (HSD) test. Encycl. Res. Des. **3**(1), 1–5 (2010)
2. Abid, M.A., et al.: Ambulance travel time estimation using spatiotemporal data. Procedia Comput. Sci. **238**, 265–272 (2024)
3. Akiba, T., et al.: Optuna: a next-generation hyperparameter optimization framework. In: Proceedings of the 25th ACM SIGKDD International Conference on Knowledge Discovery & Data Mining, pp. 2623–2631 (2019)
4. Bélanger, V., Ruiz, A., Soriano, P.: Recent optimization models and trends in location, relocation, and dispatching of emergency medical vehicles. Eur. J. Oper. Res. **272**(1), 1–23 (2019)
5. Bozorgmanesh, H., Rydén, P.: Optimal placement of ambulance stations using data-driven direct and surrogate search methods. Int. J. Med. Inf. 10579 (2025)
6. Brotcorne, L., Laporte, G., Semet, F.: Ambulance location and relocation models. Eur. J. Oper. Res. **147**(3), 451–463 (2003)
7. Daskin, M.S.: A maximum expected covering location model: formulation, properties and heuristic solution. Transp. Sci. **17**(1), 48–70 (1983)
8. Gallego, C., et al.: A metaheuristic algorithm guided by simulation for optimizing the static allocation of emergency medical vehicles. In: International Conference on Soft Computing Models in Industrial and Environmental Applications, pp. 295–305. Springer (2024). https://doi.org/10.1007/978-3-031-75013-7_28

9. Hauke, J., Kossowski, T.: Comparison of values of Pearson's and Spearman's correlation coefficients on the same sets of data. Quaestiones Geographicae **30**(2), 87–93 (2011)
10. Jagtenberg, C.J., Bhulai, S., van der Mei, R.D.: An efficient heuristic for real-time ambulance redeployment. Oper. Res. Health Care **4**, 27–35 (2015)
11. Mahdiraji, S.A., Abid, M.A., Holmgren, J.: Integrating machine learning-based ambulance travel time estimation into an emergency medical services simulation modeling framework. Procedia Comput. Sci. **251**, 479–486 (2024)
12. McCormack, R., Coates, G.: A simulation model to enable the optimization of ambulance fleet allocation and base station location for increased patient survival. Eur. J. Oper. Res. **247**(1), 294–309 (2015)
13. Ridler, S., Mason, A.J., Raith, A.: A simulation and optimisation package for emergency medical services. Eur. J. Oper. Res. **298**(3), 1101–1113 (2022)
14. Schjølberg, M.E., et al.: Comparing metaheuristic optimization algorithms for ambulance allocation: an experimental simulation study. In: Proceedings of the Genetic and Evolutionary Computation Conference, pp. 1454–1463 (2023)
15. Vecina, M.Á., et al.: A decision support tool for the static allocation of emergency vehicles to stations. In: International Conference on Hybrid Artificial Intelligence Systems, pp. 141–152. Springer (2022). https://doi.org/10.1007/978-3-031-15471-3_13
16. Yang, W., et al.: Simulation modeling and optimization for ambulance allocation considering spatiotemporal stochastic demand. J. Manage. Sci. Eng. **4**(4), 252–265 (2019)
17. Yue, Y., Marla, L., Krishnan, R.: An efficient simulationbased approach to ambulance fleet allocation and dynamic redeployment. In: Proceedings of the AAAI Conference on Artificial Intelligence, vol. 26., no. 1, pp. 398–405 (2012)
18. Zaffar, M.A., et al.: Coverage, survivability or response time: a comparative study of performance statistics used in ambulance location models via simulation-optimization. Oper. Res. Health Care **11**, 1–12 (2016)

Hyperparameter Optimization in Graph Embeddings Using Simulated Annealing and Genetic Algorithms

Pablo Zubasti(✉), Miguel A. Patricio, José M. Molina, and Antonio Berlanga

Computer Science and Engineering Department, Applied Artificial Intelligence Group, Universidad Carlos III de Madrid, Madrid, Spain
pzubasti@pa.uc3m.es, mpatrici@inf.uc3m.es, {molina,aberlan}@ia.uc3m.es
https://giaa.uc3m.es/

Abstract. This work introduces an automated hyperparameter tuning method for the *Cauchy Graph Embeddings* (CGE) algorithm, which improves upon *Laplacian Eigenmaps* (LE) by better preserving local topology in latent representations. Although CGE achieves superior performance, it involves increased complexity and sensitivity to hyperparameter changes. To address this, the optimization problems and objective functions for both techniques are formulated, enabling automatic tuning via *Simulated Annealing* and *Genetic Algorithms*.

Keywords: Graph Embeddings · Laplacian Eigenmaps · Optimization · Simulated Annealing · Genetic Algorithms

1 Introduction

Graph Embedding algorithms [4] comprise a broad set of unsupervised learning techniques aimed at computing a vector representation of a graph, commonly referred to as a latent space or embedding. This process enables the transformation of each vertex in the original graph into a δ-dimensional point within a Euclidean space of the same dimensionality. The primary motivation for performing such transformations or "projections" of graphs/topologies into Euclidean spaces lies in the need to manage the structural and semantic information of the graph in a more accessible and computationally efficient manner. This vector-based representation allows for the application of a wide range of Machine Learning techniques, both supervised and unsupervised, facilitating tasks such as node classification, link prediction, community detection, similarity analysis, and more. Essentially, embedded representations capture both local node properties (i.e., direct connections to neighbors) and global graph properties (such as structural patterns and hierarchies). This is essential for modern applications in which graphs are used to model complex data across diverse domains such as social networks, computational biology, recommender systems, transportation networks, and knowledge analysis in semantic databases.

E. Corchado et al. (Eds.): SOCO 2025, CCIS 2806, pp. 13–23, 2026.
https://doi.org/10.1007/978-3-032-19763-4_2

The information contained in a weighted graph $G = (\mathcal{V}, \mathcal{E}, \mathcal{W})$ is generally modeled through a square adjacency matrix $\mathcal{W} \equiv M_{n \times n}$, which captures the individual information of each edge $e_{ij} \in \mathcal{E}$, so that each value $m_{ij} \in M$ represents the weight (often also referred to as the cost) associated with the edge connecting the pair of vertices $v_i, v_j \in \mathcal{V}$. The adjacency matrix encodes the relationships among all possible pairs of vertices; however, handling this information directly from the matrix representation may become a significant issue, thereby rendering the need to vectorize the matrix representation (and, by extension, that of the graph) critically important. Having understood the necessity of vectorizing topological information to work with it effectively, any Graph Embedding algorithm is hereafter defined as one that implements the transformation shown in Eq. 1.

$$T : G \to M' \mid G = (\mathcal{V}, \mathcal{E}, \mathcal{W}),\ M' = \left\{\mathbf{x}_i \in \mathbb{R}^{\delta}\right\}_{i=1}^{|\mathcal{V}|} \quad (1)$$

where δ denotes the selected dimensionality for the resulting embedding (it is the fundamental hyperparameter of any Graph Embedding technique), G is the original graph, and M' is the matrix that contains the column vectors $\mathbf{x}_i$ that represent each vertex $v_i \in \mathcal{V}$. From this point on, a commonly accepted taxonomy [4] can be established for the various types of Graph Embedding techniques, based on the approach they adopt to solve the problem presented in 1:

- **Spectral Decomposition-based Methods**: These approaches leverage the spectral properties of graph-associated matrices—such as the adjacency matrix or the graph Laplacian—to derive node embeddings through eigendecomposition techniques. Foundational works in this category include classical methods like ISOMAP, LLE (Locally Linear Embedding), and Spectral Clustering [6,12,16], which laid the groundwork for manifold learning. More recent contributions such as Laplacian Eigenmaps [1] and Cauchy Graph Embedding [11] have refined the theoretical basis for low-dimensional projections. Additionally, methods like GraRep and its enhanced variant GraRep++ [2,13] aim to preserve higher-order proximity information, while LINE offers an alternative scalable formulation [17]. Analytical studies further investigate embedding configurations and their implications for downstream tasks [20].

- **Random Walk-based Methods**: Techniques in this category exploit random walk mechanisms to capture both local and global graph structures by simulating node co-occurrence patterns over sampled paths. DeepWalk introduced the use of truncated random walks combined with the Skip-Gram model for learning node embeddings [14], a methodology later generalized by node2vec, which incorporates flexible biased walks to balance breadth- and depth-first exploration strategies [5]. Subsequent analysis has provided further insights into the optimization and theoretical underpinnings of these models [15].

- **Neural Network-based Methods**: The advent of deep learning has catalyzed the development of neural architectures tailored to graph-structured data. Early neural models such as SDNE and DNGR employed deep autoencoders to capture nonlinear structures and preserve node proximities [3,18]. This was followed by the emergence of message-passing paradigms exemplified by GCNs [10], inductive frameworks like GraphSAGE [7], and attention-based mechanisms such as GAT and its extensions [8]. More recent surveys and perspectives explore the broader applicability of GNNs and identify ongoing challenges and open directions in this rapidly evolving field [9,19].

The present work explores the optimization of hyperparameters aimed at enhancing the quality of graph embeddings (employing the CGE [11] algorithm), thereby yielding representations that more faithfully reflect the information conveyed by the original topology.

2 Theoretical Foundations of Graph Embeddings

The fundamental principle underlying all Graph Embedding algorithms is the property known as *Local Topology Preserving* [11]. This property is mathematically formulated as shown in Eq. 2.

$$If\ w_{ij} \geq w_{pq} \rightarrow (x_i - x_j)^2 \leq (x_p - x_q)^2,\ \forall i, j, p, q \tag{2}$$

The expression presented in Eq. (2) states that if any graph weight w_{ij} has a higher value than another weight w_{pq}, then the Euclidean distance between the embedded points x_i and x_j must be smaller than the distance between x_p and x_q. Essentially, it implies that edges with higher weights should correspond to closer points in the embedding space, whereas lower-weight edges (indicating weaker relationships between nodes) should result in greater distances. Although this property may seem trivial, it gives rise to infeasible problems in which the representation in a latent space cannot fully preserve the original topology. Consider the example in Fig. 1.

In Fig. 1, a strong relationship is observed between the vertices A and B, as well as between B and C. However, the relationship between the vertices A and C is significantly weaker compared to the other two edges. In other words, the graph conveys that A and B should be close in the latent space, as should B and C, while C must remain simultaneously distant from A. This topology is entirely valid from a mathematical standpoint as a discrete structure but lacks a faithful Euclidean spatial representation of the original information. These types of infeasible problems go beyond the specific Graph Embedding technique used, since the technique will compute an embedding that minimizes an objective function under a constraint inherent to the graph's topology. Therefore, it must be emphasized that tuning the hyperparameters of Graph Embedding algorithms does not indefinitely improve embedding quality, as there are often pre-existing limitations stemming from the graph structure itself that cannot be resolved within Euclidean spaces.

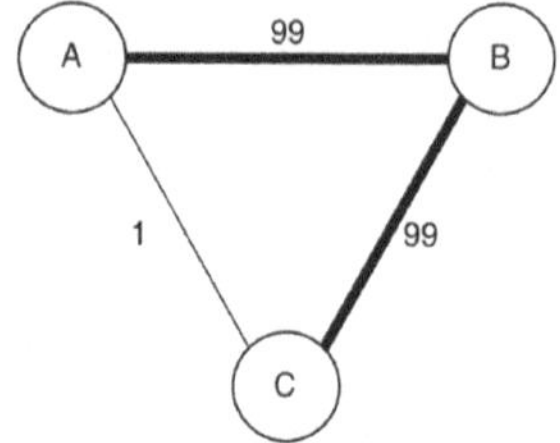

Fig. 1. Example of an infeasible problem for Graph Embedding techniques.

3 Proposed Hyperparameter Tuning Approach

3.1 Problem Modeling

The objective is to optimize the hyperparameters σ, L, γ and N on which the Cauchy Graph Embeddings (CGE) algorithm [11] relies—an improved extension of the Laplacian Eigenmaps method [1]—as these parameters are of critical importance, given their substantial impact on the solutions produced by the algorithm. This is formulated as a classical single-objective optimization problem with multiple variables comprising the objective function.

Simulated Annealing

The first technique used to solve this optimization problem is known as *Simulated Annealing*, which, although not strictly a biologically inspired algorithm, is often grouped with such methods due to its conceptual similarities. *Simulated Annealing* is inspired by the physical phenomenon of metal annealing, which involves the gradual cooling of a material that has previously been heated. The algorithm relies on a temperature function that decreases over time, influencing stochastic decisions based on the cooling process. The essence of the algorithm lies in making small probabilistic decisions, where random variations of the current state guide the search for lower-energy solutions. The method can be interpreted as an "intelligent random search" in which the final result corresponds to the best energy state encountered. More formally, the algorithm requires the definition of states, a *movement* function for the random variation of the current state, and an energy (objective) function. The procedure is as follows:

1. Initialize with a starting state.
2. Calculate the energy associated with the current state.
3. Modify the current state gradually and randomly.
4. Recalculate the energy.
5. If the new energy is lower than the previous one, return to Step 3.
6. Otherwise, compute the transition probability using:

$$P(S_{t+1}|S_t) = \begin{cases} 1 & \text{if } \Delta c \leq 0 \\ e^{-\Delta c/T} & \text{if } \Delta c > 0 \end{cases} \tag{3}$$

7. Make a stochastic decision (based on the probability of Step 6) whether to transition to the new state or stay in the current one. In either case, return to Step 3.

The algorithm executes the steps above for a defined number of iterations. In expression 3, Δc represents the increase in energy between the current state and the previous state and T represents the decaying temperature in the current iteration. With a clear understanding of the *Simulated Annealing* mechanism, the first step is to mathematically define the states over which the algorithm will explore and the *movement* function that allows transitions between states. Since the variables to be optimized are the CGE hyperparameters, states are modeled as the 4-tuple.

$$S_t = (\sigma_t,\ L_t,\ \gamma_t,\ N_t) \tag{4}$$

The movement function is defined as follows:

$$f(S_t) = \begin{bmatrix} \sigma_t \leftarrow \sigma_t + RND(\text{-}0.02,\ 0.02) \\ L_t \leftarrow L_t + RND(\text{-}0.02,\ 0.02) \\ \gamma_t \leftarrow \gamma_t + RND(\text{-}0.02,\ 0.02) \\ N_t \leftarrow N_t + RND(\text{-}1,\ 1) \end{bmatrix} \tag{5}$$

$$S_{t+1} = f(S_t) \tag{6}$$

Here, $RND(a,\ b)$ represents a function that generates continuous random values uniformly distributed within the interval $[a, b]$.

Genetic Algorithm

The second technique employed is the *Genetic Algorithm* (GA), which is fully inspired by biological processes. The classical *Genetic Algorithm* is a metaheuristic based on the principles of natural selection and genetics, mimicking evolutionary processes observed in nature. Its goal is to find optimal or near-optimal (i.e. sufficiently good) solutions for a given problem defined through an objective function. A GA can be described as follows:

1. **Initialization**: Create an initial population of N random individuals.
2. **Evaluation**: Calculate the fitness function $f(x)$ for each individual x.
3. **Genetic operators**: Apply selection, crossover, and mutation operators to generate a new population.
4. **Iteration**: Repeat steps 2 and 3 until the stopping criteria are met.

3.2 Objective Function Design Based on RMSE

The objective function must reflect and quantify how "faithful" the resulting embedding is to the structural information encoded in the original graph. The first step is the definition of the *recovered matrix* derived from the embedding, which is mathematically formulated as shown in Eq. 7.

$$R = \left\{ e_{ij} \cdot \dot{d}(\hat{E}_i^\delta,\ \hat{E}_j^\delta) \right\}_{i,j=1}^{n} \mid e_{ij} \in \{0,1\},\ \text{and } n \equiv |\mathcal{V}|$$

$$d(\hat{E}_i^\delta,\ \hat{E}_j^\delta) = \sqrt{\sum_{k=1}^{\delta} (\hat{e}_{ik} - \hat{e}_{jk})^2} \Rightarrow \dot{d}(\hat{E}_i^\delta,\ \hat{E}_j^\delta) = \frac{1}{d(\hat{E}_i^\delta,\ \hat{E}_j^\delta) + \epsilon} \tag{7}$$

Here, $\hat{E}_i^\delta$ represents the vector associated with the i-th vertex in the δ dimensional embedding, and e_{ij} indicates the presence (1) or absence (0) of an edge. Essentially, the recovered matrix assigns the value $\dot{d}(\hat{E}_i^\delta, \hat{E}_j^\delta)$ (the inverse of the distance in the embedding space, plus a very small constant ϵ to avoid division by zero) to each edge that existed in the original graph and sets all others to zero, thus avoiding the introduction of new adjacencies that could distort the evaluation. A direct comparison $m_{ij} - r_{ij}$ would yield a large error, as both values are on different scales. To correctly compare elements of matrix M with those of matrix R, a *MinMax* scaling is applied, having both matrices M and R normalized to the interval $[0,\ 1]$, allowing a proportional and meaningful comparison.

The objective function is then defined based on the *Root Mean Squared Error* (RMSE) between the original matrix scaled $\tilde{M}$ and the scaled recovered matrix $\tilde{R}$. Mathematically:

$$J(\tilde{M}; \sigma, L, \gamma, N) = \sqrt{\frac{1}{n^2} \sum_{i=1}^{n} \sum_{j=1}^{n} \left(\tilde{m}_{ij} - \tilde{r}_{ij}^{(\sigma,L,\gamma,N)} \right)^2} \mid \tilde{m}_{ij} \in \tilde{M},\ \tilde{r}_{ij}^{(\sigma,L,\gamma,N)} \in \tilde{R} \tag{8}$$

With the objective function defined in 8, all necessary components are available to feed the *Simulated Annealing* algorithm. However, for the *Genetic Algorithm*, a slight modification is required. As previously mentioned, *Simulated Annealing* solves a minimization problem by default, which in this case is:

$$\arg \min_{\sigma,L,\gamma,N} J(\tilde{M}; \sigma, L, \gamma, N)$$

In contrast, *Genetic Algorithms* are inherently designed to solve maximization problems. Since RMSE always yields values greater than or equal to zero, multiplying the objective by -1 effectively inverts the problem, allowing the maximization framework to minimize RMSE, as desired:

$$\arg \max_{\sigma,L,\gamma,N} -J(\tilde{M}; \sigma, L, \gamma, N)$$

With the objective functions now defined for both algorithms, they can be used respectively as the energy function (for *Simulated Annealing*) and the fitness function (for the *Genetic Algorithm*).

4 Experimentation and Analysis of Results

To implement the proposed problem, the Python programming language was used, along with the `simanneal` and `DEAP` libraries for integrating *Simulated Annealing* and the *Genetic Algorithm*, respectively. The experimental phase focused on testing cases where the CGE algorithm typically produces poor embeddings by default, followed by evaluating the improvements achieved through hyperparameter optimization. The simplest graph analyzed is shown in Fig. 2.

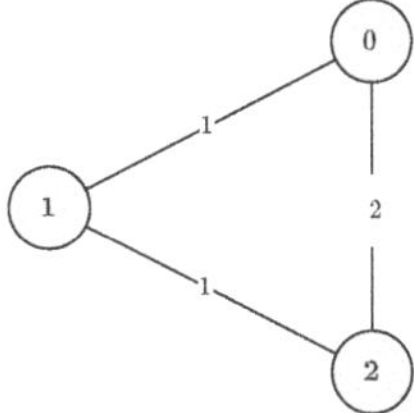

Fig. 2. Complete graph with three vertices.

The graph in Fig. 2 represents a complete graph $\mathbb{K}_3$, where there is no infeasibility in spatial representation, as previously described (see Fig. 1). Although simple in appearance, the distribution of edge weights makes the Euclidean representation of the vertices possible but not immediate. Representing a triangle with sides of lengths 1, 1, and 2 leads to an aligned configuration of the three vertices. The corresponding embedding obtained by the CGE algorithm is shown in Fig. 3. As shown in Fig. 3, the CGE algorithm does not provide a faithful representation of the original topology. This is reflected in the RMSE, calculated between the recovered and original matrices, which yields a result of **0.33**, which is significant within the $[0, 1]$ scale used. Using the *Simulated Annealing* algorithm with a minimum and maximum temperature of **1** and **100**, respectively, and a total of **128** epochs, the RMSE (energy) was reduced by 100%, reaching a final energy of $E = 0$.

$$\mathcal{S} = (\sigma \leftarrow 2.448, L \leftarrow 0.225, \gamma \leftarrow 0.923, N \leftarrow 71)$$

Figure 4 confirms that the resulting embedding matches the expected geometric configuration of the three aligned points, achieving zero RMSE. The same process was repeated using the *Genetic Algorithm*, configured with **40 individuals**, **30 generations**, a **70%** crossover probability, and a **20%** mutation probability. The mutation probability is set to a high value, as empirical observations have shown that the problem's search space is highly irregular and noisy, containing a vast number of local minima that may cause the algorithm to become trapped. Therefore, a high mutation probability favors exploration

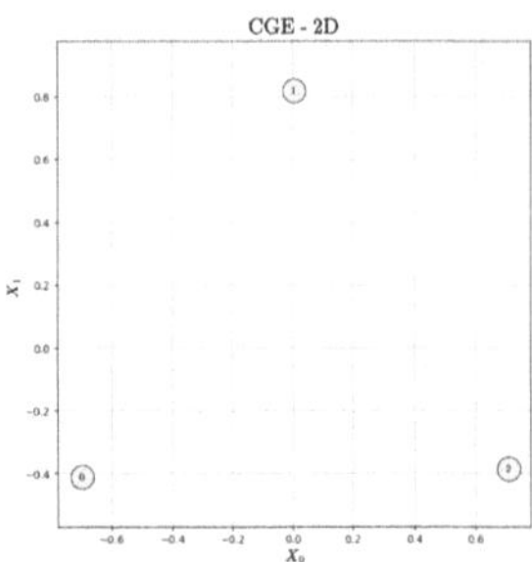

Fig. 3. Resulting embedding generated from the graph in Fig. 2.

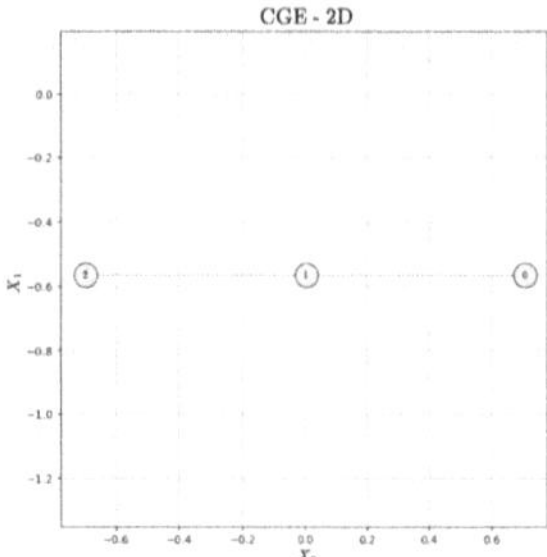

Fig. 4. Embedding resulting from optimization via *Simulated Annealing*.

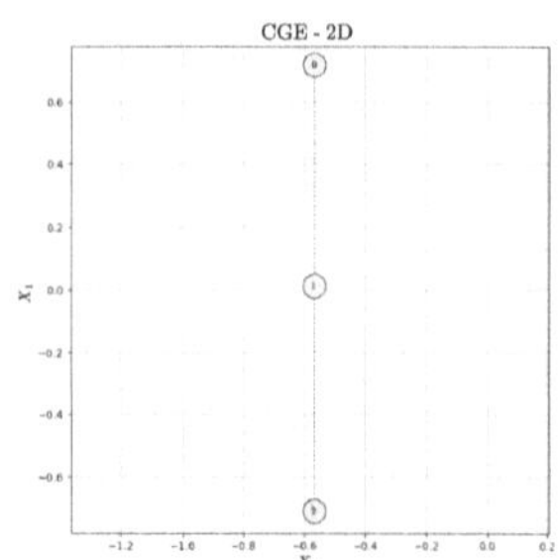

Fig. 5. Embedding resulting from optimization via the *Genetic Algorithm*.

over exploitation, leveraging the stochastic nature of the algorithm. The optimal hyperparameter values were:

$$\mathcal{S} = (\sigma \leftarrow 1.535, L \leftarrow 0.516, \gamma \leftarrow 0.989, N \leftarrow 28)$$

The hyperparameter configuration differs from that of *Simulated Annealing*, yet both reached valid solutions with zero error. The final embedding is shown in Fig. 5, obtaining a perpendicular solution to that of *Simulated Annealing*, yet equally valid under the constraints established before. This explains the different hyperparameter values: each algorithm found distinct, but optimal solutions.

Finally, the same experimental methodology applied to the trivial graph with 3 vertices was extended to fully connected graphs with 5, 10, 15, 20, and 25 vertices (corresponding to 10, 45, 105, 190, and 300 edges, respectively), assigning weights using a Gaussian pseudo-random number generator. The results of this experimentation are summarized in Table 1.

Table 1. Results obtained during experiments with randomly generated graphs using *Simulated Annealing* (SA) and the *Genetic Algorithm* (GA).

Vertices/Edges	Initial RMSE	Final RMSE (SA)	Final RMSE (GA)	Time (SA)	Time (GA)
5/10	0.348	0.293	0.289	34.4 s	46.7 s
10/45	0.314	0.254	0.233	98.5 s	113.9 s
15/105	0.358	0.234	0.256	132.9 s	374.9 s
20/190	0.301	0.239	0.212	188.3 s	628.3 s
25/300	0.302	0.208	0.207	248.6 s	758.5 s
50/808	0.226	0.182	0.173	303.8 s	836.0 s
75/1819	0.195	0.164	0.171	641.6 s	967.5 s

The results show slightly better accuracy with the *Genetic Algorithm*, at the cost of significantly higher computation time. Overall, the process is computationally intensive due to constant embedding recalculation and matrix comparisons. *Simulated Annealing* yields nearly identical results in much less time, as it avoids population-based operations like crossover and mutation. Since the graphs were randomly and fully connected, these scenarios include cases of potential infeasibility, meaning that embeddings inherently carry structural limitations. This explains the apparent performance ceiling, with the RMSE stabilizing around 0.20.

5 Conclusions

The proposed methodology for hyperparameter tuning of the CGE algorithm has demonstrated substantial effectiveness, yielding marked improvements in the fidelity of the resulting embeddings relative to the original graph topology. The empirical findings validate the initial hypothesis regarding the CGE algorithm's high sensitivity to its hyperparameters, as evidenced by the complex, non-convex search landscape characterized by numerous local optima. Furthermore, the comparative analysis reveals that while both *Simulated Annealing* and the *Genetic Algorithm* produce solutions of comparable quality, the former offers significantly reduced computational overhead. Importantly, despite focusing on the CGE algorithm, the generality of the proposed tuning framework extends to a broad class of Graph Embedding techniques, contingent on appropriate implementation of recovery matrices and RMSE-based evaluation.

6 Future Work

In future work, we intend to explore the integration of adaptive or self-tuning metaheuristics to further enhance convergence efficiency in high-dimensional search spaces. Additionally, the methodology will be extended to support non-Euclidean embedding spaces, potentially improving representation quality for graphs exhibiting complex topologies. Furthermore, we plan to evaluate the methodology using real-world graph data, such as those derived from social network analysis, to better understand its performance in practical scenarios.

Acknowledgments. This study was funded by public research projects of the Spanish Ministry of Science and Innovation PID2023-151605OB-C22 and the project under the call PEICTI 2021-2023 with the identifier TED2021-131520B-C22.

References

1. Belkin, M., Niyogi, P.: Laplacian eigenmaps for dimensionality reduction and data representation. Neural Comput. **15**(6) (2003). https://doi.org/10.1162/089976603321780317
2. Cao, S., Lu, W., Xu, Q.: GraRep: learning graph representations with global structural information. In: International Conference on Information and Knowledge Management, Proceedings (2015). https://doi.org/10.1145/2806416.2806512
3. Cao, S., Lu, W., Xu, Q.: Deep neural networks for learning graph representations. In: 30th AAAI Conference on Artificial Intelligence, AAAI 2016 (2016). https://doi.org/10.1609/aaai.v30i1.10179
4. Goyal, P., Ferrara, E.: Graph embedding techniques, applications, and performance: a survey. Knowl. Based Syst. **151** (2018). https://doi.org/10.1016/j.knosys.2018.03.022
5. Grover, A., Leskovec, J.: Node2vec: scalable feature learning for networks. In: Proceedings of the ACM SIGKDD International Conference on Knowledge Discovery and Data Mining (2016). https://doi.org/10.1145/2939672.2939754
6. Hall, K.M.: An r-dimensional quadratic placement algorithm. Manage. Sci. **17**(3) (1970). https://doi.org/10.1287/mnsc.17.3.219
7. Hamilton, W.L., Ying, R., Leskovec, J.: Inductive representation learning on large graphs. In: Advances in Neural Information Processing Systems (2017)
8. Hu, J., Cao, L., Li, T., Dong, S., Li, P.: GAT-LI: a graph attention network based learning and interpreting method for functional brain network classification. BMC Bioinf. **22**(1) (2021). https://doi.org/10.1186/s12859-021-04295-1
9. Khemani, B., Patil, S., Kotecha, K., Tanwar, S.: A review of graph neural networks: concepts, architectures, techniques, challenges, datasets, applications, and future directions. J. Big Data **11**(1) (2024). https://doi.org/10.1186/s40537-023-00876-4
10. Kipf, T.N., Welling, M.: Semi-supervised classification with graph convolutional networks. In: 5th International Conference on Learning Representations, ICLR 2017 - Conference Track Proceedings (2017)
11. Luo, D., Ding, C., Nie, F., Huang, H.: Cauchy graph embedding. In: Proceedings of the 28th International Conference on Machine Learning, ICML 2011 (2011)
12. Ng, A.Y., Jordan, M.I., Weiss, Y.: On spectral clustering: analysis and an algorithm. In: Advances in Neural Information Processing Systems (2002)
13. Ouyang, M., Zhang, Y., Xia, X., Xu, X.: GraRep++: flexible learning graph representations with weighted global structural information. IEEE Access **11** (2023). https://doi.org/10.1109/ACCESS.2023.3313411
14. Perozzi, B., Al-Rfou, R., Skiena, S.: DeepWalk: online learning of social representations. In: Proceedings of the ACM SIGKDD International Conference on Knowledge Discovery and Data Mining (2014). https://doi.org/10.1145/2623330.2623732
15. Qiu, J., Dong, Y., Ma, H., Li, J., Wang, K., Tang, J.: Network embedding as matrix factorization: unifying DeepWalk, LINE, PTE, and node2vec. In: WSDM 2018 - Proceedings of the 11th ACM International Conference on Web Search and Data Mining (2018). https://doi.org/10.1145/3159652.3159706
16. Roweis, S.T., Saul, L.K.: Nonlinear dimensionality reduction by locally linear embedding. Science **290**(5500) (2000). https://doi.org/10.1126/science.290.5500.2323
17. Tang, J., Qu, M., Wang, M., Zhang, M., Yan, J., Mei, Q.: LINE: large-scale information network embedding. In: WWW 2015 - Proceedings of the 24th International Conference on World Wide Web (2015). https://doi.org/10.1145/2736277.2741093

18. Wang, D., Cui, P., Zhu, W.: Structural deep network embedding. In: Proceedings of the ACM SIGKDD International Conference on Knowledge Discovery and Data Mining, vol. 13 (2016). https://doi.org/10.1145/2939672.2939753
19. Zhou, J., et al.: Graph neural networks: a review of methods and applications. AI Open (2020). https://doi.org/10.1016/j.aiopen.2021.01.001
20. Zhou, Z., Amini, A.A.: Analysis of spectral clustering algorithms for community detection: the general bipartite setting. J. Mach. Learn. Res. **20** (2019)

Multi-robot Path Planning Problem with Multiple Destinations Applied to Warehouse Environments

Diego Villa García, Enol García González(✉), and José R. Villar

Computer Science Department, University of Oviedo, Oviedo, Spain
{garciaenol,villarjose}@uniovi.es

Abstract. This paper addresses a variant of the Multi-Robot Path Planning (MPP) problem in which each robot must visit multiple destinations within a known environment. The problem formulation combines classical MPP with the Traveling Salesman Problem (TSP), introducing the additional challenge of determining the optimal order in which destinations are visited. To solve this, a two-phase hybrid approach is proposed. First, multiple route alternatives are generated using the A* algorithm. Then, a genetic algorithm is applied to select the best combination of routes and visiting orders. This work proposes two alternatives for the second phase: a single-objective genetic algorithm and a multiobjective genetic algorithm. The experimentation evaluates and compares the performance of both approaches with different fitness functions. We conclude from the results that the multiobjective approach based on NSGA-II is the best in all the evaluated metrics: length, number of collisions, and execution time.

Keywords: Multi Robot Path planning · real-world application · multiple destinations · A* algorithm · genetic algorithm

1 Introduction

Nowadays, using robots to automate transport tasks is increasingly common in environments such as hospitals [1,10] and large logistics centers [3,7]. Robots must operate in shared spaces without colliding in many of these contexts, leading to the Multi-robot Path Planning (MPP) problem. This problem has diverse applications, including surveillance [13], intelligent laboratories [14], and space exploration [9].

In this work, MPP is addressed in the context of warehouse logistics [3,7], where the goal is to compute safe and efficient routes for robots in a known environment with predefined obstacles and start and goal points. The particularity of this work is that it poses this problem in an extended form combining it with another well known in the literature as is the case of the Traveling Salesman Problem (TSP). This combination arises by including multiple destination

E. Corchado et al. (Eds.): SOCO 2025, CCIS 2806, pp. 24–34, 2026.
https://doi.org/10.1007/978-3-032-19763-4_3

points for each robot so that the optimization will be based on two aspects: i) the length of the paths, and ii) the order in which to visit the points.

Two main approaches dominate the literature for those problems: heuristic algorithms and metaheuristics. Heuristic algorithms are mainly represented by A* [8], and some extensions of A* for multirobot environments such as Windowed HCA* [12], DLite [11], FieldD* [5], and ThetA* [4]. Metaheuristic algorithms include methods like Differential Evolution [2], Ant Colony Optimization (ACO) [6], or the Grey Wolf Optimizer (GWO) [15]. In this work, we combine the two approaches, the heuristic using A* and the metaheuristic using genetic algorithms to solve the problem.

The structure of this paper is as follows. Section 2 describes the problem posed by combining the MPP problem and the TSP. Then, Sect. 3 describes the proposed solution to the problem. Section 4 details how the experimentation will be carried out to validate the proposed solution. The results of this experimentation are presented and discussed in Sect. 5. Finally, Sect. 6 presents the conclusions and the lines of future work to be developed.

2 The Problem

The classical approach to the MPP problem consists of establishing the set of paths that a set of robots must follow to reach a goal point from their start position. This problem is posed as an optimization problem, since the length of the paths established to reach the destination must be minimized while guaranteeing that no collisions occur.

On the other hand, the Traveling Salesman Problem (TSP) is another optimization problem in which it is posed that there is a travelling salesman who must visit a set of cities, and it is necessary to determine in which order he visits them in order to minimize the time required to complete all the cities.

This work combines both problems. The classical MPP problem approach assigns a target point to each robot in the problem. In this work, each robot will have multiple points that it must visit, and, as in the TSP, it must be determined in what order the points are visited in order to complete all of them in the shortest possible time. Formally, this combined MPP problem and TSP can be described as M, R, S, G, where:

- M is a representation of the environment, which is represented by a 2D graph in which the robots will move. This 2D environment will contain the information about the static obstacles that the routes must avoid.
- R is the set of robots in the problem $r_1, r_2, ..., r_n$.
- S is the set of initial points for each robot. $s_1, s_2, ..., s_n$, where s_i is the starting point for the robot r_i.
- G is the set of goal points that the robot must visit to complete the task. Each robot will have a number m of points to visit so the structure of this set will be: $g_1^1, g_2^1, ..., g_m^1, g_1^2, g_2^2, ..., g_m^2, ..., g_m^n$, where g_j^i is a point for robot r_i.

The solutions to this problem must establish a route for each robot in R such that, starting at its corresponding point in the set S, it passes through all the points of the corresponding subset in G, avoiding the obstacles described in M and the positions of the rest of the robots. The problem must establish the order of the points to visit in G and the routes to use in such a way that the total length of all paths used is minimized.

3 Proposal

A solution composed of two main phases has been designed to address this task. The first phase uses the A* search algorithm to pre-compute multiple alternative routes between all relevant pairs of points for each robot (starting positions and destinations). The second phase employs a genetic algorithm to choose the best combination of routes to solve the problem.

3.1 A* Algorithm

The first phase of the algorithm consists of generating a set of alternative routes between all pairs of interest for each robot. This includes routes from the robot's initial position to each of its destinations and routes between all pairs of destinations assigned to that robot. The goal is to have several viable alternatives that can be evaluated and combined later during optimization.

For this task, the A* informed search algorithm, widely used in discrete environments for its efficiency and its ability to find optimal routes in terms of cumulative cost, is employed. In this work, the cost between adjacent cells is considered constant, and the Manhattan distance has been used as a heuristic, since robots can only move in four directions (up, down, left, and right) on a two-dimensional grid.

An important modification to the standard A* algorithm is obtaining not only the optimal path but also the K best paths between each pair of points. This technique, known as k-shortest path routing, is achieved by extending the algorithm beyond the first route found: after registering a valid solution, the search continues until other suboptimal variants have been explored, avoiding cycles or redundant paths. This results in a set of routes ordered by increasing length, with greater spatial diversity.

3.2 Genetic Algorithm

The second phase of the system aims to find optimal planning for a set of robots that must visit multiple destinations without collisions, using the routes previously generated using the A* algorithm. For this phase of the proposal, a genetic algorithm is proposed since it is a clear example of combinatorial optimization.

When looking for the best solution, evaluating and optimizing based on different criteria defined below, such as the number of collisions and the length of paths, is desired. Since this problem has multiple objectives that need to be optimized, the use of NSGA-II (Non-dominated Sorting Genetic Algorithm II) is proposed. NSGA-II is a multi-objective genetic algorithm that does not seek a single optimal solution, but a set of solutions that constitute the Pareto frontier: those that cannot be improved in one of the objectives without worsening another. This approach will also be compared to a traditional genetic algorithm in which only one objective function is evaluated to determine which approach is able to achieve better results.

Representation of the Solutions. Each individual in the genetic algorithm population, represents a complete robot schedule. Since each robot must visit multiple destinations in a given order and can choose among several precomputed routes for each path, the representation of the solutions must encode both the order of visitation and the selected route variants.

For this purpose, a matrix is used, where each row corresponds to a robot, and each column represents one of its destinations. The content of each cell is a tuple (d, v), where d indicates the destination identified by its index, and v is the index of the route variant chosen among the K routes previously generated by A*.

(2,1)	(1,4)	(3,5)
(1,4)	(3,1)	(2,3)
(1,3)	(2,5)	(3,2)

Fig. 1. Example of the representation of a solution.

Figure 1 contains an example of how a solution could be represented as a matrix. In this matrix, robot one first visits destination 2 with variant route 1, destination 1 with variant 4, and destination 3 with variant 5. This scheme allows the algorithm to evaluate solutions that vary both in the order of destinations and in the physical routes used.

Fitness Functions. The genetic algorithm uses what is known as a fitness function when comparing and determining the best solution to the problem. The different fitness functions used are:

- F_c. Which represents only the number of collisions in the solution.
- F_l. Which represents the total aggregate length of all routes. That is: $F_l = \sum_i^N \sum_j^K d_{ij}$, where d_{ij} is the distance of robot i in segment j. And therefore, N is the number of robots and K is the number of points it visits.
- F_1. It is a combined function that takes into account the two criteria and is calculated as follows: $F_1 = \sum_i^N \sum_j^K d_{ij} + pc$, where p is an adjustment factor to weight the priority of one criterion over another, c are the number of collisions, andd_{ij} is the distance of robot i in segment j. And therefore, N is the number of robots and K is the number of points it visits.

In the single-objective genetic algorithm, a only one of those functions are used, while NSGA-II consider both the length and the number of collisions.

Operators. One of the most important parts of the genetical algorithm is the operators to generate new solutions and to select which individuals remains in the population between two generations. Two crossover and mutation operators have been defined, inspired by classical techniques in the literature, to generate new solutions. These two crossover and mutation operators will be the same for the two proposals, both for the experimentation with NSGA-II and the experimentation with the single-target genetic algorithm.

The crossover operator developed generates a new individual by randomly selecting rows from each of the parents. As each row is identified with a robot, the operator generates an individual by combining the information from the different robots. Figure 2 contains an example of the execution of this crossover operator, where a new individual is generated containing the first and last row of the first parent and the intermediate row of the second parent.

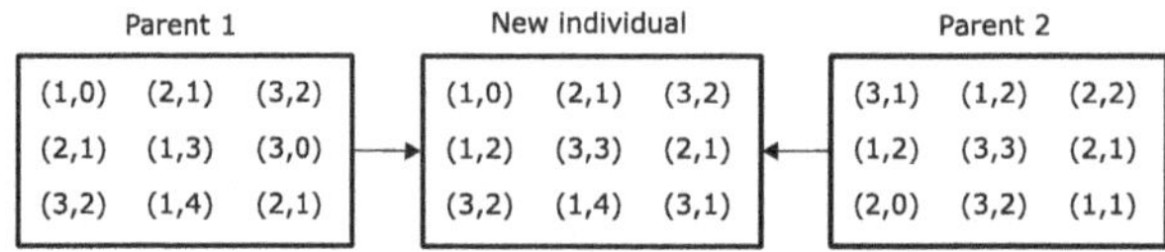

Fig. 2. Example of how the crossover operator works.

The designed mutation operator mutates the solution in two ways. On the one hand, a mutation is performed at the destination level, in which the two robot destinations are exchanged. On the other hand, a path-level mutation is performed, which consists of randomly changing the path variant used to reach a destination. Figure 3 shows an example of the two mutation strategies. In Fig. 3a, it can be seen that two tuples of the first robot are exchanged, and in Fig. 3b it can be seen that what is mutated is the path variant selected for the second destination of the first robot.

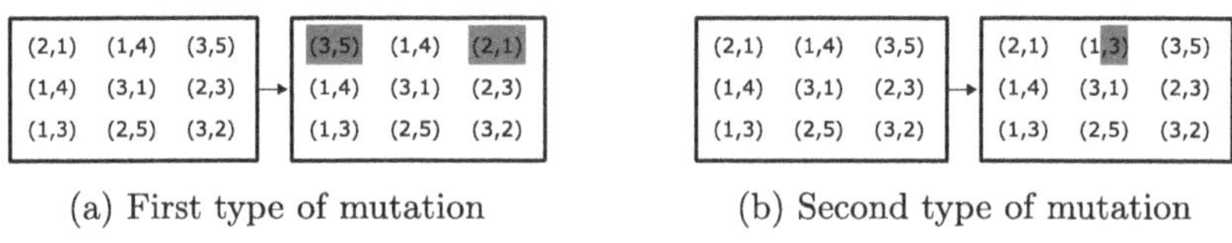

(a) First type of mutation (b) Second type of mutation

Fig. 3. Graphical examples of the different strategies for the mutation.

The last operator to consider for the genetic algorithm is selection. This is different between the single-objective genetic algorithm and the NSGA-II approach.

In the case of the genetic algorithm that works only with an objective function, it has been proposed to use a selection strategy well known in the literature, based on a tournament selection with elite. This selection consists of first separating the elite solutions, which are the best compared to fitness. Then, we perform a tournament that confront two solutions and keep the best solution.

On the other hand, in the case of the NSGA-II algorithm, which works with multiple targets simultaneously, selection is based on sorting by dominance and population diversity.

4 Experimental Setup

The experimentation to validate and compare the elaborated proposals will consist of solving different instances of the problem with different numbers of robots and scenarios. The different scenarios used for the experimentation are:

- Scenario 1 (Fig. 4a). This is a room of 10×30 squares in which there are different types of internal obstacles that generate different situations and conflicts between the robots.
- Scenario 2 (Fig. 4c). This is a set of 4 rooms of the same size as scenario 1. The 4 rooms are connected by a long narrow corridor into which only one robot wide enters.
- Scenario 2b (Fig. 4d). This scenario is a variant of the previous scenario in which the empty rooms are replaced by replicated rooms from scenario 1.
- Scenario 3 (Fig. 4b). This scenario represents a real environment, specifically a warehouse. This scenario has been included to evaluate what the performance of our proposal would be like in a physical system.

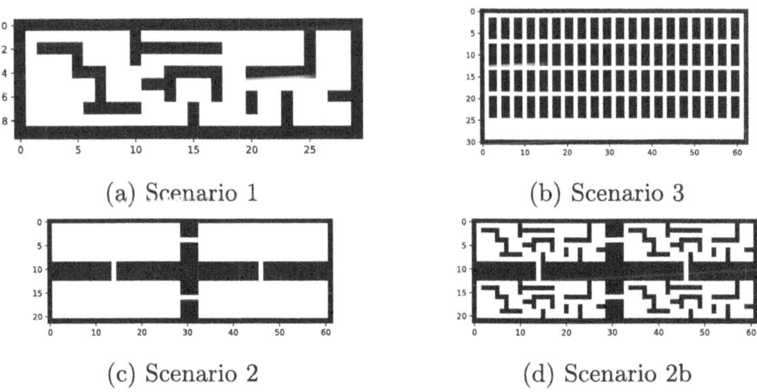

(a) Scenario 1 (b) Scenario 3

(c) Scenario 2 (d) Scenario 2b

Fig. 4. Graphical representation in the form of a matrix of the different scenarios used in the experimentation.

All scenarios have been tested with a number of robots between 3 and 7. To complete the task, each robot will have to visit 10 points. Three metrics will be extracted from each experiment: the sum of the lengths of all the paths in the best solution found. The number of collisions in the best solution found and the execution time required. Each instance of the problem will be executed 10 times.

Before running the full experimentation, a parameter search was carried out to determine which combination of parameters made the genetic algorithm perform best. In this parameter search it was established that the best population size was 100 and the best probabilities for the crossover and mutation operators were 65% and 30% respectively.

5 Results and Discussion

The execution was carried out as indicated in the previous section and the following metrics were obtained.

Table 1 shows the results of the aggregate sum of distances for each problem instance and method. This metric sums the length of all the paths of all the robots. It can be observed that the genetic algorithm obtains worse results than the NSGA-II proposal. The results of NSGA-II are better than those of the

Table 1. Mean (MN) and Standard Deviation (SD) of the aggregate length results for each method and problem instance in the experiment

		Genetic Algorithm						NSGA-II	
	Number	F_c		F_l		F_1		F_c & F_l	
	robots	MN	SD	MN	SD	MN	SD	MN	SD
Scenario 1	3	1030.70	84.30	555.60	20.87	664.90	54.06	**248.00**	13.30
	4	1273.30	63.66	726.60	26.74	1166.50	64.90	**290.90**	18.62
	5	1702.00	67.57	1032.60	27.73	1579.40	58.94	**433.90**	21.31
	6	2044.00	108.53	1245.30	44.75	1934.10	142.27	**499.30**	43.49
	7	2296.80	104.70	1433.80	51.11	2173.10	127.22	**670.90**	49.77
Scenario 2	3	1482.80	105.91	828.30	41.82	941.70	39.52	**489.20**	28.31
	4	1868.60	99.01	1147.10	72.62	1334.20	56.62	**569.80**	22.52
	5	2373.70	103.76	1539.90	52.46	2022.10	107.71	**753.00**	28.98
	6	2803.30	108.70	1842.00	55.86	2627.90	200.03	**929.60**	68.44
	7	3158.10	119.60	2167.80	83.21	2982.70	172.11	**1054.90**	50.60
Scenario 2b	3	1774.80	125.28	1026.50	43.66	1257.00	55.26	**603.20**	25.79
	4	2171.20	115.22	1353.10	46.36	1868.40	130.32	**752.00**	32.92
	5	2611.50	125.50	1797.90	77.39	2538.80	119.61	**969.40**	42.68
	6	3278.60	91.93	2309.10	67.16	3198.80	191.42	**1279.40**	49.79
	7	3835.70	186.94	2651.60	50.37	3590.30	154.43	**1440.20**	75.49
Scenario 3	3	1485.20	101.81	823.30	36.31	899.70	28.99	**473.50**	26.04
	4	2065.10	134.87	1255.90	70.77	1412.20	64.64	**653.90**	16.23
	5	2512.70	170.15	1571.60	87.75	1922.70	83.45	**776.50**	52.95
	6	2869.50	96.04	1886.30	90.13	2429.00	107.76	**925.30**	52.87
	7	3620.80	138.74	2397.90	84.68	3290.80	261.74	**1125.70**	48.72

Table 2. Mean (MN) and Standard Deviation (SD) of the number of collisions for each method and problem instance in the experiment

	Number	Genetic Algorithm						NSGA-II	
		F_c		F_l		F_1		F_c & F_l	
	robots	MN	SD	MN	SD	MN	SD	MN	SD
Scenario 1	3	0.0	0.0	4.3	2.1	0.0	0.0	0.0	0.0
	4	0.1	0.3	11.9	7.1	0.3	0.5	0.0	0.0
	5	3.4	1.6	22.7	11.0	3.0	1.2	0.0	0.0
	6	5.5	2.1	23.6	11.1	5.2	2.0	0.0	0.0
	7	9.9	3.1	41.6	9.1	11.1	2.7	0.4	0.7
Scenario 2	3	0.0	0.0	1.2	2.5	0.0	0.0	0.0	0.0
	4	0.0	0.0	5.2	4.0	0.0	0.0	0.0	0.0
	5	0.0	0.0	7.0	4.5	0.0	0.0	0.0	0.0
	6	0.0	0.0	14.9	13.4	0.1	0.3	0.0	0.0
	7	1.0	0.8	18.9	8.1	1.3	1.2	0.0	0.0
Scenario 2b	3	0.0	0.0	2.0	2.9	0.0	0.0	0.0	0.0
	4	0.0	0.0	7.5	2.6	0.0	0.0	0.0	0.0
	5	0.4	0.5	10.0	8.0	0.0	0.0	0.0	0.0
	6	3.2	1.3	15.8	5.0	2.2	1.0	0.1	0.3
	7	4.3	1.5	18.1	8.2	5.2	1.6	0.1	0.3
Scenario 3	3	0.0	0.0	0.7	0.8	0.0	0.0	0.0	0.0
	4	0.0	0.0	3.5	3.5	0.0	0.0	0.0	0.0
	5	0.0	0.0	6.9	8.6	0.0	0.0	0.0	0.0
	6	0.0	0.0	5.5	3.3	0.0	0.0	0.0	0.0
	7	0.3	0.5	13.8	9.7	0.1	0.3	0.0	0.0

single-function genetic algorithm, even when the genetic algorithm only aims to minimize the length. This could may be because NSGA-II maintain a large diversity of solutions that broadly represent the search space, which makes it a method more prone to explore a greater variety of solutions.

Table 2 shows the results on the number of collisions that remains in the best solution achieved. Again, the method that demonstrates the best performance is NSGA-II, since it usually achieves collision-free solutions. There are only three cases where the average number of collisions is not exactly 0. Even for those instances, it can be seen that the mean values over 10 runs were 0.1 and 0.3, with a considerable standard deviation. That denotes that in most of the runs, the number of collisions was 0. On the other hand, the genetic algorithm could not compete with these low collision counters even when it only had the objective function of minimizing the number of collisions.

The results regarding the average execution time are presented in Table 3. This table shows that, again, NSGA-II wins by a landslide over the genetic algorithm, since its execution time is approximately half in all the problem instances evaluated.

Analyzing the overall results of all the metrics raised and compared, NSGA-II is the best proposal to address the problem posed. NSGA-II has shown better results in both length and number of collisions, the two fundamental metrics associated with the problem. In addition, it has also stood out for its speed of execution, requiring a shorter execution time for all the evaluated instances.

Table 3. Mean (MN) and Standard Deviation (SD) of the execution time for each method and problem instance in the experiment

		Genetic Algorithm						NSGA-II	
	Number	F_c		F_l		F_1		F_c & F_l	
	robots	MN	SD	MN	SD	MN	SD	MN	SD
Scenario 1	3	7.936	0.746	5.805	0.562	6.608	1.033	3.023	0.063
	4	11.823	1.341	8.401	0.731	10.698	1.166	4.107	0.136
	5	14.108	1.551	11.226	1.332	12.518	1.281	5.069	0.041
	6	17.554	1.112	13.333	1.05	17.432	1.913	6.259	0.261
	7	20.383	1.025	14.949	1.209	19.544	1.516	7.419	0.299
Scenario 2	3	6.134	0.708	5.899	0.703	5.835	0.564	3.003	0.008
	4	8.453	0.821	7.108	0.474	7.499	0.864	4.018	0.022
	5	9.836	0.768	8.966	0.500	10.752	0.895	5.036	0.037
	6	11.946	0.789	11.284	1.004	12.170	1.417	6.118	0.176
	7	14.807	1.381	12.900	0.793	14.573	0.871	7.128	0.096
Scenario 2b	3	5.269	0.653	5.237	0.471	5.508	0.708	3.046	0.074
	4	7.189	0.644	7.057	0.551	7.565	1.134	4.061	0.076
	5	9.110	0.428	8.158	0.294	8.904	0.948	5.090	0.108
	6	10.987	0.806	10.848	0.716	10.841	0.709	6.228	0.272
	7	13.638	0.770	12.742	0.755	13.377	0.988	7.186	0.189
Scenario 3	3	5.347	0.276	4.858	0.505	5.236	0.607	3.004	0.009
	4	7.490	0.706	6.757	0.314	7.142	0.648	4.036	0.094
	5	9.736	0.593	8.860	0.559	9.874	0.788	5.000	0.094
	6	12.223	0.672	11.347	1.028	11.978	1.537	6.035	0.040
	7	13.402	0.737	12.628	0.878	13.787	1.166	7.075	0.118

6 Conclusions

In this paper, we presented a new proposal for a classic problem in the world of robots that represents the combination of two classic problems in the literature: the MPP problem and the TSP. This approach to the robot problem for task and route assignment is accompanied by two proposals based on genetic algorithms: single-objective genetic algorithm and NSGA-II. In the experimentation, both proposals were compared, and it was observed that the proposal based on the multiobjective algorithm achieved the best results in all the metrics evaluated.

As future work develops from this work, two lines of work are proposed. On the one hand, the application of other non-evolutionary metaheuristic techniques to address and solve this problem, such as GWO or ACO. On the other hand, we propose to solve this problem in different types more complex, for example, in a open scenario with freedom of movements. It is also considered to develop a study about how the size of the initial path base can affect the results of the genetic algorithm.

Acknowledgments. This research has been funded by the Spanish Research Agency –grant PID2023-146257OB-I00–. Also, by Principado de Asturias, grant IDE/2024/000734, and by the Council of Gijón through the University Institute of Industrial Technology of Asturias grants SV-25-GIJÓN-1-23.

References

1. Causse, O., Pampagnin, L.: Management of a multi-robot system in a public environment. In: Proceedings 1995 IEEE/RSJ International Conference on Intelligent Robots and Systems. Human Robot Interaction and Cooperative Robots, vol. 2, pp. 246–252 (1995). https://doi.org/10.1109/IROS.1995.526168
2. Chakraborty, J., Konar, A., Jain, L.C., Chakraborty, U.K.: Cooperative multi-robot path planning using differential evolution. J. Intell. Fuzzy Syst. **20**, 13–27 (2009). https://doi.org/10.3233/IFS-2009-0412
3. Chen, X., Li, Y., Liu, L.: A coordinated path planning algorithm for multi-robot in intelligent warehouse. In: 2019 IEEE International Conference on Robotics and Biomimetics (ROBIO), pp. 2945–2950 (2019). https://doi.org/10.1109/ROBIO49542.2019.8961586
4. Daniel, K., Nash, A., Koenig, S., Felner, A.: Theta*: any-angle path planning on grids. J. Artif. Intell. Res. **39**, 533–579 (2010). https://doi.org/10.1613/jair.2994
5. Ferguson, D., Stentz, A.: Using interpolation to improve path planning: the field d* algorithm. J. Field Robot. **23**, 79–101 (2006). https://doi.org/10.1002/rob.20109
6. Gul, F., Rahiman, W., Alhady, S.S.N., Ali, A., Mir, I., Jalil, A.: Meta-heuristic approach for solving multi-objective path planning for autonomous guided robot using PSO–GWO optimization algorithm with evolutionary programming. J. Ambient Intell. Humanize Comput. **12**, 7873–7890 (2021). https://doi.org/10.1007/s12652-020-02514-w
7. Han, S.D., Yu, J.: Effective heuristics for multi-robot path planning in warehouse environments. In: 2019 International Symposium on Multi-Robot and Multi-Agent Systems (MRS), pp. 10–12 (2019). https://doi.org/10.1109/MRS.2019.8901065
8. Hart, P.E., Nilsson, N.J., Raphael, B.: A formal basis for the heuristic determination of minimum cost paths. IEEE Trans. Syst. Sci. Cybern. **4**, 100–107 (1968). https://doi.org/10.1109/TSSC.1968.300136
9. Huang, D., Jiang, H., Yu, Z., Kang, C., Hu, C.: Leader-following cluster consensus in multi-agent systems with intermittence. Int. J. Control Autom. Syst. **16**, 437–451 (2018). https://doi.org/10.1007/s12555-017-0345-2

10. Huang, X., Cao, Q., Zhu, X.: Mixed path planning for multi-robots in structured hospital environment. J. Eng. **2019**(14), 512–516 (2019). https://doi.org/10.1049/joe.2018.9409
11. Koenig, S., Likhachev, M.: Fast replanning for navigation in unknown terrain. IEEE Trans. Rob. **21**, 354–363 (2005). https://doi.org/10.1109/TRO.2004.838026
12. Silver, D.: Cooperative pathfinding. In: Proceedings of the First AAAI Conference on Artificial Intelligence and Interactive Digital Entertainment (AIIDE 2005), p. 117–122 (2005). https://doi.org/10.1609/aiide.v1i1.18726
13. Stump, E., Michael, N.: Multi-robot persistent surveillance planning as a vehicle routing problem. In: 2011 IEEE International Conference on Automation Science and Engineering, pp. 569–575 (2011). https://doi.org/10.1109/CASE.2011.6042503
14. Tan, Q., Denojean-Mairet, M., Wang, H., Zhang, X., Pivot, F.C., Treu, R.: Toward a telepresence robot empowered smart lab. Smart Learn. Environ. **6**, 5 (2019). https://doi.org/10.1186/s40561-019-0084-3
15. Zheng, Y., Luo, Q., Wang, H., Wang, C., Chen, X.: Path planning of mobile robot based on adaptive ant colony algorithm. J. Intell. Fuzzy Syst. **39**, 5329–5338 (2020). https://doi.org/10.3233/JIFS-189018

Geometric Algebra as a Language for Physics: Application to Rigid Body Dynamics

David Muñoz Hernández, Carmen García Barceló, Antonio Jimeno-Morenilla, and Higinio Mora(✉)

Computer Science Technology and Computation, University of Alicante, Alicante, Spain
{david.mhernandez,jimeno,hmora}@ua.es, cgb72@gcloud.ua.es

Abstract. This paper explores the application of geometric algebra (or Clifford algebra) as a powerful and intuitive mathematical framework for physics, particularly in the domain of rigid body dynamics. Traditionally underutilized in physics curricula, geometric algebra, since its resurgence through the work of David Hestenes, offers a compelling alternative to vector calculus and differential geometry. We highlight its computational advantages in describing rotations using rotors and its conceptual clarity in formulating physical quantities. The paper develops the foundational elements of geometric algebra, applies these concepts to rewrite fundamental aspects of rigid body dynamics such as angular momentum and the inertia tensor, and presents an analytical and computational solution of a symmetric top using the `Clifford` Python library, demonstrating the simplicity and efficiency gained by employing this algebraic structure.

Keywords: Geometric Algebra · Clifford Algebra · Rigid Body Dynamics · Rotors · Python · Computational Physics · Computer Science

1 Introduction

The conventional mathematical toolkit of physics, heavily reliant on vector algebra as formulated by Gibbs and Heaviside, has been instrumental in our understanding of the physical world. However, geometric algebra emerges as a unifying framework that integrates scalar, vector, and higher-order entities into a single algebraic structure. Rediscovered and championed by David Hestenes, notably in his work "New Foundations for Classical Mechanics" [5], geometric algebra provides a more geometric and computationally advantageous language for various areas of physics, including classical mechanics, relativity, and quantum mechanics. Its inherent description of rotations through rotors offers significant benefits over matrix representations, finding increasing relevance in computer science and engineering. Moreover, this mathematical structure shows significant potential

E. Corchado et al. (Eds.): SOCO 2025, CCIS 2806, pp. 35–44, 2026.
https://doi.org/10.1007/978-3-032-19763-4_4

to solve complex problems in which artificial intelligence (AI) is involved, such as image processing [2] and deep learning algorithms [1].

The novel contribution of this work lies in providing an explicit computational and analytical demonstration of how geometric algebra can be applied to rigid body dynamics, going beyond traditional vector calculus approaches. By combining the theoretical formulation with an open-source Python implementation, the aim is to highlight the practical advantages and versatility of geometric algebra, especially for problems that arise in physics and AI.

1.1 Ease of Use: Motivation for Geometric Algebra

Geometric algebra distinguishes itself by its intuitive geometric interpretation of algebraic operations and entities. Unlike the somewhat ad-hoc nature of vector products (scalar and vector), the geometric product unifies these into a single, invertible operation. This product, defined as the sum of the inner (scalar) and outer (wedge) products of vectors ($ab = a{\cdot}b{+}a{\wedge}b$), naturally encodes information about both the magnitude and the relative orientation of vectors. The outer product, $a \wedge b$, directly represents the oriented plane (bivector) spanned by the vectors, offering a coordinate-free description of areas. Furthermore, rotations in geometric algebra are elegantly handled by rotors, which are elements of the algebra that act on multivectors via a double-sided transformation ($a' = RaR^{\dagger}$). This rotor-based approach is dimension independent and avoids the complexities associated with Euler angles and matrix representations, leading to simpler and more computationally efficient algorithms to handle rotations [3].

1.2 Methodology Summary

This paper begins by laying the foundations of geometric algebra, introducing the outer and geometric products, the concept of multivectors and their grades, and the properties of bivectors and the pseudoscalar in both 2 and 3 dimensions. Special emphasis is placed on rotors, their construction from reflections, and their application in performing rotations on vectors and other multivectors. Subsequently, we apply these theoretical tools to the domain of rigid body dynamics. We reformulate key quantities such as angular momentum and the inertia tensor within the geometric algebra framework. The equations of motion for a rigid body are then derived using geometric algebra, and the case of a symmetric top under torque-free motion is solved analytically. To further illustrate the power of this approach, a computational solution of the symmetric top is implemented using the `Clifford` Python library, demonstrating the ease with which rotational dynamics problems can be tackled numerically. The results of these simulations, which show phenomena like precession, are presented and compared with expected theoretical behavior.

2 Geometric Algebra Basics

The outer product ($a \wedge b$) of two vectors yields a bivector, representing the oriented area spanned by them (Fig. 1).

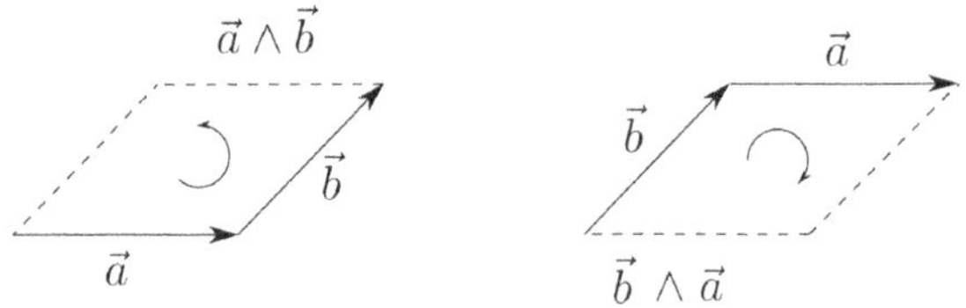

Fig. 1. Bivector representation $\boldsymbol{ab}$, as a result of the outer product of two vectors $(a \wedge b)$.

It is antisymmetric ($a \wedge b = -b \wedge a$) and is defined in any dimension. In 3D with an orthonormal basis $\{e_1, e_2, e_3\}$, a bivector can be expressed as a linear combination of $\{e_2e_3, e_3e_1, e_1e_2\}$. The geometric product ($ab = a \cdot b + a \wedge b$) combines the scalar inner product and the bivector-valued outer product into a single invertible algebraic operation. This product is associative and distributive. Multivectors are general elements of geometric algebra, formed by linear combinations of scalars (grade 0), vectors (grade 1), bivectors (grade 2), trivectors (grade 3), and so on. In 3D, the highest grade element is the pseudoscalar $I = e_1e_2e_3$ (a trivector), which has the property $I^2 = -1$ and acts similarly to the imaginary unit [8]. Furthermore, trivectors can be interpreted as oriented volumes.

2.1 Reflections

To explain rotors properly, the reflection operation has to be defined. Let a be a vector and n a unit vector ($n^2 = 1$). We can decompose the vector a as:

$$a = n^2 a = n(n \cdot a + n \wedge a) = a_{\parallel} + a_{\perp} \tag{1}$$

Then, the projection of a onto n is: $a_{\parallel} = n \cdot an$. On the other hand, the perpendicular part is: $a_{\perp} = nn \wedge a$. Since $a_{\perp}$ is perpendicular to n, then:

$$n \cdot a_{\perp} = \langle nnn \wedge a \rangle = \langle n \wedge a \rangle = 0 \tag{2}$$

where we used the fact that the scalar part of a bivector is zero.

Reflecting a in the plane orthogonal to n gives $a' = a_{\perp} - a_{\parallel}$. For spaces of any dimension:

$$\begin{aligned} a' &= a_{\perp} - a_{\parallel} = nn \wedge a - n \cdot an \\ &= -n \cdot an - n \wedge an = -nan \end{aligned}$$

3 Rotors

Rotations in geometric algebra are generated by rotors, which can be constructed from the product of two unit vectors, $R = nm = n \cdot m + n \wedge m = \cos\theta + B \sin\theta =$

$e^{-B\theta}$. Here, θ is the angle between n and m, and B is the unit bivector in the plane spanned by them.

Rotations can be seen as combinations of successive reflections (Fig. 2). If we have a plane $m \wedge n$, with $m \cdot n = \cos\theta$, then rotations in this plane of a vector a are achieved through successive reflections in hyperplanes perpendicular to m and n. The component $a_{\perp}$ is not affected by reflection, and the angle between a and its reflection a' is 2θ.

Starting from a, the first reflection is $b = -mam$. The second reflection is $c = -nbn$. Therefore, we get:

$$c = -n(-mam)n = nmamn \tag{3}$$

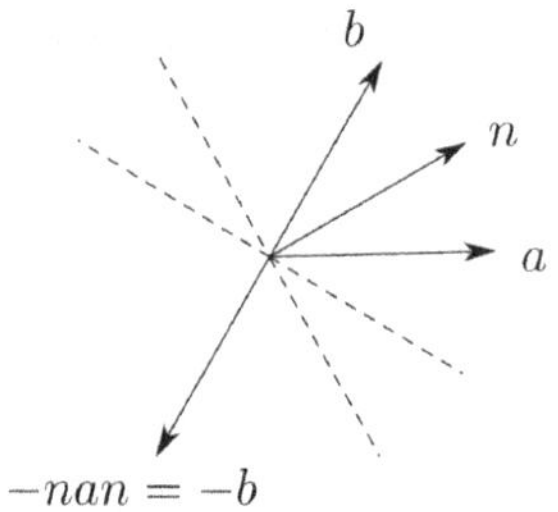

Fig. 2. Scheme for constructing a rotor to take us from a vector a to a vector b (unitary), from two successive reflections.

If we define $R = nm$, we can write the total rotation as:

$$c = RaR^{\dagger} \tag{4}$$

This transformation is a general method for performing rotations, in any dimension, and for any grade of multivector.

A rotation of a vector a by an angle 2θ in the plane B is achieved by the double-sided transformation $a' = RaR^{\dagger}$, where $R^{\dagger}$ is the reverse of the rotor. For a rotation of angle θ, the rotor is given by $R = e^{-B\theta/2}$. Composing successive rotations simply involves multiplying their corresponding rotors ($R = R_2R_1$). This rotor formalism offers a compact and efficient way to represent and manipulate rotations in any dimension, bypassing the complexities of Euler angles and potentially singular parameterizations. This is a great computational advantage, as the rotors vanish the matrix operations from Euler angles in any physical rotation.

4 Rigid Body Dynamics with Geometric Algebra

Within the geometric algebra framework, the angular momentum can be naturally represented as a bivector L, reflecting the plane and magnitude of rotation [9].

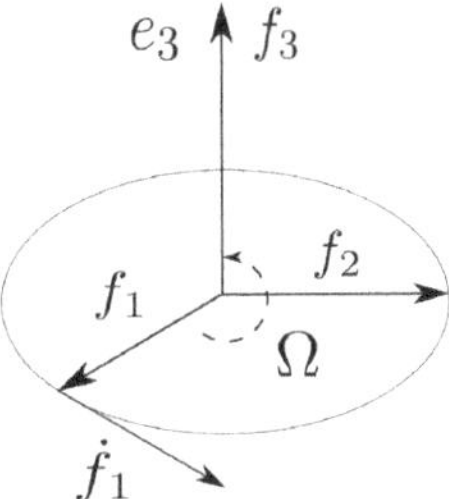

Fig. 3. Angular velocity in geometric algebra representation.

The inertia tensor $I(B)$ becomes a linear map that acts on the bivectors (angular velocity Ω_B) to produce the bivectors (angular momentum L_B in the body frame), $L_B = I(\Omega_B)$. This formalism allows a more geometrically intuitive understanding of the physical quantities involved. In Fig. 3, the angular velocity is represented as a bivector, illustrating how rotation occurs within a specific plane and how this interpretation is related to classical mechanical magnitudes. In this diagram, f_1, f_2, and f_3 represent the basis vectors of a rotating reference frame, not necessarily the principal axes of inertia. However, the moving frame vectors may be related to the body's principal axes. These axes are used throughout the analysis to express components of angular velocity and momentum. The equations of motion for a rigid body, in the absence of external torque, take the form $\dot{L} = N$ [4] [10], which translates to $I(\dot{\Omega}_B) - \Omega_B \times I(\Omega_B) = 0$ in the body frame, where $\times$ is the commutator product. For a symmetric top with principal moments of inertia, the standard notation is $i_1 \leq i_2 \leq i_3$, where the symmetric case corresponds to $i_1 = i_2 \neq i_3$ [4], the torque-free motion can be solved analytically in geometric algebra, revealing the precession of the angular velocity vector around the body's symmetry axes [6].

The inertia ellipsoid, shown in Fig. 4 (a), provides a visual representation of how the distribution of mass determines the rotational response of a rigid body in general terms. Its shape reflects the body's principal moments of inertia, with the axes scaled according to the inverse square roots of the moments.

The symmetric top to be solved analytically via the rotor equations is depicted in Fig. 4 (b). This configuration exemplifies the practical application of the geometric algebra approach, highlighting the elegance and power of rotor-based formulations for rigid body dynamics [7].

5 Computational Implementation with the Clifford Python Library

The `Clifford` Python library provides a powerful tool for performing computations in geometric algebra [11]. It allows for the direct representation of multivectors and the implementation of geometric product, outer product, and inner product with an intuitive syntax. Rotors can be easily defined and applied to

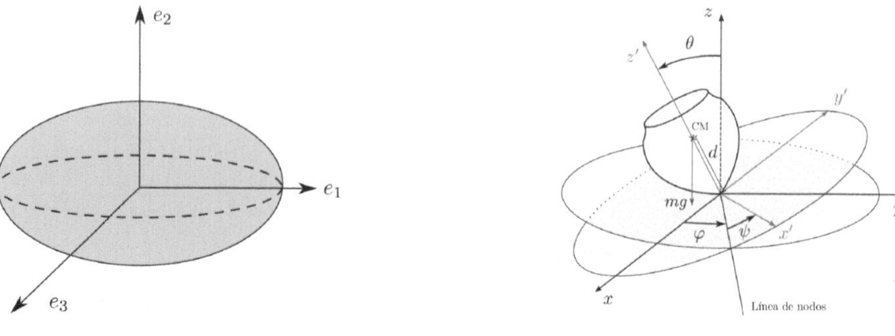

(a) Inertia ellipsoid. (b) Symmetric top system.

Fig. 4. Illustration of the inertia ellipsoid and the symmetric top system.

rotate multivectors. The library was used to simulate the motion of a symmetric top, solving the analytical expressions obtained for the angular velocity as a function of time. By varying the initial angular momentum, the precession of the top's symmetry axes was visualized, demonstrating the library's capability to handle rotational dynamics problems without resorting to solving complex differential equations in generalized coordinates. The conservation of kinetic energy was also verified through the simulation. This computational approach highlights the practical advantages of geometric algebra for simulating physical systems involving rotations.

The implementation described in this work, leveraging the Clifford Python library, can be seamlessly extended to various applications in the AI domain. For example, geometric algebra provides a natural way to represent geometric transformations, which is particularly relevant in computer vision, robotics, and geometric deep learning [12]. In such contexts, encoding and manipulating spatial relationships efficiently and robustly is critical, and the rotor-based formalism can enhance the interpretability and performance of AI models working with spatial or physical data. Our codebase and computational examples serve as a foundation for further integration of geometric algebra techniques into AI-driven research and applications.

6 Results

The simulations of the symmetric top using the `Clifford` library successfully reproduced the expected precessional motion[1]. Two experiments have been made with different initial conditions. For an initial angular momentum in the xy-plane, the principal axis of inertia I_1 exhibits sinusoidal and cosinoidal oscillations in their x and y components (Fig. 5), while the z components remain

[1] Experimentation code: https://github.com/cloudlab-aia/Geometric-Algebra-applied-to-rigid-solid-physics.

relatively constant, consistent with rotation around the z-axis (Figs. 6 (a) and 7 (b)). Nevertheless, the principal axis of Inertia I_3 remains constant, as there is no movement in the xy-plane (Figs. 6 (b) and 7 (a)). It is the case where a symmetric top is rotating around the z-axis without friction. Determining the movement of two axes of inertia, such as I_1 and I_3, is enough to know the motion of this rigid body. Another thing to be taken into account is the initial angular momentum. This physical quantity is what determines the motion of the symmetric top, as it describes where is spinning and how the movement develops. In the first experiment is spinning perpendicular to the floor in the z-axis. The reason why Fig. 7 (b) is not constant is because of the numerical error. In the y axis of the figure is shown that the scale is 10^{-12}, so it is a little significant variation, and it can be considered as a constant in zero.

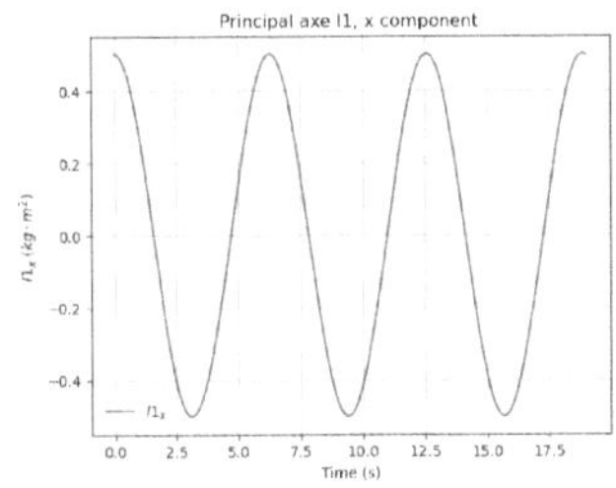

(a) Symmetric top solution to I_1x

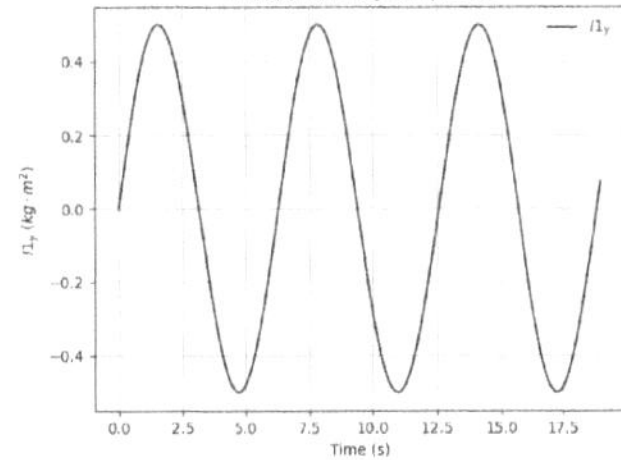

(b) Symmetric top solution to I_1y

Fig. 5. Symmetric top solutions to I_1x and I_1y for the first experiment.

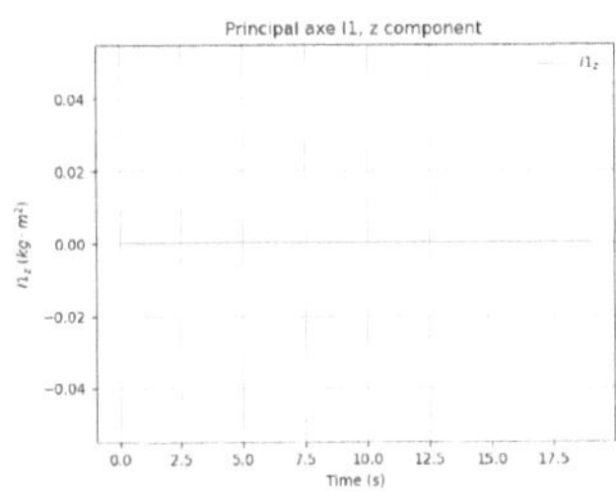

(a) Symmetric top solution to I_1z

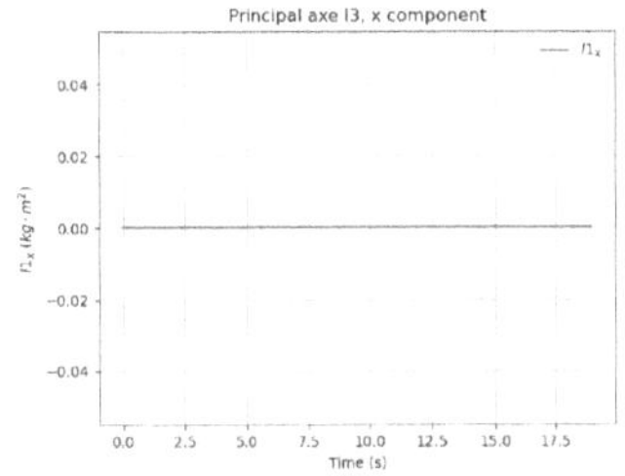

(b) Symmetric top solution to I_3x

Fig. 6. Symmetric top solutions to I_1z and I_3x for the first experiment.

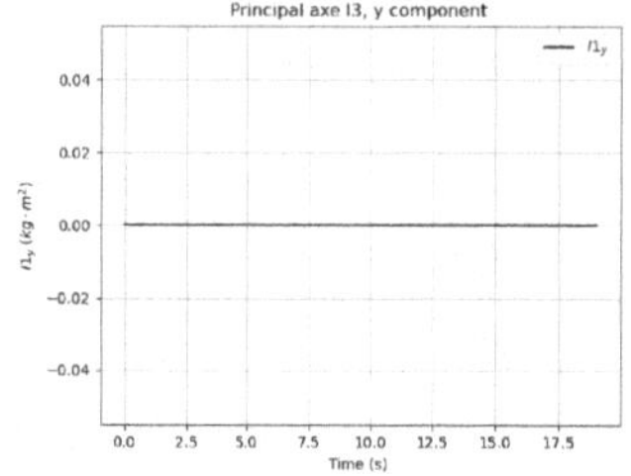

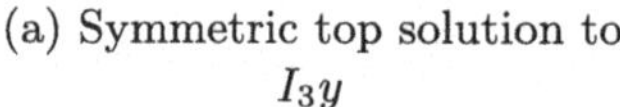
(a) Symmetric top solution to I_3y

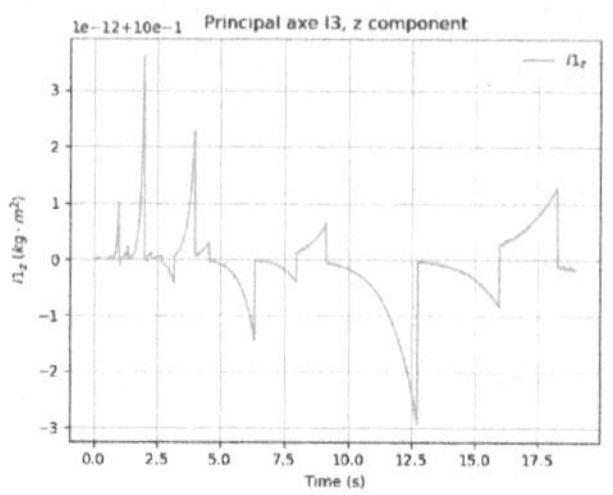

(b) Symmetric top solution to I_3z

Fig. 7. Symmetric top solutions to I_3y and I_3z for the first experiment.

In the second experiment, the rigid body is no longer spinning perpendicular to the floor. When a small yz-component is introduced to the initial angular momentum, the precession of the I_3 axis becomes evident, with its z-component no longer constant and exhibites a slow oscillation. This oscillation is shown in Fig. 8 (a). It also oscilates in the principal axis I_1 direction due to this modification (Fig. 8 (b)). In this case, I_3x and I_3y are no longer constant and start to oscilate as the top is not perpendicular to the floor and they have a component in this direction (Fig. 9). The kinetic energy remains constant throughout both simulations (Fig. 10), confirming the conservation law for torque-free motion. These results demonstrate the accuracy and ease of simulating rigid body dynamics using geometric algebra and the `Clifford` library.

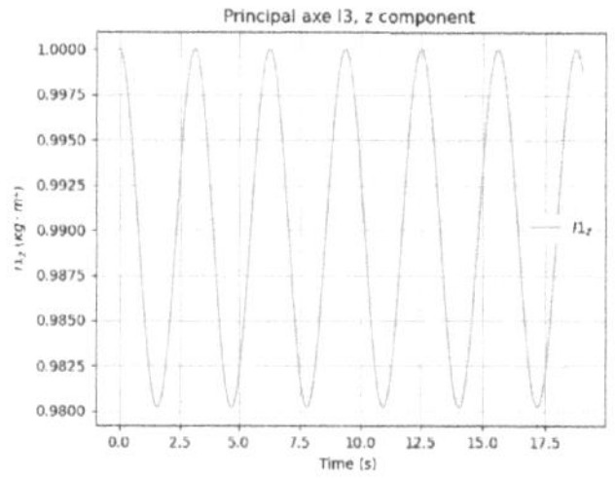

(a) Symmetric top solution to I_3z

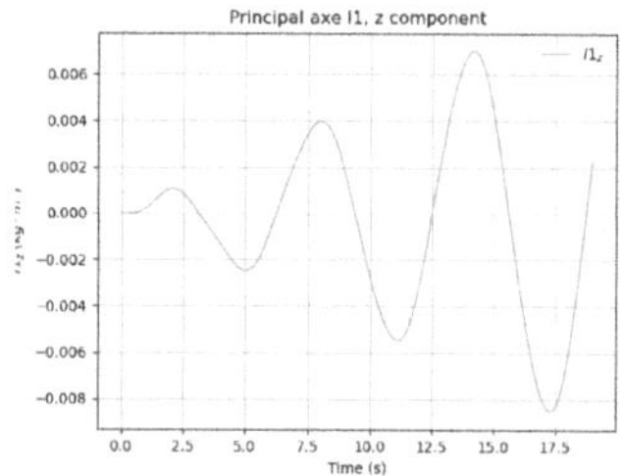

(b) Symmetric top solution to I_1z

Fig. 8. Symmetric top solutions to I_3z and I_1z for the second experiment.

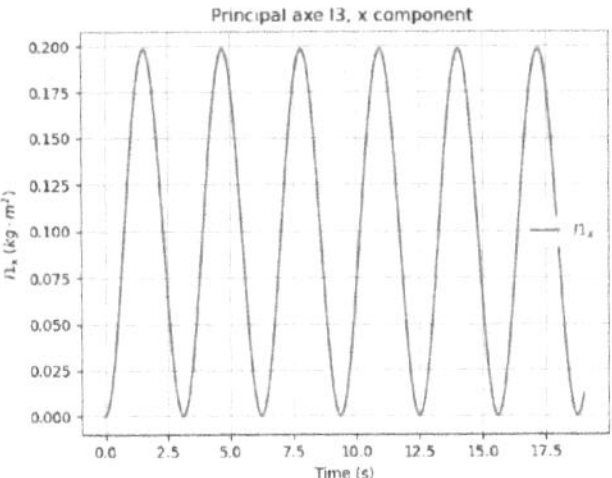

(a) Symmetric top solution to I_3x

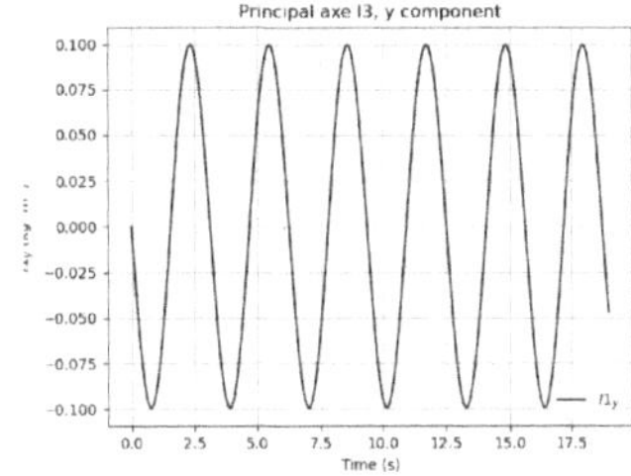

(b) Symmetric top solution to I_3y

Fig. 9. Symmetric top solutions to I_3x and I_3y for the second experiment.

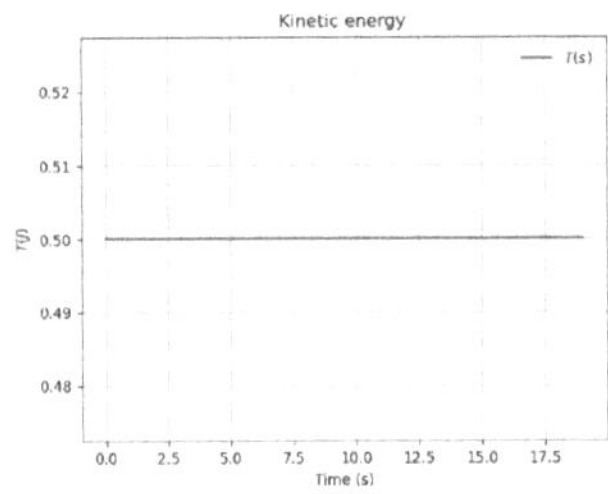

(a) Kinetic energy of the symmetric top in the first experiment.

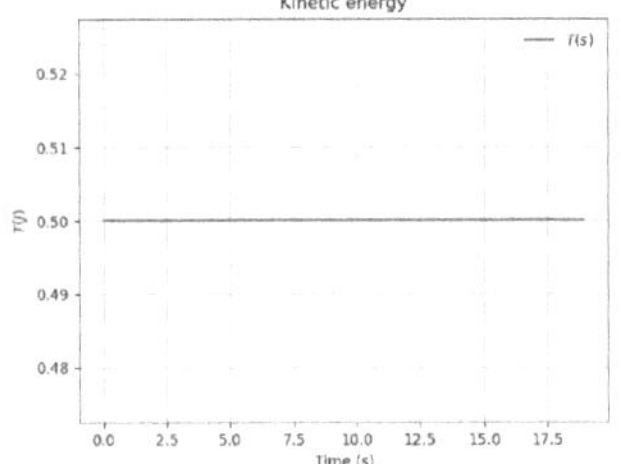

(b) Kinetic energy of the symmetric top in the second experiment.

Fig. 10. Kinetic energy of the symmetric top in both experiments.

7 Conclusion

This paper has showcased the power and elegance of geometric algebra as a language for physics, with a specific focus on its application to rigid body dynamics. The rotor formalism provides a particularly compelling advantage for handling rotations, offering both conceptual clarity and computational efficiency. The analytical and computational treatment of the symmetric top demonstrates the practical utility of geometric algebra and the `Clifford` Python library for solving complex rotational motion problems. The intuitive nature of the algebraic operations and the direct geometric interpretation of multivectors offer a promising avenue for enhancing our understanding and computational modeling of physical phenomena involving rotations.

Future work could explore the application of this framework to more complex scenarios, such as the motion of asymmetric tops or rigid bodies under external torques. Moreover, the integration of geometric algebra into deep learning provides a promising avenue for future research. Its ability to encode geometric relationships in a unified algebraic framework opens new directions for developing models that are not only more interpretable and efficient, but also inherently aligned with spatial and structural properties of data. Exploring this synergy further may lead to significant advances in AI fields such as geometric deep learning, robotics, and physics-informed machine learning.

Acknowledgments. Grant Serverless 4HPC PID2023-152804OB-I00 funded by MICIU/AEI/10.13039/501100011033 and by ERDF/EU.

References

1. Spellings, M.: Geometric algebra attention networks for small point clouds. arXiv arXiv:2110.02393 (2022)
2. Bhatti, U.A., et al.: Geometric algebra applications in geospatial artificial intelligence and remote sensing image processing. IEEE Access (2020). https://doi.org/10.1109/ACCESS.2024.3373339
3. Doran, C., Lasenby, A.: Geometric Algebra for Physicists. Cambridge University Press (2007)
4. Goldstein, H., Poole, C.P., Safko, J.L.: Classical Mechanics, 3rd edn. Addison-Wesley, pp. 184–237 (2001)
5. Hestenes, D.: New Foundations for Classical Mechanics, 2nd edn. Kluwer Academic Publishers (2002)
6. Witte, F.M.C.: Lecture Notes in Physics, pp. 3–28. University College Utrecht (2007)
7. Nearing, J.: Mechanics Contents, pp. 260–293. University of Miami (2013)
8. Doran, C.: Geometric Algebra and Its Application to Mathematical Physics. Sidney Sussex College (1994)
9. Taylor, J.R.: Mecánica Clásica, 1st edn. Editorial Reverté (2003)
10. Kittel, C., et al.: Mecánica (Berkeley Physics Course) (Vol.1), 2nd edn. Editorial Reverté (1982)
11. Hadfield, H., Wieser, E., Arsenovic, A., Kern, R.: pygae/clifford: v1.4.0 (2021). https://clifford.readthedocs.io/en/latest/tutorials/g2-quick-start.html
12. Bronstein, M.M., Bruna, J., Cohen, T., Veličković, P.: Geometric deep learning: grids, groups, graphs, geodesics, and gauges. arXiv preprint arXiv:2104.13478 (2021)

Deep Learning for Environmental, Energy and Bio-Vision Applications

From Data to Decisions: A Novel AI-Based Framework to Energy Purchase Optimization in Volatile Markets

Guillermo Dominguez Pesquera(✉), Pablo García Bringas, and Iker Pastor López

Deusto University, 48007 Bilbao, Vizcaya, Spain
guillermo.dominguez@deusto.es
http://www.deusto.es

Abstract. This paper introduces a novel artificial intelligence (AI)-based methodology aimed at optimizing electricity purchase decisions in volatile energy markets. Leveraging a hybrid machine learning approach, the model classifies future market favorability into three decision classes: (1) the spot market (OMIE, in the case of the Iberian market analyzed), (2) the derivatives market (OMIP), and (3) a neutral scenario. Validated with real-world data from the Iberian electricity market, the proposed model demonstrates superior decision-making performance compared to traditional forecasting approaches. Furthermore, it exhibits notable resilience under market stress conditions, such as renewable intermittency and limited interconnection capacity. During critical periods, the methodology proves to be more robust and reliable than conventional methods. The proposed framework is designed to support strategic energy procurement in increasingly unstable and decarbonized power systems.

Keywords: Energy volatility · Data mining · Energy management optimization · Industry 4.0 for strategic decision-making

1 Introduction

1.1 Current Context

In the context of the global energy transition, the integration of renewable energy sources into power systems and the decarbonization of the economy have become top priorities to mitigate the effects of climate change and reduce dependence on fossil fuels. The commitment of the scientific community is increasingly evident. Researchers worldwide, such as Belmar et al. [1], indicated that integration of renewable energy sources has a significant impact on the electricity market, potentially affecting electricity prices directly are working on solutions to maximize the integration of renewables, reduce greenhouse gas emissions, and ensure a reliable and affordable electricity supply. In this scenario, the development of forecasting tools in electricity markets acquires critical relevance, as it enables the optimization of energy trading decisions, encouraging greater

E. Corchado et al. (Eds.): SOCO 2025, CCIS 2806, pp. 47–59, 2026.
https://doi.org/10.1007/978-3-032-19763-4_5

renewable participation in the market and reducing the costs associated with intermittency and uncertainty, as highlighted by the European Commission in recent publications [2]. The ability to anticipate information in energy markets provides tangible benefits to energy businesses. First, it facilitates better planning of generation and consumption, allowing operators to efficiently adapt to market conditions. Second, it reduces the need to rely on backup generation based on fossil fuels, thereby lowering both emissions and costs and fostering a more competitive and transparent market environment, attracting investment in clean technologies and promoting innovation in the sector.

The growing penetration of renewable energy sources in power systems of countries with limited international interconnections has posed additional and significant challenges to the efficient management of electricity markets. These systems are characterized by volatile production and pricing of renewables and limited capacity to balance that production with international backup supplies, resulting in increased instability in the ordinary operation of wholesale markets. These instabilities have led to large-scale supply shortages affecting entire countries, such as the recent nationwide blackout event in Spain and Portugal on April 28, 2025 [3]. In this context, wholesale electricity markets (spot or continuous) and derivatives markets play a crucial role by offering tools to balance supply and demand, optimize costs, and manage risk.

In the aforementioned markets dominated by volatile renewables, traditional forecasting methodologies face well-kwon, as Matrenin et al. indicated [4], the difficulty in accurately forecasting renewable energy generation, and consequently, its impact on electricity prices, represents a fundamental challenge in energy market analysis. The new analytical approach proposed here addresses this uncertainty by laying the groundwork for the development of AI-powered Smart Business tools that can deliver energy procurement benefits across all types of electricity markets. This research focuses on the Iberian Electricity Market (MIBEL), applying data from the Spanish hub due to its high renewable penetration in the generation mix, as noted by various energy agencies [5], and its significant limitations in interconnection capacity with neighboring countries, as reported by electricity market operators [6].

The current literature includes numerous studies on electricity price forecasting. This goal is common across all reviewed articles. Price forecasting has progressively improved through advancements in analytical techniques and methods, as explained by Jedrzejewski et al. [7], from time series-based models used by Wang et al. [8] in their research of electricity prices in the DE-LU market, to fundamental models by Ghelasi & Ziel [9], who develop a fully structural merit order model for German day-ahead electricity pricing, estimating supply, demand curve parameters directly from data and outperforming both traditional fundamental and pure data-driven alternatives machine learning approaches by Castelli et al. [10], deep learning models proposed by Silva et al. [11], with studies employing random forest techniques like Magalhães et al. [12], and Bayesian models as applied by Wang [13], all contributing to the sophistication and accuracy of price forecasting, as evidenced in the publication by Lago et al. [14].

However, forecasting electricity prices in the wholesale market is not the aim of this research. The primary objective of this study is to identify the optimal timing for purchasing electricity in the spot or derivatives market to achieve greater profitability in energy trading. This approach represents a substantial advancement over traditional price forecasting methods, as it shifts from a passive predictive framework to an active strategic decision-making tool, with a direct and measurable impact on business profitability. Furthermore, it demonstrates enhanced robustness under conditions of high market volatility, and its implementation offers immediate and practical applicability in real-world operational contexts.

To achieve this, decision-support tools based on the analysis of energy market data will be used. These tools, as stated by Alsaigh et al. [15], are AI-powered, interpretable and explainable, capable of utilizing data mining techniques and optimizing procurement results through user-accessible platforms tailored for various market agents and companies managing energy purchases, all under the Smart Business philosophy.

The traditional forecasting algorithms mentioned are capable of predicting electricity prices for the next day or even the following hours within intraday markets.

This study is grounded on the following questions: Why develop a new analytical approach when multiple effective electricity price forecasting methods already exist? What can this new approach offer that has not already been achieved?

This research introduces a new analytical perspective compared to traditional approaches in electricity market forecasting both for stable and more volatile markets. It aims to serve as a gateway for developing Industry 4.0 compliant, AI-driven technological tools, which, as emphasized by Pourdaryaei et al. [16], are in high demand in the current literature. These tools, with simple interpretability, aim to optimize energy-related expenditures for stakeholders in the energy sector as well as end consumers. This paper is organized as follows: after the introduction, in Sect. 2, the methodology outlines the proposed approach in comparison with the current situation, detailing the experimental setup and the classification algorithms employed. Then, in Sect. 3, are presented the results, the findings obtained from the models, followed in Sect. 4 by a critical discussion, which explores the technical and strategic implications of the results. Subsequently, the paper offers in Sect. 5 the conclusions, summarizing the main contributions and suggesting future lines of research. Finally, a references section provides the scientific foundation supporting the theoretical and methodological framework of the study.

2 Methodology

2.1 Description of the Proposed Situation Compared to the Current One

In pursuit of developing a new analytical framework better suited to the realities of today's electricity markets, it becomes necessary to address a series of strategic questions that have thus far been insufficiently explored in the existing literature. For instance, if the electricity price predicted by current algorithms is exceptionally high, should energy purchases be suspended? Is it appropriate to disconnect certain consumer equipment from the grid? What operational disruptions might such actions generate for businesses and institutions? Present-day forecasting models, predominantly based on time-series techniques and conventional statistical methods, that are limited to generating point estimates and are not designed to provide operational guidance or strategic insight. As such, they fail to contextualize price forecasts within broader decision-making frameworks, as Lu et al. [17] emphasized the absence of data-driven models aimed at supporting strategic decision-making in energy markets. This limitation often results in reactive, suboptimal, or even disorderly responses, particularly in highly volatile energy environments. Accordingly, there is a pressing need for a novel methodological approach that moves beyond mere forecasting and integrates intelligent decision-making mechanisms capable of anticipating, evaluating, and managing the potential impacts of diverse actions in scenarios of extreme price volatility. The principal advantage of shifting from conventional methodologies to this new paradigm lies in its capacity to continuously choose between the spot and derivatives markets, thereby optimizing electricity purchase and sale decisions in alignment with both economic and operational objectives.

The following methodological workflow was considered throughout the development of the research (Fig. 1).

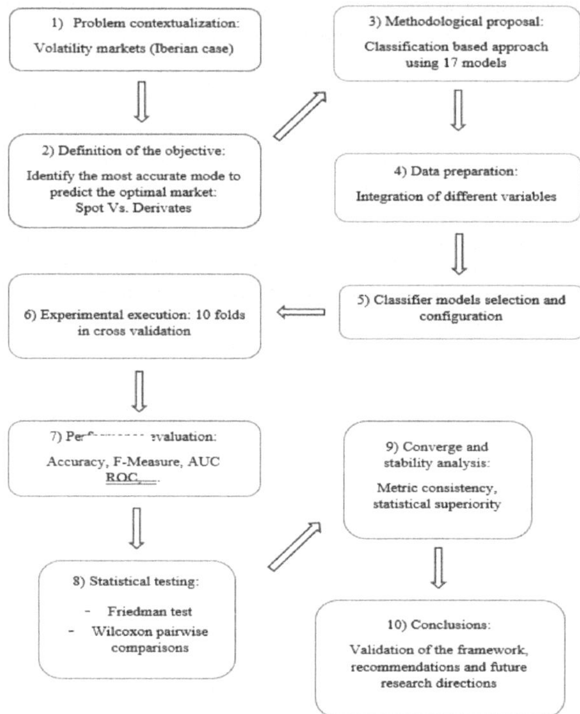

Fig. 1. The methodological workflow to archive the results of this research.

To this end, the following hypotheses will be considered, which differ from the concepts used in current methodologies:

In contrast to conventional approaches that treat electricity price forecasting as a discrete and isolated event, this study proposes a continuous-market perspective, wherein decision-making is dynamically optimized over a weekly horizon, adaptable to longer periods according to stakeholder requirements. Traditional models, as exemplified by Manfre et al. [18], commonly focus on forecasting spot or intraday prices using analytical variables such as electricity demand, weather, and fuel costs; however, they often neglect the temporal decision component that underpins optimal market participation. Instead of solely estimating prices, this work introduces the concept of "timing" as a core variable, arguing that knowing when to buy, either in the spot or derivatives market, is more valuable than the mere knowledge of price levels, as the daily electricity demand must invariably be sourced from one of these markets, a principle acknowledged since Burger et al. [19], the authors pointed out the need to establish a connection between the spot and derivatives markets in order to achieve consistent pricing, representing one of the main contributions of this work. Nonetheless, that approach has not been extensively explored. Furthermore, the study expands the temporal scope beyond day-ahead forecasts, incorporating medium-term horizons by leveraging the structure of energy derivative products available in the Spanish market, as indicated by Harkin & Liu [20] and Ziel & Steinert [21], to accommodate the strategic needs of a broader range of market actors, including IPPs and policymakers. A wide array of predictive variables and costs, often overlooked or underexplored in previous literature, is integrated into the analysis, encompassing spot and futures electricity prices (OMIE [22] and OMIP [23]), natural gas prices (MIBGAS [24]), CO_2 emission allowances, underground gas storage levels, hydrological reserves, exchange rates (EUR/USD), generation costs, technological distribution of generation, and interconnection flows with neighboring countries. Data from January 2016 to December 2024 (excluding 2022 due to regulatory intervention) was processed using seventeen distinct analytical techniques, including time series, machine learning, deep learning, Bayesian models, and hybrid approaches, to ensure methodological robustness and comprehensive validation of the proposed predictive system. The choice to use 2,740 daily records across 23 variables and 17 analysis methods stems from the complex interplay of multiple factors influencing energy prices. Fuel prices, the electricity generated by each fuel type, and how that electricity is used (e.g., for export or import consumption) all matter, as highlighted by Bernal et al. [25]. After appropriate data preprocessing, such as normalization, missing value treatment, and temporal feature extraction, plays a crucial role in improving the accuracy and robustness of electricity price forecasting models, as indicated Chai et al. [26]. As has been consistently demonstrated in previous research, including the study by Madadkhani and Ikonnikova [27], electricity price forecasting greatly benefits from the integration of advanced machine learning techniques and the consideration of key exogenous variables such as fuel and CO_2 prices. Shiri et al. [28] demonstrate that incorporating oil and natural gas prices enhances short-term electricity price forecasting accuracy in some electricity markets, and Arcos Vargas et al. [29] demonstrate a similar opinion in their research. In addition to fossil fuel prices, it is also widely recognized since several years ago that other exogenous variables, unrelated to electricity generation, have a significant impact on electricity prices. In this case, the financial conditions of the markets play a key role. Specifically, for the case studied in this research, the exchange rate between the euro, the currency of the country where the proposed methodology is tested, and the US dollar, the currency in which imported fossil fuels are typically traded, is particularly relevant, as convincingly described by Muñoz and Dickey [30] in their study in 2009.

All of this resulted in a multidimensional dataset that allows for a highly detailed analysis, including complementary variables often excluded from other studies. This has enabled highly accurate forecasts. Moreover, using diverse analytical methods mitigates the uncertainty that arises when relying on any one method alone, as suggested by Lago et al. [31].

2.2 Experimentation

To obtain the research results, daily data over 8 years of the indicated variables were considered. The principal comparison was made between the average weekly data published by the derivatives market (OMIP) during week (w) for energy delivery one week ahead (w + 1), and the data published one week later (w + 1) by the spot or wholesale market (OMIE).

Once the database was defined, the objective of the model was to forecast, week-ahead, which of the two purchasing options, spot market (OMIE) or futures market (OMIP), would be more advantageous. To this end, the prediction target was categorized into three mutually exclusive classes:

- Expensive OMIE: if the average weekly price published in week (w + 1) by the OMIE wholesale market is more than €5/MWh higher than the average weekly price published in week (w) by the OMIP derivatives market for delivery in (w + 1).
- Expensive OMIP: if the average weekly price published in week (w) by the OMIP derivatives market for delivery in (w + 1) is more than €5/MWh higher than the average weekly price published in week (w + 1) by the OMIE wholesale market.
- Moderate: all other cases.

Considering these categories, the data in the database were analyzed using the previously mentioned variables. The analysis was performed using WEKA, an open-source software developed by the University of Waikato that provides a wide range of algorithms, including classification, regression, clustering, and attribute selection models, as well as data preprocessing and model evaluation tools. The relevance of this approach has been widely acknowledged for over a decade, initially highlighted by Hall et al. [32] and more recently adopted in the field of residential electricity forecasting in the national context by Bilal et al. [33]. The robustness and variety of the models make this software a valid tool for the research presented in this paper. A noteworthy result of this study is that, in times of crisis, the system responds better than current predictions, due to the strong correlation between the indices of the spot market and the derivatives market. This implies that if the OMIE wholesale market spikes during a crisis, the OMIP derivatives market will also spike, and similarly if it collapses. In the case of the Spanish market, the correlation over the years between the average weekly values published in week (w) by the OMIP derivatives market for delivery in (w + 1), and the average weekly values published in week (w + 1) by the OMIE wholesale market is shown in the following table (Table 1).

The year 2022 has not been considered, as the wholesale market prices were regulated by the Spanish Government; this aspect was previously described in our earlier work [34]. The impact of this intervention can also be seen in early 2023, particularly in the first quarter of the year, distorting the relationship between the two indices. Before launching the prediction algorithm, and as a result of the high linear correlation obtained, the following research hypothesis was proposed:

Table 1. Lineal correlation between OMIP & OMIE (week + 1).

Year	Pearson Coefficient
2016	0,973599
2017	0,946312
2018	0,935719
2019	0,884262
2020	0,923912
2021	0,968069
2022 2023	0,682843
2024	0,854848
From 2016 to 2024	0,964901

Can this new methodology detect when the difference between the two indices is large enough to favor one market over the other?

To design this new methodology, the cross-validation technique was applied in most of the analysis methods used. In other methods, such as decision tree-based ones, results were validated by training the prediction algorithm with 70% of the data and testing it with the remaining 30%, which was not included in the training phase. The results obtained were similar to those from the cross-validation technique. The described database was subjected to statistical analysis using 17 different methods with the three specified categories as prediction targets. The analysis and classification methods were parameterized under the following conditions:

- **Ensemble Methods**.

Random Forest used 100 fully-grown trees, each trained on the entire dataset, with $\sqrt{N}$ features considered at each split and no depth limit.

Random Committee combined 10 Random Trees using the same $\sqrt{N}$ split rule, requiring minimal error improvement (0.001).

Random Subspace employed 10 REPTrees, each trained on 50% of the attributes with deep trees allowed by a low split variance threshold.

Bagging applied 10 REPTrees with bootstrap sampling, no pruning depth, and 3-fold cross-validation.

Random Tree used all attributes with no pruning, growing trees deeply when variance gain $\geq$0.001.

- **Decision Trees and Rule-Based Models**.

J48 used 25% pruning confidence, splitting only when $\geq$2 instances existed, to improve generalization.

PART followed a similar pruning strategy with balanced tree complexity.

REPTree split nodes with at least 2 instances and a 0.001 variance reduction threshold.

JRip used 3-fold internal cross-validation for pruning, a minimum of 2 instances per rule, and two rounds of rule refinement.

Decision Table selected attributes using BestFirst without cross-validation and stopped after five non-improving nodes.

Classification via Regression relied on the M5P algorithm with a minimum of 4 instances per leaf.

- **Instance-Based Learning.**

Lazy.IBK (KNN) used K = 1 with Euclidean distance and equal weights.

KStar applied a Global Blending value of 20 and handled missing data by averaging all outcomes.

Randomizable Filtered Classifier reduced attributes to 10 dimensions via Random-Projection and used 1-NN with Euclidean distance.

- **Bayesian and Meta-Learning Methods.**

Bayesian Networks employed the K2 algorithm with a maximum of one parent per node and Laplace smoothing ($\alpha = 0.5$).

Meta AttributeSelectedClassifier reduced dimensionality using Chi-squared evaluation and BestFirst search before final classification.

- **Hybrid Model.**

LMT (Logistic Model Tree) dynamically determined LogitBoost iterations, generating logistic models only if $\geq$15 instances were available per node.

3 Results

Once the different classification analytical methods were parameterized, the multidisciplinary database was analyzed with the various categories to be predicted as the main objective of the research, obtaining the following results (Table 2).

A convergence and stability analysis were conducted across the seventeen classification models to ensure the internal consistency and methodological rigor of the experimental evaluation. The performance of each classifier was assessed using eight core metrics: True Positive Rate, False Positive Rate, Precision, Recall, F-Measure, Matthews Correlation Coefficient (MCC), Area Under the ROC Curve (AUC), and Area Under the Precision-Recall Curve (PRC AUC). For each metric, the mean, standard deviation, and coefficient of variation (CV) were calculated to quantify not only the average performance but also the relative dispersion of results. Among all models, Random Forest stood out with the highest average performance and a notably low coefficient of variation (CV $\approx$ 0.408), indicating a high level of convergence across all evaluation dimensions. This suggests that its superior accuracy is consistent and not driven by isolated metric outliers. Other models such as J48 and Random Tree also demonstrated a favorable balance between accuracy and stability, with CV values below 0.41. Conversely, models like Bayesian Networks and Classification via Regression showed higher performance variability, reflected in elevated CVs, implying weaker internsistency and greater sensitivity to metric selection. These results underscore the greater reliability

Table 2. Results achieved after the analysis.

Applied method	Accuracy	Kappa Cohen Coef.	ROC Area
Random forest	96.422%	0.9260	0.991
Random committee	96.094%	0.9197	0.988
J48	94.016%	0.8935	0.961
Decision table	94.451%	0.8835	0.948
Lazy.IBK	91.566%	0.8280	0.909
Kstar	94.268%	0.8824	0.972
Random subspace	94.305%	0.8805	0.984
Part	94.122%	0.8801	0.946
Meta attributeselectedclassifier	94.049%	0.8777	0.955
LMT	93.611%	0.8693	0.947
Bagging	93.611%	0.8653	0.983
Random tree	93.173%	0.8588	0.929
Randomizable filtered classifier	92.187%	0.8401	0.913
Jrip	91.858%	0.8294	0.899
Reptree	90.435%	0.8015	0.951
Classification via regression	90.179%	0.7912	0.960
Bayesian net	83.425%	0.6676	0.901

of ensemble-based methods in high-volatility contexts such as electricity markets. To statistically validate these findings, a Friedman test was applied to seven key metrics across all classifiers, yielding significant differences ($\chi 2 = 56.48$, $p < 0.0001$). Subsequent Wilcoxon signed-rank tests comparing Random Forest with each of the other models confirmed its statistically significant superiority ($p < 0.05$), particularly in terms of F-Measure, MCC, and AUC. These results confirm Random Forest as the most reliable and robust model within the proposed framework. It is also important to note that conventional benchmarking against price forecasting models is not appropriate in this context. The objective of the proposed methodology is not to estimate future prices, but to identify the most advantageous market and moment for electricity procurement. Therefore, comparisons with regression-based price models are not meaningful in quantitative terms and should instead focus on qualitative aspects, such as decision reliability, robustness under market stress, and operational applicability.

4 Discussion of the Results

The results obtained in this study highlight significant differences in the performance of various classification algorithms used to determine the optimal electricity market for purchasing decisions (OMIE vs. OMIP). The superiority of ensemble learning methods, especially Random Forest and Random Committee, is both statistically validated and practically relevant for strategic decision-making in volatile electricity markets. Random Forest achieved an exceptional performance, with an accuracy of 96.42%, Cohen's Kappa of 0.9260, and an AUC of 0.991. These results demonstrate its strong ability to capture complex and nonlinear patterns in market behavior, an essential capability in electricity systems characterized by uncertainty, multicollinearity, and renewable intermittency. Its low variability across metrics and high Kappa value confirms it as a consistent and trustworthy decision-support tool, particularly for short-term operational planning. Random Committee, with slightly lower metrics (96.09% accuracy, 0.9197 Kappa), confirms the value of classifier aggregation in enhancing model stability. However, its slightly reduced diversity compared to Random Forest may account for its marginally lower performance. Simpler models such as J48, REPTree, and PART exhibited decent accuracy levels (90%-94%) but limited capacity to represent the complex interdependencies among variables in the energy domain. Their strengths in interpretability and computational efficiency may be outweighed by their reduced precision in high-stakes decision contexts, potentially leading to suboptimal or economically damaging outcomes. Notably, Bayesian Networks and instance-based methods such as Lazy. IBK and KStar performed poorly. The low Kappa value of Bayesian Networks (0.6676) reflects their difficulty in handling nonlinear relationships or high-dimensional data without over-constraining the conditional structure. These shortcomings raise concerns about overconfidence in probabilistic estimations if not properly regularized or guided by expert input.

The practical implications of these methodological differences are considerable. In markets like MIBEL, where price volatility, renewable integration, and interconnection constraints are prevalent, the ability to consistently identify the more favorable market for procurement can result in substantial economic advantages. In this sense, the selection of Random Forest is not merely a technical preference, but a strategic choice with direct financial implications.

From a methodological perspective, two promising lines of future work emerge:

The integration of ensemble methods with explainability techniques such as SHAP or LIME, which would enhance model transparency without sacrificing performance.

The application of robust validation strategies, including block-wise or temporal cross-validation, to ensure generalizability across varying market conditions.

Finally, it must be acknowledged that no model is immune to failure under extreme scenarios such as unexpected demand shocks, infrastructural disruptions, or abrupt regulatory changes. As such, the proposed models should be embedded within broader decision support systems, incorporating predictive analytics, domain expertise, and real-time contextual data (e.g., weather forecasts, interconnection capacity, geopolitical risks).

5 Conclusions

The results obtained in this study allow us to affirm that ensemble-based classification methods, especially Random Forest, constitute a fundamental tool within the field of data mining for solving complex classification problems in industrial and technological contexts. These algorithms' ability to capture nonlinear relationships, model multidimensional datasets, and maintain high generalization capacity makes them key elements in the digital transformation of organizations. In particular, Random Forest demonstrated the best overall performance in terms of accuracy (96.42%), model stability (Kappa coefficient of 0.9260), and discriminative ability (AUC of 0.991), significantly outperforming other algorithms, including both individual models and other ensemble approaches. This superior performance aligns with the principles of computational optimization by enabling the selection of efficient predictive strategies that minimize errors and maximize classification performance. Moreover, it maintains a high level of agreement between predictions and actual classes, which is crucial for practical applications in critical environments. In the context of Industry 4.0, where system interconnection, intelligent automation, and large-scale data analysis are central elements, the use of algorithms such as Random Forest becomes especially relevant. These models not only allow processing and learning from large volumes of data generated by cyber-physical systems, sensors, and digital platforms, but also provide effective decision-making support by delivering reliable predictions that can be integrated into real-time planning, control, and monitoring systems. Furthermore, the findings of this study highlight the need for a comprehensive approach to the selection of predictive models, valuing not only quantitative performance but also interpretability, robustness, and adaptability to the operational environment. In this regard, data mining stands out as a strategic process for extracting useful knowledge from complex data, contributing to the development of intelligent solutions aligned with modern industrial environments' efficiency, quality, and sustainability goals. In summary, the evaluated methods lead to the conclusion that ensemble techniques not only represent a quantitative improvement over traditional approaches but also signify a methodological evolution that strengthens the link between data science, artificial intelligence (AI), and the real needs of contemporary industry. The integration of these tools into decision support systems not only optimizes operational outcomes but also enhances the resilience and adaptive capacity of organizations in an environment characterized by complexity, volatility, and increasing demand for innovation. In conclusion, this work emphasizes the importance of selecting classification techniques tailored to the nature of the problem and dataset in business decision-making, highlighting the effectiveness of ensemble methods as a powerful tool to enhance data-driven decision-making. Based on the research presented, it can be confirmed that the study's proposed objective has been achieved by developing a new methodology capable of accurately predicting the optimal timing for electricity purchase and sale management. This is particularly significant in markets where traditional algorithms are less reliable and during periods of instability. The current values of the Spanish electricity market, and globally, have withstood several recent crises, such as the 2020 pandemic (which caused electricity prices to plummet, as reported by Abadie [35]), the Russia-Ukraine war in 2022 (which drove prices to historic highs, according to Inacio et al. [36]), and the present day, where electricity prices remain higher than those before 2020, as described by Gajdzik et al. [37]. Several technological advancement opportunities stem from this innovative methodology. New applications based on generative AI could be developed to interpret and simplify results, tailored to the needs of various market agents. As numerous authors, including Surathunmanun et al. [38], suggest, such tools can assist in interpreting AI-based method outputs and optimizing the intended objective. Additionally, this new approach to electricity market analysis serves as a foundation for future development of new analytical methodologies in other energy-related markets and disciplines. These could lead to substantial improvements in current prediction mechanisms by introducing innovations in variable selection and analytical techniques used.

References

1. Belmar, F., Baptista, P., Nieves, D.: Modelling renewable energy communities: assessing the impact of different configurations, technologies and types of participants. Energy Sustain. Soc. (2023)
2. European Commission: Tackling energy price volatility: A smarter approach to price forecasting. Joint Research Centre (2025). https://joint-research-centre.ec.europa.eu/jrc-news-and-updates/tackling-energy-price-volatility-smarter-approach-price-forecasting-2025-03-13_en. Accessed 02 Apr 2025
3. BBC News Mundo: ¿Qué pudo haber causado el masivo apagón eléctrico en España y Portugal? 29 April 2025. https://www.bbc.com/mundo/articles/cjr7vdjegp7o. Accessed 20 May 2025
4. Matrenin, P., Atabaeva, L.Sh., Sergeev, N.: Limitations and perspectives of short-term renewable energy generation forecasting methods. In: IEEE Region International Conference on Computational Technologies in Electrical and Electronics Engineering (2020)
5. "Renewables 2023" International Energy Agency, January 2024. https://www.iea.org/reports/renewables-2023. Accessed 12 Sept 2024
6. "Market Report 2020". European Network of Transmission System Operators for Electricity (ENTSO-E), September 2020. https://www.entsoe.eu/news/2020/06/30/2020-entso-e-market-reports. Accessed 09 Jan 2025
7. Jedrzejewski, A., Lago, J., Marcjasz, G., Weron, R.: Electricity price forecasting: the dawn of machine learning. IEEE Power Energy Mag. (2022)
8. Wang, D., Gryshova, I., Kyzym, M., Salashenko, T., Khaustova, V., Shcherbata, M.: Electricity price instability over time: time series analysis and forecasting. Sustainability (2022)
9. Ghelasi, P., Ziel, F.: A data-driven merit order: Learning a fundamental electricity price model. arXiv (2025)
10. Castelli, M., Groznik, A., Popovic, A.: Forecasting electricity prices: a machine learning approach. Algorithms (2020)
11. Rita Silva, A., Nuno Fidalgo, J., Andrade, J.R.: Easing predictors selection in electricity price forecasting with deep learning techniques. In: 19th International Conference on the European Energy Market, Lappeenranta, Finland (2023)
12. Magalhäes, B.G., Bento, P., Pombo, J., Calado, M., Mariano, S.: Short-term load forecasting based on optimized random forest and optimal feature selection. Energies (2024)
13. Wang, Y.: Electricity price forecasting based in Bayesian network and Monte Carlo simulation. In: IEEE 6th International Conference on Automation, Electronics and Electrical Engineering (AUTEEE), Shenyang, China (2023)
14. Lago, J., Ridder, F., Schutter, B.: Forecasting spot electricity prices deep learning approaches and empirical comparison of traditional algorithms. Appl. Energy **221**, 386–405 (2018)
15. Alsaigh, R., Mehmood, R., Katib, I.A.: AI explainability and governance in smart energy systems: a review. Front. Energy Res. (2022)
16. Pourdaryaei, A., et al.: Recent development in electricity price forecasting based on computational intelligence techniques in deregulated power market. Energies (2021)
17. Lu, H., Ma, X., Ma, M., Zhu, S.: Energy price prediction using data-driven models: a decade review. Comput. Sci. Rev. (2021)

18. Manfre, D., Zamudio, M., Zareipour, H., Quashie, M.: A hybrid model for multi-day-ahead electricity price forecasting considering price spikes. Forecasting (2023)
19. Burger, M., Müller, A., Schindlmayr, G.: A spot market model for pricing deratives in electricity markets. Quant. Finance **4**, 109–122 (2004)
20. Harkin, B., Liu, X.: Forecasting day-ahead electricity prices in the integrated single electricity market: addressing volatility with comparative machine learning methods. asXiv.org (2024)
21. Ziel, F., Steinert, R.: Probabilistic mid-and long-term electricity price forecasting. Renew. Sustain. Energy Rev. (2017)
22. OMIE - Operator of Iberian energy market. Spanish hub. Hourly electricity prices. https://www.omie.es/es/market-results/daily/daily-market/day-ahead-price. Accessed 26 May 2025. mibgas26
23. OMIP - Operator of Iberian energy market. Portuguese hub. Listed products. https://www.omip.pt/es/node. Accessed 23 May 2025
24. MIBGAS. Operator of Iberian natural gas market. Daily price publication. https://www.mibgas.es/es/market-results. Accessed 12 May 2025
25. Bernal, B., Molero, J., García, F.P.: Impact of fossil fuel prices on electricity prices in Mexico. J. Econ. Stud. (2019)
26. Chai, S., Li, Q., Zoynol Abedin, M., Lucey, B.M.: Forecasting electricity prices from the state-of-the-art modeling technology and the price determinant perspectives. Res. Int. Bus. Finance (2023)
27. Madadkhani, S., Ikonnikova, S.: Toward high-resolution projection of electricity prices: a machine learning approach to quantifying the effects of high fuel and CO_2 prices. Energy Econ. (2023)
28. Shiri, A., Afshar, M., Rahimi-Kian, A., Maham, B.: Electricity price forecasting using Support Vector Machines by considering oil and natural price impacts. In: IEE International Conference on Smart Energy Grid Engineering (SEGE). Oshawa-Ontario, Canada (2015)
29. Arcos Vargas, A., Nuñez Hernandez, F., Ballesteros Gallardo, J.A.: CO_2 price effects on the electricity market and greenhouse gas emissions levels: an application to the Spanish market. Clean Technol. Environ. Policy (2022)
30. Pilar Muñoz, M., Dickey, D.A.: Are electricity prices affected by the US dollar to Euro exchange rate? The Spanish case. Energy Econ. **31**(6), 857–866 (2009)
31. Lago, J., Marcjasz, G., Schutter, B., Weron, R.: Forecasting day-ahead electricity prices: a review of state-of-the-art algorithms, best practices and an open-access benchmark. Appl. Energy (2020)
32. Hall, M., Frank, E., Holmes, G., Pfahringer, B., Reutemann, P., Witten, I.: The WEKA data mining software: an update. SKDD (2009)
33. Bilal, M., Kim, H., Fayaz, M., Pawar, P.: Comparative analysis of time series forecasting approaches for household electricity consumption prediction. arXiv (2022)
34. Dominguez, G., Garcia Bringas, P.: Comprensi6n del mercado energético español 2022 con un análisis de los datos oficiales. Dyna, 357–361 (2024)
35. Abadie, L.M.: Energy market prices in times of covid-19: the case of electricity and natural gas in Spain. Energies (2021)
36. Inacio, C., Kristoufek, L., David, S.: Assessing the impact of the Russia-Ukraine war on energy prices: a dynamic cross-correlation analysis. Physica A Stat. Mech. Appl. (2023)
37. Gajdzik, B., Wolniak, R., Nagaj, R., Zuromskaité-Nagaj, B., Grebski, W.: The influence of the global energy crisis on energy efficiency: a comprehensive analysis. Energies (2024)
38. Surathunmanun, S., Ongsakul, W., Singh, J.G.: Exploring the role of generative artificial intelligence in the energy sector: a comprehensive literature review. In: 2024 International Conference on Sustainable Energy: Energy Transition and Net-Zero Climate Future (ICUE), Pattaya City, Thailand (2024)

Anomaly Detection and Predictive Maintenance of H_2 Compressors Using Machine Learning

Beatriz Gil-Arroyo[1](✉), David García[2], Carolina Gutiérrez[2], Eduardo García[3], Luis Ángel Ramos[3], Daniel Urda[1], and Álvaro Herrero[1]

[1] Grupo de Inteligencia Computacional Aplicada (GICAP), Departamento de Digitalización, Escuela Politécnica Superior, Universidad de Burgos, Av. Cantabria s/n, 09006 Burgos, Spain
{bgarroyo,durda,ahcosio}@ubu.es
[2] DGH Technological Solutions, C/ Claustrillas 1-3, 09001 Burgos, Spain
{david.garcia,carolina.gutierrez}@dgh.es
[3] HIPERBARIC ESPAÑA, C/ Condado de Treviño 6, 09001 Burgos, Spain
{l.garcia,la.ramos}@hiperbaric.com
http://www.ubu.es, http://www.dgh.es, http://www.hiperbaric.com

Abstract. In industrial environments, the early detection of anomalies in equipment, such as hydrogen compressors, is critical for ensuring operational safety and reliability. This paper proposes a novel hybrid approach that combines time series forecasting with supervised classification to predict future alarm and warning events. The study uses real-world data from hydrogen compressors provided by Hiperbaric, a company specialized in industrial equipment for high-pressure technologies. The data underwent extensive preprocessing to ensure quality and temporal consistency. Eight tailored datasets were constructed to represent different components and types of alerts. The methodology integrates a Vector Autoregressive model for forecasting endogenous variables and an XGBoost classifier for predicting binary alarm signals. To address the inherent class imbalance in the data, a comprehensive evaluation was conducted using metrics such as AU-ROC, AU-PR, F1-score, and G-mean. The results show that the proposed approach effectively captures temporal dependencies and enables the accurate early detection of alarms and warnings. This study demonstrates the practical applicability of combining forecasting and classification models for the predictive maintenance of hydrogen compression systems.

Keywords: Alarm prediction · Predictive maintenance · Hydrogen compressors · Anomaly detection · Machine learning · Performance metrics · Industrial data analysis

1 Introduction

Hydrogen compressors are essential components in the infrastructure of hydrogen technologies, widely used for pressurizing and storing hydrogen gas in applica-

E. Corchado et al. (Eds.): SOCO 2025, CCIS 2806, pp. 60–70, 2026.
https://doi.org/10.1007/978-3-032-19763-4_6

tions such as fuel cell vehicles, power generation, and chemical synthesis. The proper functioning of these compressors is critical, as failures can lead to significant operational disruptions, safety concerns, and high maintenance costs. Given the growing importance of hydrogen in energy systems, ensuring the reliability and optimal performance of compressors is of paramount importance for the sustainability and efficiency of hydrogen-based technologies [4]. Moreover, the efficiency and costs of hydrogen storage systems remain significant challenges, with ongoing research aiming to reduce these barriers and enhance overall system performance [7].

In this context, predictive maintenance and fault detection using artificial intelligence (AI) techniques, particularly Machine Learning (ML) and Deep Learning (DL), have emerged as powerful tools for monitoring and forecasting abnormal behaviors in industrial systems [5,6]. These techniques have been successfully applied to components such as pumps, turbines, and compressors [2,9], demonstrating their potential to enhance operational safety and reduce downtime. However, although some initial studies have explored the use of AI in hydrogen compressors, dedicated research in this area remains limited, especially regarding the early detection of alarms and warnings [3].

This study proposes a methodology for predicting alarms and warnings in hydrogen compressors using a time series-based machine learning approach. Real operational data from a hydrogen compressor operated by Hiperbaric—comprising both normal conditions and recorded alarm and warning events—serve as the foundation for model development. A VAR model [8] is used to forecast future endogenous variables, and the predictions from VAR are fed into an XGBoost classifier [1] to predict alarms and warnings. The challenge of class imbalance is addressed by using metrics such as AU-ROC, AU-PR, F1-score, and G-mean. The methodology will be validated with the aforementioned real-world dataset.

The rest of the paper is structured as follows: the dataset is described in Sect. 2, followed by the methodology and experimental setup (Sect. 3), the results (Sect. 4), and concluding remarks with future perspectives (Sect. 5).

2 Case Study

Owing to confidentiality agreements with Hiperbaric, specific details about the variables used in the dataset and the exact type of compressor employed in this study cannot be provided. This restriction ensures the protection of the sensitive information and commercial specifications. However, it is guaranteed that the data used are representative of the purpose of the research and comply with appropriate ethical and privacy standards.

Hyperbaric hydrogen compressors are advanced systems designed to compress hydrogen to high pressures efficiently, safely, and oil-free, ensuring high gas purity. The compressor features two piston intensifiers (Comp A and Comp B) that handle the two-stage compression process, an advanced cooling system (Cooling) and a range of auxiliary systems to ensure the safe functioning of the whole equipment. Figure 1 shows a photograph of the Hiperbaric compressor, highlighting its main components such as the piston intensifiers, the cooling system, and the auxiliary subsystems.

Fig. 1. Illustration of Hiperbaric compressor.

This study was based on a dataset obtained from the data acquisition system of a hydrogen compressor provided by Hiperbaric. This dataset underwent an extensive pre-processing procedure to ensure its quality, consistency, and relevance for the application of machine learning models. Initially, there were 82 relevant variables. The preprocessing stage involved outlier correction (e.g., clamping negative sensor values to zero), removal of non-informative and redundant features, semantic standardization through variable renaming, and binary encoding of categorical attributes. After preprocessing, the number of variables was reduced to 39, as constant features, counter variables, and those with a high degree of correlation were eliminated. The different dimensions of the subsets are due to each event belonging to a different functional block of the compressor (General, A, By Cooling), with each block being associated with a specific set of variables. Therefore, each subset contains a different number of variables and samples. Additionally, the engineered binary indicators captured the presence of general (alarm/warning status) and component-level (H01–H05) alarms and warnings.

From the fully preprocessed dataset, eight data subsets (Datasets 1–8) were generated, each focused on analyzing a specific type of event (alarm or warning) and associated with a particular component of the system (General, Component A, Component B and Cooling). Each subset included both records in which the event of interest was detected (value 1 in the corresponding field) and those without events (value 0), enabling comparative analysis and the construction of binary classification models.

A constant sampling frequency is required to apply the time-series models. Since the original data captures did not follow a uniform frequency (e.g., every 2 s or every 30 s), and gaps existed due to previous filtering, new datasets were generated with a regular time frequency adjusted to 1 min intervals. To achieve this, the timestamps were rounded to the defined frequency, and the data were aggregated using different functions depending on the variable type. Binary variables corresponding to alarms and warnings (e.g., `alarmas_H01`, `alarmas_H03`, etc.) took the maximum value within each interval, thus preserving the activation of events. The other variables were averaged to ensure smooth fluctuations and to maintain the continuity of the time series. Finally, a complete time series with regular intervals was generated to ensure temporal continuity of the dataset. Periods without data are filled with `NULL` values.

After preprocessing, the generated files were analyzed to identify the periods with and without data. Through this segmentation, continuous time ranges were identified in which data were available, as well as periods with missing records. Time intervals without missing values were selected, prioritizing those in which the class distribution between 0 and 1 was more balanced.

Table 1 lists the training and validation details of several components. The table includes information on the start and end dates of the training and validation periods, as well as the distribution of classes (number of 0's and number of 1's) in these periods. The values correspond to different components and alerts/warnings, allowing us to observe the number of predictor variables used (X) and the name of the target variable (Y) in each dataset.

Table 1. Overview of training and validation data

Component	X	Y	Start date	End date	No 0's	No 1's
General	5	Alarm_01	22/06/2023 14:45	03/07/2023 17:26	9261	6741
Comp A	14	Alarm_03	26/11/2023 00:12	28/11/2023 23:09	2872	1386
Comp B	14	Alarm_04	20/09/2023 16:46	21/09/2023 10:27	950	112
Cooling	6	Alarm_05	04/08/2023 20:25	06/08/2023 05:59	1064	951
General	5	Warning_01	22/06/2023 08:57	03/07/2023 17:26	8100	8250
Comp A	14	Warning_03	03/04/2023 11:03	04/04/2023 19:42	1364	596
Comp B	14	Warning_04	06/08/2023 06:02	09/08/2023 13:06	1786	2959
Cooling	6	Warning_05	10/04/2023 13:50	12/04/2023 07:55	1153	1373

Initial experiments were conducted using prediction horizons of 10, 20, and 30 min. Given the satisfactory performance observed at these intervals, the horizon was progressively extended up to 115 min to evaluate the model's behavior over longer-term forecasts. Notably, the prediction error tends to increase with longer horizons, which is a common trade-off in time-series forecasting. Table 2 displays the start timestamp and count of class 0 and class 1 labels for each component and alarm over the 115 min prediction window.

As seen in the two previous tables, two time periods were chosen: one for training and validation, and the other for predictions (test). This decision was made because using a single time interval, even with sufficient records, can lead to an imbalance in the test set, especially in time-series data. When splitting

Table 2. Overview of future prediction data

Component	X	Y	Start date	No 0's	No 1's
General	5	Alarm_01	09/08/2023 15:30	37	78
Comp A	14	Alarm_03	04/01/2024 06:00	83	32
Comp B	14	Alarm_04	25/09/2023 09:15	59	56
Cooling	6	Alarm_05	25/10/2023 05:25	31	84
General	5	Warning_01	09/08/2023 15:30	47	68
Comp A	14	Warning_03	12/12/2023 10:35	94	21
Comp B	14	Warning_04	04/01/2024 05:37	60	55
Cooling	6	Warning_05	12/04/2023 17:20	57	58

the dataset into training and testing, the last records of the test set tend to be imbalanced, with a predominance of one class (either 1 or 0). This does not provide an adequate representation of either class. Therefore, a second time period was selected for testing, ensuring a better representation of both classes, and it was later than the training period, guaranteeing a more balanced and representative evaluation of the model's performance under diverse conditions.

3 Methods and Experimentation

This section describes the methodology and experimental procedures used to predict alarms and warnings in hydrogen compressors using a time series-based modeling approach. Special attention is paid to capturing temporal dependencies in the data and evaluating the model performance within the context of a binary classification task with imbalanced classes. To address this challenge, we propose a hybrid approach that combines time series forecasting with classification techniques. This approach will allow us to effectively analyze and predict the target variable on the datasets previously described.

3.1 Model and Experimental Set up

This section presents the classification model used for determining alarms and warnings, along with the experimental setup designed for training and validation using historical data. Initially, the model is trained and validated on historical data to classify alarms and warnings. Following this, a forecasting model is introduced to predict future values of the exogenous variables. These predicted values are then used as inputs for the classifier to predict whether an alarm or warning will occur.

The classification model selected in this study is **Extreme Gradient Boosting (XGBoost)**, a tree-based ensemble learning method optimized for supervised tasks such as classification and regression. XGBoost is based on the gradient boosting framework, where decision trees are trained sequentially, each one aiming to correct the errors of its predecessor. Its scalability and regularization mechanisms make it particularly well suited for time series classification tasks involving multiple features.

In this study, XGBoost is used to predict the activation of alarms and warnings based on the values of several input variables. Different XGBoost models were trained for each event because they correspond to different functional blocks of the compressor (e.g., General, A, By Cooling), each with its own set of relevant variables and signal behaviors. Additionally, the datasets used for each model consist of different samples and exhibit varying class imbalance levels. For these reasons, a specific model architecture and hyperparameter configuration was defined for each case, as detailed in Table 3.

Table 3. Parameter values for alarms and warnings classification XGBoost

Alarm/Warning	Estimator	Max_depth_tree	Colsample_bytree	Learning Rate
Alarm_01	150	3	1	0.01
Alarm_03	50	3	1	0.01
Alarm_04	50	3	0.8	0.01
Alarm_05	150	5	0.8	0.20
Warning_01	150	3	0.8	0.10
Warning_03	50	5	0.8	0.01
Warning_04	50	3	18	0.01
Warning_05	50	3	0.8	0.01

The model was trained on the training/validation periods outlined in Table 1, and subsequently, a forecasting approach was implemented for future alarm prediction. Specifically, a **Vector Autoregressive (VAR)** model was used to forecast future values of the endogenous variables. These predicted variables were then used as input features for the previously trained XGBoost model to predict alarms and warnings in the test period, as shown in Table 2.

Figure 2 illustrates the overall workflow used in this study. It highlights the two main phases: (1) model training and validation using historical data, and (2) future prediction using forecasted variables obtained from the VAR model.

3.2 Performance Metrics

The performance of the model was evaluated using several common metrics, including accuracy, precision, recall, F1-score, geometric mean (G-mean), area under the precision-recall curve (AU-PR), and area under the ROC curve (AU-ROC). Accuracy measures the proportion of correct predictions, but it can be misleading in imbalanced datasets. Precision and recall focus on the model's ability to correctly identify positive cases, with precision emphasizing false positives and recall emphasizing false negatives. The F1-score provides a balance between

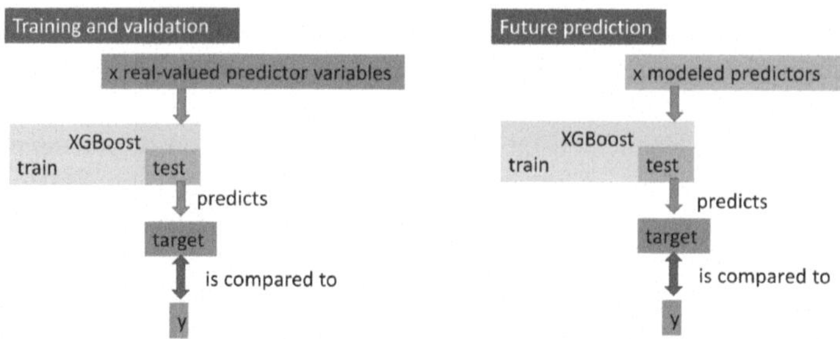

Fig. 2. Illustration of the predictive modeling pipeline, where future values of endogenous variables are generated via VAR and used to predict alarms/warnings using XGBoost.

precision and recall, while the G-mean evaluates the balance between sensitivity and specificity. AU-PR and AU-ROC assess model performance with respect to the positive class and the model's ability to distinguish between classes, respectively. These metrics are widely recognized and effective for evaluating classification tasks, especially when dealing with imbalanced data.

4 Results

Table 4 presents the performance metrics for predicting alarms and warnings across the different components, using a forecasting horizon of 115 min. The results are categorized into two main groups: alarms and warnings, with further breakdowns for each component (Component A, Component B, and Cooling). These metrics help evaluate the effectiveness of the model in classifying both alarms and warnings accurately.

Table 4. Performance metrics for alarms and warnings in a compressor

	ALARMS				WARNINGS			
Metric	General	Comp A	Comp B	Cooling	General	Comp A	Comp B	Cooling
Accuracy	0.85	0.94	0.90	**0.99**	0.97	0.18	0.88	**0.99**
Precision	0.89	**1.00**	**1.00**	0.99	0.94	0.18	0.80	0.98
Recall	0.87	0.78	0.79	**1.00**	**1.00**	**1.00**	**1.00**	**1.00**
F1 Score	0.88	0.88	0.88	**0.99**	0.97	0.31	0.89	**0.99**
Geometric Mean	0.83	0.88	0.89	0.98	0.96	0.00	0.88	**0.99**
AU-PR	0.98	0.92	0.94	0.99	**1.00**	0.59	0.90	0.99
AU-ROC	0.94	0.89	0.89	0.98	**1.00**	0.50	0.88	0.99
Thresdold	0.400	0.444	0.484	0.997	0.944	0.192	0.384	0.674

The date ranges used during the training and testing phases were not identical across each of the components and for each alarm or warning (Tables 1, 2).

In particular, time windows were chosen that contained no missing values and where the distribution between classes 0 and 1 was more balanced. Although the date ranges are not the same, the prediction horizon has been kept constant at 115 min, ensuring consistency in the evaluation of the model.

Figure 3 shows a radar plot that visually represents the performance of models trained for alarms. This plot provides an in-depth comparison of the models across various performance metrics, for four categories: General, Component A, Component B and Cooling. Similarly, Fig. 4 shows a radar plot for the warning category, following the same structure and performance metrics.

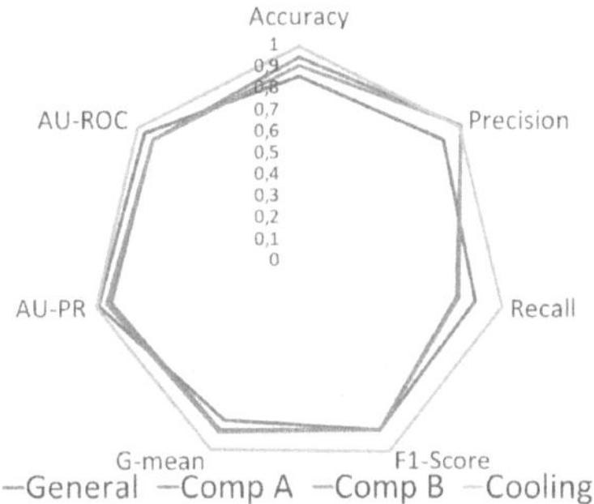

Fig. 3. Radar plot showing models performance for **alarms**.

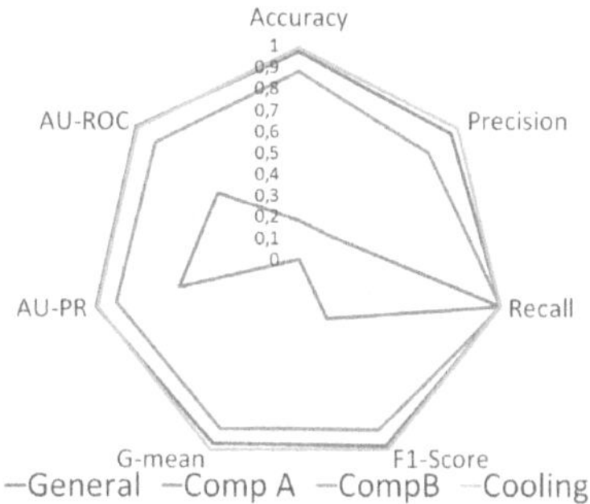

Fig. 4. Radar plot showing models performance for **warnings**.

General Component displays solid overall performance, with an accuracy of 0.85 for alarms and 0.97 for warnings. The F1 scores of 0.88 (alarms) and 0.97 (warnings) indicate a well-balanced trade-off between precision and recall. According to Tables 1 and 2, the number of training instances and the balance between classes are favorable, particularly for warnings, where the model benefits from a large and balanced dataset (over 16000 instances, nearly evenly split). In contrast, alarm data shows a mild class imbalance (9261 controls vs. 6741 cases), which may explain the slightly lower recall. The warning predictions also benefit from broader date ranges and more operational variability, potentially enhancing the model's ability to generalize. These observations suggest that General Component serves as a strong baseline, although efforts to further improve alarm recall could help align its performance more closely with that of the warning predictions.

Component A shows solid performance for alarm predictions, with an accuracy of 0.94 and an F1-Score of 0.88, but the recall of 0.78 suggests room for improvement in identifying all positive alarms. However, its performance in warning predictions is much lower, with an accuracy of 0.18 and an F1-Score of 0.31. This discrepancy can be attributed to several factors: a class imbalance in the warning dataset (1364 "0" instances vs. 596 "1" instances), a small training set with only 1960 instances, and a very short training period (April 3-4, 2023), which may have resulted in a lack of variability in the data. Additionally, the test set begins 8 months later, on December 12, 2023, which could affect the model's ability to generalize to future predictions (see Tables 1 and 2).

Component B shows acceptable performance in both types of prediction, with an accuracy of 0.90 for alarms and 0.88 for warnings. The F1-Score is also solid (0.88 for alarms and 0.89 for warnings), indicating a good balance between precision and recall. However, the recall of 0.79 in alarms suggests there is room for improvement in detecting positive cases. Considering Tables 1 and 2, the training data for alarms in Component B is notably imbalanced, with 950 control samples (label 0) and only 112 positive cases (label 1). Despite this imbalance, the model performs reasonably well, which may reflect its ability to learn distinguishing patterns even from a limited number of positive instances. In contrast, for warnings, the training data is more balanced (1786 controls vs. 2959 cases), which likely contributes to the strong and consistent metrics observed. Nonetheless, the limited number of positive alarm cases in the training set may restrict the model's ability to capture broader variability in alarm conditions, as suggested by the suboptimal recall.

Cooling is the component with the best performance in both alarms and warnings, with a model that shows outstanding metrics across all areas and hardly any differences between the two types of predictions. The accuracy close to 1.00, along with an F1-Score of 0.99, and perfect AU-PR and AU-ROC values, highlight an exceptional ability to correctly predict both alarms and warnings. The recall and F1-Score metrics reflect that the model is working perfectly in identifying both types of events. The Cooling model stands out for its near-perfect performance, supported by well-balanced class data and consistency between training and test sets. The short time gap and homogeneous behavior of the component facilitate accurate event detection. This highlights how data quality and structure directly impact predictive effectiveness.

5 Conclusions and Future Work

The objective of this study was to develop and evaluate an innovative hybrid approach for the early detection of alarm and warning signals in industrial hydrogen compressors. This approach combined time series forecasting with classification techniques, enabling the anticipation of potential failures in the key subsystems of a compressor. A real-world dataset of operational conditions was used to train and evaluate predictive models.

The results show very solid performance in alarm detection, with precision, recall, F1, and AU-ROC metrics exceeding 0.85 in most cases (Table 4). The cooling subsystem stands out particularly, with values close to 1.00 across all metrics, indicating reliable detection of critical failures in this component. Regarding warnings, the behavior is more variable: while in Compressor B and cooling, the model reaches high levels of precision and F1 (around 0.88–0.99), in Compressor A, performance is significantly lower, with a precision of 18% and an F1 of only 0.31, despite achieving a recall of 100%. From these results, it can be concluded that the model is highly effective at detecting alarms, especially those related to critical failures that precede system shutdowns. This reinforces the potential of the proposed approach as a predictive maintenance tool. However, warnings present a greater challenge, particularly in the case of Compressor A, where the high false positive rate indicates that signals preceding minor failures are more difficult to characterize or are underrepresented in the data. This behavior suggests that warnings may be associated with more diffuse or less consistent operating conditions than alarms.

It should be noted that, due to industrial confidentiality reasons, the data used in this study cannot be shared publicly. However, the proposed methodology is applicable to other industrial environments with similar data structures and prediction problems.

Future work will focus on improving the precision of warning detection by addressing the high rate of false positives observed in some components. This could involve exploring advanced machine learning techniques to reduce false positive rates without compromising recall. Additionally, although the models have shown good performance in offline analysis, their real-time implementation and deployment within the compressor system will be crucial. Real-time monitoring and alerting are essential for preventing failures and optimizing system performance. Integrating these models with a real-time data stream, as well as deploying an automated alert system, would enable faster responses to detected alarms and warnings. Furthermore, an area for improvement lies in handling missing data, as certain values were removed during preprocessing. Future work could explore methods to impute missing values, ensuring that these gaps do not negatively impact model performance or lead to bias.

Acknowledgments. This study was carried out within the framework of "VALOR H2 – Research Project on new technologies, materials, and processes associated with the hydrogen value chain", under the collaboration agreement signed between Hiperbaric, DGH Technological Solutions, and the University of Burgos. This project has been funded by the Spanish Ministry of Science and Innovation (MCIN) and the Centre for the Development of Industrial Technology (CDTI), and co-funded by the European Union through the Recovery, Transformation and Resilience Plan – NextGenerationEU.

Disclosure of Interests. The authors have no competing interests.

References

1. Chen, T., Guestrin, C.: XGBoost: a scalable tree boosting system. In: KDD 2016, Proceedings of the 22nd ACM SIGKDD International Conference on Knowledge Discovery and Data Mining, pp. 785–794. Association for Computing Machinery, New York, NY, USA (2016). https://doi.org/10.1145/2939672.2939785
2. Liu, Y., Duan, L., Yuan, Z., Wang, N., Zhao, J.: An intelligent fault diagnosis method for reciprocating compressors based on LMD and SDAE. J. Safety Ocean Eng. **102249**, 1–10 (2019)
3. Lv, Q., Yu, X., Ma, H., Ye, J., Wu, W., Wang, X.: Applications of machine learning to reciprocating compressor fault diagnosis: a review. Processes **9**(6) (2021). https://doi.org/10.3390/pr9060909

4. Palacín, I., et al.: Anomaly detection for diagnosing failures in a centrifugal compressor train. Front. Artif. Intell. Appl. **339**, 217–220 (2025). https://doi.org/10.3233/FAIA210137
5. Ren, Y.: Optimizing predictive maintenance with machine learning for reliability improvement. ASCE-ASME J. Risk Uncertainty Eng. Syst. B: Mech. Eng. **7**(3) (2021). https://doi.org/10.1115/1.4049525
6. Saravanan, S., Khare, R., Km, U., Khare, S., Gowda, B., Boopathi, S.: AI and ML adaptive smart-grid energy management systems: exploring advanced innovations. In: Principles and Applications in Speed Sensing and Energy Harvesting for Smart Roads, pp. 166–196. IGI Global (2024). https://doi.org/10.4018/978-1-6684-9214-7.ch006
7. Shin, H.K., Ha, S.K.: A review on the cost analysis of hydrogen gas storage tanks for fuel cell vehicles. Energies **16**(13), 5233 (2023). https://doi.org/10.3390/en16135233
8. Sims, C.A.: Macroeconomics and reality. Econometrica **48**(1), 1–48 (1980). https://doi.org/10.2307/1912017
9. Zhang, W., Yang, D., Wang, H.: Data-driven methods for predictive maintenance of industrial equipment: a survey. IEEE Syst. J. **13**(3), 2213–2227 (2019). https://doi.org/10.1109/JSYST.2019.2905565

Analysis of Honey Bee Drone Sperm Using Deep Learning

Miguel Santiago Gómez[1], Víctor Manuel López[1], Jose Divasón[1(✉)], Francisco Javier Martinez-de-Pison[2], Pilar Santolaria[3], and Jesús L. Yániz[3]

[1] University of La Rioja, Logroño, Spain
{miguel-santiago.gomez,victor-manuel.lopez,jose.divason}@unirioja.es
[2] Scientific Computation Research Institute (SCRIUR), University of La Rioja, Logroño, Spain
fjmartin@unirioja.es
[3] BIOFITER research group, Environmental Sciences Institute (IUCA), Department of Animal Production and Food Sciences, University of Zaragoza, Huesca, Spain
{psantola,jyaniz}@unizar.es

Abstract. Bees are vital for their pollination services, making the conservation of healthy colonies essential. Drone sperm quality directly impacts colony fitness, yet its analysis remains challenging compared to those of mammals due to the unique morphology and motion of insect sperm. This work presents a hybrid method for tracking drone sperm heads in microscopy videos by combining YOLO-based deep learning models with classical computer vision techniques. Using rotational analysis and a custom scoring function, the approach enables reliable, GPU-free head tracking. Results suggest practical applicability in low-resource laboratory settings.

Keywords: tracking · honey bee drone · sperm quality analysis

1 Introduction

Bees play a crucial role in global food security through their pollination services. Bee pollination provides excellent value to crop quality and quantity, improving global economic and dietary outcomes [5]. Pollination improves the yield of most crop species and contributes to one-third of global crop production [12], making bee conservation essential for sustainable agriculture and food systems worldwide. The analysis of drone sperm quality is vital for understanding bee reproductive health and colony fitness. Most studies on drone semen quality have assessed parameters such as sperm volume, sperm concentration and sperm plasma membrane integrity [19,29]. Drones whose reproductive competitiveness is affected by extrinsic factors may contribute dead or suboptimal sperm to a queen, which can have severe negative consequences for the colony's overall productivity and survival [20], highlighting the importance of drone sperm analysis in bee conservation efforts. Beekeepers and breeders select drones from colonies with superior traits, such as high honey production, gentleness, or disease resistance, to improve the performance of future generations [4].

E. Corchado et al. (Eds.): SOCO 2025, CCIS 2806, pp. 71–81, 2026.
https://doi.org/10.1007/978-3-032-19763-4_7

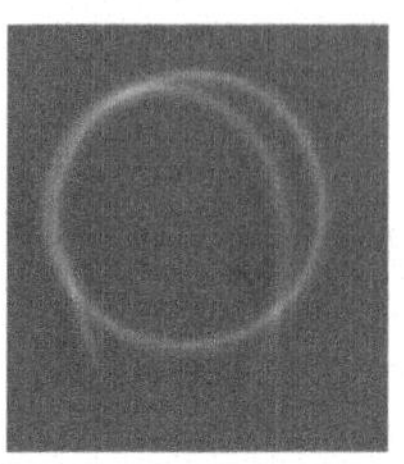

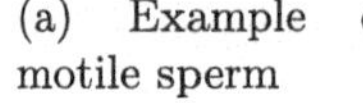

(a) Example of motile sperm

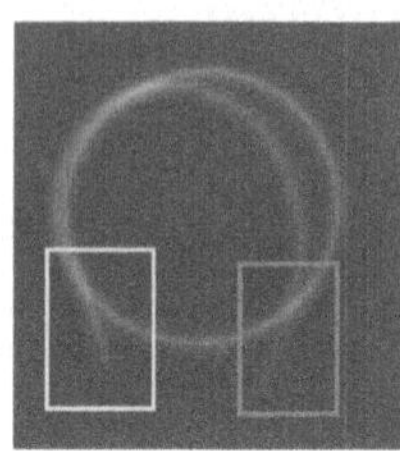

(b) Head (yellow) and tail (green)

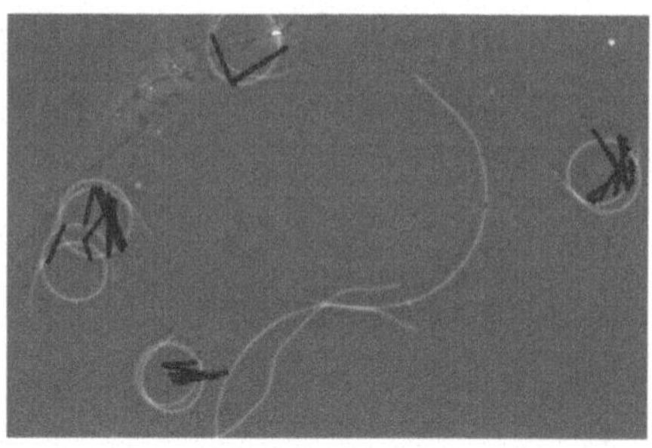

(c) Failed tracking of sperm. Blue lines show the (wrong) detected trajectory.

Fig. 1. Example of motile sperm and use of conventional trackers.

The analysis of semen using computerized tools is well-established in mammals [2]. However, the spermatozoa of insects –particularly those of honey bee drones– differ significantly from their mammalian counterparts. An example of a motile drone spermatozoon is shown in Fig. 1a. As illustrated, its morphology tends toward a more circular shape, and its movement is predominantly rotational rather than translational. Unlike mammalian sperm, distinguishing the head is notably more challenging. While the two extremities of the spermatozoon can usually be identified, determining which end corresponds to the head is not straightforward. Three main features help identify the head: it tends to be slightly thicker, exhibits a somewhat straighter shape, and the spermatozoon typically rotates in the direction of the head. In Fig. 1b, it is marked which is the head and which is the tail.

The morphological similarity between the head and tail of drone spermatozoa presents challenges for their identification using neuronal networks, particularly in scenarios involving overlapping. This ambiguity often leads to misidentification and tracking errors. As a result, conventional multi-object tracking methods, such as BoT-SORT [1], ByteTrack [30], and those included in the OpenCV Tracker API [14,23], perform poorly in this context, frequently failing to reliably distinguish and follow individual spermatozoa, see Fig. 1c.

This work presents a first approach to tracking insect sperm using a hybrid method that combines two deep learning models with classical computer vision techniques. The ultimate goal is to have a reliable and fast method, capable of inferring trajectories without the need for a graphics processing unit (GPU).

The remainder of this paper is organized as follows. Section 2 reviews the related literature. Section 3 describes the dataset and evaluation metrics used in this study. Our proposed hybrid approach is detailed in Sect. 4. The experimental results and their analysis are presented in Sects. 5 and 6, respectively. Finally, Sect. 7 outlines the conclusions.

2 Related Works

2.1 Artificial Intelligence for Beekeeping

The use of Artificial Intelligence (AI) techniques is slowly increasing in hive management, although it is still far from being widely implemented in the field. There are works on beekeeping precision systems [10], monitorization of hives [3], insect detection at hive entrance [27] and honey bee state identification [13], among others. In recent years, efforts have been made to use deep learning methods to count Varroa mites [7,24,28], an endemic parasite that causes the most destructive disease in global beekeeping (*varroosis*). As for sperm analysis, there are several programs that work very well for mammals, for instance, Open-CASA [2]. However, the only one specialized in insects is CASABee [9]: it uses classical imaging techniques (such as Hough transform, dilations and erosions) to count the number of motile and dead sperm and analyze the concentration. The program does not track the sperm heads to properly evaluate motility, though. Advanced topological data analysis techniques have also been utilized to analyze sperm drone videos [8], but again without tracking the heads.

2.2 The YOLO Family

YOLO (You Only Look Once) [21] represented a revolutionary family of real-time object detection algorithms that transformed computer vision by treating detection as a single regression problem. The original YOLO model introduced a paradigm shift from traditional two-stage approaches, with subsequent iterations showing significant improvements. These models excel in real-time applications such as autonomous driving, surveillance systems, and detection based on unmanned aerial vehicles (UAV). YOLOv8 and YOLOv11 are two well-known versions provided by Ultralytics, being very simple to use and widely applied by the community in different fields [11,25]. The current last version is YOLOv12, although it is still considered experimental and not ready for production.[1]

While YOLO dominates real-time detection, several alternative approaches offer different trade-offs between speed and accuracy. Two-stage detectors like Faster R-CNN [22] provide higher accuracy through region proposal networks but at slower inference speeds. Single-stage alternatives include RetinaNet [15], which introduced focal loss to address class imbalance. More recent transformer-based approaches such as DETR (DEtection TRansformer) [6] eliminate hand-designed components such as non-maximum suppression and anchor generation, viewing object detection as a direct set prediction problem. In any case, as one of the final objectives of this work is to be able to perform the tracking on average computers (without GPU), we chose the YOLO family (specifically versions 8, 11 and 12) for their resource-efficient real-time processing.

[1] See https://github.com/ultralytics/ultralytics/issues/19595.

3 Material and Methods

3.1 The Dataset

A dataset of 31 videos of drone sperm captured with microscope was available, which were randomly divided into 24 videos for training, 4 for validation and 3 for test. Each video consists of approximately 58 frames (about 2 s long) and 1920×1200 resolution. The dataset is not labeled, i.e., we only have the raw videos without bounding boxes of where each spermatozoon is, nor its head and tail. If we use CASABee to detect the number of motile spermatozoa, we have that there are approximately 970 in total where each video ranges between 5 and 61. Figure 1c is precisely an example of a small crop of a video.

3.2 Metrics

For object detection tasks, Average Precision (AP) is usually the predominant evaluation metric, assessing detection quality by calculating the area beneath the precision-recall curve for each individual class. This metric evaluates the model's capability to correctly identify objects by incorporating both precision (proportion of correct predictions among all positive predictions) and recall (proportion of correctly identified objects among all ground truth instances). The mean Average Precision (mAP) extends this evaluation by computing the averaged AP values across all object classes. AP computation depends on the selected Intersection over Union (IoU) threshold value, which determines whether a predicted bounding box is categorized as a true positive or false positive detection. mAP is commonly evaluated across multiple IoU threshold values, with the standard notation mAP@[0.5, 0.95, 0.05] indicating averaged AP scores computed at IoU thresholds ranging from 0.5 to 0.95 in increments of 0.05. Alternative formulations frequently employed in research literature and detection benchmarks [18] include mAP50, representing mAP calculated exclusively at an IoU threshold of 0.5. Consequently, mAP scores range between 0 and 1, with higher values approaching 1 indicating superior detector performance.

This study employs mAP as the primary benchmark for evaluating model performance; in particular, mAP50. The rationale is straightforward: our focus centers on detecting the presence and quantity of motile sperm and their heads in each frame, rather than achieving highly precise bounding box alignment; thus, IoU value of 0.50 meets our requirements.

4 Implementation

Our proposal for tracking is based on several steps summarized in Algorithm 1. The most important steps are as follows:

- **Line 1:** Detection and tracking of motile sperm
- **Line 5:** Analysis of the rotation of each motile sperm
- **Line 8:** Detection of possible head and tail of each motile sperm

Algorithm 1: Proposed tracking system

```
Data: A video of the drone sperm
Result: New video with tracked heads drawn
1  Detect and track motile sperm in all frames
2  Initialize previous positions: last_head_position
3  foreach frame do
4  |   Extract bounding boxes and tracking IDs, update tracking history, extract
   |     crop of each motile sperm and add padding
5  rotation ← PerformRotationAnalysis(crops)
6  foreach frame do
7  |   foreach crop_of_motile_sperm in frame do
8  |   |   head_candidates ← DetectHeads(crop_of_motile_sperm)
9  |   |   best_head ← SelectBestByRotation(head_candidate, rotation,
   |   |     last_head_position)
10 |   |   new_head_position ← adjusted coordinates
11 |   |   Update last_head_position and draw bounding box
12 Export processed frames to output video
```

- **Line 9:** Detection of the best suited head candidate using a score function

Lines 1 and 8 were performed using deep learning models. Next, we detail the dataset construction process for each model training. In the first case, since manually labeling all frames of all videos is an extremely tedious and manual process, we used CASABee to annotate the motile spermatozoa and then manually checked the labels. With this we get a dataset of motile spermatozoa where the training set consists of 1408 images with 34035 instances, the validation set has 228 images and 8684 instances and the test set contains 177 images with 2773 instances. In the second case, unlike the labeling of motile sperm, the annotation of their heads and tails cannot be automated, as no suitable utility exists for this task. Due to the time required for manual annotation, we limited this process to a subset of the video frames and only a single frame per video. As a result, the dataset comprises 474 images for training (with a total of 973 annotations), 109 images for validation (224 annotations), and 42 images for testing (85 annotations). Note that some images contain multiple labeled heads and/or tails: this occurs when several sperm appear within the same crop, often overlapping. The following subsections provide more details on how Lines 5 and 9 are performed.

4.1 Analysis of Rotation

Once we have a model that detects motile sperm (Line 1), we should detect the direction of rotation. To do this, we analyze each detected motile spermatozoon across video frames using classical computer vision techniques. Specifically, we extract image crops centered on the spermatozoon in each frame and apply the Scale-Invariant Feature Transform (SIFT) algorithm [16] to detect key points within each crop. We then employ a nearest-neighbor matching strategy, using

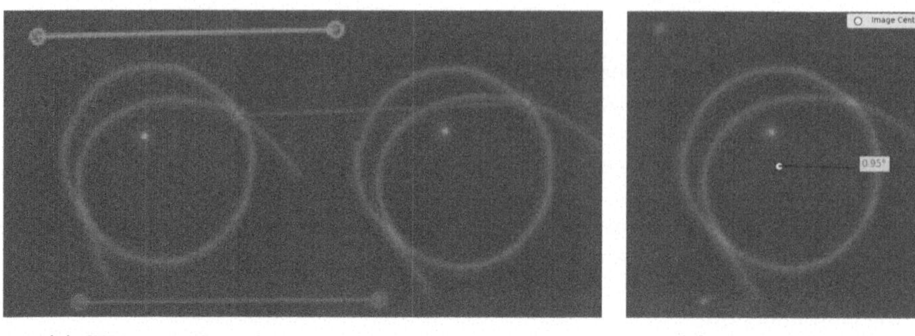

(a) SIFT matching betweem consecutive frames (b) Detected angle

Fig. 2. SIFT matching to estimate the rotation angle.

OpenCV's `FlannBasedMatcher` function to match SIFT keypoints between pairs of frames, see Fig. 2. For each matched pair of key points, we compute the vectors connecting the corresponding points and calculate their angles. These angles serve as the basis for estimating rotational movement.

To enhance robustness, this analysis is performed not only between consecutive frames but also between each frame and the initial frame. This redundancy helps mitigate issues such as insufficient keypoint detection or poor matching in individual frames. Outliers in the resulting angle data are removed using the Interquartile Range (IQR) method [26]. Finally, by counting the number of positive (counterclockwise) and negative (clockwise) rotation angles, we infer the overall direction of rotation and estimate the confidence level of the result. Note we do not mind if the computed angles are not always exact (in fact, it is very unlikely to obtain highly accurate results), we are interested in the rotation direction.

4.2 Selecting the Best Head Candidate

Once we have a model capable of distinguishing between heads and tails (Line 8) –perhaps with a low mAP– our next objective is to ensure reliable head tracking across consecutive frames. This step is critical, as the head-tail classification model may produce errors such as missed detections or misclassifications. To address this, we develop a tracking function that, given the last known position of the head and the current set of candidate detections (both heads and tails), selects the most likely continuation based on a scoring mechanism. This function estimates the direction of movement of the sperm and, using the rotation direction computed in Line 5 (Subsect. 4.1), calculates the cross product to prioritize candidates aligned with the expected rotation. More factors contribute to the scoring: higher scores are assigned to detections classified as heads and to those closer to the previous head position (encouraging spatial continuity). Candidates near the image center are penalized since they often correspond to overlaps or artifacts. The detection with the highest score above a predefined threshold is selected as the new head position. If no candidate meets the threshold, the previous head position is retained.

5 Results

Regarding the detection of motile sperm, Table 1 shows the results (in the validation set) of training different versions of YOLO (8, 11 and 12) with different model sizes (nano, small, medium, large, extralarge) and different input image sizes (640, 1200, 1920). Modern YOLO versions can handle arbitrary-sized images since they do not use fully connected (dense) layers, and the reshape is done simply through the *imgsz* parameter of the model. The neuronal networks were trained 100 epochs with an early stopping if the metric did not improve in 20 epochs. As can be seen, the results are very good in all cases, obtaining high mAP50 scores. This is due to the fact that the circular shape of the motile spermatozoa allows them to be easily detectable, even in the presence of overlapping or blurred areas in the samples. As the final model, we chose the YOLOv8 nano at 640 image size because it is a light model (therefore fast) with good results.

Table 1. mAP50 for motile sperm detection with different model configurations. Some YOLOv12 variants were not tested due to the long training time.

	Nano			Small			Medium			Large			ExtraLarge		
Input size	640	1200	1920	640	1200	1920	640	1200	1920	640	1200	1920	640	1200	1920
YOLOv8	0.964	0.915	0.983	0.961	0.936	0.978	0.954	0.963	0.970	0.947	0.954	0.979	0.947	0.961	0.970
YOLOv11	0.931	0.955	0.942	0.928	0.924	0.945	0.926	0.934	0.953	0.918	0.960	0.952	0.935	0.96	0.962
YOLOv12	0.939	0.938	0.958	0.941	0.941	0.942	–	–	–	–	–	–	–	–	–

Table 2. mAP50 score for detecting the sperm heads. Within the *heads and tails* configuration, we show the mAP50 score for detecting heads.

	Nano	Small	Medium	Large	ExtraLarge	Nano	Small	Medium	Large	ExtraLarge
	Heads					Heads and tails				
YOLOv8	0.634	0.649	0.551	0.606	0.634	0.706	0.618	0.632	0.648	0.655
YOLOv11	0.61	0.571	0.605	0.606	0.642	0.654	0.679	0.693	0.635	0.661
YOLOv12	0.611	–	–	–	–	0.64	–	–	–	–

With respect to the detection of the heads of the sperm, Table 2 shows the results obtained by different models, both predicting heads only and heads and tails. Although our primary interest lies in detecting heads, training the network to distinguish both heads and tails improves overall prediction performance.

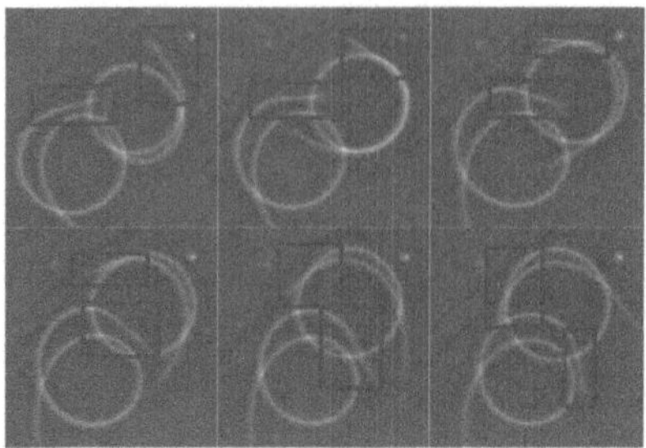

(a) Same crop of a video in six different frames with correct head detections

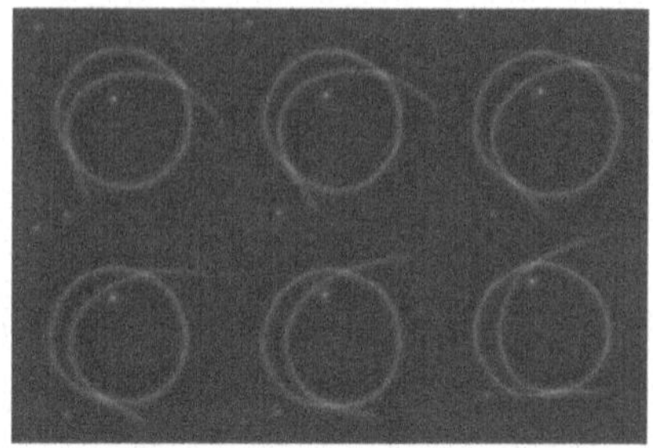

(b) Grid of six images of a sperm in a video used for RT-DETR

Fig. 3. Video analysis showing sperm tracking and detection results

This is because heads and tails are visually similar and can be easily confused, so learning to differentiate both helps the model make more accurate classifications. All models were trained with an input image size of 250 (which is approximately the crop size), for 1000 epochs, also with an early stopping criterion. Of all of them, we choose the YOLOv8 nano version (for heads and tails) for its balance between good metric and inference speed, obtaining a mAP50 0.706. To further improve this model, we modified the loss function by assigning greater weights to the classification loss component and the distribution focal loss. This adjustment prioritizes accurate class predictions over precise bounding box localization. Our primary objective is to ensure a clear distinction between heads and tails, even if it comes at the expense of slightly less accurate bounding boxes. After retraining that YOLOv8 nano-model (for heads and tails), the mAP50 score increased to 0.712 for the head class.

6 Discussion

Once all components of Algorithm 1 have been integrated, the results using our hybrid technique are good, as can be seen for example in Fig. 3a where two spermatozoa are shown close together in a video and the head detection performed in each frame with our algorithm. These detections allow trajectory detection. In addition to YOLO, we explored alternative detection approaches, including 3D convolutional neural networks (3D-CNNs) for incorporating motion information, and transformer-based models, which we hypothesized might capture higher-level temporal or spatial patterns—such as rotation direction—that standard convolutional architectures may overlook. One such attempt is shown in Fig. 3b: we constructed a 6-frame grid for each spermatozoon and fed it into a transformer model (Real-Time Detection Transformer, RT-DETR [17]), aiming to leverage the temporal sequence to estimate the head position. However, the results from this alternative method were significantly worse (mAP50 < 0.2). This poor performance is likely due to the limited amount of labeled training data available, which constrains the learning capacity of more complex architectures.

Although the results are acceptable, there are still difficulties when there are overlaps or very close spermatozoa and further research is necessary. In any case, it is something that can be minimized by entomologists by increasing the dilution of the samples and therefore decreasing the concentration in the image. As for performance, the combination of nano YOLO networks + SIFT points can be run on CPU in moderate time (the analysis needs about one minute per video, depending on the number of sperm). This makes it feasible for implementation in laboratories without the need for infrastructure or computing servers.

7 Conclusions and Further Work

This work introduces a hybrid approach for tracking drone sperm in video microscopy, combining lightweight deep learning models with classical computer vision techniques. Our method addresses the challenges posed by the morphology and motion of insect sperm—particularly the difficulty of distinguishing between head and tail—by leveraging a combination of YOLO-based object detection, SIFT-based rotation analysis, and a custom scoring mechanism for consistent head tracking. The results demonstrate the potential for practical deployment in beekeeping laboratories, offering a cost-effective tool to aid in the evaluation of drone reproductive health. As further work, we plan to develop a new CASABee program to include the detection of static sperm using deep learning and improve the current models using pseudolabelling with more videos.

Acknowledgments. This research was funded by MCIU/AEI/10.13039/501100011033 (grant: PID2023-148475OB-I00), the EU Horizon Europe (grant: 101082073), the DGA-FSE (grant: A07_23R), and La Rioja Government (grant: INICIA2023/02).

References

1. Aharon, N., Orfaig, R., Bobrovsky, B.Z.: Bot-sort: robust associations multi-pedestrian tracking. arXiv preprint arXiv:2206.14651 (2022)
2. Alquézar-Baeta, C., et al.: OpenCASA: a new open-source and scalable tool for sperm quality analysis. PLoS Comput. Biol. **15**(1), e1006691 (2019)
3. Andrijević, N., Urošević, V., Arsić, B., Herceg, D., Savić, B.: Iot monitoring and prediction modeling of honeybee activity with alarm. Electronics **11**(5), 783 (2022)
4. Bienefeld, K., Ehrhardt, K., Reinhardt, F.: Genetic evaluation in the honey bee considering queen and worker effects – a blup-animal model approach. Apidologie **38**(1), 77–85 (2006). https://doi.org/10.1051/apido:2006050
5. Calderone, N.W.: Insect pollinated crops, insect pollinators and US agriculture: trend analysis of aggregate data for the period 1992–2009. PLoS ONE **7**(5), e37235 (2012)
6. Carion, N., Massa, F., Synnaeve, G., Usunier, N., Kirillov, A., Zagoruyko, S.: End-to-end object detection with transformers. In: European Conference on Computer Vision, pp. 213–229 (2020)

7. Divasón, J., et al.: Analysis of varroa mite colony infestation level using new open software based on deep learning techniques. Sensors **24**(12), 3828 (2024)
8. Divasón, J., Romero, A., Santolaria, P., Yániz, J.L.: Zigzag persistence for image processing: new software and applications. Pattern Recogn. Lett. **184**, 111–118 (2024)
9. Divasón, J., Romero, A., Silvestre, M.A., Santolaria, P., Yániz, J.L.: In vitro maintenance of drones and development of a new software for sperm quality analysis facilitate the study of honey bee reproductive quality. J. Apic. Res. **63**(5), 1088–1095 (2024)
10. Hadjur, H., Ammar, D., Lefèvre, L.: Toward an intelligent and efficient beehive: a survey of precision beekeeping systems and services. Comput. Electron. Agric. **192**, 106604 (2022)
11. Khanam, R., Hussain, M.: Yolov11: an overview of the key architectural enhancements. arXiv preprint arXiv:2410.17725 (2024)
12. Klein, A.M., et al.: Importance of pollinators in changing landscapes for world crops. Proc. Roy. Soc. B Biol. Sci. **274**(1608), 303–313 (2006)
13. Kviesis, A., Zacepins, A.: Application of neural networks for honey bee colony state identification. In: 2016 17th International Carpathian Control Conference (ICCC), pp. 413–417. IEEE (2016)
14. Li, X., Hu, W., Shen, C., Zhang, Z., Dick, A., Hengel, A.V.D.: A survey of appearance models in visual object tracking. ACM Trans. Intell. Syst. Technol. (TIST) **4**(4), 1–48 (2013)
15. Lin, T.Y., Goyal, P., Girshick, R., He, K., Dollár, P.: Focal loss for dense object detection. In: Proceedings of the IEEE International Conference on Computer Vision, pp. 2980–2989 (2017)
16. Lowe, D.G.: Distinctive image features from scale-invariant keypoints. Int. J. Comput. Vision **60**(2), 91–110 (2004)
17. Lv, W., et al.: DETRs beat YOLOs on real-time object detection (2023)
18. Padilla, R., Passos, W.L., Dias, T.L.B., Netto, S.L., da Silva, E.A.B.: A comparative analysis of object detection metrics with a companion open-source toolkit. Electronics **10**(3), 279 (2021)
19. Paynter, E., Millar, A.H., Welch, M., Baer, B., Keller, A.: Insights into the molecular basis of long-term storage and survival of sperm in the honeybee (apis mellifera). Sci. Rep. **4**(1), 4236 (2014)
20. Rangel, J., Fisher, A.: Factors affecting the reproductive health of honey bee (apis mellifera) drones–a review. Apidologie **50**(6), 759–778 (2019)
21. Redmon, J., Divvala, S., Girshick, R., Farhadi, A.: You only look once: unified, real-time object detection. In: Proceedings of the IEEE Conference on Computer Vision and Pattern Recognition, pp. 779–788 (2016)
22. Ren, S., He, K., Girshick, R., Sun, J.: Faster R-CNN: towards real-time object detection with region proposal networks. In: Advances in Neural Information Processing Systems, vol. 28 (2015)
23. Salti, S., Cavallaro, A., Di Stefano, L.: Adaptive appearance modeling for video tracking: survey and evaluation. IEEE Trans. Image Process. **21**(10), 4334–4348 (2012)
24. Scutaru, D., et al.: An AI-based digital scanner for varroa destructor detection in beekeeping. Insects **16**(1), 75 (2025)
25. Sohan, M., Sai Ram, T., Rami Reddy, C.V.: A review on yolov8 and its advancements. In: Jacob, I.J., Piramuthu, S., Falkowski-Gilski, P. (eds.) International Conference on Data Intelligence and Cognitive Informatics, pp. 529–545. Springer, Singapore (2024). https://doi.org/10.1007/978-981-99-7962-2_39

26. Tukey, J.W., et al.: Exploratory data analysis, vol. 2. Springer (1977)
27. Vdoviak, G., Sledevič, T., Serackis, A., Plonis, D., Matuzevičius, D., Abromavičius, V.: Evaluation of deep learning models for insects detection at the hive entrance for a bee behavior recognition system. Agriculture **15**(10), 1019 (2025)
28. Yániz, J., et al.: An AI-based open-source software for varroa mite fall analysis in honeybee colonies. Agriculture **15**(9), 969 (2025)
29. Yániz, J.L., Silvestre, M.A., Santolaria, P.: Sperm quality assessment in honey bee drones. Biology **9**(7), 174 (2020)
30. Zhang, Y., et al.: Bytetrack: multi-object tracking by associating every detection box (2022)

Analyzing the Efficacy of Urban Low Emission Zones: A Machine Learning Approach to Air Quality in Madrid (Spain)

Gilberto J. Brito(✉), Manuel Curado, Stefan Rada, and Jose F. Vicent

Department of Computer Science and Artificial Intelligence, University of Alicante, Campus de San Vicente del Raspeig, Ap. Correos 99, 03080 Alicante, Spain
{gjb3,rsr65}@alu.ua.es, {manuel.curado,jvicent}@ua.es

Abstract. Urban air pollution presents a persistent and critical global health concern, significantly diminishing the quality of life in densely populated areas and contributing to a spectrum of environmental issues. Cities worldwide are increasingly adopting proactive measures to combat this, with Low Emission Zones (LEZs) and the strategic pedestrianization of urban spaces emerging as key interventions aimed at curbing vehicular emissions. This work evaluates the impact of Low Emission Zones and pedestrianization on Madrid's urban pollution mitigation program. Models trained on pre-policy data were evaluated against post-implementation measurements in three distinct Madrid districts. Furthermore, the approach was to forecast air pollutant concentration levels in multivariate time-series data leveraging deep learning models. Observed post-policy pollutant concentrations were lower than pre-policy forecasts which indicates a quantifiable positive impact of the mitigation program.

Keywords: Low Emission Zones · Time Series Analysis · Urban Pollution · LSTM · CNN · Predictive Modeling

1 Introduction

Urban air pollution is a critical global issue, due to its serious public health and environmental consequences. On the health side, it is associated with a higher risk of respiratory diseases such as ashtma, as well as cardiovascular conditions, particularly stroke [3]. Environmentally, its most significant negative effects include acid rain, climate change, and damage to soil and water quality [13].

Among the most widely adopted mitigation strategies are Low Emission Zones (LEZs) and pedestrianization policies. Analyzing the trade-offs involved is crucial, as environmental interventions often have complex and multifaceted consequences. While such policies may improve air quality, they can also result in increased economic costs and unequal social impacts, as discussed in [15,26].

E. Corchado et al. (Eds.): SOCO 2025, CCIS 2806, pp. 82–91, 2026.
https://doi.org/10.1007/978-3-032-19763-4_8

This work evaluates the impact of such policies in Madrid using a machine learning approach to forecast air pollutant concentrations. The pollutants analyzed in this study are nitrogen oxides (NO, NO_2, and NO_x) and particulate matter of different diameters ($PM_{2.5}$ and PM_{10}), as their impact on human health and the environment has been among the most extensively studied and documented.

The main contributions of this study include the use of real-world data from three distinct districts and the application of a rigorous counterfactual analysis using state-of-the-art predictive models.

2 Related Work

Various approaches have been employed to address the challenge of deducing the impact of air quality improvement programs in urban areas, particularly those involving the implementation of LEZs and pedestrianization strategies. A standard practice in such evaluations is the measurement of key pollutants (NO, NO_2, NO_x, PM_{10} and $PM_{2.5}$), given that their detrimental effects on human health have been extensively documented [14,21,25].

Also, the employed approach uses Differene-in-differences (DD) strategy [2], which estimates the effects of an applied intervention by comparing changes in the studied variable at different time points against a projected baseline. This approach is common in policy evaluation studies, some relevant ones are [6,16,17].

In contrast to [18–20] our work performs the forecasting of pollution levels using advanced deep learning models which are more powerful and capable of recognize more complex patterns than traditional machine learning models. Also, these works use only one data source (meteorological, satellite imagery, etc.), unlike the present work which integrates the various data sources (air quality, traffic intensity, and meteorological conditions). The models used are Long Short-Term Memory (LSTM) [8] and Convolutional Neural Networks (CNNs) [11]

The chosen models are adept at discerning more complex patterns from richer feature sets compared to simpler machine learning models [9,22,23].

3 Background

Low Emission Zones are designated urban areas that restricts access for high-polluting vehicles, in order to improve local air quality and public health. These regulations are often based on vehicle emission standards, each city has its own standard.

In the context of this work, Madrid's LEZs are regulated by the Sustainable Mobility Ordinance [12]. It has been applied progressively, with initial circulation prohibitions starting from January 1, 2022. The transitional regime to facilitate compliance ended on December 31, 2024. Since January 1, 2025, vehicles without an environmental label are prohibited from circulating within these LEZs.

Convolutional Neural Networks were originally developed for image processing, but are adapted for time-series analysis by treating sequences as one-dimensional (1D) signals, one example is [1]. 1D CNNs work by sliding small filters (kernels) across the input time series. Each filter is designed to detect specific local patterns or features, such as short-term trends or recurrent motifs, at different time steps. The outputs of these filters, known as feature maps, highlight where these patterns occur. Subsequent layers, often include pooling layers to summarize features and reduce dimensionality. Then, fully connected layers use these learned hierarchical features for the final forecasting task. Through this algorithm CNNs can learn relevant temporal features from the raw data.

Long Short-Term Memory are a type of Recurrent Neural Network (RNN) specifically designed to address the vanishing gradient problem, enabling them to learn long-range dependencies in sequential data . LSTMs achieve this through a sophisticated cell structure containing gates (input, forget, and output gates). These gates regulate the flow of information, allowing the network to selectively remember or forget information over extended time steps. This capability makes LSTMs particularly well-suited for modeling time series where long-term context is crucial for accurate forecasting as shown in [10,24].

Difference-in-differences is a technique widely used in econometrics, health and policy evaluation to estimate a causal effect of an specific intervention by comparing the change in outcomes over time in the exposed group to the change in the outcomes in the not exposed one, [4,5]. The core assumption is that, in the absence of the intervention, both groups would have followed parallel trends.

3.1 Evaluation Metrics

The performance of the forecasting models in this study is assessed using two standard regression metrics:

Mean Squared Error (MSE). MSE is a common metric used to measure the average squared difference between the estimated values and the actual values. It penalizes larger errors more heavily due to the squaring term, making it sensitive to outliers. For N samples, if y_i is the actual value and $\hat{y}_i$ is the predicted value, MSE is calculated as:

$$\text{MSE} = \frac{1}{N}\sum_{i=1}^{N}(y_i - \hat{y}_i)^2$$

Mean Absolute Error (MAE). MAE measures the average magnitude of the errors in a set of predictions, without considering their direction. It is the average over the test sample of the absolute differences between prediction and actual observation where all individual differences have equal weight. MAE is generally less sensitive to outliers than MSE. MAE is calculated as:

$$\text{MAE} = \frac{1}{N}\sum_{i=1}^{N}|y_i - \hat{y}_i|$$

4 Methodology

4.1 General Overview

The data employed were sourced from the open data portal of the Madrid City Council encompassing air quality data, traffic measurements along with their sensor locations, and meteorological data.

The followed approach in this work was to treat the forecasting of the studied pollutants as a multivariate time-series problem. Accordingly, various deep learning models were trained over the different sensor data. The strategy followed to measure the impact of the pollution reduction program involved using models trained exclusively on pre-intervention data to generate forecasts for the post-intervention period. This allowed for the measurement of the variation between model predictions and the actual post-intervention observations.

The selected models were chosen based on experimentation, as they demonstrated superior performance during the modeling phase on the pre-intervention dataset.

4.2 Data

The dataset employed includes air quality data, traffic measurements with their corresponding sensor locations, and meteorological data from the last three available years (2022–2025). These three distinct datasets were combined into a single, comprehensive dataset. The specific input features and target variables utilized are detailed in Table 1.

It is worth mentioning that the air quality and meteorological data were extracted from the measurements' stations data. These stations locations were grouped in each district dataset.

4.3 Models Architecture

The selection of these architectures is justified as they yielded the best performances during an iterative experimental process according to the evaluation metrics.

***CNN*.** The CNN model employs a deep one-dimensional convolutional (Conv1D) architecture composed of five stacked convolutional layers. These layers progressively reduce the feature dimensionality from 512 to 32 filters, all using the ReLU activation function. Each Conv1D layer maintains the temporal dimension (sequence length of 1) while increasing the depth of learned representations. After the final convolutional layer, a `Flatten` layer is used to convert the output into a 1D vector, which is then passed to a fully connected output layer with five units—one for each target pollutant. The model is trained using MSE loss function and optimized with the Nadam optimizer.

***LSTM*.** For the LSTM model the length of the sequence is 1 and each timestep includes 10 features. Also, it is composed of a stack of five sequential LSTM

Table 1. Variables utilized in the study

Input Features	
Variable Name	Description
Timestamp	Date and time of measurement
Wind Speed	Speed of the wind
Wind Direction	Direction of the wind
Temperature	Ambient temperature
Relative Humidity	Relative humidity level
Barometric Pressure	Atmospheric pressure
Solar Radiation	Solar radiation intensity
Precipitation	Amount of precipitation
Traffic Flow	Vehicle count at a measurement point
District Traffic Intensity	Aggregated traffic intensity within the district
Target Pollutants	
Variable Name	Description
NO	Nitric Oxide concentration
NO_2	Nitrogen Dioxide concentration
NO_x	Nitrogen Oxides (total) concentration
PM_{10}	Particulate Matter $<$ 10µm concentration
$PM_{2.5}$	Particulate Matter $<$ 2.5µm concentration

layers, with each layer progressively reducing the hidden state dimensionality from 512 to 32 units. Each LSTM layer is followed by a Dropout layer with a dropout rate of 20% to prevent overfitting, following the regularization strategy proposed by [7]. The final output layer is a fully connected layer with five units, corresponding to the predicted concentrations of the five target pollutants. The model is trained using MSE and optimized with the Nadam optimizer.

4.4 Impact Analysis Framework

The models trained exclusively with pre-intervention data, predict the levels of pollutants over the post-intervention data. This counterfactual scenario evaluates the efficacy of these interventions. The evaluation is performed using the percentage of predictions ($\hat{y}_i$) that are greater than the actual values (y_i).

Let N be the total number of observations in the post-intervention period, and $I(\hat{y}_i > y_i)$ be an indicator function that equals 1 if the predicted value $\hat{y}_i$ is greater than the actual value y_i, and 0 otherwise. The impact metric, termed "Positive Impact Percentage" (PIP), is given by:

$$\text{PIP} = \left(\frac{\sum_{i=1}^{N} I(\hat{y}_i > y_i)}{N} \right) \times 100$$

A higher PIP suggests a greater reduction in pollutant concentrations attributable to the interventions, as the observed reality is better than the no-policy-change forecast.

5 Experiments

This section details the experimental setup, which steps are data selection, data preparation, model training and evaluation of the models and the impact of the environmental policies.

5.1 Data Selection

This work only focuses on the LEZs implementation within three Madrid districts: Salamanca, Tetuán, and Villa de Vallecas, each presenting unique urban and socioeconomic characteristics.

Salamanca, centrally located, spans approximately 5.38 km^2 with a population of around 147,700, resulting in a density of about 27,400 inhabitants per km^2. The pedestrianization projects, such as the transformation of Calle Recoletos and Calle del Cid in February 2022, converted over 2,700 m^2 into pedestrian-friendly spaces.

Tetuán, situated northwest of the city center, covers 5.38 km^2 and houses approximately 155,600 residents, yielding a density of 28,888 inhabitants per km^2. The district is characterized by a mix of residential and commercial zones, the LEZs measures in Tetuán aim to mitigate traffic pollution.

Villa de Vallecas, located in the southeastern periphery, encompasses 51.59 km^2 with a population of about 122,300, leading to a lower density of approximately 2,370 inhabitants per km^2. This district has experienced significant urban development, particularly in the Ensanche de Vallecas area, and the LEZs initiatives here focus on improving air quality and promoting sustainable mobility.

5.2 Data Preparation

For the experimental phase, outliers were removed by defining the lower and upper bounds for measurements as (Q1 - 1.5 * IQR) and (Q3 + 1.5 * IQR), respectively, where Q1 and Q3 are the first and third quartiles, and IQR is the interquartile range. Furthermore, the pre-intervention data was split into training and validation sets using a 70–30 ratio. Numerical features were scaled to a [0, 1] range using Min-Max scaling prior to model training to ensure uniformity and improve convergence.

5.3 Pre-intervention Training Results

The performance of the models on the pre-intervention test set was evaluated using the MSE and MAE metrics. The obtained results are presented in Table 2.

Based on these results, the models that performed the best in this test set were chosen for the subsequent impact analysis.

The errors of NO_x and NO_2 are considerably higher either in MSE and MAE. This is due to the variability in the data, both measurements have a wide range of values and outliers compared to other pollutants.

Table 2. Pre-intervention model performance on the test set

	LSTM		**CNN**	
Pollutant	MAE	MSE	MAE	MSE
NO	2.10	9.09	2.10	8.98
NO_2	10.64	168.78	10.85	172.12
NO_x	13.15	258.21	13.45	263.68
PM_{10}	7.43	88.97	7.35	87.43
$PM_{2.5}$	2.48	15.07	2.54	14.97

5.4 Post-intervention Impact Analysis

The impact of the implemented measures is considered beneficial for air quality improvement, as the observed pollutant concentrations were predominantly lower than the counterfactual levels predicted by the models. A summary of these variations, quantified using the PIP metric, is presented in Table 3.

Salamanca is a high socioeconomic level district. There is a clear correlation between NO levels and the presence of diesel-powered vehicles, which emit larger amounts of NO. This could explain why the reduction of NO in Salamanca is significantly higher compared to other pollutants.

Tetuán, in contrast, shows a more uniform reduction across all pollutants, with similar percentage decreases observed for NO, NO_2, NO_x, and particulate matter. However, compared to Salamanca, non-NO pollutants in Tetuán experienced a more pronounced decrease. This may be due to the predominance of gasoline-powered vehicles in the district, which emit relatively lower levels of NO compared to diesel engines.

Villa de Vallecas stands out for having the lowest socioeconomic level among the districts analyzed. This factor is reflected in the older average age of vehicles circulating in the area. Prior to the implementation of traffic restrictions, older vehicles made up a significant portion of traffic and thus contributed substantially to air pollution. After the policies were enacted, a more substantial reduction in pollution levels was observed in this district compared to the others. The reduction is particularly notable for $PM_{2.5}$ concentrations, which may be attributed to the prevalence of gasoline-powered vehicles which emit fewer fine particles than diesel vehicles more common in wealthier areas.

Table 3. Positive Impact Percentage of interventions by pollutant and district

Pollutant	Salamanca	Tetuán	Villa de Vallecas
NO	82.12% ↓	62.23% ↓	76.62% ↓
NO_2	58.47% ↓	61.44% ↓	60.46% ↓
NO_x	59.14% ↓	64.69% ↓	60.94% ↓
PM_{10}	55.28% ↓	55.39% ↓	57.21% ↓
$PM_{2.5}$	56.05% ↓	56.80% ↓	79.00% ↓

6 Conclusions

To conclude, this work demonstrates the tangible effectiveness of LEZs and pedestrianization policies in the districts of Villa de Vallecas, Tetuán and Salamanca in order to improve urban air quality. Using deep learning models, a counterfactual analysis was conducted, revealing that the predicted pollutants levels—based on pre-intervention data—consistently exceeded the actual observed values. This discrepancy suggests that the implemented policies have had a significant positive impact on reducing air pollution in the studied districts.

All analyzed pollutants, including NO, NO2, NOx, PM10, P2.5 showed notable reduction in concentration. These findings support the existence of a causal relationship between the deployment of LEZs and pedestrian zones with the observed improvements in air quality.

The positive environmental impact observed underscores the value of proactive urban planning, which not only enhances air quality but also contributes to public health and citizens' well-being.

Future work could involve expanding the analysis to additional districts to determine whether pollution has been displaced to surrounding areas not covered by LEZ policies. Additionally, evaluating other mitigation strategies could help identify those with more favorable trade-offs between environmental benefit, economic cost, and social equity. Geographic characteristics could also be integrated into the models to refine predictions. Finally, incorporating socioeconomic data could allow for the development of more equitable and targeted policies, adapting interventions based on factors such as average income or vulnerability of the population in each zone.

Acknowledgements. Financial support for this research has been provided under grant CIGE/2023/52 funded by Generalitat Valenciana, Conselleria de Educación, Cultura, Universidades y Empleo (Spain).

References

1. Borovykh, A., Bohte, S., Oosterlee, C.W.: Conditional time series forecasting with convolutional neural networks (2018). https://arxiv.org/abs/1703.04691
2. Card, D., Krueger, A.B.: Minimum wages and employment: a case study of the fast-food industry in New Jersey and Pennsylvania. Am. Econ. Rev. **84**(4), 772–793 (1994). https://doi.org/10.1257/aer.84.4.772, http://www.jstor.org/stable/2118030
3. Dominski, F.H., Branco, J.H.L., Buonanno, G., Stabile, L., da Silva, M.G., Andrade, A.: Effects of air pollution on health: a mapping review of systematic reviews and meta-analyses. Environ. Res. **201**, 111487 (2021). https://doi.org/10.1016/j.envres.2021.111487, epub 2021 Jun 8
4. Feng, S., Ganguli, I., Lee, Y., Poe, J., Ryan, A., Bilinski, A.: Difference-in-differences for health policy and practice: A review of modern methods (2024). https://arxiv.org/abs/2408.04617
5. Goodman-Bacon, A.: Difference-in-differences with variation in treatment timing. Working Paper 25018, National Bureau of Economic Research (2018). https://doi.org/10.3386/w25018, http://www.nber.org/papers/w25018
6. Guo, J., Wu, X., Guo, Y., Tang, Y., Dzandu, M.D.: Spatiotemporal impact of major events on air quality based on spatial differences-in-differences model: big data analysis from China. Nat. Hazards **107**(3), 2583–2604 (2021). https://doi.org/10.1007/s11069-021-04517-y
7. Hinton, G.E., Srivastava, N., Krizhevsky, A., Sutskever, I., Salakhutdinov, R.R.: Improving neural networks by preventing co-adaptation of feature detectors (2012). https://arxiv.org/abs/1207.0580
8. Hochreiter, S., Schmidhuber, J.: Long short-term memory. Neural Comput. **9**(8), 1735–1780 (1997). https://doi.org/10.1162/neco.1997.9.8.1735
9. Jin, S.: A comparative analysis of traditional and machine learning methods in forecasting the stock markets of China and the US. Int. J. Adv. Comput. Sci. Appl. **15**(4) (2024). https://doi.org/10.14569/IJACSA.2024.0150401
10. Kong, Y., et al.: Unlocking the power of LSTM for long term time series forecasting (2025). https://arxiv.org/abs/2408.10006
11. Le Cun, Y., et al.: Handwritten digit recognition with a back-propagation network. In: Proceedings of the 3rd International Conference on Neural Information Processing Systems, pp. 396–404. NIPS'89, MIT Press, Cambridge, MA, USA (1989)
12. Madrid City Council: Ordenanza de movilidad sostenible de madrid. BOE Madrid City Council, núm. 8.267, 23 de octubre de 2018. Ver. 24/09/2023 (2018). https://sede.madrid.es/eli/es-md-01860896/odnz/2018/10/23/(1)/con/20230924/spa/pdf, changes until ANM 2023/152
13. Manisalidis, I., Stavropoulou, E., Stavropoulos, A., Bezirtzoglou, E.: Environmental and health impacts of air pollution: a review. Front. Public Health **8**, 14 (2020). https://doi.org/10.3389/fpubh.2020.00014, published 2020 Feb 20
14. Nitschke, M.: Respiratory health effects of nitrogen dioxide exposure and current guidelines. Int. J. Environ. Health Res. **9**(1), 39–53 (1999). https://doi.org/10.1080/09603129973344
15. OECD: Assessing the Economic Impacts of Environmental Policies: Evidence from a Decade of OECD Research. OECD Publishing, Paris (2021). https://doi.org/10.1787/bf2fb156-en

16. Park, D., Lim, B.I.: The effect of the ultra-low emission zone on PM2.5 concentration in Seoul, South Korea. Atmos. Environ. **340**, 120908 (2025). https://doi.org/10.1016/j.atmosenv.2024.120908, https://www.sciencedirect.com/science/article/pii/S1352231024005831
17. Prieto-Rodriguez, J., Perez-Villadoniga, M.J., Salas, R., Russo, A.: Impact of London toxicity charge and ultra low emission zone on NO2. Transport Policy **129**, 237–247 (2022). https://doi.org/10.1016/j.tranpol.2022.10.010, https://www.sciencedirect.com/science/article/pii/S0967070X22002955
18. Rahman, M.M., et al.: AirNet: predictive machine learning model for air quality forecasting using web interface. Environ. Syst. Res. **13**(1), 44 (2024). https://doi.org/10.1186/s40068-024-00378-z
19. Ravindiran, G., Hayder, G., Kanagarathinam, K., Alagumalai, A., Sonne, C.: Air quality prediction by machine learning models: a predictive study on the Indian coastal city of Visakhapatnam. Chemosphere **338**, 139518 (2023). https://doi.org/10.1016/j.chemosphere.2023.139518, https://www.sciencedirect.com/science/article/pii/S004565352301785X
20. Samad, A., Garuda, S., Vogt, U., Yang, B.: Air pollution prediction using machine learning techniques – an approach to replace existing monitoring stations with virtual monitoring stations. Atmos. Environ. **310**, 119987 (2023). https://doi.org/10.1016/j.atmosenv.2023.119987, https://www.sciencedirect.com/science/article/pii/S1352231023004132
21. Sarmiento, L., Wägner, N., Zaklan, A.: The air quality and well-being effects of low emission zones. J. Public Econ. **227**, 105014 (2023). https://doi.org/10.1016/j.jpubeco.2023.105014, https://www.sciencedirect.com/science/article/pii/S0047272723001962
22. Shobayo, O., Adeyemi-Longe, S., Popoola, O., Okoyeigbo, O.: A comparative analysis of machine learning and deep learning techniques for accurate market price forecasting. Analytics **4**(1) (2025). https://doi.org/10.3390/analytics4010005, https://www.mdpi.com/2813-2203/4/1/5
23. Siami-Namini, S., Namin, A.S.: Forecasting economics and financial time series: ARIMA vs. LSTM. CoRR abs/1803.06386 (2018). http://arxiv.org/abs/1803.06386
24. Siami-Namini, S., Tavakoli, N., Namin, A.S.: A comparative analysis of forecasting financial time series using ARIMA, LSTM, and BiLSTM (2019). https://arxiv.org/abs/1911.09512
25. Solomon, P.A.: Air pollution and health: bridging the gap from sources to health outcomes. Environ. Health Perspect. **119**(4), 156–157 (2011). https://doi.org/10.1289/ehp.1103660
26. Wang, S., et al.: The costs, health and economic impact of air pollution control strategies: a systematic review. Global Health Res. Policy **9**(1), 30 (2024). https://doi.org/10.1186/s41256-024-00373-y

Fusion of UAV Images for 3D Road Defect Analysis: A Context-Enhanced Framework for Depth and Geospatial Mapping

Paula López Álvarez(✉), Pablo Zubasti Recalde, Jesús García Herrero, and José Manuel Molina López

Department of Computer Science, Universidad Carlos III de Madrid, Colmenarejo, Spain
{palopeza,jgherrer}@inf.uc3m.es, pzubasti@pa.uc3m.es, molina@ia.uc3m.es
https://www.uc3m.es

Abstract. Efficient road maintenance requires accurate detection of surface defects. Manual inspections are labor-intensive and error-prone, limiting scalability. This study introduces a UAV-based framework integrating adaptive flight planning, hybrid segmentation, and lightweight 3D reconstruction with geospatial mapping. Defects are segmented using the Segment Anything Model (SAM) combined with adaptive thresholding, and the resulting masks are fused and interpolated to produce grayscale depth maps for surface modeling. Boundaries are georeferenced into GIS layers, enabling spatial analysis and long-term monitoring. The framework is evaluated with RMSE, MAE, PSNR, SSIM, IoU, and Dice Score, showing strong agreement between reconstructed models and expert annotations, confirming the method's effectiveness.

Keywords: UAV imagery · 3D reconstruction · Geospatial mapping

1 Introduction

Road infrastructure is critical for economic development and safety, yet it degrades over time due to traffic, construction flaws, and environmental stressors such as water infiltration and temperature changes [8]. These factors cause cracks, potholes, and rutting, which increase maintenance costs and accident risks, especially in urban areas.

Traditional inspections are manual, subjective, and hazardous in traffic-exposed zones [1]. Their limited scalability and delayed processing hinder timely maintenance [18].

Unmanned Aerial Vehicles (UAVs) offer a scalable alternative, enabling rapid and safe image acquisition [15]. Combined with deep learning, UAVs support automated defect detection [13]. However, most works rely on two-dimensional imagery, lacking depth and geospatial context essential for structural assessment [2].

E. Corchado et al. (Eds.): SOCO 2025, CCIS 2806, pp. 92–101, 2026.
https://doi.org/10.1007/978-3-032-19763-4_9

1.1 Main Contributions

This work proposes a lightweight UAV framework for 3D road defect analysis from single-view images. The main contributions are summarized as follows:

- **Path planning:** Efficient UAV trajectories via the Nearest Neighbour algorithm minimize flight time and redundancy.
- **Hybrid segmentation:** Fuses the Segment Anything Model (SAM) and adaptive thresholding to capture both contours and textures.
- **3D reconstruction:** Interpolates fused masks to estimate relative depth with low computational cost.
- **Geospatial mapping:** Projects defect contours into GIS coordinates using UAV GPS and camera data.
- **Evaluation:** Quantitative validation with RMSE, MAE, PSNR, SSIM, R^2, IoU, and Dice metrics.

2 Related Work

The detection and assessment of road infrastructure damage are essential for ensuring transportation safety and efficiency. Over the years, various methods have been developed, ranging from manual inspections to computer vision, photogrammetry, and sensor-based techniques. This section reviews key advances in UAV-based inspection, 3D reconstruction, sensor fusion, and GIS integration.

2.1 UAV-Based Road Inspection

UAVs provide faster, safer, and more scalable alternatives to traditional ground-based assessments for road surface inspection. When combined with computer vision methods (e.g., CNN-based segmentation, YOLO detection), they enable accurate detection of surface anomalies [15]. However, two-dimensional image analysis faces limitations such as occlusions, lighting variability, and scale distortion, which motivate the use of complementary three-dimensional and spatial techniques.

2.2 3D Reconstruction of Road Surfaces

Two-dimensional UAV imagery identifies visible defects but lacks geometric depth and morphological detail. Laser-based systems such as Terrestrial Laser Scanners (TLS), Mobile Laser Scanning (MLS), and LiDAR generate high-resolution 3D point clouds for accurate rut and pothole measurement [6]. However, their cost, weight, and complexity hinder scalability when deployed on UAVs.

Photogrammetry and stereo vision offer low-cost, UAV-compatible alternatives by estimating 3D geometry from overlapping images through camera pose recovery and dense matching. Nevertheless, these methods require calibrated dual-camera setups and may be sensitive to lighting and motion artifacts [7,11].

2.3 GIS Integration and Spatial Analysis

Integrating UAV data with Geographic Information Systems (GIS) enables road monitoring by linking defect information with georeferenced maps, facilitating visual inspection and spatial pattern analysis. For instance, Hoang et al. [12] mapped pavement distress using UAV-GIS fusion, while Chun et al. [2] applied machine learning within GIS environments to automate crack detection and classification. Such approaches support multi-scale diagnostics and data-driven maintenance planning.

2.4 Limitations in Existing Work

Despite recent advances, four major limitations persist in previous studies and have been considered in the design of this framework. First, predefined static flight paths often lead to redundant image capture and inefficient energy use [5]. Second, segmentation performance remains sensitive to pavement variability, as CNN-based methods typically require large annotated datasets for generalization. Third, many approaches provide limited three-dimensional insight, neglecting valuable depth-related information [4]. Finally, geospatial mapping is frequently omitted or poorly integrated with GIS platforms [17].

The present work addresses these limitations by combining efficient trajectory planning, robust hybrid segmentation, and lightweight 3D reconstruction with geospatial integration, enabling scalable UAV-based monitoring of road surface defects.

3 Materials and Methods

The proposed four-phase pipeline includes flight path planning, defect segmentation, three-dimensional reconstruction, and geospatial mapping. The approach integrates UAV-based imaging, computer vision techniques, and geospatial analysis to enhance the detection and structural assessment of pavement defects.

3.1 Study Area and Path Planning

The study was conducted in a simulated environment using AirSim (Aerial Informatics and Robotics Simulation) [16], a high-fidelity simulator for UAV research. Defect coordinates were obtained from a previous study [14], and an optimized flight path was generated using the Nearest Neighbour (NN) algorithm [9], which provides an efficient approximation to the Travelling Salesman Problem (TSP) [3] by iteratively visiting the nearest unvisited defect. Euclidean distance was used to minimize the total route length.

3.2 Segmentation and 3D Reconstruction

The segmentation process combines two complementary techniques to improve robustness. First, the Segment Anything Model (SAM), a vision transformer capable of extracting object boundaries without task-specific training, is applied to generate initial defect contours. SAM's generalization ability across visual domains makes it well suited for UAV imagery, particularly in unsupervised scenarios. Second, adaptive grayscale thresholding is performed to refine internal textures and defect shapes. These outputs are fused into a single binary mask prior to reconstruction.

To determine the optimal binarization threshold T_{opt}, a heuristic is introduced based on the cumulative count of foreground pixels (value 0) in the binary mask. The threshold is selected such that the number of zero-valued pixels closely matches a target quantile α of the distribution of foreground pixel counts across all possible thresholds.

Let $\mathcal{V} = \{P(T_i)\}_{T_i \in \{0,5,\ldots,255\}}$ denote the set of pixel counts for threshold candidates T_i sampled every 5 intensity levels in the range $[0, 255]$, where $P(T)$ is defined as:

$$P(T) = \sum_{i=0}^{H-1} \sum_{j=0}^{W-1} \begin{cases} 1, & \text{if } f_T^*(x_{ij}) = 0, \\ 0, & \text{otherwise.} \end{cases} \tag{1}$$

Then, the optimal threshold is given by:

$$T_{\text{opt}} = \arg\min_{T} |P(T) - Q_\alpha(\mathcal{V})|, \tag{2}$$

where $Q_\alpha(\mathcal{V})$ denotes the α-quantile of the distribution $\mathcal{V}$, and $\alpha \in (0, 1)$ controls the target density of foreground pixels. As discussed in Sect. 4, $\alpha = 0.4$ yielded the best segmentation performance.

After binarization, grayscale values are reassigned row-wise using a piecewise linear function centered at the midpoint $h[i]$, computed as the integer average between the leftmost ($m[i]$) and rightmost ($M[i]$) foreground pixels:

$$I(x_{ij}) = \begin{cases} \dfrac{h[i] - j}{h[i] - m[i]} I_{\max}, & \text{if } j \leq h[i], \\ \dfrac{j - h[i]}{M[i] - h[i]} I_{\max}, & \text{if } j > h[i]. \end{cases} \tag{3}$$

This symmetric interpolation creates a smooth intensity gradient from edges to center, which is later used as a proxy for surface depth. Gaussian smoothing is applied to reduce discontinuities between adjacent rows (Fig. 1).

For 3D visualization, interpolated grayscale values are assumed to vary linearly with surface depth and are normalized to the range $[0, 1]$ for Z-axis mapping, where $Z = 0$ denotes the darkest pixel (maximum depth).

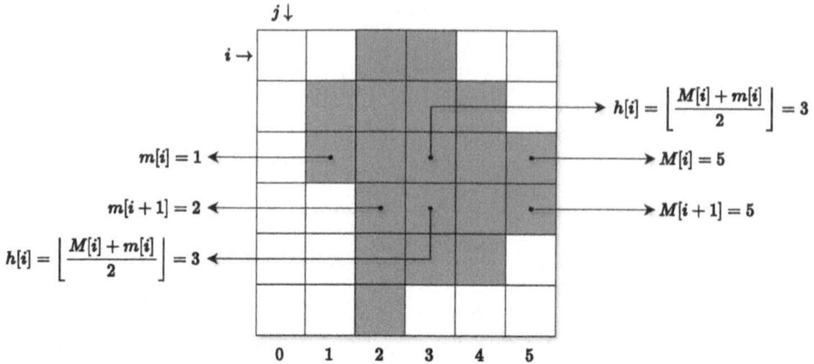

Fig. 1. Example of row-wise grayscale interpolation. Intensity decreases toward the center.

3.3 Contour Detection and Geospatial Mapping

Final contours are extracted using Canny edge detection on the binary masks. Pixels are mapped to geographic coordinates using UAV GPS data, altitude, and camera intrinsics. The process involves three main steps:

1. **Pixel normalization.** Image coordinates $(X_{\text{pixel}}, Y_{\text{pixel}})$ are centered relative to the optical center $(X_{\text{center}}, Y_{\text{center}})$ and scaled by focal lengths (f_x, f_y):

$$X_{\text{norm}} = \frac{X_{\text{pixel}} - X_{\text{center}}}{f_x}, \quad Y_{\text{norm}} = \frac{Y_{\text{pixel}} - Y_{\text{center}}}{f_y}. \tag{4}$$

2. **Metric projection.** The normalized coordinates are projected into real-world distances $(X_{\text{real}}, Y_{\text{real}})$ on the UAV's ground plane using an estimated depth Z:

$$X_{\text{real}} = X_{\text{norm}} \cdot Z, \quad Y_{\text{real}} = Y_{\text{norm}} \cdot Z. \tag{5}$$

3. **Geodetic conversion (latitude and longitude).** Using the UAV's geolocation $(Lat_{\text{center}}, Lon_{\text{center}})$ and Earth's radius R:

$$Lat = Lat_{\text{center}} + \frac{Y_{\text{real}}}{R}, \quad Lon = Lon_{\text{center}} + \frac{X_{\text{real}}}{R \cdot \cos(Lat_{\text{center}})}. \tag{6}$$

Contours are saved as Shapefiles using the EPSG:4326 reference system, enabling integration with standard GIS platforms for further spatial analysis.

4 Results and Evaluation

This section reports outcomes from the four pipeline stages: flight path optimization, segmentation, 3D reconstruction, and GIS localization, using 41 annotated defects in a simulated environment.

4.1 Flight Path Optimization

A Nearest Neighbour (NN)-based trajectory was generated to minimize travel distance while ensuring complete coverage of all defects. The UAV flew at a constant altitude and paused at each waypoint for image capture. Figure 2 illustrates the optimized route.

Fig. 2. Optimized UAV flight path with waypoints (red dots) at defect locations. The start and end points coincide, indicated by the arrow labeled "L." (Color figure online)

4.2 Segmentation Performance

Segmentation performance was evaluated against 41 expert-annotated ground truth masks using seven quantitative metrics grouped into two categories. Pixel-level error metrics include Root Mean Squared Error (RMSE), Mean Absolute Error (MAE), Peak Signal-to-Noise Ratio (PSNR) [10], and the coefficient of determination (R^2), which quantify numerical deviation between predicted and ground truth masks. Structural similarity and spatial alignment were assessed using the Structural Similarity Index Measure (SSIM) [19], Intersection over Union (IoU), and Dice Score, capturing perceptual fidelity and overlap between predicted and annotated regions.

Table 1 summarizes the quantitative results. Median IoU = 0.79 and Dice = 0.88 indicate high spatial consistency, while SSIM values above 0.99 confirm strong perceptual agreement. Outliers in R^2 and MAE correspond to shallow or low-contrast defects, which are inherently harder to delineate precisely.

The binarization parameter α, introduced in Sect. 3, was tuned via grid search. The setting $\alpha = 0.4$ achieved the best balance between contour fidelity and over-segmentation, yielding optimal SSIM, IoU, and Dice scores. This configuration was used for all reported results.

4.3 3D Reconstruction Visualizations

Grayscale masks were rendered as three-dimensional surfaces using bicubic interpolation and Gaussian smoothing. Figures 3 and 4 illustrate representative reconstructions from multiple viewpoints.

Table 1. Segmentation metrics comparing fused masks to ground truth.

Metric	Mean	Median	Std	Min	Max
RMSE	7.28	5.54	3.08	2.84	16.43
MAE	0.57	0.40	0.45	0.13	2.10
PSNR (dB)	32.47	33.26	3.53	23.82	39.07
SSIM	0.990	0.993	0.007	0.966	0.997
R^2	0.59	0.65	0.24	-0.18	0.83
IoU	0.77	0.79	0.10	0.54	0.94
Dice	0.86	0.88	0.07	0.71	0.97

(a) Ground truth mask.

(b) Fused interpolated mask.

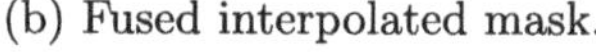

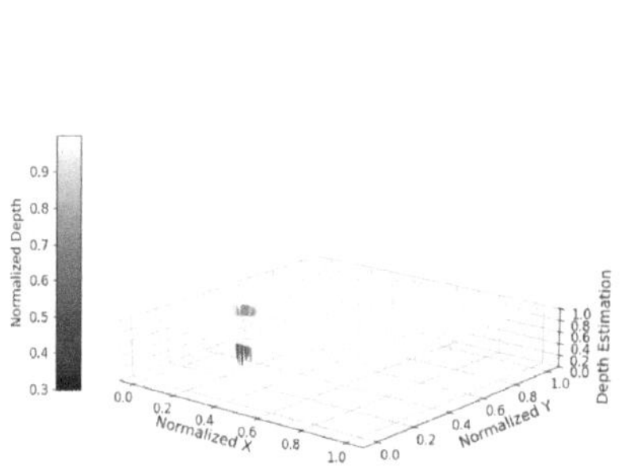

(c) 3D view at elev = 45°, azim = 135°.

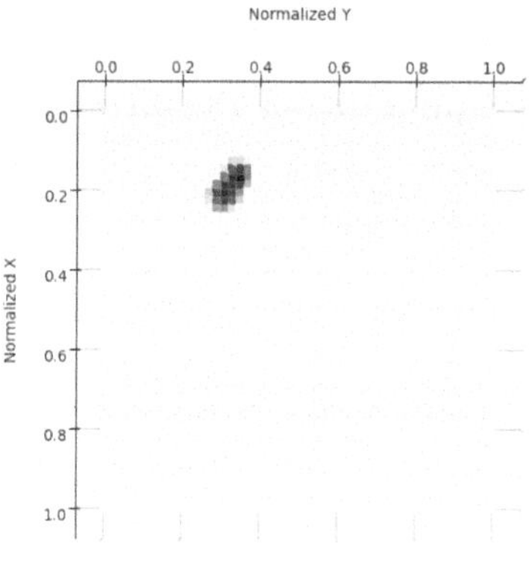

(d) Top-down view at elev = 90°, azim = 0°.

Fig. 3. Comparison of ground truth and reconstructed 3D surfaces.

The results confirm that the fusion and interpolation pipeline effectively capture both surface geometry and relative depth. The visual resemblance to ground truth supports its fidelity. An additional example is presented in Fig. 4, reinforcing the alignment between annotated masks and reconstructed surfaces.

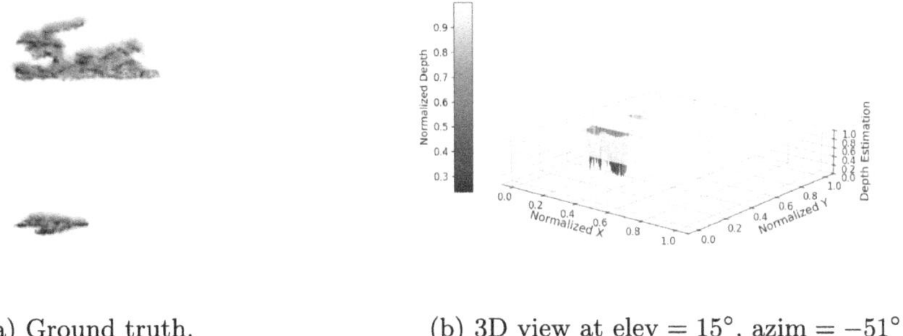

(a) Ground truth.

(b) 3D view at elev = 15°, azim = −51°.

Fig. 4. Additional example comparing ground truth and 3D reconstruction.

(a) Top-down geospatial view with geolocated defect markers.

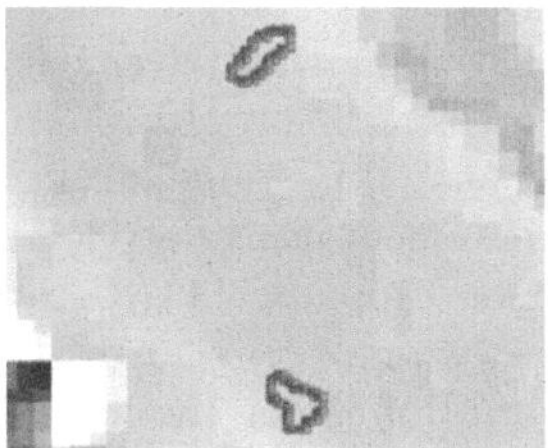

(b) Zoomed-in GIS visualization of interpolated defect contours.

Fig. 5. Geospatial representation of segmented road defects in GIS.

4.4 Geospatial Mapping

Detected defect contours were georeferenced using the EPSG:4326 (WGS 84) reference system and exported as Shapefiles for GIS integration. This step enables multi-scale spatial analysis and the transition from visual inspection to data-driven infrastructure management.

Figures 5a and 5b show the global spatial distribution of detected defects and a detailed view of individual interpolated contours, respectively.

5 Conclusions and Future Work

This paper presented a lightweight method for 3D reconstruction and geospatial mapping of road defects from single-view UAV imagery. The integration of SAM segmentation, adaptive thresholding, and grayscale interpolation enables surface profiling without stereo or LiDAR. Fused masks align with expert annotations, and 3D models preserve relative depth and geometry. Outputs are exported as EPSG:4326 Shapefiles for GIS-based maintenance planning.

Simulation results are promising, though lighting and material variations may affect grayscale-depth mapping, and real-world validation is pending. Future work will integrate monocular depth sensing, enable real-time deployment, classify defect types, and validate under field conditions.

Acknowledgments. Fun by public research projects of the Spanish Ministry of Science and Innovation under grants PID2023-151605OB-C2, PID2020-118249RB-C22, and TED2021-131520B-C22 (PEICTI 2021-2023 call).

Disclosure of Interests. The authors declare no competing interests.

References

1. Bogus, S., Song, J., Waggerman, R., Lenke, L.: Rank correlation method for evaluating manual pavement distress data variability. J. Infrastruct. Syst. **16** (2010). https://doi.org/10.1061/(ASCE)1076-0342(2010)16:1(66)
2. Chun, P.j., Yamane, T., Tsuzuki, Y.: Automatic detection of cracks in asphalt pavement using deep learning to overcome weaknesses in images and GIS visualization. Appl. Sci. **11**(3), 892 (2021). https://doi.org/10.3390/app11030892, https://www.mdpi.com/2076-3417/11/3/892, number: 3 Publisher: Multidisciplinary Digital Publishing Institute
3. Davendra, D.: Traveling Salesman Problem: Theory and Applications. BoD – Books on Demand (2010). google-Books-ID: gKWdDwAAQBAJ
4. Ellenberg, A., Kontsos, A., Moon, F., Bartoli, I.: Bridge deck delamination identification from unmanned aerial vehicle infrared imagery. Autom. Constr. **72**, 155–165 (2016). https://doi.org/10.1016/j.autcon.2016.08.024
5. Guerrero, J.A., Bestaoui, Y.: UAV path planning for structure inspection in windy environments. J. Intell. Robot. Syst. **69**(1), 297–311 (2013). https://doi.org/10.1007/s10846-012-9778-2
6. Inzerillo, L., Mino, G.D., Roberts, R.: Image-based 3D reconstruction using traditional and UAV datasets for analysis of road pavement distress. Autom. Constr. **96**, 457–469 (2018). https://doi.org/10.1016/j.autcon.2018.10.010
7. Jiang, S., Jiang, C., Jiang, W.: Efficient structure from motion for large-scale UAV images: a review and a comparison of SfM tools. ISPRS J. Photogrammetry Remote Sens. **167**, 230–251 (2020). https://doi.org/10.1016/j.isprsjprs.2020.04.016, https://www.sciencedirect.com/science/article/pii/S0924271620301131
8. Jiang, Y., Yan, H., Zhang, Y., Wu, K., Liu, R., Lin, C.: RDD-YOLOv5: road defect detection algorithm with self-attention based on unmanned aerial vehicle inspection. Sensors (Basel, Switzerland) **23**(19), 8241 (2023). https://doi.org/10.3390/s23198241
9. Kizilateş, G., Nuriyeva, F.: On the nearest neighbor algorithms for the traveling salesman problem. In: Nagamalai, D., Kumar, A., Annamalai, A. (eds.) Advances in Computational Science, Engineering and Information Technology, pp. 111–118. Springer International Publishing, Heidelberg (2013). https://doi.org/10.1007/978-3-319-00951-3_11

10. Kotevski, Z., Mitrevski, P.: Experimental comparison of PSNR and SSIM metrics for video quality estimation. ICT Innovations **2009**, 357–366 (2010). https://doi.org/10.1007/978-3-642-10781-8_37, https://www.mendeley.com/catalogue/46577954-c720-3acb-be50-cf10cf2e625c/
11. Mustafah, Y.M., Noor, R., Hasbi, H., Azma, A.W.: Stereo vision images processing for real-time object distance and size measurements. In: 2012 International Conference on Computer and Communication Engineering (ICCCE), pp. 659–663 (2012). https://doi.org/10.1109/ICCCE.2012.6271270, https://ieeexplore.ieee.org/document/6271270/
12. Nthaga, T.M.E., Nyomboi, T., Mwaniki, M.: An integrated remote sensing and GIS road condition assessment framework: applying geospatial techniques to improve pavement condition analysis. Eng. Technol. Appl. Sci. Res. **15**(2), 21021–21028 (2025). https://doi.org/10.48084/etasr.9785, https://etasr.com/index.php/ETASR/article/view/9785
13. Palacios-Rodríguez, A., Cabero-Almenara, J., Serrano-Hidalgo, M.: Educación Médica y Carga Cognitiva: Estudio de la Interacción con Objetos de Aprendizaje en Realidad Virtual y Vídeo 360º. Revista de Educación a Distancia (RED) **24**(79) (2024). https://doi.org/10.6018/red.582741, https://revistas.um.es/red/article/view/582741
14. Recalde, P.Z., Fernández, M.S., Herrero, J.G., López, J.M.M.: Computer Vision-based road surveillance system using autonomous drones and sensor fusion. In: 2024 27th International Conference on Information Fusion (FUSION), pp. 1–7 (2024). https://doi.org/10.23919/FUSION59988.2024.10706468
15. Roberts, R., Inzerillo, L., Mino, G.D.: Using UAV based 3D modelling to provide smart monitoring of road pavement conditions. Information **11**(12) (2020). https://doi.org/10.3390/info11120568
16. Shah, S., Dey, D., Lovett, C., Kapoor, A.: AirSim: high-fidelity visual and physical simulation for autonomous vehicles. In: Hutter, M., Siegwart, R. (eds.) Field and Service Robotics, pp. 621–635. Springer International Publishing, Cham (2018). https://doi.org/10.1007/978-3-319-67361-5_40
17. Siebert, S., Teizer, J.: Mobile 3D mapping for surveying earthwork projects using an Unmanned Aerial Vehicle (UAV) system. Autom. Constr. **41**, 1–14 (2014). https://doi.org/10.1016/j.autcon.2014.01.004
18. Turki, I., Al-Suleiman (Obaidat), T.: EFFECT OF HUMAN FACTOR ON VARIABILITY OF PAVEMENT CONDITION DATA (2015)
19. Wang, Z., Bovik, A.C., Sheikh, H.R., Simoncelli, E.P.: Image quality assessment: from error visibility to structural similarity. IEEE Trans. Image Process. Publ. IEEE Signal Process. Soc. **13**(4), 600–612 (2004). https://doi.org/10.1109/tip.2003.819861

On the Limitations of Neural Collapse for the Detection of Out-of-Distribution Observations: Can We Do Any Better?

Erik B. Terres-Escudero[1(✉)], Javier Del Ser[1,2], and Pablo Garcia-Bringas[3]

[1] TECNALIA, Basque Research and Technology Alliance (BRTA), 48160 Derio, Bizkaia, Spain
{erik.terres,javier.delser}@tecnalia.com
[2] University of the Basque Country (UPV/EHU), 48940 Leioa, Spain
[3] University of Deusto, 48007 Bilbao, Spain
pablo.garcia.bringas@deusto.es

Abstract. In the last decade AI systems have become increasingly popular in real-world applications. As such, ensuring their reliability and safety in open-world settings remains a critical task for the field. Within AI safety, Out-of-Distribution (OoD) detection aims to prevent AI models from making overconfident (and potentially dangerous) predictions when presented with inputs outside the domain of their training data. Recently, the Neural Collapse (NC) phenomenon has been employed to detect OoD instances in neural networks by relying solely on the similarity between the latent vectors and the weights of the classification layer, with no noticeable impact on inference speed. While NC-based OoD detection framework has shown competitive performance, we argue that several overlooked limitations restrict the robustness and broader applicability of this approach. In this paper we present a critical analysis of these limitations, identifying three core challenges: 1) the dependency on the generalization of the base model, which may lead to unreliable overconfident estimates; 2) structural dataset issues that break the NC assumption, particularly in large-scale datasets like ImageNet; and 3) the alignment between the out-of-distribution datasets used for benchmarking and the requirements of the AI model's operational design domain, which may affect the fairness of performance comparisons. As a result of our analysis, we provide learned lessons and outline workarounds to overcome these limitations in real-world scenarios, showing how minor adjustments to the NC-based OoD detection framework can potentially improve its effectiveness.

Keywords: Out-of-Distribution Detection · Neural Collapse · AI Safety

1 Introduction

In recent years, AI-based systems have become deeply embedded in a wide range of user-facing products, operating increasingly in open-world environments. This

E. Corchado et al. (Eds.): SOCO 2025, CCIS 2806, pp. 102–114, 2026.
https://doi.org/10.1007/978-3-032-19763-4_10

shift exposes models to unexpected and potentially adversarial inputs. As a consequence, guaranteeing the safety of these products has become an emergent task that requires careful consideration of model robustness and real-world reliability [7]. Among these tasks, Out-of-Distribution (OoD) detection stands as a priority, as it seeks how to protect AI models from unpredictable behaviors when presented with inputs that fall outside their training data distribution [21]. Although several tasks can result unharmed, when working in life-threatening domains, such as law or medicine, it is crucial to verify that the model operates in a solid, trustworthy, and consistent manner when processing unexpected inputs. Failing to detect such OoD stimuli can lead to severe consequences, making OoD detection not just a technical requirement, but a foundational aspect of responsible AI deployment [9].

Recently, the Neural Collapse (NC [15]) framework has gained attention as a candidate theoretical basis for developing OoD detection heuristics. Payman et al. analyzed the behavior of neural network parameters in the last classification layer and the latent space, showcasing how these parameters evolve during training towards satisfying a set of geometric properties. Building on this foundation, methods such as NECO [1] and NCI [13] have demonstrated that NC can enable robust OoD detectors, even achieving inference-level runtimes.

Following this line of work, this paper investigates the limitations that arise when using NC in OoD detection tasks, with a focus on proposing effective solutions. To this end, we conduct experiments with NCI [13], the current state-of-the-art NC-based OoD detector, and identify three key problems (P) that hinder its performance: (P1) overconfident predictions, (P2) violated NC assumptions, and (P3) feature-space similarity of semantically different samples. Since OoD detection is intended for real-world, safety-critical scenarios, we concentrate our analysis on how NC-based methods perform on specific datasets, rather than on aggregate benchmark comparisons. Our contributions are summarized as follows:

1. We propose i-NCI($\cdot$) (*improved* NCI), a modification of the original NCI method that reduces setup time and improves OoD detection accuracy.
2. We identify three key challenges in relying solely on the NC-based framework for OoD detection, and provide practical solutions to mitigate each of them.
3. We introduce a simple yet effective extension to NC-based OoD detectors that leverages a small amount of available OoD data to boost performance.

2 Preliminaries and Proposed Improved NCI

This section revisits important literature related to the NC framework and OoD Detection. We also detail the i-NCI algorithm used in the rest of this work.

Neural Collapse is a phenomenon observed during the terminal phase of neural network training, particularly in classification tasks, where the feature representations of samples from the same class converge to a common mean, and these class means become symmetrically arranged and align with the weight vectors of the final linear layer [15]. In this state, the geometry of the network exhibits a highly regular structure: features collapse to class centers (within-class variability vanishes), these centers are equidistant (forming a simplex equiangular tight frame), and the classifier's decision boundaries become maximally separated. This emergent behavior has drawn attention not only for its theoretical elegance but also for its practical utility. Applications such as model compression [20], transfer learning [5], and OoD detection can leverage NC properties. For instance, the consistency and predictability of class-wise feature alignment under NC provide a natural baseline against which anomalous inputs (those deviating from the learned geometry) can be detected. This makes NC a promising theory for developing lightweight, interpretable, and robust OoD detectors capable of identifying deviations from learned data distributions in open-world settings.

Several axioms related to NC must be defined to understand the challenges identified in this work. To properly define them, we first introduce the required notation that will be followed during this section. Let $\ell_{i,c}$ represent the i-th feature vector from class $c \in \mathcal{C}$, where $\mathcal{C}$ denotes the set of classes, taken from the last classification layer. Similarly, let $\mu_G = \mathbb{E}_{\forall c,i}[\ell_{i,c}]$ be the mean global vector from the in-distribution (ID) dataset, whereas $\mu_c = \mathbb{E}_{\forall c}[\ell_{\cdot,c}]$ denotes the mean vector of class $c \in \mathcal{C}$. For the sake of simplicity in the notation, we define $\overline{\mu} = (\mu - \mu_G)/\|\mu - \mu_G\|_2$ as the renormalized vector over the global mean μ_G. Similarly, we refer as $M = [\overline{\mu_c} : c \in \mathcal{C}]$ to the collection of class mean vectors. Finally, Using this notation, the axioms of NC state that:

NC1 (*Variability Collapse*) As the training process of the network reaches its terminal phase, the variance of activations within each class approaches zero:

$$\mathrm{Var}_{i,c}\left[(\ell_{i,c} - \mu_c)(\ell_{i,c} - \mu_c)^\top\right] \to 0. \tag{1}$$

NC2 (*Simplex Equiangular Tight Frame*) At the final training stages, the renormalized vector means converge into having the same norm:

$$\Big| \|\mu_c - \mu_G\|_2 - \|\mu_{c'} - \mu_G\|_2 \Big| \to 0 \quad \forall c, c' \in \mathcal{C}, \tag{2}$$

maximizing the pairwise angles between said vectors:

$$\langle \bar{\mu}_c, \bar{\mu}_{c'} \rangle \to \frac{C}{C-1}\delta_{c,c'} - \frac{1}{C-1} \quad \forall c, c' \in \mathcal{C}, \tag{3}$$

where $\langle \cdot, \cdot \rangle : \mathbb{R}^n \times \mathbb{R}^n \to \mathbb{R}$ denotes the dot product of two vectors of dimension n, and $\delta_{a,b}$ denotes the Kronecker's delta, which equals 1 if $a = b$ and 0 otherwise.

NC3 (*Self-Dual Alignment*) At the end of the training process, the weight matrix vectors and the mean class vectors converge up to normalization:

$$\frac{W^\top}{\|W\|_F} = \frac{M}{\|M\|_F}. \tag{4}$$

where W denotes the weight matrix and $\|\cdot\|_F$ denotes the Frobenius norm.

NC4 (*Equivalence to Nearest-Class Center*) The classifier's decision is equivalent to applying a nearest class center classification:

$$\arg\max_{c\in\mathcal{C}}\langle w_c, \ell\rangle + b_c = \arg\min_{c\in\mathcal{C}} \|\ell - \mu_c\|_2, \tag{5}$$

where w_c and b_c represent the weight vector and bias of class $c \in \mathcal{C}$.

Out-of-Distribution Detection. OoD Detection is the task of identifying whether an input sample originates from the training distribution, in order to prevent potentially unreliable or harmful predictions. Early methods, such as Maximum Softmax Probability (MSP), relied on analyzing the softmax output to distinguish ID from OoD samples by leveraging the model's tendency to be overconfident on known classes [8]. Subsequent approaches, such as ODIN, introduced temperature scaling and small input perturbations to amplify the confidence gap between ID and OoD samples [12]. More recent techniques operate by modifying the feature space directly. Among these, ReAct gained traction recently due to its high accuracy and speed, suppressing high-magnitude activations at the penultimate layer to reduce spurious responses triggered by OoD inputs [16].

Recently, the NC framework has gained traction as a theoretical foundation for developing OoD detection methods. The first work in this direction was proposed by Ben Ammar et al., introducing NECO, which leveraged the null space with respect to the classifier weights to identify OoD inputs [1]. Building on this, Liu and Qin introduced NCI, a method that employed the simplex structure of the weight vectors to measure the dissimilarity between input latents and class weights [13]. This approach demonstrated competitive accuracy and near-inference speed when compared to purely distributional baselines. Formally, the detection score is formulated as follows:

$$\mathrm{NCI}_{\mathrm{Base}}(\ell; W, \mu_G) = d_{\cos}(w_c, \ell - \mu_G) \cdot \|w_c\|_2, \tag{6}$$

where $d_{\cos}(\cdot,\cdot)$ denotes cosine distance, $\ell \in \mathbb{R}^n$ represents the latent vector previous to the last classification layer, W denotes the weight matrix of the last classification layer, and w_c being the associated weight vector of class $c \subset \mathcal{C}$. The class c is selected as the one with the highest softmax output probability.

Proposed Improved NCI (i-NCI). In addition to the critical analysis of NC-based OoD detection, in this paper we propose two incremental improvements to the original NCI implementation aimed to boosting both the speed of the method and its accuracy in detecting OoD samples:

- First, although the original implementation relies on the mathematical formulation of NC2's simplex structure, we opt to exclude the global class-mean vector μ_G. Computing this vector requires running inference over the entire training dataset to average the latent activations, which can be computationally expensive, especially for large datasets. The second reason for omitting μ_G is that, since both the global and latent vectors are obtained after the

ReLU activation, they remain in the positive orthant. As a result, the following condition $\langle \ell, \mu_c - \mu_G \rangle = \langle \ell, \mu_c \rangle - \langle \ell, \mu_G \rangle \geq \langle \ell, \mu_c \rangle$, holds. Therefore, since latent vectors are expected to align with class weight vectors rather than the global mean, removing the global vector should reduce the likelihood of OoD instances falling within a small angular proximity to it.
- Secondly, while NCI employs unnormalized weight vectors for classes with high variance, we diverge from this approach and advocate for normalizing the weights. The original motivation suggested that uncertain classes during prediction tend to develop larger weight norms. However, in our early empirical evaluations, we observed that this increased variance could inadvertently raise the ID-ness score for certain OoD samples-particularly those that lie between multiple class directions.

Our alternative implementation of the NCI detector, i-NCI($\cdot$), is given by:

$$\text{i-NCI}(x_{\text{new}}) = \max_{c \in \mathcal{C}} \frac{w_c}{||w_c||_2} \cdot \ell(x_{\text{New}}), \tag{7}$$

where we extend the notation of ℓ to $\ell(x)$ to denote the activation vector obtained from the last classification layer using x as an input. As shown later, i-NCI($\cdot$) attains improved detection with respect to the baseline NCI.

3 When Does NC Fail?

Although NC-based OoD detection algorithms have proven effective in achieving high accuracy and near inference-time detection [13], they still struggle with specific edge cases. The literature on the theoretical foundations of OoD detection has already highlighted how semantic overlap between OoD and ID datasets makes perfect accuracy unattainable. However, to better approximate this accuracy ceiling, we analyze the current limitations of NC for OoD, organizing them into three problems (P): P1) Over/underconfidence (Subsect. 3.1); P2) broken NC assumptions (Subsect. 3.2); and P3) near-OoD cases (Subsect. 3.3). Our findings are supported by the results of side experiments aimed to inform them empirically. In order to achieve a reproducible evaluation process, we employ the OpenOoD library [22] to conduct the experiments. For the sake of reproducibility and to support follow-up studies, code and results are available at https://github.com/erikberter/iNCI.

Datasets. We consider three in-distribution classification datasets: *CIFAR-10* [10] and *CIFAR-100* [10] and *ImageNet-1K* [4]. Regarding OoD datasets, we use TinyImagenet [10], Textures [3], SVHN [14], MNIST [11], Places365 [23] and CIFAR-100/10 for CIFAR-10/100, respectively, and SSB_hard [18], Ninco [2], INaturalist [17], Textures [3] and OpenImage_o [19], for Imagenet [4].

Network Architectures. We employ two convolutional neural networks, ResNet-18 for CIFAR-10 and CIFAR-100 experiments, and ResNet-50 for ImageNet-1K. All models are pretrained using the protocols defined in OpenOOD to ensure a fair comparison across methods.

i-NCI Implementation. Our method is implemented using the mathematical formulation of i-NCI from Equation (7) and the *few-shot* (FS) version of this detector (i-NCI$_{\text{FS}}$) later defined in Equation (10). We use normalized latent vectors for CIFAR-10 and CIFAR-100 and unnormalized latents for Imagenet.

Experiments on i-NCI$_{OoD}$. Each experiment in this section was conducted using independent ID/OoD dataset pairs. For each experiment, only one ID dataset and one OoD dataset were used, and the OoD data for i-NCI was collected exclusively from the corresponding OoD dataset. A total of $k \in \{10, 100, 1000\}$ OoD samples were randomly selected for each experiment. The experiments were repeated three times for CIFAR-10 and CIFAR-100, and once for ImageNet.

3.1 P1: Overconfident and Ambiguous Predictions

Overconfidence, a problem by which the calibration of networks tends to assign high probability scores that exceed the value expected based on the actual accuracy of the network, has been a well-known issue in modern deep neural networks for several years [6]. This effect arises as a natural consequence of NC1, whereby the variability collapse within the latent vectors into the weight vectors makes the final prediction align more tightly with one of the classes. As a consequence of this effect, it is not uncommon to observe OoD samples that attain high probability scores, due to the similarity between the latent vector and the weight class vectors. This incapability of classification-based networks to assign low probability scores to samples outside the training distribution is especially detrimental in NC-based OoD detection, as it makes ID and some OoD samples equally proximal to the expected class means.

A second branch within this limitation arises from the inability of models to remain close enough to multiple classes when making uncertain predictions. For example, if a sample cannot be confidently assigned to a specific class, the resulting latent may emerge in the middle space between the most semantically similar labels. However, this reduced cosine similarity with respect to the class vectors leads to lower OoD detection scores than those of overly confident OoD samples, which remain closely tied to a single class.

This disparity between predictions and OoD scores arises as a consequence of the training objective, where models are expected to solely maximize the probability of the class they believe the input belongs to. Since the NC framework relies on the assumption that class-aligned representations are separable, and that input samples fall within one of the weight vectors, having samples that align between multiple vectors remains a challenge. While NECO [1] advocates for studying the orthogonal space to W, instances that remain within the span of W, either due to being ID or due to being OoD samples, remain indistinguishable.

Learned Lesson (Confidence): NC-based methods are effective but should be combined with additional techniques to handle over and underconfident edge cases.

3.2 P2: Broken NC Assumption

One of the fundamental theoretical pillars for NC-based OoD detection relies on the validity of its axioms. Naturally, if these bases fail to hold for some instances, the empirical accuracy degrades consequently. This broken assumption is observed within the weight structure of the classifier layer, where several weight vectors fail to split properly, breaking the NC2 framework. To illustrate this, Fig. 1 depicts heatmaps representing the cosine similarity between the weight vectors w_c and $w_{c'}$ corresponding to a neural network learned on the CIFAR-100 and Imagenet datasets. As shown in these plots, several class pairs (c, c') expose a high degree of similarity, far exceeding the expected scores from Equation (3).

Fig. 1. Similarity heatmaps of the weight vectors w_c for a network learned on CIFAR-100 (left) and ImageNet (right).

The main cause of this effect lies in the semantic similarity of different classes within the ID data. One example of this similarity can be observed in the Imagenet dataset, where images from the class of *Cassette Players* semantically overlap with those from *Tape Players*. Due to this semantic proximity, we obtain a cosine similarity greater than 0.5 between the weight vectors corresponding to these two classes, far greater than the expected almost orthogonal expectation. This case, where several images can correspond to both classes, is one of the multiple cases within the dataset. CIFAR-100 also suffers from this effect to some extent, having images that remain semantically tied to each other.

Formally, a compliance score with respect to axiom NC2 can be measured by the so-called *equiangularity* over every pair of classes, defined in [1] as:

$$\text{EqAng}^{\text{Agg}}_{\text{class-means}} = \text{Agg}_{c,c'} \left| \frac{\langle \mu_c - \mu_G, \mu_{c'} - \mu_G \rangle + \frac{1}{C-1}}{\|\mu_c - \mu_G\|_2 \cdot \|\mu_{c'} - \mu_G\|_2} \right|, \tag{8}$$

where μ_G is the global mean vector, μ_c is the mean vector of class c and $c, c' \in \mathcal{C}$. The function $\text{Agg}_{c,c'}(\cdot)$ denotes an aggregation method employed during evaluation, chosen from 2 options: $\text{Avg}(\cdot)$, representing the average of the equiangularity scores; and Max(·), denoting the maximum equiangularity score. When applying Avg, the NC2 values remain relatively low and consistent across datasets (0.086 for CIFAR-10, 0.092 for CIFAR-100, and 0.10 for ImageNet). However, the Max aggregation yields markedly higher values (0.33 for CIFAR-10, 0.84 for CIFAR-100, and 0.92 for ImageNet) indicating substantially greater peak semantic similarity in CIFAR-100 and ImageNet compared to CIFAR-10.

Table 1. Comparison of algorithms on CIFAR-10, CIFAR-100, and ImageNet. For each dataset, performance is measured in terms of AUROC and FPR95 for both NearOoD and FarOoD scenarios. Arrows indicate the direction in which metrics attain better results. The experiments on CIFAR-10 and CIFAR-100 were performed 3 times, according to the number of pretrained models on OpenOoD (variance as subindex). Bold denotes the highest scoring methods.

	CIFAR-10				CIFAR-100				Imagenet			
	NearOoD		FarOoD		NearOoD		FarOoD		NearOoD		FarOoD	
Algorithm	AUROC ↑	FPR95 ↓	AUROC ↑	FPR95 ↓	AUROC ↑	FPR95 ↓	AUROC ↑	FPR95 ↓	AUROC ↑	FPR95 ↓	AUROC ↑	FPR95 ↓
MSP [8]	$88.03_{0.25}$	$48.17_{3.92}$	$90.73_{0.43}$	$31.71_{1.83}$	$80.27_{0.11}$	$\mathbf{54.80_{0.33}}$	$77.76_{0.44}$	$58.70_{1.06}$	76.02	**65.66**	85.23	51.45
ODIN [12]	$82.87_{1.85}$	$76.18_{6.04}$	$87.96_{0.61}$	$57.62_{4.25}$	$79.90_{0.11}$	$57.92_{0.48}$	$79.28_{0.21}$	$58.87_{0.80}$	74.75	72.47	89.47	43.97
React [16]	$87.11_{0.61}$	$63.56_{7.32}$	$90.42_{1.41}$	$44.90_{8.37}$	$80.77_{0.05}$	$56.39_{0.33}$	$80.39_{0.49}$	$54.20_{1.58}$	**77.38**	66.74	**93.67**	**26.31**
NCI [13]	$88.71_{0.21}$	$47.19_{1.17}$	$91.30_{0.29}$	$30.81_{0.93}$	$\mathbf{80.92_{0.09}}$	$56.25_{0.32}$	$\mathbf{81.85_{0.38}}$	$\mathbf{50.35_{1.00}}$	73.50	69.27	91.68	29.79
i-NCI	$\mathbf{90.46_{0.09}}$	$\mathbf{36.57_{1.12}}$	$\mathbf{93.12_{0.35}}$	$\mathbf{24.48_{1.11}}$	$80.83_{0.07}$	$57.29_{0.75}$	$79.14_{0.42}$	$56.46_{0.65}$	73.63	68.82	92.24	27.00

As previously mentioned, the main detrimental effect of the class semantic proximity is the reduced accuracy correlation present between CIFAR-10 (which has almost orthogonal weights), and CIFAR-100 and Imagenet. To measure this, we perform OoD detection experiments using different algorithms (including i-NCI our proposed improvement of the baseline NCI detector), yielding the results presented in Table 1. In the case of CIFAR-10, where the maximal weight vector similarity is less extreme, NC-based methods present the highest accuracy, even surpassing feature-based or probability-based methods. In contrast, in cases where the weight similarity is higher, score differences between the distinct OoD detection algorithms are reduced, with NC-based methods even underperforming in Imagenet. These results illustrate how, in cases of high semantic correlation between classes, NC-based methods perform inherently worse, as the network activations fail to achieve the theoretical geometric properties that enable their effectiveness in detecting OoD instances.

Learned Lesson (Broken NC Assumption): NC assumptions may not hold if data is not properly curated into clearly defined classes, leading to a accuracy drop compared to other OoD detectors in the presence of strong semantic overlap between categories.

3.3 P3: Feature-Space Similarity of Semantically Different Instances

The last problem identified in this work involves samples that, while not being fully detected as either ID or OoD, hold a strong resemblance to certain ID class features despite being semantically distinct. Due to this similarity in the feature space, such OoD samples often receive higher ID scores than typical OoD inputs, yet lower scores than true ID instances. One example of this phenomenon can be noticed between landscape images from the CIFAR-100 dataset and digit images from the SVHN dataset. Images of such classes often share similar color an shapes, as exemplified by the data samples depicted in Fig. 2. In this figure, the distribution of OoD scores appears to collapse toward the ID cluster, highlighting how ResNet-18 fails to isolate the feature space from that class.

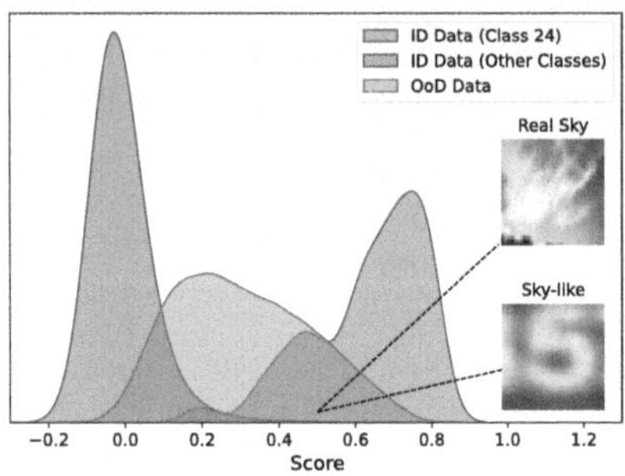

Fig. 2. OoD score of ID data (CIFAR-100), split between class 24 and the rest of the samples, and OoD data (SVHN), with respect to the weight vector of class 24. We include two examples with score near 0.5 from CIFAR-100 and SVHN.

The most restricting aspect of this observation is that, in several cases, these high average ID scores can surpass the scores obtained by ID samples that align close to multiple classes. Naturally, given the normalization of both the feature vector and the weight vector, and the simplex-like structure of the weight vector, any sample aligned with multiple classes will have reduced similarity to any weight vector, which can result in lower scores than those observed here. Therefore, as opposed to P1 and P2, this effect does not arise from a lack of model generalization capability or from the ID semantic overlap, but as a lack of training data for the model to learn how to isolate ID features from OoD ones.

One property of these OoD examples is that they are enclosed in a specific region of the space, as their feature set is similar within the OoD specific data, but distant to other ID data. Therefore, NC-based detectors fail to capture these samples due to them having similar angular distance to weight vectors than ID data. However, such samples can be found when assuming access to a small sample of OoD data. This assumption is reasonable in realistic open-world settings with a priori knowledge of the semantics of OoD samples that the model can receive when deployed.

In order to tackle this *few-shot* OoD scenario, we propose a straightforward adaptation of our i-NCI($\cdot$) algorithm to incorporate information of OoDs. Let $\mathcal{X}_{\text{OoD}} \subset \mathcal{D}_{\text{ODD}}$ be a subset extracted of the set $\mathcal{D}_{\text{OoD}}$ of OoD data (assumed to be known at inference time), and let $\ell(x_{\text{OoD}})$ denote the normalized activation vector of the input sample x_{OoD}, obtained by following the same process as with ID samples. We define the OoD-ness score i-NCI$_{\mathrm{OoD}}$ of any new sample x_{new} as:

$$\text{i-NCI}_{\mathrm{OoD}}(x_{\mathrm{new}}) = \max_{x_{\mathrm{OoD}} \in \mathcal{X}_{\mathrm{OoD}}} \ell(x_{\mathrm{new}}) \cdot \ell(x_{\mathrm{OoD}})^{\top}. \tag{9}$$

Similarly to the heuristic employed on i-NCI, this function evaluates the similarity of the new latent vector to a subset of representatives. However, in this case, given the lack of NC-based OoD weight vectors, the method relies on a procedure equivalent to a distance-based KNN OoD detector. Once the OoD-ness score is computed, we subtract it from the original score to penalize these proximal samples in the feature space. This process gives rise to the final score i-NCI$_{\mathrm{OoD}}$ of the *few-shot* adaptation of our proposed i-NCI$(\cdot)$:

$$\text{i-NCI}_{\mathrm{FS}}(x_{\mathrm{new}}) = \text{i-NCI}(x_{\mathrm{new}}) - \text{i-NCI}_{\mathrm{OoD}}(x_{\mathrm{new}}). \tag{10}$$

Remark. As pointed by other papers, the condition $\ell(x_{\mathrm{ID}}) \perp \ell(x_{OOD})$ is usually verified. As a consequence, i-NCI$_{\mathrm{FS}}(x_{\mathrm{New}})$ tends to approximate the value of i-NCI(x_{New}) for ID data. This implies that the use of a sample of OoD data does not degrade the accuracy of the detector over ID instances.

Results. As depicted in Table 2, the usage of a small set of OoD samples (with size $|\mathcal{X}_{\mathrm{OoD}}| = k$) can result in enhanced accuracy for the rest of the data. In scenarios where the problem P3 (closely aligned features) is not present, such as CIFAR-10, the reported AUROC gains are significantly lower. However, when dealing with more complex datasets, where the complexity of the data requires model to take weaker generalization over the features (CIFAR-100 and Imagenet), we can see a stronger increase in detection performance.

Table 2. AUROC differences for i-NCI $\rightarrow$ i-NCI$_{\mathrm{FS}}$ and varying k. For conciseness, we denote CIFAR-10 and CIFAR-100 as C-10 and C-100, respectively.

	i-NCI			i-NCI$_{\mathrm{FS}}$ ($k = 1000$)			i-NCI$_{\mathrm{FS}}$ ($k = 100$)			i-NCI$_{\mathrm{FS}}$ ($k = 10$)		
Dataset	C-10	C-100	Imagenet	C-10	C-100	Imagenet	C-10	C-100	Imagenet	C-10	C-100	Imagenet
Near OoD	90.46	80.83	73.63	91.18	87.22	90.81	90.54	85.55	85.89	87.13	83.34	79.96
Far OoD	93.12	79.14	92.24	95.47	94.29	96.37	95.73	93.98	94.19	94.58	92.78	89.87

Naturally, as the amount of OoD samples k is lower, the AUROC scores gradually decrease. In some specific cases, low values of k can even result in sub-optimal scores with respect to the baseline, as observed in NearOoD in CIFAR-10 or FarOoD in Imagenet. Nevertheless, this detrimental result is highly specific of the OoD dataset under consideration, and does not occur within all the datasets of each group. Instances that do not suffer from the feature-similarity problem stated in this section end up achieving only subtle accuracy increases with large values of k, ending up being detrimental for the overall value of i-NCI when used with small subsets of OoD samples.

When inspecting the improvements for every (ID,OoD) dataset combination, the most notable result occurs when comparing SVHN and CIFAR-100, where the AUROC increases from 79.04 to an average of nearly 98 across all OoD subset sizes. In this case, although there is a strong feature similarity between the OoD and ID samples, the internal consistency among the detected OoD features is stronger, making i-$\text{NCI}_{\text{FS}}(\cdot)$ achieve near-perfect detection.

Learned Lesson (Feature-space Similarity): In scenarios where OoD samples have proximal features, but distinct overall semantics, using small sets of OoD data can be beneficial to enhance precision.

4 Conclusions and Learned Lessons

NC-based methods have recently gained momentum in the literature as a theoretical foundation for developing OoD detection tools. These methods combine fast inference speeds with high accuracy across various benchmarks. However, their inherent reliance on the NC framework introduces limitations in scenarios that are poorly aligned with its geometric assumptions. In this work, we investigate edge cases where the model either fails to detect OoD instances or misclassifies ID samples.

Through this analysis, we have identified three primary sources of failure: (1) Overconfidence and inaccurate detections, stemming from the classification-based training paradigm used in these models; (2) Violations of NC assumptions, particularly when classes exhibit high semantic similarity; and (3) Feature-space overlap, occurring when OoD classes resemble ID ones at the representation level, despite their semantic differences.

To address these issues, we have proposed i-NCI, an enhanced version of the state-of-the-art NC-based OoD detector, NCI, designed to improve accuracy while reducing runtime. We further analyze the root causes of these challenges and provide insights into how they should be tackled. To specifically address the third issue, we have introduced i-$\text{NCI}_{\text{FS}}(\cdot)$, a few-shot OoD detector built upon i-$\text{NCI}(\cdot)$ that leverages the geometric structure of classifier weights and incorporates small subsets of OoD data to boost detection performance.

We envision two main lines of research to extend the practical utility of this method. First, we aim at developing a more refined methodology to extract OoD samples for i-$\text{NCI}_{\text{FS}}(\cdot)$, with the goal of improving accuracy by maximizing the representative power of this small OoD subset with respect to the real OoD distribution. Next, we plan to broaden the range of supported models to enable OoD detection in video and audio datasets, as the current formulation is limited to data without a temporal dimension. Finally, we aim to adapt the method for scenarios where the training data is non-static, enabling real-world mechanisms to maintain and improve detection performance throughout the model's lifecycle.

Acknowledgments. The authors acknowledge funding support from the Basque Government through ELKARTEK program (project KK-2023/00012) and the consolidated research groups MATHMODE (IT1456-22) and D4K-Deusto for Knowledge (IT1528).

Disclosure of Interests. The authors have no competing interests to declare that are relevant to the content of this article.

References

1. Ammar, M.B., et al.: NECO: neural collapse based out-of-distribution detection. In: ICLR (2024)
2. Bitterwolf, J., et al.: In or out? Fixing ImageNet out-of-distribution detection evaluation. arXiv:2306.00826 (2023)
3. Cimpoi, M., et al.: Describing textures in the wild. In: IEEE CVPR, pp. 3606–3613 (2014)
4. Deng, J., et al.: ImageNet: a large-scale hierarchical image database. In: IEEE CVPR, pp. 248–255 (2009)
5. Galanti, T., et al.: On the role of neural collapse in transfer learning. arXiv:2112.15121 (2021)
6. Guo, C., et al.: On calibration of modern neural networks. In: ICML, pp. 1321–1330 (2017)
7. Gyevnár, B., Kasirzadeh, A.: AI safety for everyone. Nat. Mach. Intell. 1–12 (2025)
8. Hendrycks, D., Gimpel, K.: A baseline for detecting misclassified and out-of-distribution examples in neural networks. In: ICLR (2017)
9. Herrera-Poyatos, A., et al.: Responsible artificial intelligence systems: A roadmap to society's trust through trustworthy AI, auditability, accountability, and governance. arXiv:2503.04739 (2025)
10. Krizhevsky, A.: Learning multiple layers of features from tiny images. Master's thesis, University of Tront (2009)
11. LeCun, Y., Cortes, C.: MNIST handwritten digit database (2010)
12. Liang, S., Li, Y., Srikant, R.: Enhancing the reliability of out-of-distribution image detection in neural networks. In: ICLR (2018)
13. Liu, L., Qin, Y.: Detecting out-of-distribution through the lens of neural collapse. arXiv:2311.01479 (2023)
14. Netzer, Y., et al.: Reading digits in natural images with unsupervised feature learning. In: NeurIPS Workshop on Deep Learning and Unsupervised Feature Learning. vol. 2011, p. 4 (2011)
15. Papyan, V., et al.: Prevalence of neural collapse during the terminal phase of deep learning training. Proc. Natl. Acad. Sci. **117**(40), 24652–24663 (2020)
16. Sun, Y., Guo, C., Li, Y.: ReAct: out-of-distribution detection with rectified activations. Adv. Neural. Inf. Process. Syst. **34**, 144–157 (2021)
17. Van Horn, G., et al.: The inaturalist species classification and detection dataset. In: IEEE CVPR, pp. 8769–8778 (2018)
18. Vaze, S., et al.: Open-set recognition: a good closed-set classifier is all you need? In: ICLR (2022)
19. Wang, H., et al.: ViM: out-of-distribution with virtual-logit matching. In: IEEE CVPR, pp. 4921–4930 (2022)
20. Wu, R., Papyan, V.: Linguistic collapse: neural collapse in (large) language models. Adv. Neural. Inf. Process. Syst. **37**, 137432–137473 (2024)

21. Yang, J., et al.: Generalized out-of-distribution detection: a survey. IJCV **132**(12), 5635–5662 (2024)
22. Zhang, J., et al.: OpenOOD v1.5: Enhanced benchmark for out-of-distribution detection. arXiv:2306.09301 (2023)
23. Zhou, B., et al.: Places: a 10 million image database for scene recognition. IEEE Trans. Pattern Anal. Mach. Intell. **40**(6), 1452–1464 (2017)

Anomaly Detection and Predictive Maintenance

Validating Anomaly Detection of One-Class Classifiers Using Statistical Generators

Rui Pinto[3], André Pilastri[1,3], Luís Ferreira[3], Rafael Freitas[2], and Paulo Cortez[1,3](✉)

[1] CCG/ZGDV ICT Innovation Institute, Universidade do Minho, Campus de Azurém, 4800-058 Guimarães, Portugal
andre.pilastri@ccg.pt

[2] Pibra - Industrial Solutions, S.A., 4765-420 Guimarães, Portugal
rafael.freitas@pibra.com

[3] ALGORITMI/LASI, Dep. Information Systems, Universidade do Minho, Campus de Azurém, 4800-058 Guimarães, Portugal
pg54209@alunos.uminho.pt, {luis.ferreira,pcortez}@dsi.uminho.pt

Abstract. One-Class Classification (OCC) is often used in Anomaly Detection (AD) Machine Learning tasks (e.g., industrial quality and maintenance). Since real anomalies are non-existent or extremely rare in several AD applications, in this paper we empirically compare five Statistical Data Generation (SDG) methods to generate anomalies for validation sets, allowing to tune OCC algorithms for AD. Using eight tabular AD datasets, we firstly measure the fidelity of the SDG methods to simulate real anomalies. Then, using a nested cross-validation scheme, we perform a Bayesian optimization of four OCC algorithms using distinct SDG validation sets. Our experiments show that the Bernoulli SDG method yields the highest average fidelity (0.43) but a low AD performance, while Laplace SDG delivers a superior AD, particularly when combined with a Local Outlier Factor (LOF) classifier, reaching an overall Area under the Precision-Recall Curve (AUPRC) of 0.91 and demonstrating its value for tuning AD methods.

Keywords: One-class classification · Model Selection · Synthetic data

1 Introduction

There are multiple industrial applications that require an Anomaly Detection (AD), such as detecting unsatisfactory assembled products or production maintenance issues. In several real-world AD tasks there is a lack of labelled anomalies, since abnormal data can be non-existent, rare or too costly to be collected. Within this context, One-Class Classification (OCC) is an interesting Machine Learning (ML) approach, since OCC algorithms only require normal examples (the majority class) during the training phase. Examples of popular OCC algorithms include: Local Outlier Factor (LOF) [3], Isolation Forest (IF) [12], One-Class Support Vector Machine (OC-SVM) [19], and deep Autoencoders (AE) [7].

E. Corchado et al. (Eds.): SOCO 2025, CCIS 2806, pp. 117–127, 2026.
https://doi.org/10.1007/978-3-032-19763-4_11

However, OCC algorithms often include hyperparameters that need to be tuned using labelled validation sets [7], thus requiring the availability of anomalies in order to properly perform model selection.

Aiming to solve the need of labelled validation data, in this paper, we study the effect of using five distinct Statistical Data Generation (SDG) methods (Bernoulli, Gaussian, Gamma, Laplace and Poisson) to generate synthetic anomalies for model selection purposes. Using eight public domain tabular AD datasets, we empirically measure the fidelity and AD performance of the analyzed SDG methods when combined with four OCC algorithms (LOF, OC-SVM, IF, and AE). The paper is organized as follows: Sect. 2 reviews related work; Sect. 3 details the AD datasets, algorithms, and evaluation strategy; Sect. 4 presents empirical results; and Sect. 5 discusses conclusions and future directions.

2 Related Work

Tabular AD is a commonly explored ML task. When anomalous data are scarce, OCC methods are typically used, with popular algorithms including: LOF, OC-SVM, IF and deep AE [7,11]. Since anomaly data are non-existent or rare in several AD contexts, some recent studies have proposed the creation of synthetic anomalies. For instance, Dai et al. [5] proposed a ML framework that corrupts normal data with zero-mean SDG methods (Gaussian, Laplace, Uniform, Rayleigh, Gamma, Poisson, Salt&Pepper and Bernoulli) at a specific standard deviation σ, and then trains a deep neural network that discriminates normal from anomalous instances. While it covers a broad amount of perturbation types, it only covers AD performance and does not quantify the fidelity, i.e., how closely the synthetic examples match the real anomaly distributions. Learned perturbation frameworks (e.g., PLAD [4]) have also been proposed to learn additive and multiplicative distortions via a deep learning perturbator. While interesting results were reported, these two recent works have drawbacks. They only evaluate AD performance and not the quality of the generated synthetic data. In effect, there are recently proposed fidelity measures, such as the Total Variation Distance (TVD) [15], which can be used to evaluate the quality of generated anomalies. Our study addresses these gaps by exploring five SDG methods (Bernoulli, Gaussian, Gamma, Laplace and Poisson) while evaluating them with a fidelity metric (TVD) alongside with an AD performance in eight tabular AD datasets.

3 Materials and Methods

3.1 Datasets

We used eight public tabular datasets suitable for AD, each with a extreme imbalance between normal and anomalous data. The data span across six different domains: Engineering, Finance, Healthcare, IoT, Image, and Physics. Table 1 summarizes the statistical characteristics of the datasets used in this benchmark (number of rows, features, numerical columns, categorical columns and anomalies, and percentage of anomalies) and also their application domain. We prioritized datasets with a low percentage of anomalies to simulate real-world scenarios where anomalies are typically scarce.

Table 1. Imbalanced AD Datasets for OCC Benchmarking.

Dataset	Rows	Features	Num.	Cat.	#Anomaly	%Anomaly	Domain
ALOI [9]	49,534	27	27	0	1,508	3.040	Image
CCFD [16]	284,807	29	29	0	492	0.170	Finance
GECCO [18]	139,566	9	9	0	1,726	1.230	IoT
Genesis [2]	16,220	20	20	0	50	0.003	Engineering
Mammography [20]	11,183	6	6	0	260	2.320	Healthcare
PMAI4I [14]	10,000	6	5	1	339	3.390	Engineering
Thyroid [17]	3,772	6	6	0	93	2.470	Healthcare
Waveform [21]	3,443	21	21	0	100	2.900	Physics

All AD datasets were pre-processed starting with removing missing values to prevent errors from occurring as some models do not handle them well. We applied a categorical encoding as implemented by the `cane`[1] Python module [13], with One-Hot Encoding for categorical columns with less than 10 categories and Inverse Document Frequency Encoding (IDF) otherwise. This IDF encoding involves transforming a categorical value into a single numerical value proportional to the logarithm of the ratio between the total number of instances and the frequency of that level. We also applied feature scaling via `scikit-learn`[2] `StandardScaler`, also known as z-score, to centre each feature to zero mean and scale it to a unit standard deviation. This approach ensures that all features contribute unevenly when we inject synthetic data. To reduce the computational cost to benchmark all OCC models, we decided to limit all training sets for normal instances up to 10,000 rows by performing a random subsampling, which is aligned with the procedure performed in the ADBench benchmark [10].

3.2 Machine Learning Algorithms

The OCC algorithms were trained with normal data using the Python language and the following modules: `scikit-learn` – for IF, LOF and OC-SVM; `TensorFlow`[3] – for AE. The `scikit-learn` IF, LOF and OC-SVM implementations provide a decision score that ranges from $\hat{y}_i = -1$ (highest abnormal score) to $\hat{y}_i = 1$ (highest normal score). In order to obtain an anomaly probability score ($d_i \in [0, 1]$, for an input example i), we rescale the IF, LOF, and OC-SVM scores

[1] https://pypi.org/project/cane/.
[2] https://scikit-learn.org/stable/.
[3] https://www.tensorflow.org/.

by computing $d_i = (1 - \hat{y}_i)/2$, following the same strategy as [8]. In addition, we optimize the OCC $\mathcal{H}$ hyperparameters using a Bayesian optimization with 50 trials, as implemented by the `Optuna` Python module [1].

The IF algorithm is a classical ML algorithm proposed in 2008 [12] and isolates anomalies instead of normal points: it works well in high-dimensional data and does not require much memory usage due to the absence of distance or density measures. Concretely, IF rests on two key assumptions: anomalies are a small fraction of the dataset and they are markedly different from those of normal points. The IF hyperparameters ($\mathcal{H} = 2$) are tuned using the search ranges: `n_estimators` $\in [50, 200]$; `max_samples` $\in [0.5, 1.0]$.

The LOF algorithm [3] is also a popular ML algorithm for AD. LOF is a density-based method that excels at detecting local outliers which are values that may fall within the normal range of the whole dataset but outside the normal range of its neighbours. This algorithm uses distance measures to calculate its k-nearest neighbours and, for that reason, has higher computational costs when dealing with high-dimensional data. The LOF hyperparameter ($\mathcal{H} = 1$) is searched using the range `n_neighbors` $\in [5, 20]$.

The OC-SVM algorithm [6,19] is an unsupervised method that is considered an extension of the Support Vector Machine (SVM) algorithm for unlabelled data. The hyperplane is controlled by a parameter v that controls the fraction of outliers that are allowed in the training set. The tuned hyperparameters ($\mathcal{H} = 2$) include: `kernel` $\in$ {`linear`, `rbf`}; and `gamma` $\in$ {`scale`, 0.01, 0.1, 1} (for `rbf`).

AEs are unsupervised neural networks that learn to compress and reconstruct data through a symmetric encoder–decoder architecture. We implement a fully connected Deep Feed Forward Network (DFFN) following [8], with layer sizes decreasing by half until the bottleneck size is reached and the input dimension is mirrored. The network is compiled with the Adam optimizer to minimise a chosen reconstruction loss, on a 10% validation split. At test time, we compute the reconstruction error per sample, either mean absolute error or mean squared error, as the anomaly score. The searched hyperparameters ($\mathcal{H} = 6$) include: `epochs` $\in [50, 100]$; `batch_size` $\in$ {32, 64, 128}; `encode_activation_function` $\in$ {`relu`, `tanh`, `sigmoid`}; `decode_activation_function` $\in$ {`linear`, `relu`, `tanh`}; `loss_function` $\in$ {`mae`, `mse`}; and `reconstruction_error_function` $\in$ {`mae`, `mse`}.

3.3 Statistical Data Generation Methods

To generate synthetic anomalies, we explore five SDG approaches (Bernoulli, Gamma, Gaussian, Laplace, and Poisson) using the distributional definitions provided by Dai et al. [5], as shown in Table 2. For a given dataset, we experiment three percentages of synthetic anomaly injection (A_I) examples: 3%, 5% and 7%. A synthetic anomaly is generated by randomly selecting a normal example, perturbing each individual input feature with a SDG method (e.g. Gaussian) and then changing the output label to "abnormal". Since the preprocessed data input attributes have a zero mean and one standard deviation ($\sigma = 1$), the intensity of the statistical perturbations assumes a higher standard deviation or scale of 1.5. Thus, a zero mean and 1.5 spread ($\mu = 0, \sigma = 1.5$) were used for the Gaussian and Laplace distributions.

Table 2. Parameters and offsets for the five SDG methods.

Type	Parameter	Value	Offset
Gaussian	μ, σ	$\mu = 0,\ \sigma = 1.5$	0
Laplace	μ, σ	$\mu = 0,\ \sigma = 1.5$	0
Poisson	λ	$\lambda = 1.5$	-1.5
Gamma	α, β	$\alpha = \sqrt{\frac{1}{\beta}}\,1.5,\ \beta = 1$	-1.5
Bernoulli	p	$p = 0.5$	0

For the Gamma distribution we used a shape parameter of $\beta = 1$ and a scale (offset) chosen with a value of 1.5. We injected Poisson anomalies with a mean (and variance) of $\lambda = 1.5$, then centred Poisson and Gamma by subtracting $\sigma = 1.5$ to achieve a zero mean. Additionally, for Bernoulli perturbations, we generated a probability vector with 50% chance of flipping the sign of an element. All SDG methods run in $O(1)$ time per feature, ensuring scalability for large datasets.

3.4 Evaluation

Using all data, we measure the fidelity of the SDG methods by comparing the real anomalies with the synthetic generated ones, assessing how closely they replicate the original anomalies. To measure fidelity, we first compute the TVD between the empirical distributions of the real and synthetic anomalies, calculated for univariate and bivariate marginals [15]. In particular, for each feature (univariate TVD) and each pair of features (bivariate TVD), we partition the range of observed values into 10 bins. The objective is to map continuous numerical observations into discrete categories, ensuring that each bin contains the same amount of data. Next, we compute the univariate and bivariate TVD between two discrete distributions (synthetic and real anomalies). For each SDG, we average these two TVD values to obtain $\overline{\text{TVD}}$. Finally, the fidelity metric is defined as $1 - \overline{\text{TVD}}$, ranging from 0 (highly different distributions, thus dissimilar samples) to 1 (identical empirical distributions, thus highly similar samples).

We assume the extreme but common scenario where abnormal data are unavailable during the model selection and training stages. To measure AD predictive performance, we applied a nested cross-validation scheme. The scheme uses an external stratified 5-fold Cross-Validation (CV), which splits the dataset into five folds. For each CV iteration, the OCC algorithms are optimized using normal examples from four folds (80% of the data) and then evaluated using data from the test fold (20% with unseen normal and real abnormal samples). During an external CV iteration, the available data for training is processed with an internal 3-fold CV, to generate fit and validation sets. The fit set (2/3 of the internal data) only contains normal examples and it is used to train an OCC model. The validation set (1/3) is then transformed such that a small percentage of A_I normal examples are selected to be perturbed by a SDG method, resulting in synthetic injected anomalies. Next, the AD performance on the validation set is used to guide the Bayesian search for the OCC hyperparameter tuning.

Regarding AD measures, the Area under the Receiver Operating Characteristic Curve (AUROC) is commonly used and it measures the trade-off between True Positive Rate (TPR) and False Positive Rate (FPR). Yet, it can be overly optimistic when the majority class (normal data) dominates. As suggested in [11], the Area under the Precision-Recall Curve (AUPRC) measures the precision against recall and provides a more discriminative assessment compared to AUROC in rare anomalous events. In this paper, we use AUPRC for the Bayesian search and report both AUROC and AUPRC average values on the test sets.

4 Results

Table 3 presents the obtained fidelity values when using A_I =5%. The results show that the Bernoulli SDG method yields the highest fidelity in most of the datasets, which means that it tends to better mimic the real anomalies. As for the other SDG methods, they obtain a lower and similar average fidelity (0.33/0.32).

Table 3. Fidelity values when A_I=5% (best values in **bold**).

Dataset	Bernoulli	Gamma	Gaussian	Laplace	Poisson
ALOI	**0.55**	0.17	0.19	0.19	0.16
CCFD	0.40	0.45	**0.48**	**0.48**	0.46
GECCO	**0.50**	0.48	0.49	0.49	0.49
Genesis	**0.30**	0.04	0.04	0.04	0.04
Mammography	**0.34**	0.28	0.28	0.30	0.27
PMAI4I	**0.62**	0.41	0.43	0.39	0.45
Thyroid	0.24	0.27	0.29	0.26	**0.30**
Waveform	**0.53**	0.45	0.43	0.40	0.46
Average:	**0.43**	0.32	0.33	0.32	0.33

Each OCC model was tuned using three distinct runs, assuming a particular SDG method with one of three $A_I \in \{3\%, 5\%, 7\%\}$ values. Table 4 presents the obtained AD AUPRC and AUROC results ($\mu \pm \sigma$ – mean and standard deviation – test values when considering the 3 A_I runs, the 5 SDG methods and the 5-CV procedure). Overall, LOF is the top OCC performer, producing a high quality AD, with an overall AUPRC performance of 83% and an AUROC performance of 98%. The only exception is the Genesis dataset, where OC-SVM (the second-best OCC algorithm) achieves significantly higher results. The overall high performance of LOF might be explained by its ability to easily detect local outliers in multidimensional data, due to its density-based nature [3].

Table 4. CV average AD test results (best values in **bold**).

	AUPRC				AUROC			
Dataset	AE	IF	LOF	OC-SVM	AE	IF	LOF	OC-SVM
ALOI	0.49 ± 0.04	0.37 ± 0.03	**0.83 ± 0.03**	0.69 ± 0.05	0.94 ± 0.01	0.92 ± 0.01	**0.98 ± 0.00**	0.94 ± 0.04
CCFD	0.37 ± 0.03	0.40 ± 0.03	**0.80 ± 0.05**	0.57 ± 0.05	0.90 ± 0.01	0.88 ± 0.01	**0.99 ± 0.00**	0.90 ± 0.05
GECCO	0.47 ± 0.06	0.55 ± 0.06	**0.80 ± 0.03**	0.78 ± 0.08	0.90 ± 0.01	0.92 ± 0.01	**0.98 ± 0.01**	0.97 ± 0.02
Genesis	0.76 ± 0.07	0.70 ± 0.04	0.76 ± 0.07	**0.99 ± 0.01**	0.97 ± 0.02	0.98 ± 0.00	0.99 ± 0.00	**1.00 ± 0.00**
Mammography	0.36 ± 0.05	0.42 ± 0.03	**0.92 ± 0.04**	0.82 ± 0.09	0.91 ± 0.02	0.90 ± 0.01	**0.99 ± 0.00**	0.97 ± 0.02
PMAI4I	0.52 ± 0.06	0.43 ± 0.04	**0.99 ± 0.00**	0.93 ± 0.07	0.90 ± 0.02	0.89 ± 0.01	**1.00 ± 0.00**	0.98 ± 0.03
Thyroid	0.29 ± 0.05	0.31 ± 0.05	**0.59 ± 0.06**	0.55 ± 0.07	0.85 ± 0.03	0.85 ± 0.02	**0.93 ± 0.01**	0.92 ± 0.03
Waveform	0.88 ± 0.03	0.79 ± 0.03	**0.91 ± 0.02**	0.89 ± 0.03	0.97 ± 0.01	0.93 ± 0.01	**0.98 ± 0.01**	0.97 ± 0.01
Average:	0.52 ± 0.05	0.50 ± 0.04	**0.83 ± 0.04**	0.78 ± 0.06	0.92 ± 0.02	0.91 ± 0.01	**0.98 ± 0.00**	0.96 ± 0.03

Table 5 compares the best performing OCC algorithm (LOF) under two approaches: the proposed Bayesian SDG (BS) tuned LOF and a baseline that consists of the LOF default hyperparameter values (thus without tuning and no validation sets with SDG anomalies). In the table, for BS LOF we report the $mu \pm \sigma$ test values when considering all five SDG methods, the three A_I runs and the external 5-CV, while for the baseline LOF the $\mu \pm \sigma$ test values are computed using only the external 5-CV (5 iterations). For both AUPRC and AUROC measures, BS LOF substantially outperforms the baseline, thus showing the advantage of the proposed BS OCC optimization method.

Table 5. Bayesian SDG optimized vs. Baseline LOF (best values in **bold**).

	BS LOF		Baseline LOF	
Dataset	AUPRC	AUROC	AUPRC	AUROC
ALOI	**0.83 ± 0.03**	**0.98 ± 0.00**	0.07 ± 0.01	0.69 ± 0.02
CCFD	**0.80 ± 0.05**	**0.99 ± 0.00**	0.55 ± 0.03	0.95 ± 0.01
GECCO	0.80 ± 0.03	0.98 ± 0.01	**0.82 ± 0.02**	0.99 ± 0.00
Genesis	**0.76 ± 0.07**	**0.99 ± 0.00**	0.12 ± 0.01	0.99 ± 0.00
Mammography	**0.92 ± 0.04**	**0.99 ± 0.00**	0.25 ± 0.06	0.85 ± 0.02
PMAI4I	**0.99 ± 0.00**	**1.00 ± 0.00**	0.33 ± 0.04	0.87 ± 0.01
Thyroid	0.59 ± 0.06	0.93 ± 0.01	**0.61 ± 0.09**	**0.97 ± 0.01**
Waveform	**0.91 ± 0.02**	**0.98 ± 0.01**	0.25 ± 0.10	0.76 ± 0.05
Mean Value:	**0.83 ± 0.04**	**0.98 ± 0.00**	0.38 ± 0.05	0.88 ± 0.02

Finally, Fig. 1 presents the Pareto frontier of the SDG methods, where each point denotes on the x-axis the mean AUPRC values (μ when considering a particular OCC algorithm, all 8 datasets, 3 A_I runs and 5-CV test data) and the mean Fidelity (μ when aggregating all 8 datasets and 3 A_I injection rates). In the plot, the vertical and horizontal whiskers correspond to the respective standard deviation (σ) values. The objective of plotting this figure is to understand which SDG approach achieves better performance in maximising both Fidelity and AUPRC detection performance for each OCC method evaluated (AE, IF, LOF, and OC-SVM). The Pareto curves show the SDG methods that provide the best Fidelity-AUPRC trade-offs.

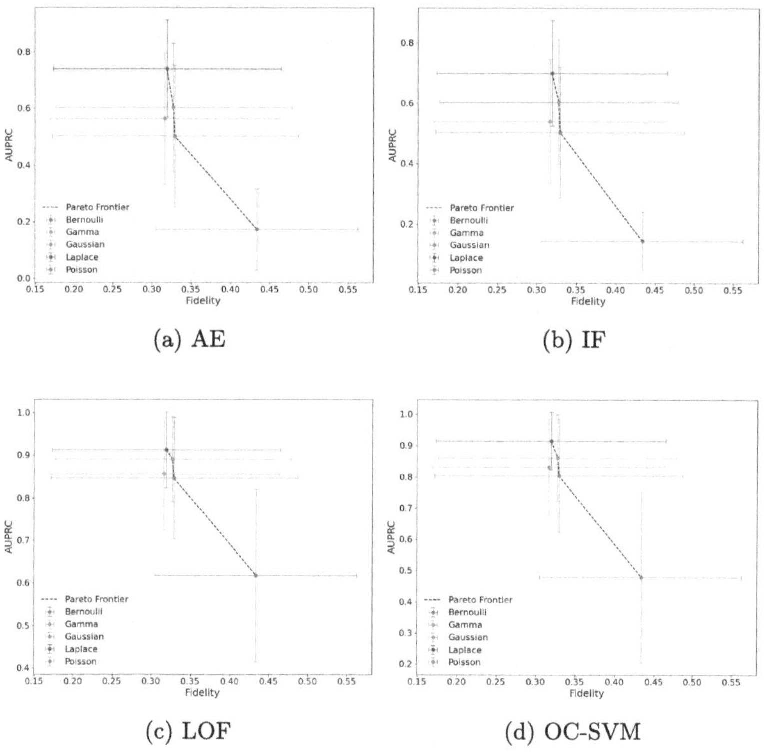

(a) AE (b) IF

(c) LOF (d) OC-SVM

Fig. 1. SDG Fidelity vs. AUPRC for AE (a), IF (b), LOF (c) and OC-SVM (d).

For all four OCC algorithms, the Gamma SDG is never included in the Pareto front. Overall, the results show a consistent and interesting pattern: the Bernoulli generator provides the highest fidelity but worst AD performance, while Laplace SDG obtains the best AD results, particularly when combined with LOF, reaching an AUPRC of 0.91, at the expense of a relatively small decay in fidelity.

5 Conclusions

This paper explores SDG methods to inject synthetic anomalies into validation sets for an AD OCC model selection. Several computational experiments were held, using eight public domain datasets, five SDG methods, three injection rates and four OCC algorithms. Overall, interesting results were achieved. The Bernoulli SDG creates anomalies that are closer to real abnormal examples, yet it also yields the lowest AD predictive performance. In contrast, the Laplace generator consistently provided the best AD results for all tested OCC algorithms, despite being associated with a lower fidelity. Furthermore, the density-based LOF method delivered the best overall performance out of the four benchmarked OCC algorithms. Indeed, the Bayesian SDG optimized LOF obtained much better AD results when compared with a simpler default LOF (with no tuning).

The obtained results favor the usage of the Laplace SDG and the LOF algorithm as an interesting AD approach when anomalies are non-existent or extremely rare during the model selection and training stage. In future work, we intend to explore other generative methods to create synthetic anomalies, such as Generative Adversarial Networks or Diffusion Models. We also plan to enlarge the empirical study by considering more AD datasets, OCC algorithms and their training times to support the goals of the green and digital transition. Additionally, we will validate the methodology in an industrial automation company with only normal operational data.

Acknowledgements. This work has been supported by the European Union under the Next Generation EU, through a grant of the Portuguese Republic's Recovery and Resilience Plan (RRP) Partnership Agreement, within the scope of the project PRODUTECH R3 – "Agenda Mobilizadora da Fileira das Tecnologias de Produção para a Reindustrialização".

References

1. Akiba, T., Sano, S., Yanase, T., Ohta, T., Koyama, M.: Optuna: a next-generation hyperparameter optimization framework. In: KDD, pp. 2623–2631. ACM (2019). https://doi.org/10.1145/3292500.3330701
2. von Birgelen, A., Niggemann, O.: Anomaly detection and localization for cyber-physical production systems with self-organizing maps. In: IMPROVE-Innovative Modelling Approaches for Production Systems to Raise Validatable Efficiency, pp. 55–71. Springer (2018). https://doi.org/10.1007/978-3-662-57805-6_4
3. Breunig, M.M., Kriegel, H., Ng, R.T., Sander, J.: LOF: identifying density-based local outliers. In: SIGMOD Conference, pp. 93–104. ACM (2000). https://doi.org/10.1145/342009.335388
4. Cai, J., Fan, J.: Perturbation learning based anomaly detection. Adv. Neural. Inf. Process. Syst. **35**, 14317–14330 (2022)
5. Dai, W., Hwang, K., Fan, J.: Unsupervised anomaly detection for tabular data using deep noise evaluation. In: Proceedings of the AAAI Conference on Artificial Intelligence. vol. 39, pp. 11553–11562 (2025)

6. Fernandes, S.E.N., Pilastri, A.L., Pereira, L.A.M., Pires, R.G., Papa, J.P.: Learning kernels for support vector machines with polynomial powers of sigmoid. In: 2014 27th SIBGRAPI Conference on Graphics, Patterns and Images, pp. 259–265 (2014). https://doi.org/10.1109/SIBGRAPI.2014.36
7. Ferreira, L., Cortez, P.: AutoOC: automated multi-objective design of deep autoencoders and one-class classifiers using grammatical evolution. Appl. Soft Comput. **144**, 110496 (2023). https://doi.org/10.1016/J.ASOC.2023.110496
8. Fontes, G., Matos, L.M., Matta, A., Pilastri, A.L., Cortez, P.: An empirical study on anomaly detection algorithms for extremely imbalanced datasets. In: AIAI (1). IFIP Advances in Information and Communication Technology, vol. 646, pp. 85–95. Springer (2022). https://doi.org/10.1007/978-3-031-08333-4_7
9. Geusebroek, J., Burghouts, G.J., Smeulders, A.W.M.: The amsterdam library of object images. Int. J. Comput. Vis. **61**(1), 103–112 (2005). https://doi.org/10.1023/B:VISI.0000042993.50813.60
10. Han, S., Hu, X., Huang, H., Jiang, M., Zhao, Y.: ADBench: anomaly detection benchmark. In: Advances in Neural Information Processing Systems 35: Annual Conference on Neural Information Processing Systems 2022, NeurIPS 2022, New Orleans, LA, USA, November 28 - December 9, 2022 (2022)
11. Li, L., Yan, J., Wang, H., Jin, Y.: Anomaly detection of time series with smoothness-inducing sequential variational auto-encoder. IEEE Trans. Neural Networks Learn. Syst. **32**(3), 1177–1191 (2021). https://doi.org/10.1109/TNNLS.2020.2980749
12. Liu, F.T., Ting, K.M., Zhou, Z.: Isolation forest. In: Proceedings of the 8th IEEE International Conference on Data Mining (ICDM 2008), December 15-19, 2008, Pisa, Italy, pp. 413–422. IEEE Computer Society (2008). https://doi.org/10.1109/ICDM.2008.17
13. Matos, L.M., Azevedo, J., Matta, A., Pilastri, A.L., Cortez, P., Mendes, R.: Categorical attribute transformation environment (CANE): a python module for categorical to numeric data preprocessing. Softw. Impacts **13**, 100359 (2022). https://doi.org/10.1016/j.simpa.2022.100359
14. Matzka, S.: Explainable artificial intelligence for predictive maintenance applications. In: AI4I, pp. 69–74. IEEE (2020). https://doi.org/10.1109/AI4I49448.2020.00023
15. Platzer, M., Reutterer, T.: Holdout-based empirical assessment of mixed-type synthetic data. Front. Big Data **4**, 679939 (2021). https://doi.org/10.3389/FDATA.2021.679939
16. Pozzolo, A.D., Caelen, O., Johnson, R.A., Bontempi, G.: Calibrating probability with undersampling for unbalanced classification. In: 2015 IEEE Symposium Series on Computational Intelligence, pp. 159–166 (2015). https://doi.org/10.1109/SSCI.2015.33
17. Quinlan, J.R., Compton, P.J., Horn, K., Lazarus, L.: Inductive knowledge acquisition: a case study. In: Proceedings of the Second Australian Conference on Applications of expert systems, pp. 137–156 (1987)
18. Ribeiro, V.H.A., Reynoso-Meza, G.: Monitoring of drinking-water quality by means of a multi-objective ensemble learning approach. In: GECCO (Companion), pp. 1–2. ACM (2019). https://doi.org/10.1145/3319619.3326745
19. Schölkopf, B., Platt, J.C., Shawe-Taylor, J., Smola, A.J., Williamson, R.C.: Estimating the support of a high-dimensional distribution. Neural Comput. **13**(7), 1443–1471 (2001). https://doi.org/10.1162/089976601750264965

20. Woods, K.S., Doss, C.C., Bowyer, K.W., Solka, J.L., Priebe, C.E., Kegelmeyer, W.P., Jr.: Comparative evaluation of pattern recognition techniques for detection of microcalcifications in mammography. Int. J. Pattern Recognit Artif Intell. **7**(06), 1417–1436 (1993). https://doi.org/10.1142/9789812797834_0011
21. Zimek, A., Gaudet, M., Campello, R.J.G.B., Sander, J.: Subsampling for efficient and effective unsupervised outlier detection ensembles. In: KDD, pp. 428–436. ACM (2013). https://doi.org/10.1145/2487575.2487676

Detection of Photovoltaic Generator Underperformance Due to Inverter Operation Modes

Leire Hernandez-Lecuona[1], Ricardo Alonso[2], Ainhoa Pereda[2], Sergio Gil[2], J. David Nuñez-Gonzalez[3(✉)], and Manuel Graña[1]

[1] Computational Intelligence Group, University of Basque Country UPV/EHU, Donostia/San Sebastián, Spain
manuel.grana@ehu.eus
[2] Tecnalia Research and Innovation, Derio, Spain
[3] Computational Intelligence Group, University of Basque Country UPV/EHU, Eibar, Spain
josedavid.nunez@ehu.eus

Abstract. The increasing deployment of photovoltaic (PV) systems demands advanced supervision tools to ensure optimal performance and minimize maintenance costs. This work presents a novel methodology for detecting and classifying inverter-related anomalies—specifically inverter clipping and thermal power derating—using synthetic SCADA data from a simulated PV plant. The proposed approach integrates linear regression, residual filtering, and Gaussian Mixture Models to identify non-MPP operating conditions, enabling improved accuracy in performance analysis. Results demonstrate high detection accuracy. The lightweight, unsupervised nature of the method makes it suitable for real-time SCADA integration and future application to diverse PV configurations, paving the way for more reliable and condition-based maintenance strategies in PV asset management.

1 Introduction

As solar Photovoltaic (PV) market grows and there are more and more PV assets installed in the field, that must be efficiently operated and maintained, suitable supervising tools are critical to avoid unnecessary preventative and unforeseen corrective maintenance actions, deploying a condition-based maintenance (CBM) strategy instead. CBM paves the way to Operational Expenditures (OPEX) reduction, Performance Ratio (PR) increase and a better and faster knowledge of real performance of PV components under different operating conditions. This last point is key to reduce risk of investment on arising PV technologies and increasing their bankability, that is another factor with a high relevance on the Levelized Cost of Energy (LCOE) of PV. When it comes to deploy CBM, it is necessary to implement a continuous supervising system based on monitoring data, available in SCADA of PV plant, in combination with

E. Corchado et al. (Eds.): SOCO 2025, CCIS 2806, pp. 128–139, 2026.
https://doi.org/10.1007/978-3-032-19763-4_12

periodic inspection techniques in the field. Regarding the supervision of DC side of PV plants based on monitoring data, most of the analyses are focused on PV panels and tracker performance. However, PV inverters play a crucial role to ensure the performance of the PV generator, making it operate at its maximum power point (MPP). Unfortunately, there are different events preventing PV inverters from this normal operation, like inverter clipping effect, when generated PV power exceeds the nominal power of the device, or power derating functionality to avoid damages due to overtemperature. In these cases, PV generator stops operating at its MPP, leading to energy losses and wrong conclusions by data analytics tools considering these energy deviations are related to PV panels or tracking system performance. Although both self-protective operating modes cited above should be reported by the PV inverter, this information is not always easily accessible in the SCADA, due to the lack of standardization of status and error codes through PV inverters and dataloggers. For this reason, in this study novel data analytics methods for automatic detection of these abnormal operating modes of PV inverter and quantification of related energy losses are proposed. This will help to determine more accurately energy losses derived from the selection and sizing of PV inverters at design stage, as well as to remove these operating points from the datasets used for subsequent PV panels and trackers performance analysis.

The paper is structured as follows: Sect. 2 gives the background of the State of the Art. Section 3 defines the experimental pipeline. Section 4 shows results. Section 5 concludes this paper.

2 State of Art

To ensure high technical and economic performance of solar PV assets, the maintenance plans for large PV plants include periodic visual inspections, measurements of the characteristic I-V curve of the PV generator, and capture and analysis of infrared or electroluminescence images of the PV modules. All of these field inspection techniques are intended to verify the proper functioning of components and systems, generally providing high resolution and reliability. Unfortunately, performing these procedures in the field is quite expensive and they can be only carried out punctually. In this context, SCADA (Supervisory Control and Data Acquisition) systems and data analytics techniques play a critical role in monitoring and optimizing the performance of solar PV plants [1]. These systems enable improved monitoring and fault detection by providing historical and real-time data, facilitating early identification of problems to minimize downtime and maximize energy production. More specifically, there are various data analytics and artificial intelligence (AI) methods in the literature for monitoring PV generator performance [2]: statistical process control to identify deviations from expected performance parameters [3]; advanced time series analysis techniques, such as ARIMA models [4], to capture seasonal variations and weather fluctuations in power generation; supervised machine learning algorithms [5], trained with historical sensor data and labeled equipment failure events; and unsupervised techniques, since labeled datasets are not always available due to the new technologies continuously arising in PV sector [6–10].

3 Experimental Pipeline

3.1 Data Generation and Preprocessing

To validate the proposed methodology, a full year of synthetic operational data from a simulated PV plant was employed, covering the entirety of 2023. Measurements were recorded at 15-minute intervals, resulting in approximately 35,000 data points per variable. Although this dataset is not sourced from a real PV installation, its synthetic nature enables precise control over failure modes, ensuring the ability to benchmark fault detection algorithms under well-defined conditions.

The simulated PV plant reflects the behavior of a realistic medium-sized system. It uses bifacial PV modules that collect irradiance from both the front and rear sides. The inverter is configured with two independent Maximum Power Point Trackers (MPPTs), a mechanism responsible for dynamically adjusting the load presented to the PV modules to maximize power extraction under varying irradiance and temperature conditions. Each MPPT is connected to a single string composed of 25 PV modules, each of them with 3 bypass diodes.

To induce the inverter clipping, the PV array peak power capacity is oversized with respect to the PV inverter nominal power capacity, leading to events of this type of failure, where the available DC power exceeds the inverter's rated power. To model thermal power derating behavior, a physical model based on energy balance is used to simulate the inverter internal temperature and a Proportional-Integral (PI) controller is employed to adjust power output in order to maintain the temperature within safe limits. The internal temperature of the inverter (T_{inv}) is estimated based on the energy balance between dissipated power and heat loss to the environment:

$$\frac{dT}{dt} = \frac{P_{dis} - \frac{T_{prev} - T_{amb}}{R_{th}}}{C_{th}} \tag{1}$$

where $P_{dis} = P_{in} - P_{out}$ is the dissipated power, T_{amb} is ambient temperature, T_{prev} is the previous internal temperature, R_{th} is the thermal resistance and C_{th} is the thermal capacitance. The temperature is updated using a discrete-time step Δt as follows:

$$T_{new} = T_{prev} + \frac{dT}{dt} \cdot \Delta t \tag{2}$$

This model simulates the thermal inertia and dynamic response of the inverter to varying input/output power conditions. The thermal resistance R_{th} and thermal capacitance C_{th} parameters were set to $0.18\,^{\circ}\mathrm{C/W}$ and $10{,}000\,\mathrm{J/^{\circ}C}$, respectively. These values were chosen to emulate the typical thermal inertia observed in commercial string inverters. Specifically, the selected parameters reflect a moderate heat dissipation capacity combined with a relatively large thermal mass, which results in a gradual temperature rise and fall in response to varying power loads and ambient conditions.

To prevent the inverter temperature from exceeding safe limits, a PI controller is implemented:

$$\Delta P = K_p e(t) + K_i \sum_{\tau=0}^{t} e(\tau) \tag{3}$$

where $e(t) = T_{target} - T_{inv}(t)$ is the temperature error, and K_p, K_i are controller gains. The control adjusts the inverter output power to mitigate thermal stress while ensuring operational continuity. When T_{inv} exceeds a defined threshold ($T_{derating}$), the controller reacts to reduce P_{out}.

To further enhance realism, shading events are simulated by means of ray-tracing computation independently for each PV submodule, since each string is assumed to be installed in different PV trackers projecting shadows to each other at sunrise and sunset. This diversity of modeled conditions makes the dataset well-suited for developing and validating fault detection methods that generalize across multiple failure modes.

Data Segmentation. Although the full year of data is available, all fault detection and model training tasks are performed using monthly time windows. This segmentation is motivated by several key considerations. First, solar irradiance and ambient temperature vary significantly over the course of the year, making a month-by-month analysis more suitable for identifying climate-dependent faults. Additionally, using one-month windows ensures statistical stability, as each segment contains approximately 3,000 data points, sufficient to train robust regression models and apply unsupervised clustering without overfitting. If daily windows were used, the algorithm would not have enough data to find anomaly trends. Finally, from a deployment perspective, it is not practical in real SCADA systems to wait an entire year to detect and diagnose faults, as this would result in extended periods of underperformance and substantial energy losses.

3.2 Anomaly Detection

In order to distinguish anomalous data from normality, it has been decided to generate a linear regression model. Because inverter clipping may occur at each MPPT input unevenly depending on string orientation, shading or temperature effects, it is critical to analyze each MPPT independently. In this case, each MPPT power (P_{MPP}) has been modeled as a linear function of global irradiance on plane of array ($GPOA$). In the case of thermal power derating detection, a linear model has been generated with the data from both MPPTs, where the current of the whole PV plant has been modeled as a linear function of the irradiance. Current has been used instead of power for detecting thermal power derating because it presents less dependence on temperature variations.

Power exhibits nonlinear behavior due to the higher PV voltage temperature coefficient, introducing distortion. This can be appreciated in the Fig. 1. The residuals (differences between predicted and observed variables) are calculated, and outliers are classified as "Other" if they lie beyond a threshold. Points within the threshold are classified as "Normal". This filtering process is repeated 5 times to iteratively refine the model and exclude anomalies.

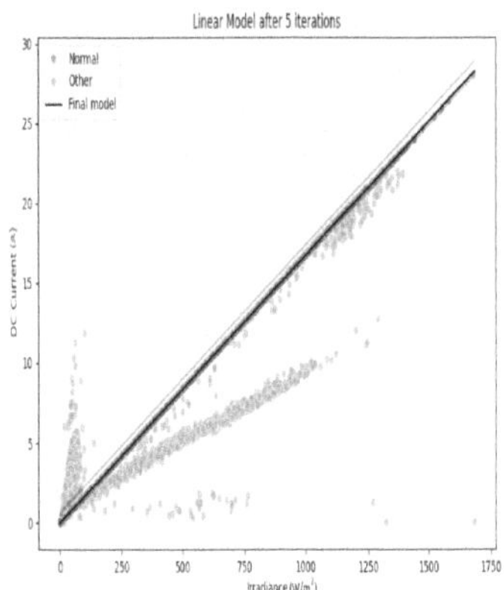

(a) Linear Regression for thermal power derating detection using DC current

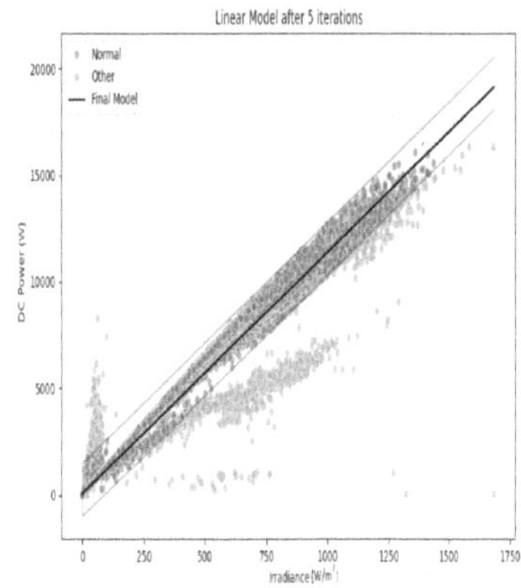

(b) Linear Regression for thermal power derating detection using DC power

Fig. 1. Comparison of the linear model using DC Current or DC Power.

3.3 Clustering with Gaussian Mixture Models

In this work, several Gaussian Mixture Models (GMM) have been used to classify the different anomalous points into different clusters. Then, each cluster has been evaluated to identify the events that are trying to be detected.

GMM are a probabilistic unsupervised learning technique used to model a dataset as a combination of multiple Gaussian distributions. Unlike k-means clustering, which assigns data points to exactly one cluster based on distance, GMM provides a soft clustering approach by estimating the probability that a data point belongs to each cluster. Mathematically, a GMM assumes that the data $\mathbf{x}$ is generated from a mixture of K Gaussian distributions:

$$p(\mathbf{x}) = \sum_{k=1}^{K} \pi_k \cdot \mathcal{N}(\mathbf{x} \mid \mu_k, \Sigma_k)$$

where π_k is the mixture weight (prior probability) of the k-th component with $\sum_k \pi_k = 1$, $\mathcal{N}(\mathbf{x} \mid \mu_k, \Sigma_k)$ is the multivariate normal distribution with mean μ_k

and covariance matrix Σ_k and K is the number of components (clusters). GMM parameters are learned using the Expectation-Maximization (EM) algorithm, which iteratively estimates the probability of each point belonging to each cluster (E-step) and updates the parameters to maximize the likelihood (M-step). To determine the optimal number of clusters, the Bayesian Information Criterion (BIC) is used. BIC penalizes model complexity while rewarding goodness of fit, favoring models that best explain the data with fewer components.

It has been decided to use GMM because, unlike K-Means which assumes spherical clusters of equal size, GMM can model clusters with varying sizes, shapes and orientations. This flexibility allows it to better capture the complexity of real-world operational data, where different system states (in our case, shading, thermal power derating and inverter clipping) may overlap or exhibit distinct statistical properties. GMM's ability to model overlapping distributions also supports more accurate clustering under non-uniform and changing conditions, such as variations in irradiance or temperature.

3.4 Inverter Clipping Detection

As mentioned above, the current signal from each MPPT is compared to the irradiance and the residuals are used to identify outliers. These outliers are clustered with GMM to detect the behavior of the inverter clipping. The signature of inverter clipping is typically observed in plots of DC power versus solar irradiance. At low irradiance levels, the relationship is linear. However, at high irradiance levels, instead of continuing the linear trend, the power levels off, forming a saturation zone. This deviation from linearity is the key feature used in this detection method. If any cluster with high average power with values close to the inverter's nominal output and low power variance at high irradiances is detected, it is labeled as "Inverter Clipping".

Although the analysis covers the entire year 2023, this section presents the January results as a representative example. In the following image, the process of outlier detection with the linear regression model and the classification of those points by clusters can be seen. Clusters 0 and 5 are labeled as Inverter Clipping by the model (Fig. 2).

Since it is possible that many data suffering from inverter clipping were classified as normal by the linear regression model, another phase of statistical expansion extends the clipping classification to nearby "Normal" data points within a defined confidence region. This expansion takes into account the characteristics of the cluster detected as inverter clipping (variance and power mean) to label data that were initially assumed to be normal (Fig. 3).

During this month, the model successfully identifies 158 instances of inverter clipping in MPPT1, out of a total of 163 true clipping events. This corresponds to 4 false positives and 1 false negative, indicating high precision and recall in the detection performance. In January, inverter clipping events were only forced on MPPT1, leaving only normality data on MPPT2 throughout the month. Consistently, the model does not detect any clusters that would suggest clipping

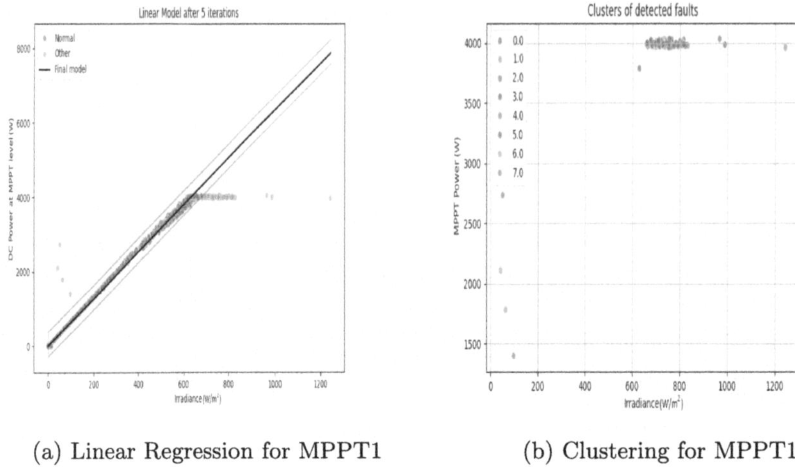

(a) Linear Regression for MPPT1 (b) Clustering for MPPT1

Fig. 2. Inverter Clipping Detection Process for MPPT1.

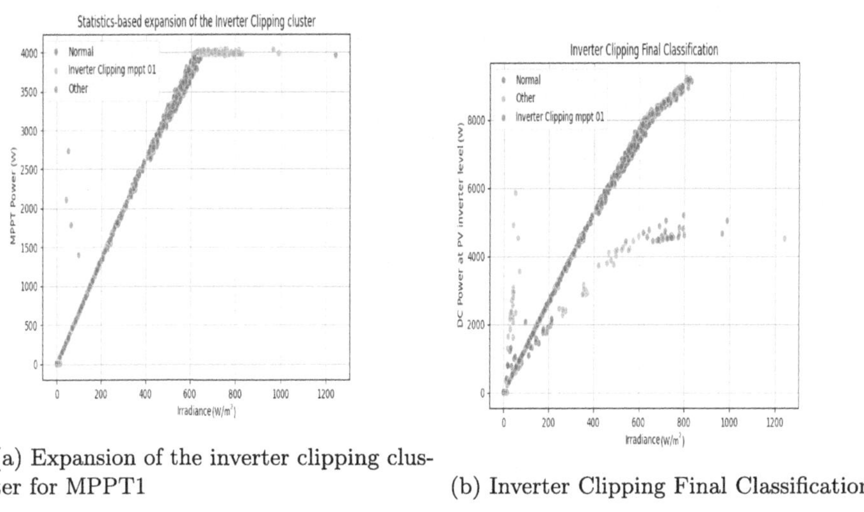

(a) Expansion of the inverter clipping cluster for MPPT1 (b) Inverter Clipping Final Classification

Fig. 3. Inverter Clipping Detection Process for MPPT1.

behavior in this channel, demonstrating its ability to distinguish between operating modes across independent MPPTs. Before performing the cluster expansion, the algorithm correctly detected 129 inverter clipping points out of the 163 points that are known to exist. Thanks to the expansion, this number grew to 163. The results section shows the confusion tables for these two cases for the whole year.

3.5 Thermal Power Derating Detection

Thermal power derating is detected by analyzing the evolution of the inverter's current and internal temperature. To better understand the detection process, the month of July is analyzed in this section. Figure 5 shows the regression model trained using accumulated data from January to July.

The DC current predicted from irradiance using the linear regression model is compared to the actual output to compute ΔI metric:

$$\Delta I = I_{\text{predicted}} - I_{\text{measured}} \tag{4}$$

Only points initially labeled as "Other" and with high temperature (T_{inv} within 20°C of the maximum observed, corresponding to a 23% margin over the total observed temperature range for the inverter) are retained for derating analysis. This ensures that only conditions where thermal derating is likely are analyzed. For these filtered data points, GMM is applied using T_{inv} and ΔI as clustering features.

It is known that when the inverter's internal temperature exceeds T_{derating}, the PI controller begins reducing output power. Due to the integral component of the PI control logic, the reduction becomes more pronounced the longer the temperature remains elevated. As a result, the ΔI vs T_{inv} curve tends to exhibit a logarithmic shape, as shown in Fig. 4b .

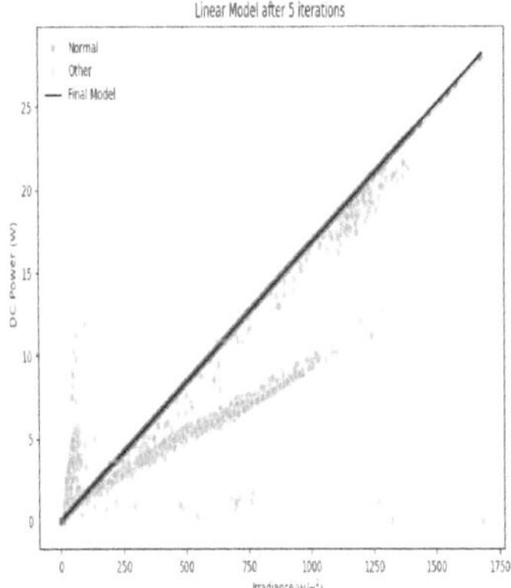

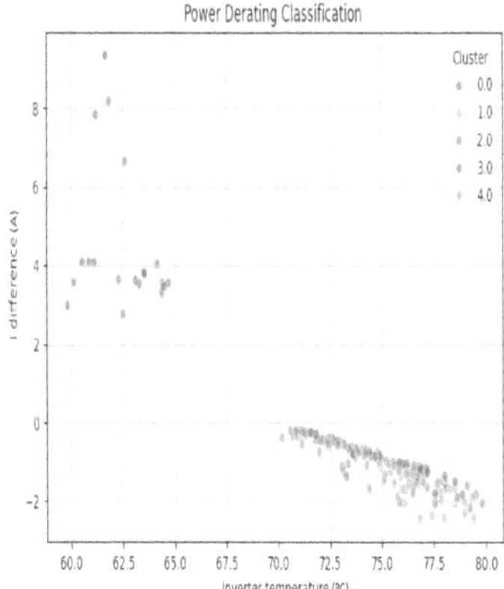

(a) Linear Regression from January to July (b) GMM Clustering on Accumulated Data

Fig. 4. Thermal Power Derating Detection Pipeline.

To capture this behavior, the correlation between $\log(\Delta I)$ and T_{inv} is calculated for each cluster. Clusters with correlation coefficients above 0.65 are labeled as thermal power derating. This value was chosen experimentally. As shown in Fig. 4b , in July all clusters except cluster 3 exceed the correlation threshold. The lowest temperature among these points is then selected as the derating threshold for that month and all points with higher inverter internal temperature are labeled as Thermal Power Derating.

When the model is trained using only data from July, the threshold temperature is 74.15° C. Figure 5 shows that Cluster 4 does not meet the correlation condition and is not labeled as derating. However, by combining information from previous months, the model adapts the threshold down to 71.54° C. This accumulated strategy improves prediction quality by using more context than just the current month. If no valid clusters are found, T_{derating} defaults to the maximum T_{inv} in the dataset.

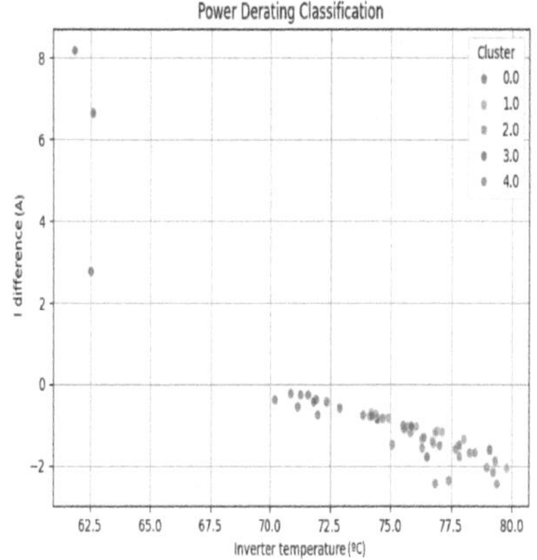

Fig. 5. Cluster Analysis for July.

In early months, the model may produce false positives by detecting derating at sub-threshold temperatures (below 70° C). To prevent this, the detection thresholds are updated monthly using historical context. If a cluster is identified as derating, its minimum temperature is included in the running average used to update T_{derating}. This adaptive approach prevents persistent misclassification.

4 Results

To evaluate the performance of the proposed anomaly detection methodology, confusion matrices have been generated for each event.

Inverter Clipping. Inverter clipping events are known to occur predominantly during the high-irradiance months of the year. It is known that there are a total of 1,918 clipping instances throughout 2023. Using the proposed detection methodology, the algorithm successfully identifies 1,902 instances. The detection algorithm produces 200 false positives and 16 false negatives during the whole year. These false negatives are associated with cases where DC power exceeds the inverter limit by a small amount and only briefly, leading to negligible energy loss. The following tables show the confusion matrix for the case of Inverter Clipping before and after the expansion of the cluster (Table 1):

Table 1. Confusion Matrix for Inverter Clipping

With Expansion

		True		
		Positive	Negative	Total
Predicted	Positive	606	0	606
	Negative	1,312	33,027	34,339
	Total	1,918	33,027	34,945

With Expansion

		True		
		Positive	Negative	Total
Predicted	Positive	1,902	200	2,102
	Negative	16	32,827	32,843
	Total	1,918	33,027	34,945

It is important to mention that the false positives are due to the fact that, at the time of the expansion, points of the normality that have the same mean and standard deviation as the points of the cluster are included. These false positives are points that do not generate much change in the estimation of the energy loss, given that they are points where, despite not having inverter clipping, they are at the limit of the inverter's nominal power.

Thermal Power Derating. Thermal power derating control is known to activate on one day in March, throughout the months of April, May, June and July, and again on one day in September. Across the entire year, there are 266 known instances of thermal power derating. Using the proposed detection technique, the algorithm correctly identifies 235 instances during April, May, June and July.

Notably, most false negatives occur due to borderline cases in the linear regression step, where some data points remain classified as normal operation. However, these missed detections are not critical in terms of energy yield: they correspond to conditions where the inverter temperature only slightly exceeds the derating threshold, and thus the control system does not significantly reduce power output.

The thermal control is designed to activate when the inverter's internal temperature exceeds 70°C. The detection model infers the onset of derating at 70.34°C at the end of the year, which aligns with the true control behavior. The following table summarizes the results obtained in a confusion matrix (Table 2):

Event-Free Dataset. To further validate the robustness of the proposed detection algorithms, both models were tested using a dataset containing only normal (event-free) operating conditions. As expected, neither algorithm identified any instances of the targeted anomalies, confirming their reliability in the absence of anomalies.

Table 2. Confusion Matrix for thermal power derating

		True Labels		
		Positive	Negative	Total
Predicted Labels	Positive	220	0	220
	Negative	46	34,679	34,725
	Total	266	34,679	34,945

5 Conclusion and Future Work

This work presents an unsupervised methodology for detecting inverter clipping and thermal power derating events in photovoltaic systems using synthetic operational data. The proposed approach identifies these two power-limiting phenomena without requiring labeled datasets or extensive domain-specific parameter tuning. This allows plant operators to optimize system design and adjust operational strategies. In the case of thermal power derating, the model also provides early insight into possible thermal stress conditions, supporting preventive maintenance and system reconfiguration.

Acknowledgment. Authors received research funds from the Basque Government as of the Grupo de Inteligencia Computacional, Universidad del Pais Vasco, UPV/EHU, from 2007 until 2025. The current code for the grant is IT1689-22. Additionally, authors participate in Elkartek projects KK-2022/00051 and KK-2021/00070. The Spanish MCIN has also granted the authors a research project under code PID2020-116346GB-I00

Research activities carried out by TECNALIA team were funded by European Commission through the Horizon Europe project "SUPERNOVA - Operation and maintenance and grid friendly tools and solutions for solar data fusion and insight explosion for reliable, bankable, circular PV plants", under Grant Agreement 101146883.

References

1. SolarPower Europe. Operation & Maintenance Best Practice Guidelines/Version 6.0 (2025)
2. Mellit, A., Tina, G.M., Kalogirou, S.A.: Fault detection and diagnosis methods for photovoltaic systems: a review. Renew. Sustain. Energy Rev. **91**, 1–17 (2018). ISSN 1364-0321, https://doi.org/10.1016/j.rser.2018.03.062
3. Atalan, Y.A., Atalan, A.: Integration of the machine learning algorithms and I-MR statistical process control for solar energy. Sustainability **15**, 13782 (2023). https://doi.org/10.3390/su151813782
4. Sapundzhi, F., Chikalov, A., Georgiev, S., Georgiev, I.: Predictive modeling of photovoltaic energy yield using an ARIMA approach. Appl. Sci. **14**, 11192 (2024). https://doi.org/10.3390/app142311192
5. Hong, Y.-Y., Pula, R.A.: Methods of photovoltaic fault detection and classification: a review. Energy Reports **8**, 5898–5929 (2022). ISSN 2352-4847, https://doi.org/10.1016/j.egyr.2022.04.043

6. Utama, C., Meske, C., Schneider, J., Schlatmann, R., Ulbrich, C.: Explainable artificial intelligence for photovoltaic fault detection: a comparison of instruments. Solar Energy (2023)
7. Chaleshtori, A.E., Aghaie, A.: A novel bearing fault diagnosis approach using the Gaussian mixture model and the weighted principal component analysis. Reliab. Eng. Syst. Saf. **242**, 109720 (2024). https://doi.org/10.1016/j.ress.2023.109720
8. Chen, J., Zhang, X., Zhang, N., Guo, K.: Fault detection for turbine engine disk using adaptive Gaussian mixture model. Proc. Inst. Mech. Eng. Part I: J. Syst. Control Eng. **231**(10), 827–835 (2017). https://doi.org/10.1177/0959651817731249
9. Hong, Y., Kim, M., Lee, H., Park, J.J., Lee, D.: Early fault diagnosis and classification of ball bearing using enhanced Kurtogram and gaussian mixture model. IEEE Trans. Instrum. Meas. **68**(12), 4746–4755 (2019). https://doi.org/10.1109/TIM.2019.2898050
10. Cao, S., Hu, Z., Luo, X., Wang, H.: Research on fault diagnosis technology of centrifugal pump blade crack based on PCA and GMM. Measurement **173**, 108558 (2021). https://doi.org/10.1016/j.measurement.2020.108558

Denial of Service in 5G Core: AMF and UPF Saturation via Virtual SIMs

Diego Narciandi-Rodríguez[1](✉), Jose Aveleira-Mata[2], Natalia Prieto-Fernandez[2], Javier Alfonso-Cendón[3], Héctor Alaiz-Moretón[2], and Isaías García-Rodríguez[2]

[1] RIASC. Research Institute of Applied Sciences in Cybersecurity, Universidad de León, MIC. Campus de Vegazana, s/n, 24071 León, Spain
dnarr@unileon.es

[2] SECOMUCI Research Group. Department of Electric, Systems and Automatics Engineering, Universidad de León, Escuela de Ingenierías. Campus de Vegazana, s/n, 24071 León, Spain
{javem,nprif,halam,isaias.garcia}@unileon.es

[3] Department of Mechanical, Computer Science and Aerospace Engineering, Universidad de León, Escuela de Ingenierías. Campus de Vegazana, s/n, 24071 León, Spain
javier.alfonso@unileon.es

Abstract. Standalone (SA) 5G networks offer improvements in speed, latency, and device density, making them suitable for industrial settings, real-time communications, and emerging IoT deployments. However, these same capabilities, combined with a virtualized architecture, also increase the attack surface. This work presents a Denial-of-Service (DoS) scenario targeting critical components of the 5G core—specifically the Access and Mobility Management Function (AMF) and the User Plane Function (UPF)—using virtual User Equipments (UEs) and simulated Subscriber Identity Modules (SIMs) generated with open-source tools such as UERANSIM. The experimental setup, based on a fully containerized 5G core, was subjected to a simulated registration flood using virtual UEs. As the load increased, the system showed clear signs of degradation, leading to consistent failures and insufficient resource errors from the UPF. The absence of mitigation mechanisms allows an attacker to cause significant disruption without specialized hardware or physical access.

Keywords: 5G · DoS · OpenAirInterface · UERANSIM · Cybersecurity

1 Introduction

5G networks have revolutionized wireless communication by enabling faster, more stable, and higher-performance connections. While much of the 5G infrastructure is virtualized, its implementation remains costly. This high cost has

E. Corchado et al. (Eds.): SOCO 2025, CCIS 2806, pp. 140–149, 2026.
https://doi.org/10.1007/978-3-032-19763-4_13

limited adoption, particularly among small and medium enterprises. Nevertheless, the demand for private 5G networks has grown rapidly, driven by their potential to improve security and operational control. Leading industrial players, such as those in the automotive sector, have embraced private 5G solutions to power smart factories and Industry 4.0 use cases [2,9].

To address the high entry cost of commercial solutions, several open-source projects have emerged as viable alternatives for deploying and experimenting with 5G networks. OpenAirInterface (OAI) [8] is a comprehensive software suite that implements both the 5G Core and Radio Access Network (RAN), allowing users to simulate or deploy end-to-end 5G systems. Open5GS [5], on the other hand, focuses specifically on the 5G Core network, offering a lightweight, modular, and standards-compliant implementation of core network functions such as the AMF, SMF, and UPF. UERANSIM [3] is a powerful UE and gNB emulator that enables the simulation of user equipment and access points for standalone (SA) and non-standalone (NSA) 5G deployments.

Together, these frameworks provide a free and flexible alternative to costly commercial platforms, requiring only minimal physical infrastructure. As a result, they have gained significant traction among researchers, startups, and academic institutions looking to explore 5G technology in a controlled and customizable environment.

However, open-source tools are accessible and easy to deploy, which makes them useful, but they can also expose vulnerabilities or be used as attack vectors. The same tools that reduce costs and barriers for legitimate experimentation can be leveraged by malicious actors to simulate large-scale attacks against critical 5G core components—without the need for physical hardware or commercial licenses.

In this paper, we explore a Denial-of-Service (DoS) attack scenario targeting the AMF and UPF Functions in a 5G Standalone (SA) core network built with OpenAirInterface. Using UERANSIM, we simulate hundreds of virtual UEs (User Equipments) with unique virtual SIMs that attempt to register with the core network.

This paper is organized as follows. In the "Related Work" section, we review existing research on security threats and DoS attacks in 5G networks. The "Methods and Materials" section describes our experimental setup, including details on the hardware, software stack, 5G network topology, and the tools used to simulate and monitor network activity. In "DoS Attack in the 5G Environment", we present the design and execution of the attack scenario in our 5G Standalone testbed, detailing how UERANSIM was used to generate massive registration attempts and saturate the AMF. The "Results" section reports on the impact of the attack, analyzing system behavior. Finally, we outline contributions of this work and suggest future directions to enhance cybersecurity in 5G networks.

2 Related Work

Denial-of-Service (DoS) attacks are among the most common and persistent threats in the field of cybersecurity. Their main objective is to render a system, service, or network inaccessible to legitimate users by generating massive amounts of requests or traffic that overwhelm available resources, thereby compromising availability. In many cases, DoS attacks are relatively simple to execute but difficult to effectively mitigate, especially in distributed and dynamic systems such as modern mobile networks [7].

In the context of 5G networks, the threat of DoS attacks takes on new dimensions due to the specific characteristics of this generation of wireless technologies. 5G networks are designed to support a high density of devices, low latency, and a wide variety of services through concepts such as Network Function Virtualization (NFV), network slicing, and dynamic resource orchestration. While these features enhance network flexibility and scalability, they also significantly broaden the attack surface [11,12].

Two key components to highlight are the Access and Mobility Management Function (AMF) and the User Plane Function (UPF). The AMF is responsible for receiving and processing Non-Access Stratum (NAS) signaling messages from user equipment (UE), as well as coordinating access to core resources [10,12]. Meanwhile, the UPF is responsible for handling user data traffic and maintaining GTP-U tunnels between the gNB and external data networks. Both channels can be exploited via DoS attacks involving the flooding of registration messages, authentication requests, creation of traffic channels.

These attacks are dangerous in experimental and low-cost environments, such as those based on open-source solutions (e.g. OpenAirInterface, Open5GS, or free5GC), since these implementations often lack advanced defense mechanisms, rate-limiting strategies, Intrusion Detection Systems (IDS) or low resources. In this context, UERANSIM—an open-source tool designed to simulate user equipment (UE) behavior and radio access network (RAN) elements— can be maliciously repurposed to simulate hundreds of virtual UEs initiating parallel registration procedures. This results in a flood of NAS messages that compromises the stability of the 5G core.

Amponis et al. developed a dataset of DoS attacks targeting the 5G core, specifically the PFCP protocol operating on the N4 interface. They implemented attacks such as unauthorized deletion and modification of PFCP sessions, as well as flooding with session establishment requests, all within an experimental environment based on Open5GS and UERANSIM. These attacks demonstrated the feasibility of disrupting user connectivity to the data plane without affecting the control plane connection, thereby complicating detection and mitigation [1].

Alharbi et al. built a 5G SA testbed using free5GC as the core and UERANSIM as the UE/RAN simulator, with the aim of evaluating the impact of DoS/DDoS attacks. In their setup, they injected malicious traffic (e.g., intermittent signaling or overload conditions) alongside benign traffic, thereby generating a new 5G traffic dataset for analysis [4].

Overall, the literature lacks comprehensive experimental studies that push the 5G core's AMF to its limits under adverse conditions in open environments. This issue reflects the need for more empirical research on DoS attacks targeting the 5G control plane, using open-source platforms to reproduce realistic attack scenarios and evaluate the effectiveness of mitigation strategies. Our work aims to help fill this gap by providing an in-depth analysis of AMF saturation attacks in an open-source 5G SA test environment, offering new perspectives in the field.

3 Methods and Materials

The experiments were conducted on a host machine running Windows 11 Home, equipped with an Intel Core i7-12700F processor, 32 GB of RAM, and an NVIDIA RTX 4060 GPU. The environment was virtualized using VirtualBox 7.1.4, with a bridged connection through an 802.11n USB Wireless LAN card. A single virtual machine (VM) running Ubuntu 22.04 hosted the entire experimental setup, including the 5G Core, RAN simulation, and UE simulation. The VM was allocated 9 GB of RAM and 4 virtual CPU, with no dedicated GPU assigned.

The 5G Core was deployed using the OpenAirInterface CN5G stack, working on the oai-cn5g-fed branch, version v2.1.0 via Docker Compose. The following network functions were implemented:

- `oai-amf` (Access and Mobility Management Function)
- `oai-smf` (Session Management Function)
- `oai-upf-vpp` (User Plane Function with VPP)
- `oai-nrf` (Network Repository Function)
- `oai-ausf` (Authentication Server Function)
- `oai-udm` (Unified Data Management)
- `oai-udr` (User Data Repository)

These functions were containerized and interconnected over Docker-defined networks, communicating via standard 5G interfaces:

- N2: UERANSIM ↔ AMF
- N3: UERANSIM ↔ UPF
- N4: SMF ↔ UPF
- N6: UPF ↔ External DN
- N10, N11, N12, N13: Internal signaling between core components

The architecture used three Docker subnets:

- **Core network:** `192.168.70.0/24`
- **Access network:** `192.168.72.0/24`
- **SGi-LAN:** `192.168.73.0/24`

UERANSIM was used to emulate both the RAN and user devices. The simulator was deployed using the `docker_support` branch, integrated via the `docker-compose-ueransim-vpp.yaml` file provided by the OAI project.

All core components were built and managed using Docker targets defined in their respective Dockerfiles. The build and runtime environment included:

- Docker Engine and Docker Compose

- Python 3.x
- Git, curl, and common networking utilities

Network forwarding was enabled at the system level, and Docker container lifecycle was managed using OAI's helper scripts, such as `core-network.py`.

To monitor the system's performance during the execution of the DoS attack, we used btop, a terminal-based resource monitor that provides real-time visualizations of CPU usage, memory consumption, network throughput, and active processes [6]. This tool allowed us to observe the system's behavior before, during, and after the attack scenario.

Diagnostic logs, including registration status, were monitored through the container logs, particularly using `docker logs ueransim`. To track system performance, btop—a terminal-based resource monitor—was used. It provides real-time visualizations of CPU usage, memory consumption, network throughput, and active processes [6].

4 DoS Attack on 5G Environment

To visually represent the Denial of Service (DoS) scenario, Fig. 1 illustrates the architecture of the simulated 5G network used in this study. The setup is entirely containerized using OpenAirInterface, with each network function—UERANSIM, AMF, and UPF—running in separate Docker containers.

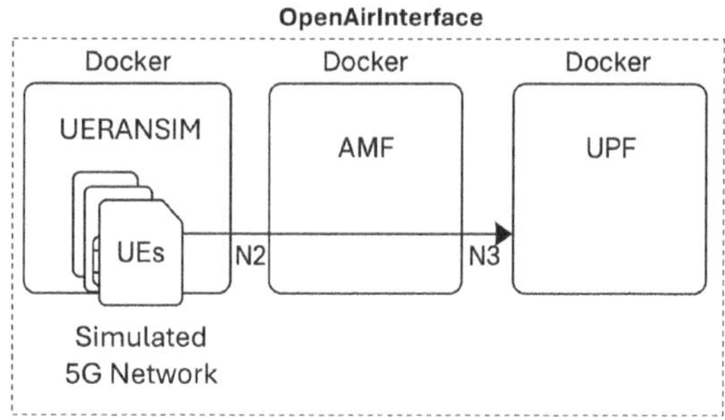

Fig. 1. Overview of the DoS attack.

To launch the Denial of Service (DoS) attack, a large number of legitimate users were added to the core database. As defined in the file `/database/oai_db2.sql`, these UEs were added consecutively, following a sequential IMSI scheme and matching the structure expected by the OpenAirInterface Core.

When the 5G Core is launched using the `docker-compose-basic-vpp-nrf.yaml` configuration, the SQL file is automatically loaded into the MySQL container. The mounting of `oai_db2.sql` into the container's initialization directory ensures that these users are injected at container startup without manual intervention.

Once the database container is initialized, the loaded entries can be verified using a direct query inside the container, executing the commands and obtaining the result shown on Fig. 2:

```
mysql> USE oai_db;
Reading table information for completion of table and column names
You can turn off this feature to get a quicker startup with -A

Database changed
mysql> SELECT COUNT(*) FROM AuthenticationSubscription;
+----------+
| COUNT(*) |
+----------+
|      170 |
+----------+
1 row in set (0.01 sec)
```

Fig. 2. Verification of UEs loaded.

The number of UEs inserted into the database determines the theoretical upper limit of UEs that can attempt to connect to the 5G Core.

Next, the parameter `NUMBER_OF_UE` was set to 170 in the file `docker-compose-ueransim-vpp.yaml`, ensuring that all provisioned UEs would attempt to register simultaneously upon deployment. This configuration also includes core network parameters and the IP addresses required for correct connectivity, as seen in Fig. 3.

```
# UE Config Parameters
- NUMBER_OF_UE=170
- IMSI=208950000000031
- KEY=0C0A34601D4F07677303652C0462535B
- OP=63bfa50ee6523365ff14c1f45f88737d
- OP_TYPE=OPC
- AMF_VALUE=8000
- IMEI=356938035643803
- IMEI_SV=0035609204079514
- GNB_IP_ADDRESS=192.168.70.141
- PDU_TYPE=IPv4
- APN=default
- SST_0=222
- SD_0=123
- SST_C-222
- SD_C=123
- SST_D=222
- SD_D=123
```

Fig. 3. UERANSIM parameters configuration.

After confirming that the 5G Core containers were running properly (via `docker ps`), the simulated UEs were launched using:

```
docker-compose -f docker-compose-ueransim-vpp.yaml up -d
```

In the output of `docker logs ueransim`, several sessions were initially established successfully (see Fig. 4). However, as the number of connection attempts

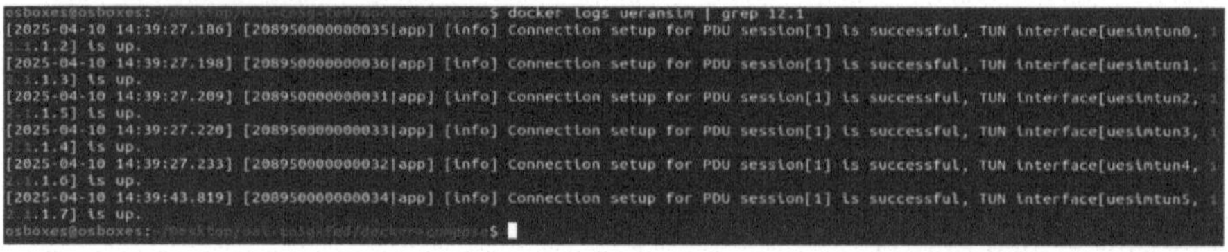

```
osboxes@osboxes:$ docker logs ueransim | grep 12.1
[2025-04-10 14:39:27.186] [208950000000035|app] [info] Connection setup for PDU session[1] is successful, TUN interface[uesimtun0,
.1.2] is up.
[2025-04-10 14:39:27.198] [208950000000036|app] [info] Connection setup for PDU session[1] is successful, TUN interface[uesimtun1,
.1.3] is up.
[2025-04-10 14:39:27.209] [208950000000031|app] [info] Connection setup for PDU session[1] is successful, TUN interface[uesimtun2,
.1.5] is up.
[2025-04-10 14:39:27.220] [208950000000033|app] [info] Connection setup for PDU session[1] is successful, TUN interface[uesimtun3,
.1.4] is up.
[2025-04-10 14:39:27.233] [208950000000032|app] [info] Connection setup for PDU session[1] is successful, TUN interface[uesimtun4,
.1.6] is up.
[2025-04-10 14:39:43.819] [208950000000034|app] [info] Connection setup for PDU session[1] is successful, TUN interface[uesimtun5,
.1.7] is up.
osboxes@osboxes:$
```

Fig. 4. Successful UEs.

increased, session failures began to occur. Eventually, the system was unable to accommodate any additional UEs.

The logs showed a series of `PDU Session Establishment Reject [INSUFFICIENT_RESOURCES]` messages, indicating that the UPF was unable to allocate further resources (Fig. 5).

```
[2025-04-07 14:32:24.767] [208950000000031|nas] [error] PDU Session Establishment Reject received [INSUFFICIENT_RESOURCES]
[2025-04-07 14:32:24.775] [208950000000067|nas] [error] PDU Session Establishment Reject received [INSUFFICIENT_RESOURCES]
[2025-04-07 14:32:24.983] [208950000000114|nas] [warning] Retransmitting PDU Session Establishment Request due to T3580 expiry
[2025-04-07 14:32:24.984] [208950000000114|nas] [debug] UAC access attempt is allowed for identity[0], category[MO_sig]
[2025-04-07 14:32:24.991] [208950000000114|nas] [error] PDU Session Establishment Reject received [INSUFFICIENT_RESOURCES]
```

Fig. 5. UPF insufficient resources.

These failures can be attributed to a number of underlying causes:

- **Session limits**: The maximum number of concurrent PDU sessions the UPF can handle was exceeded.
- **IP address exhaustion**: The /24 subnet (with 256 total IPs) may have reached its allocation limit.
- **Machine constraints**: The VM running the core was limited to 4 vCPU and 9 GB RAM, constraining overall performance.
- **Container resources**: Docker container default limits may not support high-throughput workloads unless explicitly adjusted.

It is important to emphasize that in real deployments, the feasibility and impact of this attack depends on network exposure. For instance, if the AMF interface is accessible via public or misconfigured remote O&M connections the risk increases. In contrast, proper network segmentation, internal-only exposure of critical services, and strict access control policies can reduce the attack surface. Therefore, network architecture plays a crucial role

5 Results

The Denial of Service (DoS) experiments aimed to saturate the control and user planes of the 5G core operating in standalone mode, deployed using the OpenAirInterface within a virtualized environment with limited resources. Up to 170

User Equipments (UEs) were emulated using UERANSIM, configured through Docker Compose. Each UE attempted to perform a complete registration and session establishment process with the core.

Under moderate load conditions (100 UEs), the system successfully managed registration procedures, assigning IP addresses to all UEs and establishing connections through the AMF and UPF functions. However, once the number of simultaneous UEs exceeded 170, signs of service degradation began to emerge. The main symptom was the inability of most UEs to complete the registration process—in many cases, only 20 or fewer UEs successfully connected, while the rest failed silently or remained in incomplete session states.

System monitoring using btop showed how deploying a 5G core imposes a significant load on the system, such that any simultaneous action requiring a high number of resources will cause adverse effects on the machine. In some cases, the graphical interface of the host system became unresponsive due to CPU resource exhaustion. Figure 6 demonstrates that even with a modern processor, resource allocation to the virtual machine significantly impacts the core's ability to handle large volumes of signaling traffic.

The bottom panel of the same figure shows a per-process breakdown of CPU consumption. The process consuming the most resources is vpp_main, which corresponds to the User Plane Function (UPF) implementation using Vector Packet Processing (VPP). This confirms that the UPF becomes the primary bottleneck under stress conditions. Secondary processes, such as nr-ue (emulated user equipment) and other core components like oai_nrf and oai_udr, show minimal CPU usage in comparison.

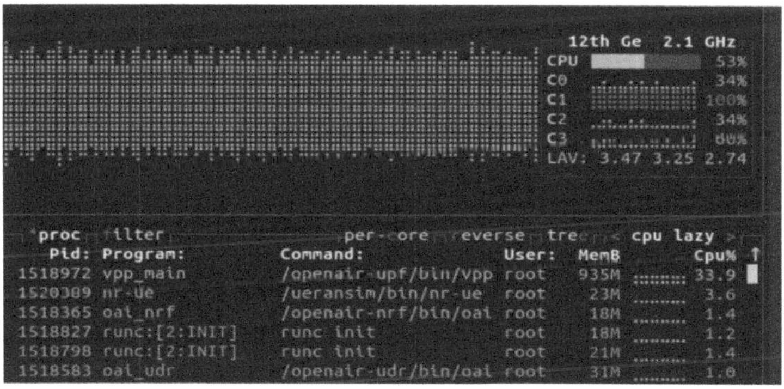

Fig. 6. CPU resource exhaustion.

To further quantify the resource impact of the attack, we examined per-process CPU usage during the overload scenario. As shown in Fig. 6, the vpp_main process, which corresponds to the User Plane Function (UPF), consumes 33.9% of total CPU, significantly more than any other process. The nr-ue emulator process accounts for 3.6%, while other core functions such as oai_nrf and oai_udr remain below 1.5%. This confirms that the UPF becomes the primary bottleneck under stress.

The system load average values (3.47, 3.25, 2.74) indicate that the system was consistently operating near or above its CPU capacity (4 vCPUs allocated to the VM), further confirming resource exhaustion. These statistics quantitatively support the qualitative findings discussed earlier and validate the UPF's role as the limiting component in this attack scenario.

From UERANSIM logs, it was observed that many UEs failed to complete the connection sequence, either due to not detecting available gNB cells or not receiving an IP address. In several instances, the logs displayed explicit errors indicating insufficient resources in the UPF, confirming the stress experienced by the User Plane Function. These messages suggest that the UPF was unable to establish the required GTP-U tunnels for each incoming UE session. Additionally, when analyzing system-level resource usage, it became evident that the UPF process exhibited higher CPU consumption compared to other core functions, highlighting it as the primary bottleneck during the attack.

Despite the attack, the 5G core was occasionally able to recover its state without manual intervention. After an overload event, retrying the connection attempts sometimes led to partial recovery, with more UEs successfully connecting than in the previous round. However, in many scenarios, it was necessary to restart the core or even the entire virtual machine to restore normal operation.

6 Conclusion and Future Work

This work has demonstrated that it is possible to carry out a Denial-of-Service (DoS) attack against a 5G Standalone (SA) core network implemented with open-source tools. The results indicate that the AMF and UPF were saturated, as the PFCP connection failed when attempting to manage 170 concurrent UEs—a situation that did not occur with 100 devices. The absence of mitigation or defensive mechanisms in the core led to all registration attempts being processed simultaneously, ultimately collapsing the function.

One of the main conclusions of this study is the paradox that solutions designed to simplify and reduce the cost of 5G network deployment—such as OpenAirInterface and UERANSIM—can be repurposed by attackers with minimal resources. The adversary does not require antennas, base stations, or specialized hardware; access to the AMF's IP address is sufficient. If this IP is publicly reachable or exposed to external networks, the associated risk is significantly increased.

As future work, we plan to replicate this attack in a more realistic environment. Specifically, we will use the Firecell Labkit 40 NSA device to evaluate the impact of the attack from an external virtual machine. This will allow us to observe the system's response when operating with dedicated resources for 5G deployment and assess the AMF's resilience under real-world conditions. In addition, we plan to extract features from the captured traffic generated during these attacks in order to build labeled datasets. These datasets will be used to train and evaluate machine learning models capable of detecting early signs of core saturation or anomalous behavior. The output of this process will serve as the basis for implementing intrusion detection systems (IDS), which represent one of the most effective forms of attack mitigation in 5G networks. Beyond IDS, we also aim to explore complementary mitigation techniques such as rate limiting, connection throttling at the AMF interface and network segmentation

Acknowledgments. This work has been funded by the Recovery, Transformation, and Resilience Plan, financed by the European Union (Next Generation) thanks to the "Internet of Things Security in Home and Business Environments in the Context of 5G-IoT Technology".

Disclosure of Interests. The authors have no competing interests to declare that are relevant to the content of this article.

References

1. Amponis, G., et al.: Threatening the 5G core via PFCP DoS attacks: the case of blocking UAV communications. Eurasip J. Wirel. Commun. Netw. **2022**, 1–27 (2022). https://doi.org/10.1186/S13638-022-02204-5/FIGURES/6
2. Frank, H., Colman-Meixner, C., Assis, K.D.R., Yan, S., Simeonidou, D.: Techno-economic analysis of 5G non-public network architectures. IEEE Access **10**, 70204–70218 (2022). https://doi.org/10.1109/ACCESS.2022.3187727
3. GÜNGÖR, A.: Documentation Open5GS (2025). https://open5gs.org
4. Khan, S., Farzaneh, B., Shahriar, N., Saha, N., Boutaba, R.: SliceSecure: impact and detection of DoS/DDoS attacks on 5G network slices. In: 2022 IEEE Future Networks World Forum (FNWF) (2022). https://www.free5gc.org/
5. Lee, S.: UERANSIM (2025). https://github.com/aligungr/UERANSIM
6. Liljenberg, J.P.: Github - btop (2025). https://github.com/aristocratos/btop
7. de Neira, A.B., Kantarci, B., Nogueira, M.: Distributed denial of service attack prediction: challenges, open issues and opportunities. Comput. Netw. **222**, 109553 (2023). https://doi.org/10.1016/J.COMNET.2022.109553
8. OpenAirInterface: OpenAirInterface (2025). https://openairinterface.org/
9. Palacios-Morocho, E., Picazo-Martinez, P., Inca, S., Monserrat, J.F.: Open source 5G-NSA network for industry 4.0 applications. In: 2021 IEEE 32nd Annual International Symposium on Personal, Indoor and Mobile Radio Communications (PIMRC) 2021-September, 1–6 (2021). https://doi.org/10.1109/PIMRC50174.2021.9569481
10. Park, S., Cho, B., Kim, D., You, I.: Machine learning based signaling DDoS detection system for 5G stand alone core network. Appl. Sci. **12** (2022). https://doi.org/10.3390/APP122312456
11. Peng, C., Zhu, Q.: Trust-aware resource management for secure and optimal network slicing in 5G mobile edge networks. INFOCOM, pp. 1–6 (2023). https://doi.org/10.1109/INFOCOMWKSHPS57453.2023.10225773
12. Peng, C., Zhu, Q.: Early detection of dos attacks in 5G core networks. In: 2024 IEEE International Conference on Advanced Networks and Telecommunications Systems (ANTS), pp. 1–6 (2024). https://doi.org/10.1109/ANTS63515.2024.10898752

Enhancing Spam Detection with Machine Learning Techniques: A Case Study

Martim Martins(✉), E. J. Solteiro Pires, and P. B. de Moura Oliveira

INESC-TEC, Department of Engineering, University of Trás-os-Montes and Alto Douro (UTAD), Vila Real, Portugal
al78681@alunos.utad.pt, {epires,oliveira}@utad.pt

Abstract. In today's digital age, email communication plays a crucial role in both personal and professional settings. However, the increasing volume of spam emails poses significant challenges, leading to wasted time, security risks, and potential financial losses. This work explores various Machine Learning models to classify emails as spam or non-spam, aiming to enhance the accuracy and efficiency of spam detection systems. The study involves data preprocessing, feature extraction, and the implementation of supervised learning techniques such as Naïve Bayes, Support Vector Machines, and Neural Networks. A comparative analysis is performed based on key performance metrics, including Accuracy, Precision, and Recall. Furthermore, the impact of feature selection and dataset balancing on classification results is examined. The findings provide valuable insights into the strengths and limitations of different models, contributing to the development of more reliable spam detection models.

Keywords: Machine Learning · Data Science · Email Classification · Spam Detection

1 Introduction

Email has become one of the most widely used forms of communication, in both personal and professional environments. Spam refers to unsolicited and often irrelevant or harmful emails sent in bulk, typically for advertising, phishing, or spreading malware (Cisco 2025). However, the growing prevalence of spam emails, which in 2024 are estimated to total 162 billion daily, accounting for approximately 46.8% of all email traffic (EmailToolTester 2024), has introduced significant challenges, including loss of productivity, exposure to phishing attacks, and increased cybersecurity risks. Spam messages often contain fraudulent content, malware, or deceptive advertisements, making their detection and filtering important for individuals and organizations alike. Traditional rule-based filtering systems, while effective to some extent, struggle to adapt to the continuously evolving nature of spam tactics (Pandian 2023). As a result, more advanced, data-driven approaches are required to increase the accuracy and efficiency of spam classification.

Machine Learning (ML) has emerged as a powerful tool in addressing the limitations of traditional spam filters. By leveraging ML models, email classification systems

E. Corchado et al. (Eds.): SOCO 2025, CCIS 2806, pp. 150–158, 2026.
https://doi.org/10.1007/978-3-032-19763-4_14

can analyze large datasets of emails to identify patterns, and automatically distinguish between spam and legitimate emails. Various supervised learning models, which are models trained on labeled datasets to learn patterns and make predictions, have been widely explored in spam detection due to their ability to learn from past data and improve over time. However, selecting the most suitable algorithm for this task requires a careful evaluation of different techniques based on their performance metrics, such as Accuracy, Precision, and Recall (Occhipinti et al. 2022).

This project aims to explore and compare different ML models for spam classification, assessing their strengths and limitations in real-world scenarios.

2 Literature Review

To ensure a comprehensive and focused literature review on the use of ML for spam detection, a systematic search strategy was implemented using the OpenAlex database (OpenAlex 2025).The initial query returned a total of 1.562 articles related to the topic using the prompt "spam Detection Machine Learning." A step-by-step screening process was then applied to refine the results and guarantee the inclusion of only the most relevant and high-quality sources. The first filtering criterion was based on the field of study, selecting only articles within Computer Science, which led to the removal of 127 articles. Next, articles were filtered by year of publication, specifically between 2020 and 2025, to ensure the inclusion of recent studies, 203 articles were excluded during this phase. Only articles written in English were retained, resulting in the exclusion of 674 entries.

Further screening considered the geographic context of the publications, focusing on works from the United States, the United Kingdom, Spain, and Italy, leading to the removal of 29 more. The selection of these countries allows for the construction of a robust, comprehensive, and globally relevant literature review. Following these criteria, the remaining 25 articles were evaluated based on title and abstract relevance, excluding 5 that were deemed either off-topic or of questionable quality. From the remaining 20 full-text articles assessed for eligibility, 11 were ultimately included in the final review. These selected studies form the foundation of the present analysis, which explores the methodologies, challenges, and outcomes associated with applying machine learning to spam classification across various data domains and contexts. The selection process used to identify and filter the most relevant studies is illustrated in the Prisma flow diagram presented in Fig. 1.

The study reported by Moutafis et al. (2023) applied ten of the most popular machine learning models across three different spam datasets. Results varied by dataset, with Neural Networks achieving the highest accuracy of 99.51% in one case and Support Vector Machines (SVM) leading in another case with 99.38%. Similarly, another work by Chen (2023) tested various models on multiple datasets, where Neural Networks again stood out with an accuracy of 98.52%, slightly outperforming SVM at 97.40%.

In addition to performance, the study (Agboola 2020) focused on the challenges posed by imbalanced datasets. In a binary classification task using AdaBoost, Naive Bayes, and Support Vector Classifier (SVC). It was observed that while models achieved high overall accuracy, they performed poorly on the minority class, which was spam.

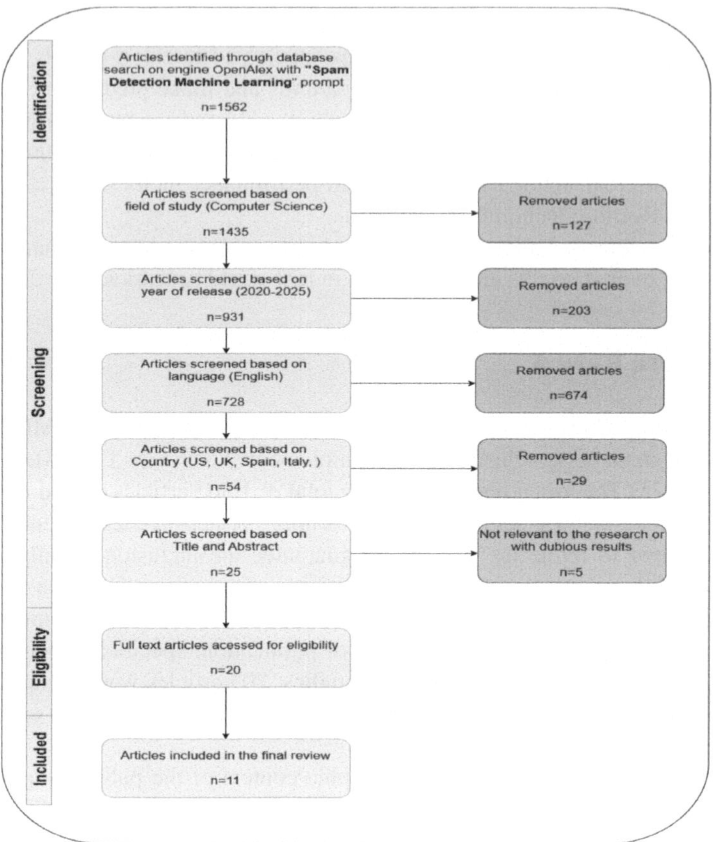

Fig. 1. Prisma flow diagram the article selection process

To address this, the authors combined AdaBoost with SVC, which significantly reduced false negatives in the spam class. A different study used logistic and linear regression across multiple datasets to estimate spam probability. While both models provided useful insights, they were limited in detecting more sophisticated spam strategies, pointing to the need for more robust solutions when dealing with modern spam tactics (Liu 2023).

Regarding spam detection in social media, Xiao and Liang (2024) addressed YouTube comments. Naive Bayes achieved a recall of 93.1%, however, its precision was lower, as it often misclassified ham as spam. On the other hand, Random Forest delivered a more balanced performance with an F1-score of 96.3%, suggesting it is a strong choice for scenarios where both false positives and false negatives must be minimized. Another approach (Binsaeed et al. 2020) targeted spam tweets on Twitter, emphasizing how spammers exploit popular domains to conceal malicious content. The study used an implementation of J48 decision tree algorithm, which successfully identified 12.939 spam tweets with a relatively low false positive count of 1.033. SMS spam classification has also seen successful application of machine learning, the study reported by Ahluwalia et al. (2023) achieved 98.44% accuracy in spam classification using Naive Bayes. ML models are suitability for text-based spam filtering in short message services. Feature extraction methods such as N-grams have proven essential for improving classification performance, using such techniques significantly enhances the accuracy of spam detection models, reinforcing the importance of proper preprocessing and feature engineering in text classification tasks (Sutta et al. 2020).

3 Material and Method

In this study, a dataset with pre-labeled emails as either spam or ham (legitimate) was used. The dataset serves as the foundation for training and evaluating machine learning models for spam classification. However, raw email data cannot be directly fed into ML models, as it often contains unstructured text, redundant information, noise and imbalanced data that can negatively impact model performance. To address this, Natural Language Processing (NLP) techniques and standard data preprocessing methods were applied. These steps include text normalization, stop-word removal, tokenization, and feature extraction techniques such as Term Frequency-Inverse Document Frequency (TF-IDF) and N-grams. By converting textual data into a structured format, these preprocessing steps enhance the ability of ML models to identify patterns and improve classification accuracy. The following sections detail the specific methodologies used for data preprocessing, model selection, and evaluation, as well as provide information about the dataset.

3.1 Database

Initially, the "Spam Email" dataset provided by Qureshi (2021) on Kaggle was selected for experimentation. However, during preliminary testing, a significant issue arose: overfitting. Although the dataset contained a substantial number of emails for model training, the distribution between the two classes, spam and ham, was highly imbalanced. Class imbalance in binary classification can severely impact model performance. This issue becomes specially evident when analyzing learning curves, which plot a model's performance on both training and validation data as the size of the training set increases (Burkov 2019). In this case, due to the overrepresentation of the ham class, the models developed a strong bias toward predicting non-spam emails, thereby reducing their ability to correctly classify spam. To address class imbalance, several techniques can be applied to improve model performance and reduce bias toward the majority class (Wainer 2018). One common approach is resampling, which involves adjusting the distribution of the dataset. For example, Synthetic Minority Oversampling Technique (SMOTE) is a popular method that creates synthetic samples of the minority class by interpolating between existing samples. This helps balance the dataset without simply duplicating data, which can lead to overfitting. Several authors in similar studies have successfully applied SMOTE to address class imbalance, demonstrating its effectiveness across various domains (e.g. see Bhatnagar et al. 2024). Another effective strategy is to use class weighting, where the learning algorithm is instructed to give more importance to misclassifications of the minority class. In libraries like scikit-learn, this is done by setting the "class_weight" parameter to "balanced", which automatically adjusts weights inversely proportional to class frequencies. With this technique, the model is penalized more when it incorrectly predicts a minority class instance, encouraging it to learn more meaningful distinctions. A third option is to increase the overall size of the dataset, particularly by collecting more examples of the underrepresented class. This can involve data augmentation or sourcing additional real-world data. In this work, it was opted to follow this approach by expanding the dataset to achieve a more balanced class distribution. To do so, additional spam emails were retrieved from similar public datasets, made public by Ozler (2019) and Noey (2025) and integrated into the main dataset, effectively boosting the representation of the minority class. Additionally, it was applied class weighting during model training to further mitigate the imbalance and therefore reduce the overfitting. After completing this process, the class distribution the dataset was as follows (Fig. 2).

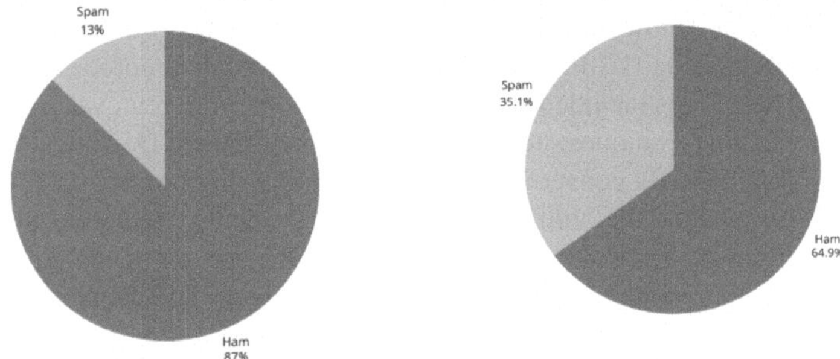

Fig. 2. Class distribution: before and after gathering more data

3.2 Data Pre-processing

Before applying machine learning models to a text classification problem like spam detection, it is crucial to process the raw text data into a format that models can understand. This is where NLP techniques come into play. NLP involves a series of steps that clean, structure, and transform human language data so it can be effectively used in computational models. Most ML models for spam detection aim to identify patterns by associating certain keywords with spam. Through training, these models learn that the presence of specific terms can strongly indicate that an email is malicious. On the other hand, common words found in everyday English phrases, known as stop words, as well as punctuation, typically have little to no value in determining whether an email is spam. This is because they appear frequently in both spam and non-spam (ham) messages. As a result, a standard and widely recommended preprocessing step in spam classification studies is to remove stop words and punctuation entirely. Doing so helps ensure that these elements do not influence the model's decision-making, either negatively or at all. Another important aspect of the preprocessing phase in email data is addressing the fact that computers cannot directly interpret text, they work only with numerical data. To overcome this issue, the text must be converted into a numerical format. This involves breaking each sentence into smaller units (tokens), typically individual words, and assigning a unique numerical representation to each token. This transformation allows the machine learning models to process the text and assign different weights to different words during classification. However, before this step, it needs to be ensured the consistency in how words are represented. For example, the model should treat "FREE", "free" and "Free" as the same word, so all text is normalized by converting it to lowercase during preprocessing. With tokenization and normalization complete, attention can now turn to the effective assignment of appropriate weights to each word. There are several methods for doing this, with two common approaches being N-grams and TF-IDF. Both techniques aim to capture the relevance of words within a document in different ways. The basic premise is to highlight terms that carry meaningful information for classification, whether by analyzing word frequency patterns (as in TF-IDF) or by capturing word sequences and context (as in N-grams). While they differ in implementation, both methods effectively contribute to distinguishing between classes, such as spam and non-spam emails.In this study, TF-IDF was chosen as the preferred method, so it will be explored in greater detail. TF-IDF (1) works by performing two main operations: Term Frequency (TF) and Inverse Document Frequency (IDF).

$$TF - IDF = TF \times IDF \tag{1}$$

TF (2) measures how often a word appears in a specific email, being calculated by dividing the number of times a word appears in a document by the total number of words in that document. This gives an indication of the word's relevance within that document.

$$TF = \frac{\text{Number of times the word "X" appears in the email}}{\text{Number of words in the email}} \tag{2}$$

IDF (3) evaluates how rare or common a word is across the entire dataset. Words that appear in many documents are considered less significant, while those that appear in only a few documents are given higher importance.

$$IDF = \log\left(\frac{\text{Number of total emails in the Dataset}}{\text{Number of emails where the word "X" appeared}}\right) \tag{3}$$

3.3 Metrics

Before presenting the performance of the ML models applied to spam detection, it is important to understand why certain evaluation metrics were chosen. Due to the inherent class imbalance in spam classification (with fewer spam emails compared to legitimate ones), metrics that fairly assess performance across both classes are necessary. Balanced Accuracy is selected because it gives equal importance to detecting spam and ham, preventing bias toward the majority class. Precision is emphasized since minimizing false positives is critical to avoid disrupting user communication. Recall for spam measures how well the model captures actual spam. Recall for ham ensures legitimate emails are correctly identified. Finally, the F1-Score provides an overall measure that reflects the trade-offs between false positives and false negatives. Together, these metrics offer a well-rounded evaluation of model effectiveness.

4 Results

This section presents the results of the various machine learning classifiers and serves to compare their performance using key evaluation metrics.

Table 1 summarizes the results for each machine learning model applied to the spam classification task.

Table 1. Testing results obtained

Classifier	BA	Pre	Re-S	Re-H	F1	TP	TN	FP	FN
SVM	0.98	0.98	0.96	0.99	0.98	2514	4782	43	100
RF	0.99	0.99	0.98	0.99	0.99	2558	4797	28	56
Naive Bayes	0.96	0.97	0.95	0.98	0.97	2474	4731	94	140
LR	0.97	0.97	0.95	0.99	0.97	2475	4758	67	139
XGBoost	0.96	0.97	0.93	0.99	0.97	2435	4794	31	179

The training and evaluation of the models were conducted using stratified 5-fold cross-validation to ensure balanced representation of both spam and ham classes in each fold, providing reliable performance estimates despite class imbalance. Hyper-parameter tuning was performed through a systematic grid search process, which explored various parameter combinations relevant to each model to optimize their predictive performance. This process used cross-validation to select the best set of hyper-parameters that maximized classification accuracy and generalization. The final results reported correspond to models trained with their respective best hyper-parameter configurations identified through this tuning approach.

The results in Table 1 demonstrate that Random Forest and SVM are the most effective models for classifying spam and ham emails. Random Forest, in particular, achieved the highest Balanced Accuracy and F1-Score, showing its ability to differentiate between legitimate and spam messages. SVM also performed strongly, especially in terms of precision and Recall-Ham. The strength of Random Forest in this spam classification task stems from its architecture. As an ensemble method, it builds multiple decision trees, each trained on random subsets of both the data and the features. This diversity among trees allows the model to capture different patterns or signals that indicate spam, such as the frequent appearance of obfuscated words, numbers, or repeated spammy phrases, while still being sensitive to the nuances of legitimate content. In the context of spam detection, where emails can vary widely in format and style, Random Forest's ability to aggregate weak learners into a robust final prediction enables it to generalize well to unseen data. For example, if one tree picks up on spam trigger words like "FREE" or "WINNER," and another focuses on suspicious formatting, their combined vote helps ensure that spam is caught even if it uses varied tactics. At the same time, the averaging process across trees reduces the likelihood of false positives, helping the model avoid flagging legitimate messages incorrectly. Moreover, Random Forest is particularly effective at handling high-dimensional feature spaces, which are common in text-based tasks. Even when many words or tokens are irrelevant, the model's ability to ignore noisy features and focus on the most informative ones contributes to its superior recall for spam (0.98) and recall for ham (0.99). In addition, the performance of the Random Forest model was further enhanced through GridSearch, a technique that systematically explores multiple combinations of hyperparameters using cross-validation to find the most effective configuration for the classification task. This tuning process helped refine the model's decision-making by selecting parameter values that maximized its accuracy and generalization. The other models (Naive Bayes, Logistic Regression and XGBoost) showed a somewhat noticable disparity between recall for spam and recall for ham, reflecting a bias toward the majority class, often leading to more missed spam (false negatives).

5 Conclusion and Future Work

This study explored the application of various supervised machine learning models for the classification of spam and ham emails, addressing the growing need for accurate, efficient, and adaptive spam detection methods. Models such as Support Vector Machine, Random Forest, Naive Bayes, Logistic Regression, and XGBoost were evaluated using key performance metrics including Balanced Accuracy, Precision, Recall for Spam and Ham, and F1-Score. The results revealed that Random Forest and Support Vector Machine were the most effective classifiers. Random Forest, in particular, achieved the highest overall scores across most metrics, benefiting from its ensemble-based structure and hyperparameter tuning through GridSearch. Support Vector Machine also demonstrated robust performance, especially in accurately classifying ham messages, a crucial aspect for minimizing the misclassification of legitimate emails. However, the other models, Naive Bayes, Logistic Regression, and XGBoost, also delivered respectable results. Machine learning presents, and is, a promising solution for spam detection, offering flexibility and strong predictive capabilities in handling imbalanced and evolving datasets. In conclusion, machine learning presents a promising solution for spam detection, offering flexibility and strong predictive capabilities in handling imbalanced and evolving datasets. Given the strong performance of Random Forest and SVM, future research may focus on integrating these models into real-time filtering systems, as well as combining them with deep learning or adaptive learning techniques to further enhance resilience against novel spam tactics.

References

Cisco: What is Spam? Cisco Security Learning Hub. https://www.cisco.com/site/us/en/learn/topics/security/what-is-spam.html. Accessed 14 Apr 2025

EmailToolTester: Spam Statistics 2025: Survey on Junk Email, AI Scams & Phishing (2024). https://www.emailtooltester.com/en/blog/spam-statistics/

Pandian, A.: Traditional Programming vs. Machine Learning: Spam Email Filtering (2023). https://arunpandianm.medium.com/traditional-programming-vs-machine-learning-spam-email-filtering-9d2a8baf37bd

Occhipinti, A., Rogers, L., Angione, C.: A pipeline and comparative study of 12 machine learning models for text classification (2022). https://arxiv.org/abs/2204.06518

OpenAlex: OpenAlex: The open catalog of the global research system. https://openalex.org/. Accessed 02 May 2025

Moutafis, I., Andreatos, A., Stefaneas, P.: Spam email detection using machine learning techniques. In: 22nd European Conference on Cyber Warfare and Security (ECCWS 2023), pp. 303–310 (2023). https://doi.org/10.34190/eccws.22.1.1208

Chen, S.: Multiple Machine Learning Algorithms for Spam Mail Detection. Highlights in Science, Engineering and Technology 39. University of Manchester, United Kingdom (2023). https://drpress.org/ojs/index.php/HSET/article/view/6698

Agboola, O.S.: Spam detection using machine learning. Comput. Eng. Intell. Syst. **11**(3), 34–39 (2020). https://iiste.org/Journals/index.php/CEIS/article/view/52548

Liu, Z.: A detection research of spams based on machine learning algorithms. ACE Appl. Comput. Eng. **17**, 81–86 (2023). https://www.ewadirect.com/proceedings/ace/volumes/vol/17/81.pdf

Xiao, A.S., Liang, Q.: Spam detection for YouTube video comments using machine learning approaches. Mach. Learn. Appl. **16**, Article no. 100550 (2024). https://doi.org/10.1016/j.mlwa.2024.100550

Binsaeed, K., Stringhini, G., Youssef, A.E.: Detecting spam in twitter microblogging services: a novel machine learning approach based on domain popularity. Int. J. Adv. Comput. Sci. Appl. **11**(11), 11–22. https://doi.org/10.14569/IJACSA.2020.0111103 (2020)

Ahluwalia, K., Gururaj, H.L., Rashmi, R., Lin, H.: Comparative analysis of various SMS spam detection methods using machine learning. In: 13th International Workshop on Computer Science and Engineering (WCSE 2023), pp. 146–155 (2023). https://doi.org/10.18178/wcse.2023.06.021

Sutta, N., Liu, Z., Zhang, X.: A study of machine learning algorithms on email spam classification. In: 35th International Conference on Computers and Their Applications (CATA 2020), EPiC Series in Computing, vol. 69, pp. 170–179 (2020). https://doi.org/10.29007/qshd

Wang, Z., Pang, Y., Lin, Y., Zhu, X.: Adaptable and reliable text classification using large language models. arXiv preprint. https://arxiv.org/abs/2405.10523 (2024)

Qureshi, F.: Spam Email. Kaggle (2021). https://www.kaggle.com/datasets/mfaisalqureshi/spam-email

Burkov, A.: The Hundred-Page Machine Learning Book, 1st edn. Andriy Burkov Publishing, Québec (2019)

Wainer, J.: An empirical evaluation of imbalanced data strategies from a practitioner's point of view. arXiv preprint. https://arxiv.org/abs/1810.07168 (2018)

Bhatnagar, P., Degadwala, S.: Efficient email spam classification with n-gram features and ensemble learning. Int. J. Sci. Res. Comput. Sci. Eng. Inf. Technol. **10**(1) (2024). https://doi.org/10.32628/CSEIT2410220

Scikit-learn developers: compute_class_weight — scikit-learn 1.7.dev0 documentation (2025). https://scikit-learn.org/dev/modules/generated/sklearn.utils.class_weight.compute_class_weight.html

Ozler, H.: Spam or Not Spam Dataset. SpamAssassin Public Datasets (2019). https://www.kaggle.com/datasets/ozlerhakan/spam-or-not-spam-dataset

Noey: Spam Emails: Identifying and Classifying Spam Emails. Kaggle (2025). https://www.kaggle.com/datasets/noeyislearning/spam-emails

Ratadiya, P., Moorthy, R.: Spam filtering on forums: a synthetic oversampling based approach for imbalanced data classification. arXiv preprint. https://arxiv.org/abs/1909.04826 (2019)

Special Session: Time Series Forecasting in Industrial and Environmental Applications

Personalized Multi-horizon Glucose Forecasting in Type 1 Diabetes via Deep Learning

Alessandro Santopaolo[1], Ilaria Basile[2], Alessandro Giuseppi[1], and Giovanna Sannino[2](✉)

[1] Department of Computer, Control, and Management Engineering Antonio Ruberti (DIAG), Sapienza University of Rome, Rome, Italy
{santopaolo,giuseppi}@diag.uniroma1.it

[2] Institute for High Performance Computing and Networking (ICAR), National Research Council of Italy (CNR), Naples, Italy
{ilaria.basile,giovanna.sannino}@icar.cnr.it

Abstract. This study presents a Deep Learning framework for forecasting blood glucose levels at a time horizon of 45 and 60 min using ElectroCardioGraphy (ECG)-derived features alongside baseline glucose readings. Using data from the D1NAMO dataset, we extracted and normalized Heart Rate (HR) and Heart Rate Variability (HRV) metrics and categorized future glucose levels into clinically meaningful bins. A feedforward neural network was trained to classify these categories. The resulting model demonstrated promising classification performance on both prediction horizons, with balanced predictive capability across most glucose categories. To enhance transparency and clinical relevance, we employed SHapley Additive exPlanations (SHAP) to assess feature importance and interpret the model's predictions. SHAP analysis revealed that HR and HRV metrics—particularly `hr_max`, `hr_min`, `hr_std`, and `nni_mean`—provide valuable complementary information to current glucose levels for predicting short-term glycemic trends. This highlights the potential of ECG-derived features as predictive biomarkers for short-term glycemic changes. Overall, the proposed framework offers a promising and explainable approach to glucose forecasting, supporting transparency, physiological insight, and trust–key factors for the adoption of predictive models in real-time clinical monitoring.

Keywords: Blood Glucose Levels Forecasting · Heart Rate Variability · Type 1 Diabetes · Deep Learning · Explainable AI

1 Introduction

Type 1 Diabetes (T1D) is a chronic autoimmune disease characterized by insulin deficiency, caused by the immune-mediated destruction of the pancreatic β-cells.

A. Santopaolo and I. Basile—Equal contribution.

E. Corchado et al. (Eds.): SOCO 2025, CCIS 2806, pp. 161–170, 2026.
https://doi.org/10.1007/978-3-032-19763-4_15

The resulting imbalance of glycidic metabolism leads to a condition of chronic hyperglycemia, which, if not properly managed, can lead to serious microvascular and macrovascular complications [1,7].

Given the lack of effective treatment for this disease, the only possible therapy is the exogenous administration of insulin in order to maintain glycaemic balance. The patient must follow this replacement therapy throughout his life, combining it with careful self-management that includes continuous monitoring of Blood Glucose Levels (BGLs) and prevention of severe complications [9].

The ability to predict glucose levels in the short term (30–60 min) has great clinical relevance, allowing the prevention of critical events and optimizing insulin therapy, by allowing timely and personalized adjustments to insulin dosing. In this context, predictive models based on Machine Learning (ML) and Deep Learning (DL) techniques have shown great potential in analyzing glucose levels. These models serve as decision support tools in closed-loop insulin delivery systems such as the artificial pancreas [15]. However, in a critical area such as health, non-transparency of DL models is a significant barrier that can hinder their adoption. In this context, the application of Explainable AI (XAI) techniques makes it possible to illustrate the rationale behind the model's predictions, furthering its comprehensibility [18,21].

In this work, we propose a DL framework for predicting glucose levels at a 45-minute and 60-minute time horizons. The model leverages BGLs provided by Continuous Glucose Monitors (CGMs) along with Heart Rate Variability (HRV) metrics and Heart Rate (HR) parameters extracted from the Electro-CardioGraphic (ECG) signal acquired by wearable devices. The introduction of ECG-derived features as an additional input aims to enable a noninvasive approach to glucose monitoring. The ease of use of wearable ECG devices, such as smartwatches or fitness bands, makes them perfectly suited to the patient's daily life, limiting the discomfort and invasiveness of glucose measurements.

Several studies have shown that fluctuations in BGLs can affect the electrical activity of the heart, particularly during hypo- or hyperglycemic events [16,22]. This is particularly relevant in patients with T1D in whom one of the serious complications that can result from vascular alterations to the Autonomic Nervous System (ANS) is the Cardiovascular Autonomic Neuropathy (CAN). Since HRV reflects the dynamic interplay between sympathetic and parasympathetic activity, its modulation by glycemic variations makes it a promising biomarker for glucose trend prediction [16,22]. Furthermore, it has been shown that in patients with T1D, hyperglycaemia and hyperinsulinaemia are associated with a high resting HR, while intensive diabetes management is associated with a lower HR [17,23]. Hence, leveraging HR and HRV data offers a valuable, non-invasive means to enhance glucose level forecasting and support diabetes management.

2 Related Work

Blood glucose level prediction has emerged as a critical research area in diabetes management, driven by the urgent need to prevent severe hypoglycemic and hyperglycemic episodes that can lead to life-threatening complications. The advancement of ML techniques has substantially enhanced glucose prediction accuracy, with Support Vector Machines (SVMs) and Random Forest algorithms successfully applied to glucose forecasting tasks [11,24]. However, these traditional approaches typically require extensive manual feature engineering and may not fully capture complex interactions between multiple physiological variables. In contrast, DL techniques have shown greater potential by automatically learning complex patterns and feature interactions without explicit programming. Artificial Neural Networks (ANNs) have demonstrated their ability to model non-linear relationships between input features and glucose levels, effectively processing multi-dimensional feature vectors containing various physiological parameters [3,19]. Moreover, several studies have shown that simple DL architectures can achieve competitive performance while maintaining computational efficiency suitable for real-time applications [20,25].

However, the "black box" nature of DL networks poses significant challenges for clinical adoption, where understanding prediction rationale is crucial for building trust and ensuring patient safety. This has led to the increasingly important application of XAI techniques in healthcare [2].

The need for explainability becomes even more crucial when integrating other physiological features in predicting BGLs to understand how these features influence the model's decision. In fact, recent studies have demonstrated that incorporating additional physiological signals beyond glucose measurements can substantially improve prediction accuracy. The integration of ECG-derived features represents a particularly promising approach [4,6]. Cordeiro et al. [6] developed a DL architecture for hyperglycemia detection using ECG-derived features, including HRV parameters in time and frequency domains, validated on a private dataset containing ECG recordings and glucose measurements from over a thousand subjects. Arbi et al. [4] created mathematical equations combining corrected QT duration, T-wave amplitude, and HR for glucose level estimation using the D1NAMO dataset, employing recurrent convolutional neural networks for ECG segmentation and achieving very low estimation errors.

Based on these promising results, this work addresses critical gaps by demonstrating the predictive value of ECG-derived features for glucose forecasting using feedforward neural networks, providing comprehensive explainability analysis using SHapley Additive exPlanations (SHAP) to understand feature contributions, and evaluating multi-horizon prediction capabilities to support different clinical decision-making timeframes, thereby bridging the gap between advanced DL techniques and interpretability requirements in diabetes management.

3 Methodology

3.1 Dataset

In this work, we used the D1NAMO dataset [10], a publicly accessible dataset containing data acquired for three weeks in real-life conditions by 29 subjects (20 healthy people and 9 people with T1D). The healthy population is composed of 16 men and 4 women with an average age of 33 years, whereas the diabetic population is composed of 6 men and 3 women with an average age of 38 years. In this study, we focused only on data from the diabetic population because our intention was to develop an approach capable of identifying critical episodes, such as hyperglycaemic or hypoglycaemic events. This dataset includes glucose measurements, food picture annotations, and ECG, breathing, and accelerometer signals. Physiological signals were acquired using the *Zephyr BioHarness 3 band*, while BGLs were measured every five minutes in diabetic subjects using the *iPro2 Professional CGM sensor*.

The dataset used for the prediction task in this paper was derived from the D1NAMO dataset, using glucose measurements and HRV parameters in both time and frequency domains calculated from the ECG signal provided by the dataset. The HRV features used in this work will be described in detail in the Sect. 3.3. Our dataset contains information from only one individual with T1D from the D1NAMO collection: patient 008. Data from the remaining diabetic patients (001, 002, 003, 004, 005, 006, 007, and 009) were excluded due to technical limitations: glucose measurements were not properly time-aligned with ECG recordings, ECG signals were too noisy due to improper band placement, or patients had insufficient observations for effective deep learning model training. Patient 008 was the only subject meeting all requirements for data quality, temporal synchronization, and adequate sample size.

3.2 Pre-processing

The ECG signal provided by the D1NAMO dataset required a pre-processing phase before the HRV features could be extracted. Firstly, due to time misalignments between the ECG signal and glucose measurements, only data points with synchronized ECG and BGLs were considered for analysis.

Additionally, due to the noisy nature of wearable ECG signals, filtering and quality assessment were also conducted prior to feature extraction. The ECG signal filtering and R-peaks identification were performed using dedicated functions from the Neurokit2 library [14]. Specifically, the filter applied to the ECG signal was a fifth-order high-pass filter with a cutoff frequency of 0.5 Hz, followed by powerline filtering at 50 Hz.

The subsequent procedure for extracting the QRS complexes and evaluating the ECG Signal Quality Index (SQI) was carried out according to the method described in [5]. At the end of this processing, the output consists of 2-minute ECG signal windows with a 1-minute overlap.

3.3 Feature Extraction

Following the pre-processing phase of the ECG signal, the features derived from HR and HRV were extracted. Given the length of the ECG segments extracted (2 min), the analysis focused only on parameters that can be evaluated over ultra-short recording periods (<5 min), calculated using the *pyHRV* toolbox [13]. In the frequency domain, this limited the analysis to parameters calculated in the

High-Frequency (HF) band (0.15–0.40 Hz) and Low-Frequency (LF) band (0.04–0.15 Hz). In the time domain, the triangular HRV index, which requires 24-hour recordings, was excluded.

Therefore, the dataset structure includes the following 21 input variables:

- `glucose_t0`: Glucose value at the baseline [mmol/L];
- `hr_mean`: Average HR computed over the window [bpm];
- `hr_max`: Maximum HR recorded in the window [bpm];
- `hr_min`: Minimum HR recorded in the window [bpm];
- `hr_std`: Standard deviation of HR over the window [bpm];
- `nni_mean`: Mean of NN intervals [ms];
- `sdnn`: Standard Deviation of NN intervals [ms];
- `rmssd`: Root Mean Square of Successive Differences [ms];
- `sdsd`: Standard Deviation of Successive Differences [ms];
- `nn50/nn20`: Number of successive intervals differing by more than 50 ms or 20 ms) [n.u.];
- `pNN50/pNN20`: Percentage of successive intervals differing by more than 50 ms or 20 ms [%];
- `peak_lf/peak_hf`: Peak frequencies in the LF band and HF band;
- `abspower_lf/abspower_hf`: Absolute powers in the LF band and HF band [ms^2];
- `logpower_lf/logpower_hf`: Logarithmic powers in the LF band and HF band [log];
- `normpower_hf/normpower_lf`: Normalized powers in the LF and HF band [n.u.].

3.4 Glycemic Class Labeling

Glucose prediction was formulated as a multiclass classification task at 45- and 60-minute horizons, with each future value assigned to one of seven glycemic categories defined according to the classification scheme proposed in [8]. This classification scheme includes two hypoglycemic, three euglycemic, and two hyperglycemic classes, allowing early detection of deviations from normoglycemia and facilitating clinical interpretability. Class labels were assigned to each sample based on the forecasted glucose value at the respective time horizon. Table 1 presents the complete classification scheme with corresponding glucose ranges and sample distribution across all glycemic classes.

3.5 Model Architecture

To model the relationship between ECG-derived features and future glycemic states, we employed a feedforward neural network (FNN) trained as a multi-class classifier. This network is particularly suitable for forecasting tasks thanks to its ability to capture dependencies in sequential data [12].

The network takes as input a 21-dimensional vector corresponding to the features described in Sect. 3.3. The architecture consists of an input layer followed

Table 1. Glycemic class definitions and number of instances used for multi-class labeling.

Class Name	Class ID	Range (mmol/L)	Range (mg/dL)	Samples
Very Low	0	<3.0	<54	44
Low	1	3.0–3.9	54–70	40
Normal (lower range)	2	3.9–5.0	70–90	132
Normal	3	5.0–7.8	90–140	262
Normal (upper range)	4	7.8–10.0	140–180	204
High	5	10.0–13.9	180–250	255
Very High	6	>13.9	>250	35

by two fully connected hidden layers. The first hidden layer comprises 64 neurons with ReLU activation, followed by a dropout layer (rate = 0.3) to reduce overfitting. The second hidden layer contains 32 neurons, also activated via ReLU. The output layer includes 7 neurons with softmax activation, corresponding to the seven glycemic classes defined in Sect. 3.4.

This architecture was identified through a grid search procedure conducted over a predefined set of hyperparameters. The search space included the number of units in the first hidden layer (32, 64), dropout rates (0.2, 0.3, 0.5) and learning rates (0.001, 0.0005). For each combination, model performance was evaluated using 5-fold cross-validation, and the average classification accuracy on the validation sets was used as the selection criterion. The configuration yielding the highest mean accuracy—64 units in the first hidden layer, a dropout rate of 0.3, and a learning rate of 0.0005—was retained for final training.

The model was trained using the Adam optimizer and a sparse categorical cross-entropy as loss function. To mitigate overfitting, an early stopping criterion was applied, terminating training when validation performance ceased to improve over a defined number of epochs.

4 Experimental Setup

We partitioned the dataset using an 80/20 train-test split. Stratified 5-fold cross-validation was applied within the training set to maintain class balance during hyperparameter optimization.

Model performance was evaluated using the classification metrics: overall accuracy, precision, recall, and F1-score for each glucose class. Confusion matrices were analyzed to identify misclassification patterns between glucose categories.

For interpretability analysis, we implemented SHAP with the DeepExplainer method. We used 100 background samples from the training data to establish baseline feature contributions, then computed SHAP values for 100 test samples. These 100 instances are drawn randomly but with the aim of mirroring the overall class distribution. Global feature importance was determined by averaging absolute SHAP values across instances.

All computations were performed on Google Colab using an NVIDIA Tesla T4 GPU (16 GB VRAM).

5 Results and Discussion

In this section, we report the performance of the proposed model evaluated at the 45-minute and 60-minute prediction horizons. Specifically, we present the classification results in terms of precision, recall, F1 score and support for each glycemic class (Tables 2 and 3). Furthermore, we include confusion matrices (Fig. 1aand Fig. 1b) to visualize misclassification patterns, along with SHAP summary graphs (Fig. 2aand Fig. 2b) to assess feature importance and model interpretability.

Table 2. Classification performance metrics for 45-minute glucose forecasting

Class	Prec.	Rec.	F1	Supp.
0	0.90	0.82	0.86	11
1	0.72	0.60	0.65	10
2	0.78	0.77	0.77	26
3	0.70	0.85	0.77	48
4	0.75	0.60	0.67	40
5	0.86	0.96	0.91	51
6	1.00	1.00	1.00	8
Acc.	0.81			
Macro	0.82	0.80	0.80	–
Weighted	0.79	0.81	0.79	–

Table 3. Classification performance metrics for 60-minute glucose forecasting

Class	Prec.	Rec.	F1	Supp.
0	0.82	0.82	0.82	11
1	0.75	0.60	0.67	10
2	0.69	0.69	0.69	26
3	0.63	0.77	0.69	48
4	0.68	0.43	0.52	40
5	0.81	0.90	0.85	51
6	1.00	1.00	1.00	8
Acc.	0.73			
Macro	0.77	0.74	0.75	–
Weighted	0.73	0.73	0.72	–

The proposed deep learning framework, tested on subject 008, demonstrated promising predictive performance across both temporal horizons, achieving 81% classification accuracy at 45 min and 73% at 60 min, with corresponding macro-averaged F1 scores of 0.80 and 0.75 respectively. It is notable that the system demonstrated improved performance for extreme glycemic states (Classes 0, 1, 5 and 6). The temporal degradation from 45 to 60 min followed expected patterns, with Class 4 experiencing the most pronounced decline in F1 score (0.67 to 0.52), reflecting the inherent uncertainty in extended physiological forecasting.

The not-balanced distribution of glycemic classes in the test set, combined with the absence of imbalance-mitigation techniques, likely inflates the performance metrics of less represent class— e.g. the F1 score achieved for Class 6.

Confusion matrices showed that misclassifications mainly occurred between adjacent glucose categories, indicating physiologically plausible errors that maintain clinical safety. This suggests the model learned meaningful decision boundaries aligned with the continuous nature of glucose dynamics.

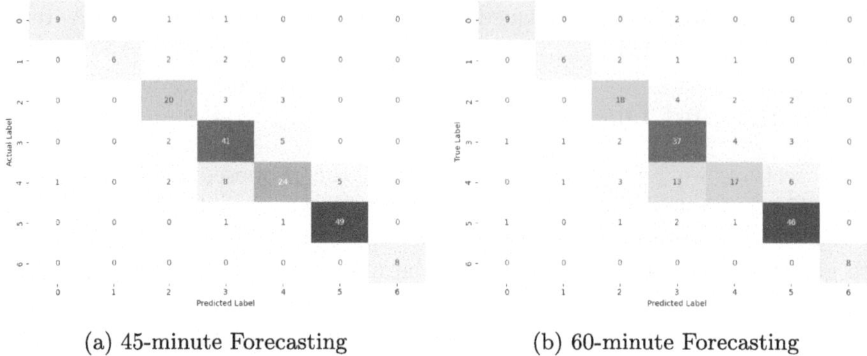

(a) 45-minute Forecasting (b) 60-minute Forecasting

Fig. 1. Confusion matrices showing classification performance across 7 glycemic classes for 45-minute (a) and 60-minute (b) forecasting.

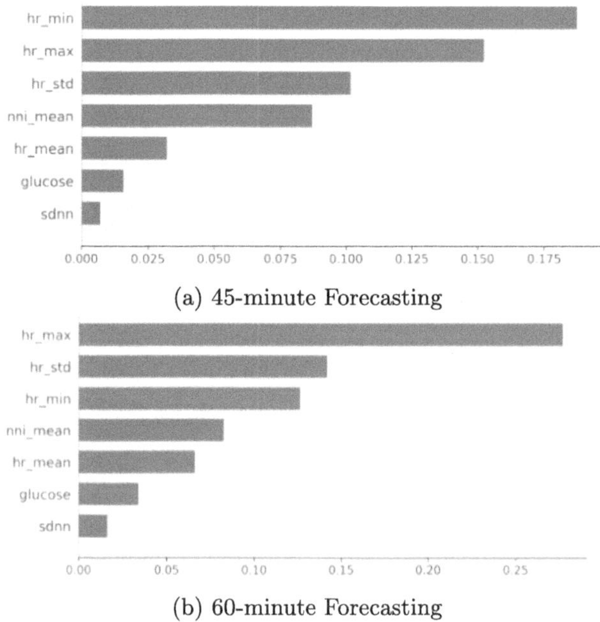

(a) 45-minute Forecasting

(b) 60-minute Forecasting

Fig. 2. Comparison of SHAP summary plots for 45-minute (a) and 60-minute (b) glycemic forecasting. The x-axis represents the mean absolute SHAP value.

SHAP analysis identified `hr_max`, `hr_min`, `hr_std`, and `nni_mean` as the primary ECG-derived contributors to prediction accuracy, alongside `glucose_t0`, the baseline glucose measurement. These ECG-derived features capture complementary information about metabolic state transitions, highlighting their utility as predictive biomarkers for short-term glycemic changes. The findings support integration of wearable ECG monitoring in diabetes management protocols.

6 Conclusions and Future Work

This study presents a DL framework for personalized glucose forecasting in T1D combining ECG-derived features with glucose levels. The model demonstrated promising classification performance for both 45-minute and 60-minute prediction horizons. SHAP analysis revealed that HR parameters and HRV metrics significantly contribute to prediction accuracy, often outweighing current glucose levels as predictive biomarkers. Despite its potential and explainability, the framework's main limitation is the use of data from only one subject with adequate ECG-glucose synchronization, restricting generalizability. Future work should focus on expanding datasets via multi-site studies and federated learning, and on integrating lifestyle factors (e.g., nutrition, insulin, physical activity, sleep) to improve accuracy and clinical relevance. Emerging tools like mobile health and Large Language Models (LLMs) could further enable real-time interpretation and personalized recommendations, aiming for a non-invasive, actionable glucose monitoring system for daily diabetes management.

Acknowledgments. The authors thank the project "Digital Driven Diagnostics, prognostics and therapeutics for sustainable Health care"-D34Health (B53C22006120001), funded by the National Complementary Plan within the National Recovery and Resilience Plan, that is part of the NextGenerationEU programme of the Italian MUR.

Disclosure of Interests. The authors have no competing interests to declare that are relevant to the content of this article.

References

1. Addissouky, T.A., Ali, M.M., El Sayed, I.E.T., Wang, Y.: Type 1 diabetes mellitus: retrospect and prospect. Bull. Nat. Res. Centre **48**(1), 42 (2024)
2. Ahmad, M.A., Eckert, C., Teredesai, A.: Interpretable machine learning in healthcare. In: Proceedings of the 2018 ACM International Conference on Bioinformatics, Computational Biology, and Health Informatics, pp. 559–560 (2018)
3. Ali, J.B., Hamdi, T., Fnaiech, N., Di Costanzo, V., Fnaiech, F., Ginoux, J.M.: Continuous blood glucose level prediction of type 1 diabetes based on artificial neural network. Biocybernetics Biomed. Eng. **38**(4), 828–840 (2018)
4. Arbi, K.F., Soulimane, S., Saffih, F.: Non-invasive method for blood glucose monitoring using ECG signal. Pol. J. Med. Phys. Eng. **29**(1), 1–9 (2023)
5. Basile, L., Sannino, G.: Blood glucose level prediction in type 1 diabetes: a comparative analysis of interpretable artificial intelligence approaches. Results Eng. **25**, 103681 (2025)
6. Cordeiro, R., Karimian, N., Park, Y.: Hyperglycemia identification using ECG in deep learning era. Sensors **21**(18), 6263 (2021)
7. Daneman, D.: Type 1 diabetes. The Lancet **367**(9513), 847–858 (2006)
8. De Falco, I., et al.: A federated learning-inspired evolutionary algorithm: application to glucose prediction. Sensors **23**(6), 2957 (2023)
9. DiMeglio, L.A., Evans-Molina, C., Oram, R.A.: Type 1 diabetes. The Lancet **391**(10138), 2449–2462 (2018)

10. Dubosson, F., Ranvier, J.E., Bromuri, S., Calbimonte, J.P., Ruiz, J., Schumacher, M.: The open d1namo dataset: a multi-modal dataset for research on non-invasive type 1 diabetes management. Inf. Med. Unlocked **13**, 92–100 (2018)
11. Georga, E.I., Protopappas, V.C., Polyzos, D., Fotiadis, D.I.: Evaluation of short-term predictors of glucose concentration in type 1 diabetes combining feature ranking with regression models. Med. Biol. Eng. Comput. **53**, 1305–1318 (2015)
12. Gil-Gamboa, A., Paneque, P., Trull, O., Troncoso, A.: Medium-term water consumption forecasting based on deep neural networks. Expert Syst. Appl. **247**, 123234 (2024)
13. Gomes, P.: pyHRV: Python toolbox for heart rate variability analysis (2021). https://github.com/rhenanbartels/pyhrv. Accessed 20 May 2025
14. Makowski, D., et al.: NeuroKit2: a python toolbox for neurophysiological signal processing. Behav. Res. Methods 1–8 (2021)
15. Martínez-Delgado, L., Munoz-Organero, M., Queipo-Alvarez, P.: Using absorption models for insulin and carbohydrates and deep leaning to improve glucose level predictions. Sensors **21**(16), 5273 (2021)
16. Olde Bekkink, M., Koeneman, M., de Galan, B.E., Bredie, S.J.: Early detection of hypoglycemia in type 1 diabetes using heart rate variability measured by a wearable device. Diabetes Care **42**(4), 689–692 (2019)
17. Paterson, A.D., et al.: The effect of intensive diabetes treatment on resting heart rate in type 1 diabetes: the diabetes control and complications trial/epidemiology of diabetes interventions and complications study. Diabetes care **30**(8), 2107–2112 (2007)
18. Pawar, U., O'shea, D., Rea, S., O'reilly, R.: Explainable ai in healthcare. In: 2020 International Conference on Cyber Situational Awareness, Data Analytics and Assessment (CyberSA), pp. 1–2. IEEE (2020)
19. Rajagopal, S., Koshy, S., Sagar, R.: A comparative study of deep learning techniques for the prediction of blood glucose level in type-1 diabetic patients
20. Rodríguez-Rodríguez, I., Chatzigiannakis, I., Rodríguez, J.V., Maranghi, M., Gentili, M., Zamora-Izquierdo, M.Á.: Utility of big data in predicting short-term blood glucose levels in type 1 diabetes mellitus through machine learning techniques. Sensors **19**(20), 4482 (2019)
21. Sindiramutty, S.R., et al.: Explainable ai in healthcare application. In: Advances in Explainable AI Applications for Smart Cities, pp. 123–176. IGI Global Scientific Publishing (2024)
22. Singh, J.P., et al.: Association of hyperglycemia with reduced heart rate variability (the framingham heart study). Am. J. Cardiol. **86**(3), 309–312 (2000)
23. Torchinsky, M.Y., Gomez, R., Rao, J., Vargas, A., Mercante, D.E., Chalew, S.A.: Poor glycemic control is associated with increased diastolic blood pressure and heart rate in children with type 1 diabetes. J. Diabetes Complications **18**(4), 220–223 (2004)
24. Zecchin, C., Facchinetti, A., Sparacino, G., De Nicolao, G., Cobelli, C.: Neural network incorporating meal information improves accuracy of short-time prediction of glucose concentration. IEEE Trans. Biomed. Eng. **59**(6), 1550–1560 (2012)
25. Zhu, T., Li, K., Herrero, P., Chen, J., Georgiou, P.: A deep learning algorithm for personalized blood glucose prediction. In: KDH@ IJCAI, pp. 64–78 (2018)

Counterfactual Simulation for Estimating Performance Loss in PV Systems: A Machine Learning Approach

José E. Sánchez-López(✉), José María Luna-Romera, Daniel Mateos-García, and F. Javier Galán-Sales

Departamento de Lenguajes y Sistemas Informáticos, Universidad de Sevilla, Avenida de la Reina Mercedes s/n, 41012 Seville, Spain
jensanlop@us.es

Abstract. Performance loss in photovoltaic systems (PV) is caused by multiple factors, such as soiling, panel degradation, and operational maintenance. Traditional estimation approaches rely on statistical methods to isolate degradation signals from environmental conditions. This study proposes a novel counterfactual simulation framework based on machine learning (ML) to address this challenge. The core idea is to train a separate regression model for each year using SCADA data from that year and then use it to simulate how the system would have performed under the environmental conditions of other years. This version of counterfactual prediction reconstructs past operational states and isolates changes in energy output that can be attributed to long-term system degradation rather than fluctuating weather. Two visual diagnostics support this analysis: a degradation envelope capturing forecast dispersion over time and a training-bias comparison that highlights how models trained on different years diverge in their forecasts. The methodology was applied to four datasets from the Desert Knowledge Australia Solar Center (DKASC), resulting in performance loss between 0.75% and 1.75% per year. These results align with expectations, as only environmental variability was decoupled while maintenance-related losses remained unmodeled.

Keywords: Counterfactual analysis · Performance Loss · Degradation Rate · Photovoltaic systems

1 Introduction and Context

Evaluation of performance in photovoltaic (PV) plants and modules is crucial because of the significant investment involved in solar energy projects and their critical role in the transition to sustainable energy. Among the key aspects of this evaluation is the energy yield degradation analysis, where the primary objective is to determine the performance loss rate (PLR).

Numerous factors influence PLR, including soiling loss, panel degradation, and maintenance and environmental conditions. Given the complexity and the multitude of influencing parameters, several purely statistical methodologies

E. Corchado et al. (Eds.): SOCO 2025, CCIS 2806, pp. 171–181, 2026.
https://doi.org/10.1007/978-3-032-19763-4_16

have been developed and applied. Notable among these are the Year-on-Year (YOY) estimation method [18], Seasonal and Trend decomposition using LOESS (STL) [2], and the Statistical Clear-Sky Fitting (SCSF) approach [13].

In ML, counterfactual explanations involve exploring hypothetical scenarios to understand outcome changes caused by different input conditions, typically by examining "what-if" questions. This proposal aims to solve this problem using counterfactual simulation. This term is a consolidated concept in causal cognition, typically describing the process of imagining alternative scenarios different from those that actually occurred in the past and subsequently simulating the potential consequences of those hypothetical scenarios.

In this study, the counterfactual simulation concept is adapted from causal cognition to ML. It is applied using yearly-specific machine learning (ML) models trained exclusively on observed historical conditions from the same PV system. It fundamentally differs from synthetic control methods, where hypothetical counterfactual scenarios are constructed by aggregating data from multiple untreated or control units. Unlike synthetic control, the counterfactual simulation presented here utilizes only historical data from a single unit. Furthermore, this method differs from general counterfactual prediction approaches, which typically predict outcomes under hypothetical future interventions or alternative scenarios that were never observed. In contrast, the counterfactual simulation method explicitly reconstructs past states to isolate system degradation independently from environmental variability.

In the literature, the performance evaluation of PV solar plants has been extensively studied over the past several decades. A wide range of approaches has been proposed, ranging from physics-based models to modern data-driven techniques. In parallel, the application of inference methods – including counterfactual analysis – has gained traction in the energy sector. However, their adoption in the specific context of PV performance evaluation remains limited.

The performance of PV plants has traditionally been assessed using physical models and empirical indicators. One of the most widely used metrics is the Performance Ratio (PR), which quantifies the quality of a PV plant independently of location and weather conditions. Various studies [8,11] have proposed improved PR formulations to account for temperature effects, inverter inefficiencies, and soil loss.

In addition to PR, simulation-based tools such as PVsyst have been widely used to model expected plant performance [17]. These models rely on detailed site-specific input data and often operate under idealized or quasi-static assumptions, which may not capture the dynamic behaviors observed in real-world PV systems [3,15].

The growing availability of SCADA and sensor data from PV installations has facilitated the use of ML techniques for performance monitoring, forecasting, and fault detection. Supervised learning models, such as random forests [16] and deep learning architectures [9,10], have been used to predict energy output, detect anomalies, or classify system operating conditions under varying environmental factors. A more recent trend has focused on interpretable ML models that estimate energy losses without requiring explicit performance indices or detailed system labeling [14].

Counterfactual analysis, a key tool in causal inference, has attracted increasing attention in the energy domain. It has been applied to estimate the impact of demand response programs [5], evaluate the effectiveness of energy efficiency interventions [19], and support decision making in policy and operations. These methods aim to infer treatment effects by comparing observed outcomes with hypothetical alternatives under different interventions or conditions [1]. Techniques such as propensity score matching, synthetic control methods, and causal forests have been adopted for this purpose. However, counterfactual analysis applications focused on PV plant performance remain scarce. Existing studies in this area often rely on simple linear models [6,12], which may not adequately capture the complex, nonlinear, and time-varying dynamics of PV systems.

Although both traditional metrics and ML approaches have contributed to improving PV performance analysis, they typically lack the causal structure needed to support robust counterfactual reasoning. Likewise, although causal inference techniques have shown promise in broader energy applications, their adaptation to the specific characteristics of PV systems, including nonstationary environmental conditions, aging effects, and maintenance events, presents ongoing challenges.

This proposal addresses these gaps by introducing a counterfactual simulation approach that offers several significant advantages. First, it exclusively uses readily available historical meteorological data, specifically irradiance and temperature, along with active power data. Additionally, the methodology is flexible and can be applied to individual modules as well as entire PV systems. Evaluating complete systems is often preferable, as it avoids the logistical and economic challenges of performing panel-by-panel evaluations, particularly given the variability of degradation rates between individual panels [7].

2 Materials and Methods

Section 2 begins by presenting the datasets and associated pre-processing procedures in Subsect. 2.1. Subsequently, Subsect. 2.2 details the methodological framework, comprising a two-stage approach: an intra-annual cross-validation phase for model selection and hyperparameter tuning, followed by a cross-year counterfactual simulation designed to quantify temporal performance loss.

2.1 Data Description

The datasets used in this study were obtained from grid-connected PV systems installed at the Desert Knowledge Australia Solar Center (DKASC) [4]. The DKASC hosts a wide range of PV arrays, allowing for comparative evaluations of different technologies, manufacturers, and mounting configurations. The four datasets analyzed in this proposal correspond to four distinct arrays at the Alice Springs site, as summarized in Table 1. The columns indicating the module manufacturer (Company), the cell technology (type, *Poly-Si* for polycrystalline silicon or *Mono-Si* for monocrystalline silicon), the installation (mounting, either *Roof* - mounted in a building or *Ground* -mounted in a freestanding structure), the nominal capacity of the array in kilowatts (Array Rating kW) and the number and model designation of panels (Count & Model).

Table 1. Summary of datasets used

Dataset	Company	Type	Mounting	Array Rating (kW)	Count & Model
1	BP Solar	Poly-Si	Roof	4.95	30 × BP 3165
2	BP Solar	Mono-Si	Ground	5.1	30 × BP 4170N
3	BP Solar	Poly-Si	Ground	4.95	30 × BP 3165N
4	Sungrid	Poly-Si	Ground	5.04	18 × SG-280P6

These datasets are publicly available on the DKASC platform. They span the years 2011 to 2021, excluding partial years with incomplete data (2008-2010 and 2022–2025). Each dataset contains time-series measurements of both environmental conditions and electrical signals. The core variables common in all datasets include the following: Global horizontal irradiance (GHI) measured in Wh/m^2, ambient temperature (°C), daily rainfall (mm) and active power (kW). Additional variables such as performance ratio, module temperature, wind speed, and wind direction are present only in some datasets.

Data Preprocessing. The raw data was subjected to several pre-processing and quality control steps before performing calculations or training models.

1. **Data Cleansing**: Filters were applied to eliminate unreliable records:
   - Elimination of hourly records with null or negative active power
   - Exclusion of years with incomplete data (2008–2010, 2022–2025).
   - Elimination of aggregated days with null GHI but positive power (physically inconsistent).
2. **Temporal Aggregation**: High-frequency measurements were aggregated to daily totals. Active power measurements were integrated over time to compute daily energy (kWh)

2.2 Methodology

The methodological framework comprises two complementary stages that aim to identify the best predictive model for each year and characterize temporal degradation. The approach begins with intra-annual cross-validation to select the optimal regression model and hyperparameters, followed by cross-year counterfactual simulation to aggregate annual forecasts and quantify performance drift.

Intra-annual Cross-Validation for Model Selection. For each calendar year $y_i \in \{2011, \ldots, 2021\}$, a grid of regression models was trained and validated using a 5-fold cross-validation applied to the aggregated daily data for that year. The objective of the models is to predict the total energy production (kWh) for each day, using aggregated meteorological data as input features. Specifically, input features include daily averaged values of global horizontal irradiance (GHI), ambient temperature, and cumulative rainfall.

The Mean Absolute Error (MAE) is used across K folds as a performance metric:

$$\mathrm{MAE} = \frac{1}{K}\sum_{f=1}^{K}\left(\frac{1}{N}\sum_{i=1}^{N}\left|y_{i,f} - \hat{y}_{i,f}\right|\right),$$

where $y_{i,f}$ is the observed daily energy (KWh) and $\hat{y}_{i,f}$ the corresponding prediction in fold f, N is the number of samples per fold and K the number of CV folds. The configuration minimizing this MAE is selected (Table 2).

Table 2. Grid of regression models and their corresponding hyperparameters evaluated in intra-annual cross-validation.

Model	Hyperparameter grid
Linear Regression	–
Ridge	$\alpha \in \{1.0, 0.1\}$
Lasso	$\alpha \in \{1.0, 0.1\}$
Random Forest	$n_{\text{estimators}} \in \{10, 50, 100, 200, 300\}$
K-Nearest Neighbors	$n_{\text{neighbors}} \in \{3, 5, 10\}$
XGBoost	$n_{\text{estimators}} \in \{10, 50, 100, 200, 300\}$, $\mathtt{lr} \in \{0.1, 0.05, 0.01\}$
AdaBoost	$n_{\text{estimators}} \in \{10, 50, 100, 200, 300\}$
MLP	hidden_layers $\in \{(20,), (50,), (100,), (50, 50), (100, 50)\}$

Cross–Year Counterfactual Simulation. To evaluate temporal generalization and isolate degradation effects, a separate Random Forest model is trained on the data for each year and used to predict daily energy for all other years, with the predictions aggregated into annual totals. (see Fig. 1). Then two diagnostic visualizations are employed:

1. **Degradation envelope:** For each training year, compute the range between minimum and maximum predicted annual outputs in all test years. This graph quantifies how the forecasts drift as the training horizon advances and allows estimation of the plant's average annual performance loss rate.
2. **Training–bias comparison:** Select a small set of representative training years and overlay their predicted annual energy curves on the observed series. This comparison illustrates how the choice of training period can introduce systematic biases in the predictions for each evaluation year.

In practical terms, the simulation can be framed as a leave-one-year-out (LOYO) cross-validation approach commonly used in time series forecasting. LOYO methods primarily aim to evaluate model robustness or predictive accuracy by systematically withholding data from one year during training and subsequently testing predictive performance in the omitted year. In contrast, the counterfactual simulation introduced here aims to build counterfactual scenarios by reconstructing past operating conditions, thereby simulating the energy production that the system would have achieved under previous conditions.

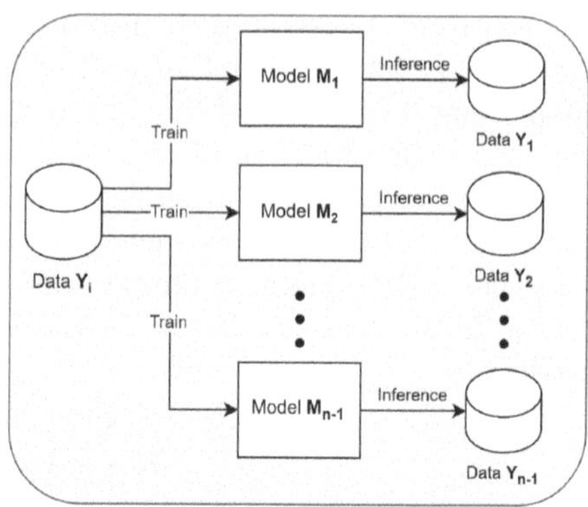

Fig. 1. Cross-year counterfactual simulation methodology

3 Experiments and Results

Based on the methodology of Sect. 2.2, the experimental evaluation is organized into two parts. The analysis begins with an intra-annual comparison of eight candidate regression models on the four DKASC datasets using 5-fold cross-validation. The corresponding mean absolute errors (MAE) are reported in Tables 3 and 4, with the best performer for each dataset highlighted. Second, temporal generalization of the selected regressor is assessed by training on progressively earlier years and evaluating on subsequent years to quantify the rate of degradation.

3.1 Models Comparison

Table 3 reports the mean absolute error (MAE) obtained by each candidate model in the four DKASC datasets. The Random Forest regressor consistently achieves the lowest error in every dataset, highlighting its ability to capture the non-linear relationships between meteorological inputs and power output. XGBoost and KNN follow as the second and third best performers, respectively, while the multilayer perceptron exhibits substantially higher errors. Linear models (Lasso, Ridge, Linear Regression) provide intermediate performance, indicating that purely linear assumptions are insufficient to model the full complexity of the response of the PV system.

The Random Forest model configured with 300 estimators is the selected configuration. Across all four DKASC datasets (Table 4), this setup yields the lowest mean MAE, outperforming the alternatives when averaged: although the 100-estimator version is superior on Datasets 1 and 3 and the 200-estimator on Dataset 2, the 300-estimator configuration minimizes the overall error and is therefore chosen. This final Random Forest (300 estimators) will serve as the model for all subsequent analyses.

Table 3. Mean absolute error (MAE) for each model across the four DKASC datasets selected

Model	Dataset 1	Dataset 2	Dataset 3	Dataset 4
AdaBoost	12.09	12.87	11.97	13.99
KNN	10.66	11.55	10.43	13.59
Lasso	13.95	13.85	13.06	15.63
Linear Regression	13.95	13.85	13.06	15.63
MLP	28.63	29.65	25.48	32.21
Random Forest	**9.09**	**9.35**	**8.69**	**11.15**
Ridge	13.95	13.85	13.06	15.63
XGB	9.22	9.69	9.14	11.35

Table 4. Mean absolute error (MAE) for Random Forest configurations across the four DKASC datasets and mean MAE

Estimators	Dataset 1	Dataset 2	Dataset 3	Dataset 4	Mean
10	9.69	9.78	9.12	11.65	10.06
50	9.20	9.50	8.81	11.38	9.72
100	**9.09**	9.38	**8.69**	11.33	9.62
200	9.14	**9.35**	8.73	11.24	9.61
300	9.12	9.40	8.72	**11.15**	**9.60**

3.2 Performance Loss Analysis

To quantify the drift in the plant's performance, a Random Forest is trained on data from each year and used to predict the total annual energy for all subsequent years. Figure 2 shows, for each of the four DKASC datasets, the envelope of min–max predicted annual energy and the mean prediction as a function of training year. The dashed line in each panel is the linear fit, annotated with the corresponding percentage loss rate per year.

All four datasets exhibit a clear downward trend in predicted annual energy as the training year advances, confirming a gradual loss of plant performance. The estimated annual degradation rates range from approximately 0.75 % to 1.75 % across the different arrays, demonstrating consistent evidence of module aging and a progressive drift in system performance over time.

To complement the aggregate envelope analysis, a year-specific training experiment was conducted across all four DKASC datasets. Random Forest models were trained on data from four representative years: 2012, 2015, 2018, and 2021, and each was used to predict the total annual energy for each evaluation year. Figure 3 presents, for each dataset, the predicted annual energy curves, color coded by training year, overlaid with the observed output shown in black.

The model trained with 2012 data (blue) systematically overestimates outputs in later years, failing to capture accumulated degradation. Models trained

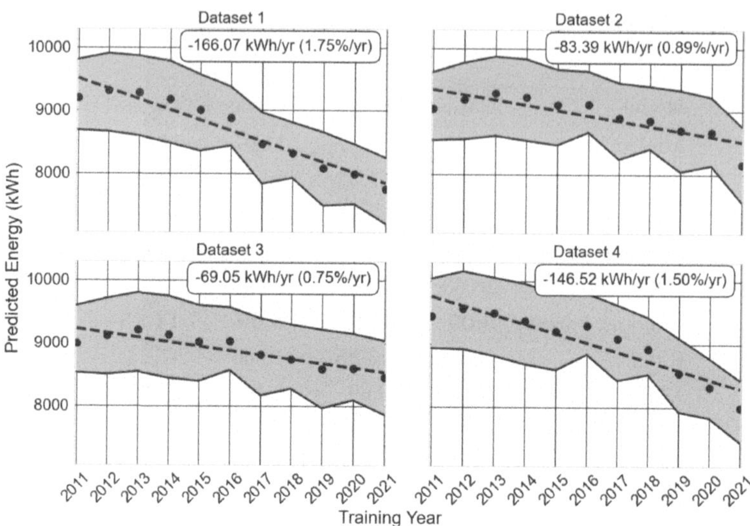

Fig. 2. Year–to–year predicted energy output for Random Forest models trained on successive years. Shaded bands span the min–max range; markers show mean predictions; dashed lines indicate linear degradation trends.

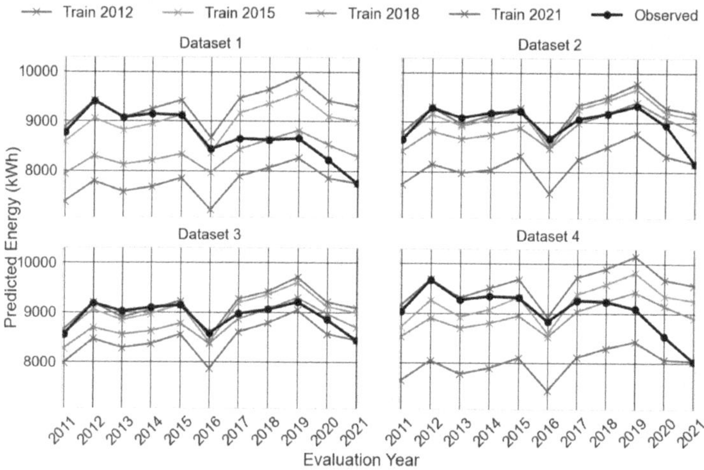

Fig. 3. Predicted vs. observed annual energy for models trained on 2012 (blue), 2015 (orange), 2018 (green), and 2021 (red) across the four DKASC datasets. Each quadrant is one dataset; the black line shows the observed output. (Color figure online)

on progressively more recent years shift downward: the 2015 and 2018 models reduce this bias, while the 2021–trained model (red) begins to underpredict earlier years but aligns more closely with late–period performance. This panel underscores the need for regular retraining to track the plant's evolving degradation.

4 Conclusions and Future Work

This study has introduced a counterfactual simulation framework that uses yearly retrained ML regressors to decouple panel degradation from meteorological variability in PV systems. When applied to four different operational PV datasets, a quasi-linear decline was obtained with degradation rates ranging from 0.75% to 1.75% per year. These values are well aligned with field measurements of crystalline-silicon panel degradation and confirm that the proposed simulation strategy can retrieve realistic degradation rates.

Three practical advantages of the framework merit emphasis. First, it relies exclusively on SCADA and meteorological signals. Second, the methodology is scale-agnostic: although demonstrated at array level, the same workflow can be extended to module strings or whole plants without major modifications. Third, the counterfactual viewpoint facilitates causal reasoning: By simulating the energy that would have been produced under unchanged equipment health, the method isolates the effects of aging that conventional statistical trends often confound.

Future work will extend the framework along four complementary axes. First, applying the workflow to different sites will test its robustness beyond the present semi-arid case study. Second, incorporating subdaily inputs will enable the capture of high-frequency phenomena such as inverter clipping or cloudiness variability. Third, a formal probabilistic treatment was left for future research. Finally, the current proposal decouples only environmental variability. Integrating maintenance-related metrics, such as performance ratio, inverter availability, and soiling indices, will allow a better approximation of the panel degradation rate.

Acknowledgements. The authors would like to thank the Spanish Ministry of Science and Innovation for the support within the projects PID2020-117954RB-C22, PID2023-146037OB-C21 funded by MICIU/AEI/ 10.13039/501100011033. It is also supported by TED2021-131311B-C21 funded by MICIU/AEI/10.13039/501100011033 and the European Union NextGenerationEU/PRTR.

References

1. Ates, E., Aksar, B., Leung, V.J., Coskun, A.K.: Counterfactual explanations for multivariate time series. In: 2021 International Conference on Applied Artificial Intelligence, ICAPAI 2021 (2021). https://doi.org/10.1109/ICAPAI49758.2021.9462056
2. Cleveland, W.S., Cleveland, R.E., McRae, J.E., Terpenning, I.: STL: a seasonal-trend decomposition procedure based on loess. J. Official Stat. **6**(1), 3–73 (1990)

3. Dahmoun, M.E.H., Bekkouche, B., Sudhakar, K., Guezgouz, M., Chenafi, A., Chaouch, A.: Performance evaluation and analysis of grid-tied large scale PV plant in Algeria. Energy Sustain. Dev. **61**, 181–195 (2021). https://doi.org/10.1016/j.esd.2021.02.004
4. Desert Knowledge Australia Solar Centre (DKASC): https://dkasolarcentre.com.au (2025). Accessed 10 June 2025
5. Islam, M.M., Sohag, K., Mariev, O.: Mineral import demand-driven solar energy generation in China: a threshold estimation using the counterfactual shock approach. Renew. Energy **221** (2024).https://doi.org/10.1016/j.renene.2023.119764
6. Krishnan, G., Thrinath, S., Reddy, M., Thukkaram, S.: Enhancing solar power generation through AC power prediction optimization in solar plants. Int. J. Appl. Power Eng. (IJAPE) **13**(3), 645–652 (2024). https://doi.org/10.11591/ijape.v13.i3.pp645-652
7. Lai, G., Wang, D., Wang, Z., Fan, F., Wang, Q., Wang, R.: Distribution-based PV module degradation model. Energy Sci. Eng. **11**(3), 1219–1228 (2023). https://doi.org/10.1002/ese3.1401
8. Li, Z., Ma, T., Zhao, J., Song, A., Cheng, Y.: Experimental study and performance analysis on solar photovoltaic panel integrated with phase change material. Energy **178**, 471–486 (2019). https://doi.org/10.1016/j.energy.2019.04.166
9. Lim, S.C., Huh, J.H., Hong, S.H., Park, C.Y., Kim, J.C.: Solar power forecasting using CNN-LSTM hybrid model. Energies **15**(21) (2022).https://doi.org/10.3390/en15218233
10. Lim, S.C., Kim, B.G., Kim, J.C.: Analysis of inverter efficiency using photovoltaic power generation element parameters. Sensors **24**(19) (2024). https://doi.org/10.3390/s24196390
11. Ma, T., Yang, H., Lu, L.: Solar photovoltaic system modeling and performance prediction. Renew. Sustain. Energy Rev. **36**, 304–315 (2014). https://doi.org/10.1016/j.rser.2014.04.057
12. Makhija, A.S., Bohra, S.S., Tiwari, V.: Investigating the performance of water-mounted solar photo-voltaic systems using different simulation tools. Energy Convers. Manage. **322**, 119116 (2024). https://doi.org/10.1016/j.enconman.2024.119116
13. Meyers, B., Hansen, C., Underwood, J.: Statistical clear-sky fitting algorithm. arXiv preprint arXiv:1907.08279 (2019). https://github.com/bmeyers/StatisticalClearSky
14. Meyers, B., Deceglie, M.: Automatic loss factor modeling and attribution on unlabeled PV energy data. In: 2024 IEEE 52nd Photovoltaic Specialist Conference (PVSC), pp. 1134–1141 (2024). https://doi.org/10.1109/PVSC57443.2024.10748940
15. Mishra, P.R., Rathore, S., Jain, V.: PVSyst enabled real time evaluation of grid connected solar photovoltaic system. Int. J. Inform. Technol. (Singapore) **16**(2), 745–752 (2024). https://doi.org/10.1007/s41870-023-01677-x
16. Rafati, A., Joorabian, M., Mashhour, E., Shaker, H.R.: High dimensional very short-term solar power forecasting based on a data-driven heuristic method. Energy **219** (2021). https://doi.org/10.1016/j.energy.2020.119647
17. Sharma, R., Sharma, S., Tiwari, S.: Design optimization of solar PV water pumping system. In: Materials Today: Proceedings. vol. 21, pp. 1673–1679 (2020). https://doi.org/10.1016/j.matpr.2019.11.322
18. Teichmann, H., Köhler, B., Ellert, D., Becker, K.: Performance analysis and degradation of a large fleet of PV systems. IEEE J. Photovoltaics **11**(5), 1316–1325 (2021). https://doi.org/10.1109/JPHOTOV.2021.3093049

19. Wang, Y., Li, J., O'Leary, N., Shao, J.: Banding: a game changer in the renewables obligation scheme in the United Kingdom. Energy Econ. **130** (2024). https://doi.org/10.1016/j.eneco.2024.107331

Multi-horizon Forecasting of Air Pollutants Using an Explainable Multi-task Attention-Based LSTM

Naeem Ullah[1(✉)], Ivanoe De Falco[2], and Giovanna Sannino[1,2]

[1] Department of Electrical Engineering and Information Technology, University of Naples Federico II, via Claudio 21, 80125 Naples, Italy
naeem.ullah@unina.it

[2] Institute for High Performance Computing and Networking (ICAR) of the National Research Council (CNR) of Italy, via P. Castellino 111, 80131 Naples, Italy
{ivanoe.defalco,giovanna.sannino}@icar.cnr.it

Abstract. Precise prediction of air pollutants is of crucial importance in urban environmental management and public health. In this article, we propose a multi-task and multi-horizon deep learning framework to forecast nitrogen dioxide (NO_2) and ozone (O_3) concentrations up to 12 h ahead using multivariate time series data. The model, based on an attention-enhanced Long Short-Term Memory (LSTM) network, learns temporal interactions between meteorological and pollutant variables from different monitoring stations. Multi-horizon forecasting is supported by a common architecture that makes simultaneous predictions of NO_2 and O_3 for 1, 3, 6, and 12-hour forecasting horizons. Optuna is used to tune hyperparameters, and model explainability is achieved through integrated gradients and attention visualization. Experimental findings demonstrate that the proposed model achieved high predictive performance, with explainability tools offering valuable understanding into temporal and feature-level importance. This approach shows the potential of explainable deep learning for high-fidelity air quality forecasting in urban areas.

Keywords: air pollution · deep learning · explainable artificial intelligence · integrated gradients · multi-task LSTM

1 Introduction

Air pollution is a key environmental and public health concern, primarily in urban and sub-urban regions where emissions from transport, industry, and households are concentrated [4]. Of all atmospheric pollutants, Nitrogen Dioxide (NO_2) and ground-level Ozone (O_3) are of particular concern due to their toxic effect on human health, their role in generating photochemical smog, and their link to respiratory and cardiovascular disease [11]. Accurate forecasting of these pollutants is crucial for environmental policy, early warning, and health risk mitigation.

E. Corchado et al. (Eds.): SOCO 2025, CCIS 2806, pp. 182–192, 2026.
https://doi.org/10.1007/978-3-032-19763-4_17

Traditional statistical methods, such as linear regression and ARIMA, have been used very widely in air quality forecasting but are prone to fail in capturing the nonlinear behavior and temporal relationships that are inherent in environmental time series data [6]. Recent advances in Artificial Intelligence (AI), particularly Deep Learning (DL) [8–10], Recurrent Neural Networks (RNNs) [7] and their variants, such as Long-Short-Term Memory (LSTM) networks [12], have been found to perform more effectively in capturing complex temporal patterns. Such models find special utility when dealing with multivariate time-series data motivated by multiple interacting meteorological and pollutant variables.

Some recent studies have explored diverse LSTM-based structures and hybrid architectures for enhancing forecast performance. As an example, an extendable LSTM-based framework has been presented for forecasting air pollutant levels from both sensor network observations and government open datasets in Bengaluru, India, demonstrating the potential to integrate real-time data across diverse sources [2]. Furthermore, a PSO-LSTM model–combining Particle Swarm Optimization (PSO) with Long Short-Term Memory (LSTM) networks–was shown to outperform both random forest and baseline LSTM models in predicting the concentration levels of fine particulate matter ($PM_{2.5}$, particles with diameters $\leq 2.5\mu m$), coarse particulate matter (PM_{10}, particles with diameters $\leq 10\mu m$), and ozone (O_3). This highlights the potential of automated hyperparameter tuning to enhance predictive performance [3]. These efforts reflect the growing applications of deep learning techniques and the hybrid models to air quality prediction since they have the capability of integrating spatial-temporal complexity as well as local environmental heterogeneity to learn.

Although deep learning models are powerful, challenges persist–including limited multi-step forecasting, lack of cross-pollutant modeling despite shared atmospheric dynamics, and poor explainability due to their black-box nature, which reduces trust among domain experts.

To address the aforementioned challenges, we propose a novel deep learning method for multi-horizon, multi-task air quality forecasting using an attention enhanced LSTM model. The forecasting horizons (1, 3, 6, and 12 h) are selected based on operational and regulatory requirements: the 1-hour and 3-hour horizons support real-time health advisories and air quality monitoring, while the 6-hour and 12-hour horizons assist in long-term environmental planning and regulatory compliance. Our model predicts future NO_2 and O_3 concentrations using multivariate input features, including meteorological conditions (e.g., temperature, wind speed) and historical pollutant levels. Leveraging a multi-task learning framework, the model learns shared representations between NO_2 and O_3 to enhance generalization and efficiency.

Additionally, we used an attention mechanism to dynamically focus on relevant temporal contexts and utilized Optuna [1] for hyperparameter tuning to optimize model performance. Explainability is enhanced through integrated gradients and attention visualizations, providing insights into the feature and temporal importance for each prediction. The model is trained and evaluated on a dataset collected from five monitoring stations in Spain between 2015 and 2023 at hourly intervals. Our main contributions are summarized as follows:

1. We develop a multi-step and multi-task LSTM model with an attention mechanism to predict NO_2 and O_3 levels at multiple future horizons.
2. We develop a joint multi-horizon training strategy under which the model can predict up to 12 h in advance accurately.
3. We perform hyperparameter tuning using Optuna to achieve more effective hyperparameter tuning, thereby leading to improved model performance.
4. We use integrated gradients and attention visualizations to enhance the explainability of the model by providing information about feature and temporal significance.

2 Methodology

The overall methodology, summarized in Fig. 1, includes data collection, preprocessing, attention-based LSTM modeling, hyperparameter tuning, prediction, explainability, and evaluation.

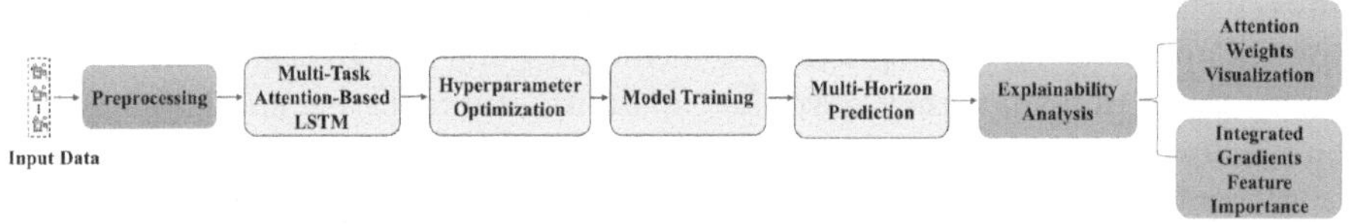

Fig. 1. Overview of the proposed methodology for multi-task, multi-horizon air pollutant forecasting.

2.1 Data Preprocessing

Before training, data preprocessing steps are used to ensure consistency, remove noise, and prepare the dataset for model training. Firstly, missing or corrupt datetime entries are handled by reindexing the time series to enforce a regular hourly frequency, followed by linear interpolation to impute missing values and preserve temporal consistency for model training. Secondly, features and target variables are standardized using z-score normalization to ensure equal contribution during training and stable gradient updates. The data is split into training (80%) and test (20%) sets to preserve temporal dependencies.

2.2 Multi-horizon Sequence Generation

The preprocessed data are structured into input-output pairs using a sliding window approach. At each timestamp t, the model receives a sequence of $l = 24$ historical time steps:

$$\mathbf{X}_t = [\mathbf{x}_{t-l}, \ldots, \mathbf{x}_{t-1}] \in \mathbb{R}^{l \times d} \tag{1}$$

where $\mathbf{x}_{t-i} \in \mathbb{R}^d$ is the feature vector at time $t-i$, l is the sequence length, and d is the number of input features. The targets are the NO_2 and O_3 concentrations at multiple future horizons $hr \in HR = \{1, 3, 6, 12\}$, producing two sets of output vectors:

$$\mathbf{y}_t^{(\mathrm{NO_2})} = \left[y_{t+h}^{(\mathrm{NO_2})}\right]_{hr \in HR}, \quad \mathbf{y}_t^{(\mathrm{O_3})} = \left[y_{t+h}^{(\mathrm{O_3})}\right]_{hr \in HR} \tag{2}$$

Here, $y_{t+hr}^{(\mathrm{NO_2})}$ and $y_{t+hr}^{(\mathrm{O_3})}$ denote the NO_2 and O_3 concentrations at future horizon hr, respectively.

2.3 Model Architecture

We design a multi-task LSTM-based architecture enhanced with an attention mechanism for improved temporal feature learning. The model consists of the following components:

- **LSTM Encoder:** A stack of n LSTM layers encodes the input sequence $\mathbf{X}_t \in \mathbb{R}^{l \times d}$ into a sequence of hidden states $\mathbf{H}_t = [\mathbf{h}_1^{(s)}, \ldots, \mathbf{h}_l^{(s)}] \in \mathbb{R}^{l \times h_s}$, where h_s is the hidden state dimension.
- **Attention Layer:** A soft attention mechanism computes a context vector $\mathbf{c}_t$ as a weighted sum of the LSTM hidden states:

$$\alpha_i = \frac{\exp(\mathbf{w}^\top \mathbf{h}_i^{(s)})}{\sum_{j=1}^{l} \exp(\mathbf{w}^\top \mathbf{h}_j^{(s)})}, \quad \mathbf{c}_t = \sum_{i=1}^{l} \alpha_i \mathbf{h}_i^{(s)} \tag{3}$$

  where $\mathbf{w} \in \mathbb{R}^{h_s}$ is a learnable attention weight vector and α_i is the attention weight for timestep i.
- **Multi-Task Heads:** Two separate fully connected layers generate predictions for NO_2 and O_3 concentrations:

$$\hat{y}_t^{(\mathrm{NO_2})} = \mathbf{W}_{\mathrm{NO_2}} \mathbf{c}_t + b_{\mathrm{NO_2}}, \quad \hat{y}_t^{(\mathrm{O_3})} = \mathbf{W}_{\mathrm{O_3}} \mathbf{c}_t + b_{\mathrm{O_3}} \tag{4}$$

  where $\mathbf{W}_{\mathrm{NO_2}}, \mathbf{W}_{\mathrm{O_3}} \in \mathbb{R}^{|H| \times h_s}$ and $b_{\mathrm{NO_2}}, b_{\mathrm{O_3}} \in \mathbb{R}^{|H|}$ are learnable parameters, and $|H|$ is the number of forecasting horizons.
- **Training Strategy and Loss Function** The model is trained using the mean squared error (MSE) loss across all prediction horizons. Model optimization is performed using the Adam optimizer.

2.4 Hyperparameter Optimization

Hyperparameters are optimized using Bayesian optimization with Optuna, focusing on model performance for NO2 prediction. The hyperparameter search space is summarized in Table 1.

2.5 Explainability via Attention and Integrated Gradients

Attention weights α_i are visualized to assess the temporal relevance of input time steps. Also, to quantify the contribution of each input feature to model outputs, we apply Integrated Gradients (IG) [13].

Table 1. Hyperparameter search space used in Optuna optimization.

Hyperparameter	Type	Search Space
Hidden Dimension (hd)	Categorical	$\{32, 64, 128\}$
Number of LSTM Layers (n)	Integer	$[1, 3]$
Learning Rate (`lr`)	Log-Uniform	$[1\times10^{-4}, 1\times10^{-2}]$
Batch Size (`batch_size`)	Categorical	$\{32, 64, 128\}$

3 Results and Discussion

3.1 Datasets Description

Hourly O_3 multivariate time series data has been collected yearly from 2015 to 2023. The information was provided from five monitoring stations in Spain. These monitoring sites were classified according to the primary pollutants (Background(B), Traffic(T)) and the type of region i.e., Urban (U) or Suburban (S). The analyzed sites included Aljarafe (S-B), Bermejales (U-B), Torneo (U-T) in Seville, Asomadilla (S-B) in Cordova, and Ronda del Valle (U-B) in Jaen. Each

Table 2. Table with datasets' relevant properties employed in all areas.

Property	Aljarafe	Asomadilla	Bermejales	Ronda del Valle	Torneo
Location	Seville	Cordova	Seville	Jaen	Seville
Area type	S	S	U	U	U
Emission source	B	B	B	B	T
Features	WD, CO, TMP, WS, NO_2, PM_{10}, and O_3				
Ozone unit	$\mu g/m^3$				

station contains 78,889 hourly data instances, comprehensively representing ozone variations across different environmental conditions. Temperature (TMP), wind direction (WD), and wind speed (WS) were among the atmospheric conditions. Carbon monoxide (CO), Nitrogen Dioxide (NO_2), and particles of 10μ or less (PM_{10}) were the local variables that were taken into account. Table 2 summarizes the properties of the datasets employed.

3.2 Implementation Details

The entire pipeline is implemented using PyTorch and trained on a computer with 16 GB of RAM and a 500 GB SSD. Windows 11 Pro is used as an Operating system. Captum library is used for integrated gradients. Visualization is conducted using Matplotlib. The model's predictive performance is assessed for each forecast horizon ($h \in \{1, 3, 6, 12\}$) and each pollutant (NO_2, O_3) using three metrics: i.e., Root Mean Square Error (RMSE), Mean Squared Error (MSE), and Mean Absolute Error (MAE).

3.3 Station-Wise and Horizon-Wise Forecasting Evaluation

For all the datasets (Tables 3), the model trained efficiently with acceptable training times, with the most challenging setup being for Ronda Del Valle station, employing a deeper network (3 layers, 128 units) and longer training duration (729.97 s). Inference times remained low across datasets, highlighting the model's real-world applicability.

Table 3. Multi-Horizon Multi-Task Model Configuration and Training Details.

Dataset	Hidden Dim	Layers	Learning Rate	Batch Size	Train Time (s)	Inference Time (s)
Aljarafe	32	1	0.0019	64	182.91	0.15
Asomadilla	128	1	0.0008	64	349.05	1.39
Bormejales	64	1	0.0033	32	217.01	0.29
Ronda Del Valle	128	3	0.0013	128	729.97	3.43
Torneo	32	1	0.0045	32	210.49	0.18

Throughout all stations and pollutants (Table 4), performance generally declines as the forecast horizon increases, as might be expected from the growing uncertainty in the longer-term forecasts. For example, for Aljarafe station, the RMSE for NO_2 increases from 0.6943 (1 step) to 0.7454 (12 steps), and the same trend is observed for O_3, from 0.6499 to 0.7029. Although this trend is found at all stations, Aljare shows especially large errors for O_3 every horizon and less consistent growth of predictive error with time. However, even at the 12-step prediction horizon, the model maintained stable performance across all datasets, highlighting its ability to capture long-term temporal dependencies.

Table 4. Forecasting Performance Metrics Across Horizons for NO_2 and O_3.

Dataset	Horizon	Pollutant	RMSE	MSE	MAE
Aljarafe	1-step	NO_2	0.6943	0.4821	0.4314
		O_3	0.6499	0.4224	0.5171
	3-step	NO_2	0.7009	0.4912	0.4408
		O_3	0.6745	0.4549	0.5372
	6-step	NO_2	0.7145	0.5106	0.4601
		O_3	0.6911	0.4776	0.5478
	12-step	NO_2	0.7454	0.5556	0.4939
		O_3	0.7029	0.4940	0.5613
Asomadilla	1-step	NO_2	0.6015	0.3618	0.4132
		O_3	0.5156	0.2658	0.3987
	3-step	NO_2	0.6541	0.4278	0.4469
		O_3	0.5415	0.2932	0.4180
	6-step	NO_2	0.6731	0.4531	0.4617
		O_3	0.5643	0.3185	0.4397
	12-step	NO_2	0.6739	0.4541	0.4754
		O_3	0.5716	0.3267	0.4493
Bermejales	1-step	NO_2	0.6580	0.4330	0.4704
		O_3	0.6384	0.4075	0.5074
	3-step	NO_2	0.7056	0.4979	0.5032
		O_3	0.6614	0.4374	0.5224
	6-step	NO_2	0.7265	0.5278	0.5317
		O_3	0.6680	0.4462	0.5265
	12-step	NO_2	0.7294	0.5321	0.5350
		O_3	0.6739	0.4542	0.5385
Ronda Del Valle	1-step	NO_2	0.5790	0.3352	0.3606
		O_3	0.5924	0.3509	0.4531
	3-step	NO_2	0.6408	0.4106	0.4061
		O_3	0.6314	0.3987	0.4832
	6-step	NO_2	0.6626	0.4390	0.4146
		O_3	0.6527	0.4261	0.4998
	12-step	NO_2	0.6899	0.4759	0.4386
		O_3	0.6725	0.4522	0.5126
Torneo	1-step	NO_2	0.5933	0.3520	0.4538
		O_3	0.6245	0.3900	0.4960
	3-step	NO_2	0.6555	0.4297	0.5014
		O_3	0.6478	0.4196	0.5102
	6-step	NO_2	0.6646	0.4418	0.5089
		O_3	0.6750	0.4556	0.5333
	12-step	NO_2	0.6868	0.4717	0.5305
		O_3	0.6885	0.4740	0.5446

Station-specific patterns reveal that Aljarafe consistently shows the greatest error levels, especially for NO_2, with 12-step RMSE = 0.7454, suggesting more complex dynamics or data variability at this station or less expressive capacity of its smaller model configuration (32 hidden units, 1 layer). Among the stations, Asomadilla showed the lowest error for O_3, with a 1-step RMSE of 0.5156. At the same time, Ronda Del Valle achieved the best short-term performance for NO_2 with a 1-step RMSE of 0.5790, benefiting from its deeper architecture (128 hidden units, 3 layers), which enhances its capacity to capture complex temporal dynamics. Bermejales and Torneo show moderate error growth over longer horizons but generally relatively stable performance. These suggest a possible link between model capacity and station-specific forecasting performance.

In terms of pollutants, NO_2 has generally exhibited slightly higher RMSE and MAE than O_3, especially with datasets such as Torneo and Asomadilla, which may be due to its more nonlinear photochemical nature, which is harder to model. Nonetheless, the proposed framework managed to capture useful patterns for both pollutants simultaneously, guaranteeing the appropriateness of the multi-task learning approach.

Overall, these results verify the model's capacity to generalize across various urban environments and forecasting horizons. The application of attention mechanisms likely played a major role in increased learning efficiency and long-term dependencies, and the multi-tasking design facilitated mutual learning benefits between NO_2 and O_3.

3.4 Explainability Analysis

We conducted an explainability analysis across multiple stations; results for Aljarafe are shown as a representative example. Two complementary explainability methods are employed: attention weight visualization and Integrated Gradients (IG) feature importance analysis.

Figure 2 shows the attention weights across timesteps for a single prediction. More recent timesteps received larger attention weights, in line with the model's importance of recent observations when predicting NO_2 and O_3 levels.

Figure 3 presents IG feature importance for NO_2 and O_3 across 1, 3, 6, and 12-step horizons. For forecasting NO_2, the feature "DD" (wind direction) is always the most important feature at all horizons, reflecting the dominant role of temperature in the dynamics of nitrogen dioxide concentration. Wind features such as "TMP Media" (mean temperature) and "VA" (wind speed) are of negligible or negative importance in general, indicating a comparatively small contribution to NO_2 forecasting in these data.

In the prediction O_3, "PM10" is also at the head of the ranks for the most essential features for the predictions in 1, 3, 6 and 12-step. However, at the 3, 6, and 12-step forecast horizons, a slight increase in the relative importance of temperature and wind speed features is observed, suggesting that meteorological dispersion effects may play a growing role for longer forecast steps. Overall, temperature is the primary driver, followed by wind-related factors.

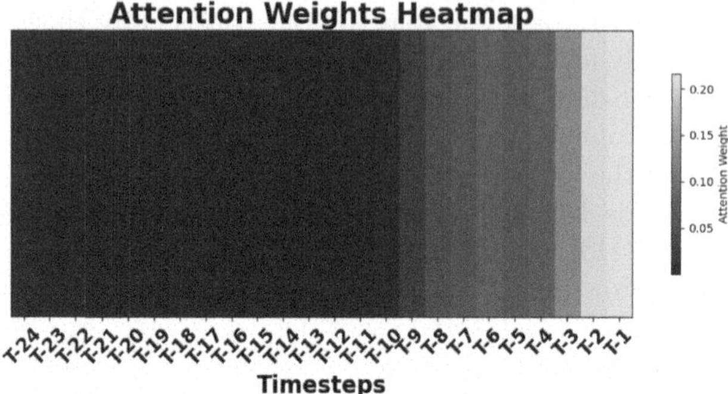

Fig. 2. Attention weight heatmap for a sample from the Aljarafe station. Higher weights correspond to timesteps with greater influence on the model's predictions.

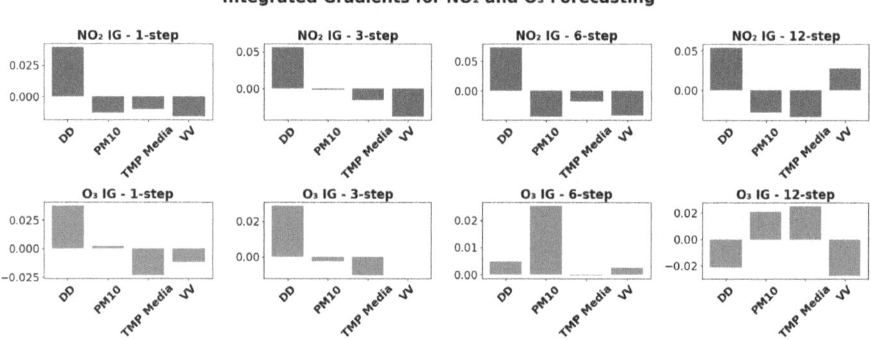

Fig. 3. Integrated Gradients feature importance scores for the Aljarafe station across different prediction horizons (1, 3, 6, and 12 steps ahead) for NO_2 and O_3 forecasting.

3.5 Comparison with Existing Work on the Same Dataset

This section aims to compare the performance of the proposed approach with an existing study that has utilized the same dataset [5]. In that existing research, a temporal LSTM model with the inclusion of a temporal selection layer (TSL) had been used to forecast ozone. For ozone forecasting, the reported RMSE values of their model were 14.9 for Aljarafe, 14.9 for Asomadilla, 16.9 for Bermejales, 17.0 for Ronda del Valle, and 13.9 for Torneo. The proposed model achieves significantly lower RMSE values across all stations.

Unlike the prior work, which focused only on ozone, our model performs multi-task forecasting of NO_2 and O_3. In addition, our model incorporates explainability using attention visualization and integrated gradients, which reveal key time steps and environmental features contributing to predictions. Thus, the proposed approach improves on prior work in both performance and interpretability.

4 Conclusion

In this study, we presented a multi-horizon forecasting model for air pollutant concentration (NO_2 and O_3) with a multi-task attention-based LSTM model. With broad experimentation on five urban datasets, including the Aljarafe station, the method showed satisfactory performance for all the forecasting horizons.

Explainability is a primary interest of this study. Attention weights highlighted the importance of recent time steps, reflecting pollutant temporal dynamics. Analysis using Integrated Gradients provided additional insights by quantifying the impact of different meteorological and environmental input features. The findings showed the critical importance of variables such as temperature and wind features in determining NO_2 and O_3 levels at different forecast steps.

Despite the promising results, some limitations still exist. First, attention analysis is performed on representative samples instead of aggregating across the whole test set, which could constrain generalizability. Second, though the model is successful when trained and tested on urban datasets, its applicability to other areas with varied climatic and emission characteristics needs to be confirmed.

Third, the current strategy did not deal with exogenous parameters such as traffic flow, industrial operations, or sudden meteorological fluctuations, which could unpredictably impact pollutant levels. The exclusion of these exogenous variables may limit the model's ability to capture real-world scenarios where such factors significantly influence air quality. In future iterations, we aim to incorporate traffic and industrial activity data, possibly through collaboration with real-time data providers, or by integrating additional sensors into the forecasting pipeline. This would provide a more comprehensive understanding of the variables driving pollutant levels and improve the robustness of the model.

Future work will aim to counter such limitations by investigating ensemble attention summarization across sequences of samples towards more robust explainability, by looking at other sources of exogenous data, and model generalization to untrained cities by looking into transfer learning strategies.

Acknowledgments. This research received no external funding.

Disclosure of Interests. The authors declare no conflicts of interest.

References

1. Akiba, T., Sano, S., Yanase, T., Ohta, T., Koyama, M.: Optuna: a next-generation hyperparameter optimization framework. In: Proceedings of the 25th ACM SIGKDD International Conference on Knowledge Discovery & Data Mining, pp. 2623–2631 (2019)
2. Belavadi, S.V., Rajagopal, S., Mohan, R., et al.: Air quality forecasting using LSTM RNN and wireless sensor networks. Procedia Comput. Sci. **170**, 241–248 (2020)
3. Chen, M., Xu, P., Liu, Z., Liu, F., Zhang, H., Miao, S.: Air pollution prediction based on optimized deep learning neural networks: PSO-LSTM. Atmos. Pollut. Res. **16**(3), 102413 (2025)

4. Crippa, M., et al.: Global anthropogenic emissions in urban areas: patterns, trends, and challenges. Environ. Res. Lett. **16**(7), 074033 (2021)
5. Jiménez-Navarro, M.J., Martínez-Ballesteros, M., Martínez-Álvarez, F., Asencio-Cortés, G.: Explaining deep learning models for ozone pollution prediction via embedded feature selection. Appl. Soft Comput. **157**, 111504 (2024)
6. Mani, G., Viswanadhapalli, J.K., et al.: Prediction and forecasting of air quality index in Chennai using regression and ARIMA time series models. J. Eng. Res. **10**(2A), 179–194 (2022)
7. Pande, C.B., et al.: Daily scale air quality index forecasting using bidirectional recurrent neural networks: case study of Delhi, India. Environ. Pollut. **351**, 124040 (2024)
8. Ullah, N., Javed, A.: Deep features comparative analysis for covid-19 detection from the chest radiograph images. In: 2021 International Conference on Frontiers of Information Technology (FIT), pp. 258–263. IEEE (2021)
9. Ullah, N., et al.: ChestCovidNet: an effective dl-based approach for covid-19, lung opacity, and pneumonia detection using chest radiographs images. Biochem. Cell Biol. (ja) (2024)
10. Ullah, N., Khan, J.A., De Falco, I., Sannino, G.: Bridging clinical gaps: multi-dataset integration for reliable multi-class lung disease classification with DeepCRINet and occlusion sensitivity. In: 2024 IEEE Symposium on Computers and Communications (ISCC), pp. 1–6. IEEE (2024)
11. World Health Organization: WHO global air quality guidelines: particulate matter (PM2.5 and PM10), ozone, nitrogen dioxide, sulfur dioxide and carbon monoxide. World Health Organization (2021)
12. Wu, Z., Tian, Y., Li, M., Wang, B., Quan, Y., Liu, J.: Prediction of air pollutant concentrations based on the long short-term memory neural network. J. Hazard. Mater. **465**, 133099 (2024)
13. Zhuo, Y., Ge, Z.: IG2: Integrated gradient on iterative gradient path for feature attribution. IEEE Transactions on Pattern Analysis and Machine Intelligence (2024)

Deep Learning for Robust Soil Moisture Forecasting Using LSTM and Transformer

Raúl Aguilar[1(✉)], Miguel A. Patricio[1], José M. Molina[1], Antonio Berlanga[1], and Sergio Zubelzu[2]

[1] Computer Science and Engineering Department. Applied Artificial Intelligence Group, Universidad Carlos III de Madrid, Madrid, Spain
raulagarr@gmail.com, mpatrici@inf.uc3m.es, {molina,aberlan}@ia.uc3m.es
[2] Departamento de Ingeniería Agroforestal, ETSI Agronómica, Alimentaria y Biosistemas, Universidad Politécnica de Madrid, 28040 Madrid, Spain
sergio.zubelzu@upm.es
https://ror.org/03ths8210 , https://giaa.uc3m.es/

Abstract. Accurate forecasting of soil moisture is essential for applications in agriculture and hydrology; however, real-world sensor networks frequently present challenges such as noise, missing data, and hardware limitations. This study proposes a robust preprocessing pipeline designed to extract high-quality, gap-free intervals from a complex in-situ soil moisture dataset collected in the Duero basin (Spain). State-of-the-art deep learning models, namely Long Short-Term Memory (LSTM) networks and Transformer architectures, were evaluated alongside a simple baseline model. All models were trained on carefully curated data segments. The results indicate that both the LSTM and Transformer models substantially outperform the baseline across medium- and long-term prediction horizons, demonstrating strong generalization capabilities to previously unseen intervals and sensors. These findings underscore the importance of rigorous data preparation for reliable forecasting in noisy environmental contexts and highlight the potential of deep learning techniques for robust soil moisture prediction in real-world scenarios.

Keywords: Soil moisture prediction · Time series forecasting · Hydrology

1 Introduction

Soil moisture prediction is a critical challenge in hydrology, agriculture, and climate science, as it supports efficient irrigation strategies, drought monitoring, and yield forecasting [3,7]. Recent advances in machine learning and deep learning have enabled substantial progress in this field, with numerous studies using large-scale, high-quality datasets from sources such as the International Soil Moisture Network (ISMN) [8–10] and satellite products such as European Space Agency's Climate Change Initiative [4]. These works typically assume that

E. Corchado et al. (Eds.): SOCO 2025, CCIS 2806, pp. 193–202, 2026.
https://doi.org/10.1007/978-3-032-19763-4_18

the input data is continuous, homogeneous, and largely free from sensor errors or significant gaps. For example, ISMN provides dense and reliable in-situ measurements, while ESA CCI integrates multiple remote sensing sources to produce spatially continuous time series. When necessary, data gaps are often addressed using auxiliary datasets (ERA5-Land, GLEAM, SRTM, OpenLandMap) and applying interpolation or composition schemes [4].

Despite these methodological advances, the preprocessing pipeline often receives limited attention. Most studies employ standard procedures such as resampling, normalization, and occasional gap-filling, but generally begin with datasets that have already been carefully curated or enhanced [2]. Even when working with raw sensor data, as in Bandaru et al. [1], preprocessing is typically restricted to subsampling and smoothing to reduce diurnal variations and sensor noise, under the assumption of regular and well-behaved data streams.

However, in practice, the quality and reliability of environmental sensor networks can vary significantly. Although many previous studies rely on well-maintained networks or satellite-derived products with exhaustive quality control, practical deployments—especially those that involve cost-effective, densely distributed or custom-built sensors, such as ours [6]—are often affected by hardware faults, calibration drift, missing intervals, and elevated measurement noise. Only a handful of works have addressed the challenge of extracting reliable and informative segments from such noisy, irregular, and incomplete datasets prior to model development.

This study addresses a critical gap by proposing a robust preprocessing pipeline adapted to the constraints of challenging sensor networks. The methodology is specifically designed to identify, clean, and select the most informative and contiguous segments from a highly noisy in-situ sensor dataset deployed in the Duero basin (Spain), ensuring that the intervals used for modeling are of high quality and free from significant gaps, thus minimizing the reliance on artificial data imputation. Subsequently, the predictive performance of state-of-the-art deep learning models, namely long short-term memory (LSTM) networks and transformer architectures, is evaluated on these curated intervals, with robustness and accuracy benchmarked against standard baseline models.

The results demonstrate that, when supported by rigorous preprocessing and careful data selection, deep learning models are capable of producing robust and accurate soil moisture predictions, even under highly challenging data conditions. This work helps bridge the gap between theoretical advancements in machine learning and practical constraints of environmental monitoring in real-world field settings.

2 Dataset

2.1 Data Description

The primary dataset contains soil moisture measurements collected by a custom network of in-situ sensors distributed across the Duero basin (Spain). Although technical documentation reported the deployment of 55 devices [6], the raw data revealed 91 unique device identifiers. This discrepancy is attributed to device resets, which cause changes in identifiers over time.

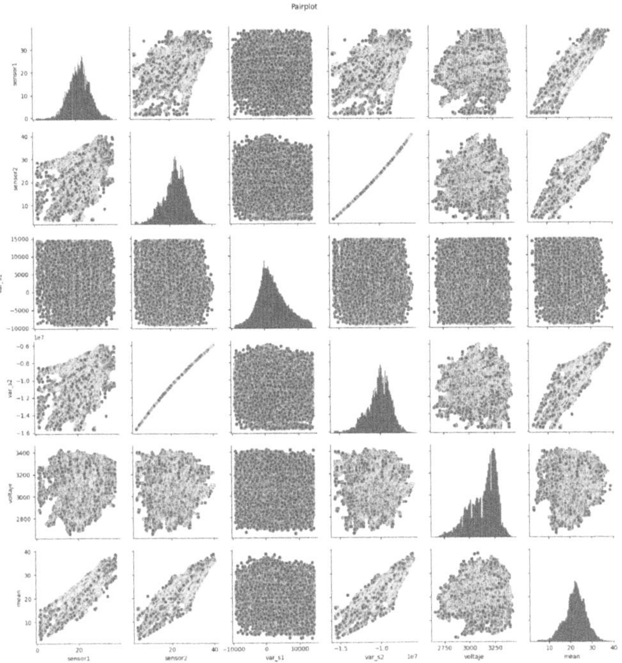

Fig. 1. Pairplot of key sensor variables.

Each device is equipped with two independent soil moisture sensors, which sample at irregular intervals and store data locally until upload. On average, each device records around 70 measurements per day, yielding approximately 1,000 daily readings in aggregate. However, the dataset displays considerable heterogeneity in sensor behavior, including:

- **Frequent outliers:** Some correspond to rare but real events (e.g., heavy rainfall), while others result from sensor malfunction.
- **Marked variance disparities:** Sensor 1 and Sensor 2 within the same device exhibit differences of several orders of magnitude in variance, likely due to hardware or floating-point issues.
- **Irregular sampling and data gaps:** Sensor outages are common, resulting in unevenly spaced time series and missing periods ranging from minutes to weeks.

To further illustrate these issues, Fig. 1 presents a pairplot of the main sensor variables, including readings, variances, supply voltage, and mean values. In particular, the variance of Sensor 1 (`var_s1`) forms a distinctive cross-shaped noise pattern, indicating a lack of relationship with its own measurements or any variable. In contrast, Sensor 2 (`var_s2`) shows a nearly linear dependency between its variance and value, suggesting systematic artifacts. Based on these findings, Sensor 1 was prioritized for subsequent analyses.

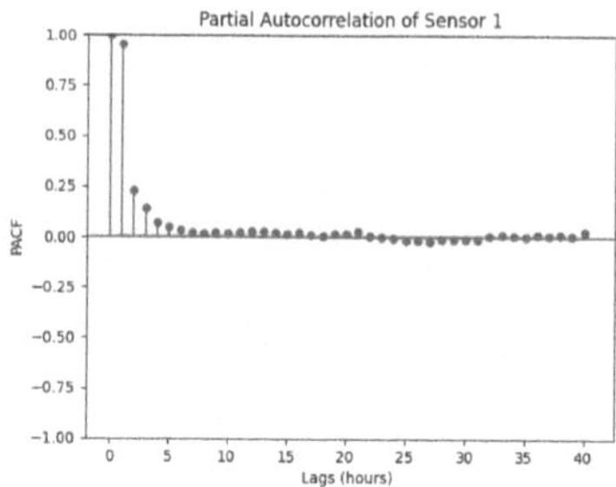

Fig. 2. Partial autocorrelation function (PACF) of soil moisture readings for a representative sensor.

The dataset exhibits very limited temporal memory: the partial autocorrelation function (PACF) reveals strong correlation only for lags 1 and 2, dropping rapidly to near zero beyond that. Thus, only the most recent values are useful for forecasting (at least when considering only the sensor values). Figure 2 illustrates this rapid decay in temporal dependency for a representative sensor.

To enhance the predictive context, the hourly meteorological variables (ambient humidity, air temperature, and precipitation) were integrated with the sensor data. Sensor and weather station records were temporally aligned by downsampling all series to a uniform hourly resolution. The resulting merged dataset offers a synchronized view of sensor measurements and environmental conditions.

Correlation analysis confirms only weak linear relationships between meteorological variables and soil moisture: precipitation shows a correlation coefficient of approximately 0.13, and ambient humidity approximately 0.23. These low values highlight the complexity and non-linearity of the interactions, indicating that weather alone does not provide strong predictive signals for soil moisture on the hourly scale.

2.2 Data Preparation

Given the heterogeneity and noise in the raw sensor data, a multi-stage data preparation pipeline was developed to ensure reliability and maximize the informative value of the training data. The process included device-level filtering, interval segmentation, and final scoring and selection, as outlined below.

1. **Device Filtering.** Only devices with at least the average number of valid measurements (1,391) were retained. To further ensure internal consistency, only devices that exhibit a minimum Pearson correlation of 0.7 between their

two sensors were considered. Devices with abrupt spikes (changes exceeding 10% within a two-hour window) were also excluded as likely to be faulty.

2. **Interval Segmentation.** For each device, time series were partitioned into continuous measurement intervals, with any gap exceeding five days marking a boundary. Intervals with fewer than 200 consecutive points were discarded.
3. **Interval Scoring and Selection.** Each candidate interval was assigned a custom score, favoring long, densely populated, and reliable segments, while penalizing intervals with excessive outliers or persistently low soil moisture. The score is defined as follows:

$$\text{Score} = \left((t_{\text{end}} - t_{\text{start}})^{w_1} \cdot N^{w_2} \cdot |\text{corr}_{12}|^{w_3} \cdot \max\left(0,\ 1 - \frac{\text{fo}_1 + \text{fo}_2}{2}\right)\right)^{w_4} \times \max\left(0,\ 1 - w_{\text{low}} \cdot \frac{1}{N}\sum_{i=1}^{N} \mathbb{I}\left(\frac{4095 - \text{Sensor1}_i}{4095} < \theta_{\text{low}}\right)\right)$$

where:

- t_{end}, t_{start}: timestamps of the last and first samples in the interval.
- N: number of samples in the interval.
- corr_{12}: absolute Pearson correlation between Sensor 1 and Sensor 2.
- fo_j $(j = 1, 2)$: fraction of outliers for Sensor j, where outliers are values outside $[Q_1 - 1.5\,\text{IQR}, Q_3 + 1.5\,\text{IQR}]$.
- $w_1, w_2, w_3, w_4, w_{\text{low}}$: hyperparameter weights.
- $\mathbb{I}(\cdot)$: indicator function (1 if the condition is true, 0 otherwise).
- θ_{low}: threshold for low normalized soil moisture (default: 0.02).

The intervals were ranked according to this score, and the top five—each corresponding to a unique device and densely populated period—were selected for model training and evaluation (27 – 38 d, 641 – 908 hourly readings each). Outliers were marked as NaNs before segmentation, so most were excluded during interval selection; remaining missing values within selected intervals were imputed by linear interpolation. Exploring more advanced imputation methods remains as future work (Table 1).

Table 1. Summary of selected device-intervals used for model training and evaluation.

Interval	Device ID	Start Date	End Date	Days	Points	Imputed
1	89882280666613233841	2024-09-17	2024-10-25	38	908	0%
2	89882280666608761716	2024-10-06	2024-11-08	34	734	8.1%
3	89882280666613233818	2024-09-17	2024-10-24	38	889	0.8%
4	89882280666613233780	2025-02-10	2025-03-09	27	641	0%
5	89882280666605450596	2024-09-17	2024-10-23	37	874	0.5%

Figure 3 and Fig. 4 illustrate the interval selection process for a representative device. The first panel shows the complete time series, with the selected interval highlighted in green. The second panel zooms in on the selected interval after post-processing and normalization, demonstrating the improved temporal density and quality achieved by the pipeline.

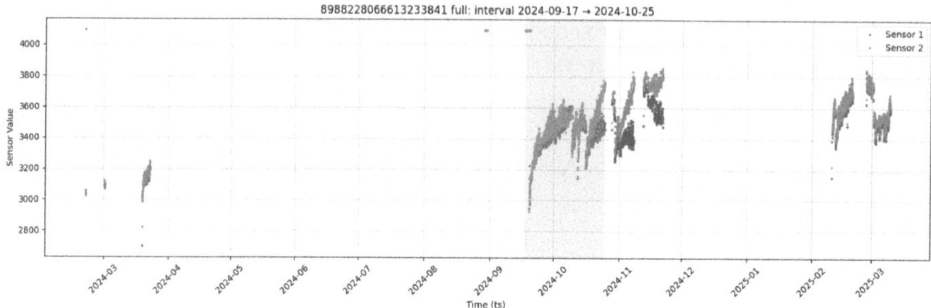

Fig. 3. Full dataset for device 8988228066613233841. The selected interval is highlighted in green. (Color figure online)

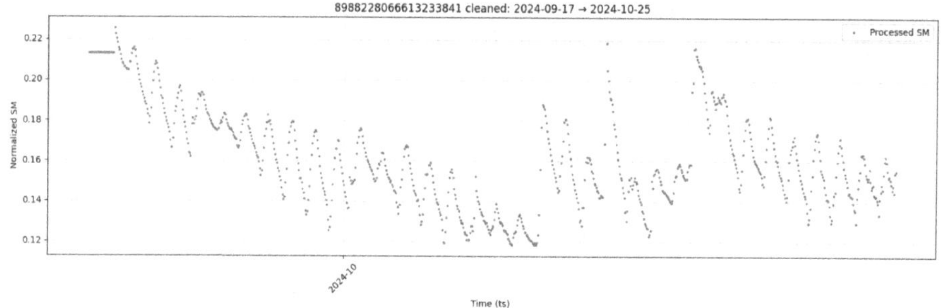

Fig. 4. Selected interval (normalized soil moisture, post-processing) for the same device (Interval 1).

This process ensures that only the most reliable and information-rich segments are used for downstream modeling, maximizing the robustness and generalizability of the results.

3 Algorithms

3.1 Model Selection

Building on recent advances in time series modeling for soil moisture prediction [10], two state-of-the-art deep learning architectures were selected: the Long

Short-Term Memory (LSTM) network and the Transformer. Both architectures are designed to capture complex, nonlinear temporal dependencies and have demonstrated competitive performance across a range of environmental forecasting tasks.

The LSTM architecture is widely recognized for its ability to retain information over long input sequences, effectively modeling time-lagged relationships common in environmental sensor data. The Transformer, originally developed for natural language processing, has recently demonstrated strong results in time series prediction [10] due to its parallelizable attention mechanism, which can flexibly learn dependencies at multiple temporal scales.

To benchmark the added value of deep learning approaches, a simple baseline model ("dummy") is also included. This model predicts future values using the mean of the most recent observations and serves as a naive reference commonly used in forecasting tasks [5]. Its inclusion enables a direct comparison of the informational gain achieved by more advanced models.

3.2 Model Implementation and Training

After the data pre-processing described in Sect. 2.2, each model is trained and evaluated on the contiguous high-quality data segments identified from the sensor network. Input features include past values of soil moisture, as well as synchronized meteorological variables (temperature, humidity, precipitation) for each time step.

For both LSTM and Transformer models, the input "lookback" window (number of past time steps) is treated as a tunable hyperparameter, allowing adaptation to the short temporal memory characteristic of the dataset. Model architecture, hyperparameters (such as number of layers, hidden units, attention heads, and dropout rates), and training procedures (optimizer, learning rate schedules, batch size) are selected based on Bayesian optimization (using Optuna) and cross-validation on the selected intervals.

Table 2. Predictive performance (RMSE, R^2) of each model across different prediction horizons. Results are averaged over all selected intervals.

Model	Horizon (h)	RMSE (%)	R^2
LSTM	1	0.29	0.98
Transformer	1	0.37	0.96
Baseline	1	0.48	0.93
LSTM	8	0.84	0.74
Transformer	8	0.69	0.83
Baseline	8	1.44	0.25
LSTM	24	1.27	0.31
Transformer	24	1.10	0.43
Baseline	24	1.68	-0.46

Each model is trained using only one high-quality data segment (interval), as identified in Sect. 2.2, with all tuning and validation performed within that interval via cross-validation. To ensure a robust generalization assessment, the final evaluation is also performed at additional intervals, specifically data segments from other sensors or time periods that the model never saw during either training or validation. This setup ensures that the reported results reflect not only within-interval performance but also the ability to generalize to unseen real-world conditions.

All models are evaluated in multiple prediction horizons (1, 8, and 24 h in advance), with the performance assessed using the root mean squared error (RMSE) and the coefficient R^2. Each model is re-trained independently for each forecasting horizon, ensuring a fair comparison.

4 Results and Discussion

4.1 Quantitative Results

The performance of the models evaluated (LSTM, Transformer, and baseline) was assessed across multiple prediction horizons (1, 8, and 24 h ahead) using the root mean squared error (RMSE) and the coefficient R^2. Table 2 summarizes the results obtained in the main evaluation intervals.

As shown in Table 2, both deep learning models consistently outperform the baseline for all prediction horizons. The Transformer achieves the best overall performance, attaining the lowest RMSE and the highest R^2 at both horizons. Notably, the baseline model's accuracy degrades rapidly with increasing horizon. These results highlight the added value of deep temporal modeling.

4.2 Hyperparameter Optimization

For each model and prediction horizon, the hyperparameters were tuned using Bayesian optimization using the Optuna framework. Table 3 summarizes the best configuration obtained for a representative run of the LSTM model at an 8-hour prediction horizon.

The reported results correspond to the models with these best-found settings. Similar tuning was performed for the Transformer and for each prediction horizon.

4.3 Qualitative Analysis and Visualizations

To further illustrate the predictive capacity and limitations of the models, Fig. 5 shows an example of Transformer predictions for a representative interval. Importantly, the interval depicted here was **never used during training or validation**: the model was trained and tuned exclusively on a different data segment and here evaluated on an entirely unseen interval to assess its true generalization capacity, as outlined in Sect. 3.2.

Table 3. Hyperparameters for LSTM and Transformer (8h horizon, Interval 1).

Hyperparameter	LSTM	Transformer
Learning Rate (LR)	0.00157	7.84×10^{-5}
Weight Decay	2.46×10^{-5}	9.08×10^{-5}
Dropout Rate	0.18	0.22
Batch Size	16	32
Lookback Window	29	23
Max Epochs	490	534
Early Stopping Patience ($PATIENCE_{ES}$)	40	57
Gradient Clipping ($CLIP\_GRAD\_NORM$)	2.76	3.36
Embedding size (d_model)	-	128
Attention heads	-	4
Layers	-	3
Feedforward size (d_ff)	-	256

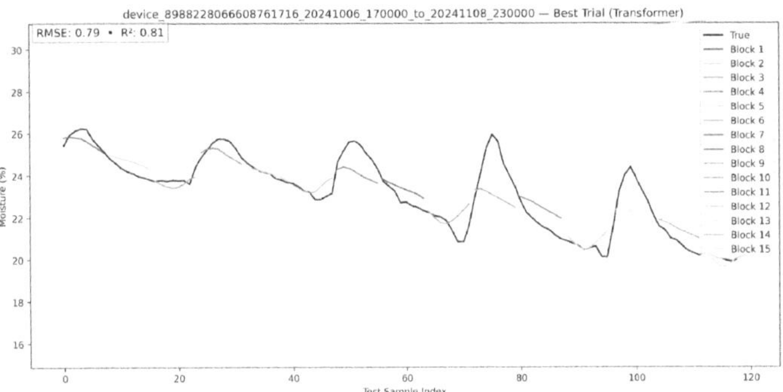

Fig. 5. Multi-step predictions (block forecasts) for Transformer on interval 2. RMSE and R^2 for the interval are indicated.

The model predicts in "blocks": for each test point t, it receives data from t – lookback to t and generates a multi-step forecast up to t + horizon in a single inference pass. The resulting predicted segments are shown as colored lines, superimposed on the ground truth.

5 Conclusions

This study addressed the challenge of soil moisture prediction in the presence of noisy and incomplete datasets by developing a robust preprocessing pipeline to extract high-quality, gap-free intervals. The evaluation demonstrated that, when trained on carefully selected data segments, deep learning models exhibit strong predictive performance over medium- and long-term horizons. The results confirm that rigorous data preparation is essential for reliable forecasting and highlight the capacity of deep learning methods to deliver robust and generalizable predictions, even under highly irregular and incomplete data conditions.

Acknowledgments. This study was funded by public research projects of the Spanish Ministry of Science and Innovation PID2023-151605OB-C22 and projects under the call PEICTI 2021–2023 with the identifiers TED2021-131520B-C21 and TED2021-131520B-C22. The source code and data used to reproduce the experiments described in this paper are publicly available on GitHub

Disclosure of Interests. The authors have no competing interests to declare that are relevant to the content of this article.

References

1. Bandaru, L., Irigireddy, B.C., Davis, B.: DeepQC: a deep learning system for automatic quality control of in-situ soil moisture sensor time series data (2023). https://doi.org/10.48550/ARXIV.2311.06735, https://arxiv.org/abs/2311.06735
2. Basak, A., Schmidt, K.M., Mengshoel, O.J.: From data to interpretable models: machine learning for soil moisture forecasting. Int. J. Data Sci. Anal. **15**(1), 9–32 (2022). https://doi.org/10.1007/s41060-022-00347-8
3. Costa, V., Rodrigues, A., Fernandes, W., De Mello, C., Dong, Q.: Droughts in a changing climate: advances in modeling, forecasting and strategies for adaptation. Front. Water **7**, 1585531 (2025)
4. Duan, Y., et al.: A deep learning framework for long-term soil moisture-based drought assessment across the major basins in China. Remote Sens. **17**(6), 1000 (2025). https://doi.org/10.3390/rs17061000
5. Kühnert, C., Gonuguntla, N.M., Krieg, H., Nowak, D., Thomas, J.A.: Application of LSTM networks for water demand prediction in optimal pump control. Water **13**(5), 644 (2021). https://doi.org/10.3390/w13050644
6. Morante Hernández, O.A.: Despliegue de un sistema IoT en la cuenca del parque forestal de Valdebebas (Madrid): Desarrollo y análisis de sensores de humedad. Master's thesis, Universidad Carlos III de Madrid (2024), tutor: Jose Manuel Molina López
7. Pan, Z., Xu, L., Chen, N.: Combining graph neural network and convolutional LSTM network for multistep soil moisture spatiotemporal prediction. J. Hydrol. **651**, 132572 (2025). https://doi.org/10.1016/j.jhydrol.2024.132572, https://www.sciencedirect.com/science/article/pii/S0022169424019681
8. Singh, A., Gaurav, K.: Deep learning and data fusion to estimate surface soil moisture from multi-sensor satellite images. Sci. Rep. **13**(1) (2023). https://doi.org/10.1038/s41598-023-28939-9,
9. Skulovich, O., Gentine, P.: A long-term consistent artificial intelligence and remote sensing-based soil moisture dataset. Sci. Data **10**(1) (2023). https://doi.org/10.1038/s41597-023-02053-x
10. Wang, Y., Shi, L., Hu, Y., Hu, X., Song, W., Wang, L.: A comprehensive study of deep learning for soil moisture prediction. Hydrol. Earth Syst. Sci. **28**(4), 917–943 (2024). https://doi.org/10.5194/hess-28-917-2024

Special Session: Efficiency and Explainability in Machine Learning and Soft Computing

Generalized Fuzzy Logical Operators: A Behavioral Study

Paolo Fosci(✉) and Giuseppe Psaila

University of Bergamo - DIGIP, Viale Marconi 5, 24044 Dalmine, (BG), Italy
{paolo.fosci,giuseppe.psaila}@unibg.it

Abstract. The paper addresses the problem of defining t-norm/t-conorm operators in the area of fuzzy logic, in order to generalize the classical logical operators.

Among the large variety of proposals for t-norm/t-conorm operators, the Vector p-norm approach is based on an elegant formalization that allows for varying the behavior of the operators by modifying a few parameters. In particular, it is possible to enhance or reduce the AND-ness/OR-ness of the operators, to obtain a stricter or weaker AND-ness/OR-ness, depending on the needs.

The paper studies the behavior of the p-norm operators using charts that plot in a 3-D format their behavior: this visual approach will help the reader understand how the operators actually behave and possibly practically exploit them.

Keywords: Fuzzy Logic · Generalized Logical Operators · Behavioral Study

1 Introduction

Since their introduction by Zadeh [17], Fuzzy Sets have proven to be a powerful tool for a large variety of tasks that are characterized by imprecise concepts. Indeed, the idea that an item can belong to a fuzzy set only partially, with a membership degree in the range $[0, 1]$ is crucial in many practical problems, such as decision making and soft querying.

A major issue when translating classical Boolean logical operators to a fuzzy interpretation is how to combine membership degrees. This is the problem of defining *t-norm/t-conorm* operators (corresponding to the `AND` and `OR` operators). In this respect, a plethora of proposals are available in the literature. One proposal is the Vector p-norm approach [3]: it is a formally simple, yet elegant and comprehensive proposal, in which it is possible to tune the degree of `AND`-ness/`OR`-ness of the operators, which includes importance weights as well.

Is it possible to practically exploit Vector p-norm operators for, e.g., soft querying? Yes, it is. The authors are investigating their practical exploitation within *J-CO-QL*$^+$, the query language of the *J-CO* Framework [11]. While applying p-norm operators, it has emerged that understanding Vector p-norm operators is not easy and could be an obstacle to their practical exploitation.

E. Corchado et al. (Eds.): SOCO 2025, CCIS 2806, pp. 205–216, 2026.
https://doi.org/10.1007/978-3-032-19763-4_19

In this paper, the authors address the problem of understanding how AND/OR operators (in the p-norm version) actually behave, depending on the settings of their parameters.

The paper is organized as follows. Section 2 introduces the notion of fuzzy aggregator and provides the Vector p-norm definitions (Sect. 2.1). Section 3 lets the reader understand the rationale of the p-norm operators. Section 4 analyzes the behavior of the operators AND and OR in the p-norm approach. Finally, Sect. 5 draws the conclusions and sketches future work.

2 Fuzzy Aggregators

Fuzzy aggregators have been defined to cope with the belonging of items (entities) to multiple fuzzy sets at the same time. For example, consider two sets of people, i.e., the fuzzy sets *Rich People* and *Popular People*, which correspond to the linguistic predicates "Rich Person" and "Popular Person", respectively. So, in the fuzzy context, what is the meaning of the following condition?

"Rich Person" AND "Popular Person"

Clearly, it is necessary to aggregate somehow the membership degrees of an item to both fuzzy sets. The classical approach is to consider the minimum degree for the AND operator (t-norm), and the maximum membership degree for the OR operator (t-conorm). Nonetheless, many proposals exist in the literature (see [2,6,7]). Another interesting aspect that was considered by works in the literature related with t-norm/t-conorm operators, is also the possibility to express preferences over aggregated fuzzy sets, i.e., preferences on linguistic predicates that are represented by fuzzy sets (see [1,5,13]). Among these proposals, some like the Hamacher (see [6]) and Aczél-Alsina (see [13]) families rely on algebraic formulations, while others, such as the Vector p-norm approach, are grounded in geometric interpretations based on Minkowski distance.

2.1 The Vector p-Norm Approach

In the literature, the "Vector p-norm" was proposed (see [3]) as an elegant mathematical formalization to cope with the above-mentioned issue, i.e., defining generalized t-norm/t-conorm operators. The general formulas are reported below.

Consider two linguistic predicates lp_1 and lp_2 and the corresponding membership degrees μ_1 and μ_2 of an item x to two fuzzy sets fs_1 and fs_2 (that correspond to the linguistic predicates lp_1 and lp_2). Consider also two "importance degrees" i_1 and i_2 (positive real numbers) for the predicates lp_1 and lp_2, respectively. Finally, consider a positive integer number p.

The membership degree computed by the operator AND in the p-norm version is defined by Eq. 1.

$$\mu_{\texttt{AND}}(\mu_1, \mu_2) = 1 - \sqrt[p]{\frac{(i_1)^p \times (1 - \mu_1)^p + (i_2)^p \times (1 - \mu_2)^p}{(i_1)^p + (i_2)^p}} \tag{1}$$

With the same premises, the operator OR in the p-norm version is defined as in Eq. 2.

$$\mu_{\texttt{OR}}(\mu_1, \mu_2) = \sqrt[p]{\frac{(i_1)^p \times (\mu_1)^p + (i_2)^p \times (\mu_2)^p}{(i_1)^p + (i_2)^p}} \quad (2)$$

The reason why these operators are named AND and OR, can be understood considering the limit for $p \to \infty$.

In the case of the operator AND, it is $lim_{p\to\infty}\mu_{\texttt{AND}}(\mu_1, \mu_2) = min(\mu_1, \mu_2)$.

In the case of the operator OR, it is $lim_{p\to\infty}\mu_{\texttt{OR}}(\mu_1, \mu_2) = max(\mu_1, \mu_2)$.

Consequently, the p-norm version encompasses the classical semantics for AND and OR, when $p \to \infty$.

Equations 1 and 2 rely on the "Minkowski distance" [14], one of the concepts of the "Minkowski geometry" [16]. These concepts are widely used in Mathematics (see, for example, [4]) and in various applied sciences [15].

3 Understanding the Vector p-Norm

The goal of this section is to let the reader understand the rationale behind the definitions of the Vector p-norm operators. To clearly illustrate, the case of $p = 2$ is considered, because it corresponds to the classical Euclidean space.

Let us start with the case $i_1 = i_2 = 1$ (no preferences). Since $p = 2$, the roots in Eqs. 1 and 2 are square roots. In Eq. 1, the square root represents the Minkowski distance [14] between the point (μ_1, μ_2) and the point $(1, 1)$ (see Fig. 1). Consequently, the final membership degree is maximized when the Minkowski distance [14] between the two above-mentioned points is minimized. In practice, the points (μ_1, μ_2) that are equidistant from the point $(1, 1)$ have the same final membership degree ("iso-membership" line), because they are on the same circle that is centered at the point $(1, 1)$ (the black solid line in Fig. 1 is an example).

In the case of the operator OR, Eq. 2 is the distance between the point (μ_1, μ_2) and the point $(0, 0)$. Consequently, the equidistant points from $(0, 0)$ obtain the same resulting membership degree; this is illustrated in Fig. 1 by the dashed circle.

In the case of different importance, things slightly change. To illustrate, let us consider the case $i_1 = 2$, $i_2 = 1$, i.e., μ_1 is twice as important as μ_2; the resulting iso-membership line for the operator AND is the blue solid line in Fig. 1. All the points on this line represent pairs (μ_1, μ_2) yielding the same final membership degree $\mu_{\texttt{AND}}(\mu_1, \mu_2) = 0.8$; the curve is no longer a circle, because a small variation of μ_1 must be compensated by a much greater variation of μ_2, to obtain the same final membership degree.

Exchanging the importance degrees, i.e., $i_1 = 1$ and $i_2 = 2$, the red solid line in Fig. 1 is obtained, for which μ_1 has half the importance of μ_2.

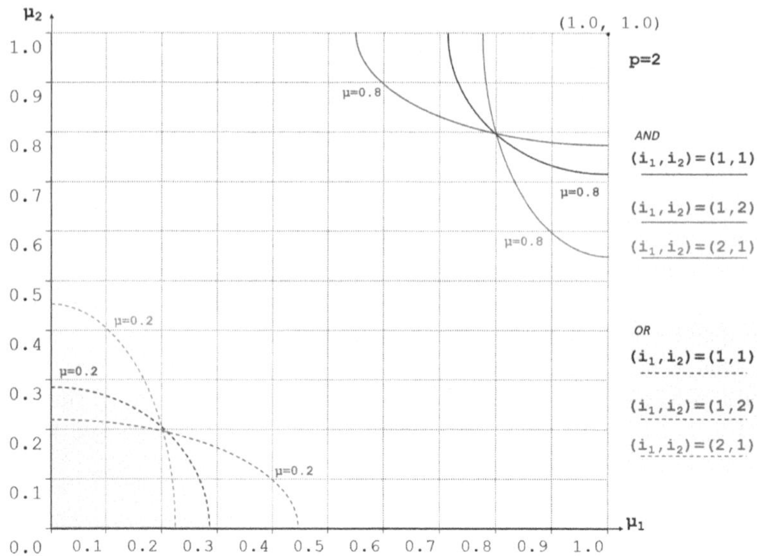

Fig. 1. Iso-membership lines for the p-norm AND (solid lines) and OR (dashed lines).

The reader can notice that the three solid lines determine the same final membership degree $\mu_{\text{AND}}(\mu_1, \mu_2) = 0.8$; furthermore, the three solid lines cross on the main diagonal; meaning that points on the main diagonal have the same final membership degree, independently of the importance degrees.

In the case of the operator OR with unequal importance, a similar result (as for the operator AND) is obtained, as illustrated by the dashed lines in Fig. 1. Indeed, the three dashed lines correspond to a final membership degree $\mu\text{OR}(\mu_1, \mu_2) = 0.2$.

Specifically, the blue dashed line corresponds to the case $i_1 = 2$ and $i_2 = 1$; again, the shape of the curve deploys vertically, because small variations of μ_1 must be compensated by greater variations of μ_2.

Exchanging the importance weights, i.e., $i_1 = 1$ and $i_2 = 2$, the red dashed line is obtained, which deploys horizontally, consequently.

Again, points on the main diagonal have the same final $\mu\text{OR}(\mu_1, \mu_2)$, independently of the importance values.

4 Behavioral Analysis

This section provides a detailed visual analysis of the behavior of the fuzzy logical operators AND and OR in the Vector p-norm version.

4.1 Operator AND

Figures 2 and 3 present a series of 3-D plots that illustrate the behavior of the operator AND. Specifically, the surfaces depict how the membership degree

$\mu_{\texttt{AND}}(\mu_1, \mu_2)$ varies, based on the variations of memberships μ_1 and μ_2 to the input fuzzy sets *fuzzysetA* and *fuzzysetB*, respectively. Different charts are obtained for different configurations of the parameters p, i_1, and i_2 in Eq. 1. Specifically, the x-axis (resp., y-axis) reports the membership degrees to *fuzzysetA* (resp., *fuzzysetB*), while the z-axis reports the aggregated membership degree. Charts are generated by incrementing input degrees in steps of 0.01 (i.e., the range $[0, 1]$ is divided into 100 intervals).

Each figure is labeled to indicate the specific configuration of the parameter p and the pair of importance weights i_1 and i_2 used to generate the plot. For example, `p=2, i`$_1$`=1, i`$_2$`=2` indicates that the computation of the membership $\mu_{\texttt{AND}}(\mu_1, \mu_2)$ (denoted as `pNormAND` in the charts) is performed with $p = 2$ and importance weights $i_1 = 1$ $i_2 = 2$.

The remainder of this section analyzes the behavior by discussing the charts in Figs. 2 and 3.

Equal Importance. Figures 2.a, 2.b, and 2.c are built with equal importance weights ($i_1 = 1$ $i_2 = 1$), with values 1, 2, and 100 for the parameter p; in particular, $p = 100$ is useful as a placeholder for $p = \infty$.

- When $p = 1$ (Fig. 2.a), the shape of the surface is a plane. Indeed, none of the input membership degrees μ_1 and μ_2 prevails on the other, because Eq. 1 reduces to a linear combination, with $p = 1$. Thus, it is possible to say that the `AND` nature is lost.
- With $p = 2$ (Fig. 2.b), the `AND` nature of the operator emerges again and the shape of the surface becomes pyramidal: this happens because in Eq. 1 the contribution of $min(\mu_1, \mu_2)$ tends to prevail (although partially) on $max(\mu_1, \mu_2)$. This explains why the shape is not a perfect pyramid.
  In particular, notice how the corner from the point $(0, 0, 0)$ to the point $(1, 1, 1)$ has emerged.
- With $p = 100$ (Fig. 2.c), the trend that the case $p = 2$ lets emerge is overstressed: the shape of the surface is definitely a pyramid, in which the z-value is substantially the minimum value between the x-value and the y-value.
  In particular, notice how the corner from the point $(0, 0, 0)$ to the point $(1, 1, 1)$ has become a straight line.

It is possible to say that small values of p, yet greater than 1, can be used to obtain an `AND` behavior that is weaker than the classical minimum degree, but it is still an `AND`, in which the minimum value tends to prevail.

Unequal Importance. Figures 2.d, 2.e, and 2.f depict the surfaces obtained in the case of unequal importance: specifically, $i_1 = 1$ and $i_2 = 2$, so as to give significantly-greater importance to μ_2, with respect to μ_1.

- In the case $p = 1$ (Fig. 2.d), the shape of the surface is still a plane, as in the case with equal importance (see Fig. 2.a). The angles have changed, in such a way μ_2 clearly dominates the contribution to the final membership degree.
- With $p = 2$ (Fig. 2.e), the shape of the surface becomes pyramidal again. However, the shape is distorted, to emphasize the contribution of μ_2.

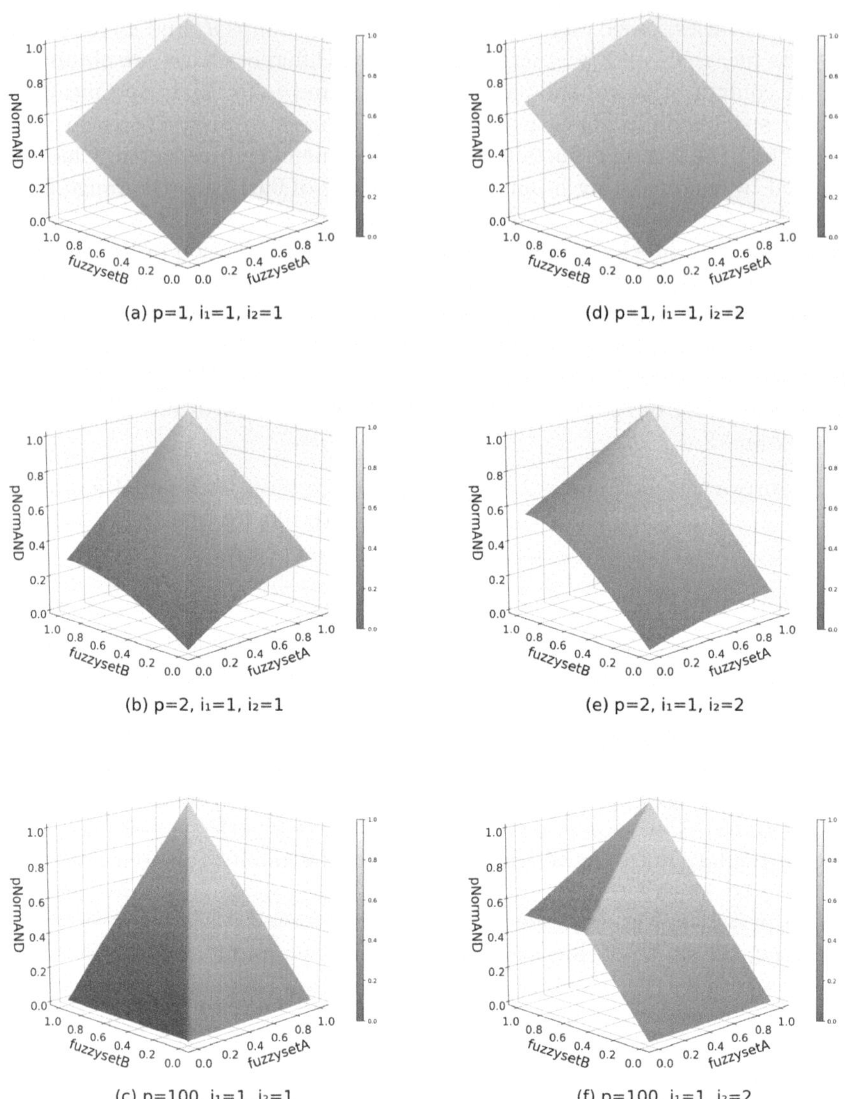

(a) p=1, i₁=1, i₂=1

(d) p=1, i₁=1, i₂=2

(b) p=2, i₁=1, i₂=1

(e) p=2, i₁=1, i₂=2

(c) p=100, i₁=1, i₂=1

(f) p=100, i₁=1, i₂=2

Fig. 2. Behavior of p-norm `AND` with importance weights $(1, 1)$ and $(1, 2)$.

- When $p = 100$ (Fig. 2.f), the above-described transformation is further amplified: lines become straight, but the shape is substantially the same.

Effects of Unequal Importance. Figure 3 reports the surfaces that can be obtained with different settings for importance weights. Specifically, Figs. 3.a, 3.b, and 3.c are obtained for the setting $i_1 = 1$ and $i_2 = 3$; Fig. 3.d, 3.e, and 3.f are obtained for the setting $i_1 = 1$ and $i_2 = 7$. This way, it is possible to visualize

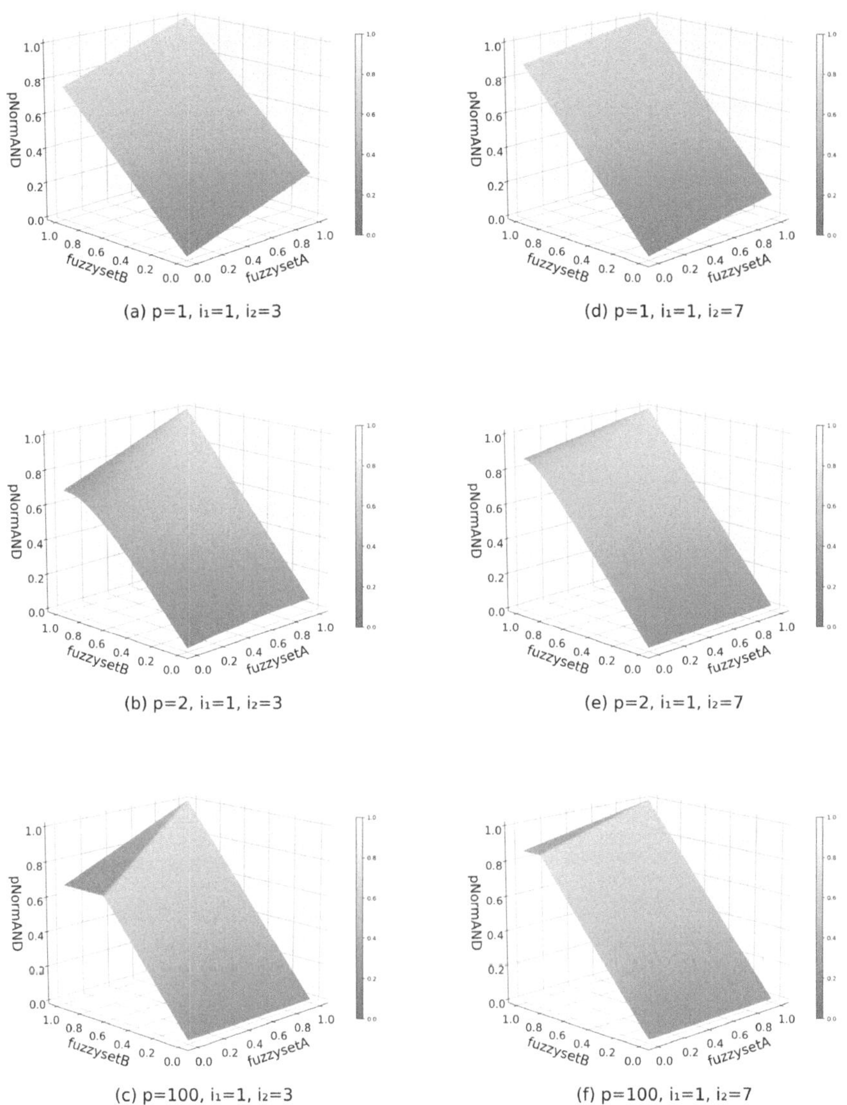

(a) $p=1$, $i_1=1$, $i_2=3$ (d) $p=1$, $i_1=1$, $i_2=7$

(b) $p=2$, $i_1=1$, $i_2=3$ (e) $p=2$, $i_1=1$, $i_2=7$

(c) $p=100$, $i_1=1$, $i_2=3$ (f) $p=100$, $i_1=1$, $i_2=7$

Fig. 3. Behavior of p-norm AND with importance weights $(1, 3)$ and $(1, 7)$.

how the behavior changes on the basis of the different settings for importance weights.

From a formal point of view, it is important to understand how the ratio $k = i_2/i_1$ influences the behavior. Equation 1 can be rewritten as in Eq. 3.

$$\mu_{\texttt{AND}}(\mu_1, \mu_2) = 1 - \sqrt[p]{\frac{(i_1)^p \times (1-\mu_1)^p + (k \times i_1)^p \times (1-\mu_2)^p}{(i_1)^p + (k \times i_1)^p}} \tag{3}$$

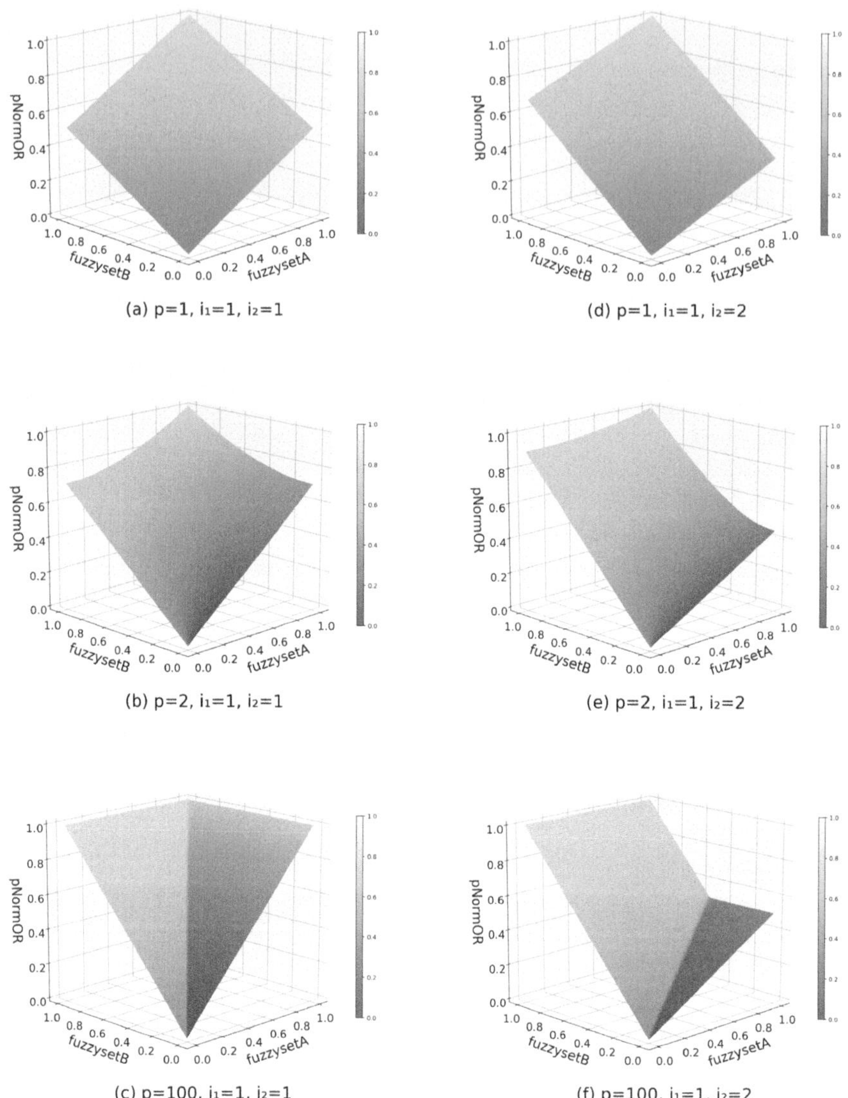

Fig. 4. Behavior of p-norm OR with importance weights $(1, 1)$ and $(1, 2)$.

Factorizing and simplifying $(i_1)^p$, Eq. 3 becomes Eq. 4.

$$\mu_{\text{AND}}(\mu_1, \mu_2) = 1 - \sqrt[p]{\frac{(1 - \mu_1)^p + k^p \times (1 - \mu_2)^p}{1 + k^p}} \tag{4}$$

The reader can see that the only factor that determines the behavior (apart from p) is the ratio $k = i_2/i_1$. Consequently, the case $i_1 = 1, i_2 = 2$ behaves exactly as $i_1 = 3, i_2 = 6$, and so on and so forth.

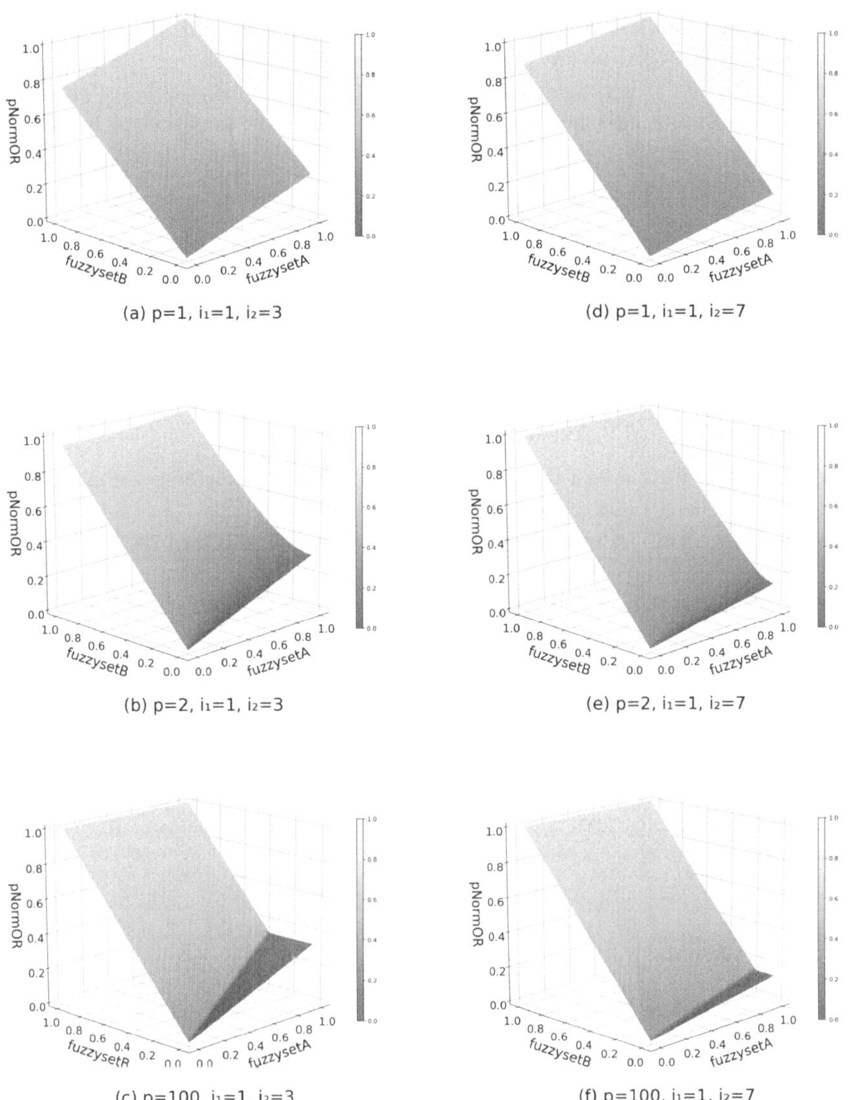

Fig. 5. Behavior of p-norm `OR` with importance weights $(1, 3)$ and $(1, 7)$.

- In the case $p = 1$ (Figs. 3.a and 3.d), the shape of the surface is still a plane. So, notice how in 3.d it is less oblique, because $k = 7$ and μ_2 heavily dominates.
- With $p = 2$ (Figs. 3.b and 3.e), the pyramid is strongly stretched, thus prizing μ_2; in particular, the bottom border that was parallel to the y-axis becomes very short and moves upward; thus, a novel oblique border appears, so as to

connect the small horizontal border with the border that is parallel to the x-axis. For $k = 7$, this effect is further stressed.
- When $p = 100$ (Figs. 3.c and 3.f), the shapes of the surfaces are consistent with the previous case; the only difference is that now borders are much more definite, approximating the generalized AND in the classical membership with different importance weights.

4.2 Operator OR

The behavior of the operator OR in the p-norm version can be studied in the same way as for the operator AND. The resulting plots are depicted in Figs. 4 and 5; the same settings for the parameters p, i_1, and i_2 are considered. The remainder of the section analyzes them in detail.

- When $p = 1$ (Figs. 4.a, 4.d, 5.a, and 5.d) the shape of surfaces remain planar; moreover, these planes coincide with those in Figs. 2.a, 2.d, 3.a, and 3.d, respectively. This fact confirms that the nature of AND and OR is lost, when $p = 1$.
- When $p = 2$ and $p = 100$ with equal importance weights (Figs. 4.b and 4.c) the shape becomes pyramidal, but with respect to the AND case, the pyramid is reversed and folded in the opposite way.
  When $p = 100$, lines become straight and the border from the point $(0, 0, 0)$ to the point $(1, 1, 1)$ is definitely evident.
- With unequal importance weights (Figs. 4.d, 4.e, 4.f, and Fig. 5), the shape of pyramids is stretched, as for the AND case, but on the opposite side, i.e., to the bottom and towards the x-axis (lower importance). Again, higher ratios k determine stronger stretching.

5 Conclusions

The contribution of the paper is the study of the behavior of t-norm/t-conorm operators based on the Vector p-norm approach. Specifically, the classical operators AND and OR, for which in the world of fuzzy logic a large variety of proposals are available.

Typically, the classical interpretation of operators AND and OR is to consider the minimum/maximum membership degree. The advantage of the p-norm approach is that this interpretation is considered as the limit behavior for $p \rightarrow \infty$ (see Eqs. 1 and 2); thus, varying the parameter p, it is possible to gradually modify the behavior of the operators, by increasing or reducing the prevalence of the minimum/maximum membership degree on the final aggregated membership degree. Furthermore, the presence of importance weights in the definitions allows either for promoting or for penalizing one of the operands.

Through the surfaces that are depicted in Figs. 2, 3, 4, and 5, the reader can visualize how the operators vary their behavior, so as to choose the proper configuration of the parameters p, i_1, and i_2.

As future work, the authors will pursue the investigation on how to exploit the p-norm version of the logical operators. Indeed, the p-norm formulation has already been applied for tasks of "Soft Web Intelligence" [10,11]. This is a specific research line within the development of the *J-CO* Framework, which is a software framework for soft querying (possibly-)large *JSON* datasets [8]. The authors are investigating several aspects of fuzzy querying, such as spatial querying [9], and multi-grade models for fuzzy sets [12], for which the p-norm operators could play a crucial role.

Acknowledgments. This study was funded by the European Union - *NextGenerationEU, in the framework of the GRINS - Growing Resilient, INclusive and Sustainable project (GRINS PE00000018 CUP F83C22001720001).* The views and opinions expressed are solely those of the authors and do not necessarily reflect those of the European Union, nor can the European Union be held responsible for them.

References

1. Abir, B.K., Amel, G.T.: Towards fuzzy querying of noSQL document-oriented databases. DBKDA **2015**, 163 (2015)
2. Boixader, D., Recasens, J.: Vague and fuzzy t-norms and t-conorms. Fuzzy Sets Syst. **433**, 156–175 (2022)
3. Bordogna, G., Psaila, G.: Soft aggregation in flexible databases querying based on the vector p-norm. Int. J. Uncertainty Fuzziness Knowl. Based Syst. **17**(supp01), 25–40 (2009)
4. Crasta, G., Malusa, A.: The distance function from the boundary in a Minkowski space. Trans. Am. Math. Soc. **359**(12), 5725–5759 (2007)
5. Deb, N., Sarkar, A., Biswas, A.: Linguistic q-rung orthopair fuzzy prioritized aggregation operators based on hamacher t-norm and t-conorm and their applications to multicriteria group decision making. Arch. Control Sci. 451–484 (2022)
6. Dong, H., Ali, Z., Mahmood, T., Liu, P.: Power aggregation operators based on Hamacher T-norm and T-conorm for complex intuitionistic fuzzy information and their application in decision-making problems. J. Intell. Fuzzy Syst. **45**(5), 8383–8403 (2023)
7. Farahbod, F., Eftekhari, M.: Comparison of different T-norm operators in classification problems. arXiv:1208.1955 (2012)
8. Fosci, P., Psaila, G.: J-CO, a framework for fuzzy querying collections of JSON documents. In: Flexible Query Answering Systems: 14th International Conference, FQAS 2021, Bratislava, Slovakia, September 19–24, 2021, Proceedings 14, pp. 142–153. Springer International Publishing (2021)
9. Fosci, P., Psaila, G.: Soft integration of geo-tagged data sets in J-CO-QL+. ISPRS Int. J. Geo-Information **11**(9), 484 (2022)
10. Fosci, P., Psaila, G.: Towards soft web intelligence by collecting and processing JSON data sets from web sources. In: Proceedings of the 18th International Conference on Web Information Systems and Technologies (2022)
11. Fosci, P., Psaila, G.: Soft querying JSON datasets with personalized preferences and aggregations. In: WEBIST 2024, pp. 153–164 (2024)
12. Fosci, P., Psaila, G.: A unified view of multi-grade fuzzy-set models in J-CO-QL+. Neurocomputing **565**, 126968 (2024)

13. Liu, P., Ali, Z., Mahmood, T., Geng, Y.: Prioritized aggregation operators for complex intuitionistic fuzzy sets based on Aczel-Alsina T-norm and T-conorm and their applications in decision-making. Int. J. Fuzzy Syst. **25**(7), 2590–2608 (2023)
14. Martini, H., Swanepoel, K.J., Weiß, G.: The geometry of Minkowski spaces—a survey. part I. Expositiones mathematicae **19**(2), 97–142 (2001)
15. Merigó, J.M., Gil-Lafuente, A.M.: Using the OWA operator in the Minkowski distance. Int. J. Econ. Manag. Eng. **2**(9), 1032–1040 (2008)
16. Thompson, A.C.: Minkowski geometry. Cambridge University Press (1996)
17. Zadeh, L.A.: Fuzzy sets. Inf. Control **8**(3), 338–353 (1965)

Exploring Zero-Shot Learning for Time Series Forecasting Using Large Language Models

José Ignacio Meseguer-Gómez, Pablo Reina-Jiménez, Manuel Jesús Jiménez-Navarro, María Martínez-Ballesteros(✉), and José C. Riquelme-Santos

Department of Computer Languages and Systems, University of Seville, 41012 Seville, Spain
{jmeseguer,prjimenez,mjimenez3,mariamartinez,riquelme}@us.es

Abstract. The ever-growing importance of forecasting in real-world applications is indisputable. Time series analysis plays a vital role in decision-making in areas such as energy, finance, or healthcare. The emergence of Large Language Models has led to a revolution in the creation of foundation models for time series forecasting. Dividing a time series into patches enables the use of a Large Language Model to predict subsequent patches instead of discrete numerical values. This approach continues to improve as long as the underlying architectures evolve. This paper shows that Chronos Bolt, a time series forecasting model developed by Amazon, can achieve competitive and, in many cases, superior performance compared to traditional statistical and machine learning models in metrics such as SMAPE and MAE, even when making zero-shot predictions without prior training. It also matches the performance of other foundation models recently developed, while achieving better computational efficiency in the process.

Keywords: Large Language Model · Zero-Shot · Time Series Forecasting · Transformers

1 Introduction

In a world driven by data, accurate time series forecasting has become indispensable for domains as varied as energy management and meteorology [7] [13]. While the volume and velocity of data grow, so does the need for reliable data-driven decision making. However, conventional forecasting methods, whether classical statistical models or machine learning (ML) approaches, often depend on iterative, dataset-specific training. This requirement not only incurs significant computational cost and energy consumption, but also inspired the emergence of Green Artificial Intelligence, which seeks to balance predictive performance with environmental sustainability [1]. Moreover, many of these techniques rely on exogenous covariates that may be unavailable or costly to obtain, limiting their scalability and generalizability across different forecasting contexts.

E. Corchado et al. (Eds.): SOCO 2025, CCIS 2806, pp. 217–226, 2026.
https://doi.org/10.1007/978-3-032-19763-4_20

The emergence of foundation models in time series forecasting offers a solution to both of these issues [5]. If a model generalizes well across the data, the computational cost of training will be avoided. This paradigm is known as Zero-Shot Learning, where models make predictions on data unseen before [15]. This approach refers to the ability of a pretrained model to perform inference on entirely new tasks or datasets without any additional fine-tuning. In the context of time series forecasting, a zero-shot model has already absorbed a broad range of temporal patterns during its pretraining phase, such as seasonality, trend shifts, and irregular fluctuations, and can apply that knowledge directly to novel series. Rather than updating model parameters to fit each specific dataset, zero-shot forecasting simply feeds the raw series into the model and obtains predictions immediately. This paradigm not only eliminates the overhead of training on every new dataset, but also ensures consistent performance even when exogenous covariates or labeled examples are unavailable. By leveraging the diversity and scale of its pretraining data, a zero-shot model effectively "recognizes" recurring patterns and dynamics, enabling rapid, reliable forecasts.

This paper explores the potential of zero-shot models for time series forecasting, focusing on their ability to provide accurate and efficient predictions. In particular, the analysis focuses on Chronos Bolt, a foundation model developed by Amazon based on a patch-based approach in which the model predicts the next segment of a time series instead of individual values. Its performance is evaluated without any fine-tuning. Our results show that Chronos Bolt achieves accurate and efficient forecasts, performing competitively against both traditional models and other foundation models.

The remainder of the paper is structured as follows. Section 2 reviews selected related works in time series forecasting. Section 3 describes the datasets used in the experiments, the models included in the benchmark, and the evaluation criteria. Section 4 discusses the results, and the final Sect. 5 presents the conclusions drawn from these findings, along with potential directions for future research.

2 Related Work

Traditional time series forecasting has often required training a dedicated model per dataset or even per time series. In recent years, global modeling approaches have emerged to train a single model across many time series, improving generalization by sharing patterns across related series. For example, DeepAR [11] is a global Deep Learning (DL) model that learn from multiple time series in a dataset. These global models yielded strong performance across diverse forecasting tasks, but they still needed re-training or fine-tuning on each new dataset. This limitation has motivated research into zero-shot forecasting, where a model pretrained on broad data can directly forecast new, unseen time series without any task-specific training. The advent of large pretrained models in Natural Language Processing (NLP) has heavily influenced this direction. The goal is to develop foundation models for time series, analogous to NLP's foundation models, that capture generic temporal patterns from massive data and can be applied to new forecasting tasks.

Transformer networks, which revolutionized NLP, have been adapted for time-series forecasting. The Time Series Transformer (TST) introduced the transformer self-attention mechanism [12] to multivariate prediction and representation learning, achieving competitive results. Subsequent work tackled time series-specific challenges, particularly long-term dependencies and varying scales, most notably PatchTST [9]. This method advances long-horizon forecasting by splitting series into patches of consecutive observations as input tokens and using a channel-independent architecture that processes each variable separately through shared transformer layers. This preserves local temporal patterns and shortens the sequences seen by attention, yielding state-of-the-art accuracy on long-range benchmarks.

Inspired by the success of Large Language Models (LLM), researchers in 2023 explored using existing LLMs directly for time series forecasting. LLM-Time [8] demonstrated that GPT-style models can produce surprisingly good forecasts when numeric time series data are converted into a textual format. In this prompting approach, past values are fed to a pretrained LLM (like GPT-3) as a prompt, as comma-separated numbers or natural language descriptions, and the model is asked to generate future values. This study reported that such LLMs possess a degree of zero-shot forecasting ability across various benchmarks.

For enabling zero-shot forecasting, foundation models were proposed. These models are trained on a wide corpus of time series data (potentially spanning many domains and frequencies) with the aim of capturing universal patterns. Early steps in this direction included TimesNet [14], a CNN-based time series foundation model. TimesNet introduced a novel temporal 2D variation block to capture multi-scale patterns by transforming 1D time series into a 2D representation. Trained on a large collection of time series, TimesNet achieved state-of-the-art results on several tasks (forecasting, classification, imputation, etc.) out-of-the-box, highlighting that a well-designed deep model can serve as a general time series backbone.

Building on the foundation of pretraining exemplified by TimesNet, Chronos Bolt [2] introduces several innovations that elevate both its flexibility and efficiency. Instead of generating forecasts one patch at a time, it employs a direct multistep decoding head that outputs an entire forecast horizon in a single forward pass, eliminating autoregressive loops and dramatically accelerating inference while achieving substantial memory savings. To accommodate diverse deployment scenarios, Chronos Bolt is offered in multiple model sizes, from lightweight variants suited for edge or embedded systems to larger configurations optimized for maximum accuracy, allowing users to tailor the trade-off between throughput and predictive performance.

3 Materials and Methodology

This section is divided into four subsections. Section 3.1 describes the characteristics of the datasets and the training partitions. Section 3.2 presents the models included in the study, while Sect. 3.3 summarizes the hyperparameters settings. The final subsection, Sect. 3.4, outlines the evaluation procedure.

3.1 Dataset

For this study, four datasets have been selected and grouped into two categories: two focused on air quality prediction [4] and two on transformer condition monitoring [16]. Note that all datasets have an hourly temporal resolution.

The first two datasets (Asomadilla and Bermejales), contain meteorological data collected from different regions in Seville and Jaén, two cities in southern Spain. Their primary objective is to forecast ground-level ozone (O_3) concentration. Although these datasets include several exogenous variables, such as temperature and nitrogen dioxide (NO_2) levels, our analysis focuses exclusively on the target variable O_3.

In contrast, the remaining two datasets are related to electrical transformer monitoring. They aim to prevent transformer failures by predicting the oil temperature (OT) in two Chinese power transformers. In this second group, OT serves as the target variable.

Table 1 summarizes the partitioning details and the number of blocks used for prediction, which will be explained in Sect. 3.4. Additionally, Fig. 1 illustrates the last 750 values of each time series. This helps highlight recent patterns, tendency or variability in the time series analyzed.

Table 1. Dataset sizes and partition information.

Dataset	Total	Train Length	Test Length	Blocks
Asomadilla	157776	133328	24448	693
Bermejales	166536	140236	26300	730
ETTh1	17420	16984	436	72
ETTh2	17420	16984	436	72

3.2 Models

Chronos Bolt is an Amazon foundation model for time series forecasting that segments input series into patches to efficiently capture temporal patterns. Pretrained via masked patch prediction on diverse datasets, it delivers fast, accurate, and energy-efficient forecasts without requiring fine-tuning. This model presents different variants which will be introduced in Sect. 3.3. Given that Chronos-Bolt is the main focus of this study, a diverse set of baseline models was selected to enable a fair comparison.

In the experimentation, the Naive model was included as a simple statistical model serving as a baseline benchmark. This is the simplest model, which predicts the next 48 values by simply repeating the last observed value of the series.

PatchTST, DeepAR and TFT [6] were also selected as they were competitors in the original Chronos paper, achieving comparable results. However, it is

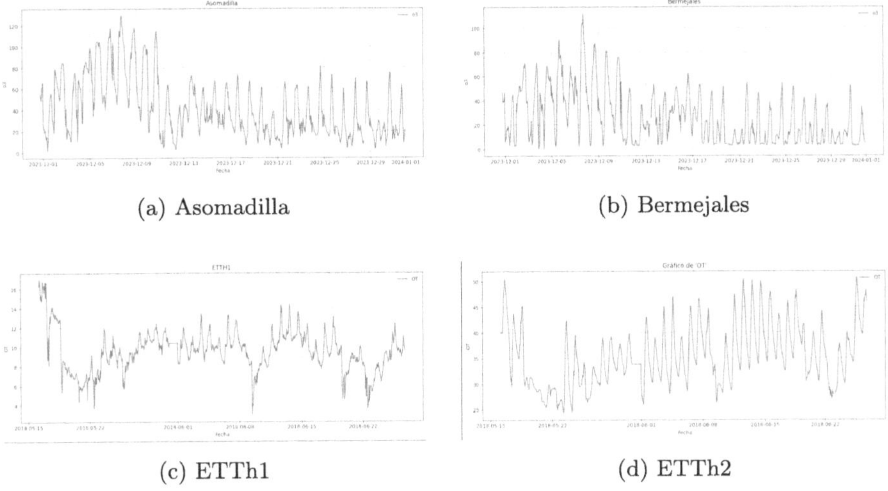

(a) Asomadilla (b) Bermejales

(c) ETTh1 (d) ETTh2

Fig. 1. Representation of the last 750 values of each time series.

important to note that these models are task-specific and typically require fine-tuning in the time series they are intended to forecast. PatchTST and TFT are also models based on Transformers, whereas DeepAR is based on a Recurrent Neural Network (RNN) architecture.

Random Forest and Gradient Boosting were also included as representative non-DL models. Both models rely on an ensemble of regression trees for estimating their outputs.

3.3 Experimental Setup

AutoGluon [3] was used as the main framework for time series forecasting, enabling the application of advanced models such as Chronos Bolt or PatchTST. Each dataset was transformed into a suitable format for forecasting, and models were trained accordingly.

On the other hand, the ML models, Random Forest and Gradient Boosting, were trained using the scikit-learn [10] library. The training partitions were created manually, ensuring consistency with the partitioning process used by the AutoGluon library, with training sets matching the length of the *context_length* hyperparameter. This hyperparameter defines the number of past time steps used as input for prediction.

Table 2 presents the models used in the experimental analysis, along with the corresponding hyperparameters. In the experimental configuration, the forecast horizon is 48 (corresponding to 48 h). This choice represents a suitable trade-off between predictive complexity and informational richness, as the model receives one week of historical data (through the hyperparameter *context_length*). A 48-hour forecast window is also practically relevant, providing enough lead time

Table 2. Hyperparameters used for each AutoGluon and Scikit model.

Model	Parameter	Value
Naive	-	-
DeepAR	epochs	50
	learning_rate	5e-4
	context_length	168
PatchTST	epochs	50
	learning rate	5e-4
	patch_len	16
	stride	8
	d_model	64
	context_length	168
TemporalFusionTransformer	context_length	168
RandomForestRegressor	n_estimators	100
HitsGradientBoostingRegressor	-	-
Chronos Bolt Tiny	model_path	`"bolt_tiny"`
	context_length	168
Chronos Bolt Mini	model_path	`"bolt_mini"`
	context_length	168
Chronos Bolt Small	model_path	`"bolt_small"`
	context_length	168
Chronos Bolt Base	model_path	`"bolt_base"`
	context_length	168

for decision-making while still maintaining a strong connection to recent observations.

Other hyperparameters not yet discussed include *epochs*, the number of passes over the training data, and *learning_rate*, which is the step size used to update model weights during training. These hyperparameters can be found in the PatchTST and DeepAR models. The main hyperparameters of PatchTST are *patch_len*, *stride*, and *d_model*, which correspond to the length of each patch, the step size of the patch window, and the embedding dimension, respectively. In RandomForestRegressor, *n_estimators* specifies the number of trees in the ensemble. Finally, four different variants of Chronos Bolt models were incorporated. These variants differ in computational complexity, and in some cases, the simpler models may outperform the more complex ones on simpler datasets. The *model_path* hyperparameter allows us to select the model.

All experiments were conducted on the same machine, equipped with an AMD Ryzen 7 5800X and an NVIDIA GeForce RTX 3080 TI with 12GB of RAM GPU.

3.4 Evaluation Procedure

A data-splitting strategy was applied in the experimental setup. Both the meteorological and transformer condition datasets were divided using an 80/20 split for training and testing, respectively. The meteorological datasets presented greater complexity due to a higher proportion of missing values, approximately 5,500 in total. To address this, linear interpolation was applied to fill in the missing values and ensure the continuity of the time series.

The datasets were divided into training and test data. Following the standard approach for time series data, the split was performed chronologically: the initial portion (80 percent) was used for training, while the most recent values (20 percent) were reserved for testing. Before model training, the training data was scaled using a Min-Max scaler. The same transformation was applied to the test data using the parameters fitted on the training set.

The training data was appropriately formatted, and a forecasting model was trained using this setup. Each model was fitted at an hourly cadence, forecasting 48 steps ahead, and optimized to minimize the SMAPE. A separate forecasting instance was created for each approach evaluated in our earlier analysis.

To ensure a fair evaluation, the assessment was not based on a single test prediction. Instead, the test set was divided into consecutive 48-length blocks. For each block, a forecast was generated using the previous 7 d (168 input steps) as context. The goodness of the prediction was evaluated using two metrics: SMAPE and MAE. As Fig. 2 shows, after each prediction, the real values of the predicted block are appended to the context window, allowing the model to generate the next prediction. To maintain a fixed input length, the oldest values in the context window are simultaneously removed. This rolling-window approach continues until the end of the test series. Finally, the metrics were averaged across all blocks. This provides us with a comprehensive view of each model's performance throughout the series. The average inference and training time for each model were also recorded, as computational efficiency plays a key role in this study.

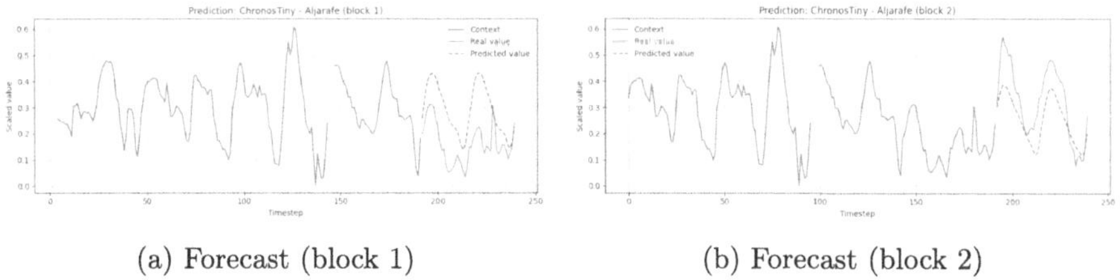

(a) Forecast (block 1)

(b) Forecast (block 2)

Fig. 2. The prediction block in the first iteration is appended to the context window in order to generate the next prediction.

4 Results and Discussion

This section reports the experimental findings regarding prediction accuracy and computational performance.

The models will be compared using the evaluation metrics discussed in 3.4, along their respective training time. The training time (TT) refers to the total time required to train the model, measured by seconds. The results for the meteorological and energy datasets are presented in Table 3.

Table 3. Comparison of the models across the Asomadilla, Bermejales, ETTh1, and ETTh2 datasets using SMAPE, MAE, and TT.

Setting	Model	Asomadilla			Bermejales			ETTh1			ETTh2		
		SMAPE	MAE	TT	SMAPE	MAE	TT	SMAPE	MAE	TT	SMAPE	MAE	TT
Zero-Shot	CBTiny	24.095	0.062	0	40.228	0.058	0	15.403	0.034	0	13.015	0.054	0
Zero-Shot	CBMini	23.919	0.062	0	39.941	0.058	0	15.377	0.034	0	13.324	0.056	0
Zero-Shot	CBSmall	23.913	0.062	0	40.254	0.058	0	14.990	0.033	0	12.986	0.054	0
Zero-Shot	CBBase	24.058	0.062	0	40.142	0.059	0	15.405	0.034	0	12.649	0.052	0
Zero-Shot	Naive	41.232	0.114	0	62.736	0.103	0	17.816	0.041	0	25.943	0.119	0
Fine-Tuned	TFT	23.710	0.061	97.182	**38.553**	**0.055**	118.605	16.310	0.036	68.947	**12.172**	**0.051**	69.523
Fine-Tuned	PatchTST	23.664	0.061	94.742	40.126	0.058	86.328	14.628	0.032	38.595	13.524	0.056	23.771
Fine-Tuned	DeepAR	26.353	0.071	96.276	40.302	0.060	62.316	25.803	0.067	28.989	20.651	0.094	28.895
Fine-Tuned	RFR	23.612	0.608	2102.137	39.492	0.058	2219.019	**14.124**	**0.031**	154.354	14.060	0.059	152.238
Fine-Tuned	HGBR	**23.221**	**0.060**	117.323	38.756	0.056	123.911	14.499	0.032	22.016	13.237	0.055	22.440

This table shows that Chronos Bolt is never the top-performing model, however, its efficiency relative to training time is notably high. For instance, on the ETTh1 dataset, the SMAPE of the Chronos Bolt Small model is less than one point higher than that of the best-performing model (Random Forest), while its training time is approximately 166 times shorter.

In certain scenarios, this level of efficiency may be more valuable than achieving the best possible prediction. Work environments that handle multiple datasets and do not require top-tier accuracy can benefit significantly from the advantages of zero-shot learning. As shown in our experiments, the difference in accuracy is often negligible, less than one point, while training times are drastically reduced.

A detailed comparison between the best-performing Chronos model and the best overall model for each dataset is presented in Table 4. The metrics reported include the difference in SMAPE, MAE, and training time (TT), denoted as *Diff*, along with the relative percentage increase in error.

These differences are computed, using SMAPE as example, as:

$$\mathrm{SMAPE}_{\mathrm{BestModel}} - \mathrm{SMAPE}_{\mathrm{BestChronosModel}}$$

As non-Chronos models achieve the best performance in all cases, these differences are negative. The relative percentage increase in SMAPE and MAE metrics of the best Chronos model compared to the best model is calculated as:

$$\frac{\mathrm{SMAPE}_{\mathrm{BestChronosModel}} - \mathrm{SMAPE}_{\mathrm{BestModel}}}{\mathrm{SMAPE}_{\mathrm{BestModel}}} \times 100$$

Table 4. Comparison of Chronos vs. Non-Chronos models on each dataset.

Dataset	Best Chronos	Best Model	SMAPE		MAE		TT
			Diff.	% Inc.	Diff.	% Inc.	Diff.
Asomadilla	CBSmall	HGBR	−0.692	2.89%	−0.002	3.33%	117.323
Bermejales	CBMini	TFT	−1.338	3.60%	−0.003	5.45%	118.605
ETTh1	CBSmall	RFR	−0.866	6.13%	−0.002	6.45%	154.354
ETTh2	CBBase	TFT	−0.466	3.83%	−0.001	1.96%	69.523

Across all datasets, the performance degradation of Chronos models remains consistently small, reinforcing the efficiency–accuracy trade-off they achieve and suggesting that their computational advantage does not come at the cost of unpredictable or highly variable accuracy losses.

5 Conclusions and Future Works

Although Chronos Bolt was never the top-performing model in terms of accuracy, the analysis shows that its results were consistently competitive, often within a one point margin in SMAPE, the metric minimized during training, while achieving dramatically lower training times. This makes it especially suitable for scenarios where computational efficiency is critical, optimal accuracy is not essential, or data is limited.

Chronos Bolt offers a compelling trade-off between accuracy and efficiency, which can be crucial for operational decision making. Moreover, it aligns well with the principles of Green AI, as its inference phase consumes minimal energy once the model has been trained.

Several directions could be pursued in future research. First, it would be valuable to investigate which types of time series benefit the most from zero-shot learning. In this study, the forecast horizon was chosen to balance context length and forecasting difficulty, requiring the models to predict two days ahead using one week of historical data. It would be insightful to explore how varying both the context window and the prediction horizon affects performance metrics, as well as to expand the evaluation with additional metrics such as memory usage and inference time. Additionally, incorporating exogenous covariates into the forecasting process may lead to further improvements and provide deeper insights into model behavior.

Acknowledgements. This research has been supported by the grant PID2020-117954 RB-C22, PID2023-146037OB-C21 funded by MICIU/AEI/10.13039/501100011033 and TED2021-131311B-C21, funded by MICIU/AEI/10.13039/501100011033 and by the "European Union NextGenerationEU/PRTR". The authors would like to thank the Dirección General de Sostenibilidad Ambiental y Cambio Climático, Consejería de Sostenibilidad, Medio Ambiente y Economía Azul for providing part of the data used in this study.

References

1. Alzoubi, Y.I., Mishra, A.: Green artificial intelligence initiatives: potentials and challenges. J. Cleaner Prod. 143090 (2024)
2. Ansari, A.F., et al.: Chronos: learning the language of time series. Trans. Mach. Learn. Res. (2024). https://openreview.net/forum?id=gerNCVqqtR
3. Erickson, N., et al.: AutoGluon-Tabular: Robust and Accurate AutoML for Structured Data (2020). https://arxiv.org/abs/2003.06505
4. Jiménez-Navarro, M.J., Martínez-Ballesteros, M., Martínez-Álvarez, F., Asencio-Cortés, G.: Explaining deep learning models for ozone pollution prediction via embedded feature selection. Appl. Soft Comput. **157**, 111504 (2024)
5. Liang, Y., et al.: Foundation models for time series analysis: a tutorial and survey. In: Proceedings of the 30th ACM SIGKDD Conference on Knowledge Discovery and Data Mining, pp. 6555–6565 (2024)
6. Lim, B., Arık, S.O., Loeff, N., Pfister, T.: Temporal Fusion Transformers for interpretable multi-horizon time series forecasting. Int. J. Forecast. **37**(4), 1748–1764 (2021)
7. Lim, B., Zohren, S.: Time-series forecasting with deep learning: a survey. Phil. Trans. R. Soc. A **379**(2194), 20200209 (2021)
8. Gruver, N., Finzi, S.Q.M., Wilson, A.G.: Large language models are zero shot time series forecasters. In: Advances in Neural Information Processing Systems (2023)
9. Nie, Y., Nguyen, H., Sinthong, P., Kalagnanam, J.: A time series is worth 64 words: long-term forecasting with transformers. In: Proceedings of International Conference on Learning Representations (2023)
10. Pedregosa, F., et al.: Scikit-learn: machine learning in Python. J. Mach. Learn. Res. **12**, 2825–2830 (2011)
11. Salinas, D., Flunkert, V., Gasthaus, J., Januschowski, T.: DeepAR: probabilistic forecasting with autoregressive recurrent networks. Int. J. Forecast. **36**(3), 1181–1191 (2020)
12. Vaswani, A., et al.: Attention is all you need. Adv. Neural Inf. Process. Syst. **30** (2017)
13. Weron, R.: Electricity price forecasting: a review of the state-of-the-art with a look into the future. Int. J. Forecast. **30**(4), 1030–1081 (2014)
14. Wu, H., et al.: TimesNet: temporal 2D-variation modeling for general time series analysis. arXiv preprint arXiv:2210.02186 (2022)
15. Xian, Y., Lampert, C.H., Schiele, B., Akata, Z.: Zero-shot learning–a comprehensive evaluation of the good, the bad and the ugly. IEEE Trans. Pattern Anal. Mach. Intell. **41**(9), 2251–2265 (2018)
16. Zhou, H., et al.: Informer: beyond efficient transformer for long sequence time-series forecasting. In: Proceedings of The Thirty-Fifth AAAI Conference on Artificial Intelligence, AAAI 2021, Virtual Conference, vol. 35, pp. 11106–11115. AAAI Press (2021)

Explainable Graph Neural Networks for Omics-Based Cancer Classification

Javier Hiruelo-Pérez, Paula Herrero-Míguez, David Gutiérrez-Avilés, Manuel J. Jiménez-Navarro, and María Martínez-Ballesteros(✉)

Department of Computer Languages and Systems, University of Seville, 41012 Seville, Spain
{jhiruelo1,dgutierrez3,mjimenez3,mariamartinez}@us.es, paulaherrero02@hotmail.com

Abstract. Cancer is one of the leading causes of death in the world, notable for its biological complexity and high mortality rate. Although recent advances in artificial intelligence provide promising solutions, challenges still remain in the context of deep learning's interpretability. In this paper, we propose a methodology to explain graph neural network outcomes in cancer classification tasks. For this purpose, we collected RNA-seq data to predict five cancer types: low-grade brain glioma, lung adenocarcinoma, cutaneous melanoma, stomach adenocarcinoma, and thyroid carcinoma. To explain the model's predictions, we combine explainable artificial intelligence methods, specifically GNNExplainer, with functional over-representation analysis. Our results show that this methodology not only achieves outstanding predictive performance (average weighted F1-score of 98.68%), but also facilitates biological interpretation by identifying key genes and pathways involved in these pathologies.

Keywords: graph neural networks · cancer · classification · explainability · gene expression · over-representation analysis

1 Introduction

Cancer was the second leading cause of death globally, accounting for 14.57% of total mortality in 2021. Despite modest declines in age-standardized death rates, continuous progress in this field depends on primary and secondary preventive measures, such as tobacco control, dietary improvements, early detection, equitable access to treatments, and investments in novel research techniques [17].

Advances in Artificial Intelligence (AI) present promising solutions for cancer care, enhancing diagnosis speed and accuracy, enabling personalized treatments, and supporting research. Fields like radiology and genomics particularly benefit from AI's ability to perform detailed analysis and big data processing [6].

J. Hiruelo-Pérez and P. Herrero-Míguez—These authors contributed equally to this work.

E. Corchado et al. (Eds.): SOCO 2025, CCIS 2806, pp. 227–236, 2026.
https://doi.org/10.1007/978-3-032-19763-4_21

However, the main challenge in applying AI to healthcare is ensuring the safety and efficacy of these systems. An essential step in achieving this is explainability, which is why eXplainable Artificial Intelligence (XAI) is a rapidly expanding field of AI. The use of XAI is crucial for the implementation of models in important fields, such as oncology, as it is strictly necessary that the results are interpretable and explainable [21].

Considering the inherent biological complexity of cancer, it is necessary to integrate multimodal data, including radiomic, metabolomic, transcriptomic, genomic, and clinical information, for a precise tumor characterization [5]. Although databases such as Linkedomics, STRING, Gene Expression Omnibus, and The Cancer Genome Atlas (TGCA) facilitate data integration, challenges remain in effectively consolidating multimodal explanations [8].

In this context, Graph Neural Networks (GNNs) represent significant advances in handling biological network data, enhancing predictive accuracy compared to traditional Deep Learning (DL) models. GNNs surpass conventional DL models in performance, particularly when dealing with partially structured and multimodal data and where interpretability is a key requirement [3].

This paper presents the application of an efficient spectral graph neural network model for cancer-type classification, emphasizes the importance of explainability, and proposes a structured implementation framework to explain the decisions made by GNNs in cancer-type classification of RNA-seq data. The remainder of the paper is organized as follows: Sect. 2 presents selected related studies relevant to this work, Sect. 3 describes the data and methodology, Sect. 4 discusses the results obtained and Sect. 5 provides the conclusions and outlines future research directions.

2 Related Works

The application of GNNs in the field of genomics is expanding rapidly due to their ability to combine experimental data, such as microarray or RNA sequencing, with prior biological knowledge. For example, Rhee et al. [13] applied Graph Convolutional Networks (GCNs) to classify protein-protein interaction networks in breast cancer using STRING and TCGA data. Subsequently, R. Ramírez et al. [11] extended this approach to all 33 cancer types in TCGA. Other models, such as scDeepSort [15], graph-sc [1], and scDMG [19], focus on classifying single-cell RNA sequencing data, each with distinct approaches.

However, a significant challenge to the practical application of these models is the need for both interpretability and explainability. Current efforts in XAI emphasize post hoc techniques such as model-agnostic and widely applicable LIME [14] and SHAP [9] methods, which treat models as black boxes and infer reasoning by perturbing inputs. For example, SHAP was used to interpret the results of a random forest classifier applied to RNA-seq data in [12], while LIME was applied to explain neural network predictions of CTCF binding sites in [16].

In this context, adaptations of existing methodologies and the development of novel approaches have been implemented for GNNs. In particular, methods such as GNNExplainer [20] and GraphLIME [4] exemplify this advancement. As discussed in Sect. 1, this domain is experiencing rapid growth, driven by the growing interest in the use of AI within increasingly complex and sensitive realms.

2.1 ChebNet Graph Neural Networks

A GNN is a class of neural networks designed to create embeddings from graph-structured data. A graph can be denoted as $G = (V, E)$, where V is a set of vertices or nodes, and E is a set of edges. The nodes can also have assigned features that are denoted as $X \in \mathbb{R}^{|V| \times |F|}$, where $|V|$ denotes the number of nodes of the graph and $|F|$ represents the number of features for each node. In our case, the values of the matrix X are the gene expression values.

ChebNet [2] is a spectral-based GNN that computes the graph convolution by approximating a spectral filter using Chebyshev polynomials. Formally, considering a graph G with an adjacency matrix A, a diagonal degree matrix D, where $D_{ii} = \sum_{j=1}^{|V|} A_{ij}$, the normalized graph Laplacian can be defined as:

$$L = I - D^{-\frac{1}{2}} A D^{-\frac{1}{2}}, L \in \mathbb{R}^{|V| \times |V|} \tag{1}$$

where I is the identity matrix of dimensions $|V| \times |V|$. The Laplacian has certain properties that allow its decomposition into a diagonal matrix of eigenvalues, $\Lambda = diag[\lambda_1, \ldots, \lambda_{|V|}]$, and eigenvectors, U, expressed as $L = U \Lambda U^T$. Going into the spectral domain allows us to compute the convolution of the GNN with the use of spectral filters, h, as follows:

$$Z = h(L)X = h(U \Lambda U^T)X = U h(\Lambda) U^T X \tag{2}$$

where Z denotes the node embeddings matrix of the input graph. The problem lies in the fact that the filter complexity has a polynomial relation with respect to the input dimension. To solve this, a polynomial expansion of h is computed as:

$$h_\theta(\Lambda) = \sum_{k=0}^{K-1} \theta_k T_k(\hat{\Lambda}) \tag{3}$$

where $\theta \in \mathbb{R}^K$ is a vector of Chebyshev coefficients, $T_k(\hat{\Lambda}) \in \mathbb{R}^{|V| \times |V|}$ is a Chebyshev polynomial of order k evaluated at $\hat{\Lambda} = \frac{2\Lambda}{\lambda_{max}} - I$, which is a diagonal matrix of scaled eigenvalues in the range $[-1, 1]$. Also, for a generic matrix B, the values of $T_k(B) = 2BT_{k-1}(B) - T_{k-2}(B)$ with $T_0(B) = 1$ and $T_1(B) = B$. After computing θ, the ChebNet convolution can be efficiently computed as:

$$Z = \theta(I - D^{-\frac{1}{2}} A D^{-\frac{1}{2}})X \tag{4}$$

All this process only gives us the embeddings of the graphs, Z, so after the model, a fully connected output layer is needed for classification tasks.

2.2 GNNExplainer

GNNExplainer [20] formulates the importance of a given node, $v \in V$, in a prediction, $\hat{y}$, using mutual information. The goal is to identify a subgraph $G' \subseteq G$ and associated characteristics $X' = \{x'|v' \in G'\}$ that are important for the prediction $\hat{y}$. This is formalized using an optimization framework as follows:

$$\max_{G'} MI(\hat{Y}, (G', X')) = H(\hat{Y}) - H(\hat{Y}|G = G_S, X = X_S) \quad (5)$$

where $H(\hat{Y})$ represents the entropy of the prediction label distribution, $\hat{Y}$ and MI the mutual information function. The term MI measures the change in confidence of the GNN in predicting $\hat{y}$ when the input graph, G, is limited to G' and its characteristics to X'.

The intuition of this approach is to find the largest subgraph G' by removing nodes and edges. So, if removing them makes a strong difference in the prediction, it provides a counterfactual explanation of the importance of the node.

The resulting subgraph G' is thus the smallest structure whose removal causes the greatest drop in confidence of the model for $\hat{y}$, providing an explanation of which connections are truly critical for this prediction.

3 Materials and Methods

This section describes the dataset collection, preprocessing methods, the proposed GNN model architecture, the explainability technique applied, and the method applied for biological interpretation of the results.

3.1 Data Description

Following the methodology proposed in [10], this study collected the TCGA Pan-Cancer (PANCAN) dataset [18], which includes 11,060 RNA-seq samples across diverse cancer types.

In order to explore the practical application and importance of explainability methods in GNNs, the analysis is focused on a representative subset of five cancer types from the PANCAN dataset. Specifically, brain lower grade glioma, lung adenocarcinoma, skin cutaneous melanoma, stomach adenocarcinoma, and thyroid carcinoma are selected. These cancer types were chosen due to their almost balanced number of samples, which can be seen in Table 1, and their biological diversity, both of which contribute to more interpretable and robust experimental results, as further discussed in Sect. 4.

3.2 Data Preprocessing

In order to construct a curated and dimensionally reduced dataset, the following preprocessing pipeline was applied:

1. **Data Cleaning**: Genes with missing data, as well as those with low mean and variance ($mean < 0.5$ and $std < 0.8$), were excluded from the dataset. This procedure is based on the assumption that genes lacking expression and, therefore, information, typically have low variance and exhibit lower expression levels. This practice is prevalent in RNA-seq and microarray analyses to effectively reduce the dimensionality of the dataset.
2. **Scaling**: The samples were normalized to the $[0, 1]$ interval using *Min-Max scaling*. This approach was chosen to ensure that each sample has equivalent significance while allowing each gene to preserve its relevance in relation to the others.
3. **Graph Generation**: The co-expression graph was constructed by calculating the Spearman correlation and p-values between genes. By evaluating both measurements, the dataset was refined to retain only the edges connecting genes with significant associations, $p - value \leq 0.05$ and Spearman correlations ≥ 0.6.

As a final remark, the Spearman correlation value range is $[-1, 1]$, where values near the extremes indicate strong inverse or direct correlations. We decided to focus on direct correlations based on methods from [10], but alternative options include examining inverse correlations or a combination of inverse and direct correlations.

3.3 Graph Neural Network Model

For this paper, a ChebNet model was implemented for the task of cancer-type classification. The model architecture was relatively small in terms of parameters due to the use of only two ChebConv modules with only 64 channels with a Chebyshev polynomial order of 5 and a fully connected classifier.

This architecture was selected for two main reasons: the first being that, due to the simplicity of the data, where the nodes only have one feature, a complex model quickly overfits, and the second being that as the number of modules and Chebyshev polynomial order increases, the risk of oversmoothing does too. Additionally, this architectural decision contributed to computational efficiency without sacrificing expressive power.

3.4 Explainability

To explain the model's predictions, we applied GNNExplainer, which identifies compact subgraph structures and subsets of node features influential in the prediction process [20]. However, in this case, only the first functionality is applicable, as each node contains a single feature, which is its expression value.

3.5 Over-Representation Analysis

Using the explanations generated by the GNNExplainer method, the main objective is to identify the nodes and edges that had the greatest influence on the model's final prediction. By mapping the node identifiers provided by the explainer back to their corresponding gene symbols, one can determine which genes and their interactions the model considers most relevant during the decision-making process.

Identifying only the most influential genes offers limited biological insight. A more informative approach is to determine which metabolic pathways, composed of sequences of reactions carried out by gene-encoded proteins, are most influential. This provides a clearer understanding of the biological processes in which the key genes are involved. To achieve this, Over-Representation Analysis (ORA), a method commonly applied in microarray and RNA-Seq gene expression studies, can be used [7]. These methods work as follows:

1. Starting from a background set, defined as a list of reference genes, an interest set is derived based on a given criterion. In this study, the background set corresponds to all genes or nodes in the graph, while the interest set consists of those genes or nodes to which the model assigns the highest attention when making a prediction.
2. Then the number of genes in our interest set and background set belonging to each metabolic pathway are counted and tested for over-representation. Pathways for which our interest set is more enriched than what would be expected by pure chance are said to be over-represented.
3. For those pathways selected as over-represented a multiple testing correction is performed to adjust the confidence p-values obtained. The false discovery rate (FDR) method was used in this case.

Integration of the trained GNN model, GNNExplainer, and ORA allows the extraction of the most relevant genes for each cancer type and facilitates examination of the reasoning behind the model's decisions.

4 Results and Discussion

This section presents the performance results of our model and provide an interpretation of its predictions through explainability analysis.

4.1 Model Performance

The model described in Sect. 3.3 was implemented and evaluated using stratified 5-fold cross-validation, with 60% of the data used for training, 20% for validation, and the remaining 20% for testing. The resulting models achieved the performance scores reported in Table 1, demonstrating high predictive confidence across classes and low variability between cancer types.

Additionally, it can be observed that the model exhibited greater predictive confidence for thyroid carcinoma and lower-grade brain glioma compared to other cancer types. Possible reasons include a lack of relevant genes to differentiate these tumor types or limited data availability.

Table 1. Results of the model, in percentage, evaluated using stratified k-fold (k = 5).

Cancer Type	Class Distribution	Precision	Recall	F1-Score
Brain Lower Grade Glioma	**530**	99.63 ± 0.51	99.81 ± 0.42	99.72 ± 0.26
Lung Adenocarcinoma	**576**	98.25 ± 1.83	96.87 ± 3.17	97.54 ± 2.10
Skin Cutaneous Melanoma	**474**	97.71 ± 1.99	98.11 ± 1.73	97.90 ± 1.52
Stomach Adenocarcinoma	**450**	97.87 ± 3.07	99.33 ± 0.99	98.59 ± 1.97
Thyroid Carcinoma	**572**	99.83 ± 0.39	99.47 ± 0.78	99.65 ± 0.37
Macro Averages		98.66 ± 1.13	98.72 ± 1.04	98.68 ± 1.10
Weighted Averages		98.71 ± 1.05	98.69 ± 1.07	98.69 ± 1.07

4.2 Explainability Results

For each analyzed cancer type, the GNNExplainer algorithm returns a set of subgraphs representing genes and their relationships, which the model identifies as most influential for making predictions. Figure 1 illustrates an example of such subgraphs for the brain lower grade glioma class. In this visualization, node size reflects the gene's betweenness centrality within the subgraph, highlighting its structural importance, while node communities were detected using the Leiden algorithm to reveal modular organization.

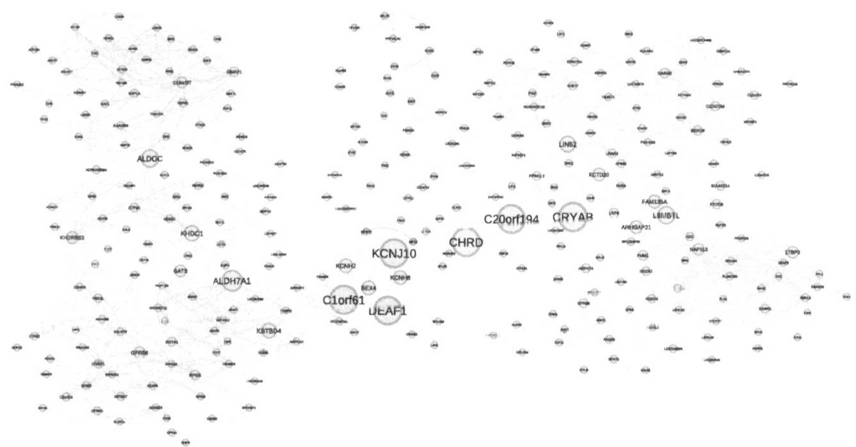

Fig. 1. Example of important subgraphs identified by GNNExplainer for classifying brain lower grade glioma samples.

However, as discussed in Sect. 3.5, identifying the most influential genes and their interactions alone does not provide meaningful biological insight. For instance, inspecting only the gene names in Fig. 1 does not yield sufficient biological information. A more informative approach involves analyzing the biological processes associated with these genes. For example, genes represented by pink nodes are linked to terms such as cerebellum and hippocampus, while green node genes are associated with terms like nervous system development and cerebral cortex.

Based on this analysis, the explanation for all the graphs in our test set was computed using our best performing model from the k-fold and averaged the importance over genes and cancer types. This procedure left us with an explanation for each cancer type for which the genes with an importance greater than 0.725 were extracted; those genes defined our interest set and all the genes in our graphs defined the background set for the ORA algorithm. Table 2 summarizes some of the most relevant biological terms identified for each cancer type.

Table 2. Three relevant terms found for each type of cancer with its associated adjusted p-value. The multiple testing correction method used for these p-values was FDR.

	Brain	Lung	Skin	Stomach	Thyroid
Relevant Term	Cerebellum	Lung	Soft tissue 2; fibroblasts[High]	Stomach 1; glandular cells [≥Medium]	Thyroid gland; glandular cells [≥Low]
P-Value	6.60×10^{-5}	8.92×10^{-5}	4.77×10^{-2}	1.20×10^{-3}	1.26×10^{-13}
Relevant Term	Cerebral cortex	Lung; macrophages [≥Medium]	Soft tissue 2; fibroblasts [≥Medium]	Stomach 2; glandular cells [≥Medium]	Thyroid gland
P-Value	1.41×10^{-4}	1.18×10^{-4}	4.77×10^{-2}	1.37×10^{-3}	1.26×10^{-13}
Relevant Term	Hippocampus	Nasopharynx; respiratory epithelial cells[≥Low]	Soft tissue 2; fibroblasts [≥Low]	Duodenum; glandular cells [High]	Thyroid gland; glandular cells [≥Medium]
P-Value	1.59×10^{-4}	1.25×10^{-4}	4.77×10^{-2}	1.47×10^{-3}	6.63×10^{-13}

As described in Table 2, our model demonstrates a high level of precision in detecting relevant genetic expressions in different cancers. Specifically, it highlights key genes in lower-grade brain glioma, lung adenocarcinoma and thyroid carcinoma, supported by strong confidence in p-value. However, the performance of the model is less robust for stomach adenocarcinoma and skin cutaneous melanoma, reflecting lower relevance and confidence in the identified terms.

5 Conclusions and Future Works

In this paper, a novel methodology is presented for interpreting the underlying reasoning of GNNs in cancer type classification tasks. The experimental results obtained show that the proposed model, in addition to having excellent predictive ability with a weighted F1-Score close to 99%, is interpretable by combining techniques such as GNNExplainer and ORA. This combined approach enables deeper insights into the biological processes influencing model predictions, thus facilitating the identification of key genes and metabolic pathways involved in cancer pathology.

Although this research applied the proposed methodology specifically to genomic data, it is adaptable to other types of gene networks. Furthermore, this methodology facilitates auditing model predictions, thus enabling the identification and rectification of weaknesses, such as those observed in the classification of cutaneous melanoma and stomach adenocarcinoma samples using the presented model. This iterative process enables the development of progressively more accurate models, ultimately enhancing patient care.

Future lines of research could explore the construction of gene co-expression graphs using both direct and inverse correlations, potentially leveraging alternative correlation metrics or incorporating external biological knowledge bases such as STRING. In addition, further research could explore the effects of feature selection methods for reducing graph dimensionality and optimizing hyperparameters to enhance cancer classification models.

Acknowledgements. This research has been supported by the grant PID2020-117954 RB-C22, PID2023-146037OB-C21 funded by MICIU/AEI/ 10.13039/501100011033 and TED2021-131311B-C21 funded by MICIU/AEI/10.13039/501100011033 and by the European Union NextGenerationEU/PRTR.

References

1. Ciortan, M., Defrance, M.: GNN-based embedding for clustering scRNA-seq data. Bioinformatics **38**(4), 1037–1044 (2021). https://doi.org/10.1093/bioinformatics/btab787
2. Defferrard, M., Bresson, X., Vandergheynst, P.: Convolutional neural networks on graphs with fast localized spectral filtering. In: Proceedings of the 30th International Conference on Neural Information Processing Systems, pp. 3844–3852. NIPS'16, Curran Associates Inc., Red Hook, NY, USA (2016). https://doi.org/10.5555/3157382.3157527
3. Gogoshin, G., Rodin, A.S.: Graph neural networks in cancer and oncology research: emerging and future trends. Cancers **15**(24) (2023). https://doi.org/10.3390/cancers15245858
4. Huang, Q., Yamada, M., Tian, Y., Singh, D., Chang, Y.: GraphLIME: local interpretable model explanations for graph neural networks. IEEE Trans. Knowl. Data Eng. **35**(7), 6968–6972 (2023). https://doi.org/10.1109/TKDE.2022.3187455
5. Hunter, B., Hindocha, S., Lee, R.W.: The role of artificial intelligence in early cancer diagnosis. Cancers **14**(6) (2022). https://doi.org/10.3390/cancers14061524
6. Jiang, F., et al.: Artificial intelligence in healthcare: past, present and future. Stroke Vasc. Neurol. **2**(4), 230–243 (2017). https://doi.org/10.1136/svn-2017-000101
7. Khatri, P., Sirota, M., Butte, A.J.: Ten years of pathway analysis: current approaches and outstanding challenges. PLOS Comput. Biol. **8**(2), 1–10 (2012). https://doi.org/10.1371/journal.pcbi.1002375

8. Klauschen, F., et al.: Toward explainable artificial intelligence for precision pathology. Annu. Rev. Pathol. **19**, 541–570 (2024). https://doi.org/10.1146/annurev-pathmechdis-051222-113147
9. Lundberg, S.M., Lee, S.I.: A unified approach to interpreting model predictions. In: Proceedings of the 31st International Conference on Neural Information Processing Systems, pp. 4768–4777. NIPS'17, Curran Associates Inc., Red Hook, NY, USA (2017)
10. Mostavi, M., Chiu, Y.C., Huang, Y., Chen, Y.: Convolutional neural network models for cancer type prediction based on gene expression. BMC Med. Genomics **13**(S5), 44 (2020). https://doi.org/10.1186/s12920-020-0677-2
11. Ramirez, R., et al.: Classification of cancer types using graph convolutional neural networks. Front. Phys. **8** (2020). https://doi.org/10.3389/fphy.2020.00203
12. Ramírez-Mena, A., et al.: Explainable artificial intelligence to predict and identify prostate cancer tissue by gene expression. Comput. Methods Programs Biomed. **240**, 107719 (2023). https://doi.org/10.1016/j.cmpb.2023.107719
13. Rhee, S., Seo, S., Kim, S.: Hybrid approach of relation network and localized graph convolutional filtering for breast cancer subtype classification. In: Proceedings of the 27th International Joint Conference on Artificial Intelligence, pp. 3527–3534. IJCAI'18, AAAI Press (2018)
14. Ribeiro, M.T., Singh, S., Guestrin, C.: "Why should i trust you?": explaining the predictions of any classifier. In: Proceedings of the 22nd ACM SIGKDD International Conference on Knowledge Discovery and Data Mining, pp. 1135–1144. KDD '16, Association for Computing Machinery, New York, NY, USA (2016). https://doi.org/10.1145/2939672.2939778
15. Shao, X., et al.: scDeepSort: a pre-trained cell-type annotation method for single-cell transcriptomics using deep learning with a weighted graph neural network. Nucleic Acids Res. **49**(21), e122–e122 (2021). https://doi.org/10.1093/nar/gkab775
16. Vanhaeren, T., et al.: Application of XAI to the prediction of CTCF binding sites. Results Eng. **25**, 103776 (2025). https://doi.org/10.1016/j.rineng.2024.103776
17. Wu, Z., Xia, F., Lin, R.: Global burden of cancer and associated risk factors in 204 countries and territories, 1980–2021: a systematic analysis for the GBD 2021. J. Hematol. Oncol. **17**, 119 (2024). https://doi.org/10.1186/s13045-024-01640-8
18. Xena, U.: UCSC Xena Datasets. https://xenabrowser.net/datapages/
19. Xu, L., Li, Z., Ren, J., Liu, S., Xu, Y.: Single-cell RNA sequencing data analysis utilizing multi-type graph neural networks. Comput. Biol. Med. **179**, 108921 (2024). https://doi.org/10.1016/j.compbiomed.2024.108921
20. Ying, R., Bourgeois, D., You, J., Zitnik, M., Leskovec, J.: GNNExplainer: generating explanations for graph neural networks. Adv. Neural. Inf. Process. Syst. **32**, 9240–9251 (2019)
21. Zhang, Y., Weng, Y., Lund, J.: Applications of explainable artificial intelligence in diagnosis and surgery. Diagnostics **12**(2) (2022). https://doi.org/10.3390/diagnostics12020237

Stress Detection with a Lightweight and Explainable Model for IoMT Devices

José L. López Ruiz[1(✉)], Carlos Montoya Peña[1], David Díaz Jiménez[1], Juan F. Gaitán Guerrero[1], Pedro A. Palomino Moral[2], and Macarena Espinilla Estévez[1]

[1] Department of Computer Science, University of Jaén, Jaén, Spain
llopez@ujaen.es
[2] Department of Nursing, University of Jaén, Jaén, Spain

Abstract. Stress is an increasingly serious public health problem, as prolonged exposure to stress is associated with cognitive, cardiovascular, and emotional disorders. However, despite its impact, stress frequently goes undetected in everyday life, emphasising the necessity for objective and continuous monitoring systems. The present paper proposes a lightweight and interpretable stress detection model based solely on physiological signals. The proposed solution is based on a deep Multilayer Perceptron architecture optimized for execution on resource-constrained devices, typically used in medical Internet of Things environments. The model was trained using the SWELL-KW dataset, and it has been demonstrated to achieve high performance in the classification of stress. In order to enhance explainability, SHAP values were calculated to identify the most relevant input features. This process enabled the development of a simplified version of the model using only 9 features. It has been demonstrated that both the full and reduced models achieve excellent results and demonstrate the feasibility of implementing accurate, explainable deep learning solutions in real-time stress monitoring systems for clinical and wearable environments.

Keywords: Health monitoring · Internet of Medical Things · Stress detection · Explainable AI

1 Introduction

According to the World Health Organization (WHO), stress [24] can be defined as a state of mental tension or worry caused by a difficult situation. We experience stress every day, as it is an innate reaction of the body to threatening or demanding situations.

This result has been partially supported by grant PID2021-127275OB-I00 funded by MICIU/AEI/10.13039/501100011033 and by "ERDF A way of making Europe" and grant PDC2023-145863-I00 funded by MICIU/AEI/10.13039/501100011033 and by "European Union NextGenerationEU/PRTR".

E. Corchado et al. (Eds.): SOCO 2025, CCIS 2806, pp. 237–246, 2026.
https://doi.org/10.1007/978-3-032-19763-4_22

Low levels of stress can be useful in helping us cope with everyday activities. However, when stress levels are excessive and sustained over time, they can trigger adverse physical and psychological effects [25]. People who are continuously exposed to stressful environments are at greater risk of central nervous system disorders, which affect the brain, memory, and cognitive functions. Similarly, negative effects on the immune, cardiovascular, and endocrine systems can also be observed.

Workplace environments significantly influence stress levels, which can vary depending on the nature of the job, the context in which professional activities take place, and the socioeconomic status of workers [21]. Several studies indicate that workers in the social and healthcare sector [2] are among the most exposed, especially those working in emergency services, intensive care units, or emergency vehicles. Similarly, caregivers of patients with cognitive impairment, including family members, are often under constant emotional strain, placing them in a situation of chronic stress that affects their physical and mental health.

Although stress can be mitigated through individual and organizational strategies, both in daily life and in the workplace, we are often not fully conscious of the specific levels of stress to which we are exposed, which makes it difficult to manage effectively and can delay the adoption of preventive or corrective measures.

Thanks to advances in artificial intelligence (AI) and the progressive reduction in the cost of microchips, the integration of AI and the Internet of Things (IoT) in the social and healthcare sector [15] represents a technological revolution with profound implications for diagnosis, treatment, and personalized care [4]. In this context, IoT has evolved into the Internet of Medical Things (IoMT), encompassing devices and systems exclusively for healthcare environments that combine large-scale data processing with intelligent automation, supporting healthcare professionals and enhancing evidence-based clinical decision-making [3]. AI plays a central role in precision medicine by enabling the analysis of clinical, genetic, and lifestyle data to tailor individualized treatments [9], while IoMT-connected devices [11,21] facilitate continuous remote patient monitoring, allowing earlier interventions and reduced healthcare costs, particularly for chronic conditions and long-term care.

These technologies take on strategic value in the context of stress. Given that individuals are often not fully aware of their stress levels, an intelligent monitoring system based on IoMT can be essential for detecting physiological patterns associated with this problem, facilitating objective measurement in real time and enabling active alerts to be sent to the person being monitored or clinical information to be provided to healthcare personnel, allowing them to assess the person's state of health and provide individualized treatment.

This paper presents a prototype designed to classify physiological signal samples and determine whether a person is in a state of stress. The proposed solution is based on a deep Multilayer Perceptron (MLP) model, optimized for execution on devices with limited resources, such as those used in IoMT-based environments. In addition, the model incorporates explainability mechanisms that allow its predictions to be interpreted and facilitate its validation in clinical contexts.

The novelties of the work are as follows: the design of a lightweight and efficient model based on deep MLP, specifically adapted to operate in IoMT environments with limited computational resources; the exclusive use of physiological signals, deliberately discarding multimodal data sources to enhance real applicability in wearable devices; and the application of the SHAP method as a model explainability technique, enabling analysis of the individual impact of each variable on the prediction and reinforcing system transparency.

This article is organized as follows. Section 2 reviews related work on stress monitoring using IoMT-based intelligent systems. Section 3 describes the dataset used, the objectives of the proposal, and the preprocessing and architecture of the developed model. Next, Sect. 4 presents the details of the experimentation and the analysis of the results obtained. Finally, Sect. 5 presents the conclusions of the work and proposes possible lines of future work.

2 Related Works

Elevated levels of stress have become a widespread public health problem, with significant effects on both mental and physical health. Even if a person appears to be physically healthy, undetected internal stress can trigger chronic disorders and seriously affect emotional stability [25]. This problem has driven AI and IoMT-based solutions that enable continuous monitoring of physiological parameters to objectively assess stress levels.

Recent advances in IoMT have revolutionized health monitoring, especially through the use of wearable devices that enable remote, real-time recording of critical physiological variables [17]. These sensor-enabled devices, integrated with IoMT capabilities, enable the collection of critical data that can be analyzed to identify signs of stress in the body [1,22].

Currently, one of the main challenges in IoMT-based stress monitoring systems is to develop solutions that adapt to end devices with limited resources in terms of processing power, memory, and energy consumption [8]. Despite these technical limitations, the potential of this technology to provide affordable and accessible health monitoring, as well as to promote well-being and disease prevention, makes it a field with great potential for future innovation [1].

In order to assess a person's stress level, IoMT-based systems focus on detecting variations in physiological signals, which tend to change during episodes of stress. These signals provide objective information that, after being analyzed by algorithms, can be used to determine the person's stress level. Among the most commonly used variables are galvanic skin response (GSR), heart rate (HR), skin temperature, and breathing patterns.

GSR is one of the most widely used indicators in stress measurement, as it directly reflects the activity of the nervous system and the degree of emotional arousal of the person being monitored [7,16,19]. HR and HR variability, obtained using photoplethysmography (PPG) or electrocardiography (ECG) sensors, are also commonly incorporated, as they provide a clear insight into how the cardiovascular system responds to stressful situations.

On the other hand, variations in skin temperature and breathing patterns, registered by sensors located in the chest or abdominal area, or even blood oxygen saturation (SpO2) together with the basic physiological signals mentioned above, create a more complete framework for stress monitoring [13,23].

However, some of the physiological variables mentioned above require sensors that can be intrusive or uncomfortable in everyday use. Therefore, despite the robustness of many existing models, it is essential to move towards solutions that are discreet, accessible, and applicable in a variety of contexts, allowing continuous stress monitoring without interfering with people's daily activities.

Regarding the methodologies used for stress detection in recent years, machine learning algorithms have been fundamental to the development of these systems [18]. With the emergence of deep learning-based techniques [6], more approaches have been designed. Highly used datasets such as WASED [20] and SWELL-KW [10] have been used for training and testing these types of methodologies.

However, the solution to this problem is still in constant development, and some researchers propose what an ideal IoMT-based system for stress or epileptic attack detection would look like [5].

3 Materials and Methods

In this section, the methodological approach followed in the development of the proposed system for stress detection based on physiological signals is detailed. First, the dataset used is described, including its main characteristics. Subsequently, the objective of the proposed model is specified, followed by the preprocessing procedure applied to the dataset. Finally, the architecture of the proposed MLP model is presented, as well as the mechanisms used for its optimization and explainability.

3.1 Purpose

Our main proposal is to design a lightweight prototype that can classify physiological samples and determine whether a person has experienced moderate or high levels of stress over a certain period of time. In addition, we want the model to provide explainable results, an aspect that is increasingly valued in the specialized literature. This feature allows healthcare professionals, in this case, to trust the model and validate its performance.

Although there are studies that distinguish multiple levels of stress, this study employs a binary classification: stressed or not stressed, considering that both medium and high stress, if prolonged, represent a significant risk to a person's health.

3.2 Material

The SWELL-KW dataset [10] was used to test the effectiveness of the proposal. This dataset were collected to facilitate research on stress and is based on an experiment in which 25 participants performed office tasks under normal conditions and also in stressful situations for three hours. Factors such as interruptions via email and time pressure were introduced to induce stress.

The data obtained is diverse and sourced from multimodal sources, including:

- User interactions with the computer: mouse clicks, scrolling, switching between applications, and typing patterns.
- Video recordings of the face and upper body, used to analyze facial activation, head orientation, global facial features, and emotional estimates.
- Body postures recorded using a Kinect depth camera.
- Physiological signals such as HR and its variability, obtained using ECG and skin conductance sensors.

Although all of these data can provide valuable information, physiological signals are particularly useful in real-world settings with IoMT-based systems. Therefore, we have chosen to use a reduced and preprocessed version of the dataset, developed by Nkurikiyeyezu et al. [14]. This version contains 29 features aggregated by one-minute intervals. More details on these variables can be found in [14].

The dataset has a total of 410,322 samples, of which 222,240 correspond to non-stress situations and 188,082 to induced stress conditions.

3.3 Proposal

The proposal is detailed below. The procedure is described in detail, starting with preprocessing, the model architecture, and the subsequent feature filtering and architecture remodeling process.

Preprocessing. First, the initial stage of preprocessing the input data is described. As a reminder, the dataset has already been processed, generating new features derived from two fundamental features: HR and the interval between two peaks of the ECG signal (RR). In addition, these features are aggregated into one-minute periods.

On the other hand, as all are numerical variables, the features are normalized using the Min-Max scaling defined by the following formula.

$$x_{scaled} = \frac{x - x_{\min}}{x_{\max} - x_{\min}}, \quad x_{scaled} \in [0, 1] \subset \mathbb{R}$$

This normalization process is applied once the dataset has been divided into three subsets: training, validation, and testing. Thus, scaling is performed using only the values in the training set, and then the same transformer is applied to the validation and testing sets, ensuring consistency.

The original dataset includes three different classes: non-stressed, stressed by interruption, and stressed by time pressure. In this proposal, both conditions associated with stress are grouped under a single class, which helps simplify the classification problem and improves the balance between classes.

Architecture. The proposed architecture is based on a deep MLP with a sequential structure, designed to operate in a single processing flow. This choice responds to the need to implement the model on devices with limited computational resources, such as single-board computers or microcontrollers, which are common in IoMT environments. Unlike more complex architectures such as convolutional networks, this configuration significantly reduces computation time, facilitating real-time execution on these types of devices.

The model consists of three consecutive dense blocks, with a decreasing number of neurons per layer, which allows complex patterns to be captured first and the representation to be progressively refined. Each block includes a PReLU activation layer, followed by batch normalization and a regularization layer using *dropout*, whose value is higher in the first block in order to mitigate overfitting in the early stages of learning.

Figure 1 shows the complete architecture.

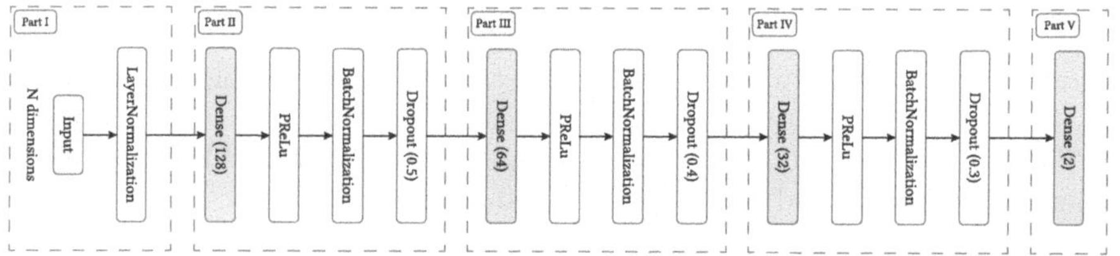

Fig. 1. Architecture of the proposed deep MLP model.

Explainability and Simplification. Finally, in order to provide the model with explicability, the SHapley Additive exPlanations (SHAP) method has been incorporated, an explainability technique based on cooperative game theory. This method allows for the accurate and consistent interpretation of the contribution of each individual feature to the model output by calculating Shapley values that quantify their impact on each prediction. In recent years, SHAP has proven to be particularly effective in complex models such as deep neural networks, as it allows the model's decision to be broken down into components that are understandable to experts, facilitating its validation by professionals and promoting trust in AI-based systems [12].

In this work, SHAP has been used not only for interpretative purposes, but also as an analysis tool to identify the physiological variables with the highest influence on predictions. This information has allowed us to generate a simplified version of the model, optimized to run in environments with limited computational resources, retaining only the most relevant features.

4 Experiments and Results

This section details the experimentation process and the results obtained from the generated models.

4.1 Experimentation

As indicated in the previous section, a reduced version of the SWELL-KW dataset [14] was used for the experimentation phase, focusing exclusively on the physiological signals recorded in the participants. After eliminating the descriptive variables that did not provide useful information for training the model, 27 of the 29 available features were used.

The dataset was randomly divided into three subsets: training (70%), validation (15%), and testing (15%), thus ensuring a balanced distribution of classes in each partition. Next, the model described in Sect. 3.3 was trained. The training process was limited to a maximum of 300 epochs, also applying an early stopping criterion with a patience of ten epochs on the accuracy value in the validation set, in order to avoid overfitting and favor the model's generalization performance.

Once the model was generated, the SHAP values were calculated to estimate the relative importance of each feature. Based on these results, the nine variables with the greatest impact on the predictions were selected, and a reduced subset of the dataset was constructed. The model was then re-trained using only these relevant features in order to validate a simplified version of the model, reducing both the complexity of the architecture and the dimensionality of the input data.

Table 1 below shows all the results obtained, and Fig. 2 shows the most relevant features when applying SHAP.

Table 1. Performance of the model using the full and optimized feature sets. Abbreviations: Ep. (Epochs), Train. (Training), Val. (Validation), Acc. (Accuracy), Prec. (Precision), Rec. (Recall) and F1 (F1-score).

Model configuration	Ep.	Train.	Val.	Test			
		Acc.	Acc.	Acc.	Prec.	Recall	F1
Full dataset (27 features)	58	0.9792	**0.9991**	**0.9995**	**0.9995**	**0.9995**	**0.9995**
Optimized dataset (9 features)	101	0.9382	0.9886	0.9887	0.9886	0.9886	0.9886

4.2 Results

The results obtained during the experimental phase demonstrate the effectiveness of the proposed model in both its complete and optimized versions. First, the model trained with the 27 features from the SWELL-KW dataset achieved

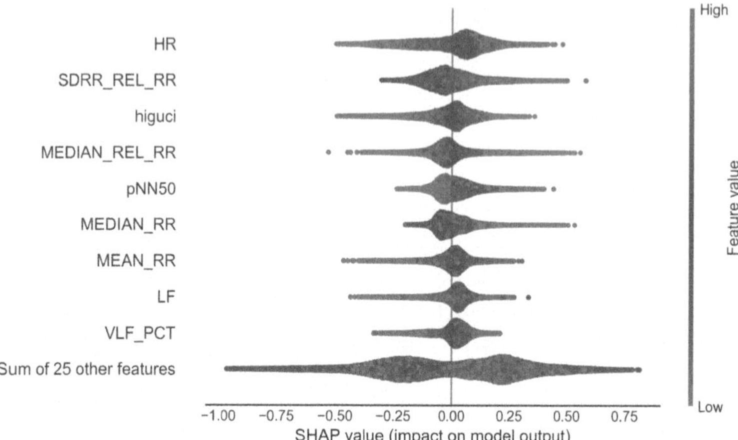

Fig. 2. Feature importance values based on SHAP analysis. The top ten variables with the greatest impact on model predictions are shown.

perfect accuracy on the test set and very high values on the training and validation subsets, indicating adequate generalization ability with no significant evidence of overfitting.

On the other hand, the simplified model, based solely on the nine features selected through SHAP analysis, also performed well, with an accuracy of 0.99 on the validation and test sets and 0.94 on the training set. The slight reduction in training set accuracy, compared to validation or test accuracy, may reflect the effect of regularization and early stopping, which contribute to improved generalization and help prevent overfitting.

In addition, all complementary evaluation metrics, such as accuracy, recall, and F1-score, achieved values of 1.0 or 0.99, reinforcing the robustness of the proposed approach. This behavior suggests that the reduced subset preserves most of the discriminative information necessary to detect stress states, which is especially valuable in contexts where computational resources are limited, such as in IoMT environments.

5 Conclusions and Future Work

The experimental results show that the proposed model, both in its full and optimized versions, performs excellently, with precision, completeness, and F1-score values above 99%. These findings demonstrate that it is possible to design deep learning models that maintain high performance even under resource constraints, which is key for their integration into IoMT devices aimed at continuous health status monitoring.

On the other hand, variable selection using SHAP has not only resulted in a lighter and more efficient architecture, but has also contributed to the interpretability of the model, providing healthcare professionals with a reliable, transparent, and viable tool for informed decision-making.

Future lines of research include extending the evaluation of the model in real-world settings using data collected from wearable sensors in real time, which would allow its performance to be validated under uncontrolled conditions. The possibility of incorporating incremental or federated learning techniques is also being considered, which would allow the model to be adapted to the end user without compromising privacy or requiring intensive centralized training.

References

1. Al-Atawi, A.A., et al.: Stress monitoring using machine learning, IOT and wearable sensors. Sensors **23**(21), 8875 (2023)
2. Alawage, H.M.A., Zaidi, U.: Traumatic stress, psychological well-being, and sociodemographic correlates in high-stress environments among healthcare professionals. Open Psychology J. **17**(1) (2024)
3. Bajwa, J., Munir, U., Nori, A., Williams, B.: Artificial intelligence in healthcare: transforming the practice of medicine. Future Healthc. J. **8**(2), e188–e194 (2021)
4. Belbase, P., Bhusal, R., Ghimire, S.S., Sharma, S., Banskota, B.: Assuring assistance to healthcare and medicine: internet of things, artificial intelligence, and artificial intelligence of things. Front. Artif. Intell. **7** (2024)
5. Bhatt, M.W., Sharma, S.: An IOMT-based approach for real-time monitoring using wearable neuro-sensors. J. Healthc. Eng. **2023**(1) (Jan 2023)
6. Gupta, A., Kansal, D., Gupta, V., Shetty, M.K., Girish, M.P., Gupta, M.D.: X-ecgnet: an interpretable dl model for stress detection using ECG in covid-19 healthcare workers. In: 2021 4th International Conference on Bio-Engineering for Smart Technologies (BioSMART). IEEE (2021)
7. Hadhri, S., Hadiji, M., Labidi, W.: An intelligent stress detection and monitoring system using the IOT environment. In: 2022 International Conference on Technology Innovations for Healthcare (ICTIH), pp. 20–25. IEEE (2022)
8. Jiang, S., Firouzi, F., Chakrabarty, K., Elbogen, E.B.: A resilient and hierarchical IOT-based solution for stress monitoring in everyday settings. IEEE Internet Things J. **9**(12), 10224–10243 (2022)
9. Johnson, K.B., et al.: Precision medicine, ai, and the future of personalized health care. Clin. Transl. Sci. **14**(1), 86–93 (2020)
10. Koldijk, S., Sappelli, M., Verberne, S., Neerincx, M.A., Kraaij, W.: The swell knowledge work dataset for stress and user modeling research. In: Proceedings of the 16th International Conference on Multimodal Interaction. ICMI '14, ACM (2014)
11. Lu, Z.x., et al.: Application of ai and IOT in clinical medicine: summary and challenges. Current Med. Sci. **41**(6), 1134–1150 (2021)
12. Lundberg, S.M., Lee, S.I.: A unified approach to interpreting model predictions. In: Proceedings of the 31st International Conference on Neural Information Processing Systems, pp. 4768–4777. NIPS'17, Curran Associates Inc., Red Hook, NY, USA (2017)
13. Mozafari, M., Firouzi, F., Farahani, B.: Towards IOT-enabled multimodal mental stress monitoring. In: 2020 International Conference on Omni-layer Intelligent Systems (COINS), pp. 1–8. IEEE (2020)

14. Nkurikiyeyezu, K., Shoji, K., Yokokubo, A., Lopez, G.: Thermal comfort and stress recognition in office environment. In: Proceedings of the 12th International Joint Conference on Biomedical Engineering Systems and Technologies. SCITEPRESS - Science and Technology Publications (2019)
15. Pise, A.A., et al.: Enabling artificial intelligence of things (AIOT) healthcare architectures and listing security issues. Comput. Intell. Neurosci. **2022**, 1–14 (2022)
16. Raju, A.R., Ramadevi, R., Babu, P.R., D.V.: Galvanic skin response based stress detection system using machine learning and IOT. In: 2023 Second International Conference on Augmented Intelligence and Sustainable Systems (ICAISS), pp. 709–714. IEEE (2023)
17. Rashid, N., Mortlock, T., Faruque, M.A.A.: Stress detection using context-aware sensor fusion from wearable devices. IEEE Internet Things J. **10**(16), 14114–14127 (2023)
18. Rawat, S.K.: Stress detection using machine learning and deep learning. Int. J. Res. Appl. Sci. Eng. Technol. **12**(12), 1399–1405 (2024)
19. Scherz, W.D., Baun, J., Seepold, R., Madrid, N.M., Ortega, J.A.: A portable ECG for recording and flexible development of algorithms and stress detection. Proc. Comput. Sci. **176**, 2886–2893 (2020)
20. Schmidt, P., Reiss, A., Duerichen, R., Marberger, C., Van Laerhoven, K.: Introducing wesad, a multimodal dataset for wearable stress and affect detection. In: Proceedings of the 20th ACM International Conference on Multimodal Interaction, pp. 400–408. ICMI '18, ACM (2018)
21. Smith, M.D., Wesselbaum, D.: Global evidence on the prevalence of and risk factors associated with stress. J. Affect. Disord. **374**, 179–183 (2025)
22. Sun, W., et al.: A review of recent advances in vital signals monitoring of sports and health via flexible wearable sensors. Sensors **22**(20), 7784 (2022)
23. Tooolkar, S., Kampali, A., Bhagwat, S., Katre, K., Tahalyani, J.: An IOT-based stress level detection system for emergency medical attention personnel. In: 2024 International Conference on Intelligent Systems and Advanced Applications (ICISAA), pp. 1–4. IEEE (Oct 2024)
24. World health organization: stress (2023). https://www.who.int/en/news-room/questions-and-answers/item/stress. Accessed 30 Oct 2025
25. Yaribeygi, H., Panahi, Y., Sahraei, H., Johnston, T.P., Sahebkar, A.: The impact of stress on body function: a review. EXCLI Journal (2017)

From Black Box to Understanding Deep Learning Hyperparameters: An Explainable Analysis

M. Olías, A.M. Chacón-Maldonado, J. F. Torres, G. Asencio-Cortés, and A. Troncoso(✉)

Universidad Pablo de Olavide, Seville, Spain
{molilop,amchamal,jftormal,guaasecor,atrolor}@upo.es

Abstract. Deep Learning models, in particular Feed-Forward Neural Networks, are highly sensitive to the configuration of their hyperparameters. However, identifying the individual contribution of each hyperparameter remains a complex and often opaque process. This study presents an explainability-driven approach to evaluate hyperparameter relevance using SHapley Additive exPlanations (SHAP). A synthetic dataset was constructed from a time series regression task in the energy domain, encompassing a wide range of hyperparameter configurations. For each configuration, a Feed-Forward Neural Network was trained and its performance evaluated on a validation set. The corresponding Mean Squared Error values were then used to train an XGBoost regressor. SHAP was subsequently employed to interpret the influence of each hyperparameter, both globally and locally. The analysis reveals that the choice of optimizer, with RMSProp having a particularly significant effect, along with the batch size, are the most impactful factors on model performance. In contrast, other parameters such as L1 regularization and specific activation functions showed minimal influence. The proposed methodology provides a transparent and interpretable framework for hyperparameter analysis, supporting more effective tuning strategies in Deep Learning models.

Keywords: XAI · deep learning · hyperparameter relevance

1 Introduction

Artificial Intelligence (AI), particularly techniques based on Deep Learning, has experienced remarkable progress in recent years, achieving promising results across diverse domains and practical applications, including image recognition, natural language processing, medical diagnostics, and predictive analytics. Feed Forward Neural Networks (FFNNs) represent one of the most widely adopted architectures, known for their flexibility, scalability, and effectiveness in a variety of prediction and classification tasks.

Nevertheless, the inherent complexity of these deep neural network models poses considerable challenges, especially regarding the optimal selection and

E. Corchado et al. (Eds.): SOCO 2025, CCIS 2806, pp. 247–256, 2026.
https://doi.org/10.1007/978-3-032-19763-4_23

tuning of hyperparameters. Some of them, such as learning rate, number of hidden layers, number of neurons per layer, activation functions and regularization parameters, substantially influence model performance. The traditional approach to hyperparameter optimization often involves heuristic methods [14], exhaustive grid [1] or random searches [21]. These methods typically demand substantial computational resources and frequently fail to provide clear explanations regarding how each individual hyperparameter contributes to the overall performance of the model.

In response to these challenges, Explainable Artificial Intelligence (XAI) has emerged as a pivotal research area. XAI aims to enhance transparency and interpretability in complex models that are usually treated as "black boxes" [10]. Employing XAI techniques facilitates deeper insights into the internal mechanisms and decision-making processes of deep neural networks, thereby enabling researchers to identify the factors most significantly influencing performance [20].

The present study applies XAI techniques specifically to identify the most influential hyperparameters within Feed Forward Neural Networks. For this purpose, a synthetic dataset has been constructed explicitly to evaluate various hyperparameter combinations and their effects on deep neural network performance metrics applied to a time series dataset. By analyzing this dataset through XAI methods, this research offers insights into the relative importance and interaction effects of hyperparameters. Such insights enable more informed and efficient hyperparameter selection processes, potentially reducing computational costs and improving the robustness and interpretability of deep neural network models.

The remainder of this paper is structured as follows: Sect. 2 reviews related literature, highlighting recent developments and current trends in hyperparameter optimization and explainability techniques in deep neural networks. Section 3 presents a detailed description of the methodology employed in this research. Section 4 provides a comprehensive analysis of experimental results along with their interpretation and implications. Finally, Sect. 5 concludes the study and suggests future research directions based on the obtained findings.

2 Related Work

Hyperparameter optimization (HPO) is critical for deep learning model performance. Traditional methods like random and grid search are often inefficient for complex search spaces, leading to the adoption of advanced techniques such as Bayesian optimization and evolutionary algorithms. For instance, genetic algorithms have improved energy consumption prediction for electric vehicles [11], and Ant Colony Optimization has enhanced hyperparameter tuning in lung cancer prediction models [4].

Further HPO advancements include using Large Language Models (LLMs) as assistants, demonstrating strong performance in tasks like object detection [8]. Automated methods such as AutoML and Neural Architecture Search (NAS) autonomously optimize both hyperparameters and architectures, reducing manual effort and computational costs [9]. Despite these advances, HPO via learning curve prediction faces challenges due to curve variability [3]. Hybrid metaheuristic and statistical approaches offer efficient HPO [5], while fine-tuning deep neural networks through advanced HPO has yielded superior results in domains like wind energy prediction [6]. Concurrently, Explainable AI (XAI) is essential for building trust and transparency in deep learning, especially in critical applications. Techniques like SHAP, Grad-CAM, and Occlusion Sensitivity are widely employed to interpret complex model decisions. For example, Grad-CAM and Occlusion Sensitivity have elucidated CNN model predictions in load disaggregation by visualizing key features [13], and Deep SHAP has provided global interpretations for multimodal network traffic classifiers [15]. In specific domains, XAI combined with transfer learning has improved the understanding and validation of deep neural networks in medical image analysis [2,7] and offered insights into factors driving accurate predictions in advanced glaucoma diagnosis [16]. In finance, XAI has clarified influential variables in stock price prediction models [17]. Furthermore, in industrial contexts, XAI has facilitated the identification and understanding of thermal errors in machinery, thereby enhancing the trustworthiness of intelligent systems [19].

3 Methodology

The methodology employed integrates dataset preparation, neural network training, hyperparameter optimization, and explainability analysis, as depicted in Fig. 1. Initially, an electricity consumption time series dataset is partitioned into training and test subsets. The training set is used for training multiple FFNN's, each with distinct hyperparameter configurations. These models are subsequently evaluated using the test set by calculating the error for each configuration.

The collected results, comprising hyperparameter configurations and corresponding performance metrics, form a structured dataset analyzed through XAI techniques. Further methodological details regarding dataset specifics, hyperparameter definition, dataset generation and XAI methods will be elaborated upon in the following subsections.

3.1 Benchmark Dataset

The dataset used in this study is a high-frequency (10-minute intervals) electricity consumption time series from Spain, spanning from January 2007 to June 2016, totaling 497832 samples. Following a preprocessing methodology previously proposed by other authors [22], the series is structured into a supervised dataset composed of input sequences of past values used to predict future values. The dataset was divided into 70% as training set and the remaining as test set.

3.2 Hyperparameter Search Space Definition

The definition of the hyperparameter search space is critical for an effective exploration and optimization of FFNN models. The considered hyperparameters

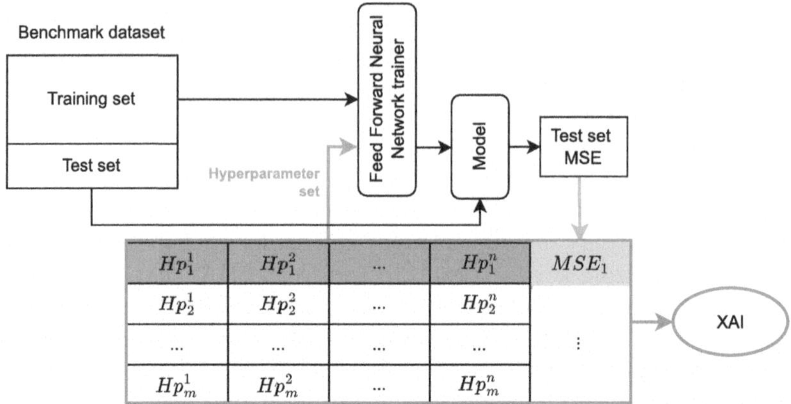

Fig. 1. Flow chart of the proposed methodology.

include both numerical ranges and categorical choices. Specifically, 2000 random hyperparameter combinations were generated for this study. The explored hyperparameters, along with their respective ranges or categorical options, are detailed in Table 1. This comprehensive definition allows for thorough exploration and optimal selection of hyperparameters to enhance the predictive performance of the neural network models.

Table 1. Hyperparameter search space.

Hyperparameter	Data type	Range/Values
Number of layers	Integer	[1, 9]. Upper bound may vary (Poisson)
Neurons per layer	Integer	[5, 75], step 5. Upper bound may vary (Poisson)
Batch size	Integer	$[2^1, 2^{15}]$. Upper bound may vary (Gaussian)
Epochs	Integer	[10, 500], step 10
Learning rate	Real	$[1\times10^{-5}, 1\times10^{-1}]$
Momentum	Real	[0.5, 0.99]
Dropout	Real	[0.2, 0.5]
L1	Real	[0, 0.1]
L2	Real	$[1\times10^{-5}, 1\times10^{-3}]$
Activation	Categorical	Sigmoid, Tanh, Relu, Softmax
Optimizer	Categorical	SDG, RMSProp, Adagrad, Adam, Adadelta, Adamw

3.3 Dataset Generation

For each of the 2000 randomly generated hyperparameter combinations depicted in Sect. 1, a FFNN was independently trained. Each trained network was evaluated using the Mean Squared Error (MSE) metric on the test set. This MSE value was then appended as a new attribute to the original hyperparameter dataset, resulting in a comprehensive dataset ready for subsequent XAI analysis.

To handle categorical variables such as the activation function and optimizer, one-hot encoding is applied. This technique transforms each category into a new binary variable, avoiding the imposition of arbitrary order among categories and allowing the model to process them effectively. Continuous hyperparameters are used directly, without additional normalization, since tree-based models can naturally handle differences in scale.

3.4 Explainability

For the explainability phase, the post-hoc SHAP (SHapley Additive exPlanations) technique [12] is used, based on Shapley values [18], which decompose the model's prediction by attributing to each feature its average marginal contribution, considering all possible combinations of features. This technique allows not only the identification of the most relevant hyperparameters, but also the estimation of the extent to which they influence the outcome. SHAP values are computed for each training configuration to estimate the marginal effect of each hyperparameter on the MSE, both globally and locally.

To this end, the XGBoost algorithm is employed as the base model for the analysis due to its strong predictive performance, computational efficiency, and compatibility with the TreeExplainer, which enables fast and accurate estimation of SHAP values. Compared to other algorithms, XGBoost is particularly well-suited for this type of tabular data and supports a structured interpretation of the influence of each hyperparameter on model performance.

4 Experimentation and Results

This section presents the experimental results obtained when SHAP is applied to the dataset to obtain the most relevant hyperparameters. Section 4.1 displays the configuration of the XGBoost algorithm. Section 4.2 shows the explainability of the model through different graphs. Finally, Sect. 4.3 discusses the results obtained.

4.1 Experimental Settings

This section reports the parameter values that have been used for the XGBoost algorithm. The XGBoost algorithm is trained with 100 trees and a fixed seed of 0 to ensure result reproducibility are used. The learning rate has been set to 0.1, a value that balances convergence speed and generalization capacity.

The maximum tree depth has been fixed at 5, as greater depths increase model complexity and may lead to overfitting. Similarly, all features have been used in each tree, avoiding restrictions in attribute selection during training. No penalties have been imposed on tree expansion. Tree growth has also been regulated by establishing a minimum requirement on the sum of instance weights at each leaf node. Furthermore, L2 regularization has been applied without additional constraints, ensuring a balanced penalization of the coefficients. The algorithm has been configured to minimize MSE.

4.2 Explainability Results

The SHAP XAI technique was used to enhance the model interpretability, focusing on identifying key hyperparameters influencing Mean Squared Error (MSE).

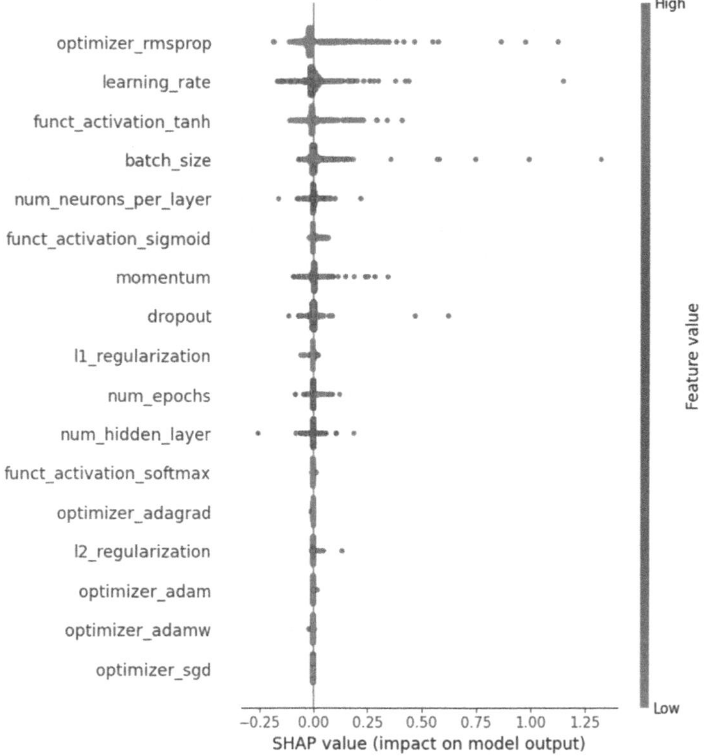

Fig. 2. SHAP summary plot obtained from XGBoost model.

Figure 2 (SHAP summary plot) illustrates global feature influence on MSE. The optimizer, particularly optimizer_rmsprop, and learning_rate were identified as the most impactful hyperparameters. Low values of these generally reduce

MSE, improving performance. However, optimizer_rmsprop also shows configurations where its presence increases MSE, indicating a potentially detrimental effect.

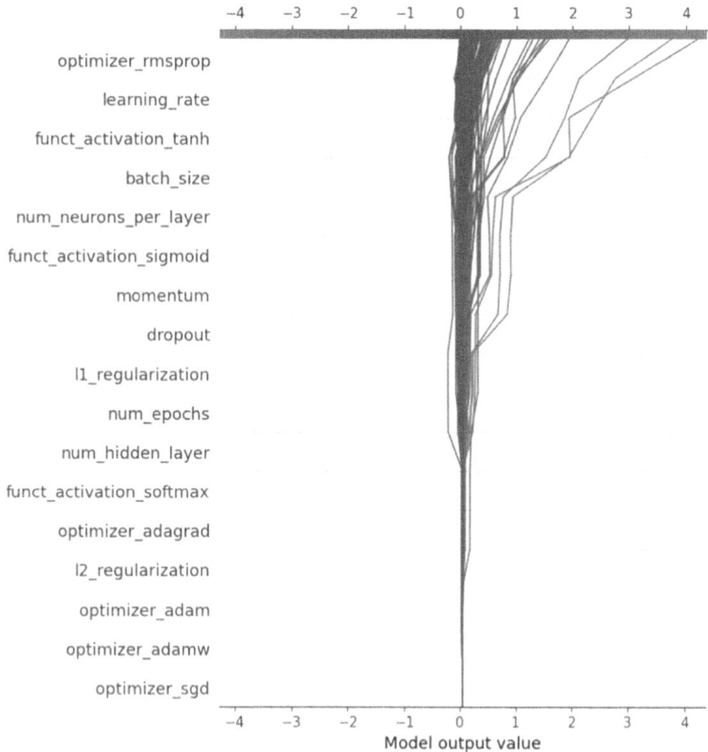

Fig. 3. SHAP decision plot obtained from XGBoost model.

Furthermore, Fig. 3 (SHAP decision plot) displays the cumulative impact of features on individual predictions, showing how each hyperparameter contributes to the final MSE value from a base value. The analysis indicates optimizer_rmsprop as the most influential hyperparameter, often increasing MSE. High learning_rate values substantially increase MSE, whereas low values reduce it. The funct_activation_tanh activation also tends to increase MSE. Large batch_size values show a modest error increase, with smaller batches having a lesser, error-reducing effect. In contrast, optimizer_sgd, optimizer_adamw, funct_activation_adam, and L2 regularization exhibited marginal impact on MSE variability within the evaluated range.

Figures 4aand 4bpresent SHAP waterfall plots for specific instances with the highest and lowest MSE, respectively. For the highest MSE instance (Fig. 4a), batch_size was the primary contributor, followed by learning_rate and optimizer_rmsprop. For the lowest MSE instance (Fig. 4b), a favorable

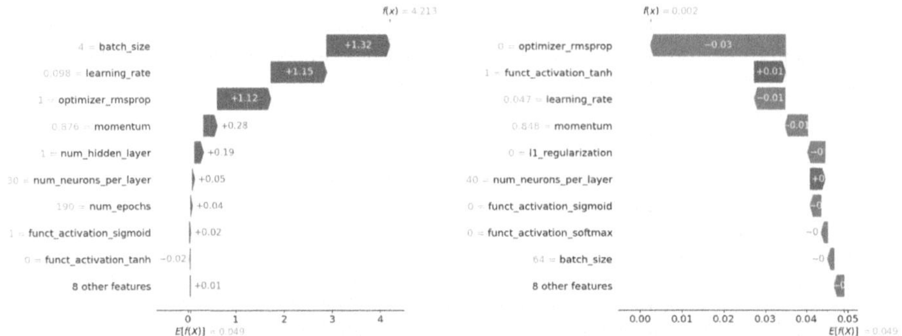

(a) Instance with the highest metric. (b) Instance with the lowest metric.

Fig. 4. SHAP waterfall charts obtained from the XGBoost model.

optimizer configuration, learning_rate, and momentum were key. Analysis of these instances suggests optimizer_rmsprop's effect is context-dependent, varying with hyperparameter interactions. batch_size was also identified as a critical factor whose impact varied significantly with context. learning_rate and MOMENTUM sensitivity was noted, where moderate values yielded better MSE outcomes compared to higher values.

4.3 Discussion of Results

The SHAP analysis confirms that the choice of optimizer (especially RMSprop) is the main driver of model error, and it also identifies the batch size and learning rate as crucial parameters. At a high level, configurations using RMSprop often raise the predicted MSE, while the opposite is true when using other optimizers.

Besides, high learning rates tend to increase error and low rates tend to decrease it, small batch sizes are usually to blame for higher error while large batches compensate for it. At a local level, it can still observe that the instance with the lowest MSE obtains it mainly because it doesn't use RMSprop, has a low learning rate and moderate momentum, while the instance with the highest MSE manages to do so because it uses both a small batch size and RMSprop, as well as a large learning rate.

Other hyperparameters (e.g. activation functions except for tanh or softmax or regularization terms) have considerably less impact. In combination with these results, these findings also imply that for the predictive performance of deep feedforward neural networks, a careful selection of the optimization algorithm, the learning rate and the batch size, considering also their interactions, is crucial. Overall, this interpretability analysis highlights the critical role of certain optimization and architectural decisions and supports their prioritization in model selection and hyperparameter tuning strategies.

5 Conclusions

In conclusion, the performed SHAP analysis indicates that the optimizer, particularly RMSprop, is the most influential factor on predictive error in feed-forward neural networks, followed by learning rate and batch size. RMSprop tends to increase Mean Squared Error (MSE), while high learning rates and small batch sizes also amplify error; other optimizers generally reduce MSE. Other hyperparameters show marginal impact. Thus, the joint selection of optimizer, learning rate, and batch size is crucial for optimal performance in time series regression, given their decisive interaction. SHAP's explanations can guide more efficient hyperparameter searches, and future applications may further validate this approach and promote transparent AI, especially when extended to other neural architectures such as convolutional or recurrent networks to assess their impact on the results.

Disclosure of Interests. The authors have no competing interests to declare that are relevant to the content of this article.

References

1. Anwar, I.: Improved grid search algorithm for optimal LSTM performance: a case study on australian electricity price forecasting. In: 2024 IEEE Power & Energy Society General Meeting (PESGM), pp. 1–5 (2024)
2. Bas, H., Kuijf, H.J., Kenneth, G.G., Max, A.V.: Explainable artificial intelligence (xai) in deep learning-based medical image analysis. Med. Image Anal. **79**, 102470 (2022)
3. Choi, D., Cho, H., Rhee, W.: On the difficulty of DNN hyperparameter optimization using learning curve prediction. In: TENCON 2018 - 2018 IEEE Region 10 Conference, pp. 0651–0656 (2018)
4. Dilip, K., Ghantasala, G.S.P., Rao, D.N., Rathee, M., Bathla, P.: ACO-based hyperparameter tuning of a dl model for lung cancer prediction. In: 2024 IEEE International Conference on Computing, Power and Communication Technologies (IC2PCT). vol. 5, pp. 883–887 (2024)
5. Gutiérrez-Avilés, D., Jiménez-Navarro, M.J., Torres, J.F., Martínez-Álvarez, F.: Metagen: a framework for metaheuristic development and hyperparameter optimization in machine and deep learning. Neurocomputing **637**, 130046 (2025)
6. Hanifi, S., Cammarono, A., Zare-Behtash, H.: Advanced hyperparameter optimization of deep learning models for wind power prediction. Renewable Energy **221**, 119700 (2024)
7. Hao, J.: Deep learning-based medical image analysis with explainable transfer learning. In: 2023 International Conference on Computer Engineering and Distance Learning (CEDL), pp. 106–109 (2023)
8. Hu, Y., Wang, Q.: Hyperparameter optimization of object detection networks based on large language models. In: IECON 2024 - 50th Annual Conference of the IEEE Industrial Electronics Society, pp. 1–5 (2024)
9. Jagadeesan, D., S.B., Purushotham, B., Kumar, S.N., Asha, G.: Auto ml and neural architecture search for deep learning model optimization. In: 2023 International Conference on Innovative Computing, Intelligent Communication and Smart Electrical Systems (ICSES), pp. 1–6 (2023)

10. Jiménez-Navarro, M.J., Troncoso-García, A.R., Troncoso, A., Martínez-Álvarez, F., Martínez-Ballesteros, M.: Explainable deep learning with embedded feature selection for electricity demand forecasting. In: 2024 International Conference on Smart Systems and Technologies (SST), pp. 153–158 (2024). https://doi.org/10.1109/SST61991.2024.10755283
11. Jlifi, B., Ferjani, S., Duvallet, C.: A genetic algorithm based three hyperparameter optimization of deep long short term memory (ga3p-dlstm) for predicting electric vehicles energy consumption. Comput. Electr. Eng. **123**, 110185 (2025)
12. Lundberg, S.M., Lee, S.: A unified approach to interpreting model predictions. In: Proceedings of the International Conference on Neural Information Processing Systems. vol. 30, pp. 4765–4774 (2017)
13. Machlev, R., Malka, A., Perl, M., Levron, Y., Belikov, J.: Explaining the decisions of deep learning models for load disaggregation (NILM) based on XAI. In: 2022 IEEE Power & Energy Society General Meeting (PESGM), pp. 1–5 (2022)
14. Martínez-Álvarez, F., et al.: Coronavirus optimization algorithm: a bioinspired metaheuristic based on the covid-19 propagation model. Big Data **8**(4), 308–322 (2020)
15. Nascita, A., Montieri, A., Aceto, G., Ciuonzo, D., Persico, V., Pescapè, A.: Unveiling mimetic: interpreting deep learning traffic classifiers via xai techniques. In: 2021 IEEE International Conference on Cyber Security and Resilience (CSR), pp. 455–460 (2021)
16. Nimmy, S.F., Hussain, O.K., Chakrabortty, R.K., Saha, S.: Explainable artificial intelligence (XAI) in glaucoma assessment: advancing the frontiers of machine learning algorithms. Knowl.-Based Syst. **316**, 113333 (2025)
17. Sasikumar, M., Raj, E.D.: Investigating explainability of deep learning models for sequential data on stock price prediction. Procedia Comput. Sci. **258**, 4190–4201 (2025)
18. Shapley, L.S.: A value for n-person games. In: Kuhn, H.W., Tucker, A.W. (eds.) Contributions to the Theory of Games II, pp. 307–317. Princeton University Press (1953)
19. Shicun, A., Sitong, X., Jianguo, Y.: A hyperparameter optimization-assisted deep learning method towards thermal error modeling of spindles. ISA Trans. **156**, 434–445 (2025)
20. Song, H., Kim, S.: Explainable artificial intelligence (XAI): how to make image analysis deep learning models transparent. In: 2022 22nd International Conference on Control, Automation and Systems (ICCAS), pp. 1595–1598 (2022)
21. Torres, J.F., Gutiérrez-Avilés, D., Troncoso, A., Martínez-Álvarez, F.: Random hyper-parameter search-based deep neural network for power consumption forecasting. In: Advances in Computational Intelligence, pp. 259–269. Springer International Publishing, Cham (2019)
22. Torres, J., Galicia, A., Troncoso, A., Martínez-Álvarez, F.: A scalable approach based on deep learning for big data time series forecasting. Integr. Comput. Aided Eng. **25**(4), 335–348 (2018)

Impact of Expert and Auto-generated Labels on Deep Learning Models for Neuronal Image Segmentation

Gerard Villarroya-Piqué[2(✉)], Víctor M. González[2,4], Esther Serrano-Pertierra[3,5], Antonello Novelli[4,5,6], M. Teresa Fernández-Sánchez[3,4,5], and Angel Rio-Alvarez[1,4]

[1] Computer Sciences Department, University of Oviedo, Oviedo, Spain
rioangel@uniovi.es

[2] Electrical Engineering Department, University of Oviedo, Oviedo, Spain
{uo244753,vmsuarez}@uniovi.es

[3] University Institute of Biotechnology of Asturias (IUBA), University of Oviedo, Oviedo, Spain

[4] Biomedical Engineering Center (BME), University of Oviedo, Oviedo, Spain

[5] Biochemistry and Molecular Biology Department, University of Oviedo, Oviedo, Spain

[6] Psychology Department, University of Oviedo, Oviedo, Spain

Abstract. Assessing neuronal viability is essential in neurotoxicity research. Traditional methods, which rely on fluorescence staining and manual image analysis, are both cytotoxic and time-consuming. To address these limitations, this study investigates how the type of annotation used in training deep learning models influences their performance, with the goal of automating viability analysis.

We compare two YOLOv11-based models, trained with different annotation strategies: expert-generated labels and algorithmically derived masks. The latter leverages geometric filtering and morphological operations to generate reliable masks from fluorescence images, separating clustered neurons. Both models were fine-tuned using transfer learning and evaluated on complementary test sets to assess generalizability.

The model trained on algorithmically generated masks showed improved generalizability and greater stability, likely due to its exposure to abstract, shape-based features during training.

Despite being imperfect, algorithm generated annotations provide a scalable and less-biased alternative for training deep learning models in biomedical imaging. These findings highlight a promising direction toward developing fully automated and non-invasive tools for neuronal viability assessment.

This research has been partially funded by the Council of Gijón through the University Institute of Industrial Technology of Asturias (IUTA) grants SV-25-GIJON-1-14, SV-25-GIJON-1-02, SV-24-GIJON-1-05, SV-24-GIJON-1-18, SV-24-GIJON-1-16, SV-23-GIJON-1-09, SV-22-GIJON-1-19, and SV-21-GIJON-1-19, and by Principado de Asturias, grant SV-PA-21-AYUD/2021/50994.

E. Corchado et al. (Eds.): SOCO 2025, CCIS 2806, pp. 257–266, 2026.
https://doi.org/10.1007/978-3-032-19763-4_24

Keywords: Deep learning · Neuron segmentation · Fluorescence microscopy · Phase-contrast imaging · Non-invasive imaging · Automated annotation · Computer vision · Neurotoxicity · Neurodegenerative disease research

1 Introduction

Viability assays play a critical role in assessing the neurotoxicity of various compounds. Traditionally, researchers manually analyze images of neuronal cultures to determine cell viability. This process is time-consuming and prone to human error. Researchers use phase-contrast microscopy as the technique to generate the images for the analysis [1,3]. This technique provides a non-destructive method to visualize the culture. The analysis is particularly challenging due to the high variability and complexity of the images. Images of neuron cultures have several specific features that pose challenges. The images show not only neurons but also glial cells. These cells have the task of maintaining a healthy environment for the neurons to thrive. Neurons form connections through elongated extensions—neurites—that enable intercellular signaling. These neurites also add features to the images that make identifying neurons more challenging. Besides, neurons may appear grouped into clusters. Clustered neuron instances are difficult to separate since they lose their clear edges.

To simplify the task of neuron viability assessment, researchers commonly use a classic staining technique [5] in which only live neurons are highlighted under the microscope by fluorescence. While this enhances image clarity by eliminating interference from dead cells and debris [13], the staining process itself is cytotoxic, rendering the culture unusable for further experiments.

We compare two different label-sources aimed to train a Deep Learning (DL) model based on the YOLOv11 architecture. In the first, experienced researchers manually label phase-contrast images, representing the current gold standard. In the second, masks are automatically generated using fluorescence images and an algorithm designed to separate clustered neurons more effectively. Despite the limitations of fluorescence imaging, their structural clarity enables more reliable mask creation.

2 Related Work

The complexity of the features of the images have motivated the use of computer vision techniques. For instance, various tools for images with cleaner features have been developed, and are freely available [2,9]. These tools don't suit the needs of the analysis due to the fact that the images that we analyze don't show the same degree of cleanliness. Research has shown that machine learning techniques are suitable to analyze complex systems [14]. Promising alternatives are those based on DL techniques [11,12]. Once the models are trained, they allow for the automated analysis of large complex datasets.

To date, the main supervised strategies in use consist of segmentation and object detection methods. Unsupervised training excels in grouping pixels that share common characteristics, but don't generate an efficient separation of individual instances. On the other hand, supervised models must be trained with annotated images. The specifics of the data sets motivate the leverage of transfer learning to fine-tune pretrained models in computer vision [6].

This task demands simultaneous object identification and segmentation. Models that focus only on identifying neurons or segmenting their shape don't include metrics for both in the training phase. This research requires not only the identification of the areas with neurons, but also separating each instance. Our recent work [7] has shown that a model whose architecture includes both metrics in its training performs at a higher level. This architecture is realized in the YOLOv11 DL model [4], pretrained to perform computer vision tasks successfully. Our study leverages this architecture, together with transfer learning, to generate two models that successfully identify and separate neurons in phase-contrast images. Fluorescence images are used exclusively for generating training masks and are not required during model training and inference. Once the models are trained, only the non-destructive phase-contrast imaging is needed.

Our research builds upon the mentioned work, by specifying it to our dataset and measuring the performance each model achieves on new data. The following section describes the methods from which the results are obtained.

3 Materials and Methods

3.1 Dataset Description

The dataset used in this study comprises microscopy images of neuronal cultures obtained from the cerebellum of 6–7-day-old mice. The cerebellar neurons were placed into in-vitro culture, were allowed a period for growth and stabilization. After the culture is in a mature healthy stage, it is subjected to various experimental treatments, including oxidative stress and micro-plastic exposure. This ensures enough variability in the characteristics of the images, so they form a faithful representation of this type of cultures.

All images were acquired using a standard phase-contrast microscope equipped with an Olympus IMT-2 digital camera and a Sony SPT-M308CE video camera. For each selected field of view, both, phase-contrast and fluorescence images, were taken, to highlight specific cellular features such as stained neuron somas. This dual-modality approach ensures each field has a corresponding pair of images, enabling precise alignment between the structural ground-truth and the labeled data for training the segmentation algorithms.

A total of 746 images were collected, from experiments spanning over 2 years, and divided into two equal subsets of 373 images:

- Subset A was annotated by human experts to serve as the ground truth.
- Subset B was labeled by the algorithm described at Subsect. 3.2, which outlines out previous work, currently under journal revision.

Each subset was split into training, validation, and test partitions using a standard 70%/20%/10% ratio. These splits were applied consistently across both expert and algorithm-labeled sets to enable robust model evaluation and comparison.

The ground truth annotations were represented as binary masks, where foreground pixels correspond to neuron somas and background pixels represent all other structures. These masks were converted into a format compatible with YOLOv11, a modern object detection and segmentation framework. This format facilitates integration into deep learning pipelines for neuron detection, classification, and segmentation tasks.

In summary, this dataset offers a diverse and representative sample of neuronal morphology under various experimental conditions, captured in both structural and labeled imaging formats. It supports the training and evaluation of segmentation algorithms in a biologically and technically rigorous setting.

3.2 Label-Generating Algorithm

Generating accurate binary masks from fluorescence microscopy images is a critical step in preparing our training data for deep learning-based neuron segmentation. These masks serve as ground truth labels, capturing the spatial structure of neuron somas while excluding background and irrelevant features. The process involves a carefully designed pipeline composed of normalization, enhancement, shape-fitting, and post-processing steps.

The workflow begins with intensity normalization, which adjusts the dynamic range of pixel values to reduce inter-image variability caused by differences in staining or imaging conditions. This ensures a consistent baseline for subsequent operations. To suppress high-frequency noise while preserving cellular structures, a Gaussian blur is applied, which smooths the image and facilitates threshold separation of the foreground and background. This way, using Otsu's thresholding, the algorithm determines a global threshold value that best separates foreground (neurons) from background.

After thresholding, maxima in brightness are detected within the image. The fluorescence stain highlights live cells, but not all highlighted areas represent valid neuron somas. Small connected components that are likely to be noise or non-neuronal structures are removed based on shape-fitting the surrounding bright pixels to an ellipse. These will provide a way to exclude non-neurons from the mask since neurons present homogeneous values for the main axes and the total area. To exclude them, a minimum and maximum area and size thresholds are set based on the dispersion of the values true for neurons. This filtering step ensures that only biologically relevant structures are retained. Fitting to ellipses is not enough to capture the finer details of each neuron, so an erosion step is performed to fit the actual edge more precisely. This last step only introduces slight corrections to the overall shape, which remains elliptical.

To further refine the shapes of detected neurons, morphological operations are applied. Specifically, erosion followed by dilation is used to remove small protrusions, smooth object boundaries, and separate closely located structures without significantly altering the shape or size of valid neuron somas. The correct sequencing of these operations is crucial: dilation destroys the instance separation accomplished by the previous step.

The output of this algorithm was validated against the same images annotated by experts. Once the generated masks were proven to be equivalent, they were used to generate a large dataset fast and reliably. For that we used 3 metrics, standard to asses the similarity in classification problems. These metrics were 'Intersection over union' (IoU), 'Matthew's Correlation Coefficient' (MCC) and the 'Jaccard index' (Jac). The comparison was done by pairing each mask with the corresponding ground-truth provided by the experts.

The first two indices are computed at a pixel level, so to asses the segmentation capabilities of the algorithm, whereas the Jac index is computed at the successful instance identification level. That is, for each neuron in one mask, if the other mask has a corresponding neuron with IoU > 0.3, it is counted as a successful neuron identification. The value is chosen to be 0.3 because the segmentation accuracy is measured by the other 2 metrics. The Jac metric indicates whether the two masks have the same corresponding neuron in the appropriate region of the image. The confidence metrics reach the values of 0.82 ± 0.04 and 0.71 ± 0.05, for the MCC and IoU metrics respectively. At the instance level the Jac metric reaches 0.88 ± 0.05. All the values are larger than the standard threshold values for a successful identification, which are MCC > 0.7, IoU > 0.5 and Jac > 0.5 (Fig. 1).

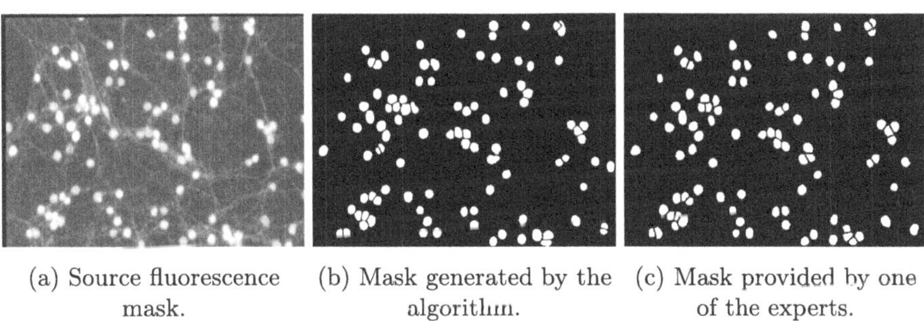

(a) Source fluorescence mask. (b) Mask generated by the algorithm. (c) Mask provided by one of the experts.

Fig. 1. The algorithm uses the source in fluorescence (a) to generate the mask (b), which closely resembles that provided by the expert (c). This allows to quickly and reliably generate masks.

The final result is a clean, smoothed binary mask where each white region corresponds to a neuron soma, and background or irrelevant structures are blacked out. Each neuron is clearly separated from other neurons by a thin line of black pixels. Clusters of neurons are reliably separated, which is the most relevant test to the algorithm's effectiveness. These masks are suitable for use in supervised learning pipelines, where they provide ground truth labels to train and validate deep learning models for automatic neuron segmentation.

3.3 Model Training

Studies have been conducted to analyze the impact of faulty labels, known as label noise [10]. Expert labels are defined to be noiseless, while masks generated by the algorithm may introduce a degree of noise. Models have a known tolerance of noise, and the generated masks fall within the limits. After training, the models should have learned the essential features that identify the neurons in the culture and not the specifics of each neuron, avoiding overfitting. Essential shapes also play a role, and the influence of the labels on the model's generalizability has been studied [8]. In fact, relying on essential shapes like ellipses helps the models to identify the fundamental neurons' features.

We used the YOLOv11 large segmentation architecture (27.6M parameters), a state-of-the-art framework for real-time object detection and instance segmentation. It was initialized with pretrained weights and fine-tuned for neuronal soma segmentation.

Training was conducted separately on the two labeled datasets described earlier—one annotated by human experts and the other by the proposed algorithm. In both cases, the models were trained using the same strategy. The training process minimizes both object detection loss and segmentation loss. The main performance metric applied was the Precision-Recall Area Under the Curve (P-R AUC), which offers a comprehensive view of model's precision and recall across threshold values. Additional metrics (e.g., training and validation loss) were used to monitor training stability and detect signs of overfitting.

The model obtained from training with the labels generated by the experts, Subset A, and the model generated by Subset B will be referred to as EL-

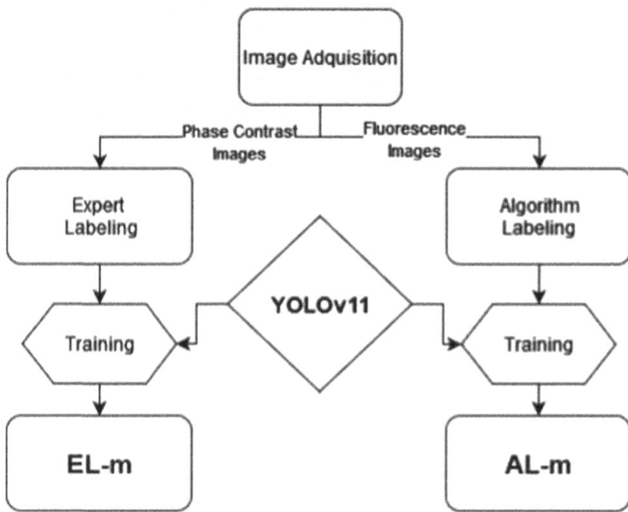

Fig. 2. Flow of the pipeline that generates the two models; EL-m and AL-m.

m (Expert-Labeled model) and AL-m (Algorithm-Labeled model), respectively (Fig. 2).

Empirically, EL-m exhibited earlier convergence, stabilizing around 100 epochs, prompting an early stopping strategy to prevent overfitting. Conversely, AL-m continued to improve and stabilize around 250 epochs, justifying full-length training. To assess generalizability and minimize dataset bias while augmenting the test dataset, a cross-evaluation strategy was implemented. Subset A was used to train the EL-m and served as the test set for AL-m; meanwhile subset B was used to train Al-m and served as the test set for EL-m.

This approach ensures that each model is evaluated on data from the complementary subset, providing an unbiased estimate of performance across both expert and algorithm-generated annotations.

4 Results and Discussion

In this section, we evaluate the performance of our two models trained using Subset A and B as described before. Figure 3 shows the performance of the models on the validation set after completing the training. The two curves at each subgraph correspond to the two metrics used to evaluate the models i.e., the segmentation of the neurons, and the individual identification of each one. Optimal training should approach the top-right margins. The numeric value stands for the area under the curve, where 1 represents perfect performance, and 0.5 or less represents random classification (Table 1).

Table 1. Each pair of values represents the mean average precision at 0.5 (`mAP@0.5`) at the segmentation and box levels, in that order. The values for EL-m show less stability than those obtained for the AL-m.

	EL-m	AL-m
Training	0.907/0.949	0.847/0.866
New culture	0.734/0.794	0.787/0.829

The performance of the models is high, with the metrics showing both of them have trained successfully. To avoid the propagation of common characteristics from one subset into the training, we validate the model against the complementary subset. Figure 4 shows the results. Compared to models trained on manually annotated data, our approach exhibits notably greater stability. The performance level shifts and the model trained on Subset A performs worse than the model trained on Subset B. We hypothesize that this improved generalizability stems from the nature of the generated masks, which tend to emphasize primitive geometric structures such as elongated cell bodies and distinct boundaries. By focusing on these essential shape cues, the model learns more robust and transferable features, enabling it to better handle variations across datasets and imaging conditions.

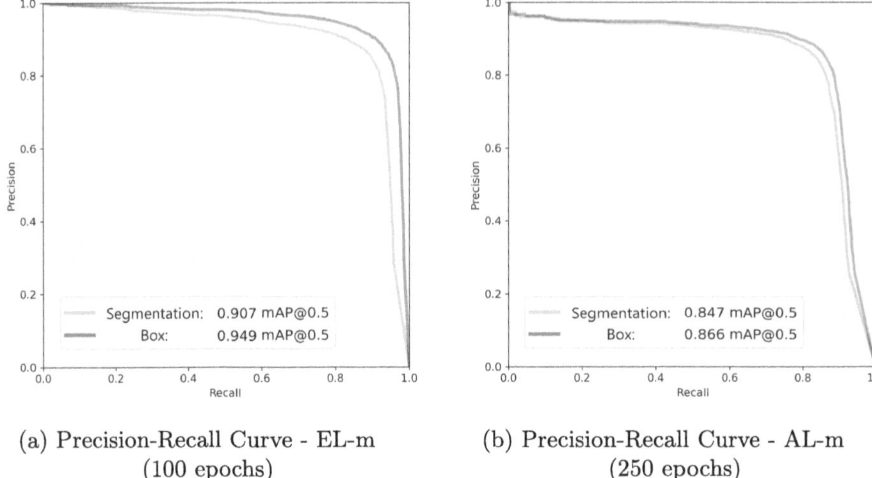

(a) Precision-Recall Curve - EL-m (100 epochs)

(b) Precision-Recall Curve - AL-m (250 epochs)

Fig. 3. Subplot (a) shows the area under the curve in a precision-recall graph metrics after training EL-m for 100 epochs, while subplot (b) shows the same metrics after training AL-m for 250 epochs. The blue lines shows the metric that evaluates the individual identification of neurons, and the orange ones evaluates the segmentation of the soma of the neurons. EL-m reaches higher metrics after training. (Color figure online)

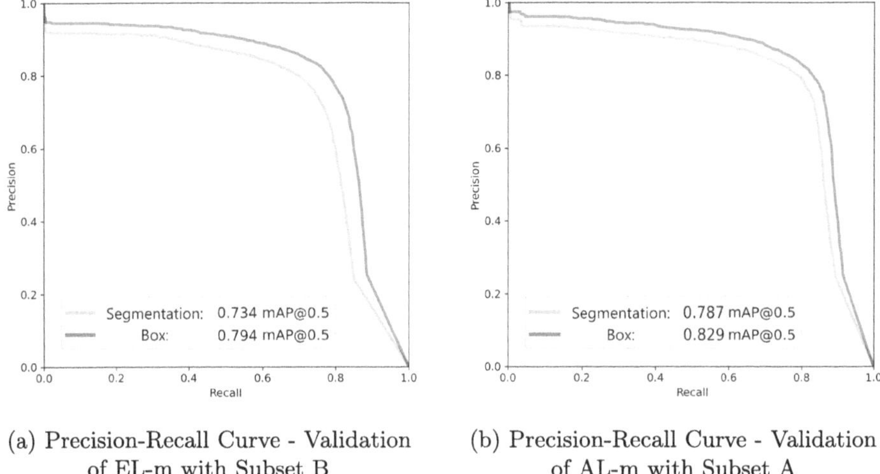

(a) Precision-Recall Curve - Validation of EL-m with Subset B

(b) Precision-Recall Curve - Validation of AL-m with Subset A

Fig. 4. Precision-recall curves for (a) EL-m validated on Subset B and (b) AL-m validated on Subset A. AL-m shows greater stability when faced with unknown images.

Interestingly, the expert-annotated model performs better during validation after the training phase, but fails to generalize effectively to data coming from different cultures. The model trained on automatically generated masks, although initially less accurate, demonstrates stronger generalization across the new set of images. We hypothesize that this is due to the simplified and abstract nature of the auto-generated masks, which guide the model towards learning more robust features. Ultimately, our findings suggest that, despite their imperfections, algorithmically derived labels may offer a scalable and generalizable alternative for training AI systems in viability assessment.

5 Conclusions and Future Work

Results show that the generalizability and stability of the model trained with generated masks is larger, although additional testing is necessary to fully assess the risk of overfitting. The current dataset, though sufficient for initial evaluation, limits the scope of our conclusions. Expanding the study to include a broader and more diverse set of images will be essential to validate the observed performance trends and to ensure that the model's robustness holds across varying experimental conditions. Future work incorporating larger-scale datasets will not only strengthen the reliability of these findings but also provide deeper insights into the impact of mask structure on model generalization.

Future work will involve a more comprehensive analysis of the performance of the algorithm to ensure the robustness and safety of future models. This includes investigating potential biases introduced by the mask generation process, as well as identifying any systematic errors that may arise across different training strategies. By closely examining failure cases and performance across subgroups of the data, we aim to optimize the method. Such an in-depth evaluation will be crucial to refining the training pipeline and building confidence in the model's application for automated, non-destructive viability assessments in neuroscience.

References

1. Baričević, Z., Ayar, Z., Leitao, S.M., Mladinic, M., Fantner, G.E., Ban, J.: Label-free long-term methods for live cell imaging of neurons: new opportunities. Biosensors **13**(3), 404 (2023). https://doi.org/10.3390/bios13030404
2. Bouchet, A., Montes, S., Ballarin, V., Díaz, I.: Intuitionistic fuzzy set and fuzzy mathematical morphology applied to color leukocytes segmentation. SIViP **14**, 557–564 (2020). https://doi.org/10.1007/s11760-019-01586-2
3. García, D.C.: Estudio toxicológico y electrofisiológico de ficotoxinas marinas: mecanismo excitotóxicos y apoptóticos en cultivos primarios de neuronas. Ph. D. Thesis. Universidad de Oviedo (2017)
4. Jocher, G., Qiu, J., Chaurasia, A.: Ultralytics YOLO (2023). https://github.com/ultralytics/ultralytics
5. Jones, K.H., Senft, J.A.: An improved method to determine cell viability by simultaneous staining with fluorescein diacetate-propidium iodide. J. Histochem. Cytochem. **33**, 77–79 (1985). https://doi.org/10.1177/33.1.2578146

6. Pu, S., Zhao, K., Zhang, H.: Learning objectness transfer networks for visual tracking. IEEE Access **7**, 148706–148717 (2019). https://doi.org/10.1109/ACCESS.2019.2946921
7. Rio-Alvarez, A., et al.: Evaluating deep learning techniques for optimal neurons counting and characterization in complex neuronal cultures. Med. Biol. Eng. Comput. **63**, 545–560 (2025). https://doi.org/10.1007/s11517-024-03202-z
8. Rüter, J., Durak, U., Dauer, J.C.: Investigating the sim-to-real generalizability of deep learning object detection models. J. Imaging **10**(10), 259 (2024). https://doi.org/10.3390/jimaging10100259
9. Seong, H., Moon, W., Lee, S., Heo, J.: Leveraging hidden positives for unsupervised semantic segmentation. In: 2023 IEEE/CVF Conference on Computer Vision and Pattern Recognition (CVPR), pp. 19540–19549. IEEE Computer Society (2023). https://doi.org/10.1109/CVPR52729.2023.01872
10. Song, H., Kim, M., Park, D., Shin, Y., Lee, J.G.: Learning from noisy labels with deep neural networks: a survey. IEEE Trans. Neural Networks Learn. Syst. (2020). https://doi.org/10.1109/TNNLS.2022.3152527
11. Stringer, C., Wang, T., Michaelos, M., Pachitariu, M.: Cellpose: a generalist algorithm for cellular segmentation. Nat. Methods **18**, 100–106 (2021). https://doi.org/10.1038/s41592-020-01018-x
12. Zargari, A., et al.: Deepsea is an efficient deep-learning model for single-cell segmentation and tracking in time-lapse microscopy. Cell Rep. Methods **3**, 100500 (2023). https://doi.org/10.1016/j.crmeth.2023.100500
13. Zhang, M., Zhao, J., Hoshino, Y.: Deep learning-based high-throughput detection of in vitro germination to assess pollen viability from microscopic images. J. Exp. Bot. **74**(21), 6551–6562 (2023). https://doi.org/10.1093/jxb/erad315
14. Zhu, X., et al.: Development of a novel noninvasive quantitative method to monitor Siraitia grosvenorii cell growth and browning degree using an integrated computer-aided vision technology and machine learning. Biotechnol. Bioeng. **118**(10), 4092–4104 (2021). https://doi.org/10.1002/bit.27886

Special Session: Intelligent Techniques Applied to Modelling and Control in Engineering Focused on Marine Energy Systems and Robotics

Detecting Yaw Anomalies in Offshore Wind Turbines Using Wind Multiple Features

Bassel Weiss[1(✉)], Segundo Esteban[2], and Matilde Santos[3]

[1] Facultad de Informática, Universidad Complutense de Madrid, 28040 Madrid, Spain
bweiss@ucm.es

[2] Departamento de Arquitectura de Computadores y Automática, Universidad Complutense de Madrid, 28040 Madrid, Spain
segundo@ucm.es

[3] Institute of Knowledge Technology, Universidad Complutense de Madrid, 28040 Madrid, Spain
msantos@ucm.es

Abstract. In this work we present a data-driven framework for early fault detection using SCADA data from the Alpha Ventus offshore wind farm. Ten-minute averages of wind speed, wind direction, and output power real data are available and are used as inputs and output variables. They are previously normalized and filtered. Support Vector Machines (SVM) and k-Nearest Neighbors (KNN) classifiers are trained to distinguish normal from anomalous operating values. Two different cases were compared, with and without wind direction included in the features dataset, together with wind speed. In general, K-NN outperforms SVM (accuracy 91–95% vs. 89–93%; F1-score 89–95% vs. 87–92%). Including wind direction also improves around 2–3% recall and F1. Spatial analysis under prevailing SW winds shows downstream turbines have reduced recall (52–55%) compared to upstream edges (74%). This approach enhances fault detection and informs wake-mitigation and maintenance strategies.

Keywords: Offshore wind turbines · anomaly detection · Support Vector Machines · k-Nearest Neighbors · Wakes · Yaw misalignment

1 Introduction

Offshore wind energy is rapidly expanding, but turbine performance is sensitive to control errors and environmental factors [1]. Early detection of anomalies is essential for predictive maintenance and to maximize wind energy capture [2].

In particular, yaw misalignment (imperfect orientation of the rotor to the wind) can cause significant power losses and mechanical stress. Even small static yaw offsets degrade performance: first principles analysis shows that a few degrees of misalignment can reduce captured energy by a significant fraction, up to 20% AEP [3]. These faults can be detected knowing the expected power that should be produced by the wind turbine according to its characteristics and checking if there is any difference.

E. Corchado et al. (Eds.): SOCO 2025, CCIS 2806, pp. 269–277, 2026.
https://doi.org/10.1007/978-3-032-19763-4_25

Traditional SCADA-based monitoring methods often lacks standardized processing and fails to isolate yaw-related faults [4]. Indeed, standard condition-monitoring systems often miss direct yaw-fault indicators: controller-based SCADA signals tend to mask misalignments by driving the nacelle to the (potentially wrong) windvane setpoint [5].

Recent research has therefore shifted towards data-driven fault diagnosis using SCADA data [6]. Machine learning (ML) approaches have been applied to wind turbine faults with promising results [7]. For example, SVM classifiers have successfully detected various turbine faults in vibration and electrical signals, and Gaussian-process models have been used to flag SCADA anomalies for condition monitoring. In the yaw-control domain specifically, Gao and Hong [8] used regression models on power vs. wind data to correct systematic yaw offsets, while Pandit et al. [9] incorporated rotor speed and pitch into a Gaussian process to improve yaw-fault detection. These studies demonstrate the feasibility of ML-based yaw monitoring but typically focus on single-turbine or corrective-control scenarios.

In contrast, our work extends the ML approach to a wind-farm context, detecting yaw-related anomalies across multiple turbines and accounting for wake effects. It presents a unified, data-driven framework using machine learning to detect such anomalies from real SCADA data. The machine learning technique here applied to obtain the model of the power curve of each wind turbine are Support Vector Machines (SVM) and K-Nearest Neighbors (K-NN).

2 Materials and Methods

2.1 Materials: The Data

SCADA data were obtained from the AlphaVentus offshore wind farm [10], in the North Sea, Germany, which has two parallel rows of six Senvion (now Repower) 5MW jacket offshore turbines (AV01-AV06 in the northern block) and six tripod Adwen (formerly AREVA Wind) AD 5MW-116 wind turbines (AV07-AV012) spaced at ≈700 m (≈5 × D) intervals.

In Fig. 1 the layout of the wind farm is shown. Each point is labeled by its turbine ID (AV1–AV12), with longitude on the x-axis and latitude on the y-axis.

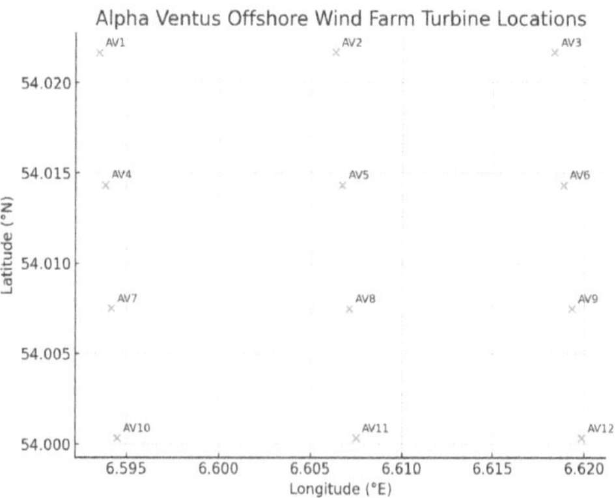

Fig. 1. Alpha Ventus offshore wind farm layout, showing all 12 turbines at their positions.

The data covers 10-minute measurements from this offshore wind farm. We are going to study the influence of the wind speed and wind direction on the classification of anomalies. Fig. 2 show the correlation between power and wind speed by a heatmap for the different wind turbines of the wind farm.

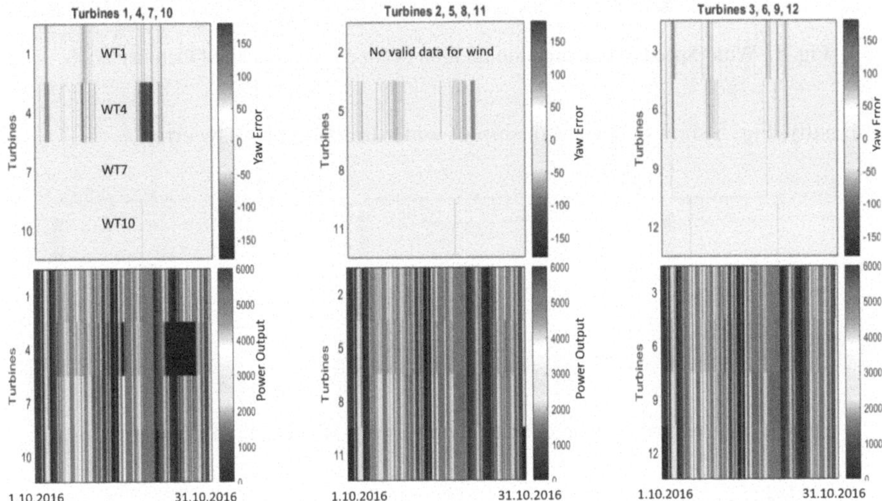

Fig. 2. Heatmap of power and wind speed along the AlphaVentus wind farm in October 2016

We focus on turbines WT4, WT5 and WT6, which belong to the same brand (Senvion) and their location will allow us to analyze different situations. Fig. 2 shows the cut off time in turbine 4 and a lot of yaw errors in turbine 5. Turbines 7–12 have different yaw controller than turbines 1–6. Which explain the correlation between yaw error and power output across turbines using a heatmap. This visualizes typical turbine response curves and helps detect deviations due to possible faults.

In Fig. 3 we can see the wind direction for those wind turbines and the relationship with the power output between the days 5th and 9th of October 2016.

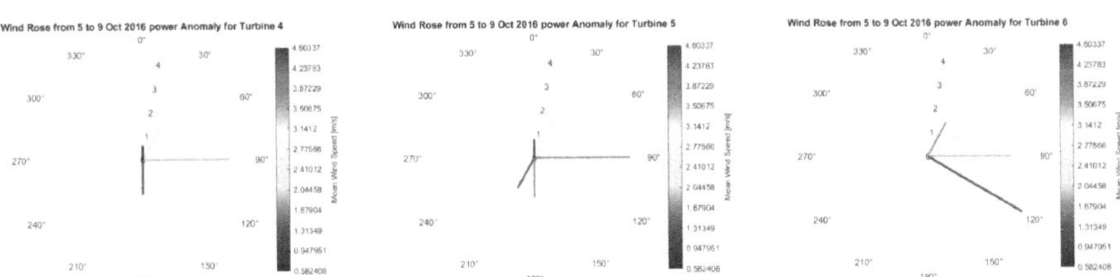

Fig. 3. Power-Wind direction relationship between 5th and 9th of October 2016

The Fig. 4 shows the relationship of wind direction and wind speed for those days.

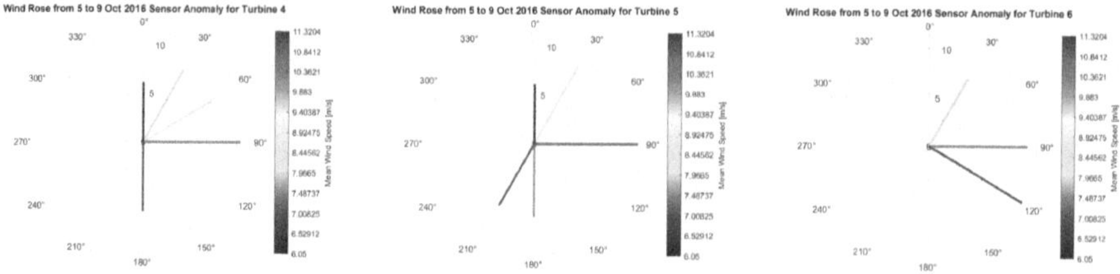

Fig. 4. Wind Speed-Wind direction relation between 5th and 9th of October 2016

Finally, Fig. 5 shows the relationship of wind direction and yaw error.

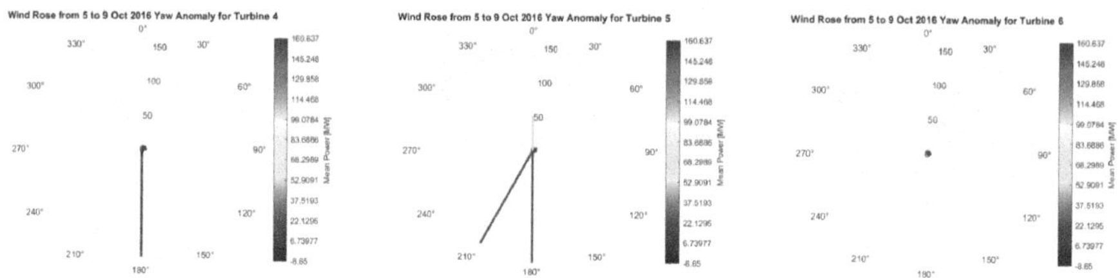

Fig. 5. Yaw Error-Wind direction relation between 5th and 9th of October 2016

Figures 3 and 4 reveal how wind direction affects power output and wind speed. Lastly, Fig. 5 illustrates how yaw errors vary with wind direction, highlighting potential systematic misalignment in specific directions. Peaks or patterns in yaw error at certain directions may suggest that turbines experience systematic misalignment when winds arrive from specific angles, possibly due to control strategy limitations or sensor biases. Understanding this relationship is key for targeted fault diagnosis.

Prior to the application of the machine learning techniques, data are cleaned by removing invalid entries (sensor errors). In addition, to homogenize data across turbines, we normalize wind speed (using the cut-in and nominal wind velocity), wind direction (circular range) and power (dividing by the rated power).

2.2 Methods: SVM and K-NN

Two classifiers have been applied to the wind farm data, Support Vector Machines (SVM) and K-Nearest Neighbors (K-NN). They perform a binary classification: normal (0) vs anomalous (1).

Labeling is semi-automated: we fit a baseline wind turbine power curve for the type of wind turbine. The configuration of these techniques is as follows. SVM works with a Radial Basis Function (RBF) kernel and hyperparameters tuned by cross-validation. K-NN has been used with $k = 5$.

For the applications of these machine learning techniques, we split data by turbine: 80% for training, 20% for testing. Then we obtained the corresponding models of the power curve of each wind turbine. The main characteristics of the Senvion 5MW are the following: cut-in wind speed: 3.5 m/s; rated wind speed: 12.5 m/s, and cut-out wind speed: 30 m/s [11].

Figure 6 compares the anomaly detection results for wind turbine 5 using SVM (left) and K-NN (right). Both plots show normalized power (y-axis) versus normalized wind speed (x-axis). Blue dots represent normal operating points, while red dots indicate points classified as anomalies. These plots illustrate how each classifier models the turbine's power curve and distinguishes deviations. SVM forms a smoother, global decision boundary, while K-NN, being a local method, adapts more closely to variations in the data. As seen, both techniques successfully identify outliers, although K-NN appears slightly better at capturing anomalies around mid-range wind speeds.

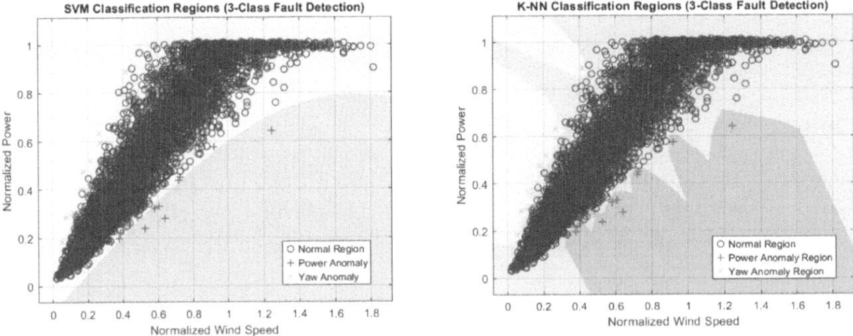

Fig. 6. Normalized power curve of wind turbine AV05 and detection of anomalies with SVM (left) and K-NN (right).

Performance is assessed via accuracy, precision, recall, and F1-score, with confusion matrices for error analysis. We selected SVM and K-NN for their interpretability, low training data requirements, and established success in anomaly detection tasks using SCADA data [6, 7]. Both are well-suited for modeling turbine power curves and are computationally efficient. While methods like Random Forests and neural networks (e.g., multilayer perceptrons, LSTMs) are promising, they typically require more extensive hyperparameter tuning, larger datasets, and may obscure decision-making transparency.

3 Discussion of the Results

First, we present in Table 1 the results when both ML techniques, SVM and K-NN, are applied to detect anomalies using the baseline power curve of the wind turbine. The features considered as inputs are wind speed (WS), wind direction (WD).

Table 1. Classification performance across periods, feature sets, and models.

Period	Feature Set	Model	Accuracy	Precision	Recall	F1-score
Dec 2016	WS + WD	SVM	0.88–0.92	≥0.85	≥0.83	≥0.85
		KNN	0.90–0.94	≥0.85	≥0.94	≥0.86
Dec 2016	WS only	SVM	0.88–0.90	≥0.85	≥0.83	≥0.85
		KNN	0.90–0.92	≥0.85	≥0.96	≥0.90
Nov 2016	WS + WD	SVM	0.86–0.90	≥0.84	≥0.82	≥0.86
		KNN	0.88–0.92	≥0.95	≥0.94	≥0.97
Nov 2016	WS only	SVM	0.84–0.88	≥0.98	≥0.81	≥0.89
		KNN	0.86–0.90	≥0.99	≥0.97	≥0.98
Full 2016	WS + WD	SVM	0.89–0.93	≥0.90	≥0.89	≥0.92
		KNN	0.91–0.95	≥0.97	≥0.92	≥0.95
Full 2016	WS only	SVM	0.87–0.90	≥0.90	≥0.89	≥0.94
		KNN	0.89–0.92	≥0.98	≥0.96	≥0.98

As it is possible to see, most models achieve high scores (≥0.85) in all metrics (Accuracy, Precision, Recall, F1), indicating robust performance for the classification.

KNN consistently outperforms SVM in accuracy and F1 score, especially when both, wind speed and wind direction are considered in comparison to only take into account wind speed. For example, in Nov 2016, KNN's F1 reaches ≥0.97 vs. SVM's ≥0.86.

Regarding Recall and F1-score, the same trend is kept. K-NN excels (e.g., ≥0.96 with "WS only" in Dec 2016), suggesting it captures true positives more effectively than SVM (which struggles with Recall ≤0.89 in most cases).

In general, when the feature set includes both inputs, wind speed WS and wind direction WD, accuracy and precision improve for both models vs. WS only. As an example, see accuracy SVM 0.89–0.93 vs. 0.87–0.90 for "Full 2016"; and precision ≥0.90 vs. ≥0.85 in Dec 2016. It is remarkable that K-NN achieves near-perfect precision (≥0.98) with "WS only" in Nov/Full 2016.

It is true that K-NN Recall drops slightly with WS + WD (e.g., Dec 2016: ≥0.94 vs. ≥0.96 for "WS only"), but F1 remains high.

There is not a clear analysis regarding the period of time considered in the evaluation of the classifiers. F1 values may suggest that November data may have clearer patterns, but December has higher accuracy ranges for both models in comparison to November.

As expected, both models give the best overall results when the full 2016 dataset is used likely due to more training data.

Despite overall high performance, classifier limitations are observed in some cases. Downstream turbines, affected by wake turbulence, show lower recall rates, as their power output patterns deviate more from the norm. Moreover, the semi-automated labeling process may introduce noise, especially in borderline cases. Seasonal variability in wind patterns could also reduce generalization when training on limited data. Lastly, excluding wind direction as a feature can lead to underperformance, particularly in detecting yaw-related anomalies.

3.1 Impact of the Location of the Wind Turbine on the Wind Farm on the Anomaly's Detection

On the other hand, we have also analyzed the rate of anomaly detection depending on the position on each wind turbine in the wind farm. This is something that some authors have begun to consider relevant for maintenance strategies, since the position of a turbine influences its wear due to the wakes of the other turbines [12, 13].

We can draw some interesting remarks from Table 2, that shows fault/event rates (in percentages) for six turbines (AV01–AV06) across three time periods (Dec 2016, Nov 2016, Full 2016). The turbines are positioned in a way that wind direction (NW → SE) creates distinct wake effects.

Table 2. Anomaly rates by turbine and period.

Turbine	Dec 2016 Rate (%)	Nov 2016 Rate (%)	Full 2016 Rate (%)
AV01	0.1–0.2	0.2	0.5
AV02	0.2	0.2	0.2
AV03	0.4	0.5	1.0
AV04	0.2	0.3	1.5
AV05	0.2	0.3	0.5
AV06	0.2	0.2	0.3

Incorporating the wind farm layout and prevailing wind dynamics in the detection of the anomalies, it seems that upstream wind turbines (AV01/AV04) are unaffected by wakes (undisturbed flow). The WT located in the middle (AV02/AV05) are partially sheltered (wake onset) and downstream turbines (AV03/AV06) strongly experience wakes (lower wind speed, higher turbulence).

That results in the following. First, all turbines show increased rates in the full-year dataset vs. monthly snapshots. That may be due to the cumulative exposure to wake effects and operational wear over time that raises fault likelihood. Nevertheless, November vs. December 2016 presents minor variations (e.g., AV04/AV05 at 0.3% in Nov vs. 0.2% in Dec). The possible reason is because seasonal wind speed/direction shifts affecting wake intensity.

Regarding the wake effects, downstream turbines are most vulnerable. An illustrative example is AV03, the northern downstream turbine, which fares worse than the southern one (AV06), suggesting the NW wind disproportionately impacts the northern row. The consistently high anomalies rates align with wake-induced turbulence, which strains components (e.g., bearings, blades) and endangers turbine health. On the contrary, upstream turbines (AV01/AV04) present low rates (0.1–0.5%), reflecting minimal mechanical stress due to undisturbed flow. Finally, center turbines (AV05/AV02) also show moderate wake effects.

Overall, Table 2 reveals a clear link between turbine position (upstream vs. downstream) and fault rates, with downstream turbines (especially AV03) suffering the most. The asymmetry between north/south rows suggests the prevailing NW wind exacerbates wakes in the northern sector. Further analysis of wind data and turbine loading is recommended to refine mitigation strategies [14].

4 Conclusions and Future Works

In this work we present a machine-learning framework to detect anomalies in offshore wind turbines from SCADA data. Regarding the two machine learning techniques applied to the detection of anomalies, SVM and K-NN, overall KNN is superior for Recall/F1, while SVM is more consistent but less exceptional. This would suggest the use of K-NN for detecting rare events, that is, if anomalies are scarce.

In addition, considering both features, wind speed and wind direction, enhances accuracy, but "WS only" can yield higher precision (especially for K-NN). Domain-specific needs (e.g., minimizing false positives) should guide feature selection.

Even more, the analysis of the anomalies considering the position of each wind turbine in the wind farm suggest the necessity of prioritize maintenance for downstream turbines that need more frequent inspections due to wake-driven wear.

I68566As future work, we propose exploring real-time implementation and deep-learning approaches for further robustness.

Beyond wind speed and wind direction, other data—such as air pressure, rotor vibration, or acoustic measurements—could further enhance model sensitivity and specificity. Integrating such sensor data would enable detection of mechanical faults that may not manifest in SCADA power curves. Future extensions of this framework will explore multi-modal input to improve fault classification granularity and robustness.

Acknowledgements. This work has been partially supported by the Spanish Ministry of Science and Innovation under the MCIU/AEI/FEDER project number PID2021-123543OBC21.

Data statement. Data was made available by the RAVE (Research at Alpha Ventus) initiative, which was funded by the German Federal Ministry of Climate Action based on a decision by the German Bundestag and coordinated by Fraunhofer IWES see: www.rave-offshore.de for more information on data access via: https://rave-offshore.de/en/data.html.

References

1. Muñoz-Palomeque, E., Sierra-García, J.E., Santos, M.: Intelligent control techniques for maximum power point tracking in wind turbines. Revista Iberoamericana de Automática e Informática industrial **21**(3), 193–204 (2024)
2. Pandit, R., Astolfi, D., Hong, J., Infield, D., Santos, M.: SCADA data for wind turbine data-driven condition/performance monitoring: a review on state-of-art, challenges and future trends. Wind Eng. **47**(2), 422–441 (2023)

3. Ulazia, A., Ibarra-Berastegi, G., Sáenz, J., Carreno-Madinabetia, S., Elosegui, U.: Wind farm analysis using SailoR diagram-based diagnostics to quantify yaw misalignment correction. Energy Convers. Manag. **X**, 100890 (2025)
4. Papatzimos, A., Lin, Y., et al.: A decision support framework for wind turbine condition monitoring. J. Mar. Sci. Eng. **11**(10), 1855 (2023)
5. Sacie, M., Santos, M., López, R., Pandit, R.: Use of state-of-art machine learning technologies for forecasting offshore wind speed, wave and misalignment to improve wind turbine performance. J. Mar. Sci. Eng. **10**(7), 938 (2022)
6. Wen, W., Liu, Y., Sun, R., Liu, Y.: Research on anomaly detection of wind farm SCADA wind speed data. Energies **15**(16), 5869 (2022)
7. Vásquez-Rodríguez, G., Maldonado-Correa, J.: Anomaly-based fault detection in wind turbines using unsupervised learning: a comparative study. In: IOP Conference Series: Earth and Environmental Science, vol. 1370, no. 1, p. 012005. IOP Publishing, July 2024
8. Gao, L., Hong, J.: Data-driven yaw misalignment correction for utility-scale wind turbines. J. Renew. Sustain. Energy **13**(6), 063301 (2021)
9. Pandit, R., David, I., Santos, M.: Accounting for environmental conditions in data-driven wind turbine power models. IEEE Trans. Sustain. Energy **14**(1), 168–177 (2022)
10. RAVE. Research at Alpha Ventus. https://rave-offshore.de/en/start.html. Accessed 2025 May
11. REpower (Senvion) 5M offshore wind turbine specifications. https://en.wind-turbine-models.com/turbines/1-repower-5m-offshore. Accessed May 2025
12. Pettas, V., Kretschmer, M., Clifton, A., Cheng, P.W.: On the effects of inter-farm interactions at the offshore wind farm Alpha Ventus. Wind Energy Sci. **6**(6), 1455–1472 (2021)
13. Li, Y., Shen, X.: Anomaly detection and classification method for wind speed data of wind turbines using spatiotemporal dependency structure. IEEE Trans. Sustain. Energy **14**(4), 2417–2431 (2023)
14. Villoslada, D., Santos, M., Tomás-Rodríguez, M.: TMD stroke limiting influence on barge-type floating wind turbines. Ocean Eng. **248**, 110781 (2022)

Neuro Fuzzy Control System for Angular Generator Speed in Floating Offshore Wind Turbines

Eduardo Muñoz-Palomeque[1](✉), Matilde Santos[2], and Jesús Enrique Sierra-García[3]

[1] Department of Computer Architecture and Automatic Control, Complutense University of Madrid, Madrid, Spain
edumun04@ucm.es

[2] Institute of Knowledge Technology, Complutense University of Madrid, Madrid, Spain
msantos@ucm.es

[3] Digitalization Department, University of Burgos, Burgos, Spain
jesierra@ubu.es

Abstract. This work presents the application of Adaptive Neuro-Fuzzy Inference System (ANFIS) controllers in the maximum power point tracking (MPPT) region of wind turbines, with the objective of mimicking the control behavior of classical MPPT strategies. The ANFIS controllers are trained to replicate the dynamic responses of two well-established control methods, simple speed-torque control and Indirect Speed Control (ISC), used for tracking maximum power extraction as two independent cases. By learning from these reference techniques, the ANFIS models are evaluated for their ability to accurately reproduce their control actions within the MPPT region. The controllers use variables like the generator speed as input and output the electromagnetic torque to drive the turbine operation. Simulation results on a high-fidelity model of a floating offshore wind turbine demonstrate that the ANFIS controllers successfully emulate the classical control strategies with high precision, showing effective adaptation to varying wind conditions and confirming their suitability for intelligent MPPT control.

Keywords: ANFIS · intelligent control · MPPT · soft computing · wind turbine

1 Introduction

Offshore wind turbine power generation has become one of the main sources of renewable energy (Mrabet et al. 2024). These systems present control challenges associated with wind variability (Apata and Oyedokun 2020). Also, there are factors associated with the non-linear dynamics of the device itself and the marine environment in the case of offshores wind turbines, and the need to operate efficiently over a wide range of wind speeds. A wind turbine (WT) operates in different regions depending on the wind speed. When the wind is below the rated speed, the WT is working in the MPPT (Maximum Power Point Tracking) operating region, and the objective is to adjust the generator torque to extract the maximum available power from the wind.

E. Corchado et al. (Eds.): SOCO 2025, CCIS 2806, pp. 278–288, 2026.
https://doi.org/10.1007/978-3-032-19763-4_26

Traditionally, linear control techniques and heuristic algorithms have been used for this task. However, these approaches can struggle to adapt to system nonlinearity and uncertainty. In this context, soft computing methods are robust options for improving wind turbine control (Muñoz-Palomeque et al. 2024a, 2024b) and other types of complex problems and systems (Santos et al. 2006; Sierra-García and Santos 2021). Among them, adaptive neuro-fuzzy inference systems (ANFIS) emerge as a promising alternative, combining the learning capabilities of neural networks with the fuzzy representation of expert knowledge.

In literature, the use of these strategies has been developed to improve the performance of the wind turbine in different aspects. For instance, an ANFIS controller is designed in (Ramana and Rosalina 2025) to improve power quality in wind energy systems connected to weak grids, while it addresses challenges related to the mitigation of voltage fluctuations and harmonic distortions by adjusting reactive power compensation. In (Griche et al. 2022), a neuro-fuzzy controller is proposed to enhance the stability and voltage regulation of a wind power system, demonstrating improved dynamic response and oscillation damping under fault conditions. Similarly, studying the electrical effects, in (Ouhssain et al. 2024), an ANFIS-based Indirect Vector Control (IVC) system for DFIG wind turbines is proposed to enhance MPPT efficiency, grid stability, and dynamic response under variable wind conditions. The authors in (Khajeh et al. 2023) propose an ANFIS-based sliding mode controller for a wind turbine using an indirect matrix converter (IMC), resulting in improved dynamic performance, reduced steady-state error, and better power quality compared to traditional PI control.

Multiple ANFIS applications are also proposed. For instance, in (Hermassi, M. et al. 2024), the authors describe the use of two ANFIS structures designed to estimate wind variations and deal with the MPPT objective in a wind power device. Additionally, in the study of grid-connected photovoltaic and wind energy systems, ANFIS controllers are presented as a solution in (Babu et al. 2024), where the controller can dynamically adjust both power signals, thus providing efficient grid connection.

Fuzzy logic and ANFIS-based MPPT controllers are also compared in (Muthu 2021), where the author uses a 250W wind power system, demonstrating through simulations that ANFIS achieves higher efficiency and faster response under varying wind speeds, with a feasible performance in maximizing energy extraction.

Focusing on the MPPT problem, a critical aspect of wind turbine operation, this study explores the use of the ANFIS technique to manage power extraction by controlling the electromagnetic torque based on wind speed and the response to the generator's angular speed. In line with this objective, an ANFIS control system was initially implemented in this work to establish a control law that balances these variables and moves toward the behavior that maximizes power output in the sub-nominal region. The controller is trained on data obtained from classical control approaches, enabling it to learn patterns related to the behavior and interconnection among system variables. This ability to replicate classical controller behavior, even without explicit knowledge of their internal control decisions, provides a solid foundation for enhancing the intelligent system and improving its response to address the limitations of traditional methods. The main contribution of this work lies in the design and validation of that control strategy using limited initial data, that provides a framework that can be further optimized and integrated into more advanced control.

The feasibility and performance of the controller are validated through simulation using a high-fidelity model of a floating offshore wind turbine, implemented with OpenFast software. The system's dynamic response and the controller's ability to replicate desired behavior are analyzed. Results demonstrate that ANFIS control is an efficient solution for this specific application, showing strong potential for further improvement. It enables quick adaptation to wind variations while power is maximized.

This paper is structured as follows. Section 2 presents the main concepts related to the MPPT operating region of a wind turbine. In Sect. 3, the design of the intelligent control in the power maximization region of the wind turbine is explained. Section 4 discusses the results of the control proposal. Lastly, Sect. 5 presents the conclusions and future work.

2 MPPT in the Partial Load Operating Region of a Wind Turbine

In the study of a wind power system, the amount of energy that is finally generated from the wind is crucial to determine the performance of this device. In this matter, the produced energy depends on the wind energy that the wind device can capture. This energy can be calculated using (1).

$$\mathrm{P} = \frac{1}{2}\rho\pi R^2 {V_w}^3 C_p(\lambda, \beta) \tag{1}$$

In this expression, some parameters and variables related to the wind condition and the wind turbine design have a significant influence. Here, ρ (Kg/m^3) is the air density, R (m) is the radius of the rotor, V_w (m/s) is the wind speed, and C_p is the power coefficient which is a function of the tip speed ratio, λ, and the pitch angle, β (deg).

In the MPPT region, the pitch angle is constant, fixed at the position where the blades capture the most wind. Thus, the generator torque must be regulated to maintain the tip speed ratio (2) at the optimal value $\lambda = \lambda_{opt}$, which maximizes the power coefficient.

$$\lambda = \frac{R\omega_r}{V_w} \tag{2}$$

The tip speed ratio denotes a speed relation between the linear speed at the tip of the blades and the wind speed that hits the blades. This linear velocity is calculated in terms of the angular rotor speed, ω_r. When these parameters achieve an optimal value, the wind turbine captures the maximum power from the wind and tracks the greatest efficiency. A classical controller can be built based on this concept that approximates an optimal mathematical description of the wind turbine model. This traditional speed controller is described with expressions (3) and (4):

$$T_{e_{ref}} = K_{opt} \cdot {\omega_r}^2 \tag{3}$$

$$K_{opt} = \frac{1}{2}\frac{\rho\pi R^5 C_{p\_opt}}{{\lambda_{opt}}^3} \tag{4}$$

where T_{eref} (Nm) is the reference for the electromagnetic torque, λ_{opt} is the optimal tip speed ratio of the wind turbine, and C_{p_opt} is the optimal power coefficient. K_{opt} denotes the optimal constant that describes the relationship between speed and torque.

This optimal relationship is difficult to establish due to the complexity of the wind power system and its changing dynamics. The ideal parameters and behavior of the wind turbine are also subjected to variations in time because of mechanical wear, blade contamination, change in air density, or wind turbulence. Additionally, sometimes the control law is unknown, and it makes it complicated to adapt it to the real scenario.

The presence of these complex phenomena motivates the use of adaptive techniques to regulate the response of the wind power system.

3 Neuro-Fuzzy Control Design for the MPPT Control of the Wind Turbine

In the partial load region of the wind turbine, where the wind speed is between the cut-in and rated values, the system must be controlled to keep an equilibrium between the angular generator speed and the operational torque of the electrical machine. The better this relationship is maintained over time at different wind speeds, the better the energy conversion process is carried out. This necessary condition allows the wind power system to track the maximum power point despite the changes produced in the environment and the dynamics of the wind device.

In this line, the control solutions based on the use of soft computing techniques are a promising alternative. Neural networks and fuzzy logic approaches are well-known examples that can be applied in control solutions. Moreover, the combination of both techniques, taking advantage of their powerful capabilities, gives place to the Adaptive Neuro-Fuzzy Inference System approach. With a controller based on this method, it behaves as a strong system able to reproduce complex, unprecise and nonlinear relationships between input and output variables. Thus, it is feasible to apply in a system that shares those characteristics, such as a floating offshore wind turbine.

This controller is implemented in the under rated wind speed region as a proposal to manage the maximum power transmission goal. The ANFIS controller is designed according to Fig. 1, where the operational flow of the controller is shown. Here, the ANFIS controller is a neural network structure with five layers based on a Sugeno fuzzy inference system. The controller receives the input variables that are the basis for the regulation of the wind device. For these purposes, some variables that contribute directly to the performance of the wind turbine and play an essential role in the behavior and control of the system in the MPPT region, are chosen. The generator speed, ω_g, and the wind speed, V_w, are defined as the two input variables that feed the neural network. The output signal that is estimated by the controller is the electromagnetic torque, $T_{e_{ref}}$, which is responsible for regulating the generator speed.

The controller is configured with three Gaussian membership functions for each input variable, which convert the numerical variables into a degree of membership to be processed. Thus, these six functions correspond to the structure of the first internal layer in the ANFIS architecture.

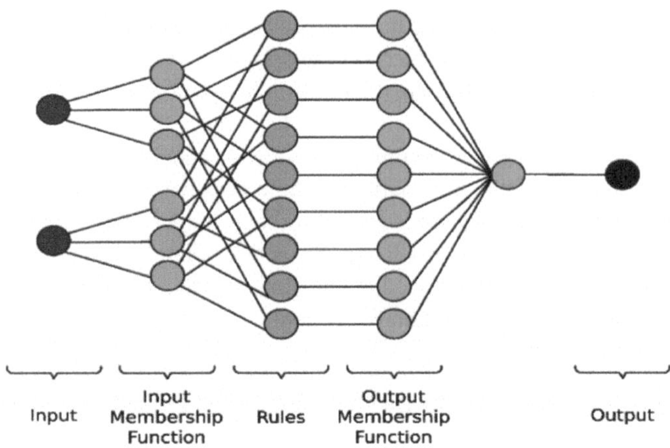

Fig. 1. Neuro-fuzzy control structure.

The ANFIS controller is trained using 50 epochs and the hybrid algorithm that combines the least-squares and backpropagation gradient descent methods. The dataset for training the controller has been obtained through simulation from the application of an unknown speed-torque tracking curve of the wind turbine embedded in the OpenFast software while operating in the sub-nominal region.

As a second and independent case, the ANFIS controller is also trained with data obtained from the classical ISC controller using the mathematical definition of the wind turbine behavior (3) with optimal Cp and tip speed ratio of 0.48 and 7.6, respectively. For this controller, the input layer is reduced to a single variable: the generator speed that received the influence of the external wind conditions.

The configuration of the neuro-fuzzy system is shown in Table 1.

The implementation of this controller is tested in the MPPT region of a floating offshore wind turbine of 5 MW. The wind speed considered changed between 8.8 m/s and 11 m/s, as shown in Fig. 2. The electromagnetic torque for optimal power tracking is then estimated, which is a variable that relates to the output power and allows it to regulate the rotational speed of the generator and the turbine.

Table 1. Configuration of the ANFIS controller.

Parameter	Value
Type of Fuzzy system	Sugeno fuzzy inference system
Input membership functions	Gaussian
Connection method	"AND" - Product of the input variables after fuzzification
Method for defuzzification	Average of the output values of the rules
Output membership function	Constant

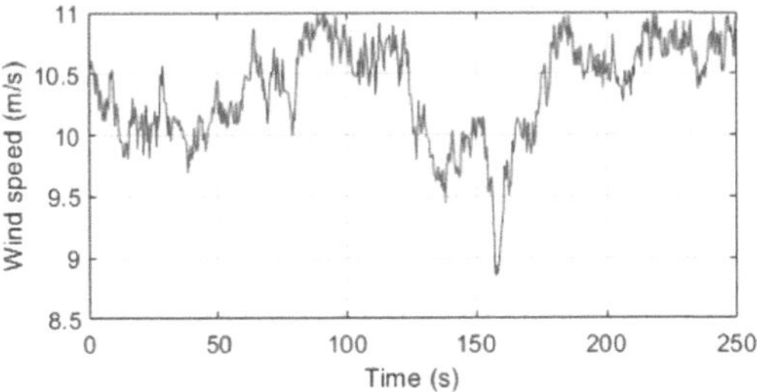

Fig. 2. Input wind speed profile.

4 Results and Discussion

The performance of the proposed ANFIS controller was evaluated through simulation using NREL OpenFAST and Matlab/Simulink software. Results show that the ANFIS controller can operate in the partial load region of operation of a wind turbine, mimicking classical controllers for tracking the maximum power curve.

The first case, using ISC data for training purposes, is tested. In Fig. 3, the power produced by the wind turbine is shown. The ANFIS controller quickly changes in response to the random wind speed input, while it tries to adjust the generator speed (Fig. 4) and output electromagnetic torque (Fig. 5).

The ANFIS control response makes the system change the electromagnetic torque in a way that the rotational speed reacts naturally to the tendency of the ISC control response produced by the wind input over time. The ANFIS response for the generator speed adapts to the variations in the wind and mimics effectively the trained patterns.

The torque response of the wind turbine in this study quickly reacts to the sudden environmental variations during the MPPT operation. This makes the speed change softly and allows the required energy levels to track the optimal efficiency, like the classical control method does. This behavior is proportional to the wind changes to achieve an equilibrium with the wind and generator speed while keeping maximum power efficiency.

For the second case, the ANFIS controller is tested using data from the application of the embedded speed-torque control in OpenFast as a technique with unknown mathematical information of the relationship between the involved variables.

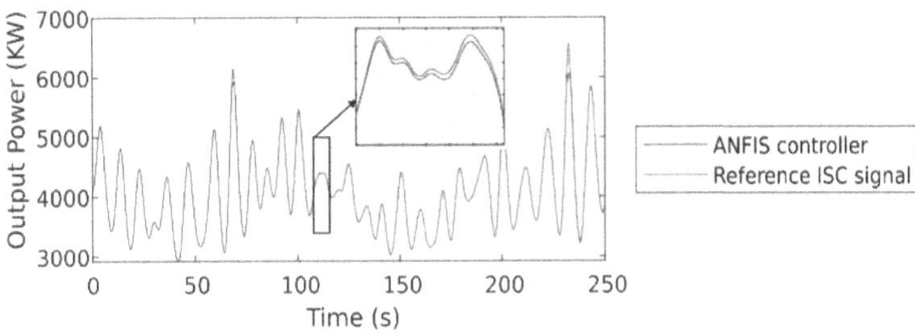

Fig. 3. Generator output power signal obtained with ANFIS trained with ISC control data.

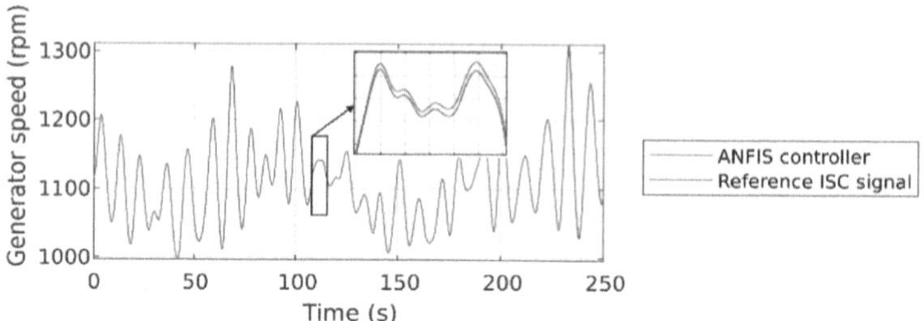

Fig. 4. Generator speed signal obtained with ANFIS trained with ISC control data.

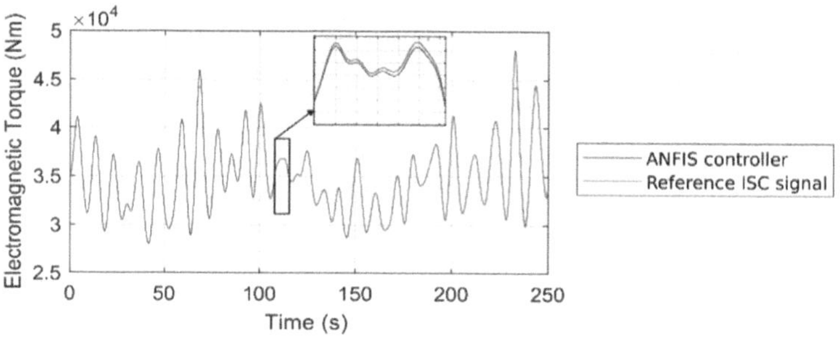

Fig. 5. Output electromagnetic torque obtained with ANFIS trained with ISC control data.

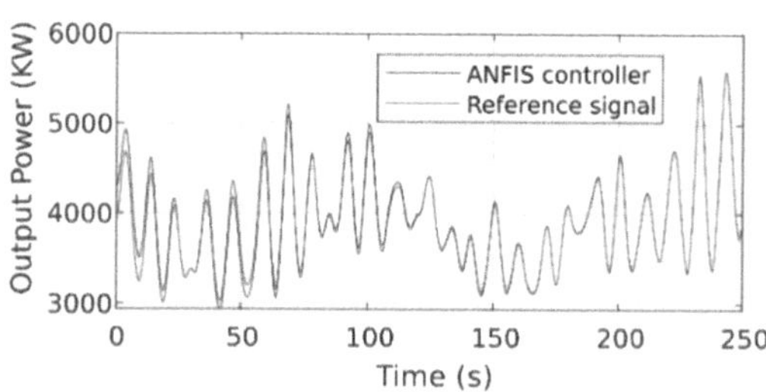

Fig. 6. Generator output power signal obtained with ANFIS trained with classical speed-torque control (controller embedded in openFast) data.

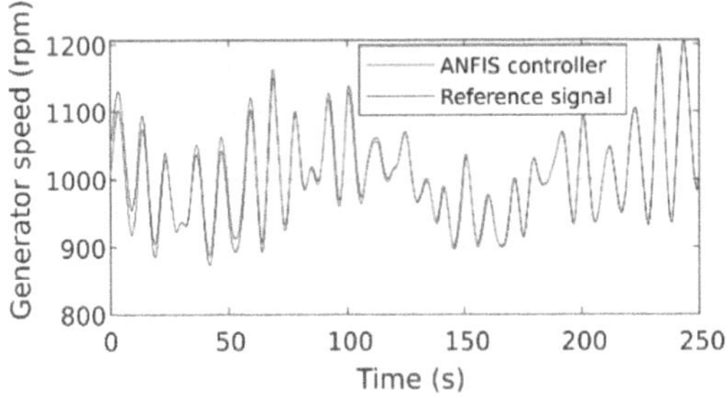

Fig. 7. Generator speed signal obtained with ANFIS trained with classical speed-torque control (controller embedded in openFast) data.

Similarly to the previous case, results indicate that the neuro-fuzzy controller can reproduce this control action efficiently. Generated power, generator speed and electromagnetic torque signals are compared in Fig. 6, Fig. 7, and Fig. 8, respectively.

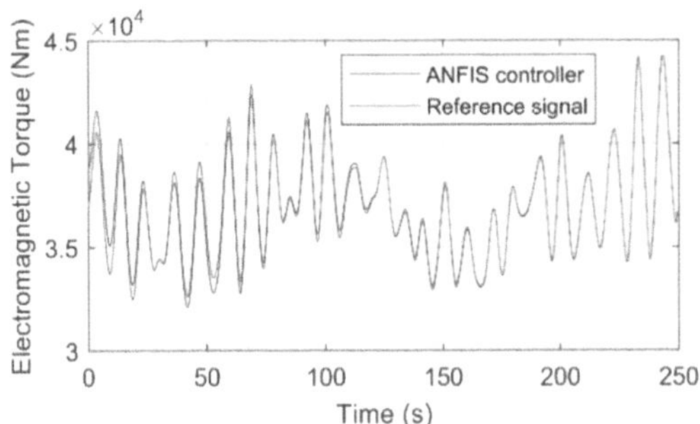

Fig. 8. Output electromagnetic torque signal obtained with ANFIS trained with classical speed-torque control (controller embedded in openFast) data.

The control response of the ANFIS approach is effective in identifying control decisions from the data and accurately mimicking control actions. A slightly larger error is observed in the control responses, especially when rapid oscillations in the wind are present. However, this small variation does not significantly affect the results, and the ability of the ANFIS technique to mimic this classical controller can be verified.

To validate the control response and capability of reproducing the behavior of other controllers, the Mean Absolute Percentage Error (MAPE) of the power signal is calculated. With the use of the neuro-fuzzy controller, a value of 0.14% is obtained with respect to the classic ISC reference controller, and 1.5% against the classical speed-torque controller. To deepen the analysis of the power control response, Table 2 summarizes the average power values calculated every 50 s of operation to verify the response of the controllers to the different wind behaviors in their input profile.

As can be seen, the performance of the control action is appropriate to manage the power tracking goal in the wind turbine. At different wind changes, including quick wind decrements, quick wind increases, and quick but small wind changes, the ANFIS approach provides a control action according to those dynamics, allowing the adjustment of the rotor speed that allows operation in optimal conditions, tracking the training behavior. This technique replies to complex control behavior with the configuration obtained after automatically training the controller rules and identifying patterns to build the control law.

Tests indicate that the ANFIS controller, even when trained with a limited dataset, successfully captures essential system dynamics relevant to MPPT in offshore wind turbines. However, since more complex behaviors can arise during operation, a narrow training set may limit the controller's ability to generalize under varied conditions. Expanding the training dataset to include a broader range of system responses and environmental variations would improve the adaptability and robustness of the controller. The current approach, nonetheless, offers a solid foundation with promising performance that can be further enhanced through extended training data to cover more operational scenarios.

Table 2. Average power (MW) of the MPPT response at different time blocks.

Time range	ISC training data		Open Fast Controller training data	
	ANFIS	Reference	ANFIS	Reference
0–50	3.9120	3.9121	3.7970	3.7395
50–100	4.2285	4.2299	3.9677	4.0088
100–150	3.9591	3.9601	3.8901	3.8792
150–200	3.8324	3.8319	3.6800	3.6835
200–250	4.3619	4.3694	4.1675	4.1647

Regarding computational burden, the study assessed elapsed time as a key metric during both training and simulation (Table 3). The ANFIS controllers were trained offline using MATLAB, and training times per epoch for both ISC and OpenFAST based datasets with 600001 samples each were comparable, differing by only about 0.1 s. Similarly, simulation times per sample showed minimal variation, with the OpenFAST-trained controller requiring approximately 0.05 ms more inference time per sample. These small differences in the context of the full wind turbine simulation confirm that the proposed control strategy maintains a low computational cost, making it suitable for practical implementation scenarios.

Table 3. Computational time with the implementation of the proposed approaches.

	ISC database	OpenFAST database
ANFIS Training time per epoch (s)	3.52	3.62
ANFIS inference time per sample (ms)	1.14	1.19

5 Conclusions and Future Work

In this work, a neuro-fuzzy control is tested through simulation in a floating offshore wind turbine for the MPPT operation region. The proposed controller is applied as a promising solution to mimic the pattern behaviors of the relationship between variables involved in the wind turbine control problem. The ANFIS approach provides a control response that quickly moves the system to acquire a state that responds to the macro variations of the wind speed profile and the reaction in the rotational speed. Therefore, the control system acts according to the patterns in the training dataset.

In future work, this control technique must be enhanced, first improving the training process with a more complete dataset that covers a wide range of relevant patterns, for high and small changes in the wind speed and in the dynamics of the wind device. Then, it is important to optimize the controller to enhance its adaptability when environmental conditions change, thereby maximizing power generation. Additionally, a comparison analysis with other control methods and experimental tests in physical models using these control approaches can be helpful to validate the performance of the proposal.

Acknowledgements. This work has been partially supported by the Spanish Ministry of Science and Innovation under the project MCI/AEI/FEDER number PID2021-123543OB-C21.

References

Apata, O., Oyedokun, D.T.O.: An overview of control techniques for wind turbine systems. Sci. Afr. **10** (2020). https://doi.org/10.1016/j.sciaf.2020.e00566

Babu, T.M., et al.: Intelligent control strategies for grid-connected photovoltaic wind hybrid energy systems using ANFIS. Int. J. Adv. Appl. Sci. **13**(3), 497 (2024). https://doi.org/10.11591/ijaas.v13.i3.pp497-506

Griche, I., et al.: A new adaptive neuro-fuzzy inference system (ANFIS) controller to control the power system equipped by wind turbine. In: ITM Web of Conferences, vol. 42, p. 01011 (2022). https://doi.org/10.1051/itmconf/20224201011

Hermassi, M., et al.: Wind speed estimation and maximum power point tracking using neuro-fuzzy systems for variable-speed wind generator. Wind Eng. **48**(6), 1088–1104 (2024). https://doi.org/10.1177/0309524x241247231

Khajeh, A., Torabi, H., Siahkaldeh, Z.S.: ANFIS based sliding mode control of a DFIG wind turbine excited by an indirect matrix converter. Int. J. Power Electron. Drive Syst. (IJPEDS) **14**(3), 1739 (2023). https://doi.org/10.11591/ijpeds.v14.i3.pp1739-1747

Mrabet, N., et al.: Enhancing wind energy conversion efficiency: a novel MPPT approach using P&O with ADRC controllers versus pi controllers with KP and Ki optimization via genetic algorithm and ant colony optimization. Clean. Energy Syst. **9**, 100159 (2024). https://doi.org/10.1016/j.cles.2024.100159

Muñoz-Palomeque, E., Sierra-García, J.E., Santos, M.: Enhancing offshore wind turbines performance with hybrid control strategies using neural networks and conventional controllers. J. Comput. Des. Eng. **12**(3), 80–97 (2024a). https://doi.org/10.1093/jcde/qwae103

Muñoz-Palomeque, E., Sierra-García, J.E., Santos, M.: Intelligent control techniques for maximum power point tracking in wind turbines. Revista Iberoamericana de Automática e Informática Industrial **21**, 193–204 (2024b). https://doi.org/10.4995/riai.2024.21097

Muthu, A.: Comparative analysis of fuzzy and ANFIS based MPPT controller for wind power generation system. Int. J. Appl. Power Eng. (IJAPE) **10**(4), 355 (2021). https://doi.org/10.11591/ijape.v10.i4.pp355-363

Ouhssain, S., et al.: Enhancing the performance of a wind turbine based DFIG generation system using an effective ANFIS control technique. Int. J. Robot. Control Syst. **4**(4), 1617–1640 (2024). https://doi.org/10.31763/ijrcs.v4i4.1451

Ramana, P.V., Rosalina, K.M.: Optimizing weak grid integrated wind energy systems using ANFIS-SRF controlled DSTATCOM. Sci. Rep. **15**(1) (2025). https://doi.org/10.1038/s41598-025-98872-6

Santos, M., López, R., de la Cruz, J.M.: A neuro-fuzzy approach to fast ferry vertical motion modelling. Eng. Appl. Artif. Intell. **19**(3), 313–321 (2006). https://doi.org/10.1016/j.engappai.2005.09.002

Sierra-García, J.E., Santos, M.: Switched learning adaptive neuro-control strategy. Neurocomputing **452**, 450–464 (2021). https://doi.org/10.1016/j.neucom.2019.12.139

Comparative Study of Machine Learning Techniques for Inverse Kinematics in Robotics: XGBoost Versus MLP Neural Networks

M. Peñacoba-Yagüe[1](✉), J. E. Sierra-García[1], and M. Santos-Peñas[2]

[1] Department of Digitalization, University of Burgos, 09006 Burgos, Spain
{mpenacoba,jesierra}@ubu.es

[2] Institute of Knowledge Technology, Complutense University of Madrid, 28040 Madrid, Spain
msantos@ucm.es

Abstract. In the field of robotics, achieving real-time response in control systems is a key requirement for the deployment of high-performance applications. Conventional iterative methods used to solve inverse kinematics are computationally demanding and often introduce delays that hinder system responsiveness. This study presents a comparative analysis between two machine learning-based approaches to accelerate inverse kinematics computation: multilayer perceptron (MLP) neural networks and XGBoost, a technique known for its efficiency in regression tasks. Both models were trained on the same dataset generated via direct kinematics and evaluated in terms of prediction accuracy and computation time. The results show that although both models maintain millimetric precision, the XGBoost-based approach achieves a dramatically faster inference speed—2.84 μs per sample versus 150 μs with MLP. These findings confirm that both models are valid for inverse kinematics computation, but the substantially faster inference speed achieved by XGBoost makes it a more suitable choice for real-time robotic applications.

Keywords: Inverse Kinematics · Robotics · XGBoost · Multi-Layer Perceptron (MLP) Neural Network · Machine Learning · Computational Efficiency

1 Introduction

Robotics has established itself as a key discipline in the automation of processes and the advanced interaction with complex environments. Its presence is increasingly common across diverse sectors such as industry, medicine, logistics, and services, where precise and rapid control is essential to ensure the proper functioning of systems [1, 2]. As applications become more dynamic and demanding, the need to implement more efficient control techniques becomes evident [3].

However, the effective control of robotic systems presents significant challenges due to the complexity of their movements and the necessity to respond quickly and accurately in dynamic environments. To achieve precise positional control, two fundamental approaches are commonly used: direct kinematics (DK) and inverse kinematics (IK) [4].

E. Corchado et al. (Eds.): SOCO 2025, CCIS 2806, pp. 289–299, 2026.
https://doi.org/10.1007/978-3-032-19763-4_27

The former allows obtaining the position and orientation of the end-effector from the robot's joint values through a straightforward and computationally light calculation [5]. In contrast, inverse kinematics—which seeks to determine the necessary joint values to reach a target position and orientation—often involves the use of iterative methods. Although effective in many cases, these procedures tend to be computationally expensive and may suffer from convergence or stability issues in certain robot configurations [6]. For this reason, inverse kinematics computation is considered one of the most demanding tasks in robotic control and remains a subject of active research aimed at improving its efficiency and robustness [7].

In this context, the use of machine learning (ML) and regression techniques emerges as a promising alternative to address the IK problem [8]. These strategies allow approximating the inverse function of the kinematic model, drastically reducing inference times without compromising the precision required for real-time applications.

This work presents a comparative study between two machine learning-based approaches for solving inverse kinematics: a model based on multilayer perceptron (MLP) networks and an approach based on the XGBoost algorithm, widely used in regression tasks for its speed and accuracy. Both models were trained on the same dataset, generated from the robot's direct kinematic model, and evaluated in terms of prediction accuracy and inference time [9].

The results show that although both approaches maintain millimetric precision, the XGBoost-based model significantly improves inference time, making it a particularly suitable alternative for environments where real-time control is a priority. This study is aligned with current industrial efforts to develop smarter and more flexible robotic systems, such as those pursued by the MANiBOT project, which focuses on modular robotic platforms for inspection and manipulation in complex environments [10].

The article is organized into five main sections: Sect. 2 describes the fundamentals of inverse kinematics and the algorithms used; Sect. 3 details the data generation and model training processes; Sect. 4 presents the obtained results and their comparative analysis; finally, Sect. 5 draws conclusions and proposes future lines of work.

2 Theoretical Background

In robotic systems, the transformation between control commands and the robot's physical motion relies on two fundamental processes: direct kinematics and inverse kinematics [11]. These formulations enable, respectively, the computation of the end-effector's position from the joint coordinates and the determination of the joint values necessary to reach a desired position in Cartesian space. Both processes are critical for the design of closed-loop controllers, as depicted in Fig. 1, where their integration within complete control architecture, together with the system dynamics, is illustrated.

2.1 Direct Kinematics

Direct kinematics allows determining the position and orientation of the robot's end-effector from the current joint values. This operation is based on the geometric description of the manipulator, typically formulated through homogeneous transformation matrices.

One of the most widely used methods for this purpose is the Denavit-Hartenberg (D-H) convention, which systematizes the assignment of reference frames to each joint using four parameters: link length, link twist, link offset, and joint angle [12].

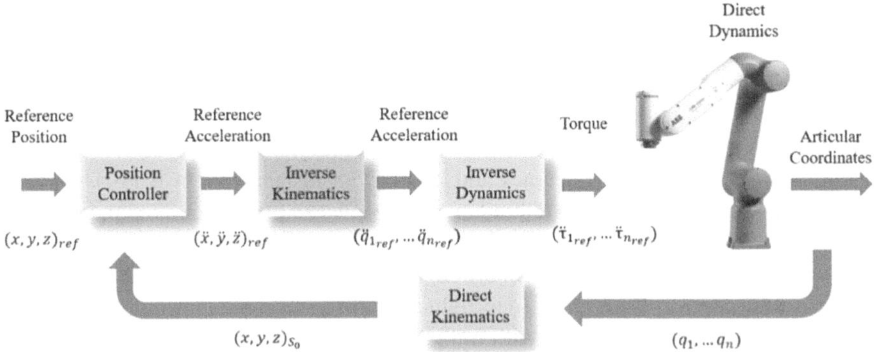

Fig. 1. Control loop of a robot manipulator scheme [9].

For a general 6 DoF robot, the direct kinematic model establishes the following relationships between the joint variables $[q_1, q_2, q_3, q_4, q_5, q_6]$ and the Cartesian position and orientation $[x, y, z, \phi, \theta, \psi]$ as depicted in Eq. 1 [9].

$$v = f_v(q_1, q_2, q_3, q_4, q_5, q_6) \tag{1}$$

Where v denotes the set of the state variables $[x, y, z, \phi, \theta, \psi]$

Thanks to its low computational cost, the direct kinematic model is used not only to control the trajectory of the end-effector but also to verify the validity of joint configurations generated by planning or learning algorithms. In this work, the model is used to generate the dataset used to train machine learning models, serving as the reference to learning the inverse relationship between Cartesian positions and joint values.

2.2 Inverse Kinematics

Inverse kinematics computation consists of finding a joint configuration that allows the end-effector to reach a desired position and orientation in three-dimensional space. This task heavily depends on the mechanical design of the robot, as each architecture presents different degrees of freedom, geometric constraints, and potential singularities.

Various approaches have been proposed to address this problem. Geometric and algebraic methods provide closed-form solutions in simple cases; however, their applicability drastically diminishes as the complexity of the manipulator increases. Consequently, iterative methods based on Jacobian matrix, such as the pseudoinverse or transpose methods, are frequently employed. These techniques provide approximate solutions through successive refinements but typically require a considerable number of iterations and may encounter convergence issues, particularly near singularities or boundary configurations [13].

Although a general analytical solution is not available for most robotic architectures, the inverse kinematics problem can conceptually be expressed as a set of nonlinear functions mapping the Cartesian position and orientation $[x, y, z, \phi, \theta, \psi]$ to the joint variables $[q_1, q_2, q_3, q_4, q_5, q_6]$ as illustrated in Eq. (2).

$$q_i = g_i(x, y, z, \phi, \theta, \psi) \quad (2)$$

Where $i = 1 \ldots 6$

To overcome the limitations of traditional methods, machine learning-based solutions have been proposed. Recent research confirms that ML models can outperform iterative numerical solvers in both accuracy and latency. For instance, Vu et al. [14] achieved $\approx$32µs joint-space solutions for a redundant KUKA iiwa by learning redundancy parameters with a small feed-forward network; Sepahvand et al. [15] mapped camera images directly to joint angles of a continuum arm using a modified VGG-16, enabling vision-based real-time control; Tsui et al. [16] introduced a diffusion-model formulation (IKDP) that generates collision-free poses with millimetric error. These advances motivate our comparison between XGBoost and MLP as data-driven alternatives for fast, reliable inverse-kinematics computation.

3 Robot Inverse Kinematics Based on Machine Learning

The traditional approach to inverse kinematics, based on iterative numerical methods, presents significant limitations when fast and reliable responses are required, particularly in systems that must operate in real time. As an alternative, this work proposes solving the problem with ML techniques trained on data generated by the robot's direct kinematic model. Specifically, a comparative study was conducted between two strategies widely used in regression problems: MLP networks and the XGBoost algorithm.

Both models were designed to take as input the Cartesian position and orientation of the end-effector, expressed through coordinates $[x, y, z]$ and Euler angles $[\phi, \theta, \psi]$, and to output the corresponding set of joint coordinates. This structure allows formulating the inverse kinematics problem as a multivariable regression task.

3.1 Dataset

A synthetic dataset was generated using the *ikpy* library in Python 3.12 to train the MLP network and the XGBoost algorithm. This library enables the computation of direct kinematics based on the Denavit-Hartenberg (D-H) parameters of the robot, allowing the automatic calculation of Cartesian coordinates and end-effector orientations from predefined joint configurations. The robotic model considered corresponds to the collaborative robot CRB15000 (GoFa 5 kg), whose structural description was implemented through its URDF file.

The dataset focused on exploring the last three degrees of freedom of the robot. Full sweeps were performed within a range of $-90°$ to $90°$ for each of these joints, with increments of $1°$. For each fixed value of the fourth joint, all possible values of the fifth joint were evaluated, and for each combination, a complete sweep was conducted over the sixth joint. To simplify the input space without compromising the study's objectives, the first three joints were kept fixed at zero. This sampling strategy generated a large and representative dataset covering a significant portion of the robot's workspace.

Each joint configuration was transformed using *ikpy* into a sample composed of spatial coordinates $[x, y, z]$ and orientation angles $[\phi, \theta, \psi]$, which were used as input features for both ML models. The corresponding outputs consist of the six joint variables $[q_1, q_2, q_3, q_4, q_5, q_6]$, representing the solution to the inverse kinematics problem.

The selection of the explored range was based on a compromise between achieving sufficient workspace coverage and maintaining computational feasibility. Expanding the sweep to a wider range would result in exponential growth of the dataset size, significantly increasing the training time and computational cost, particularly for the MLP model. Nevertheless, the generated dataset provides a reliable basis for evaluating the performance and generalization capabilities of the proposed approaches. Future studies may consider extending this dataset to cover broader trajectories or incorporate different operational modes of the robot.

3.2 MLP Based Model

The first evaluated model is a MLP network, implemented in *Keras* with TensorFlow as the backend. Its architecture consists of an input layer with six neurons, corresponding to the Cartesian and angular components, followed by 4 hidden layers with 512 neurons each. LeakyReLU activation functions were used, along with regularization techniques such as batch normalization and dropout (set at 30%) to prevent overfitting.

The output layer has 6 neurons, one for each joint of the robot. The network was trained using the Adam optimizer, with the mean squared error (MSE) as loss function.

3.3 XGBoost Based Model

The second approach considered for solving the inverse kinematics problem is XGBoost, a machine learning technique based on decision trees that implements the gradient boosting method. The same dataset generated from direct kinematics was used for training, after being normalized through standardization by centering each variable and scaling it according to its standard deviation. The data were divided into a training set (80%) and a test set (20%), using a fixed random seed to ensure reproducibility.

The model configuration was carefully tuned to maximize accuracy without compromising inference speed. Specifically, a maximum tree depth (*max_depth*) of 10 and a *min_child_weight* of 5 were set to promote generalization and prevent overfitting. The learning rate (*eta*) was set to 0.1, and regularization strategies were applied with a penalty of 1 for the L2 term (*lambda*) and 0.1 for the L1 term (*alpha*). High values were assigned to *subsample* and *colsample_bytree* (both set to 0.9), enhancing model robustness by introducing variability among trees. The training objective was the minimization of the mean squared error (*reg:squarederror*), with model performance evaluated using the root mean squared error (RMSE) metric.

Training was performed using the *train()* function from the Python xgboost library, with a maximum of 100 boosting rounds and early stopping enabled with a patience of 10 rounds to terminate training if no substantial improvement was observed on the validation set. Both methodologies are illustrated in Fig. 2a and Fig. 2b.

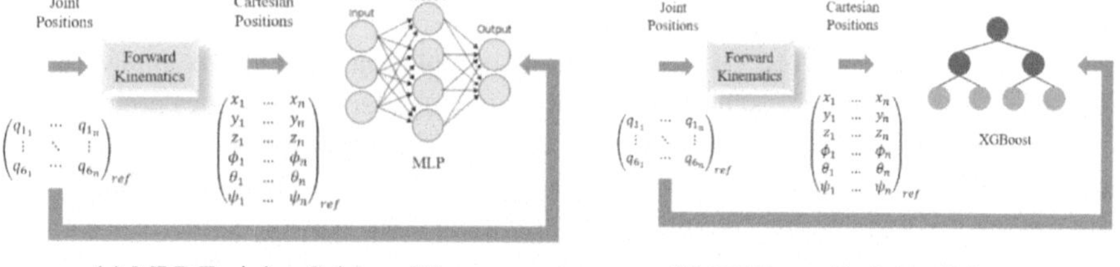

(a) MLP Training Scheme [9]. (b) XGBoost Training Scheme.

Fig. 2. MLP and XGBoost Training Schemes.

3.4 Validation Process

A validation trajectory of 1000 samples was used to evaluate the models. Each sample was generated by applying a sine function to the 3 final joints of the robot, Eq. (3), where $A = 45°$ and t is a time vector uniformly distributed between $-\pi$ and π.

$$q_3(t) = q_4(t) = q_5(t) = q_6(t) = A\sin(t) \tag{3}$$

The Cartesian positions and orientations $[x, y, z, \phi, \theta, \psi]$ corresponding to each joint configuration were obtained using the DK model. These values are the inputs to the trained neural network. Subsequently, the intelligent system predicted the associated joint values, which were then reconverted to Cartesian coordinates using the DK, obtaining the set denoted as $PREDICTEDML_{coordinate}$.

For comparison purposes, the same Cartesian positions were also solved using the conventional iterative inverse kinematics method available in the *ikpy* library, generating the reference set $IK_{coordinate}$.

The accuracy of the predictions is quantified using two metrics calculated in Cartesian space: the mean absolute error (MAE) and the root mean squared error (RMSE), defined respectively by Eqs. (4) and (5), where n is the number of samples.

$$MAE = \frac{1}{n}\sum\nolimits_{i=1}^{n} |IK_{coordinate} - PredictedML_{coordinate}| \tag{4}$$

$$RMSE = \frac{1}{n}\sum\nolimits_{i=1}^{n} \sqrt{\frac{(IK_{coordinate} - PredictedML_{coordinate})^2}{n}} \tag{5}$$

4 Analysis of the Results

This section presents the evaluation results of the two machine learning approaches applied to the IK problem: the MLP network and the XGBoost algorithm.

Figure 3 shows the comparison between the real Cartesian coordinates and those predicted by both models for the X, Y, and Z axes, respectively. It can be observed that both models achieve a high degree of accuracy, with predictions closely following the reference trajectories across the entire set of 1000 samples. The small oscillations visible in the figure are mainly a consequence of having trained with a finite-resolution dataset built at 1° increments. A possible way to reduce these oscillations is to enlarge the dataset and increase its resolution.

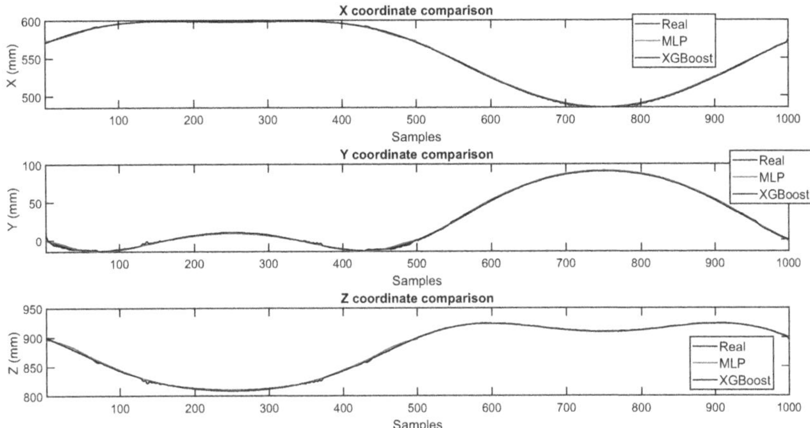

Fig. 3. Cartesian Coordinates Comparison

To further analyze the prediction errors, Fig. 4 presents histograms of the residuals for each coordinate axis. These histograms reveal that the errors obtained with XGBoost are more tightly distributed around zero compared to those of the MLP, particularly in the Y and Z axes. This suggests that XGBoost offers not only a lower mean error but also greater consistency across the samples.

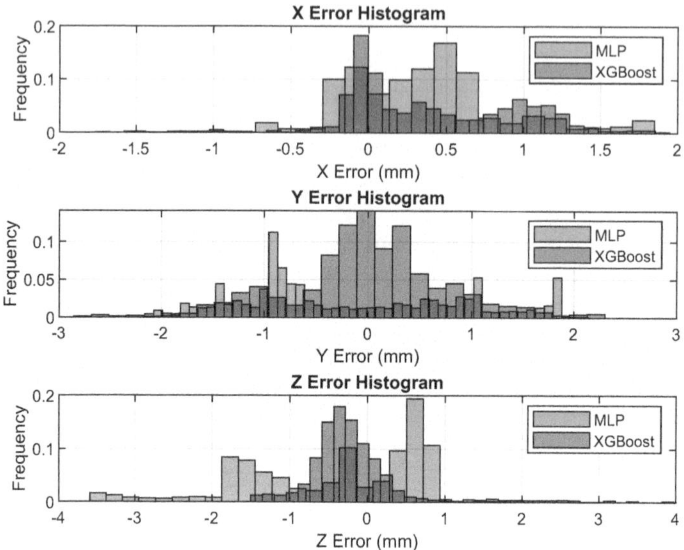

Fig. 4. Cartesian Coordinates Comparison

Quantitative results are summarized in Tables 1 and 2. Table 1 reports the mean absolute error (MAE) and root mean squared error (RMSE) for each coordinate axis. XGBoost consistently outperforms MLP in all metrics, achieving significantly lower error values, particularly notable in the Y and Z axes.

Table 1. Cartesian coordinates error comparison.

Metric	X (mm)	Y (mm)	Z (mm)
MLP - MAE	0,4824	0,9763	0,9654
MLP - RMSE	0,6603	1,0928	1,2388
XGBoost - MAE	0,5006	0,5716	0,5018
XGBoost - RMSE	0,6911	0,8072	0,7688

In addition to accuracy, the mean inference time per sample was measured in an Intel NUC 11 (Core i7-1185G7) for both models, as summarized in Table 2. XGBoost demonstrated a dramatically faster prediction time, achieving an inference speed of 2.84 μs per sample, compared to 150 μs for the MLP model. This substantial difference highlights XGBoost's potential advantages for real-time robotic applications.

Table 2. Mean prediction time comparison.

Metric	XGBoost	MLP
Mean prediction Time per Sample	2,84 μs	150 μs

Overall, both models provided millimetric prediction accuracy; however, XGBoost offered superior performance in terms of both precision and computational efficiency, making it a highly competitive option for real-time inverse kinematics applications.

5 Conclusions and Future Works

This study has presented a detailed comparative analysis between two machine learning approaches—a MLP network and the XGBoost algorithm—applied to the inverse kinematics problem in robotics. Both models were trained on a dataset generated through direct kinematics and evaluated in terms of prediction accuracy and computational efficiency.

The experimental results demonstrate that both approaches achieve millimetric accuracy in predicting the Cartesian coordinates of the end-effector, making them valid solutions for control tasks requiring high precision. However, the XGBoost-based model exhibited a significant advantage in inference speed, achieving an average prediction time of 2.84 μs per sample, compared to 150 μs for the MLP model. This substantial difference positions XGBoost as a highly competitive option for real-time applications, particularly in scenarios where computational resources or cycle times are constrained.

Moreover, the error histograms indicate that XGBoost produces error distributions more tightly centered around zero, especially along the Y and Z axes, suggesting greater robustness to input variations. The trajectory comparison plots confirm that both models are capable of accurately tracking smooth paths with high fidelity.

Several directions for future work are proposed:

- **Expand the training dataset** by including a broader range of joint combinations and wider angular spans, allowing the validation of model generalization over complex trajectories and workspace boundaries.
- **Explore hybrid architectures**, such as combining neural networks with boosting techniques (e.g., DeepGBM) or incorporating attention mechanisms.
- **Implement and validate the models** in a real robotic environment, integrating the machine learning predictions within the robot's real-time control loop.
- **Evaluate additional multivariable regression algorithms**, such as convolutional neural networks (CNNs) or transformer-based models, which may further improve accuracy and generalization capabilities.
- **Robustness to Noise and Uncertainty**: Evaluate model performance under sensor noise, model mismatches, and unpredictable operating conditions to ensure reliability in practical deployments.
- **Dynamics and Continuous-Trajectory Evaluation**: Assess the models on continuous trajectories within the real-time control loop, monitoring potential jitter or instability (as hinted in Fig. 3) and proposing mitigation strategies such as smoothing or feedback compensation.

The results obtained confirm that machine learning not only provides a viable solution but also offers a highly efficient approach to addressing the inverse kinematics problem, delivering fast, accurate, and scalable solutions for present and future robotic systems.

Acknowledgments. This work has been partially supported by European Commission project MANIBOT grant 101120823 and the Spanish Ministry of Science and Innovation under the project MCI/AEI/FEDER number PID2021-123543OB-C21.

References

1. Peñacoba, M., Sierra-García, J.E., Santos, M., Mariolis, I.: Path optimization using metaheuristic techniques for a surveillance robot. Appl. Sci. **13**(20), 11182 (2023)
2. Peñacoba, M., Bayona, E., Sierra-García, J.E., Santos, M.: Route optimization for UVC disinfection robot using bio-inspired metaheuristic techniques. Biomimetics **9**(12), 744 (2024). https://doi.org/10.3390/biomimetics9120744
3. Tinoco, V., Silva, M.F., Santos, F.N., Morais, R., Magalhães, S.A., Oliveira, P.M.: A review of advanced controller methodologies for robotic manipulators. Int. J. Dyn. Control **13**(1), 1–17 (2025)
4. Bouzid, R., Narayan, J., Gritli, H.: Investigating neural network hyperparameter variations in robotic arm inverse kinematics for different arm lengths. In: 2024 Third International Conference on Power, Control and Computing Technologies (ICPC2T), pp. 351–356. IEEE, January 2024
5. Ayazbay, A.A., Balabyev, G., Orazaliyeva, S., Gromaszek, K., Zhauyt, A.: Trajectory planning, kinematics, and experimental validation of a 3D-printed delta robot manipulator. Int. J. Mech. Eng. Robot. Res **13**, 113–125 (2024)
6. Jleilaty, S., Ammounah, A., Abdulmalek, G., Nouveliere, L., Su, H., Alfayad, S.: Distributed real-time control architecture for electrohydraulic humanoid robots. Robot. Intell. Autom. **44**(4), 607–620 (2024)
7. Hughes, N., et al.: Foundations of spatial perception for robotics: hierarchical representations and real-time systems. Int. J. Robot. Res. **43**(10), 1457–1505 (2024)
8. Sierra-García, J.E., Santos, M.: Combining reinforcement learning and conventional control to improve automatic guided vehicles tracking of complex trajectories. Expert. Syst. **41**(2), e13076 (2024)
9. Peñacoba-Yagüe, M., Sierra-García, J.E., Santos-Peñas, M., Ruano, A.: Enhancing robotic control efficiency with MLP-based inverse kinematics: first approach. In: Aguiar, A.P., Rocha Malonek, P., Pinto, V.H., Fontes, F.A.C.C., Chertovskih, R. (eds.) CONTROLO 2024. CONTROLO 2024. Lecture Notes in Electrical Engineering, vol. 1325, pp. 574–582. Springer, Cham (2025). https://doi.org/10.1007/978-3-031-81724-3_51
10. MANiBOT. MANiBOT project (2023). https://manibot-project.eu/
11. Xie, G., Yang, H., Chen, G.: A framework for formal verification of robot kinematics. J. Log. Algebraic Methods Program. **139**, 100972 (2024)
12. Rascón, R., Flores-Mendoza, A., Moreno-Valenzuela, J., Aguilar-Avelar, C.: Control para seguimiento de trayectorias cartesianas en robots manipuladores. Revista Iberoamericana de Automática e Informática industrial **21**(3), 252–261 (2024)
13. Lu, J., Zou, T., Jiang, X.: A neural network based approach to inverse kinematics problem for general six-axis robots. Sensors **22**(22), 8909 (2022)
14. Vu, M.N., Beck, F., Schwegel, M., Hartl-Nesic, C., Nguyen, A., Kugi, A.: Machine learning-based framework for optimally solving the analytical inverse kinematics for redundant manipulators. Mechatronics **91**, 102970 (2023)

15. Sepahvand, S., Wang, G., Janabi-Sharifi, F.: Image-to-joint inverse kinematic of a supportive continuum arm using deep learning. arXiv preprint arXiv:2405.20248 (2024)
16. Tsui, H.T., Tuan, Y.R., Shuai, H.H.: IKDP: inverse kinematics through diffusion process. arXiv preprint arXiv:2410.15341 (2024)

Low Resource Edge-Based Vision for Quadruped Robots: A Comparative Evaluation of Classical HSV and YOLOv5s Under Varying Illumination Conditions

Anabel Díaz-Labrador[1,3](✉), Ángel Delgado[3], Héctor J. Pérez-Iglesias[3], Óscar Fontenla-Romero[2], and José Luis Calvo-Rolle[1]

[1] CTC, CITIC, Department of Industrial Engineering, University of A Coruña, 15071 Ferrol, A Coruña, Spain
anabel.dlabrador@udc.es
[2] LIDIA, CITIC, Faculty of Computer Science, University of A Coruña, 15071 A Coruña, Spain
[3] Fundación Instituto Tecnológico de Galicia, A Coruña, Spain

Abstract. Quadruped robots are gaining traction due to their superior mobility over uneven terrain compared to wheeled platforms. However, their computational constraints pose a challenge when it comes to deploying high-performance models for vision-based navigation and object detection on edge computing platforms. This research work presents a comparative evaluation of a classical HSV colour segmentation approach and a deep learning-based method using YOLOv5s for detecting uniformly coloured objects under varying illumination conditions. Experiments were conducted on a Unitree Go1 quadruped robot equipped with enhanced embedded hardware (Jetson Nano and Raspberry Pi CM4). The test scenario involved detecting four coloured balls (pink, blue, yellow and red) at various distances and under two defined lighting setups: uniform ambient light and focused direct light. The results show that HSV segmentation achieves high detection accuracy at close ranges under stable lighting but performs poorly with illumination variation and darker colours. YOLOv5s offers greater robustness and generalisation to lighting changes, achieving better detection consistency, but at the expense of increased computational load and lower frame rates. The trade-off between computational efficiency and detection robustness is central to choosing a method for edge-based vision systems. Future work will explore hybrid models and lightweight neural architectures tailored to dynamic, resource-constrained robotic environments.

Keywords: Quadruped robots · Edge computing · HSV segmentation · YOLOv5s · Object detection · Illumination variation

1 Introduction

In recent years, quadrupedal robots have become increasingly important as they have demonstrated superior mobility compared to wheeled mobile robots [2].

E. Corchado et al. (Eds.): SOCO 2025, CCIS 2806, pp. 300–311, 2026.
https://doi.org/10.1007/978-3-032-19763-4_28

Robots are often compared to animals on account of their movement and morphology [11], but this particular one has the ability to adapt to changing situations without being in a controlled environment. This makes it useful for a variety of applications, such as patrolling and controlling the security of almost any type of land area.

Over time, many sensors have been used in robotics. These sensors have been used to locate robots in space and for area mapping, also known as Simultaneous Localization and Mapping (SLAM). These sensors were typically LiDAR, ultrasound, inertial measurement units (IMUs), cameras and odometry [12]. However, recent advances in cloud computing and reduction in network latency have driven the development of deep learning based techniques [3]. These techniques make it possible to provide visual context through RGB images under varying lighting conditions, thus mimicking human visual perception. In particular, significant progress has been made in the real-time dynamic detection of targets, from the simplest to the most complex. Among these techniques, convolutional neural networks (CNN) have played a pivotal role, given their powerful feature extraction capabilities and adaptability to various computer vision tasks, including object detection and video prediction [10]. However, these techniques have the disadvantage of requiring high computational power [8].

On the other hand, traditional image processing methods based on colour processing have been widely used to detect specific targets, although they tend to be less accurate [1]. In general, all methods based on visible-spectrum vision suffer from significant limitations in the face of illumination variations, which can significantly affect their performance.

The motivation of this research work is to evaluate the cost-benefit trade-off between a deep learning-based CNN model for machine vision and a classical colour evaluation-based method implemented on the Edge, particularly focused on the detection of a simple object with uniform colour. Due to limitations in the computational capacity of the robot, the implemented CNN model must necessarily be compact and efficient.

To define the structure of the document, the case study is presented below, describing the robot hardware. The vision methods that will be compared are then presented. The experiment setup is then presented, describing how the two experiments were configured. Finally, the results are analysed, and the contributions are concluded, along with proposals for future research.

2 Case Study

A Unitree Go1 quadruped robot, Edu version, was used for this study. The general architecture of the acquired system is shown in Figure refunitree-go1-architecture.

The Go1 model was selected for its stable locomotion, its ability to move nimbly over uneven terrain and its open design, which facilitates both hardware and software modifications. This flexibility makes it an ideal platform for the implementation and evaluation of different systems related to robotics and, as

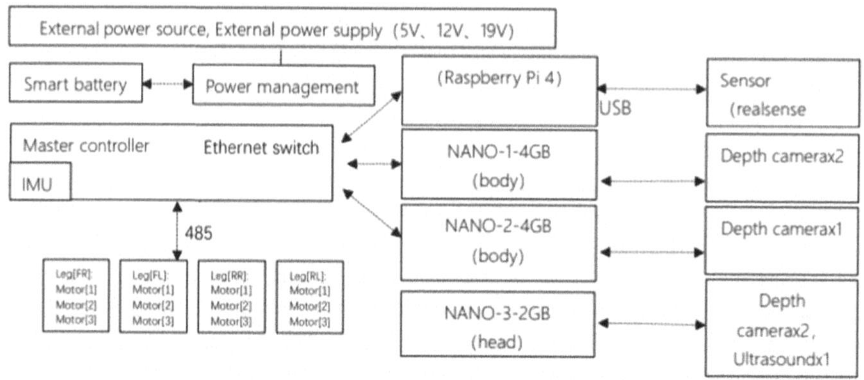

Fig. 1. General architecture of the Unitree Go1 robot.

in this case, machine vision. The availability of the robot at the Instituto Tecnológico de Galicia (ITG) also had a positive influence on its choice, facilitating access and integration in the research work.

In order to improve its performance in artificial intelligence tasks, the following hardware upgrades were made:

- The Jetson Nano integrated into the robot head was replaced by a version with 4GB of RAM.
- The 8GB Jetson Xavier NX was replaced with the 16GB version.
- An NVMe disk was added to allow storage and use of larger models, both computer vision (CV) and natural language processing (NLP).
- The SBC (Session Border Controller) for motor control and communications was upgraded from a Raspberry Pi CM4 with 2GB of RAM and microSD, to a Raspberry Pi CM4 with 8GB of RAM and 32GB of eMMC storage.

The resulting modular architecture allows the computational load to be distributed between the Jetson Nano, responsible for vision and language processing, and the Raspberry Pi CM4, responsible for motor control and communications. This separation favours a balanced distribution of computational resources, which is essential given the constraints of the embedded environment.

As shown in the Fig. 1, the robot has five depth cameras with fisheye optics distributed throughout its structure. These cameras provide three-dimensional spatial perception, which is indispensable for navigation and obstacle detection tasks. However, the optical distortion they introduce makes it difficult to perform precision vision tasks, such as object-specific detection, and their image quality is limited for certain purposes.

For this reason, an additional undistorted fisheye camera was added, connected to the Jetson Nano 2. The Logitech HD Pro Webcam C920 was chosen, with the following technical specifications:

- Max Resolution: 1080 p/30 fps - 720p/ 30 fps

- Camera mega pixel: 3
- Focus type: Autofocus
- Lens type: Glass
- Built-in mic: Stereo
- Mic range: Up to 1 m
- Diagonal field of view (dFoV): 78°
- Tripod-ready universal mounting clip fits laptops, LCD or monitors

Despite these improvements, the system still faces constraints related to power consumption, heat dissipation, autonomy and computational capacity. These limitations directly condition the size and complexity of the deep learning models that can be run in real time on the device.

Experimental tests were carried out in a semi-controlled environment, with variations in lighting conditions, which allowed the robustness of the considered vision methods to light changes to be evaluated.

3 Methods

This research work conceptually compares two main computer vision approaches applied to the detection of objects with uniform colour: a classical approach based on colour segmentation using the HSV (Hue, Saturation, Value) colour model, and a modern deep-learning approach utilizing convolutional neural networks, specifically the YOLO network family [6]. Both methodologies are evaluated conceptually and technically to determine their suitability for practical implementation in embedded systems, notably the Go1 quadruped robot.

3.1 HSV-Based Segmentation

The classical approach employs segmentation in the HSV colour space [7], advantageous due to its robustness against illumination variations compared to the RGB colour model. HSV space represents colour information in a cylindrical coordinate system, separating chromatic content (hue and saturation) from luminance (value):

- **Hue (H)** represents the dominant wavelength, defining the type of colour. It is theoretically measured in degrees ($H \in [0^\circ, 360^\circ]$), though practical implementations, such as OpenCV [5], scale it to an integer range of $[0, 179]$ for 8-bit representation.
- **Saturation (S)** quantifies colour purity, ranging from grayscale ($S = 0$) to fully saturated colours ($S = 1$), thus $S \in [0, 1]$. In practical applications like OpenCV, saturation is represented as an integer in the range $[0, 255]$.
- **Value (V)** corresponds to brightness, ranging from black ($V = 0$) to maximum intensity ($V = 1$). OpenCV represents this component with integer values in the range $[0, 255]$.

Conversion from RGB to HSV space is non-linear and involves computing the maximum (C_{max}) and minimum (C_{min}) intensities among RGB channels. Given a pixel with normalised RGB values (R, G, B) in the range $[0, 1]$, the transformation equations are:

$$V = C_{max}, \quad S = \begin{cases} 0 & \text{if } C_{max} = 0 \\ \frac{C_{max} - C_{min}}{C_{max}} & \text{otherwise} \end{cases} \tag{1}$$

$$H = \begin{cases} 0 & \text{if } C_{max} = C_{min} \\ 60^\circ \cdot \frac{G - B}{C_{max} - C_{min}} & \text{if } C_{max} = R \\ 60^\circ \cdot \frac{B - R}{C_{max} - C_{min}} + 120^\circ & \text{if } C_{max} = G \\ 60^\circ \cdot \frac{R - G}{C_{max} - C_{min}} + 240^\circ & \text{if } C_{max} = B \end{cases} \tag{2}$$

However, practical implementations such as OpenCV use adapted scales:

- RGB values are directly used without explicit normalisation, operating within the range $[0, 255]$.
- Hue values are scaled to $[0, 179]$ instead of $[0^\circ, 360^\circ]$ to accommodate 8-bit integer representation.
- Saturation and Value components are similarly scaled to integer ranges of $[0, 255]$.

Despite differences in scale, this adaptation maintains conceptual equivalence to the theoretical HSV model.

Segmentation in HSV space involves thresholding the H, S, and V channels separately to isolate regions of interest, producing binary masks that highlight potential detection areas. Although this method is computationally efficient and easily adjustable through empirical threshold tuning, its performance may degrade significantly under conditions of variable illumination, shadows, or when target objects exhibit chromatic similarity to the background.

3.2 YOLOv5 Object Detection

The deep-learning approach selected for comparison is YOLOv5s [9], belonging to the *You Only Look Once* (YOLO) family, specifically chosen due to compatibility constraints with the Jetson Nano's Ubuntu 18.04 OS and Python 3.6 environment. YOLOv5s balances computational efficiency and accuracy, making it suitable for real-time applications on resource-constrained devices typical of edge computing contexts.

YOLOv5 is a single-stage detection model that simultaneously regresses bounding boxes and class probabilities from image feature maps, structured into three primary components [4]:

- **Backbone**: CSPDarknet53, employing Cross Stage Partial (CSP) connections to enhance gradient flow and computational efficiency.

- **Neck**: Path Aggregation Network (PANet), integrating multi-scale features for improved detection accuracy across various object sizes.
- **Head**: Generates bounding box coordinates (x, y, w, h), objectivity scores, and class probabilities at multiple scales.

Formally, given an input image $I \in \mathbb{R}^{H \times W \times 3}$, YOLOv5s outputs predictions at multiple scales through feature maps derived from a convolutional backbone and PANet neck. Each output layer encodes bounding box information using anchor-based predictions. These outputs are typically structured as tensors of the form $S \times S \times (B \times (5 + C))$, where B is the number of anchors per cell, and C the number of classes.

In addition to architectural design, YOLOv5s leverages advanced training strategies. Among them, data augmentation techniques—such as mosaic augmentation—play a crucial role in improving robustness and generalization by exposing the model to diverse object scales and spatial configurations during training.

YOLOv5 [9] training utilizes a composite loss function:

$$Loss = \lambda_1 \cdot L_{cls} + \lambda_2 \cdot L_{obj} + \lambda_3 \cdot L_{loc} \tag{3}$$

where L_{cls} is the Binary Cross-Entropy (BCE) loss for class prediction, L_{obj} is the BCE loss for objectness, and L_{loc} is the Complete Intersection over Union (CIoU) loss for localization.

YOLOv5 includes five variants: n, s, m, l, and x, which trade off between inference speed and accuracy. YOLOv5s, with 7.2 million parameters, was selected for its suitability in CPU-limited environments such as the Jetson Nano.

Despite its efficiency, YOLOv5s may produce irrelevant detections when target object classes are underrepresented during training. This highlights a common limitation in object detection pipelines reliant on pre-trained models, especially in domain-specific scenarios. Nonetheless, the model's pretraining on large-scale datasets and use of advanced augmentation strategies, such as mosaic augmentation, contribute to robust generalization across diverse visual conditions.

4 Experiments Setup

The experimental scenario was designed to simulate typical operating conditions of the Unitree Go1 quadruped robot in indoor environments, considering factors such as variations in lighting and distances, which significantly affect the efficiency of the machine vision algorithms. The robot was equipped with a Logitech 1080p camera and a Jetson Nano module. The central objective was to evaluate and compare the performance of two proposed methods in detecting four balls of specific colours (pink, blue, yellow and red), under different controlled lighting conditions and at given distances.

Two clearly defined lighting conditions were established to ensure the reproducibility of the experiments: general office light, representative of uniformly illuminated indoor environments, and focused direct light, which generates high contrast and sharp shadows on the objects.

Seven different distances (1 m, 2 m, 3 m, 4 m, 4.5 m, 5 m, 5.5 m and 6 m) representing practical ranges for navigation and object search applications were selected. For each combination of light condition and distance, three consecutive images were captured with the robot static and pointing directly at the balls, resulting in a total of 48 captures, as seen in Table 1.

Table 1. Experimental configuration overview

Parameter	Value
Distances tested	1.0 m, 2.0 m, 3.0 m, 4.0 m, 4.5 m, 5.0 m, 5.5 m, 6.0 m
Lighting conditions	Uniform ambient light, Direct focused light
Objects	4 colored balls (pink, blue, yellow, red)
Images per config	3 captures per light – distance combination
Total tests	8 distances × 2 lights × 3 samples per config = **48** tests
Success criterion	Detection within ±5 cm of ball center

The detection was considered successful when the ball was correctly identified within the image frame and the centroid of the detected area coincided with the actual position of the ball, within an allowed margin of ±5 cm. In the specific case of the YOLOv5s method, only those detections whose probability of confidence exceeded 30% were accepted.

Each method was implemented under a specific workflow. The HSV method involved the conversion of the RGB colour space to HSV, followed by the application of pre-calibrated chromatic thresholds, Gaussian blur filtering, morphological operations (opening and closing) and finally the selection of the largest detected contour. On the other hand, the YOLOv5s method consisted of a pre-trained Ultralytics model, running GPU-accelerated inferences with CUDA to extract the class, bounding box coordinates and confidence of each detection.

Finally, specific metrics were used to evaluate and compare the methods:

- Detection accuracy, defined as the percentage of images in which the ball was correctly identified
- False negative rate, indicating the percentage of images where the ball was present but not detected
- Computational performance, including the processing rate in FPS and the average CPU/GPU resource usage

The computational performance was recorded using monitoring tools such as jtop on Ubuntu, which provides detailed information about CPU/GPU usage, temperature, power consumption and memory usage during execution. These computational metrics correspond to the averages obtained from three images for each combination of distance and illumination.

5 Results

The results obtained allow us to compare the performance of the HSV and YOLOv5s methods in the task of detecting coloured objects under different lighting conditions and distance ranges. In all cases, three samples per distance and illumination combination were considered, and the success rate per colour was recorded, considering as valid detection the one that met the criteria established in the methodology.

Under mixed artificial light conditions (uniform illumination), both methods showed an overall positive performance at short and medium distances, as seen in Table 2. The HSV approach achieved 100% detection for all colours at 1.2 m, and maintained high detection rates up to 3 m for most colours, especially blue and yellow. From 4.5 m onwards, accuracy decreased markedly for darker colours, notably pink and red, although yellow remained relatively stable even at longer distances. This highlights HSV's vulnerability in distinguishing tones at longer distances. Additionally, false positives appeared in the blue channel from 3.5 m onwards, indicating its sensitivity to reflections and chromatic noise, potentially limiting its applicability in variable lighting conditions.

Table 2. Detection success rate (%) under uniform illumination

Dist (m)	Pink		Blue		Yellow		Red	
	HSV	YOLO	HSV	YOLO	HSV	YOLO	HSV	YOLO
1.0	100	83	100	80	100	78	100	83
2.0	80	83	100	80	100	78	100	83
3.0	80	69	100	78	100	70	100	81
4.0	27	23	33	53	100	63	93	53
4.5	0	0	0	41	100	59	90	39
5.0	0	0	0	0	100	32	100	0
5.5	0	0	0	0	100	16	100	0
6.0	0	0	0	0	100	0	100	0

Under direct focused artificial light conditions, the performance of the HSV method deteriorated rapidly, as shown in Table 3. Only blue achieved consistent detection at the closest distances (100 and 200 cm), while the other colours had very low or null detection rates from 2 m and beyond. This clearly demonstrates HSV's strong susceptibility to brightness variations and shadows, severely limiting its reliability under non-uniform lighting conditions. Figure 2a illustrates this issue at a 100 cm distance, where detections occurred but bounding boxes were incomplete.

The YOLOv5s method demonstrated a more robust performance under adverse lighting conditions compared to HSV (Tables 3 and 2). Under mixed

Table 3. Detection success rate (%) under direct focused illumination

Dist (m)	Pink		Blue		Yellow		Red	
	HSV	YOLO	HSV	YOLO	HSV	YOLO	HSV	YOLO
1.0	26	88	100	83	16	53	1	67
2.0	0	84	100	58	0	0	0	67
3.0	0	0	0	0	0	0	0	0
4.0	0	0	0	0	0	0	0	0
4.5	0	0	0	0	0	0	0	0
5.0	0	0	0	0	0	0	0	0
5.5	0	0	0	0	0	0	0	0
6.0	0	0	0	0	0	0	0	0

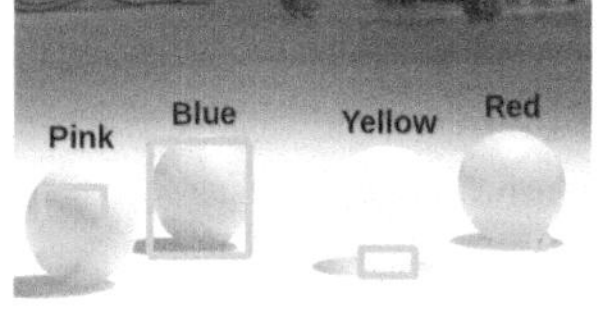

(a) HSV detection at 100 cm under direct light.

(b) HSV detection at 200 cm under mixed artificial light.

Fig. 2. Sample results of HSV-based object detection under different lighting conditions.

artificial illumination, it consistently detected all colours up to 3.5 m, achieving detection rates around or above 80% at short distances. At 4.5 m, detection became partial but remained effective for yellow and blue. However, YOLOv5s performance notably deteriorated beyond 5.5 m, with no detections recorded. Under direct illumination, YOLOv5s maintained acceptable detection rates at close distances (1.2 and 2 m), though less accurate than under uniform lighting conditions, emphasizing its robustness relative to HSV but still illustrating limitations under harsh illumination at longer distances. Figure 3a shows a representative detection at 100 cm, highlighting the relatively high confidence scores despite challenging lighting.

The HSV method showed some false positives, mainly in the blue channel, suggesting it could struggle significantly in environments with reflections or colour noise. YOLOv5s, on the other hand, did not produce false positives but incurred false negatives due to stringent confidence thresholds. This indicates a trade-off between avoiding incorrect detections and potentially missing valid ones, which should be considered carefully according to the application's tolerance for detection errors.

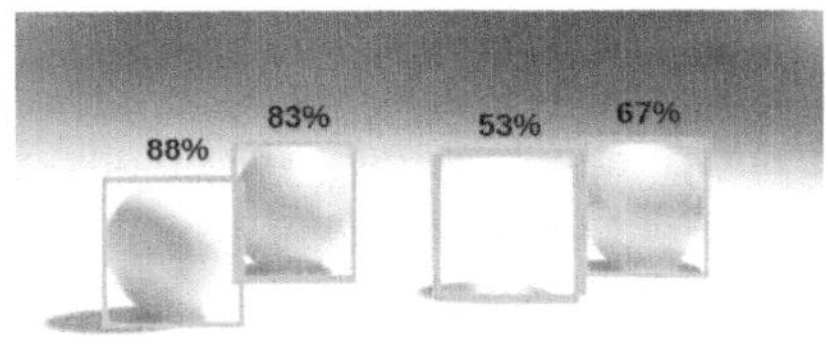

(a) YOLOv5 detection at 100 cm under direct light.

(b) YOLOv5 detection at 300 cm under mixed artificial light.

Fig. 3. Sample results of YOLOv5s object detection under different lighting conditions.

In terms of computational performance, clear differences were reflected in the data between the two approaches. The HSV method achieves an average of 15 – 20 FPS, with stable but high CPU usage (70 – 80% across all cores) and RAM consumption of 2.9 GB (48% swap), while maintaining temperatures of 58 – 65°C. In contrast, YOLOv5s is limited to operating below 5 FPS on the Jetson Nano, with CPU usage of 85–93%, GPU usage of 100% (921 MHz) and VRAM usage of 516 MB, resulting in CPU temperatures above 66°C and GPU temperatures above 61°C, as can be seen in Table 4.

Table 4. Computational performance metrics of HSV and YOLOv5s methods on the Jetson Nano.

Metric	HSV Segmentation	YOLOv5s
Average FPS	15–20	<=5
CPU Usage	70–80%	85–93%
GPU Usage	0%	100% (921 MHz)
RAM Usage	~2.9 GB	~3.2 GB
Swap Usage	48%	62%
VRAM Usage	0%	516 MB
CPU Temperature	58–65°C	66–70 °C
GPU Temperature	N/A	60–65 ÂžC

6 Conclusions and Future Work

This study enabled a comprehensive comparison between HSV colour segmentation and YOLOv5s neural network detection for colour object recognition across varied lighting and distance scenarios, using a Unitree Go1 quadruped with a Jetson Nano onboard. The findings provide actionable insights into the performance, limitations, and practical trade-offs of these two approaches for embedded robotic perception.

The results highlight a clear trade-off between the two methods. HSV segmentation stands out for its high speed (30 fps) and low resource consumption, which makes it suitable for very controlled environments, where the colour of the object to be detected is clearly distinguishable from the environment and is not repeated in other parts of the scene. However, its performance degrades significantly when faced with lighting variations or backgrounds with similar colours. In contrast, YOLOv5s is more robust to varying environmental conditions and allows detections at moderate distances. However, its computational demands limit its use in real time, especially on resource-constrained platforms such as the Jetson Nano, where it achieves only 5 fps. Although with some tolerance it could be considered feasible, it is important to note that currently only one target is detected. If the number of targets is increased, performance could deteriorate further. These findings are particularly relevant for the design of perception modules in mobile robotic systems, where a balance between accuracy, speed and resource availability must be found.

Given the observed strengths and weaknesses of each method, future work should explore hybrid strategies that leverage the speed of HSV segmentation and the robustness of deep learning. For example, HSV can serve as a fast prefiltering step to reduce the search space, followed by more accurate validation using optimised neural networks—thus balancing efficiency and accuracy.

To address the computational limitations observed with YOLOv5s on Jetson Nano, future work should focus on model optimisation techniques such as pruning, quantization, and knowledge distillation. Using deployment frameworks like TensorRT or ONNX can help significantly improve inference speed and efficiency on edge devices.

Another promising direction is the evaluation of lightweight architectures such as MobileNet-SSD or EfficientDet-Lite. These models, combined with transfer learning, can be adapted more efficiently to the specific perception needs of quadruped robots.

Future evaluations should also address more demanding dynamic conditions, including multiple simultaneous objects, frequent occlusions, continuous motion of both the robot and the objects, and more complex backgrounds. These scenarios would improve the ecological validity and generalisation of the findings.

Finally, it is important to investigate the energy consumption and autonomy of the robotic platform under different configurations. Optimising the trade-off between computational performance and battery life is key to enabling practical, long-duration deployment of autonomous robotic systems.

Acknowledgements. Anabel Díaz Labrador's research was supported by the Xunta de Galicia (Regional Government of Galicia in Spain), through grants to industrial PhD (http://gain.xunta.gal/), under the "Doutoramento Industrial 2024" grant with reference: 02_IN606D_2024_3100897.

CITIC, as a Research Center of the University System of Galicia, is funded by Consellería de Educación, Universidade e Formación Profesional of the Xunta de Galicia through the European Regional Development Fund (ERDF) and the Secretaría Xeral de Universidades (Ref. ED431G 2019/01).

Xunta de Galicia. Grants for the consolidation and structuring of competitive research units, GPC (ED431B 2023/49).

References

1. Al-Hetar, A.M., Rassam, M.A., Shormani, O., Salem, A.S.A., Al-Yousofi, H.: Color-based object categorization model using fuzzy HSV inference system. In: 2019 First International Conference of Intelligent Computing and Engineering (ICOICE), pp. 1–6 (2019). https://doi.org/10.1109/ICOICE48418.2019.9035173
2. Biswal, P., Mohanty, P.K.: Development of quadruped walking robots: a review. Ain Shams Eng. J. **12**(2), 2017–2031 (2021)
3. Chan, K.Y., et al.: Deep neural networks in the cloud: review, applications, challenges and research directions. Neurocomputing **545**, 126327 (2023). https://doi.org/10.1016/j.neucom.2023.126327. https://www.sciencedirect.com/science/article/pii/S0925231223004502
4. Khanam, R., Hussain, M.: What is yolov5: a deep look into the internal features of the popular object detector (2024). https://arxiv.org/abs/2407.20892
5. OpenCV Development Team: Open source computer vision library, version 4.11.0 (2025). https://opencv.org/. Stable release 4.11.0 – 9 January 2025
6. Redmon, J., Divvala, S., Girshick, R., Farhadi, A.: You only look once: unified, real-time object detection. In: Proceedings of the IEEE Conference on Computer Vision and Pattern Recognition (CVPR) (2016)
7. Shaik, K.B., Ganesan, P., Kalist, V., Sathish, B., Jenitha, J.M.M.: Comparative study of skin color detection and segmentation in HSV and YCBCR color space. Proc. Comput. Sci. **57**, 41–48 (2015)
8. Sultana, S., Alam, M.M., Su'ud, M.M., Mustapha, J.C., Prasad, M.: A deep dive into robot vision - an integrative systematic literature review methodologies and research endeavor practices. ACM Comput. Surv. **56**(9) (2024). https://doi.org/10.1145/3648357
9. Ultralytics: Yolov5 by ultralytics (2020). https://github.com/ultralytics/yolov5. Accessed: 2025-07-07
10. Zhao, X., Wang, L., Zhang, Y., Han, X., Deveci, M., Parmar, M.: A review of convolutional neural networks in computer vision. Artif. Intell. Rev. **57**(4), 99 (2024). https://doi.org/10.1007/s10462-024-10721-6
11. Zhou, Z., et al.: Progresses of animal robots: a historical review and perspectiveness. Heliyon **8**(11), e11499 (2022)
12. Zhu, J., Li, H., Zhang, T.: Camera, lidar, and IMU based multi-sensor fusion slam: a survey. Tsinghua Sci. Technol. **29**(2), 415–429 (2024). https://doi.org/10.26599/TST.2023.9010010

Special Session: Quantum Computing

Task Allocation on a Quantum Computer for a Two-Machine Manufacturing Cell

Wojciech Bożejko[1(✉)], Sergii Trotskyi[1], Mariusz Uchroński[1,2], and Mieczysław Wodecki[1]

[1] Wrocław University of Science and Technology, Janiszewskiego 11–17, 50–372 Wrocław, Poland
{wojciech.bozejko,sergii.trotskyi,mariusz.uchronski,mieczyslaw.wodecki}@pwr.edu.pl

[2] Wroclaw Centre for Networking and Supercomputing, Wybrzeże Wyspiańskiego 27, 50–370 Wrocław, Poland

Abstract. The paper presents an algorithm for solving the NP-hard problem of task allocation in a two-machine work cell. The calculations were performed on a D-wave quantum computer. The results obtained were compared with the results of the Gurobi package and the simulated annealing algorithm. Taking these into account and the rapid technological progress, it can be assumed that in the near future it will be possible to solve problems on quantum computers that today pose a challenge to algorithms run on classical computers. The research currently conducted provides some knowledge about the computational capabilities of modern quantum computers. They are a necessary step on the way to a breakthrough, which will be the solution of problems on a quantum computer in the future that are considered impossible to solve today.

1 Introduction

Many problems generated by industry and business related to the practice of planning, control, production scheduling, transport, design, management, etc. belong to the class of NP-hard problems. The discreteness of problems and the large size of practical examples mean that it is not possible to determine optimal solutions on classical computers today. Significant progress in algorithm construction methods (e.g., elimination properties of blocks Grabowski, Wodecki [10,11], Bożejko et al. [2]) has increased the expectations of practitioners. New technologies such as parallel computing (multiprocessor computers, clusters), distributed computing in networks, and cloud computing have not resulted in significant improvement. The size of examples that can be solved in an acceptable time has increased by only a few or a dozen units. In practice, therefore, approximate algorithms are used to solve large examples. These are mainly metaheuristics, published en masse since the 1990 s. The most universal and most frequently used are: tabu search, simulated annealing, and genetic algorithm. For many examples, the solutions they determine differ only by a few percent

E. Corchado et al. (Eds.): SOCO 2025, CCIS 2806, pp. 315–325, 2026.
https://doi.org/10.1007/978-3-032-19763-4_29

from the optimal values. A comprehensive review of metaheuristics is included in Siarra's monograph [14].

Great hope is associated with the possibility of performing calculations on quantum computers. Despite the fact that research on quantum computing has been conducted for several decades, it is only in the last few years that the possibility of performing practical calculations has appeared. However, solving discrete optimization problems on a quantum machine is currently significantly limited, among others, due to low availability, a small number of qubits, and the lack of a high-level programming language. Currently, quantum computers are not competitive with classical computers in solving optimization problems.

Today, the main barrier is the number of too few qubits. However, significant technological progress can be observed. Computer companies such as IBM, Rigetti, IQM, D-Wave plan to provide computers with thousands of qubits in the next three years: up to 4000 for the quantum gate model (Rigetti) and over 7000 for quantum annealers (D-Wave). This allows us to assume that in the near future it will be possible to solve problems that today are a challenge for algorithms run on classical computers. Currently conducted research on calculations on modern quantum computers is therefore a certain way to significantly increase the size of solvable examples.

In many complex production processes, we deal with the flow of manufactured parts in a technological sequence. Individual operations are performed in cells equipped with machines. Process scheduling consists in determining the order of flow of parts and in each cell assigning individual operations to machines in order to optimize a certain criterion. Usually, this criterion is to minimize the total execution time of all parts. The process optimization algorithm therefore requires determining:

1. the order of flow of parts through individual cells,
2. each cell, assigning operations to machines and determining the order of their execution on the machines.

This problem (Flexible Multi-machine Production Cell) is one of the most difficult, classic multi-machine scheduling problems today. Different variants of the problem, methods and algorithms for solving them are, among others, presented in the works: Bozejko et al. [6], Pezzella et al. [13], Terkaj et al. [15], Janin et al. [12] and Weckenborg et al. [16].

In the further part of the work we consider the problem of uniform allocation of operations to machines in Flexible Multi-machine Production Cell (Step 2) for the case when the nest contains two identical machines. We present the algorithm that was run on a D-Wave quantum computer. The results obtained were compared with the results of the Gurobi package and the simulated algorithm. Despite the fact that the calculations performed on the D-Wave quantum computer in the process of quantum annealing do not guarantee optimality, the obtained results indicate great potential capabilities of quantum computers. Especially, for example, taking into account the computation times. They also provide some knowledge about the computational capabilities of modern quantum computers.

The work is a continuation of research on methods and constructions of algorithms to solve discrete optimization problems on quantum computers. The results to date have been published in the works of Bożejko et al. [3–5].

The article is structured as follows: In Sect. 2 we define the problem considered in the work and present the method of its solution. In next, Sect. 3 we describe the conditions that the mathematical model of the problem must meet in order to be able to solve it on the D-Wave quantum computer. Section 4 contains a description of the computational experiments performed. The last section contains a brief summary.

2 Problem Formulation and Solution Method

The problem of minimizing the working time of the machining cell (abbreviated as M2C) considered in this work is defined as follows.
Data:
$\mathcal{J} = \{1, 2, \ldots, n\}$ – set of tasks,
$\mathcal{M} = \{M1, M2\}$ – set of machines,
p_i – task $i \in \mathcal{J}$ execution time on a machine.
Tasks must be assigned to machines in such a way that the execution time is minimized (i.e., the work of the machines is balanced - with minimal difference in working times).

We introduce binary variables

$$x_i = \begin{cases} 1, \text{ if task } i \text{ is performed on machine M1}, \\ 0, \text{ otherwise, i.e., on machine M2}, \end{cases} \tag{1}$$

for $i = 1, 2, \ldots, n$. Let Ω denote the set of all binary sequences of length n. For a binary sequence (solution to the M2C problem)

$$\mathbf{x} = (x_1, x_2, \ldots, x_n), \tag{2}$$

$$S_{M1}(\mathbf{x}) = \sum_{i=1}^{n} x_i p_i \text{ and } S_{M2}(\mathbf{x}) = \sum_{i=1}^{n} (1 - x_i) p_i \tag{3}$$

are, respectively, the execution times of tasks on machine M1 and M2. If $S_{sum} = \sum_{i=1}^{n} p_i$, then

$$S_{M2}(\mathbf{x}) = S_{sum} - S_{M1}(\mathbf{x}). \tag{4}$$

Let‘s

$$R(\mathbf{x}) = S_1(\mathbf{x}) - S_2(\mathbf{x}). \tag{5}$$

The solution to the problem involves determining the sequence $\mathbf{x}^\star \in \Omega$ such as

$$R^2(\mathbf{x}^\star) = \min\{R^2(\mathbf{x}) : \mathbf{x} \in \Omega\}. \tag{6}$$

The makespan $C_{max}(\mathbf{x})$ for machines M1 and M2 can be defined as

$$C_{max}(\mathbf{x}) = \max\{S_{M1}(\mathbf{x}), S_{M1}(\mathbf{x})\}.$$

A higher load imbalance $|S_{M1}(\mathbf{x}) - S_{M2}(\mathbf{x})|$ leads to an increase in idle time on one of the machines, which in turn increases $C_{max}(\mathbf{x})$.

To determine the solution to (6) on a D-Wave quantum computer, this task must be formulated as a Quadratic Unconstrained Binary Optimization problem (QUBO), see Glover [9].

3 Running a Task on a D-Wave Quantum Computer

Quantum annealing is a promising computational method that uses quantum phenomena to solve particularly difficult computational tasks. The rapid development of this type of computing is caused by the commercially available quantum annealing device from D-Wave Systems, as well as by the work of the Japanese corporation NEC. The quantum machine performing quantum annealing supports a certain limited class of optimization problems. In order to solve the problem on the D-Wave quantum annealer, the problem under consideration should be formulated as the Quadratic Unconstrained Binary Optimization (QUBO) problem. Currently, intensive searches are being carried out to apply this technology in practice, to learn its possibilities and limitations. The quantum annealer, through the continuous evolution of the quantum system, searches for the minimum energy of the Ising Hamiltonian (see Ajagekar et al. [1], Denkena et al. [8]).

In practice, the problems formulated for a quantum machine performing quantum annealing take the form of the Ising or QUBO model, and translation of problems between these models is trivial. The Ising model is used in statistical mechanics, and the criterion function is:

$$E_{\text{Ising}}(s) = \sum_{i=1}^{N} h_i s_i + \sum_{i=1}^{N} \sum_{j=i+1}^{N} J_{i,j} s_i s_j, \tag{7}$$

where s_i, $i = 1, 2, \ldots, N$ represent spins with values $+1$ and -1, while the linear coefficients corresponding to the qubit deviations are h_i, and the quadratic coefficients corresponding to the coupling forces are J_i, j. In the QUBO model, the function to be minimized has the form

$$f(x) = \sum_{i=1}^{N} Q_{i,i} x_i + \sum_{i=1}^{N} \sum_{j=i+1}^{N} Q_{i,j} x_i x_j, \tag{8}$$

where Q is an upper diagonal matrix with the dimension $N \times N$ of real weights, x is a vector of binary variables.

QUBO is an unconstrained model, which means that in practice all problem constraints must be included in the objective function. Some of the computation models (solvers) *Leap Hybrid* can handle constraints natively – for them, the translation of the constrained problem to the uncostrained problem is done inside the solver. This work is dedicated to such a model – specifically *LeapHybridCQMSampler* (Constrained Quadratic Model, CQM). The formulation of the problem for the CQM model takes the form of minimization:

$$\sum_{i=1}^{N} a_i x_i + \sum_{i=1}^{N} \sum_{j=i+1}^{N} b_{i,j} x_i x_j + c, \tag{9}$$

with constraints:

$$\sum_{i=1}^{N} a_i^{(m)} x_i + \sum_{i=1}^{N} \sum_{j=i+1}^{N} b_{i,j}^{(m)} x_i x_j + c^{(m)} \propto 0, \qquad \mathrm{m} = 1, 2, \ldots, M \tag{10}$$

where x_i, $i = 1, 2, \ldots, N$ can be binary or integer variables, $a_i, b_{i,j}, c$, $i, j = 1, 2, \ldots, N$, are real values, the relation $\propto \in \{\geq, \leq, =\}$ and M is the total number of constraints.

From the definition of the function R^2, the formula (6)

$$R^2(\mathbf{x}) = (S_1(\mathbf{x}) - S_2(\mathbf{x}))^2 = (S_1(\mathbf{x}) - S + S_1(\mathbf{x}))^2 = (2S_1(\mathbf{x}) - S)^2 =$$

$$= (2\sum_{i=1}^{n} x_i p_i - S)^2 = 4(\sum_{i=1}^{n} x_i p_i)^2 - 4S \sum_{i=1}^{n} x_i p_i + S^2. \tag{11}$$

Note that

$$(\sum_{i=1}^{n} x_i p_i)^2 = \sum_{i=1}^{n} x_i^2 p_i^2 + 2 \sum_{i=1}^{n} \sum_{j=i+1}^{n} x_i x_j p_i p_j.$$

Since for binary variables $x_i^2 = x_i$, so finally the expression (11) takes the form

$$R^2(\mathbf{x}) = 4(\sum_{i=1}^{n} x_i p_i^2 + 2 \sum_{i=1}^{n} \sum_{j=i+1}^{n} x_i x_j p_i p_j) - 4S \sum_{i=1}^{n} x_i p_i + S^2. \tag{12}$$

Determining the optimal solution of $\mathbf{x}^* \in \Omega$ of the M2C problem considered in the paper, therefore, comes down to determining

$$\min_{(x_i)_{1 \leq i \leq n} \in 2^{\{0,1\}}} \left(\sum_{i=1}^{n} x_i (p_i^2 - S p_i) + 2 \sum_{i=1}^{n} \sum_{j=i+1}^{n} x_i x_j p_i p_j \right). \tag{13}$$

The above formulation, importantly, does not contain any constraints; it has the form of Quadratic Unconstrained Binary Optimization (QUBO) problem. The calculations can therefore be performed natively on the D-Wave quantum computer.

4 Calculations Performed

The purpose of the experiments conducted was to assess the quality of the solution on the D-Wave quantum computer with `Advantage_system6.4` architecture possessing 5612 qubits. The D-Wave results were compared with the results obtained using the Gurobi optimization package and the Simulated Annealing (SA) algorithm.

As a first reference method, we used the Gurobi solver 2.3.1. As a second reference method, we used SA with an initial $T_{\mathrm{SA}} = 1000$, a cooling rate $= 0.995$, where the initial solution is randomly selected, new solutions are generated using Hamming distance $= 1$, and acceptance follows the Boltzmann criterion:

$$P = \exp\left(-\frac{\mathbf{x_c}^T Q \mathbf{x_c} - \mathbf{x_n}^T Q \mathbf{x_n}}{T_{\mathrm{SA}}},\right)$$

where $\mathbf{x_c}$ and $\mathbf{x_n}$ are current and neighboring solutions.

Task execution times (task processing times) were generated randomly using a uniform distribution. The calculations were performed in the Google Colab environment, which has a total RAM of approximately 12.67 GB.

For solutions $\mathbf{x} \in \Omega$, to measure whether jobs are equally distributed between machines M1 ($S_{M1}(\mathbf{x})$) and M2 ($S_{M2}(\mathbf{x})$), the load imbalance ratio (LIR) was calculated as:

$$LIR(\mathbf{x}) = \frac{|S_{M1}(\mathbf{x}) - S_{M2}(\mathbf{x})|}{S_{M1}(\mathbf{x}) + S_{M2}(\mathbf{x})} = \frac{|2S_{M1}(\mathbf{x}) - S_{sum}|}{S_{sum}}. \tag{14}$$

The LIR helps to express unnormalized values in a normalized format, within the range of $[0, 1]$. The results obtained are presented in Tables 1,2,3,4 and 5 and visualized in two charts 1 –2. Let the set $\Theta = \{$D-Wave, SA, Gurobi$\}$. The other columns of the tables contain, respectively:

- N – number of variables,
- Q – a QUBO matrix size,
- T_A – execution times in seconds (for D-Wave preparation QPU and post-processing [7]),
- LIR_A – load imbalance ratios, for $A \in \Omega$.

The results for a small synthetic dataset (with a number of binary variables ranging from 5 to 25) are presented in Table 1. The execution times are shown in Fig. 1.

Considering the nature of the LIR score—where a value of 0 indicates that tasks are equally distributed across machines, and a value of 1 means that all tasks are scheduled on a single machine with none on the other—and given that our objective in M2C is to achieve a balanced task distribution, multiplying the LIR by 100 provides the percentage deviation $DEV_{OPT,\%}$ from the optimal solution. The $DEV_{OPT,\%}$ for the D-Wave solver on the medium-size dataset typically remains within the ∼0.1–3% range (based on experiments 1–3 for the medium-size dataset). Some outliers were observed in experiment 1 (not reproducible in experiments 2–3) for Q matrices of size 50×50 and 150×150. For the small

Table 1. Execution Times and LIRs for small-size synthetic dataset.

N	Q	T_{Gurobi}	$T_{\text{D-Wave Total}}$	T_{SA}	LIR_{Gurobi}	$LIR_{\text{D-Wave}}$	LIR_{SA}
5	5×5	0.012628	1.081710	0.015881	0.095070	0.095070	0.095070
6	6×6	0.011909	2.229923	0.043143	0.006494	0.006494	0.025974
7	7×7	0.029402	2.176490	0.034325	0.007812	0.007812	0.007812
8	8×8	0.024059	0.856229	0.035630	0.005076	0.005076	0.015228
9	9×9	0.032048	0.880409	0.051442	0.000000	0.009174	0.009174
10	10×10	0.038595	1.756136	0.067533	0.002865	0.008596	0.014327
15	15×15	0.494875	1.145211	0.119920	0.000815	0.000815	0.000815
20	20×20	24.172904	2.026091	0.218571	0.000601	0.088394	0.000601
25	25×25	545.880668	4.389437	0.342394	0.000000	0.090703	0.000000

data set for the 5×5 instance, for all solvers, a higher LIR was observed, the value reaching approximately $\sim$0.09 ($DEV_{OPT,\%}$ reaching approximately $\sim$9%), suggesting that it is a result of the input data.

Table 2. Experiment #1. Execution Times and LIRs for middle-size synthetic dataset.

N	Q	$T_{\text{D-Wave Total}}$	T_{SA}	$LIR_{\text{D-Wave}}$	LIR_{SA}	$DEV_{\text{OPT,D-Wave}}$	$DEV_{\text{OPT,SA}}$
50	50×50	21.256417	1.335597	0.228927	0.002113	$\sim$ 22.9%	$\sim$ 0.2%
75	75×75	41.199345	3.014241	0.000615	0.000923	$\sim$ 0.1%	$\sim$ 0.1%
100	100×100	272.336952	5.472846	0.000798	0.001255	$\sim$ 0.1%	$\sim$ 0.1%
125	125×125	534.657800	8.447932	0.000673	0.000096	$\sim$ 0.1%	$\sim$ 0.0%
150	150×150	538.048275	11.946800	0.063488	0.000227	$\sim$ 6.3%	$\sim$ 0.0%

The results presented in Tables 1, 2, 3, and 4 (with the data in Tables 2, 3 and 4 corresponding to three re-runs on medium-sized datasets) show that Gurobi efficiently provides optimal solutions for small-sized problems (up to 15 jobs). For medium-sized datasets, Gurobi was excluded from evaluation due to its long execution times (without setting a time limit). As the problem size increases, the execution time also increases. In contrast, Simulated Annealing consistently delivered competitive results with relatively low execution times. The data further suggests that, across all experiments, D-Wave's execution times were generally longer than those of Simulated Annealing for problem sizes up to 150 jobs. Advanced D-Wave timing information like (in μs) for Experiment 1 (middle-size dataset):

- $T_{\text{D-Wave Total}}$ – are the execution times for D-Wave (preparation, QPU and post-processing [7]),
- ST_{QPU} – Sampling Time,
- $ATPS_{\text{QPU}}$ – Anneal Time Per Sample,

Table 3. Experiment #2. Execution Times and LIRs for middle-size synthetic dataset.

N	Q	$T_{\text{D-Wave Total}}$	T_{SA}	$LIR_{\text{D-Wave}}$	LIR_{SA}	$DEV_{\text{OPT,D-Wave}}$	$DEV_{\text{OPT,SA}}$
50	50×50	14.883785	0.992645	0.006888	0.002152	~ 0.7%	~ 0.2%
75	75×75	93.333812	2.124789	0.016206	0.003001	~ 1.6%	~ 0.3%
100	100×100	64.795899	3.683603	0.011306	0.000231	~ 1.1%	~ 0.0%
125	125×125	482.616506	5.770808	0.002523	0.000776	~ 0.2%	~ 0.1%
150	150×150	350.977091	8.474834	0.038751	0.000076	~ 3.9%	~ 0.0%

- $RTPS_{\text{QPU}}$ – Readout Time Per Sample,
- AT_{QPU} – Access Time,
- AOT_{QPU} – Access Overhead Time,
- PT_{QPU} – Programming Time,
- $DTPS_{\text{QPU}}$ – Delay Time Per Sample

are presented in the columns of Table 5.

Table 4. Experiment #3. Execution Times, Energies, and LIRs for middle-size synthetic dataset.

N	Q	$T_{\text{D-Wave Total}}$	T_{SA}	$LIR_{\text{D-Wave}}$	LIR_{SA}	$DEV_{\text{OPT,D-Wave}}$	$DEV_{\text{OPT,SA}}$
50	50×50	10.840082	0.945648	0.007862	0.000491	~ 0.8%	~ 0.0%
75	75×75	61.559180	2.101507	0.017975	0.001892	~ 1.8%	~ 0.2%
100	100×100	63.447477	3.713767	0.019544	0.000116	~ 1.9%	~ 0.0%
125	125×125	243.048601	5.814203	0.030899	0.000197	~ 3.1%	~ 0.0%
150	150×150	846.365334	8.400427	0.003687	0.000885	~ 0.4%	~ 0.1%

Table 5. D-Wave Advanced information for Experiment 1. Execution Times and LIRs for middle-size synthetic dataset.

N	Q	$T_{\text{D-Wave Total}}$	$LIR_{\text{D-Wave}}$	ST_{QPU}	$RTPS_{\text{QPU}}$	AT_{QPU}	AOT_{QPU}	PT_{QPU}	$DTPS_{\text{QPU}}$
50	50×50	21.256417	0.228927	1958.4	155.26	17720.36	1075.64	15761.96	20.58
75	75×75	41.199345	0.000615	1948.8	154.3	17710.36	1325.64	15761.56	20.58
100	100×100	272.336952	0.000798	2304.4	189.86	18067.56	933.44	15763.16	20.58
125	125×125	534.657800	0.000673	2381.6	197.58	21034.2	1877.8	18652.6	20.58
150	150×150	538.048275	0.063488	2613.4	220.76	18381.36	2356.64	15767.96	20.58

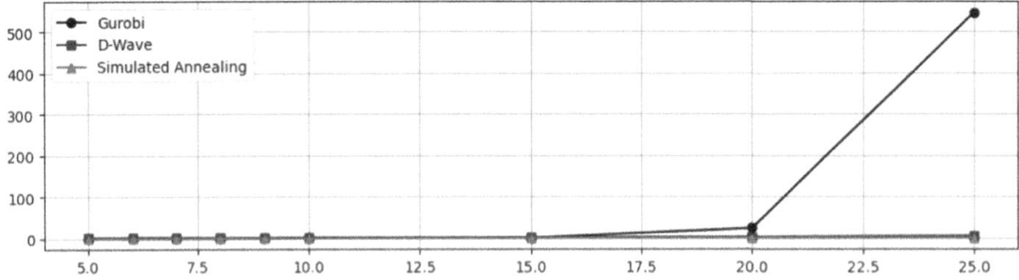

Fig. 1. Execution Times for Gurobi, D-Wave and Simulated Annealing solvers for small-size dataset (up to 25 jobs)

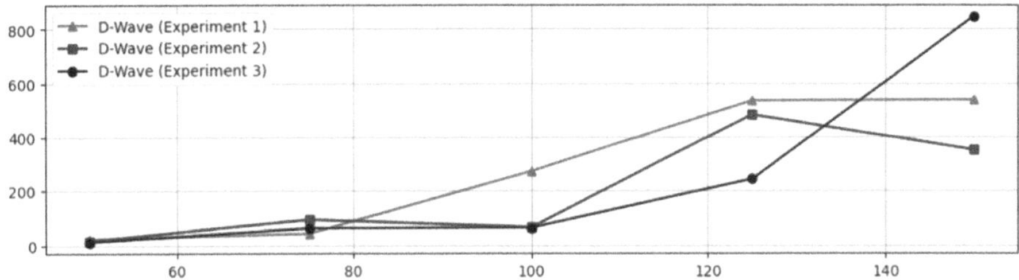

Fig. 2. Execution Times for D-Wave Pegasus solver for medium-size dataset (50–150 jobs). Experiment # 1, 2, 3

5 Conclusion

In this study, we formulated the two-machine job scheduling problem (not including into consideration job sequencing) as a QUBO model to tackle the NP-hard task allocation problem within a two-machine work cell. The model's performance was checked using different solvers, including D-Wave's quantum annealer (Pegasus), and compared against classical methods such as the Gurobi optimizer and the simulated annealing.

We compared execution times as well as LIR and $DEV_{\mathrm{OPT},\%}$ metrics across both small and medium-sized datasets. Gurobi performed well on smaller datasets but required significantly more time as problem size increased. Meanwhile, simulated quantum annealing produced great results, indicating its potential scalability.

The current scalability limits of quantum machines, especially quantum annealers, allow in practice to solve only small-sized examples of NP-hard problems. Depending on whether the problem under consideration has constraints or not, it is possible to solve an instance of a size from several tasks (e.g., for the job-shop problem, which has relatively many constraints) to several hundred (for problems with a small number of constraints, such as the knapsack problem, or completely without constraints, such as the set partition problem). The binary representation of the initial problem is particularly useful here. Additionally, the connectivity of the topology of connections between qubits in quantum machines is gradually improving (e.g. the Advantage2 model by D-Wave). In summary, adiabatic quantum computers should soon be able to solve complex optimization problems, the solution of which is currently time-consuming in the case of classical approaches.

References

1. Ajagekar, A., Humble, T., You, F.: Quantum computing based hybrid solution strategies for large-scale discrete-continuous optimization problems. Comput. Chem. Eng. **132**, 106630 (2020)
2. Bożejko, W., Grabowski, J., Wodecki, M.: Block approach tabu search algorithm for single machine total weighted tardiness problem. Comput. Ind. Eng. **50**(1/2), 1–14 (2006)
3. Bożejko, W., et al.: Optimal solving of a scheduling problem using quantum annealing metaheuristics on the d-wave quantum solver. IEEE Trans. Syst. Man Cybern. Syst. **55**(1), 196–208 (2024)
4. Bożejko, W., Pempera, J., Uchroński, M., Wodecki, M.: Quantum annealing-driven branch and bound for the single machine total weighted number of tardy jobs scheduling problem. Futur. Gener. Comput. Syst. **155**, 245–255 (2024)
5. Bożejko, W., Trotskyi, S., Uchroński, M., Wodecki. M.: Comparison of d-wave quantum computing environment solvers for a two-machine jobs scheduling problem. In: Quintian, H., et al. (eds.) SOCO 2024, vol. 888 of Lecture Notes in Networks and Systems, pp. 68–76 (2024)
6. Bożejko, W., Uchroński, M., Wodecki, M.: Multi-GPU tabu search metaheuristic for the flexible job shop scheduling problem. In: Klempous, R., Nikodem, J., Chaczko, Z. (eds.) Advanced Methods and Applications in Computational Intelligence, vol. 6 of Topics in Intelligent Engineering and Informatics, pp. 43–60. Springer (2014)
7. D-Wave Systems Inc. QPU timing: understanding quantum processing unit timing(2024). https://docs.dwavesys.com/docs/latest/c_qpu_timing.html
8. Denkena, B., Schinkel, F., Pirnay, J., Wilmsmeier, S.: Quantum algorithms for process parallel flexible job shop scheduling. CIRP J. Manuf. Sci. Technol. **33**, 100–114 (2021)
9. Glover, F., Kochenberger, G., Du, Y.: A tutorial on formulating and using qubo models (2019). https://arxiv.org/abs/1811.11538
10. Grabowski, J., Wodecki, M.: A very fast tabu search algorithm for the permutation flow shop problem with makespan criterion. Comput. Oper. Res. **31**, 1891–1909 (2004)
11. Grabowski, J., Wodecki, M.: A very fast tabu search algorithm for the job shop problem. In: Rego, C., Alidaee, B. (eds.) Adaptive memory and evolution; tabu search and scatter search, pp. 117–144. Kluwer Academic Publishers, Dordrecht (2005)
12. Jain, A., Jain, P.K., Chan, F.T.S., Singh, S.: A review on manufacturing flexibility. Int. J. Prod. Res. **51**(19), 5946–5970 (2013)
13. Pezzella, F., Morganti, G., Ciaschetti, G.: A genetic algorithm for the flexible job-shop scheduling problem. Comput. Oper. Res. **35**(10), 3202–3212 (2008)
14. Siarry, P.: Metaheuristics. Springer International Publishing (2016)
15. Terkaj, W., Tolio, T., Valente, A.: A review on manufacturing flexibility. In: Tolio, T. (ed.) Design of Flexible Production Systems. Springer, Berlin, Heidelberg (2009)

16. Weckenborg, Ch., Schumacher, P., Thies, Ch.: Flexibility in manufacturing system design: a review of recent approaches from operations research. Eur. J. Oper. Res. **315**(2), 413–441 (2024)

Solving Capacitated Vehicle Routing Problem with Equal Demand on a D-Wave Quantum Machine

Jan Gromiec, Michał Kapica, Michał Nowicki, Mariusz Uchroński(✉), and Mateusz Wasilewski

Wrocław University of Science and Technology, Wyb. Wyspiańskiego 27, 50-370 Wrocław, Poland
mariusz.uchronski@pwr.edu.pl

Abstract. The article presents an analysis of equal demand capacitated vehicle routing problem quantum-ready formulations. Analysed formulations were based on different approaches for constraints definition, such as remaining vehicle capacity between customers, remaining vehicle capacity at the customer, a truck serving customer binary decision variables, and only goal function without constraints. Considered formulations were evaluated on classical computer and quantum computer. Computational experiments confirm that the equal demand capacitated vehicle routing problem can be solved on a quantum computer, and results strongly depend on the problem formulation and its size.

Keywords: capacitated vehicle routing problem · equal demand · D-Wave

1 Introduction

Capacitated Vehicle Routing Problem has numerous applications in many fields of science and industry such as logistics and urban mobility. As outlined by Sebastian Feld et al. [5], CVRP is a NP-hard optimization problem, where every vehicle has a constrained capacity. Classical solution methods, while powerful, often struggle with scalability in real-time or large-scale settings. The goal of this problem is to find the most efficient route between customers. Eneko Osaba et al. [8] stated that the main objective of real-world routing problems is to present a solution with realistic instances while maintaining the restrictions of the original real-world problem, and he advanced his research in another study et al. [9]. In both articles it was demonstrated that hybrid approaches are a promising alternative to traditional computing resources. According to Michał Borowski et al. [3] hybrid solutions for solving VRP and CVRP can also give effective and realistically optimised results. Salvatore Sinno et al. [12] demonstrated that using a D-Wave CQM solver QUBO formulation is able to find feasible solutions with maximum of 15 nodes in CVRP. Anthony Palmieri [10] stated that the most efficient route is achieved by minimizing the total distance travelled by vehicles or

E. Corchado et al. (Eds.): SOCO 2025, CCIS 2806, pp. 326–336, 2026.
https://doi.org/10.1007/978-3-032-19763-4_30

minimizing the overall cost. David Simchi-Levi et al. [11] introduced the Equal Demand Capacitated Vehicle Routing Problem (EDCVRP). There are a lot of various CVRP variants however, due to the limitations of quantum computing, EDCVRP provides a computationally approachable yet non-trivial test base to estimate the capabilities and limitations of such technology. Ilan Karpas et al. [7] compared methods of solving EDCVRP using different formulations and highlighted the superiority of quantum-inspired platform LightSolver over classical solver Gurobi. Our goal is to make a comparison between classical, quantum and hybrid approaches to EDCVRP, where each of them was implemented using the following four formulations based on: remaining vehicle capacity between customers (F1), remaining vehicle capacity at customer (F2), truck serving customer binary decision variables (F3) and with truck serving customer binary decision variables without constraints (F4). In our study, the ILP (Integer Linear Programming) models were implemented according to two established formulations: one proposed by Deleplanque et al. [4], and another derived from the models of Godinho et al. [6] and Borcinowa [2]. Each formulation was considered and evaluated on CPU using Gurobi solver and on QPU with D-Wave samplers – `LeapHybridCQMSampler` and `DWaveSampler`. We are currently in the so-called NISQ (noisy intermediate scale quantum) era, which means that quantum technology is neither completely reliable nor capable of natively solving medium or large-scale problems. For that reason, this paper focuses on formulation efficiency and quantum readiness, not real-world instance optimisation.

2 Problem Formulation

Capacitated vehicle routing problem with equal demand can be defined as follows. There is set of n customers which should be served (by delivering some goods) by m vehicles with limited capacity Q. Each vehicle starts and finish its delivery route in depot. Between any two locations (two customers or the depot and a recipient) the distance is d_{ij}. Demand for all customers is equal 1 ($D = 1$) which means that each vehicle visit Q customers. The goal is to assign orders to the vehicles and plan routes for all of vehicles, while minimizing the total distance traveled by all vehicles.

Considered capacitated vehicle routing problem with equal demand can be mathematically formulated using some graph representation. Let $G = (V, E, d)$ be a complete directed graph, where $V = \{0, 1, 2, \ldots, n\}$ is a set of nodes and $E = \{(i, j) : i, j \in n, i \neq j\}$ is a set of edges, where node 0 represents the depot for m vehicles with the same capacity Q. Remaining n nodes represents customers. The number of customers is determined by truck capacity and node count according to this equation $n = m * Q$. Each customer has equal demand $D = 1$. With each edge $(i, j) \in E$ of graph G some non negative travel distance d_{ij} is associated.

2.1 Formulation Based on Remaining Vehicle Capacity Between Customers (F1)

Let there be following variables:

- x_{ij} – binary decision variable equals 1 if vehicle travels directly from customer i to the customer j, 0 otherwise,
- ϱ_{ij} – positive integer variable represents the remaining capacity of the vehicle when it travels from customer i to customer j.

The mixed integer linear programming model for capacitated vehicle routing problem with equal demand can be formulated as follows.
Minimize:

$$\sum_{i=1}^{n}\sum_{j=1}^{n} x_{ij}d_{ij} \tag{1}$$

Subject to:

$$\sum_{i=0}^{n} x_{ij} = 1,\ j = 1, \ldots, n \tag{2}$$

$$\sum_{j=0}^{n} x_{ij} = 1,\ i = 1, \ldots, n \tag{3}$$

$$\sum_{j=1}^{n} x_{0j} = m \tag{4}$$

$$\sum_{i=1}^{n} x_{i0} = m \tag{5}$$

$$0 \leq \varrho_{ij} \leq Qx_{ij},\ i, j = 1, \ldots, n, \tag{6}$$

$$\sum_{i=0}^{n} \varrho_{ik} - \sum_{j=0}^{n} \varrho_{kj} = 1,\ k = 1, \ldots, n \tag{7}$$

The objective function (1) minimizes total travel distance for all vehicles. Constraints (2)–(3) guarantee that each recipients is visited only once. Constraints (4)–(5) guarantee that each vehicle leaves depot and each vehicle is returning to the depot. Vehicles capacity constraints (6) – remaining capacity of the vehicle when travels from customer i to the customer j is less than or equal to the capacity of the vehicle Q. Constraint (7) ensures that the capacity of truck decreases by 1 after each visit and tracks the path of the vehicles.

2.2 Formulation Based on Remaining Vehicle Capacity at Customer (F2)

Let there be following variables:

- x_{ij} – binary decision variable equals 1 if vehicle travels directly from customer i to the customer j, 0 otherwise,
- u_i – positive integer variable represents the remaining capacity of the vehicle at customer i.

The mixed integer linear programming model for capacitated vehicle routing problem with equal demand can be formulated as follows.
Minimize:

$$\sum_{i=1}^{n}\sum_{j=1}^{n} x_{ij}d_{ij} \tag{8}$$

Subject to:

$$\sum_{i=0}^{n} x_{ij} = 1, \; j = 1, \ldots, n \tag{9}$$

$$\sum_{j=0}^{n} x_{ij} = 1, \; i = 1, \ldots, n \tag{10}$$

$$\sum_{j=1}^{n} x_{0j} = m \tag{11}$$

$$\sum_{i=1}^{n} x_{i0} = m \tag{12}$$

$$u_i - u_j + Qx_{ij} \leq Q - 1, \; i, j = 1, \ldots, n \tag{13}$$

The objective function (8) minimizes total travel distance for all vehicles. Constraints (9)–(10) guarantee that each recipients is visited only once. Constraints (11)–(12) guarantees that each vehicle leaves depot and each vehicle is returning to the depot. Constraint (13) ensures that the capacity of truck decreases by 1 after each visit.

2.3 Formulation Based on Serving Customer Binary Decision Variable (F3)

Considered problem can be mathematically formulated using serving customer binary decision variable. Let x_{kij} be a binary decision variable that is equal to 1 if vehicle k at i-th site is serving customer j. The mixed integer linear programming model for capacitated vehicle routing problem with equal demand can be formulated as follows.

Minimize:

$$\sum_{k=1}^{m}\sum_{j=1}^{n} d_{0j}x_{k1j} + \sum_{k=1}^{m}\sum_{j=1}^{n} d_{j0}x_{kQj} + \sum_{k=1}^{m}\sum_{i=1}^{Q-1}\sum_{n1=1}^{n}\sum_{n2=1}^{n} d_{n1n1}x_{kin1}x_{k(i+1)n2} \tag{14}$$

Subject to:

$$\sum_{j=1}^{n} x_{kij} = 1,\ k = 1, \ldots, m,\ i = 1, \ldots, Q, \tag{15}$$

$$\sum_{k=1}^{m}\sum_{i=1}^{Q} x_{kij} = 1,\ j = 1, \ldots, n \tag{16}$$

The objective function (1) minimizes total travel distance for all vehicles. The first sum corresponds to the first part of the vehicle route – leaving the depot. Similarly, the second sum corresponds to the last part of the vehicle route – return to the depot. The last sum corresponds to the intermediate part of the vehicle route. Constraint (15) guarantee that when vehicle k visits i-th site, it serves exactly one customer. Constraint (16) guarantees that every customer j is served once by one vehicle.

2.4 Formulation Based on Serving Customer Binary Decision Variable – QUBO (F4)

For QUBO model formulation all decision variables must be binary, all constraints must be in equality form and all constraints must be moved to the goal function. In formulation F3, all variables are binary and all constraints have equality form. Constraints (15)–(16) can be moved to the goal function in the following form:

$$\sum_{k=1}^{m}\sum_{i=1}^{Q}(\sum_{j=1}^{n} x_{kij} - 1)^2 = 0 \tag{17}$$

$$\sum_{j=1}^{n}(\sum_{k=1}^{m}\sum_{i=1}^{Q} x_{kij} - 1)^2 = 0 \tag{18}$$

Finally the objective function minimizes total travel distance for all vehicles can be defined as follows:

$$F(x) = F_{cost}(x) + \lambda F_{constraints}(x), \tag{19}$$

where λ is sufficiently large coefficient equals $n \cdot 10^3$ and

$$F_{cost}(x) = \sum_{k=1}^{m}\sum_{j=1}^{n} d_{0j}x_{k1j} + \sum_{k=1}^{m}\sum_{j=1}^{n} d_{j0}x_{kQj} + \sum_{k=1}^{m}\sum_{i=1}^{Q-1}\sum_{n1=1}^{n}\sum_{n2=1}^{n} d_{n1n2}x_{kin1}x_{k(i+1)n2} \tag{20}$$

and

$$F_{constraints}(x) = \sum_{j=1}^{n}(\sum_{k=1}^{m}\sum_{i=1}^{Q} x_{kij} - 1)^2 + \sum_{k=1}^{m}\sum_{i=1}^{Q}(\sum_{j=1}^{n} x_{kij} - 1)^2 \tag{21}$$

3 Computational Experiments

Presented mathematical models of capacitated vehicle routing problem with equal demand was implemented and then executed on CPU using Gurobi solver and also executed on QPU D-Wave quantum computer. Implementation for Gurobi solver was prepared using Python library `gurobipy` version 12.0.1. Similarly implementation for D-Wave quantum computer was prepared using Python `dimod` library version 8.1.0. Computation experiments for Gurobi solver was executed on the server with Intel(R) Core(TM) i7-7700 3.60 GHz CPU working under Fedora 41 operating system. Computational experiments using `DWaveSampler` and `LeapHybridCQMSampler` was executed on D-Wave quantum computer trough Leap Quantum Application Environment with `Advantage_system4.1` solver. `DWaveSampler` use quantum solver executed directly on QPU. Before executing sampling problem was automatically converted to QUBO (if necessarily) and then mapped (embedded) to the physical architecture of quantum computer. `LeapHybridCQMSampler` use quantum–classical hybrid solver where QPU is used for guiding multithreaded classical heuristic module to the more promising areas of the search space or to improve existing solutions.

Test instances were constructed as follows. For each instance, customer locations were randomly generated with uniform distribution on a two-dimensional integer grid of size 100×100. The depot was deterministically placed at the centre of the grid, at coordinates $(50, 50)$. Given the total number of customers $n = m \cdot Q$, where m denotes the number of vehicles and Q the number of customers per vehicle. Customer coordinates $(x_i,\ x_j)$ were sampled uniformly at random from the range $[1, 100]$. The complete set of $n + 1$ locations (including the depot) was used to construct a distance matrix D, where each entry d_{ij} represents the Euclidean distance between location i and location j.

$$d_{ij} = \left\lceil \sqrt{(x_i - x_j)^2 + (y_i - y_j)^2} \right\rceil, \ \forall\ i \neq j = 0, 1, \ldots, n. \tag{22}$$

The depot was always assigned to index 0 of the matrix, and the remaining indices were mapped to randomly generated customer locations. The test instances used in this study were randomly generated to reflect realistic and challenging scenarios. The random placement ensures comparability across different formulations. It was deliberately avoided to use real-life scenarios to maintain the theoretical focus of this work. The primary goal was to assess the behaviour and limitations of the D-Wave quantum computer under clean, synthetic input conditions, rather than solving applied logistics problems.

Computational experiments results – goal functions values – was presented in Tables 1, 2 and 3. Particular columns in tables denote:

- m – number of vehicles,
- Q – vehicle capacity,
- n – number of customers,
- F1 – goal function value for model with remaining vehicle capacity between customers,
- F2 – goal function value for model with remaining vehicle capacity at customer,
- F3 – goal function value for model with truck serving customer binary decision variables,
- F4 – goal function value for QUBO model with truck serving customer binary decision variables.

Table 1. Goal functions values for Gurobi solver with 600 s time limit.

m	Q	n	F1	F2	F3	F4
2	4	8	153	153	153	153
2	5	10	391	391	391	391
2	10	20	442	442	443	459
2	15	30	496	496	519	508
2	20	40	581	581	613	615
4	5	20	601	601	601	601
4	10	40	616	616	756	655
4	15	60	741	742	835	920
4	20	80	894	923	1485	1240
8	5	40	977	982	979	994
8	10	80	1185	1226	1346	1373
8	15	120	1265	1331	6380	1976
8	20	160	–	–	8833	2732
10	5	50	1179	1191	1187	1216
10	10	100	1422	1493	1640	1884
10	15	150	–	–	7926	2744
10	20	200	–	–	11069	4990

Table 2. Goal functions values for D-Wave `DwaveSampler`

m	Q	n	F1	F2	F3	F4
2	4	8	153	153	153	153
2	5	10	458	391	391	402
2	10	20	1044	544	443	811
2	15	30	–	699	1182	1206
2	20	40	–	–	605	1717
4	5	20	859	677	601	834
4	10	40	–	–	671	1616
4	15	60	–	–	870	2708
4	20	80	–	–	1115	3632
8	5	40	–	–	992	1840
8	10	80	–	–	1443	3962
8	15	120	–	–	1555	5293
8	20	160	–	–	2220	7865
10	5	50	–	–	1214	2448
10	10	100	–	–	1595	4878
10	15	150	–	–	2074	7383
10	20	200	–	–	2561	9638

Table 3. Goal functions values for D-Wave `LeapHybridCQMSampler`

m	Q	n	F1	F2	F3	F4
2	4	8	–	–	192	187
2	5	10	–	–	449	584

Figure 1 presents visual comparison of the vehicles routes obtained on D-Wave quantum computer for each analysed formulation for $n = 20$ customers. There are noticeable differences in route efficiency and structure between used formulations for solving considered problem. The F3 formulation clearly yields the most spatially efficient routes, characterized by minimal overlaps and relatively short total path lengths. This aligns with properties typically seen in near-optimal solutions to the Euclidean Traveling Salesman Problem, where routes tend to follow compact, non-intersecting loops. The F4 formulation, with slightly less efficient route also demonstrates a well-structured routing pattern. This reflects the effective enforcement of capacity constraints, despite the additional complexity introduced by the QUBO-based representation. On the other hand, routes produced by F2 are visibly more overlapping with each other, with a high number of intersections. The F1 formulation generates the least efficient route with the most route crossings. This lack of locality likely contributes to longer

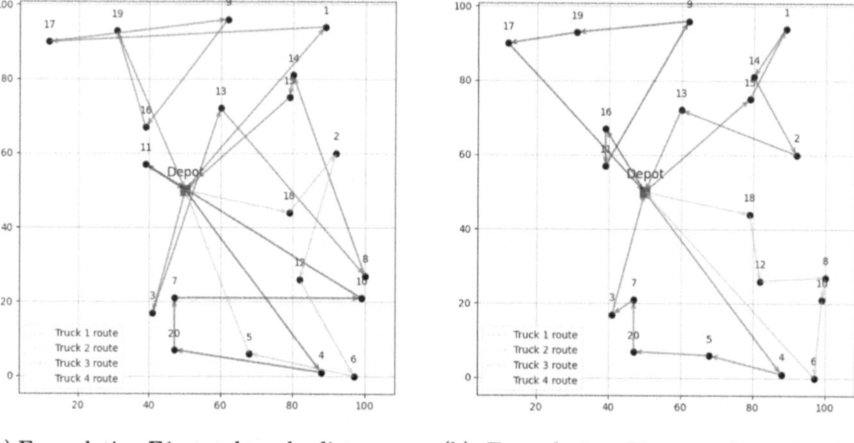

(a) Formulation F1 - total trucks distance: 989

(b) Formulation F2 - total trucks distance: 885

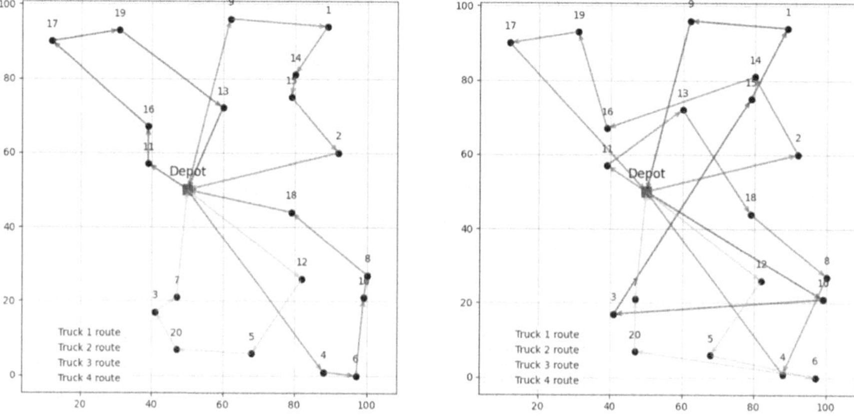

(c) Formulation F3 - total trucks distance: 601

(d) Formulation F4 - total trucks distance: 655

Fig. 1. Solution visualisation for test instance with $n = 20$ customers obtained on D-Wave quantum computer.

routes and reduced efficiency, as trucks frequently traverse the same regions rather than servicing spatially grouped customers.

Based on computation experiments results on D-Wave quantum computer it can be noticed that the F3 formulation is the most effective. For the model with truck serving customer binary decision variables goal function values are the lowest out of the other considered problem formulations. The second most efficient result was generated using F4 formulation. There is a slight difference in the value of goal function for formulations based on serving customer binary decision variable one that was for QUBO model and the other one that was formulated differently.

Formulations F1 and F2 have limited possibilities for solving considered problem with limited execution time. For Gurobi solver using formulation F1 and F2 only problems with up to $n = 120$ can be solved with 600 s time limit. Using F1 and F2 formulation on D-Wave quantum computer with `LeapHybridCQMSampler` only problems with up to $n = 20$ can be solved. Furthermore, using the `DWaveSampler`, no feasible solution was obtained for any instance, regardless of size. Formulation F1 tracks the paths traversed by the vehicles. The need to account for the vehicle's capacity at each step of the route introduces additional variables, making the optimization process more difficult. In contrast, formulation F2 tracks capacity at each customer but does not require the path between customers to be managed. As a result, formulation F2 is less complex than formulation F1 and it simplifies the problem by focusing more on capacity constraints without tracking every individual part of the vehicle journey. This reflects a trade-off between model expressiveness and computational feasibility. Formulations F3 and F4.

For Gurobi solver and `LeapHybridCQMSampler` using formulation F3 i F4 all tested problem instances can be solved (up to $n = 200$). Also using formulations F3 i F4 some small instances ($n = 8$ and $n = 10$) can be solved directly on quantum computer using `DWaveSampler`. Generally obtained cost functions values for formulation F3 and F4 are the same or worst than cost function values for formulation F1 and F2 but problems with greater sizes can be solved.

Computational times on a quantum computer using `LeapHybridCQMSampler` or `DWaveSampler` are constant. Total run time, including problem serialisation, deserialisation, and execution on QPU, for the considered problem is around 5 s. For `LeapHybridCQMSampler`, only QPU access time is reported. For the considered experiments, it was around 30–70 μs. For `DWaveSampler`, the reported QPU time depends on the number of samples (reads), and for the considered experiments, it was 100–200 ms. QPU anneal time per sample is always equal to 20 μs. More details about D-Wave quantum computer operation and timing can be found in the documentation [1]. In summary, computation time on D-Wave quantum computer are very small in comparison with classical methods, but only small size problems can be solved.

4 Conclusions

This paper presents an analysis of equal demand capacitated vehicle routing problem formulations and its readiness for solving on D-Wave quantum computer. Based on the computation experiments results formulations based on truck serving customer binary decision variables can be successfully solved on D-Wave quantum computer for all considered in this paper problem sizes with hybrid approach – `LeapHybridCQMSampler`. So far problem sizes which can be solved on D-Wave quantum computer are limited to $n = 10$ when computations are directly performed on quantum hardware using `DWaveSampler`. Generally computational experiments confirm that the equal demand capacitated vehicle routing problem can be solved on a quantum computer, and results strongly depend on the problem formulation and its size. This proof quantum readiness for solving equal demand capacitated vehicle routing problem with mathematical formulation based on truck serving customer binary decision variables.

References

1. D-wave quantum computer operation and timing (2025). https://docs.dwavequantum.com/en/latest/quantum_research/operation_timing.html
2. Borcinova, Z.: Two models of the capacitated vehicle routing problem. Croatian Oper. Res. Rev. 463–469 (2017)
3. Borowski, M., et al.: New hybrid quantum annealing algorithms for solving vehicle routing problem. In: International Conference on Computational Science, pp. 546–561. Springer (2020). https://doi.org/10.1007/978-3-030-50433-5_42
4. Deleplanque, S., Yaagoubi, A.E., Gras, T., Descamps, F., Mourrier, J., Lefebvre, I.: Applying analog quantum computing to solve optimisation problems: case studies on max-cut and CVRP. Int. J. Syst. Sci. Oper. Logistics **12**(1), 2463530 (2025)
5. Feld, S., et al.: A hybrid solution method for the capacitated vehicle routing problem using a quantum annealer. Front. ICT **6**, 13 (2019)
6. Godinho, M.T., Gouveia, L., Magnanti, T.L., Pesneau, P., Pires, J.: On the unit demand vehicle routing problem: flow based inequalities implied by a time dependent formulation (2009)
7. Karpas, I., et al.: Lightsolver: new platform challenges the vehicle routing problem
8. Osaba, E., Villar-Rodriguez, E., Asla, A.: Solving a real-world package delivery routing problem using quantum annealers. arXiv:2403.15114 (2024)
9. Osaba, E., Villar-Rodriguez, E., Miranda-Rodriguez, P., Asla, A.: Optimizing package delivery with quantum annealers: addressing time-windows and simultaneous pickup and delivery. arXiv:2504.01560 (2025)
10. Palmieri, A.: Quantum integer programming for the capacitated vehicle routing problem (2023)
11. Simchi-Levi, D., Chen, X., Bramel, J., Simchi-Levi, D., Chen, X., Bramel, J.: The capacitated VRP with equal demands. In: The Logic of Logistics: Theory, Algorithms, and Applications for Logistics Management, pp. 301–312. Springer (2014). https://doi.org/10.1007/978-1-4614-9149-1_16
12. Sinno, S., et al.: Performance of commercial quantum annealing solvers for the capacitated vehicle routing problem. arXiv:2309.05564 (2023)

A Comparative Study of Hybrid Classical–Quantum Transfer Learning

D. Martín-Pérez, F. Rodríguez-Díaz, A. Troncoso, and F. Martínez-Álvarez[(✉)]

Data Science and Big Data Lab, Pablo de Olavide University, 41013 Seville, Spain
{dmarper2,froddia,atrolor,fmaralv}@upo.es

Abstract. The combination of classical deep neural networks with quantum circuits has recently attracted attention as a promising paradigm for the current era of noisy intermediate-scale quantum (NISQ) technology. In this research, we perform a detailed comparative analysis of three classical Convolutional Neural Network architectures (ResNet18, VGG16, and MobileNetV2) combined with two distinct variational quantum circuits (VQC) acting as classifiers. To carefully assess the quantum contribution, we also introduce a purely classical baseline with an equivalent parameter budget. Each model is trained on the Hymenoptera dataset, and performance is checked based on validation accuracy and training time. Our findings show the important trade-offs between computational cost and model complexity, offering practical guidance for future hybrid classical-quantum applications and clarifying the performance benefits of different quantum circuit designs.

Keywords: Quantum Machine Learning · Transfer Learning · Pennylane · Hybrid Models

1 Introduction

Transfer learning in deep neural networks uses representations learned from large-scale datasets to solve tasks with relatively sparse labeled data [2,8]. By reusing low and middle level feature maps from pre-trained models on well-established benchmarks, practitioners can often get higher performance and faster convergence compared to training models from scratch. Convolutional neural network (CNN) architectures, in particular, have consistently shown their effectiveness in capturing meaningful visual features when pre-trained on large image repositories such as ImageNet. These learned representations can be adapted to a wide variety of subsequent tasks, such as classification, object detection and segmentation, thus reducing both the computational cost and the need for a large amount of domain-specific data.

Recently, the developing field of quantum machine learning has suggested combining well-known CNN frameworks with quantum circuits, giving rise to the concept of hybrid classical-quantum architectures. This approach, called quantum transfer learning (QTL), uses the strengths of the classical and quantum

E. Corchado et al. (Eds.): SOCO 2025, CCIS 2806, pp. 337–346, 2026.
https://doi.org/10.1007/978-3-032-19763-4_31

paradigms. On one hand, a pre-trained CNN is used as a robust feature extractor, taking advantage of its ability to distill high-level abstractions from raw image data. On the other hand, a quantum circuit processes these extracted features to perform classification.

In this paper, we present a comparative study based on the hybrid architecture put forward in [5]. Our main goal is to check the effectiveness of the QTL approach when combined with different classical convolutional neural networks as feature extractors. Thus, we explore the combination of ResNet architectures with quantum circuits and experiment with a broader set of classical architectures. Specifically, we check five hybrid classical-quantum models, each of which uses a different pre-trained CNN to extract features from image data.

In addition to this architectural comparison, we introduce a key modification to the hybrid model, aimed at significantly reducing the computational cost. While the runtimes reported in [5] were around 400 s per experiment, our modified approach gets similar or better accuracy with runtimes close to 50 s, a reduction of almost 87.5%. This substantial improvement shows the potential of optimised hybrid designs for more practical and scalable quantum machine learning applications.

By methodically comparing these architectures and optimising the hybrid design, we aim to determine whether some CNNs provide more compatible or beneficial features for quantum classification. This check offers a wider and more efficient perspective on the role of classical components in hybrid models. Thus, our study adds to the ongoing search for more efficient and scalable hybrid models in quantum machine learning, while pointing to the effect of classical architecture choices and system design on overall performance.

The main contributions of this paper can be summarized as follows:

1. We check the effectiveness of hybrid classical-quantum architectures, combining different pretrained Convolutional Neural Networks (CNNs) (ResNet18, VGG16, MobileNetV2) as feature extractors with a variational quantum circuit as the classifier.
2. We present modifications to previously suggested hybrid quantum models, significantly reducing computational costs by approximately 87.5% compared to earlier approaches, getting comparable or superior accuracy.
3. We provide insights into the balance between computational cost and model accuracy, showing that simpler CNN backbones (ResNet18 and MobileNetV2) give high validation accuracy with considerably less computational expense compared to more complex architectures.
4. We show through empirical experiments on the Hymenoptera dataset the practicality and effectiveness of quantum transfer learning, particularly pointing to the efficiency and rapid convergence of simpler architectures.
5. We offer practical guidelines for selecting classical CNN architectures in hybrid classical-quantum models based on computational resources and desired accuracy, helping with scalable quantum machine learning applications.

The rest of the paper is structured as follows. Section 2 provides a review of related work in the field of QTL. Section 3 introduces the suggested methodology. Section 4 presents the experimental results. Finally, Sect. 5 concludes the paper with a summary of our results.

2 Related Work

Classical transfer learning approaches based on CNNs are extensively documented and validated in the literature [3,10,11]. At the same time, quantum machine learning has presented novel hybrid architectures combining classical neural networks with quantum elements, including variational quantum classifiers [5]. Notably, recent research by Azevedo et al. applied quantum transfer learning successfully in medical imaging for breast cancer detection [1]. Additionally, cutting-edge classical architectures like MobileNetV2 [10] have significantly advanced the efficiency and accuracy of traditional vision models. Our study expands this research by methodically comparing these diverse CNN architectures within a unified hybrid classical-quantum framework.

In [4], the authors present transfer learning as an effective strategy for taking advantage of small quantum convolutional neural networks (QCNNs) in the noisy intermediate-scale quantum (NISQ) era. Their method combines a pre-trained classical convolutional neural network (CNN) with a QCNN to enhance classification tasks without the need for large-scale quantum circuits. Through numerical simulations on the MNIST dataset, where a classical CNN is first trained on Fashion MNIST data, they show that transferring learning from a classical to a quantum framework performs better than purely classical transfer learning under comparable training conditions. Their findings point to the potential of hybrid quantum-classical architectures to improve classification accuracy while mitigating the limitations of current quantum hardware.

The study in [7] suggests a quantum transfer learning strategy for image classification by combining a classical pre trained feature extractor with a variational quantum circuit classifier. The model's effectiveness is assessed on three tasks: classifying organic and recyclable waste, detecting tuberculosis (TB) using chest X-ray images, and identifying cracks on concrete surfaces. To determine the best architecture for each task, several classical networks, including VGG19, DenseNet169, and AlexNet are checked as feature extractors. The results show that DenseNet169 performs best for waste classification, AlexNet is most suited for TB detection, and VGG19 leads in crack detection, pointing out that the choice of model is influenced by factors like dataset size, training-test split, and learning rate.

In [12], the authors describe a classical to quantum transfer learning system for improving synthetic speech detection. Their approach uses a large-scale unsupervised pre-trained model WavLM Large to extract feature maps from speech signals, which are subsequently transformed into low-dimensional embedding vectors by classical network components. A variational quantum circuit (VQC) is then added to fine-tune both the pre-trained model and the classical layers.

Checked on the ASVspoof 2021 DF task, simulations of the quantum circuit show that this quantum transfer learning strategy performs better than conventional classical transfer learning baselines, pointing to its potential to enhance speech detection performance.

The authors in [9] compared Qiskit and Pennylane as frameworks for implementing hybrid quantum-classical support vector machines with quantum kernels. They found that Qiskit required less theoretical background and consistently got higher classification accuracy, while Pennylane showed faster execution times. Despite these differences, both frameworks kept stable performance up to 20 qubits, indicating their potential for practical quantum machine learning applications.

Finally, the work in [6] presents details on a hybrid quantum transfer learning strategy for Diabetic Retinopathy (DR) detection, aiming to overcome the drawbacks of classical transfer learning models in terms of maintenance costs and detection performance. Their approach uses the APTOS 2019 Blindness Detection dataset from Kaggle, combining InceptionV3 as a pre-trained classical neural network for feature extraction with a Variational Quantum Classifier for stratification. The model is trained and checked across various quantum simulation platforms, including Pennylane's default device, IBM Qiskit's BasicAer, and a Google Cirq Simulator using the PyTorch machine learning library. The hybrid quantum model gets an accuracy ranging from 93% to 96%, significantly better than the 85% accuracy of its classical counterpart.

3 Methodology

Figure 1 shows our hybrid architecture. A pre-trained CNN acts as a feature extractor. A trainable linear layer then reduces the feature dimensionality to match the input of a variational quantum circuit (VQC), which performs the final classification.

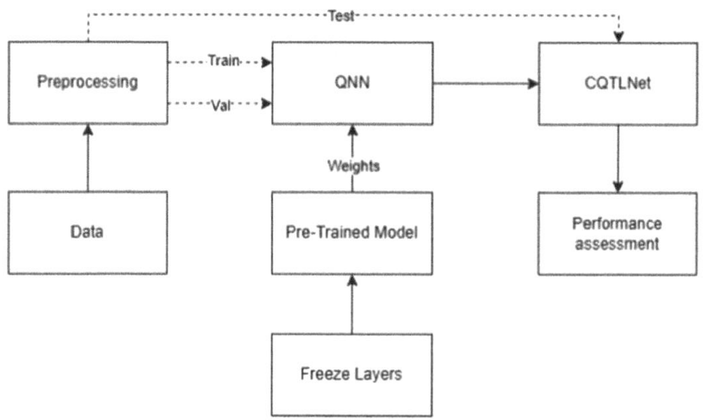

Fig. 1. Diagram of the Hybrid Quantum Transfer Learning workflow.

The hybrid models constructed in this study have two primary components: a pretrained CNN serving as a fixed feature extractor, and a variational quantum classifier. The CNN extracts high-level features from input images, which are subsequently condensed into a lower-dimensional space using a trainable linear layer matching the quantum circuit input size (four qubits). These reduced features are then encoded into a quantum circuit using angle encoding. The circuit, implemented in PennyLane, gives expectation values that act as logits for classification.

3.1 Quantum Classifier Architectures

We compare two quantum classifiers, HQM1 and HQM2, whose specifications are given in Table 1. They differ greatly in their structure, depth, and number of trainable parameters, allowing us to study the effect of circuit complexity on performance.

Table 1. Specifications of the Variational Quantum Classifiers (VQC). Both circuits use 4 qubits and angle encoding for data input.

Model	Variational Layers	Parameters	Depth	CNOTs
HQM1	BasicEntanglerLayers	8	7	8
HQM2	StronglyEntanglingLayers	72	31	24

To further clarify the structural differences between the two quantum classifiers, we provide a conceptual pseudocode for each in Algorithms 1 and 2. HQM1 is built upon a simple entangling structure, whereas HQM2 uses a more complex one with a greater number of rotations and a different connectivity pattern. This architectural choice is the primary reason for the large difference in parameter count and circuit depth detailed in Table 1.

Algorithm 1. HQM1 Circuit Structure (Conceptual)

1: **Input:** classical features x, weights w
2: AngleEncode(x) on all qubits
3: **for** each layer l in depth D **do**
4: Apply RY(w_l) to all qubits
5: Apply CNOT chain: CNOT(0,1), CNOT(1,2), ...
6: **end for**
7: **return** Expectation values

Algorithm 2. HQM2 Circuit Structure (Conceptual)

1: **Input:** classical features x, weights w
2: AngleEncode(x) on all qubits
3: **for** each layer l in depth D **do**
4: Apply RX($w_{l,1}$), RY($w_{l,2}$), RZ($w_{l,3}$) to all qubits
5: Apply CNOT ring: CNOT(0,1), CNOT(1,2), ..., CNOT($n-1$,0)
6: **end for**
7: **return** Expectation values

3.2 Classical Baseline

To carefully separate the quantum contribution, we also introduce a purely classical baseline. This baseline consists of a fully connected neural network that replaces the variational quantum circuit. Crucially, this classical classifier is constrained to have a similar number of trainable parameters as its quantum counterpart (HQM1) and is fed the identical feature vectors from the CNN backbone. This allows for a fair comparison of whether performance gains come from quantum processing or simply from model capacity.

During training, only the parameters of the classifier (either quantum or classical) and the linear reduction layer are optimized using the Adam optimizer with cross-entropy loss. Each model is trained for 10 epochs on the Hymenoptera dataset.

4 Results and Discussion

The experimental setup is described in Sect. 4.1. Section 4.2 describes the dataset used. Finally, the results obtained are reported and discussed in Sect. 4.3.

4.1 Experimental Setup

We trained three classical backbones (ResNet18, VGG16, MobileNetV2) each paired with three different classifiers: the simple HQM1, the complex HQM2, and a classical fully-connected (FC) network as a baseline. The CNN backbones remained frozen. We monitored validation accuracy and recorded training time for each of the resulting nine models.

4.2 Dataset Description

The Hymenoptera dataset used for this study consists of approximately 240 training images and 150 validation images distributed equally between two classes: ants and bees. Standard ImageNet normalization and augmentation were applied.

4.3 Discussion

The comparative results for all models are shown in Table 2. The data shows several key insights into the trade-offs between classical backbones, quantum circuit complexity, and performance.

Table 2. Test accuracy and training time (30 epochs) for each model configuration.

Model	Accuracy (%)	Training Time (s)
ResNet18 + Classical FC	77.22	483.2
ResNet18 + HQM1	94.77	52.5
ResNet18 + HQM2	95.42	434.0
VGG16 + Classical FC	70.89	501.5
VGG16 + HQM1	92.16	61.9
VGG16 + HQM2	94.12	443.3
MobileNetV2 + Classical FC	93.67	440.9
MobileNetV2 + HQM1	96.08	53.1
MobileNetV2 + HQM2	92.16	432.9

First, the choice of quantum classifier has a large effect on training time. The simple HQM1 circuit trains very fast (50–60 s), while the complex HQM2 requires much more time (around 430 s), likely due to its 9× increase in parameters and deeper circuit. Interestingly, the classical baseline, despite being fast, did not consistently perform better than the quantum models.

The MobileNetV2 and ResNet18 backbones proved to be excellent feature extractors for this task. The **MobileNetV2 + HQM1** model obtained the highest accuracy (96.08%) with a very low training time (53.1 s), showing an outstanding balance of performance and efficiency. The **ResNet18 + HQM1** also performed very well (94.77% accuracy in 52.5 s). In contrast, the VGG16 backbone produced consistently lower accuracy across all classifiers, indicating its feature representation was less optimal for this dataset.

When comparing quantum classifiers to the classical baseline, the quantum models almost universally got higher accuracy. For instance, ResNet18 + HQM1 performed better than its classical counterpart by over 17% points. This indicates that for this task, the variational quantum circuits were able to find more effective decision boundaries than a simple fully connected layer with a similar parameter count.

The training dynamics, shown in Figs. 2, 3, and 4, show important differences between the CNN backbones when paired with HQM2. The models using ResNet18 and MobileNetV2 (Figs. 2 and 4) show rapid and stable convergence, with validation loss decreasing smoothly and accuracy climbing consistently. In contrast, the VGG16-based model (Fig. 3) shows considerable instability, with large oscillations in both loss and accuracy. This indicates that the feature space

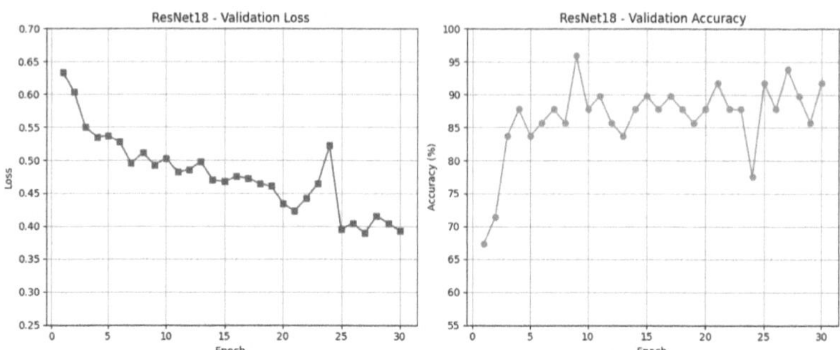

Fig. 2. Validation accuracy and loss per epoch for ResNet18 + HQM2.

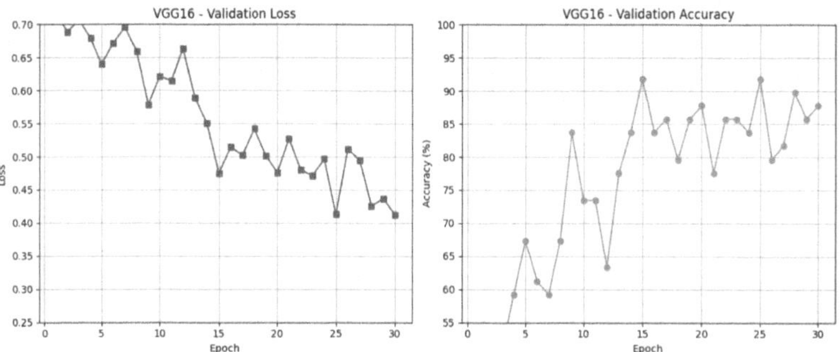

Fig. 3. Validation accuracy and loss per epoch for VGG16 + HQM2.

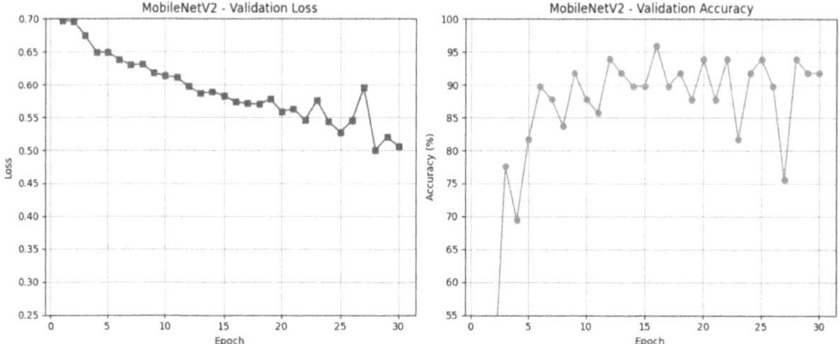

Fig. 4. Validation accuracy and loss per epoch for MobileNetV2 + HQM2.

from VGG16 is more challenging for the quantum classifier to generalize from, which points to the important role of the classical backbone choice in hybrid model performance.

5 Conclusions

In this work, we performed a methodical comparison of hybrid classical-quantum models, changing both the classical CNN backbone and the complexity of the quantum classifier. Our results show that lightweight backbones like MobileNetV2 and ResNet18, when paired with a simple quantum circuit (HQM1), provide an excellent combination of high accuracy and computational efficiency. These hybrid models had better results than their parameter-equivalent classical counterparts on the Hymenoptera dataset, indicating that variational quantum circuits can provide a tangible advantage in classification tasks. The choice of both the classical feature extractor and the quantum circuit architecture is important, as a more complex quantum circuit (HQM2) did not always give better results and incurred a substantial training cost. These findings give practical guidance for designing efficient and effective hybrid models, preparing for their application in real-world scenarios. Future work could look into adaptive quantum circuit designs and apply these models to more complex datasets.

Acknowledgments. The authors would like to thank the Spanish Ministry of Science and Innovation for the support within the projects PID2020-117954RB-C21, PID2023-146037OB-C22, and TED2021-131311B-C22.

References

1. Azevedo, V., Silva, C., Dutra, I.: Quantum transfer learning for breast cancer detection. Quan. Mach. Intell. **4** (2022)
2. Gil-Gamboa, A., Torres, J.F., Martínez-Álvarez, F., Troncoso, A.: Energy-efficient transfer learning for water consumption forecasting. Sustain. Comput. Inform. Syst. **46**, 101130 (2025)
3. He, K., Zhang, X., Ren, S., Sun, J.: Deep residual learning for image recognition. In: Proceedings of the IEEE Conference on Computer Vision and Pattern Recognition (CVPR), pp. 770–778 (2016)
4. Kim, J., Huh, J., Park, D.K.: Classical-to-quantum convolutional neural network transfer learning. Neurocomputing **555** (2023)
5. Mari, A., Bromley, T., Izaac, J., Schuld, M., Killoran, N.: Transfer learning in hybrid classical-quantum neural networks. Quantum **4**, 340 (2020)
6. Mir, A., Yasin, U., Khan, S.N., Athar, A., Jabeen, R., Aslam, S.: Diabetic retinopathy detection using classical-quantum transfer learning approach and probability model. Comput. Mater. Continua **71**, 3733–3746 (2021)
7. Mogalapalli, H., Abburi, M., Nithya, B., Vamsi Bandreddi, S.K.: Classical–Quantum Transfer Learning for Image Classification. SN Comput. Sci. **3** (2021)
8. Molina, M.A., Jiménez-Navarro, M.J., Arjona, R., Mártinez-Álvarez, F., Asencio-Cortés, G.: DIAFAN-TL: an instance weighting-based transfer learning algorithm with application to phenology forecasting. Knowl.-Based Syst. **254**, 109644 (2022)
9. Rodríguez-Díaz, F., Torres, J.F., Gutiérrez-Avilés, D., Troncoso, A., Martínez-Álvarez, F.: An experimental comparison of Qiskit and Pennylane for hybrid quantum-classical support vector machines. Lecture Notes Artif. Intell. **14640**, 121–130 (2024)

10. Sandler, M., Howard, A.G., Zhu, M., Zhmoginov, A., Chen, L.: Mobilenetv2: inverted residuals and linear bottlenecks. In: Proceedings of the IEEE Conference on Computer Vision and Pattern Recognition (CVPR), pp. 4510–4520 (2018)
11. Simonyan, K., Zisserman, A.: Very deep convolutional networks for large-scale image recognition. arXiv preprint arXiv:1409.1556 (2014)
12. Wang, R., Du, J., Gao, T.: Quantum transfer learning using the large-scale unsupervised pre-trained model WAVLM-large for synthetic speech detection. In: Proceedings of the IEEE International Conference on Acoustics, Speech and Signal Processing (ICASSP) (2023)

A Practical Implementation of Quantum LSTM Using Qiskit and PyTorch

I. Rojas-García[1], F. Rodríguez-Díaz[1], D. Gutiérrez-Avilés[2], A. Troncoso[1], and F. Martinez-Álvarez[1](✉)

[1] Data Science and Big Data Lab, Pablo de Olavide University, 41013 Seville, Spain
{irojgar,froddia,atrolor,fmaralv}@upo.es

[2] Department of Computer Science, University of Seville, 41012 Seville, Spain
dgutierrez3@us.es

Abstract. Quantum computing is emerging as a promising tool to enhance classical machine learning models, especially in areas that involve complex temporal dynamics and high-dimensional data. This paper presents a practical implementation of a Quantum Long Short-Term Memory network. This hybrid architecture integrates parameterized quantum circuits into the gating mechanisms of a classical Long Short-Term Memory network. Built using PyTorch and Qiskit, our model enables end-to-end simulation, training, and analysis of quantum-enhanced recurrent networks. We design modular quantum circuits that interact with the Long Short-Term Memory's internal states and evaluate their performance in processing synthetic sequential data. The architecture includes visualization and diagnostic tools to explore hidden state evolution, final memory values, and gate behaviors. Furthermore, we introduce an automated performance optimization framework that compares multiple Quantum Long Short-Term Memory network configurations using varying key quantum parameters such as qubit count and entanglement depth. Our findings demonstrate the feasibility of constructing interpretable, tunable quantum-classical sequence models using widely available tools. This work contributes a fully reproducible platform for experimentation and lays the groundwork for future research in scalable quantum recurrent architectures.

Keywords: Quantum Machine Learning · Quantum Long Short-Term Memory · Qiskit

1 Introduction

Long Short-Term Memory (LSTM) networks are a widely used type of recurrent neural network, particularly effective in handling data sequences over time. Thanks to their memory cells and gating mechanisms, LSTMs have been successful in tasks ranging from speech recognition to time-series prediction and natural language processing. However, they still operate within the limitations of classical computing. As our models grow in complexity and data becomes more

E. Corchado et al. (Eds.): SOCO 2025, CCIS 2806, pp. 347–356, 2026.
https://doi.org/10.1007/978-3-032-19763-4_32

demanding, researchers are beginning to explore whether emerging technologies, like quantum computing, can offer new advantages.

Quantum computing [10] has gained significant interest in recent years due to its potential to perform certain types of computation more efficiently than classical methods. In particular, hybrid approaches that combine classical neural networks with quantum circuits are being actively studied. These models make use of parameterized quantum circuits, which can represent complex and highly entangled transformations. Incorporating such circuits into neural architectures could lead to more expressive models or reduce the number of parameters required to achieve strong performance.

This paper explores that idea through the implementation of a Quantum-enhanced LSTM (QLSTM). In our design, the traditional LSTM gates responsible for deciding what information to forget, remember, or update are influenced by outputs from a quantum circuit. These circuits are built using Qiskit [11], IBM's quantum computing framework, and are integrated into a PyTorch-based model that handles the classical parts of the network. The result is a hybrid system where quantum and classical components work together in a single recurrent architecture.

To support this model, we have also developed a set of tools for analyzing its behavior and performance. Thus, this work also includes visualizations of hidden state evolution, printed summaries of internal values, and metrics like memory usage and inference time. In addition, we implemented an optimization routine that automatically tests different model configurations, such as varying the number of qubits, quantum layers, and hidden size, to find combinations that balance accuracy, speed, and resource consumption.

The main contributions of this work are:

1. A working QLSTM architecture that combines classical deep learning with quantum circuit execution.
2. A modular quantum circuit design capable of learning from input sequences and interacting with LSTM-like gate structures.
3. Tools for visualization and performance analysis to help interpret the model's inner workings.
4. An automated method for exploring and comparing different QLSTM configurations.
5. A fully reproducible code base that can be extended or adapted for further research in quantum machine learning.

The rest of the paper is structured as follows. Section 2 overviews recent and related works. The proposed QLSTM architecture is introduced in Sect. 3. Its evaluation and associated experimental results are reported in Sect. 4. Finally, the conclusions drawn are summarized in Sect. 5.

2 Related Work

The integration of quantum computing into classical machine learning architectures has gained attention, particularly in developing hybrid models. Among these, QLSTM networks have emerged as a promising approach for enhancing sequential data modeling tasks.

Chen et al. [3] introduced one of the earliest QLSTM models, replacing classical LSTM gates with variational quantum circuits. Their findings indicated that such hybrid models could offer improved learning capabilities for time-series data, demonstrating faster convergence and better accuracy in certain scenarios compared to classical counterparts.

One of the notable theoretical innovations is the Quantum Kernel-Based LSTM (QK-LSTM), which integrates quantum kernel methods within the LSTM structure to enhance the modeling of complex temporal patterns [4]. The authors show that this approach allows for compact, non-linear mappings that outperform classical LSTMs in sequential classification tasks under constrained resources.

Qi et al. proposed an enhanced QLSTM structure using a bidirectional ring variational quantum circuit [7]. This circuit design improves the entanglement capabilities and sequence encoding, which translates to better convergence and generalization across benchmark datasets. The model also addresses issues of vanishing gradients, often encountered in classical deep learning.

In another contribution, Cao et al. introduced a linear-layer-enhanced QLSTM aimed at financial forecasting applications [1]. By embedding classical linear layers into the quantum sequence model, the architecture benefits from both interpretability and improved optimization, achieving state-of-the-art results in carbon price prediction.

A complementary optimization approach is explored by Li et al., who integrate a modified Electric Eel Foraging Optimization algorithm with QLSTM to optimize model parameters in civil engineering contexts [6]. Their work demonstrates that bio-inspired metaheuristics can be effectively combined with quantum architectures to address real-world predictive challenges.

From the application perspective, Schuld et al. present a QLSTM model designed for financial time series forecasting [8]. Their results show a marked improvement in predictive performance and training stability, suggesting that quantum sequence models may be especially suited for high-frequency, volatile data streams.

Tripathi et al. propose a QLSTM-based framework for cyber threat detection in virtualized environments [9]. The model achieves high detection accuracy in identifying malicious behavior from system call sequences, showcasing the viability of QLSTM in security-critical applications.

In the renewable energy domain, Khan et al. conduct a comparative study between classical LSTM and QLSTM models for solar power forecasting [5]. Their findings reveal that QLSTM models not only match but, in some cases, surpass the predictive power of their classical counterparts, especially when modeling irregular or noisy time series.

Lastly, Chehimi et al. introduce the Federated Quantum Long Short-Term Memory (FedQLSTM) architecture, which combines quantum recurrent learning with federated computing principles [2]. The model supports decentralized learning over non-IID datasets and shows robustness under limited communication bandwidth, thus addressing practical deployment challenges in distributed quantum AI systems.

After this analysis, it can be concluded that what distinguishes our work from previous contributions is its focus on reproducibility, interpretability, and empirical evaluation of a QLSTM model developed using accessible tools such as Qiskit and PyTorch. While earlier studies often concentrate on theoretical models or limited application areas, our implementation offers a modular framework that includes tools for visualizing internal LSTM behavior, assessing resource consumption in terms of parameter count, memory usage, and inference time, and performing automated exploration of key quantum parameters including the number of qubits and circuit depth.

3 QLSTM Architecture

The proposed QLSTM model is a hybrid neural network that combines classical LSTM mechanisms with quantum circuit computations. It retains the standard LSTM structure while incorporating quantum-enhanced components into its gate operations, offering potential improvements in expressivity and parameter efficiency.

3.1 Classical LSTM Overview

Traditional LSTM networks maintain two internal states: a hidden state h_t and a cell state c_t, which are updated through three gates: forget, input, and output. These gates control the information flow over time steps, enabling the network to learn long-term dependencies in sequential data.

3.2 Hybrid Integration with Quantum Circuits

In the QLSTM, each gate is partially driven by the output of a parameterized quantum circuit. These circuits are initialized with classical input features and evolved using learnable quantum gates. The results, extracted as expectation values from measured quantum states, are incorporated into the LSTM gate activations.

Later, the quantum circuit is constructed by three main steps: (1) encoding the input via R_y rotations, (2) applying multiple layers of parameterized R_y gates followed by CNOT entanglement in a ring topology, and (3) extracting normalized outputs through measurement.

Finally, classical inputs are concatenated with the previous hidden state and passed through a linear layer. This output modulates each LSTM gate—forget, input, candidate, and output—where quantum circuit outputs influence the final gate activations via classical-quantum fusion.

3.3 Forward Pass and Output Modes

At each time step, the model performs the following operations:

1. Concatenates the current input x_t with h_{t-1}.
2. Applies classical transformation via a projection layer.
3. Feeds the result into a quantum circuit to produce expectation values.
4. Updates the gates using both classical and quantum-influenced values.
5. Computes new states c_t and h_t.

The architecture supports both sequence-to-sequence and sequence-to-one predictions. Figure 1 illustrates the architecture of the QLSTM model. At each time step, the classical input x_t and the previous hidden state h_{t-1} are concatenated and transformed via a classical projection layer. The result is simultaneously passed to a parameterized quantum circuit, which applies a sequence of quantum gates and returns expectation values after measurement. These quantum-derived outputs, together with classical activations, influence the LSTM's forget, input, and output gates. The updated states c_t and h_t are then used in the subsequent time step or directly for prediction σ.

3.4 Configurable Parameters

The model exposes configurable parameters, including:

1. Input size or dimensionality of the input vector.
2. Hidden size or number of LSTM units.
3. Number of qubits and quantum layers, determining the circuit complexity.
4. Batching and sequence output flags (PyTorch-compatible I/O flexibility).

4 Evaluation and Experimental Results

The QLSTM model was evaluated on synthetic time-series data to analyze its computational behavior, state dynamics, and performance across multiple configurations. This section combines visual inspection of internal states with quantitative metrics from structured experiments.

4.1 Data Description

To evaluate the performance of different QLSTM configurations under controlled conditions, synthetic input data were generated using the *torch.randn*() function. Specifically, each input sequence was created as a three-dimensional tensor of $shape(1, 10, 3)$, representing a batch of one sequence with ten time steps and three features per step. The values were sampled from a standard normal distribution (mean zero and unit variance), ensuring randomness while preserving statistical consistency across trials.

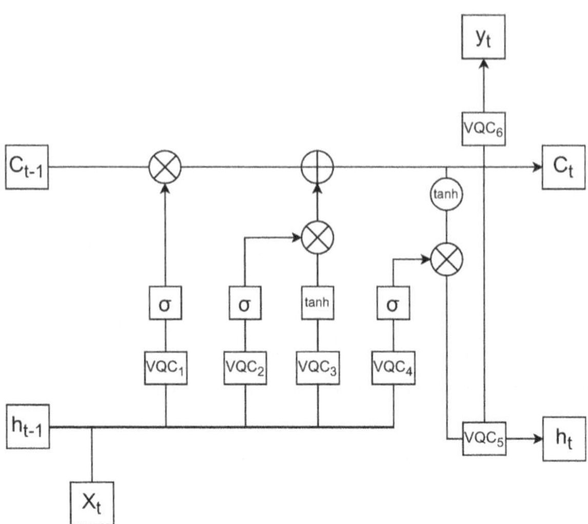

Fig. 1. Block diagram of the QLSTM architecture. The classical input and previous hidden state are combined and passed through both a classical projection layer and a parameterized quantum circuit. The quantum outputs influence the LSTM gates before computing updated hidden and cell states.

This approach allows for systematic testing of the QLSTM model's behavior independent of specific real-world datasets. By keeping the sequence structure and feature dimensionality fixed, and varying only the model configuration, such as the number of hidden units, qubits, and quantum layers, it becomes possible to isolate and analyze the effects of architectural changes on performance metrics like inference time, memory usage, and hidden state dynamics.

Although this data does not correspond to a specific application domain, it is sufficient for evaluating internal consistency, computational efficiency, and the functional stability of different QLSTM designs.

4.2 Input and Hidden State Evolution

Figure 2 displays the 10-step input sequence used in the evaluation. Each feature evolves differently over time, providing a diverse challenge for the recurrent model.

The resulting hidden state activations across time are shown in Fig. 3. Each unit within the hidden state vector captures different aspects of the temporal structure. The evolution is stable, with moderate variability and clear responsiveness to the input fluctuations.

4.3 Final State Comparison

Figure 4 compares the values of the final hidden state and cell state after processing the input sequence. The difference between these states reflects the LSTM's

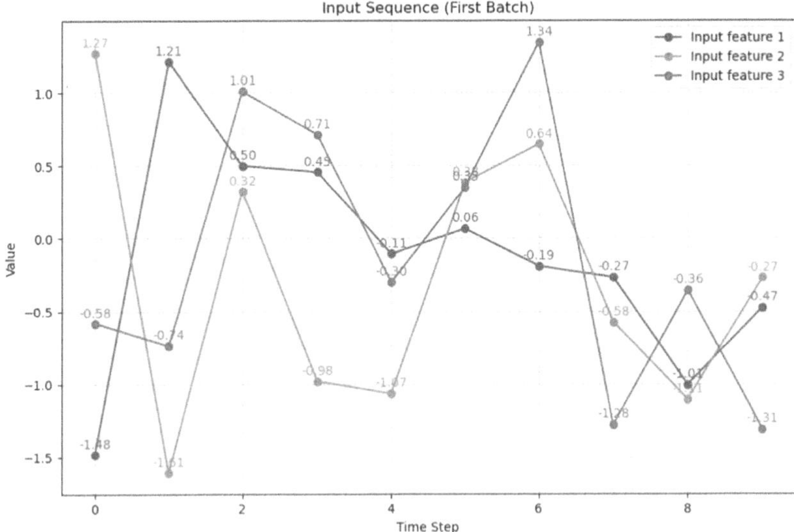

Fig. 2. Input sequence with three features over 10 time steps.

memory gating behavior, with the cell state retaining accumulated information while the hidden state emphasizes immediate outputs.

4.4 Architecture Optimization Trials

To investigate the impact of architecture size and quantum depth, we tested five QLSTM configurations varying in hidden size, number of qubits, and number of quantum layers. Performance metrics included inference time, memory usage, and hidden state stability. These results are shown in Table 1.

Table 1. Performance across multiple QLSTM configurations.

Trial	Hidden	Qubits	QLayers	Params	Time (ms)	Mem (MB)	H Mean	H Std	Output Range
1	4	2	2	80	1.63	0.0003	−0.0259	0.1952	[−0.3840, 0.2932]
2	4	2	2	80	2.13	0.0003	−0.0587	0.1595	[−0.2520, 0.2091]
3	16	4	2	432	3.01	0.0016	0.0297	0.1289	[−0.1830, 0.2457]
4	16	4	3	448	3.55	0.0017	0.0456	0.1438	[−0.2699, 0.4025]
5	16	4	1	416	3.40	0.0016	−0.0275	0.1514	[−0.2620, 0.2849]

The table summarizes the results of five distinct QLSTM configurations, each combining different values for hidden layer size, number of qubits, and depth of quantum layers. These trials were designed to evaluate the trade-offs between model complexity, runtime, memory consumption, and output stability. Despite

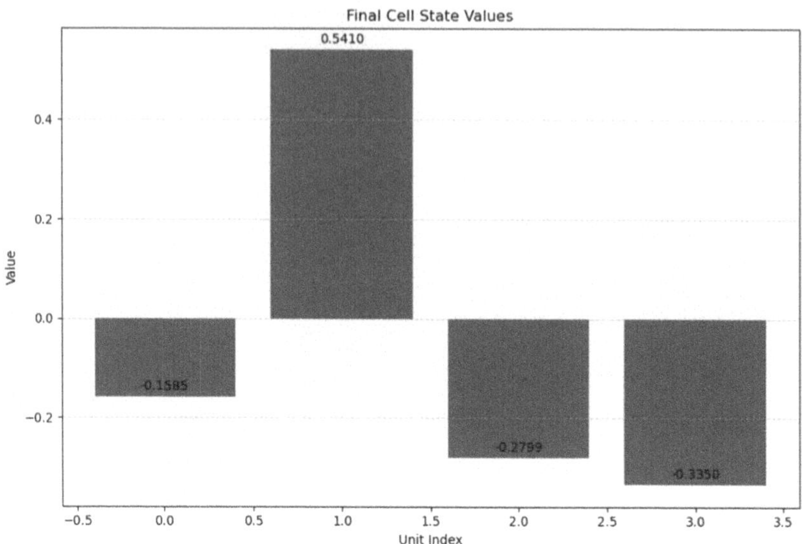

Fig. 3. Evolution of hidden state activations over time.

the differences in architectural parameters, all configurations were tested under consistent experimental conditions, ensuring comparability across performance indicators.

Thus, Trial 1, which used a minimal configuration of 4 hidden units, 2 qubits, and 2 quantum layers, exhibited the lowest inference time (1.63 ms) and the smallest memory footprint (0.0003 MB). However, its hidden state variability was higher than that of some more complex models. In contrast, Trial 3, with 16 hidden units and 4 qubits, achieved the most stable hidden state distribution (standard deviation of 0.1289), suggesting that increased capacity can enhance representational stability without necessarily leading to excessive computational cost.

These results demonstrate that careful tuning of architectural parameters can significantly influence the efficiency and robustness of QLSTM models. Configurations with moderate complexity, such as Trial 3, appear to offer a balanced compromise between performance and resource usage. These results confirm that the QLSTM can be tuned to prioritize low latency, low memory usage, or stable state evolution, making it adaptable for a range of practical applications.

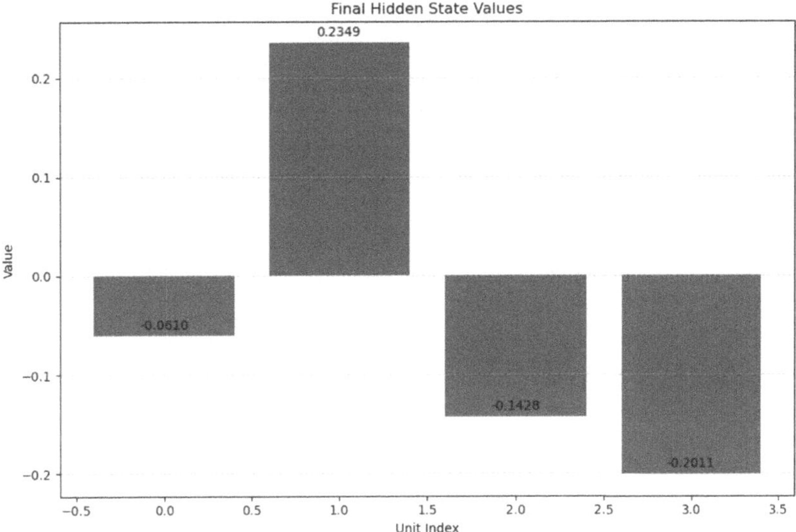

Fig. 4. Bar chart comparison of final hidden state and cell state values.

5 Conclusions

In this work, we introduced a QLSTM, a hybrid approach that combines the strengths of classical deep learning and quantum computing. By embedding parameterized quantum circuits into the gating mechanisms of a standard LSTM, we aimed to explore how quantum operations might improve the flexibility and expressiveness of recurrent models.

We implemented the QLSTM using widely adopted tools: PyTorch for the classical components and Qiskit for simulating the quantum parts. The result is a fully functional and reproducible model that is easy to experiment with, even on standard hardware. Alongside the architecture, we developed practical tools to visualize its internal behavior, evaluate performance metrics, and search for the best configuration through automated testing.

Our experiments showed that the QLSTM is capable of producing stable and interpretable results, even with very compact models. In particular, we found that it is possible to strike a balance between speed, memory use, and model stability by adjusting the number of qubits, quantum layers, and hidden units.

This work shows that quantum-enhanced neural networks are not just theoretical concepts since they can be implemented, tested, and understood using today's tools. While our tests were performed on simulated quantum circuits, the same code could be extended to run on actual quantum hardware as it becomes more accessible.

Looking ahead, we see many possibilities for future work. These include training the model on real-world datasets, exploring deeper or multi-layer QLSTM variants, and studying how quantum noise might affect performance on real devices. We hope this work can serve as a foundation for others interested in building and learning from hybrid quantum-classical models.

Acknowledgments. The authors would like to thank the Spanish Ministry of Science and Innovation for the support within the projects PID2020-117954RB-C21 and PID2023-146037OB-C22.

Disclosure of Interests. The authors declare that they have no known competing financial interests or personal relationships that could have appeared to influence the work reported in this paper.

References

1. Cao, Y., Zhou, X., Fei, X., Zhao, H., Liu, W., Zhao, J.: Linear-layer-enhanced quantum long short-term memory for carbon price forecasting. Quan. Mach. Intell. **5**(2), 26 (2023)
2. Chehimi, M., Chen, S.Y.-C., Saad, W., Yoo, S.: Federated quantum long short-term memory (FedQLSTM). Quan. Mach. Intell. **6**(1), 43 (2024)
3. Chen, K., Wang, Z., Zhang, Z., Qiu, X.: Quan. Long Short-Term Memory. arXiv **2009**(01783), 1–27 (2020)
4. Yu-Chao Hsu, Tai-Yu Li, and Chen, K.-C.: Quantum kernel-based long short-term memory. Quan. Mach. Intell. (2024)
5. Khan, S.Z., et al.: Quantum long short-term memory (QLSTM) vs. Classical LSTM in time series forecasting: a comparative study in solar power forecasting. Front. Phy. **12**, 1439180 (2024)
6. Li, X., Wang, Y., Zhang, Z., Liu, W.: Optimized quantum LSTM using modified electric eel foraging optimization for civil engineering applications. Eng. Appl. Artif. Intell. **122**, 105857 (2024)
7. Qi, H., Ma, H., Song, J., Liu, L., Lu, J.: EQLSTM: enhanced quantum long short-term memory for time series forecasting. J. Supercomput. **81**, 117 (2025)
8. Schuld, M., Sinayskiy, I., Petruccione, F.: Quantum long short-term memory neural network for financial time series prediction. Quan. Inf. Process. **23**(2) (2024)
9. Tripathi, S., Upadhyay, H., Soni, J.: A quantum LSTM-based approach to cyber threat detection in virtual environments. J. Supercomput. **81**, 142 (2025)
10. Rodríguez-Díaz, F., Gutiérrez-Avilés, D., Troncoso, A., Martínez-Álvarez, F.: A survey of quantum machine learning: foundations, algorithms, frameworks, data and applications. ACM Comput. Surv. **58**, 1–35 (2026)
11. Rodríguez-Díaz, F., Torres, J.F., Gutiérrez-Avilés, D., Troncoso, A., Martínez-Álvarez, F.: An experimental comparison of Qiskit and Pennylane for hybrid quantum-classical support vector machines. Lecture Notes Artif. Intell. **14640**, 121–130 (2024)

Solving Capacitated Vehicle Routing Problem with Equal Demand on a D-Wave Quantum Machine

Jan Gromiec, Michał Kapica, Michał Nowicki, Mariusz Uchroński(✉), and Mateusz Wasilewski

Wrocław University of Science and Technology, Wyb. Wyspiańskiego 27, 50-370 Wrocław, Poland
mariusz.uchronski@pwr.edu.pl

Abstract. The article presents an analysis of equal demand capacitated vehicle routing problem quantum-ready formulations. Analysed formulations were based on different approaches for constraints definition, such as remaining vehicle capacity between customers, remaining vehicle capacity at the customer, a truck serving customer binary decision variables, and only goal function without constraints. Considered formulations were evaluated on classical computer and quantum computer. Computational experiments confirm that the equal demand capacitated vehicle routing problem can be solved on a quantum computer, and results strongly depend on the problem formulation and its size.

Keywords: capacitated vehicle routing problem · equal demand · D-Wave

1 Introduction

Capacitated Vehicle Routing Problem has numerous applications in many fields of science and industry such as logistics and urban mobility. As outlined by Sebastian Feld et al. [5], CVRP is a NP-hard optimization problem, where every vehicle has a constrained capacity. Classical solution methods, while powerful, often struggle with scalability in real-time or large-scale settings. The goal of this problem is to find the most efficient route between customers. Eneko Osaba et al. [8] stated that the main objective of real-world routing problems is to present a solution with realistic instances while maintaining the restrictions of the original real-world problem, and he advanced his research in another study et al. [9]. In both articles it was demonstrated that hybrid approaches are a promising alternative to traditional computing resources. According to Michał Borowski et al. [3] hybrid solutions for solving VRP and CVRP can also give effective and realistically optimised results. Salvatore Sinno et al. [12] demonstrated that using a D-Wave CQM solver QUBO formulation is able to find feasible solutions with maximum of 15 nodes in CVRP. Anthony Palmieri [10] stated that the most efficient route is achieved by minimizing the total distance travelled by vehicles or

E. Corchado et al. (Eds.): SOCO 2025, CCIS 2806, pp. 357–367, 2026.
https://doi.org/10.1007/978-3-032-19763-4_33

minimizing the overall cost. David Simchi-Levi et al. [11] introduced the Equal Demand Capacitated Vehicle Routing Problem (EDCVRP). There are a lot of various CVRP variants however, due to the limitations of quantum computing, EDCVRP provides a computationally approachable yet non-trivial test base to estimate the capabilities and limitations of such technology. Ilan Karpas et al. [7] compared methods of solving EDCVRP using different formulations and highlighted the superiority of quantum-inspired platform LightSolver over classical solver Gurobi. Our goal is to make a comparison between classical, quantum and hybrid approaches to EDCVRP, where each of them was implemented using the following four formulations based on: remaining vehicle capacity between customers (F1), remaining vehicle capacity at customer (F2), truck serving customer binary decision variables (F3) and with truck serving customer binary decision variables without constraints (F4). In our study, the ILP (Integer Linear Programming) models were implemented according to two established formulations: one proposed by Deleplanque et al. [4], and another derived from the models of Godinho et al. [6] and Borcinowa [2]. Each formulation was considered and evaluated on CPU using Gurobi solver and on QPU with D-Wave samplers – `LeapHybridCQMSampler` and `DWaveSampler`. We are currently in the so-called NISQ (noisy intermediate scale quantum) era, which means that quantum technology is neither completely reliable nor capable of natively solving medium or large-scale problems. For that reason, this paper focuses on formulation efficiency and quantum readiness, not real-world instance optimisation.

2 Problem Formulation

Capacitated vehicle routing problem with equal demand can be defined as follows. There is set of n customers which should be served (by delivering some goods) by m vehicles with limited capacity Q. Each vehicle starts and finish its delivery route in depot. Between any two locations (two customers or the depot and a recipient) the distance is d_{ij}. Demand for all customers is equal 1 ($D = 1$) which means that each vehicle visit Q customers. The goal is to assign orders to the vehicles and plan routes for all of vehicles, while minimizing the total distance traveled by all vehicles.

Considered capacitated vehicle routing problem with equal demand can be mathematically formulated using some graph representation. Let $G = (V, E, d)$ be a complete directed graph, where $V = \{0, 1, 2, \ldots, n\}$ is a set of nodes and $E = \{(i, j) : i, j \in n, i \neq j\}$ is a set of edges, where node 0 represents the depot for m vehicles with the the same capacity Q. Remaining n nodes represents customers. The number of customers is determined by truck capacity and node count according to this equation $n = m * Q$. Each customer has equal demand $D = 1$. With each edge $(i, j) \in E$ of graph G some non negative travel distance d_{ij} is associated.

2.1 Formulation Based on Remaining Vehicle Capacity Between Customers (F1)

Let there be following variables:

- x_{ij} – binary decision variable equals 1 if vehicle travels directly from customer i to the customer j, 0 otherwise,
- ϱ_{ij} – positive integer variable represents the remaining capacity of the vehicle when it travels from customer i to customer j.

The mixed integer linear programming model for capacitated vehicle routing problem with equal demand can be formulated as follows.
Minimize:

$$\sum_{i=1}^{n}\sum_{j=1}^{n} x_{ij}d_{ij} \tag{1}$$

Subject to:

$$\sum_{i=0}^{n} x_{ij} = 1,\ j = 1,\ldots,n \tag{2}$$

$$\sum_{j=0}^{n} x_{ij} = 1,\ i = 1,\ldots,n \tag{3}$$

$$\sum_{j=1}^{n} x_{0j} = m \tag{4}$$

$$\sum_{i=1}^{n} x_{i0} = m \tag{5}$$

$$0 \leq \varrho_{ij} \leq Qx_{ij},\ i,j = 1,\ldots,n, \tag{6}$$

$$\sum_{i=0}^{n} \varrho_{ik} - \sum_{j=0}^{n} \varrho_{kj} = 1,\ k = 1,\ldots,n \tag{7}$$

The objective function (1) minimizes total travel distance for all vehicles. Constraints (2)–(3) guarantee that each recipients is visited only once. Constraints (4)–(5) guarantee that each vehicle leaves depot and each vehicle is returning to the depot. Vehicles capacity constraints (6) – remaining capacity of the vehicle when travels from customer i to the customer j is less than or equal to the capacity of the vehicle Q. Constraint (7) ensures that the capacity of truck decreases by 1 after each visit and tracks the path of the vehicles.

2.2 Formulation Based on Remaining Vehicle Capacity at Customer (F2)

Let there be following variables:

- x_{ij} – binary decision variable equals 1 if vehicle travels directly from customer i to the customer j, 0 otherwise,
- u_i – positive integer variable represents the remaining capacity of the vehicle at customer i.

The mixed integer linear programming model for capacitated vehicle routing problem with equal demand can be formulated as follows.
Minimize:

$$\sum_{i=1}^{n}\sum_{j=1}^{n} x_{ij}d_{ij} \tag{8}$$

Subject to:

$$\sum_{i=0}^{n} x_{ij} = 1,\ j = 1, \dots, n \tag{9}$$

$$\sum_{j=0}^{n} x_{ij} = 1,\ i = 1, \dots, n \tag{10}$$

$$\sum_{j=1}^{n} x_{0j} = m \tag{11}$$

$$\sum_{i=1}^{n} x_{i0} = m \tag{12}$$

$$u_i - u_j + Qx_{ij} \leq Q - 1,\ i, j = 1, \dots, n \tag{13}$$

The objective function (8) minimizes total travel distance for all vehicles. Constraints (9)–(10) guarantee that each recipients is visited only once. Constraints (11)–(12) guarantees that each vehicle leaves depot and each vehicle is returning to the depot. Constraint (13) ensures that the capacity of truck decreases by 1 after each visit.

2.3 Formulation Based on Serving Customer Binary Decision Variable (F3)

Considered problem can be mathematically formulated using serving customer binary decision variable. Let x_{kij} be a binary decision variable that is equal to 1 if vehicle k at i-th site is serving customer j. The mixed integer linear programming model for capacitated vehicle routing problem with equal demand can be formulated as follows.

Minimize:

$$\sum_{k=1}^{m}\sum_{j=1}^{n} d_{0j}x_{k1j} + \sum_{k=1}^{m}\sum_{j=1}^{n} d_{j0}x_{kQj} + \sum_{k=1}^{m}\sum_{i=1}^{Q-1}\sum_{n1=1}^{n}\sum_{n2=1}^{n} d_{n1n2}x_{kin1}x_{k(i+1)n2} \quad (14)$$

Subject to:

$$\sum_{j=1}^{n} x_{kij} = 1,\ k = 1, \ldots, m,\ i = 1, \ldots, Q, \quad (15)$$

$$\sum_{k=1}^{m}\sum_{i=1}^{Q} x_{kij} = 1,\ j = 1, \ldots, n \quad (16)$$

The objective function (1) minimizes total travel distance for all vehicles. The first sum corresponds to the first part of the vehicle route – leaving the depot. Similarly, the second sum corresponds to the last part of the vehicle route – return to the depot. The last sum corresponds to the intermediate part of the vehicle route. Constraint (15) guarantee that when vehicle k visits i-th site, it serves exactly one customer. Constraint (16) guarantees that every customer j is served once by one vehicle.

2.4 Formulation Based on Serving Customer Binary Decision Variable – QUBO (F4)

For QUBO model formulation all decision variables must be binary, all constraints must be in equality form and all constraints must be moved to the goal function. In formulation F3, all variables are binary and all constraints have equality form. Constraints (15)–(16) can be moved to the goal function in the following form:

$$\sum_{k=1}^{m}\sum_{i=1}^{Q}(\sum_{j=1}^{n} x_{kij} - 1)^2 = 0 \quad (17)$$

$$\sum_{j=1}^{n}(\sum_{k=1}^{m}\sum_{i=1}^{Q} x_{kij} - 1)^2 = 0 \quad (18)$$

Finally the objective function minimizes total travel distance for all vehicles can be defined as follows:

$$F(x) = F_{cost}(x) + \lambda F_{constraints}(x), \quad (19)$$

where λ is sufficiently large coefficient equals $n \cdot 10^3$ and

$$F_{cost}(x) = \sum_{k=1}^{m}\sum_{j=1}^{n} d_{0j}x_{k1j} + \sum_{k=1}^{m}\sum_{j=1}^{n} d_{j0}x_{kQj} + \sum_{k=1}^{m}\sum_{i=1}^{Q-1}\sum_{n1=1}^{n}\sum_{n2=1}^{n} d_{n1n1}x_{kin1}x_{k(i+1)n2} \quad (20)$$

and

$$F_{constraints}(x) = \sum_{j=1}^{n}(\sum_{k=1}^{m}\sum_{i=1}^{Q} x_{kij} - 1)^2 + \sum_{k=1}^{m}\sum_{i=1}^{Q}(\sum_{j=1}^{n} x_{kij} - 1)^2 \quad (21)$$

3 Computational Experiments

Presented mathematical models of capacitated vehicle routing problem with equal demand was implemented and then executed on CPU using Gurobi solver and also executed on QPU D-Wave quantum computer. Implementation for Gurobi solver was prepared using Python library `gurobipy` version 12.0.1. Similarly implementation for D-Wave quantum computer was prepared using Python `dimod` library version 8.1.0. Computation experiments for Gurobi solver was executed on the server with Intel(R) Core(TM) i7-7700 3.60GHz CPU working under Fedora 41 operating system. Computational experiments using `DWaveSampler` and `LeapHybridCQMSampler` was executed on D-Wave quantum computer trough Leap Quantum Application Environment with `Advantage_system4.1` solver. `DWaveSampler` use quantum solver executed directly on QPU. Before executing sampling problem was automatically converted to QUBO (if necessarily) and then mapped (embedded) to the physical architecture of quantum computer. `LeapHybridCQMSampler` use quantum–classical hybrid solver where QPU is used for guiding multithreaded classical heuristic module to the more promising areas of the search space or to improve existing solutions.

Test instances were constructed as follows. For each instance, customer locations were randomly generated with uniform distribution on a two-dimensional integer grid of size 100×100. The depot was deterministically placed at the centre of the grid, at coordinates $(50, 50)$. Given the total number of customers $n = m \cdot Q$, where m denotes the number of vehicles and Q the number of customers per vehicle. Customer coordinates $(x_i,\ x_j)$ were sampled uniformly at random from the range $[1, 100]$. The complete set of $n + 1$ locations (including the depot) was used to construct a distance matrix D, where each entry d_{ij} represents the Euclidean distance between location i and location j.

$$d_{ij} = \left\lceil \sqrt{(x_i - x_j)^2 + (y_i - y_j)^2} \right\rceil,\ \forall\, i \neq j = 0, 1, \ldots, n. \quad (22)$$

The depot was always assigned to index 0 of the matrix, and the remaining indices were mapped to randomly generated customer locations. The test instances used in this study were randomly generated to reflect realistic and challenging scenarios. The random placement ensures comparability across different formulations. It was deliberately avoided to use real-life scenarios to maintain the theoretical focus of this work. The primary goal was to assess the behaviour and limitations of the D-Wave quantum computer under clean, synthetic input conditions, rather than solving applied logistics problems.

Computational experiments results – goal functions values – was presented in Tables 1, 2 and 3. Particular columns in tables denote:

- m – number of vehicles,
- Q – vehicle capacity,
- n – number of customers,
- F1 – goal function value for model with remaining vehicle capacity between customers,
- F2 – goal function value for model with remaining vehicle capacity at customer,
- F3 – goal function value for model with truck serving customer binary decision variables,
- F4 – goal function value for QUBO model with truck serving customer binary decision variables.

Table 1. Goal functions values for Gurobi solver with 600s time limit.

m	Q	n	F1	F2	F3	F4
2	4	8	153	153	153	153
2	5	10	391	391	391	391
2	10	20	442	442	443	459
2	15	30	496	496	519	508
2	20	40	581	581	613	615
4	5	20	601	601	601	601
4	10	40	616	616	756	655
4	15	60	741	742	835	920
4	20	80	894	923	1485	1240
8	5	40	977	982	979	994
8	10	80	1185	1226	1346	1373
8	15	120	1265	1331	6380	1976
8	20	160	-	-	8833	2732
10	5	50	1179	1191	1187	1216
10	10	100	1422	1493	1640	1884
10	15	150	-	-	7926	2744
10	20	200	-	-	11069	4990

Figure 1 presents visual comparison of the vehicles routes obtained on D-Wave quantum computer for each analysed formulation for $n = 20$ customers. There are noticeable differences in route efficiency and structure between used formulations for solving considered problem. The F3 formulation clearly yields the most spatially efficient routes, characterized by minimal overlaps and relatively short total path lengths. This aligns with properties typically seen in near-optimal solutions to the Euclidean Traveling Salesman Problem, where routes tend to follow compact, non-intersecting loops. The F4 formulation, with slightly

Table 2. Goal functions values for D-Wave `DwaveSampler`

m	Q	n	F1	F2	F3	F4
2	4	8	153	153	153	153
2	5	10	458	391	391	402
2	10	20	1044	544	443	811
2	15	30	-	699	1182	1206
2	20	40	-	-	605	1717
4	5	20	859	677	601	834
4	10	40	-	-	671	1616
4	15	60	-	-	870	2708
4	20	80	-	-	1115	3632
8	5	40	-	-	992	1840
8	10	80	-	-	1443	3962
8	15	120	-	-	1555	5293
8	20	160	-	-	2220	7865
10	5	50	-	-	1214	2448
10	10	100	-	-	1595	4878
10	15	150	-	-	2074	7383
10	20	200	-	-	2561	9638

Table 3. Goal functions values for D-Wave `LeapHybridCQMSampler`

m	Q	n	F1	F2	F3	F4
2	4	8	-	-	192	187
2	5	10	-	-	449	584

less efficient route also demonstrates a well-structured routing pattern. This reflects the effective enforcement of capacity constraints, despite the additional complexity introduced by the QUBO-based representation. On the other hand, routes produced by F2 are visibly more overlapping with each other, with a high number of intersections. The F1 formulation generates the least efficient route with the most route crossings. This lack of locality likely contributes to longer routes and reduced efficiency, as trucks frequently traverse the same regions rather than servicing spatially grouped customers (Table 3).

Based on computation experiments results on D-Wave quantum computer it can be noticed that the F3 formulation is the most effective. For the model with truck serving customer binary decision variables goal function values are the lowest out of the other considered problem formulations. The second most efficient result was generated using F4 formulation. There is a slight difference in the value of goal function for formulations based on serving customer binary

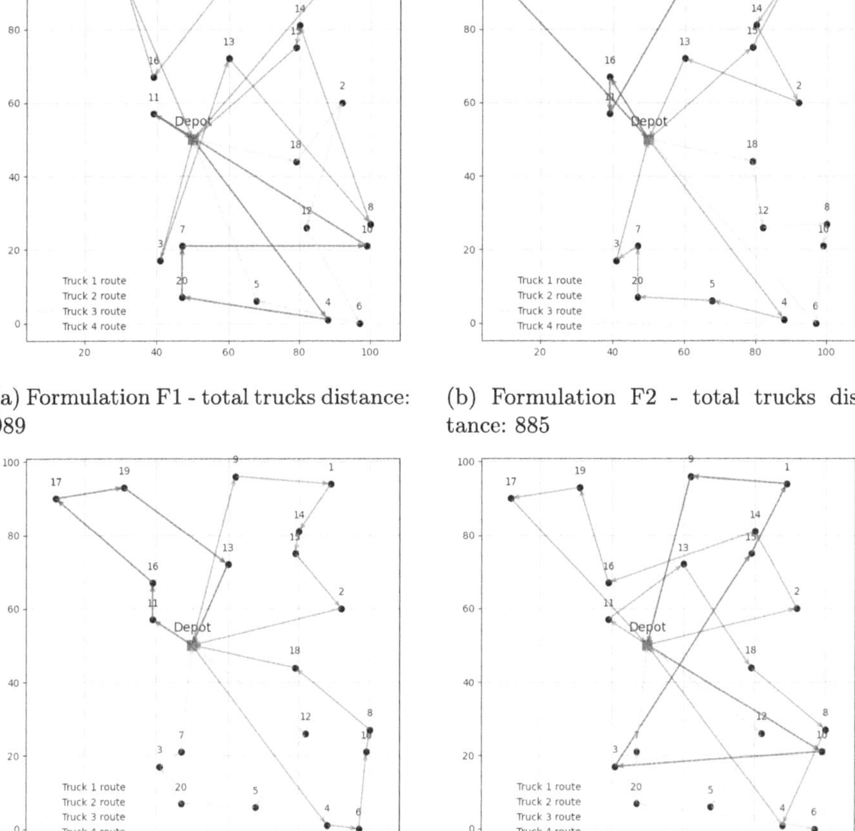

(a) Formulation F1 - total trucks distance: 989

(b) Formulation F2 - total trucks distance: 885

(c) Formulation F3 - total trucks distance: 601

(d) Formulation F4 - total trucks distance: 655

Fig. 1. Solution visualisation for test instance with $n = 20$ customers obtained on D-Wave quantum computer.

decision variable one that was for QUBO model and the other one that was formulated differently.

Formulations F1 and F2 have limited possibilities for solving considered problem with limited execution time. For Gurobi solver using formulation F1 and F2 only problems with up to $n = 120$ can be solved with $600s$ time limit. Using F1 and F2 formulation on D-Wave quantum computer with `LeapHybridCQMSampler` only problems with up to $n = 20$ can be solved. Furthermore, using the `DWaveSampler`, no feasible solution was obtained for any instance, regardless of size. Formulation F1 tracks the paths traversed by the vehicles. The need to account for the vehicle's capacity at each step of the route introduces additional variables, making the optimization process more difficult. In contrast, formulation F2 tracks capacity at each customer but does not require the path between customers to be managed. As a result, formulation F2 is less complex than formulation F1 and it simplifies the problem by focusing more on capacity constraints without tracking every individual part of the vehicle journey. This reflects a trade-off between model expressiveness and computational feasibility. Formulations F3 and F4

For Gurobi solver and `LeapHybridCQMSampler` using formulation F3 i F4 all tested problem instances can be solved (up to $n = 200$). Also using formulations F3 i F4 some small instances ($n = 8$ and $n = 10$) can be solved directly on quantum computer using `DWaveSampler`. Generally obtained cost functions values for formulation F3 and F4 are the same or worst than cost function values for formulation F1 and F2 but problems with greater sizes can be solved.

Computational times on a quantum computer using `LeapHybridCQMSampler` or `DWaveSampler` are constant. Total run time, including problem serialisation, deserialisation, and execution on QPU, for the considered problem is around 5 s. For `LeapHybridCQMSampler`, only QPU access time is reported. For the considered experiments, it was around 30–70µs. For `DWaveSampler`, the reported QPU time depends on the number of samples (reads), and for the considered experiments, it was 100–200 ms. QPU anneal time per sample is always equal to 20 µs. More details about D-Wave quantum computer operation and timing can be found in the documentation [1]. In summary, computation time on D-Wave quantum computer are very small in comparison with classical methods, but only small size problems can be solved.

4 Conclusions

This paper presents an analysis of equal demand capacitated vehicle routing problem formulations and its readiness for solving on D-Wave quantum computer. Based on the computation experiments results formulations based on truck serving customer binary decision variables can be successfully solved on D-Wave quantum computer for all considered in this paper problem sizes with hybrid approach – `LeapHybridCQMSampler`. So far problem sizes which can be solved on D-Wave quantum computer are limited to $n = 10$ when computations are directly performed on quantum hardware using `DWaveSampler`. Generally computational experiments confirm that the equal demand capacitated vehicle routing problem can be solved on a quantum computer, and results strongly depend on the problem formulation and its size. This proof quantum readiness for solving equal demand capacitated vehicle routing problem with mathematical formulation based on truck serving customer binary decision variables.

References

1. D-wave quantum computer operation and timing (2025). https://docs.dwavequantum.com/en/latest/quantum_research/operation_timing.html
2. Borcinova, Z.: Two models of the capacitated vehicle routing problem. Croatian Oper. Res. Rev. 463–469 (2017)
3. Borowski, M., et al.: New hybrid quantum annealing algorithms for solving vehicle routing problem. In: International Conference on Computational Science, pp. 546–561. Springer (2020)
4. Deleplanque, S., Yaagoubi, A.E., Gras, T., Descamps, F., Mourrier, J., I.L.: Applying analog quantum computing to solve optimisation problems: case studies on max-cut and CVRP. Int. J. Syst. Sci. Oper. Logistics **12**(1), 2463530 (2025)
5. Feld, S., et al.: A hybrid solution method for the capacitated vehicle routing problem using a quantum annealer. Front. ICT **6**, 13 (2019)
6. Godinho, M.T., Gouveia, L., Magnanti, T.L., Pesneau, P., Pires, J.: On the unit demand vehicle routing problem: flow based inequalities implied by a time dependent formulation (2009)
7. Karpas, I., Kaneet al.: Lightsolver: new platform challenges the vehicle routing problem
8. Osaba, E., Villar-Rodriguez, E., Asla, A.: Solving a real-world package delivery routing problem using quantum annealers (2024). https://arxiv.org/abs/2403.15114
9. Osaba, E., Villar-Rodriguez, E., Miranda-Rodriguez, P., Asla, A.: Optimizing package delivery with quantum annealers: addressing time-windows and simultaneous pickup and delivery (2025). https://arxiv.org/abs/2504.01560
10. Palmieri, A.: Quantum integer programming for the capacitated vehicle routing problem (2023)
11. Simchi-Levi, D., Chen, X., Bramel, J., Simchi-Levi, D., Chen, X., Bramel, J.: The capacitated VRP with equal demands. Logic Logistics: Theory, Algorithms, Appl. Logistics Manage. 301–312 (2014)
12. Sinno, S., Groß, T., Mott, A., Sahoo, A., Honnalli, D., Thuravakkath, S., Bhalgamiya, B.: Performance of commercial quantum annealing solvers for the capacitated vehicle routing problem (2023). https://arxiv.org/abs/2309.05564

Special Session: EUropean Transnational Educational

Enhancing Programming Education Through a Blended Methodology with ProgTutor

Javier Ortega-Morla[1(✉)], Antonio Leis[2], Alma María Mallo Casdelo[2], Laura Morán-Fernández[1], and Noelia Sánchez-Maroño[1]

[1] LIDIA Laboratory, CITIC, Universidade da Coruña, A Coruña, Spain
{javier.ortega1,laura.moranf,noelia.sanchez}@udc.es
[2] GII, CITIC, Universidade da Coruña, A Coruña, Spain
{antonio.leis,alma.mallo}@udc.es

Abstract. This article presents a novel blended learning methodology designed for classroom integration of ProgTutor, an Intelligent Tutoring System (ITS) for programming education. The proposed methodology combines self paced learning through the ITS with structured teacher led instruction in a recurring five phase cycle. This structure aims to reinforce conceptual understanding while maintaining teacher involvement and promoting active learning. To evaluate its effectiveness, the methodology was applied over one academic year in formal education settings and compared against two alternative approaches: traditional instruction without ProgTutor and traditional instruction using ProgTutor as a standalone tool. Results show that the blended approach led to deeper student understanding, greater progress through advanced topics, and significantly better performance in complex programming constructs, particularly loops.

Keywords: Intelligent Tutoring Systems (ITS) · Adaptive Learning · Programming · Blended Learning · Educational Technology

1 Introduction and Scope

The pursuit of effective personalized learning through Artificial Intelligence (AI) tools is well established in the field of AI in education (AIEd) [11]. Systems such as Adaptive Learning Systems [3], AI based Learning Assistants [5], and Intelligent Tutoring Systems (ITS) [6] aim to optimize individual learning paths based on student preferences and pace, enhancing both student outcomes and teacher effectiveness. Although AI driven systems have shown success in subjects such as mathematics and science [12], their integration into formal education remains limited due to regulatory and evaluative challenges [7]. One of the key aspects for such integration is the teacher's acceptance of the learning methodology, that is, how to introduce these AI powered tools into existing practices while keeping educators at the center of the learning process.

E. Corchado et al. (Eds.): SOCO 2025, CCIS 2806, pp. 371–380, 2026.
https://doi.org/10.1007/978-3-032-19763-4_34

A particularly active area for AI integration is programming education, where ITSs have advanced significantly in recent years. For instance, iSnap [8] offers contextual code hints using Abstract Syntax Trees, supporting novice learners autonomously. A recent study introduced TinyBERT as a tool to answer Python questions, with positive effects on student progress and comprehension [9]. Sharma and Harkishan [10] employed case-based reasoning and optimization algorithms to generate feedback in multiple programming languages with high success. However, these systems still lack key features necessary for adoption in formal education, such as reliable personalized learning paths, structured curricula, and effective integration with teachers. A broader review [2] identifies major challenges, such as low model transparency and the absence of public datasets, which further complicate real-world classroom deployment.

To address these challenges, this work proposes a blended teaching methodology supported by ProgTutor [4], an ITS designed for programming instruction. ProgTutor integrates a traditional ITS architecture with a 3D robotics simulator to promote a "learning by doing" approach. Its core design principle is to personalize learning experiences while keeping educators in the loop, enabling them to monitor and effectively support student progress. The presented methodology includes, after theoretical instruction, programming exercises, followed by practicing with a 3D robotics simulator (using a plugin in Visual Studio Code, see Fig. 1), and finally interaction with the ProgTutor ITS, allowing students to test functionality and evaluate their results in a realistic environment.

Fig. 1. 3D robotics simulator interface

This paper presents a comparison of the outcomes of applying this methodology over the course of one academic year with two alternative approaches: (1) a traditional methodology without ProgTutor, and (2) a traditional methodology using ProgTutor as a standalone tool. The study offers practical insights and conclusions on how AI driven systems can be effectively integrated into formal education settings.

2 ProgTutor Teaching Methodology

To support student learning, programming and robotics professors at the University of A Coruña designed a series of Python based exercises that integrate four core programming topics input/output, conditionals, loops, and functions with robotics elements to enhance practical understanding (Table 1). Each topic is structured into three difficulty levels to promote progressive learning and reinforce prior knowledge. While robotics is used to contextualize tasks, the primary objective of ProgTutor is to teach programming. The exercises are aligned with Level 5 of Bloom's revised taxonomy [1], requiring students to evaluate and develop solutions to complex problems.

Table 1. Programming skills and robotics integration across levels

Topic	Level	Programming focus	Robot sensors and actuators
Input/Output	Beginner	Basic input and output without type conversion, using simple expressions (formulas) to calculate values, focused on basic variables.	Simple actuators to move the robot (timed wheel movements) and express itself (text to speech, sounds, and LED activation).
	Intermediate	Input with type conversion, handling more complex mathematical expressions and outputting different data types. Multiple variable types are used.	
	Advanced	Extends level 2 with the inclusion of more complex algorithms or expressions.	
Conditionals	Beginner	Simple conditional statements using `if else` or `if elif else` with basic logical operators (`and`/`or`).	Actuators control the robot's face to show different emotions, with sensors detecting face position.
	Intermediate	Multiple `elif` statements with complex conditions involving multiple logical operators.	
	Advanced	Nested conditional structures with increased algorithmic complexity.	
Loops	Beginner	Use of loops to validate input data and perform iterations within algorithms.	Actuators enable movement without time control; sensors detect wheel positions, orientation, and blobs.
	Intermediate	Loops combined with conditional statements to manage algorithm flow.	
	Advanced	Use of loops to validate input data, integrated with nested loops or tasks involving higher algorithmic complexity.	
Functions	Beginner	Exercises build on those of intermediate level loops, where the task description clearly defines the functions to implement, along with their input and output parameters.	Advanced sensors support detection of QR codes and objects, enhancing functional integration.
	Intermediate	The number of functions is given, but parameters and implementation details are left to the student.	
	Advanced	Students must structure programs with appropriate functions independently; evaluation checks for a minimum number of functions.	

For each exercise, students can make multiple attempts including Python coding and testing with the robot simulator before submitting their solution for automated feedback. Once submitted, the ITS performs several checks to verify the solution's correctness. If it is correct, ProgTutor automatically assigns a new exercise to the student. If it is not, it displays a screen with the identified errors and allows the student to try again. Once the ITS considers the student's achievement is adequate, it move on to the next level.

A teaching methodology for the use of ProgTutor in a classroom is proposed here. It emphasizes a blended learning approach, which combines self paced learning facilitated by the ITS with structured teacher led learning. The proposed session schedule is shown in the left part of Fig. 2 "ProgTutor methodology". The first two sessions are preparatory to introduce students to a basic knowledge of the Python programming language and the tools they will use during the course. The next sessions follow a cyclical pattern with five main goals: a topic lecture, Python exercises without the simulator, Python exercises with the simulator, sessions with the ProgTutor ITS, and an exam. The cycle starts with a lecture where the teacher explains one of the programming topics established in Table 1. After introducing a topic, practical sessions focus on fundamental Python programming using console based exercises. In the next session, the teacher guides the students through exercises using the robot simulator, encouraging participation rather than lecturing. The final cycle consists of sessions where students complete ProgTutor exercises, with the teacher facilitating and providing support.

In addition to this blended methodology, traditional programming teaching without ProgTutor was used by another student group at the same educational center. The session schedule for traditional teaching is shown in the right part of Fig. 2 "Traditional methodology". The first session introduces students to the tools they will use during the course. The next sessions follow a cyclical pattern to teach Python in the following way: a topic lecture followed by practical sessions with console based exercises. Optionally, several sessions using the ProgTutor ITS can be included after those teaching sessions, as shown in Fig. 2 "Traditional methodology". Finally, evaluation exams can be performed.

3 Classroom Validation

To validate the proposed blended methodology in conjunction with the features of the ITS, the approach was implemented with several groups of students from two different high schools in the northwest of Spain. The students were between 16 and 17 years old, with no previous training on programming fundamentals. Consequently, the results obtained in this study could be applied to similar groups of students, like those in the first year of University degrees or VET education.

For privacy reasons, each institution is referred to as SS1 and SS2. At SS1, there are two groups: one that used the blended methodology with ProgTutor, and a control group that followed a traditional methodology without ProgTutor. In contrast, SS2 consists of a single group taught using a non-blended methodology. In this approach, all theoretical content from input/output to functions

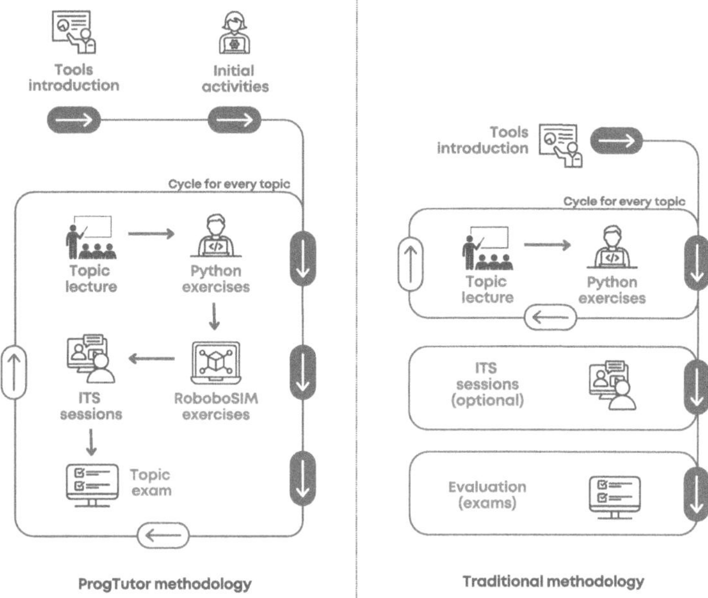

Fig. 2. Different methodologies employed

was first delivered through traditional lectures. Students began using ProgTutor after completing the entire theoretical instruction, starting from the beginning of the curriculum. Notably, no theoretical topics were revisited once the practical use of ProgTutor commenced. This methodology is illustrated in Fig. 2 as the "Traditional methodology". Each group consists of between 16 and 19 participants.

The following sections compare the three methodologies to identify which is most suitable for students and which best promotes effective student learning, so which is the most adequate to introduce an ITS in formal education.

3.1 Comparison Between Traditional Methodology Without ProgTutor and Blended Methodology Using ProgTutor

First, the results from the control and ProgTutor groups at SS1 were compared. It is worth noting that ProgTutor students come from STEM backgrounds, unlike the control group—even though none had prior programming experience. As a result, the control group was unable to complete any functions exercises. Therefore, both groups completed three exams at different stages of the course: the first exam assessed input/output concepts, the second focused on conditional statements, and the third on loops. Boxplots of the scores obtained by each group in these exams are shown in Fig. 3.

Based on these results, a statistical test was conducted to assess whether the ProgTutor group outperformed the traditionally taught group in the exams. After revealing that the data did not follow a normal distribution with the Shapiro–Wilk test, the non-parametric Mann–Whitney U test was applied to evaluate differences between the groups. Table 2 shows the p-values for each topic, where values below 0.05 indicate a significant difference between groups. It can be concluded that for introductory topics with low complexity, both the traditional and the ProgTutor methodologies yield comparable learning outcomes. However, as the complexity of the subject increases, the ProgTutor approach demonstrates a clear advantage over the non-blended methodology. This may be attributed to its practical orientation and the use of more realistic examples, which help students better understand complex programming concepts such as multi-branch conditionals and nested loops.

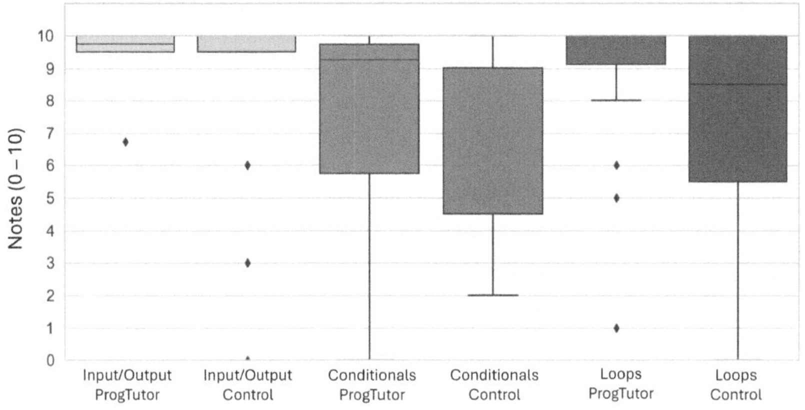

Fig. 3. Distribution of scores by programming topic and group

3.2 Comparison Between Blended Methodology Using ProgTutor and Traditional Methodology with ProgTutor

After confirming that the use of ProgTutor enhances learning outcomes, the two methodologies incorporating ProgTutor were analyzed to identify which is more effective for instructional purposes. Figure 4 shows the distribution of hours dedicated to theoretical and practical sessions at both institutions. At SS2, over twice as much time was devoted to theory, with only half as much allocated to ProgTutor based practice. Nevertheless, the total number of instructional hours was comparable across both schools.

An analysis of the different instructional approaches was conducted using a set of three metrics collected throughout the course. These metrics were extracted from each exercise completed by every student in both groups. Similarly to Sect. 3.1, statistical tests were employed. After confirming non-normal

Table 2. Results of the Mann–Whitney U Test

Topic Assessed	p-value	Hypothesis Result
Input/Output	0.358	Not significant
Conditionals	0.100	Trend, not significant
Loops	0.033	Significant in favor of ProgTutor

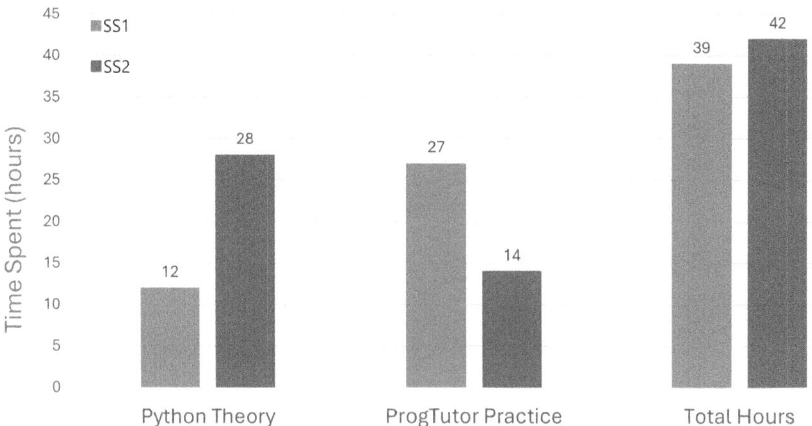

Fig. 4. Distribution of Hours Throughout the Course

data distributions using the Shapiro-Wilk test ($p < .001$), the Mann-Whitney U test was applied for further analysis.

The first metric, shown in Fig. 5 "Distribution of Python Errors Across Institutions", presents the number of Python errors students made before receiving automated feedback. The data show that students at SS1 made significantly more errors than those at SS2. This reflects differences in instructional methodology: at SS1, the blended approach with ProgTutor emphasizes hands-on exploration, encouraging students to learn through practice. As a result, more early mistakes are expected. However, these initial errors are part of an active learning process that promotes deeper understanding over time. In contrast, the more traditional approach at SS2 tends to produce fewer errors, likely due to stronger emphasis on guided instruction and correctness from the outset.

The second metric analyzed was the distribution of errors during the automated feedback phase of the exercises completed in ProgTutor. As shown in Fig. 5 "Distribution of Evaluation Errors Across Institutions", the results reveal no statistically significant differences between the two groups. This suggests that both instructional approaches were similarly effective in helping students reach a comparable level of competence by the time they engaged with the ITS. A possible explanation is that, although students in the blended model made more errors during early exploration, structured support and iterative practice helped consolidate their understanding by the time formal evaluation occurred.

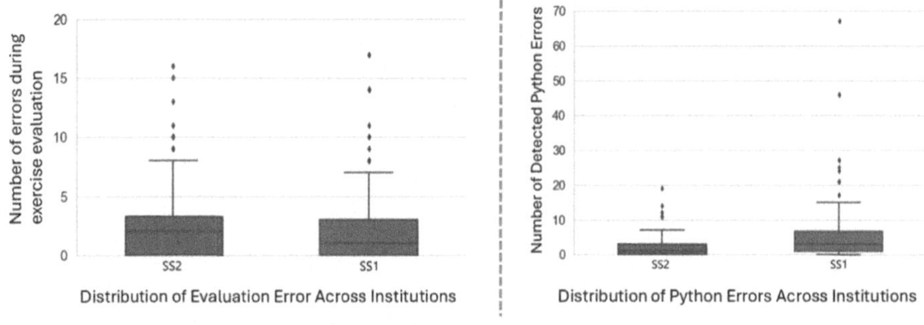

Fig. 5. Distribution of Errors Across Institutions

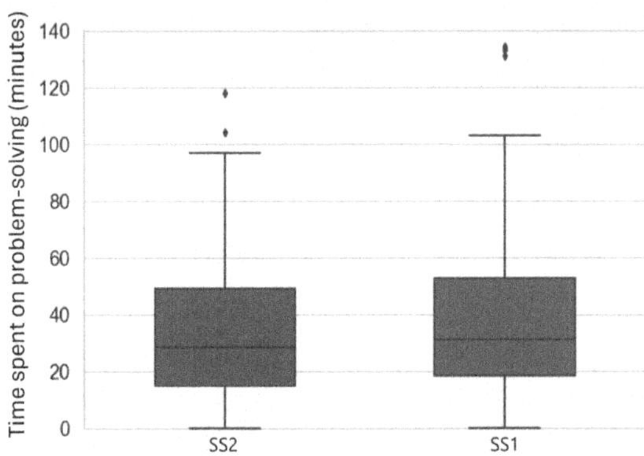

Fig. 6. Distribution of Time Resolution Across Institutions

As the final metric, the time students spent solving each of the various exercises was analyzed. Figure 6 presents the distribution of completion times (in minutes). The results indicate that there is no statistically significant difference between the two institutions. This suggests that, from the students' perspective, both teaching methodologies perform similarly in terms of exercise completion time.

Additionally, the extent of student progress throughout the course was examined by assessing the topic and level each student had reached by the end of the course. This analysis is presented in Fig. 7. The results show that students from SS1 made greater progress, with most of them reaching the advanced level in loops (see Table 1). Specifically, 63.2% of students from SS1 reached or surpassed the topic Intermediate Loops, whereas only 37.5% of students from SS2 reached that topic, highlighting a marked difference in progression between the two groups. The greater progress could be partly attributed to the higher number of practical hours (see Fig. 4). However, the key takeaway is that reducing theoretical instruction and increasing hands on practice, while allowing students to advance at their own pace, leads to better performance by the end of the course.

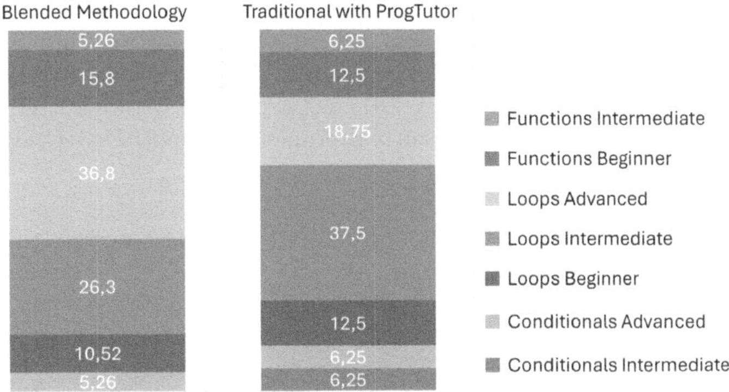

Fig. 7. Distribution of Topics Reached

4 Conclusions and Future Work

The main conclusion of this validation is that the blended methodology, where traditional and ITS based sessions are interleaved throughout the course, proved to be the most effective. Although students following this methodology made more programming errors during the learning process, these mistakes are a natural part of hands on learning and did not hinder their overall performance. On the contrary, they showed fewer errors during challenge evaluations and achieved similar exercise completion times compared to their peers using a compact approach.

Most notably, the blended methodology enabled students to progress further through the course content, with the majority reaching higher levels in advanced topics. This result highlights that fewer traditional and more theoretical classes do not necessarily lead to reduced learning. On the contrary, allowing students to advance at their own pace appears to facilitate deeper and more enduring knowledge acquisition.

In a more specific realm, findings of this study show that ProgTutor provides advantages over traditional instruction, particularly for more complex topics such as loops. While learning outcomes for introductory topics were comparable between the two approaches, ProgTutor's interactive and practice-oriented methodology led to significantly better performance as topic complexity increased.

While the results are promising, it is important to acknowledge certain considerations. The study involved a limited number of groups from only two high schools, which may influence the extent to which the results can be generalized. Additionally, although students reportedly had no prior programming experience, potential differences in initial knowledge or motivation were not systematically assessed. Future research should aim to replicate these results with a larger number of students and across different educational institutions.

Acknowledgments. The TED2021-131172B-I00 grant was funded by MCIN/AEI and the European Union NextGenerationEU/PRTR. Support was also received from the Galician Research Center (CITIC), funded by the Government of Galicia and the European Union (FEDER GALICIA 2014–2020 program), through grant ED431G 2019/01, as well as from the Galician Government group (ED431C 2022/44) with FEDER funds.

Disclosure of Interests. The authors have no competing interests to declare that are relevant to the content of this article.

References

1. Anderson, L.W., Krathwohl, D.R.: A taxonomy for learning, teaching, and assessing: a revision of bloom's taxonomy of educational objectives (2001)
2. Francisco, R.E., de Oliveira Silva, F.: Intelligent tutoring system for computer science education and the use of artificial intelligence: a literature review. In: Proceedings of the 14th International Conference on Computer Supported Education (CSEDU 2022), pp. 338–345. SCITEPRESS (2022). https://doi.org/10.5220/0011084400003182
3. Kabudi, T., Pappas, I., Olsen, D.H.: AI-enabled adaptive learning systems: a systematic mapping of the literature. Comput. Educ. Artif. Intell. **2**, 100017 (2021)
4. Leis, A., Mallo, A., Guerreiro-Santalla, S., Paz-López, A., Bellas, F.: Personalized learning of programming fundamentals through robotic simulations. AI in Educ. Educ. Res. **2519** (2025)
5. Maedche, A., et al.: AI-based digital assistants: opportunities, threats, and research perspectives. Bus. Inf. Syst. Eng. **61**, 535–544 (2019)
6. Mousavinasab, E., Zarifsanaiey, N., Kalhori, S.R.N., Rakhshan, M., Keikha, L., Saeedi, M.G.: Intelligent tutoring systems: a systematic review of characteristics, applications, and evaluation methods. Interact. Learn. Environ. **29**(1), 142–163 (2021). https://doi.org/10.1080/10494820.2018.1558257
7. Panigutti, C., et al.: The role of explainable AI in the context of the ai act. In: Proceedings of the 2023 ACM Conference on Fairness, Accountability, and Transparency, pp. 1139–1150 (2023)
8. Price, T.W., Dong, Y., Lipovac, D.: ISNAP: towards intelligent tutoring in novice programming environments. In: Proceedings of the 2017 ACM SIGCSE Technical Symposium on Computer Science Education (SIGCSE '17), pp. 483–488. ACM, Seattle (2017). https://doi.org/10.1145/3017680.3017762
9. Roldán-Álvarez, D., Mesa, F.J.: Intelligent deep-learning tutoring system to assist instructors in programming courses. IEEE Trans. Educ. **67**(1), 153–160 (2024). https://doi.org/10.1109/TE.2023.3331055
10. Sharma, P., Harkishan, M.: Designing an intelligent tutoring system for computer programming in the pacific. Educ. Inf. Technol. **27**(5), 6197–6209 (2022). https://doi.org/10.1007/s10639-021-10882-9
11. Shemshack, A., Spector, J.M.: A systematic literature review of personalized learning terms. Smart Learn. Environ. **7**(1), 33 (2020)
12. Zhang, L., Basham, J.D., Yang, S.: Understanding the implementation of personalized learning: a research synthesis. Educ. Res. Rev. **31**, 100339 (2020)

Generative AI Use Scale Indicators in Higher Education

Elena Del Val(✉) and Joaquin Taverner

Valencian Research Institute for Artificial Intelligence (VRAIN), Universitat Politècnica de València, Valencia, Spain
edelval@dsic.upv.es

Abstract. The integration of generative artificial intelligence (GAI) in higher education offers promising opportunities for personalized learning, enhanced assessment, and improved teaching practices. However, the ethical, pedagogical, and practical challenges posed by GAI require structured frameworks to guide its responsible use. This paper reviews five scales of GAI use in educational contexts and identifies their strengths and limitations. Based on this analysis, we propose a new framework called SCALE-E (Structured Competency-Aligned Levels for Ethical AI in Education), which focuses on aligning AI use with the development of key transversal competencies such as critical thinking, creativity, problem solving, and autonomy. SCALE-E advances current approaches by transforming AI interaction into a formative and ethical learning experience through five progressive levels of AI integration. We illustrate the application of SCALE-E in a programming course, demonstrating its potential to enhance technical skills and transversal competencies.

Keywords: Generative AI · Higher Education · Scale · Computer Science

1 Introduction

The incorporation of generative artificial intelligence (GAI) in education provides new opportunities to personalize learning, improve assessment, and facilitate teaching, especially in higher education [2,5]. However, the deployment of these technologies also poses significant challenges related to ethics, privacy, fairness and transparency [3]. Particularly, evaluating the responsible use of GAI in educational environments requires a framework that considers both the pedagogical impact and the social and ethical implications [13]. The automation of educational processes with GAI must ensure that fundamental values of learning, such as learner autonomy and academic integrity, are not compromised [12].

In this scenario, various scales and guides emerge that seek to regulate, orient and evaluate the use of GAI in higher education, particularly in the area of learning assessment. These scales guide teachers and students and standardize institutional practices, promoting equity and ensuring academic integrity [3,15].

E. Corchado et al. (Eds.): SOCO 2025, CCIS 2806, pp. 381–392, 2026.
https://doi.org/10.1007/978-3-032-19763-4_35

The diversity of proposals reflects different pedagogical philosophies: from normative approaches focused on control to formative perspectives that promote critical literacy in GAI.

This paper provides an analysis of the main current proposals on scales for the use of GAI in educational contexts, with a special focus on higher education. Based on this analysis, a new framework called SCALE-E is proposed, which seeks to integrate the use of generative artificial intelligence with the development of key transversal competencies in higher education. This proposal aims to provide a formative and ethical vision, oriented towards a pedagogical transformation that enhances critical, creative and strategic skills in students. Finally, the practical application of this scale is illustrated in a real case within the subject of Programming demonstrating its potential to facilitate a responsible and enriching use of GAI in academic environments.

The remainder of this paper is structured as follows. Section 2 provides a review of relevant existing scales and frameworks for GAI use in education. Section 3 presents the SCALE-E framework. Section 4 illustrates the practical application of SCALE-E within a university programming course as a concrete use case. Finally, Sect. 5 summarizes the main conclusions and outlines directions for future research.

2 Use Scales

The rapid evolution of GAI has resulted in a growing amount of research focused on its integration in higher education, particularly with regard to the assessment of learning. In this context, various theoretical and practical proposals have emerged that seek to establish frameworks, ethical principles and operational criteria for a responsible and effective use of these technologies. One of the first proposals that has gained visibility in this field is the AIAS (Artificial Intelligence Assessment Scale) [7]. This scale was developed with the purpose of providing an ethical and operational framework for the integration of GAI in academic assessment tasks. The AIAS scale establishes five clearly defined levels: Level 0 (No AI (Artificial Intelligence)), Level 1 (Limited AI), Level 2 (Moderate AI), Level 3 (High AI) and Level 4 (Full AI). These levels reflect the proportion and nature of AI use permitted by the student when performing an evaluative task, ranging from complete prohibition to active co-authorship with AI systems. Each level includes specific examples of permitted use and recommendations for implementation. Initially, the scale was presented with a traffic light-like color structure (red to green), but was redesigned with a pastel color palette, in order to avoid symbolic associations of judgment or sanction, and to emphasize its guiding rather than restrictive intent [8]. The fundamental objectives of the scale are to assist educators in adapting their assessments in light of GAI tools, to clarify for students the permissible use of AI in their work, and to support the completion of assessments in accordance with the principles of academic integrity. In this sense, the scale seeks to promote an ethical use of GAI tools, while fostering both the development of academic knowledge and practical skills related to the critical and effective use of these technologies. Its main strength lies in its normative clarity and its transversal applicability to different educational levels and disciplines. The AIAS has been widely adapted in many different contexts, both K-12 and higher education [1,11].

In [10], the AIAS scale was adapted for English language teaching contexts for academic purposes. This adaptation, denoted EAP-AIAS (English for Academic Purposes - Artificial Intelligence Assessment Scale), retains the five-level structure of the original model, but redefines each level in terms of specific language tasks, such as writing argumentative essays, writing critical summaries, and preparing oral presentations. In addition, it offers detailed examples and realistic scenarios that illustrate the potential use of GAI tools in learning academic English. It also provides differentiated guidance according to students' level of language proficiency. In contrast to the general AIAS scale, which is more normative, the EAP-AIAS incorporates a formative component that recognizes the value of AI as a mediator in language learning processes and explicitly links each level to well-established approaches in teaching English for academic purposes.

The CAIAF (Comprehensive AI Integration Assessment Framework) scale represents a significant evolution in the design of frameworks for the integration of AI in educational contexts [4]. Unlike more normative models such as AIAS, the CAIAF introduces six progressive levels of AI use, accompanied by a gradient-based visual system that allows for nuancing the intensity and risk associated with each level. Level 0 corresponds to the total absence of AI in academic production. Level 1 allows minimal use, such as spelling corrections or superficial aids. In Level 2, AI is used functionally as a support tool, without intervening directly in the authorship of the content. Level 3 contemplates a collaborative use, in which the AI actively contributes to the generation of content under the critical supervision of the learner. At Level 4, the AI can act more autonomously, generating substantive parts of the work, but requires the student to perform a reflective review. Finally, Level 5 allows for advanced, personalized use of AI, with dynamic real-time adaptations and an emphasis on ethical co-authorship. This structure recognizes the diversity of possible interactions between learners and generative technologies. It also incorporates ethical principles such as transparency, equity, pedagogical alignment and data protection. In addition, its design contemplates the possibility of adaptation to different educational levels (primary, secondary and higher education), making CAIAF a flexible, expansive and future-oriented tool for the design and evaluation of AI-mediated educational tasks.

The scale proposed in [14] is adapted for K-12/secondary education and highlights practical classroom application, teacher support, and age-appropriate considerations, including legal compliance (GDPR, parental consent). It departs from the tiered approach of previous proposals and instead offers an open-ended, formative guide for integrating AI into teaching and assessment. This guide emphasizes the need for critical reflection, co-creation, and student agency. Its approach is not intended to categorize the use of AI in a normative way, but to provide a repertoire of best practices, ethical principles, and examples of assignments that foster meaningful integration of GAI. It aims to guide students in the responsible use of artificial intelligence in teaching and learning processes and to clarify the extent to which AI can be used. The strength of this guide lies in its flexibility and its ability to adapt to changing contexts. It is particularly useful for institutions seeking to promote digital literacy and critical competence.

Finally, the AI Creative Framework Indicator (ACFI) is a framework with short codes that creators can use to declare, in a transparent and standardized way, the level of involvement of artificial intelligence (AI) in their creative works (texts, images, videos, etc.) [6]. The system is based on two-letter labels representing different uses of AI in the creative process. These labels are placed on a standard line, where each code indicates a specific type of AI intervention. Among the main codes are: IG (Idea Generation), when AI is used for idea generation or concept brainstorming; GE (Generative Editing), which indicates that AI has directly produced textual, visual, or auditory content; and AE (AI Editing), referring to the use of AI for editing, proofreading, or rewriting. Other relevant codes include RD (Research/Development Support), which implies a use of AI to gather or synthesize information, and IN (Inspiration), which identifies a purely inspirational use, with no direct intervention in the final product. Finally, the NA code (No AI Used) certifies the total absence of AI in the process. For instance, ACFI: IG, AE, means that AI was used to generate ideas and also to edit the final text.

Table 1 presents a clear and structured overview of the main scales and frameworks developed to guide the use of generative artificial intelligence in educational contexts. It can be seen that, although all proposals share an interest in integrating AI in an ethically and pedagogically meaningful way, they differ markedly in their approach, levels of structuring, and scope of application. For example, both the AIAS and the EAP-AIAS propose progressive levels of AI use, focused on the validity of the assessment and the development of competencies, the latter being a specific adaptation to the field of academic English. In contrast, CAIAF adopts a more ethical and evaluative approach, moving away from formal levels and prioritizing risk analysis and alignment with learning outcomes. The AI Assessment and Teaching Guide, while also structuring the use of AI into five levels, is distinguished by its practical orientation, designed by and for teachers, and by its explicit intention not to hierarchize the use of AI, but to promote contextualized pedagogical decisions. Finally, the ACFI focuses on introducing a concise, code-based system that does not prescribe levels of use. It classifies the type of AI participation in a creative or academic product, thus encouraging learner self-awareness, transparency and responsible authorship.

While each framework offers valuable insights, they also involve certain considerations that condition their practical application. For example, the AIAS scale offers a clear normative structure, but its effectiveness depends on sufficiently trained faculty and may require adaptation to address the nuances of highly specialized tasks. Similarly, the EAP-AIAS focuses specifically on academic English, so it largely conforms to language acquisition principles, but its applicability beyond this domain would benefit from further empirical validation.

Table 1. Comparison of GAI use scales in education

Scale	Purpose	AI Use	Pedagogical Approach	Application
AIAS [7]	Ethical integration of GenAI into educational assessments	5 levels: No AI, Limited AI, Structured AI Use, Creative AI Use, AI Exploration	Social constructivism, emphasis on assessment validity	General education (K12 and higher education)
EAP-AIAS [10]	Adaptation of AIAS to English for Academic Purposes (EAP) instruction	5 adapted levels: No AI, Awareness, Light AI Use, Moderate AI Use, Full AI Use	Development of language skills and academic acculturation	English for Academic Purposes (EAP) teaching
CAIAF [4]	Ethical and effective integration of AI in assessment and learning	6 levels: No AI, Minimal AI Use, Functional AI Use, Collaborative AI Use, Autonomous AI with Oversight, Advanced and Personalized AI Use	Ethics-centered pedagogy, authentic learning, adaptable to diverse educational levels	General and specialized education (primary, secondary, and higher education)
AI Assessment and Teaching Guide [14]	To guide pedagogical decisions regarding AI use in educational tasks	5 levels: No AI, AI for Feedback, AI as a Tool, AI as a Co-Creator, AI as an Evaluator	Teacher reflection, alignment with learning objectives, contextual decision-making	Secondary education (adaptable to other contexts)
ACFI [6]	To promote transparency in AI involvement in creative/academic work	Code-based: IG, GE, AE, RD, IN, NA	Digital literacy, critical reflection, authorship accountability	Creative writing, media production, student self-reporting

The CAIAF model provides an ethically robust and future-proof structure, yet its methodological complexity may pose challenges in under-resourced contexts or where AI literacy is still emerging. The AI Assessment and Teaching Guide provides flexibility and encourages critical thinking but may lack the clarity some educators require in regulated settings. The ACFI, by contrast, does not function as a normative scale but rather as a complementary transparency tool;

it empowers students to explicitly declare how AI was involved in their work using a simple coding system. This can be particularly powerful in formative contexts, promoting ethical reflection, digital literacy, and academic integrity through learner-centered practices. However, its effectiveness relies on students' understanding of the codes and teachers' ability to interpret them meaningfully within assessment criteria.

Taken together, these considerations illustrate the growing need for diversified frameworks that respond to the multiple dimensions (i.e., ethical, pedagogical, and institutional) of AI integration in education. Rather than seeking a universal solution, a more productive path may lie in adopting complementary tools, such as combining normative scales like AIAS or CAIAF with reflective instruments like ACFI, to foster both regulation and agency in the use of GAI.

3 Proposal

In response to the growing integration of generative artificial intelligence in educational environments, we present the SCALE-E (Structured Competency-Aligned Levels for Ethical AI in Education) framework, a new scale designed to guide its use in higher education. This proposal focuses on how the application of GAI can enhance the development of transversal competencies essential for the integral formation of students. These competencies include critical thinking, professional ethics, effective communication, autonomous learning, creativity, and teamwork. By integrating the ethical and critical use of AI with these skills, SCALE-E aligns with the principles of European digital competence frameworks such as DigCompEdu [9], which aim to empower educators and facilitate the development of students' digital competence in the digital age. The SCALE-E framework aims to transform that use into a deliberate and measurable formative component. To achieve this, the scale establishes progressive levels of AI integration aligned with the promotion of transversal competencies across diverse learning activities. The framework is structured into five levels of integration. These levels are designed as *progressive steps* in terms of the complexity of AI interaction, the degree of student autonomy, and the depth of required critical reflection (see Table 2). They are not hierarchical in the sense that a higher level is inherently better or more desirable than a lower one. The appropriate level is determined by the specific learning objectives and pedagogical context of the task. Each level explicitly incorporates one or more transversal competencies that students are expected to develop through their interaction with AI:

- *Level 1: Assisted Exploration*: At this level, the use of GAI is limited to consultation or informational support tasks, such as obtaining basic explanations, definitions, or summaries. The student does not use GAI to produce final content, but rather to enrich their understanding and access different perspectives before developing their own ideas. This initial interaction with AI fosters competencies such as information management, basic critical thinking, and the ethical use of digital tools, laying the foundation for autonomous and informed learning.

- *Level 2: Guided Elaboration*: Here, the student uses AI as an active companion in constructing academic work, for example, by generating drafts, structural suggestions, or preliminary ideas that are later reviewed and reworked. AI becomes a cognitive support tool that facilitates idea organization and process planning. The objective of this level is to promote competencies such as written communication, analytical capacity, strategic planning, and eval-

Table 2. SCALE-E Proposal.

Level	Description	Developed Competencies	Example in Programming
Level 1: Assisted Exploration	Use of AI to obtain basic explanations, definitions, or summaries, without generating final content.	Information management, basic critical thinking, ethical use of digital tools.	The student queries AI to understand what a recursive function is or how a `for` loop is structured.
Level 2: Guided Elaboration	AI helps construct initial ideas such as outlines or drafts that the student later critically revises.	Strategic planning, written communication, analysis, and evaluative judgment.	The student asks AI for a solution outline for a sorting algorithm and adapts it, modifying its logic.
Level 3: Reflective Co-creation	Collaboration with AI to create products that integrate creativity and personal judgment.	Creativity, informed decision-making, complex problem solving.	The student generates an initial graphical interface with AI, then redesigns and customizes both design and functionality.
Level 4: Critical and Ethical Evaluation	The student analyzes AI outputs, detecting errors, biases, or limitations.	Advanced critical thinking, digital literacy, professional ethics.	After reviewing AI-generated code, the student identifies security vulnerabilities and proposes justified improvements.
Level 5: Metareflection and Design of AI Interactions	The student designs strategies for AI use in academic or professional contexts.	Strategic thinking, innovation, ethical leadership, advanced digital competence.	Develops a guide for implementing AI in collaborative software development, considering ethical and technical implications.

Table 3. Competency Assessment Rubric Using the SCALE-E Scale

Competency	Level 1	Level 2	Level 3	Level 4	Level 5
Information Management	Consults AI for basic explanations.	Selects useful suggestions to organize work.	Integrates relevant AI information into creative solutions.	Identifies limitations in AI-generated information.	Critically evaluates the appropriateness of using AI in different contexts.
Critical Thinking	Recognizes simple concepts suggested by AI.	Evaluates proposed ideas, accepting or rejecting them with justification.	Makes informed decisions by combining AI inputs with own reasoning.	Analyzes biases, errors, and limitations in generated content.	Designs strategies for responsible and effective AI use.
Communication and Planning	Understands basic structures with AI help.	Uses AI to outline solutions and organize tasks.	Refines planning by integrating creativity and functionality.	Proposes organizational improvements and argues their impact.	Leads strategic design processes involving AI.
Creativity and Problem Solving	Uses simple examples suggested by AI.	Modifies AI-proposed examples to adapt them.	Creates original products through interactions with AI.	Reformulates solutions to complex problems generated by AI.	Designs new humanAI interaction methods that innovate the process.
Ethics and Digital Literacy	Uses AI with teacher guidance and basic awareness of its role.	Recognizes the need to review and validate AI output.	Acts responsibly when integrating AI into decisions.	Highlights ethical implications and proposes actions to mitigate them.	Leads critical reflection on the social impact of AI use.

uative judgment, as the student must critically engage with the system's suggestions and build upon them.

- *Level 3: Reflective Co-creation*: At this level, the student collaborates with AI in a generative process that requires judgment, adaptation, and creativity. It is not merely about accepting or rejecting what AI produces, but about integrating it thoughtfully into the development of original solutions or outputs. This symmetrical interaction activates competencies such as creativity, informed decision-making, complex problem-solving, and autonomous learning, as the student becomes a designer and critical editor of the AI's contributions. While Level 2 focuses on using AI for planning and structuring initial ideas, Level 3 moves towards a more generative partnership where AI contributes substantively to the creative output, requiring the student to act as a critical co-designer and integrator of AI-generated content into an original work.

- *Level 4: Critical and Ethical Evaluation*: This level involves a shift in focus: the student not only uses AI, but also analyzes its outputs, functioning, and limitations. They are expected to identify errors, biases, or inconsistencies in AI-generated content and propose more rigorous or responsible alternatives. Such tasks promote a deep understanding of the technology and develop key competencies such as advanced critical thinking, digital literacy, professional ethics, and argumentative reasoning, preparing students for conscious and socially responsible use of AI in their field. This level builds upon the critical understanding and evaluation skills developed in Level 4, extending them to a meta-cognitive and strategic level regarding the role and integration of AI in broader contexts. It requires not just using or evaluating AI outputs, but critically analyzing the very process of AI interaction and its systemic implications.
- *Level 5: Metareflection and Design of AI Interactions*: This level represents the highest degree of student agency and strategic thinking regarding AI integration. It positions the student not just as a user, but as a designer and strategist of AI use. Students are expected to critically reflect on how and when AI should be integrated into academic or professional processes and even propose new forms of human-AI collaboration that maximize added value and minimize risks. This level activates competencies such as strategic thinking, innovation, ethical leadership, and advanced digital competence, equipping students to lead the transformative integration of AI in real-world contexts.

SCALE-E builds on a competency-based framework that directly aligns with learning outcome-oriented university curricula, ensuring that each level of AI integration corresponds to specific learning objectives. SCALE-E places learner development at the center, illustrating how each task (with its varying degrees of autonomy, complexity and reflection) contributes to the growth of transversal competencies. To make this process transparent and practical, each level is supported by clear, observable indicators that teachers can use to assess mastery and provide specific feedback during real academic activities.

4 Use Case: Programming Course

The Programming course of the Industrial Informatics and Robotics Degree of the Universitat Politècnica de Valncia aims to provide students with a solid foundation in programming logic, data structures, and computational problem-solving. In this context, the incorporation of GAI can be a powerful tool to enrich the learning process and the development of transversal competencies, provided its use is aligned with clear pedagogical goals.

The following example illustrates how the SCALE-E can be applied in this course to guide and enhance the formative use of AI, linking different levels of interaction with the progressive development of skills such as critical thinking, creativity, professional ethics, and self-assessment (see Table 3). This approach seeks to optimize the technical learning of programming and prepare students for a conscious and responsible use of emerging technologies in their future professional careers.

- *Level 1: Assisted Exploration*: At this level, the student uses AI to obtain simple explanations of concepts such as variables, control structures, or data types. For example, they may ask the AI to explain what a `for` loop is or to provide simple C++ code examples. Here, AI acts as a supplementary resource to clarify doubts and support autonomous learning without replacing the student's own coding process.
- *Level 2: Guided Elaboration*: While developing a small program, the student uses AI to suggest the general structure of the code or to generate initial fragments that they then manually adapt and correct. For instance, they might ask for help creating pseudocode for a search algorithm and then convert it into functional code, reviewing and debugging the final result. This level works on competencies related to planning, analysis, and technical communication.
- *Level 3: Reflective Co-creation*: The student collaborates with AI to design and develop more complex solutions, such as implementing recursive functions or basic data structures (linked lists, stacks). AI may suggest alternative implementations that the student must evaluate, modify, and justify based on criteria like efficiency or readability. This stage fosters creativity and applied critical thinking in software development.
- *Level 4: Critical and Ethical Evaluation*: At this stage, the student analyzes AI-generated or suggested code to identify potential logical errors, security vulnerabilities, or poor programming practices. They also reflect on the ethical implications of automating certain programming tasks or potential bias in algorithm selection. For example, the student might detect if AI proposes solutions that fail to handle exception cases properly or could lead to failures in industrial systems.
- *Level 5: Metareflection and Design of AI Interactions*: Finally, the student designs a small project or methodology in which AI is integrated responsibly and efficiently into the software development process, such as automating unit tests or generating technical documentation. Additionally, they reflect on AI's limitations in industrial programming and propose strategies to supervise or audit its contributions, promoting ethical leadership and innovation.

5 Conclusions

In this paper, we have conducted a critical analysis of five existing scales for evaluating the use of GAI in education. The examined scales present diverse approaches, ranging from measuring the degree of AI involvement to offering ethical and pedagogical frameworks for educators. Building on the strengths and limitations identified, we propose a new scale: SCALE-E. This scale emphasizes the integration of AI with the development of transversal competencies such as creativity, critical thinking, problem solving, and autonomy. SCALE-E represents a meaningful advancement beyond previous models by explicitly linking GAI use to the cultivation of essential transversal skills in higher education. This approach supports ethical and pedagogical decision making regarding the use of AI and transforms the interaction with AI into a holistic formative experience. Additionally, we have presented an instantiation of the scale applied to a first-year university programming course, illustrating how SCALE-E can guide and enhance the formative use of AI in concrete academic settings. While SCALE-E provides a general framework, its application in various disciplines such as the humanities, medical sciences, or others will require specific adaptations. Each field presents unique pedagogical goals, assessment practices, and potential ethical challenges related to the use of AI. Therefore, effective implementation of SCALE-E will require adapting the levels, examples, and alignment of competencies to the specific context of each academic area. Future work will focus on empirically validating the scale's effectiveness and exploring tailored adaptations for specific academic fields.

Acknowledgements. Work supported by PIME/24–25/429 project funded by Instituto de Ciencias de la Educación (ICE) of the Universitat Politècnica de València.

References

1. Furze, L.: The AI assessment scale in action: examples from k-12 and higher education across the world (2024). https://leonfurze.com/2024/05/20/the-ai-assessment-scale-in-action-examples-from-k-12-and-higher-education-across-the-world/
2. Holmes, W., Bialik, M., Fadel, C.: Artificial intelligence in education: promises and implications for teaching and learning. Boston: Center for Curriculum Redesign (2019). https://curriculumredesign.org/wp-content/uploads/AI-in-Education-Promises-and-Implications-for-Teaching-and-Learning-Holmes-Bialik-Fadel.pdf
3. Holmes, W., Bialik, M., Fadel, C.: Ethics of AI in education: towards a typology of risks and opportunities. Int. J. Artif. Intell. Educ. **31**, 209–233 (2021). https://doi.org/10.1007/s40593-021-00236-6
4. Kılınç, S.: Comprehensive AI assessment framework: Enhancing educational evaluation with ethical ai integration. J. Educ. Technol. Online Learn. **7**(4-ICETOL 2024 Special Issue), 521–540 (2024)
5. Luckin, R., Holmes, W., Griffiths, M., Forcier, L.B.: Intelligence unleashed: an argument for AI in education. Pearson Educ. (2016). https://www.pearson.com/content/dam/one-dot-com/one-dot-com/global/Files/about-pearson/innovation/Intelligence-Unleashed-Publication.pdf
6. McGee, N.J.: AI creative framework indicator (ACFI) (2024). https://www.linkedin.com/posts/nnekamcgee_i-am-a-hopeful-creator-i-have-been-seeking-activity-7197608265013223424-4x7V, posted on LinkedIn. A framework for disclosing AI involvement in creative work using standardized codes
7. Perkins, M., Furze, L., Roe, J., MacVaugh, J.: The artificial intelligence assessment scale (AIAS): a framework for ethical integration of generative AI in educational assessment. J. Univ. Teach. Learn. Pract. **21**(6), 49–66 (2024)

8. Perkins, M., Roe, J., Furze, L.: The AI assessment scale revisited: a framework for educational assessment. arXiv preprint arXiv:2412.09029 (2024)
9. Redecker, C.: Digcompedu. Beurteilung der Digitalen Kompetenz Lehrende. Online verfügbar unter https://joint-research-centre.ec.europa.eu/system/files/2018-09/digcompedu/_leaflet/_de/_2018-01.pdf (abgerufen am: 20.11. 2020) (2017)
10. Roe, J., Perkins, M., Tregubova, Y.: The EAP-AIAS: adapting the AI assessment scale for english for academic purposes. arXiv:2408.01075 (2024)
11. Rostan, J.: Leon furze's "AI assessment scale": a critique and two alternative options (2023). https://www.jeremierostan.com/home/leon-furzes-ai-assessment-scale-a-critique-and-two-alternative-options, publicado en el blog personal de Jeremie Rostan
12. Selwyn, N.: Should robots replace teachers? AI and the future of education, Polity (2019)
13. Williamson, B.: Education governance and datafication. Learn. Media Technol. **45**(1), 4–15 (2020). https://doi.org/10.1080/17439884.2019.1649690
14. Wulgaert, R.: AI assessment and teaching guide. https://www.robbewulgaert.be/education/ai-assessment-scale (2024), accedido el 27 de mayo de 2025
15. Zawacki-Richter, O., Marín, V.I., Bond, M., Gouverneur, B.: Systematic review of research on artificial intelligence applications in higher education – where are the educators? Int. J. Educ. Technol. High. Educ. **16**(1), 39 (2019). https://doi.org/10.1186/s41239-019-0171-0

Semantic Systematic Review of Artificial Intelligence-Based Maturity Models for Teaching Feedback in Higher Education

Cristian Valdes Perez[1(✉)], Angel Arroyo Puente[2(✉)], and Jose Manuel Galán Ordax[3(✉)]

[1] Facultad de Ingeniería, Universidad San Sebastián, Bellavista 7, Recoleta, Chile
cristian.valdes@uss.cl

[2] Grupo de Inteligencia Computacional Aplicada (GICAP), Departamento de Digitalización, Escuela Politécnica Superior, Universidad de Burgos, Av. Cantabria s/n, 09006 Burgos, Spain
aarroyop@ubu.es

[3] Departamento de Ingeniería de Organización, Escuela Politécnica Superior, Universidad de Burgos, Ed. A1, Avda, Cantabria S/N, 09006 Burgos, Spain
jmgalan@ubu.es

Abstract. This study presents a semantic systematic review of the literature on Artificial Intelligence enhanced maturity models for improving teaching feedback in higher education. Using PRISMA methodology and semantic analysis, 159 relevant publications were identified and examined. The review process included database searches, multi-stage screening, and natural language processing techniques to cluster thematic patterns and identify re-search gaps. The findings reveal a predominance of conceptual models with limited empirical validation and a fragmented integration of Artificial Intelligence technologies, particularly and machine learning into structured institutional frameworks. The review highlights the need for models that prioritize contextualized implementation, faculty engagement, ethical transparency, and actionable analytics, or data-driven insights that directly inform decision making and pedagogical improvement. It provides a foundation for developing integrated AI-powered maturity models that support continuous improvement in teaching quality.

Keywords: Artificial Intelligence · Teaching Evaluation · Maturity Models · Systematic Review · Semantic Analysis · Higher Education

1 Introduction

In recent years, higher education institutions have increasingly adopted data-driven approaches to improve teaching quality. Among these, Student Evaluations Teaching (SET) [1] have become one of the most used mechanisms for measuring academic performance and identifying opportunities for pedagogical improvement [2].

This gap has led to the search for more structured and intelligent systems capable of processing and interpreting information in a meaningful way. In this context, Artificial

E. Corchado et al. (Eds.): SOCO 2025, CCIS 2806, pp. 393–401, 2026.
https://doi.org/10.1007/978-3-032-19763-4_36

Intelligence (AI) offers promising capabilities to transform raw feedback into relevant and individualized insights [3]. AI enables the application of data mining, semantic analysis, and predictive modeling techniques that enhance feedback cycles and contribute to teaching quality improvement [4].

Maturity models, on the other hand, provide structured frameworks to assess and improve institutional processes. A maturity model typically defines a series of development stages that describes how an organization can evolve from an initial, "ad hoc" state to a more optimized and more effective one. When applied to educational management, these models help diagnose current institutional states, define improvement pathways, and monitor their evolution over time [5, 6].

However, the existing literature appears fragmented across disciplines such as educational technology, data science, and institutional research. This article aims to synthesize this literature through a semantic systematic review, identifying studies that propose or apply AI-based maturity models focused on faculty feedback.

The structure of this article is as follows: Sect. 2 presents a review of relevant literature on AI-based maturity models and their application in teaching evaluation. Section 3 details the research objectives and describes systematic review methodology, including data collection and semantic analysis. Section 4 reports the main results, including thematic patterns and model characteristics. Section 5 offers a discussion of the findings and identifies future research opportunities. Finally, Sect. 6 presents the conclusions and outlines the implications for institutional practice and educational research.

2 Theoretical Framework

Maturity Models are structured frameworks that characterize the progressive development of organizational capabilities or processes across defined levels. These models are typically used to evaluate current performance, identify improvement areas, and guide organizations through systematic advancement. According to Becker et al. [7], "a maturity model defines a sequence of levels (or stages) that characterize the evolution of an organization's practices or processes from an ad hoc, immature state to a mature, disciplined state."

In the context of education, several maturity models have been developed to assess institutional capabilities. For instance, the HELA-CMM model (Higher Education Learning Analytics Capability Maturity Model) proposes five levels of institutional maturity in the adoption of learning analytics, ranging from basic data collection to advanced adaptive feedback systems [3]. Another example is the e-Learning Maturity model (eMM) developed by Marshall and Mitchell, which evaluates e-learning process across five dimensions: learning, development, support, evaluation and organization. These models provide valuable roadmaps but are often focused on system-level infrastructure or general digital strategy, rather than directly enhancing the pedagogical feedback loop based on student evaluation.

Simultaneously, the use of AI in higher education has grown rapidly. A notable example is Assessment's, an intelligent tutoring system designed for middle school math, which combines machine learning with human-in-the-loop feedback, track student progress and provides teachers with real-time data dashboard [3, 6].

The integration of maturity models and AI provides dynamic and contextualized information that is useful for teaching quality enhancement. Despite this potential, few studies systematically explore this integration for faculty development. The intended model in this study seeks to bridge this gap by focusing on the feedback ecosystem – understood as the interconnected processes, tools, and actors involved in collecting and analyzing to teaching evaluations-, incorporating AI capabilities not only for data analysis but also for delivering personalized, actionable insights and aligning them with Institutional faculty development strategies.

3 Methodology

This study follows the PRISMA (Preferred Reporting Items for Systematic Reviews and Meta-Analyses) methodology [6]. The PRISMA methodology is a widely adopted framework for conducting and reporting systematic reviews. It provides a structured approach to ensure transparency, reproducibility, and completeness in the identification, selection, and synthesis of research evidence. PRISMA includes a four-phase flow diagram: (1) identification, (2) screening, (3) eligibility, and (4) inclusion. PRISMA also provides a checklist that guides researchers through rigorous review processes. The aim was to identify and analyze academic studies that explore the use of AI-enhanced maturity models for faculty feedback in higher education.

The search strategy was designed to identify studies at the intersection of three concepts: (1) maturity models or capability frameworks, (2) artificial intelligence or machine learning techniques and (3) applications in higher education, in the context of teaching evaluation or feedback. The Boolean structure was used: ("maturity model" OR "capability framework") AND ("artificial intelligence" OR "machine learning") AND ("higher education" OR "teaching evaluation" OR "instructional feedback").

The search was conducted in two major databases: Scopus and Web of Science, including publications from 2010 to June 2025. A total of 1,959 records were retrieved. The selection process included two main filtering stages:

1. Initial screening: Titles and abstracts were reviewed to remove unrelated studies. During this stage, exclusion criteria were applied, including: (1) Articles not available in English or Spanish, (2) studies without full text-access, (3) works outside the educational domain or unrelated to teaching evaluation and feedback, (4) Publications that mentioned maturity models of AI only superficially and (5) duplicate records from both databases.
2. Eligibility assessment: Full-text articles were analyzed based on the following combined inclusion criteria:

   - Use or proposal of a maturity model.
   - Integration of AI or machine learning techniques.
   - Application in higher education and teaching evaluation context.

These criteria were applied jointly to ensure the relevance of each article to the scope of the review.

To support this process, we used the systematic platform Rayyan [8], which enabled a semi-automated screening. This tool enabled blind screening, collaborative inclusion/exclusion tagging, and conflict resolution between reviewers, ensuring consistency and reducing selection bias.

The final sample consisted of 159 articles. Additionally, semantic analysis techniques were applied to abstracts and keywords using Bibliometrix. Tools such as term frequency analysis, topic clustering, and conceptual-mapping were employed to identify patterns and gaps in literature.

Figure 1 presents the PRISMA flow diagram corresponding to the methodology described earlier. This figure summarizes the systematic process applied to select the studies included in this review.

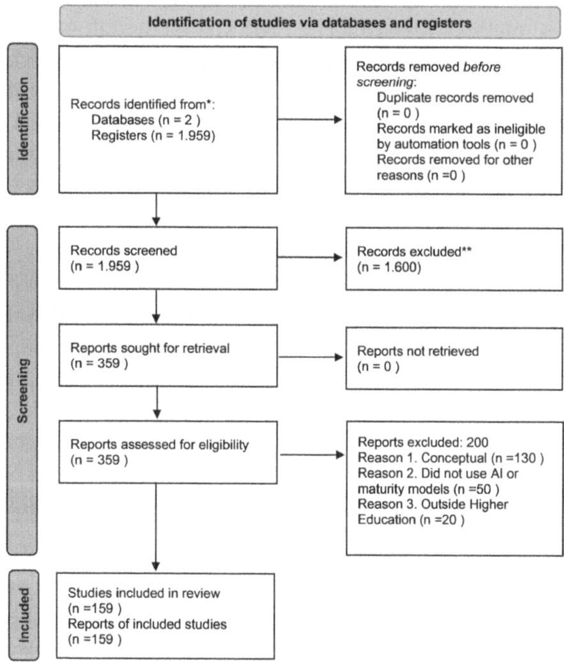

Fig. 1. PRISMA Flow Diagram

During the identification phase, a total of 1,959 records were retrieved from Scopus and Web of Science, covering publications from 2010 to June 2025. In the screening phase, 189 duplicates records were removed. The remaining 1,770 titles and abstract were retrieved using predefined exclusion criteria, including language restrictions (English and Spanish only), lack of full-text Access, irrelevance to the educational domain or teaching evaluation, and superficial mention of maturity models or AI without methodological application.

The eligibility stage involved full-text analysis of 93 potentially relevant articles. After careful evaluation, 159 studies met all inclusion criteria: use or proposal of a maturity model, integration of artificial intelligence or machine learning techniques, and application in higher education teaching evaluation contexts.

4 Results

The final sample of 159 articles covered a wide temporal and disciplinary range. Most studies were published between 2018 and 2025, with a significant increase in recent years reflecting growing interest in AI applications in education. The majority of contributions came from institutions in Europe (34%), North America (28%), and Asia (22%), with notable representation from Latin America (10%).

For illustrate the scope and methodological nature of the select studies, the Figs. 2 and 3 present the classification of the literature based in two key aspects: the type of study conducted, and the AI technologies employed:

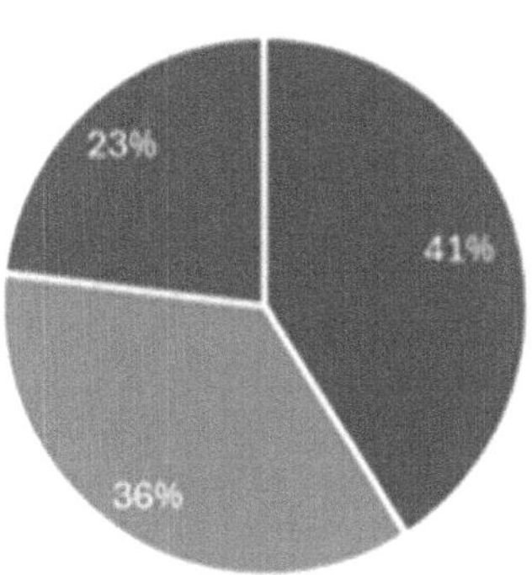

Fig. 2. Distribution of study types.

Characteristics of Maturity Models: Most models adopted a five-level structure, often inspired by frameworks such as CMMI (Capability Maturity Model Integration) or quality assurance models. CMMI, originally developed by the Software Engineering Institute, provides a structured model for evaluating and improving processes [4]. In the context of AI-enhanced maturity models in education, the most frequently assessed domains are summarized in table 1. These domains align with maturity models previously described in AI and education literature [2, 5], see Table 1.

In parallel, a semantic analysis of abstract and keywords using NLP techniques revealed four recurring thematic clusters across the literature. The findings are summarized in Table 2.

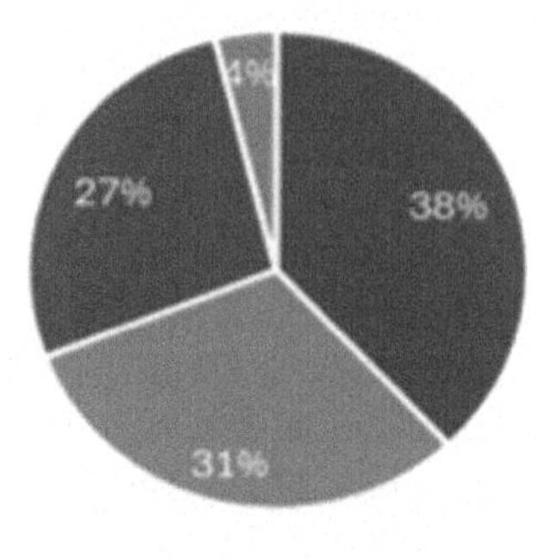

Fig. 3. AI technologies employed

Table 1. Common domains Assessed in AI-based maturity models.

Domain Assessed	Description
Institutional Data Infrastructure	Evaluation of systems, platforms and data available across the institution
Teaching Quality and Evaluation Practices	Assessment of teaching methodologies and feedback collection mechanism
Feedback Cycles and personalization	Degree of automation, customization and feedback loop integration
Governance and strategic alignment	Integration of feedback models into institutional planning and leadership

These themes reflect not only the technological orientation of literature, but also institutional concerns about ethics, scalability, and user acceptance.

Only less than 10% of the studies reported full implementation or longitudinal validation of maturity models in real-world educational environments. This high-lights a gap between research and practice and underscores the opportunity for future empirical investigations that validate the effectiveness and sustainability of AI-based feedback models in institutional settings [3, 9].

Table 2. Thematic Clusters identified through semantic analysis.

Thematic cluster	Interpretation
Personalized feedback and continuous improvement	Emphasis on adaptative and iterative improvement mechanism for teaching
Institutional Analytics and decision-making	Use of AI-generated insights for inform educational leadership
AI Integration into quality assurance frameworks	Connection between technical systems and institutional quality models
Ethics, transparency and faculty engagement	Issues of fairness, explainability and academic staff participation

5 Discussion

The results of this semantic systematic review underscore both the growing academic interest and the methodological fragmentation in the development and application of AI-based maturity models for teaching feedback in higher education. The predom-inance of theoretical contributions—often lacking empirical validation—reveals a persistent gap between conceptual frameworks and their practical implementation in real educational settings [3, 9].

Although techniques such as natural NLP and ML are increasingly applied to analyze student evaluations [3], few studies propose integrated frameworks that effec-tively align these tools with institutional feedback systems via maturity models. This reflects a missed opportunity to operationalize data-driven models that could enhance teach-ing quality through structured, iterative improvement mechanisms [3].

When compared to traditional SET approaches, AI-powered maturity models demonstrate a clear potential to foster adaptive feedback cycles, enabling more per-sonalized, timely, and actionable guidance for faculty. However, key issues such as ethical transparency, faculty engagement referring to the active participation of academic staff in feedback processes and model adoption-, and alignment with insti-tutional strategies are rarely addressed in depth [3, 9].

These findings align partially with existing research in educational data mining and quality assurance, which emphasize the importance of converting student-generated data into actionable analytics [6]. In this context, the added value of ma-turity models lies in their ability to contextualize AI-driven insights within broader trajectories of institutional development and faculty professionalization.

In comparison with established maturity models such as CMMI and the e-learning Maturity Model (eMM), the AI-Based frameworks reviewed in this study tend to emphasize technological capabilities over institutional alignment or pedagogical usability. Few models explicitly address how AI-generated feedback can be integrat-ed into ongoing faculty development programs or quality assurance mechanisms. Bridging this gap requires further discussion between AI research and educational policy.

Additionally, ethical considerations remain underexplored. While some studies mention fairness or transparency, very few provide explicit mechanisms to mitigate algorithmic bias or explain model decisions to academic stakeholders.

To move beyond conceptual contributions, researchers should also prioritize real-world validation. Case study designs involving controlled implementation of AI-enhanced maturity models in diverse educational contexts would provide empirical insights into feasibility. Pilot programs co-designed with faculty could test feedback personalization mechanisms, monitor their impact on teaching practices, and inform iterative refinement of the models.

The thematic clusters identified in the semantic analysis suggest that the field is structured around four core dimensions: Personalized feedback and continuous improvement, Institutional analytics and decision-making support, AI integration into quality assurance and ethics, transparency, and faculty involvement.

The present study aims to advance towards a maturity model specifically de-signed for the domain of teaching feedback. This model seeks not only to incorpo-rate AI tools for data analysis, but also to integrate the resulting insights into effec-tive pedagogical improvement processes and faculty development initiatives. The proposed approach addresses the identified gaps by offering a structured, actionable and contextualized framework that supports the institutional evolution from descrip-tive feedback system towards predictive, generative and formative mechanism.

6 Conclusions and Recommendations

This study presents a semantic systematic review of the literature on AI-enhanced maturity models aimed at improving teaching feedback in higher education. Through a PRISMA-guided methodology and semantic analysis, 159 relevant publications were identified, analyzed, and categorized.

The review reveals that despite growing conceptual interest, few validated and operational implementations exist. Only 9% of the 159 analyzed studies reported real-world deployment of longitudinal validation of AI-based maturity models in higher education institutions. AI technologies, particularly NLP and machine learn-ing, are employed to analyze student evaluations, but their integration into struc-tured maturity models remains limited.

Findings suggest that future research should:

- Prioritize empirical validation of proposed models in diverse educational con-texts.
- Incorporate faculty engagement and ethical transparency as design principles.
- Use AI not only for analysis, but for generating contextualized, actionable improvement pathways.

In conclusion, this review provides a solid foundation for the development of integrated AI-powered maturity models that support sustained teaching improvement processes. These models have the potential to transform evaluation data into a pow-erful tool for institutional learning and academic development. Beyond higher edu-cation, the methodology used in this review -particularly the integration of semantic analysis and maturity model frameworks-may also be applicable to other domains where feedback, data-driven improvement, and institutional development are rele-vant. For example, healthcare organizations, public administration, and corporate training environments increasingly seek structured models for assessing perfor-mance and guiding digital transformation.

References

1. Spooren, M., Brockx, B., Mortelmans, D.: On the validity of student evaluation of teaching: the state of the art. Rev. Educ. Res. **77**(4), 507–545 (2007)
2. Roe, J., Perkins, M., Ruelle, D.: Understanding Student and Academic Staff Perceptions of AI Use in Assessment and Feedback, arXiv (2024)
3. Zhan, Y., Boud, D., Dawson, P., Yan, Z.: Generative artificial intelligence as an enabler of student feedback engagement: a framework Higher Education Research and Development (2025)
4. Walter, Y.: Embracing the future of artificial intelligence in the classroom: the relevance of AI literacy, prompt engineering, and critical thinking in modern education. Int. J. Educ. Technol. High. Educ. **21** (2024)
5. Sadiq, R.B., Safie, N., Abd Rahman, A.H., Goudarzi, S.: Artificial intelligence maturity model: a systematic literature review. PeerJ Comput. Sci. **7**, 661 (2021)
6. Langer, B.: Understanding data & analytics maturity: a systematic review of maturity model composition. Schmalenbach J. Bus. Res. (2025)
7. Becker, J., Knackstedt, R., Pöppelbuß, J.: Developing maturity models for IT management. Bus. Inf. Syst. Eng. **1**(3), 213–222 (2009)
8. Ouzzani, M., Hammady, H., Fedorowicz, Z., Elmagarmid, A.: Rayyan—a web and mobile app for systematic reviews. Syst. Rev. **5** (2016)
9. Benitez, A., Contreras, J., Reinoso, M.: HELA-CMM: capability maturity model for adoption of learning analytics. Educ. Inf. Technol. (2025)

From Didactic Proposal to Practical Implementation: Results of an Industrial Robotics Training Course Within the EAGLE Project

Mario Peñacoba(✉), Eduardo Bayona, Jesus Enrique Sierra-Garcia, and Bruno Baruque-Zanon

Universidad de Burgos, Burgos, Spain
{mpenacoba,ebayona,jsierra,bbaruque}@ubu.es

Abstract. This paper presents the results of a practice-based training initiative in industrial robotics developed within the EAGLE project. The course was designed to address the digital skill gap in small and medium-sized enterprises (SMEs) by providing professionals with hands-on experience in programming, simulation, and safe operation of robotic systems. The training combined theoretical content with guided exercises using ABB RobotStudio and physical robot manipulators. Evaluation of participant performance was conducted through a technical questionnaire and a simulation-based challenge. The results show strong competence acquisition and high satisfaction rates among trainees, particularly regarding applicability and teaching methodology. Additionally, pre- and post-course surveys revealed positive shifts in perceived ease of use, intention to adopt, and confidence in applying robotics in the workplace. These findings support the scalability of the course and its contribution to workforce digitalization in European SMEs.

Keywords: Robotics · Workplace Learning · SMEs · Digitalization · ICT

1 Introduction

As industrial settings evolve towards digital ecosystems, robotics emerges as a pivotal enabler of smarter, more efficient production systems. In the context of Industry 4.0, robotic automation not only optimizes resources and enhances flexibility but also serves as a catalyst for competitiveness in global markets [5]. Nevertheless, a significant portion of the workforce, particularly in small and medium-sized enterprises (SMEs), lacks the required training to fully leverage these technologies [3,8].

While robotic platforms and simulation environments are increasingly accessible, the gap between available technologies and workforce competencies remains too large [6]. Many SMEs struggle to adopt robotic solutions due to the absence

E. Corchado et al. (Eds.): SOCO 2025, CCIS 2806, pp. 402–411, 2026.
https://doi.org/10.1007/978-3-032-19763-4_37

of trained professionals who can manage the deployment, programming, and maintenance of such systems [4,9]. This situation highlights the pressing need for tailored, practice-oriented educational initiatives that facilitate the adoption of robotics in industrial settings [7].

While large industries advance in automation and digital transformation, SMEs remain behind in accessing advanced robotics education. EU-funded projects like the Digital Innovation Hubs and EIT Manufacturing academies have sought to address this gap through flexible and modular approaches. However, many of these efforts remain conceptual or are embedded within broader industrial strategies, lacking replicable, lightweight, and outcome-oriented training methodologies. Although the value of blended learning—combining simulation with hands-on experimentation—has been recognized, a significant gap persists in the full implementation of training programs that incorporate evaluation frameworks aligned with technology acceptance models [11].

The European project EAGLE (CovEring the trAining Gap in digitaL skills for European SMEs manpower) addresses this need by developing specialized courses that target essential digital skills. One of its key deliverables is a hands-on training course in industrial robotics designed with the operational needs of SMEs in mind [10]. This paper reports on the initial implementation of this course, presenting its pedagogical approach and structure, and analyzing results from the first participants to evaluate its effectiveness and potential for scalability.

This study builds upon these foundations and aims to contribute a transferable and validated framework for robotics training in SMEs, supported by a rigorous methodology that includes technical performance evaluation, pre-post comparative analysis, and user satisfaction metrics.

The rest of this paper is organized as follows: Sect. 1 describes the EAGLE project and the consortium structure. Section 2 presents the robotics training course, detailing its objectives, structure, and pedagogical design. Section 3.2 shows the results of the implementation in terms of participant performance and satisfaction. Conclusions and future work are summarized in Sect. 5.

2 The EAGLE Project

The EAGLE project is a European initiative co-funded by the Digital Europe Programme (grant agreement No. 101100660), whose main goal is to reduce the digital skill gaps observed in the SME workforce across Europe [2]. The project develops and delivers specialized training programs in emerging technologies such as industrial robotics, artificial intelligence, and cybersecurity.

The consortium brings together a diverse set of institutions: universities, research centers, training providers, industry associations, and specialized SMEs. It is coordinated by the Universidad de Burgos (Spain) and includes partners from Slovakia, the United Kingdom, Finland, Czech Republic, Ireland, Lithuania, and Cyprus. This diversity ensures a wide applicability of the training programs and fosters transnational cooperation in digital upskilling.

The project's implementation plan is structured around four interrelated Work Packages:

- WP1: Project Management and Coordination
- WP2: Courses Design
- WP3: Course Delivery
- WP4: Communication and Dissemination

After successfully completing the design phase, the project has entered its course delivery phase, where pilot courses are being implemented and monitored. One of the most mature and impactful actions is the Industrial Robotics Course developed under WP2 and delivered in multiple contexts within WP3 [2]. The dissemination of project activities is carried out through the official website and social media channels [1].

3 Industrial Robotics Course

The Industrial Robotics Course was designed to address the practical training needs of SME technicians and professionals, aiming to provide them with the competencies needed to work with industrial robotic systems in both virtual and real environments. The course is structured to combine theoretical concepts, practical applications, and problem-solving tasks, thereby offering a holistic training experience.

The course follows a blended learning model, with 8 h of in-person sessions and 8 h of autonomous work. The theoretical content is supported by instructional videos, while the practical activities are centered around the use of simulation tools (notably ABB RobotStudio) and, when available, physical robots. The in-person sessions focus on hands-on tasks using real robot manipulators and teach pendants, whereas the autonomous portion emphasizes self-paced learning and simulation-based exercises. The pilot course included 36 participants. Table 1 summarizes the participant profiles.

Table 1. Participant Demographics Summary

Attribute	Categories	Count	Percentage (%)
Age group	20–35/36–50/51+	18/12/6	50/33/17
Gender	Male/Female	28/8	78/22
Experience in Robotics	None/Basic/Advanced	10/20/6	28/56/16
Role	Technician/Engineer/Manager	20/12/4	56/33/11

The course is built around the acquisition of both specific and transversal competences:

Specific Competences:

- Identification and classification of industrial robots by structure, control mode, and application domain.
- Selection and use of robotic components including controllers, manipulators, end-effectors, and sensors.
- Programming and execution of robotic tasks using teach pendants and simulation tools.
- Implementation of safety measures and risk mitigation strategies according to ISO standards.

Transversal Competences:

- Logical reasoning and troubleshooting skills.
- Effective information search and management.
- Application of problem-solving strategies in realistic industrial scenarios.

3.1 Course Content

The course content is organized into three progressive blocks: Transversal Concepts, Core Content and Advanced Content. Each block has a different number of theoretical and practical training hours. The first number in the bracket denotes the theoretical hours and the second one the hours dedicated to practical content.

- Transversal Concepts (1 h/0.5 h): Introduction to robotics, etymology and evolution of the term, classifications, applications, and success cases.
- Core Content (3.5 h/3 h): Robot components, controller interfaces, manipulators, mobile bases, end-effectors, sensors. Practical training in robot selection, basic motion commands, and RAPID programming in RobotStudio.
- Advanced Content (1.5 h/1.5 h): Kinematic modeling, workspace analysis, integration of safety devices, and compliance with ISO 10218–1/2, ISO/TS 15066.

Each block combines theory with guided practice to ensure the applicability of acquired skills.

3.2 Evaluation Methodology

Competence acquisition is assessed through several complementary mechanisms:

- A practical challenge, where students must develop and simulate a robotized sealing station for parcels using an IRB120 robot. It is evaluated with a test-type questionnaire designed to measure the understanding of the challenge, the understanding of the robot's operating principles and the use of the ABB RobotStudio tool to develop the application.

- A technical multiple-choice questionnaire focused on the understanding of robotic operations, programming logic, and safety considerations.
- A multiple choice questionnaire to measure the level of satisfaction of the student.
- A pre-post test to measure the improvement in the level of technology acceptance. To do it, the Unified Theory of Acceptance and Use of Technology (UTAUT) has been used [12].

In the following section, we present the results obtained during the first implementations of the course, based on participant responses, performance data, and qualitative feedback.

4 Results

This section presents the results gathered from the first iterations of the Industrial Robotics Course, focusing on the evaluation of participant performance, task completion, and knowledge acquisition. The data are based on the practical exercise involving the simulation of a robotic sealing station using RobotStudio and on the answers to a technical questionnaire. The analysis also includes qualitative feedback to assess the perceived usefulness and applicability of the training.

Figure 1 displays a violin plot that simultaneously illustrates the density distribution and individual dispersion of the students' final exam grades. The highest concentration of students lies within the 8 to 9.5 grade range, indicating a generally positive performance trend.

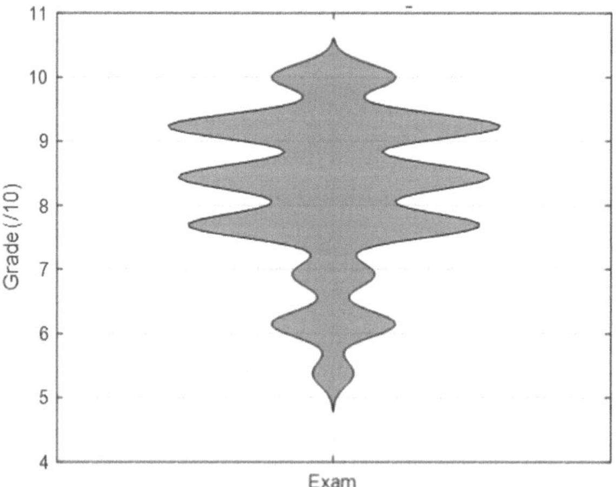

Fig. 1. Distribution of exam grades using a violin plot.

Figure 2 complements the analysis with a normalized density histogram overlaid with a kernel density estimation (KDE) curve. This visualization confirms the presence of several density peaks around higher grade values, suggesting that a significant portion of students achieved a satisfactory level of mastery in the evaluated content.

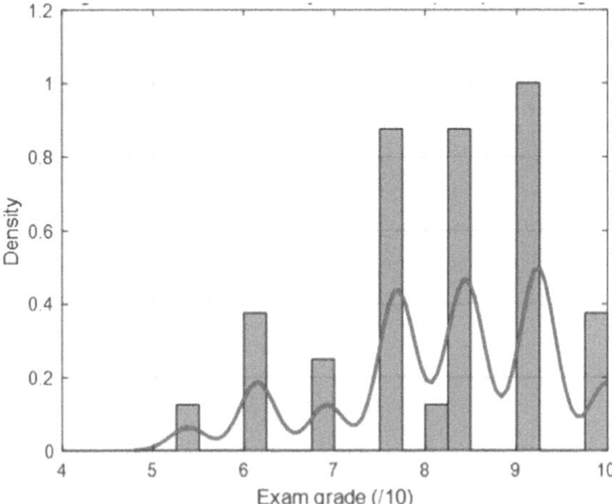

Fig. 2. Histogram with kernel density estimation (KDE) of exam grades.

Table 2. Correspondence between chart labels and full questionnaire items.

Label in Figure	Full Questionnaire Item
Accreditation	What importance do you place on this type of course having an official university accreditation (or from another higher education institution) when enrolling?
Applicability	Has this course helped you acquire new skills or knowledge that you can directly apply in your job or within your professional sector?
Quality	Are you satisfied with the overall quality of the course?
Recommendation	Would you recommend this course to other prospective students?
Tech Impact	Has this course helped you better understand the impact of the studied technology in your sector and how to use it more effectively in your company?
Content	Did the course content meet your learning expectations?
Method	Are you satisfied with the teaching method (quality and clarity of explanations, balance between theory and practice, etc.) used in this course?

In addition to the technical assessment, the level of general satisfaction has been also evaluated. Table 2 shows the items of the questionary. A stacked bar chart illustrating the distribution of student responses for each evaluation item, rated on a Likert scale from 1 ("strongly disagree") to 5 ("strongly agree"), is presented in Fig. 3. The labels on the x-axis correspond to the summary names of the evaluation dimensions, which are described in full in Table 2. The results reveal a clear predominance of positive ratings (4 and 5) across all categories, with 'Applicability', 'Method', and 'Tech impact' showing particularly high proportions of scores in the upper range. Negative ratings (1 and 2) were minimal, suggesting a generally favorable perception of the course structure, content, and teaching approach.

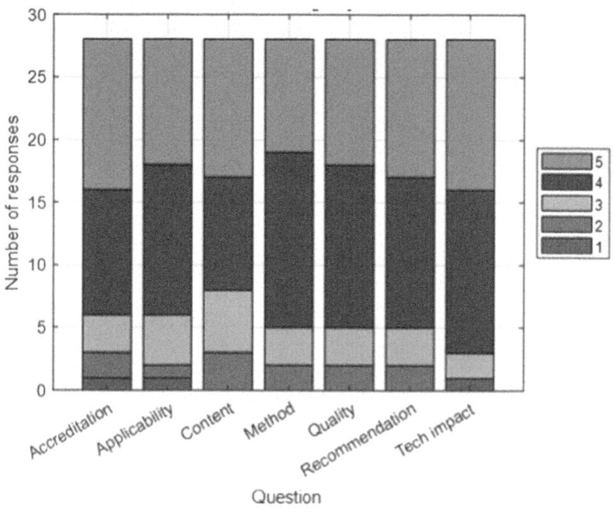

Fig. 3. Distribution of student ratings by evaluation item

Finally, the acceptance of robotics is measured before and after the course based on the UTAUT model. This point is evaluated with the questionnaire described in Table 3 divided into 9 categories. Figure 4 displays a radar chart comparing the average scores across nine analytical dimensions derived from the initial and final questionnaires (pre and post tests). These dimensions were constructed by grouping conceptually related items, as detailed in Table 3, which links each category with the specific questions used in both the pre- and post-course surveys.

Results show improvements in all categories after completing the course, with the most notable increases observed in "Support", "Habit", "Usefulness", and "Ease of Use". These findings suggest that the training program positively influenced participants' perceptions not only regarding the utility and enjoyment of the studied technology (robotics), but also in terms of their perceived competence, habitual integration, and access to support. The changes observed give some indication that the course contributes to both practical and attitudinal preparation for technology adoption in the workplace.

Table 3. Questionnaire items associated with each analytical group.

Group	Associated questionnaire items
Intention to use	I intend to use the studied technology (robotics) at work.
	I will try to use the studied technology (robotics) at work.
	I plan to frequently use the studied technology at work.
Social influence	People important to me at work think I should use the studied technology.
	People who influence my behavior at work think I should use the studied technology.
	People whose professional opinions I value think I should use the studied technology.
Perceived ease of use	Learning to use the studied technology for my job is easy.
	It is easy for me to use the studied technology for my job.
	It is easy for me to become skilled in using the studied technology at work.
Resources and knowledge	I have the resources needed to use the studied technology.
	I have the knowledge needed to use the studied technology at work.
	The studied technology is compatible with other technologies I use at work.
External support	I can get help from others when I have difficulties using the studied technology at work.
Hedonic value	Using the studied technology at work is fun.
	Using the studied technology at work is pleasant.
	Using the studied technology at work is very entertaining.
Habit	Using the studied technology at work has become a habit for me.
	I am used to using the studied technology at work.
	Using the studied technology is a routine part of my workflow.
Perceived usefulness	I find the studied technology useful for my job.
	Using the studied technology helps me get things done at work more quickly.
	Using the studied technology increases my productivity at work.
Economic evaluation	The studied technology offers good value for money.
	At its current price, the studied technology offers good value for money.
	The studied technology is reasonably priced.

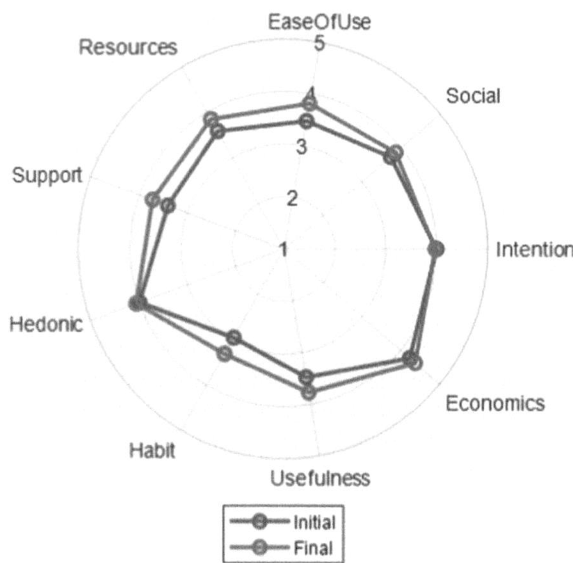

Fig. 4. Group-level comparison: Initial vs Final Questionnaire

Cronbach's alpha has been used to evaluate the reliability of the UTAUT questionaries. It is possible to conclude that some categories are quite well represented as Economics (α pre 0.47, α post 0.31), Hedonic (α pre 0.28, α post 0.42) and Intention (α pre 0.35, α post 0.65). However, the values in other categories are low and the results should be taken with caution.

5 Conclusions

The implementation of the Industrial Robotics Course within the EAGLE project confirms the effectiveness of a hands-on, SME-oriented training model. Participants demonstrated solid acquisition of key competencies, particularly in simulation-based programming, trajectory control, and safety integration. The combination of practical and conceptual evaluation proved to be a reliable method for assessing outcomes.

The successful completion of the practical task and the high performance on conceptual questions indicate that the training approach was well-received and impactful. Feedback highlighted the value of real-world scenarios, simulation environments, and the accessibility of materials.

These findings support the broader application of the course and its potential for adaptation in other contexts. Future work will focus on scaling its implementation, refining content, and evaluating long-term impact.

Future work will focus on expanding the delivery of the course across partner institutions, adapting it to specific industrial sectors, and conducting longitudinal studies to assess the long-term impact on workforce capabilities and SME digitalization. In addition, statistical studies will be developed with a larger population sample to improve the statistical reliability of the results.

In pedagogical terms, this work offers a transferable model that integrates theory, simulation, and real-world interaction in a balanced structure. Its scalability, low infrastructure requirements, and modular assessment mechanisms make it highly applicable across SME contexts and regions. Next editions of the course will include follow-up studies to evaluate knowledge retention and mid-term adoption in work environments.

Acknowledgments. This work was partially supported by the European Union's Digital Europe Programme (DIGITAL) under grant agreement No 101100660.

References

1. Eagle news. https://www.projecteagle.eu/news/ (2025)
2. Eagle project official website. https://www.projecteagle.eu/ (2025)
3. Buerkle, A.e.a.: Towards industrial robots as a service (IRAAS): flexibility, usability, safety and business models. Robot. Comput. Integrated Manufact. **81**, 102484 (2023). https://doi.org/10.1016/j.rcim.2022.102484
4. Hulla, M., Herstätter, P., Wolf, M., Ramsauer, C.: Towards digitalization in production in smes – a qualitative study of challenges, competencies and requirements for trainings. Procedia CIRP **104**, 887–892 (2021)
5. Javaid, M., Haleem, A., Singh, R.P., Suman, R.: Substantial capabilities of robotics in enhancing industry 4.0 implementation. Cogn. Robot. **1**(1–2), 58–75 (2021). https://doi.org/10.1016/j.cogr.2021.06.001
6. Jennes, P., Minin, A.D.: Cobots in smes: implementation processes, challenges, and success factors. In: 2023 IEEE International Conference on Technology and Entrepreneurship (ICTE). IEEE (2023). https://doi.org/10.1109/ICTE58739.2023.10488658
7. Shmatko, N., Volkova, G.: Bridging the skill gap in robotics: global and national environment. SAGE Open **10**(3) (2020). https://doi.org/10.1177/2158244020958736
8. Telukdarie, A., Dube, T., Matjuta, P., Philbin, S.: The opportunities and challenges of digitalization for SME's. Procedia Comput. Sci. **217**, 689–698 (2023)
9. Thrassou, A., Uzunboylu, N., Vrontis, D., Christofi, M.: Digitalization of SMEs: a review of opportunities and challenges. In: The Changing Role of SMEs in Global Business: Volume II, pp. 179–200 (2020)
10. Urrea, C., Kern, J.: Recent advances and challenges in industrial robotics: a systematic review of technological trends and emerging applications. Processes **13**(3), 832 (2025). https://doi.org/10.3390/pr13030832
11. Warnhoff, K.e.a.: Learning factories as innovative training locations for SMEs: qualitative analysis of concepts and cooperations. Industry 4.0 Sci. **40**(4), 32–41 (2024)
12. Williams, M.D., Rana, N.P., Dwivedi, Y.K.: The unified theory of acceptance and use of technology (UTAUT): a literature review. J. Enterp. Inf. Manag. **28**(3), 443–488 (2015)

WhoSQLWho: A Gamified Web Application for Learning SQL

Bernat Costa[1], Juan M. Alberola[2](✉), and Victor Sánchez-Anguix[3]

[1] IES Antonio Machado, C. Arroyo, 80, 41008 Sevilla, Spain
bernatcosta@iesamachado.org

[2] Valencian Research Institute for Artificial Intelligence, Universitat Politècnica de València, Camino de Vera, s/n, 46022 Valencia, Spain
jalberola@dsic.upv.es

[3] Instituto Tecnológico de Informática, Grupo de Sistemas de Optimización Aplicada, Ciudad Politécnica de la Innovación, Universitat Politècnica de València, Camino de Vera, s/n, 46022 Valencia, Spain
vsanchez@eio.upv.es

Abstract. Learning Structured Query Language (SQL) remains a persistent challenge for students due to its abstract nature and syntactic complexity. Traditional pedagogical approaches often rely on generic business-oriented database examples, which may hinder student engagement and limit conceptual understanding. In this paper, we present *WhoSQLWho*, a web-based gamified application designed to make SQL learning more accessible, motivating, and interactive. An evaluation conducted with 97 students and 14 instructors revealed strong perceived effectiveness in supporting SQL learning. By combining playful mechanics with pedagogical rigor, *WhoSQLWho* offers a compelling alternative to traditional SQL instruction and demonstrates the potential of game-based learning environments in technical education.

Keywords: SQL learning · gamification · educational technology · web application

1 Introduction

Learning database systems and Structured Query Language (SQL) presents significant challenges for students across a wide range of educational contexts. SQL is a declarative language used to interact with relational databases and represents a core component in most computer science and information systems curricula. However, mastering its syntax and logic is often difficult, especially for students with limited programming or data modeling experience [3]. In addition to these technical difficulties, the abstract nature of SQL may reduce student engagement and motivation [7]. Therefore, it is essential to design learning environments that are not only effective but also motivating [4].

In recent years, gamification has emerged as one of the most effective techniques to enhance motivation [9,11,12]. Its adoption in education has increased,

E. Corchado et al. (Eds.): SOCO 2025, CCIS 2806, pp. 412–421, 2026.
https://doi.org/10.1007/978-3-032-19763-4_38

as an strategy to improve motivation and foster a more dynamic learning experience. By incorporating game-like elements such as competition, progress tracking, and rewards, educators aim to promote deeper engagement and active participation among learners. Several studies have shown that gamified environments can lead to increased student interest and improved academic outcomes. Notably, [2] reported that gamification significantly boosted both student involvement and performance, providing strong evidence of its pedagogical benefits in technical disciplines.

In the specific context of SQL instruction, various innovative approaches have been proposed to make learning more attractive through game-based mechanics. For example, [8] introduced a block-based learning tool designed to simplify SQL query writing for secondary school students. Similarly, [5] presented a gamified platform where students compete by solving SQL queries. [1] proposed a gamification framework for teaching NoSQL concepts, while [10] highlighted the positive impact of game elements on student performance in database-related topics. More recently, [6] developed *SQLValidator*, an interactive web-based tool focused on SQL query practice, aimed at improving learner proficiency through hands-on engagement.

In this work, we present a web-based application designed to support SQL learning through a gamified experience inspired by the classic game *Guess Who?*. In this application, students formulate SQL queries to identify a hidden character based on database attributes, thereby reinforcing their understanding of SQL syntax and logic through an engaging and iterative process. The tool is intended for use both in the classroom (encouraging competition based on accuracy and speed) and for independent study, enabling students to review and practice content at their own pace.

Beyond its primary learning purpose, the application was developed with three complementary educational goals: to provide students with a playful and engaging alternative to traditional instructional methods; to foster motivation through the integration of core database concepts into a game-based environment; and to serve as a support tool for reviewing course material at home. Additionally, by publishing the complete development process—including source code, database diagrams, and deployment instructions—the tool offers students a real-world example of a software project, encouraging technical skills and promoting a deeper understanding of software engineering practices.

The remainder of this paper is structured as follows: Sect. 2 describes the proposed application and its implementation. Section 3 presents the results of an evaluation conducted with both students and instructors. Finally, Sect. 4 offers conclusions and outlines possible directions for future work.

2 Description of the Proposal

This project *WhoSQLWho*[1] is a gamified web application inspired by the classic game "Guess Who?", reimagined to facilitate the learning of SQL query language.

[1] https://whosqlwho.org/.

The core objective is to transform the abstract and often demotivating experience of learning database concepts into an engaging and interactive process.

Traditional teaching approaches in database courses often rely on generic business examples (e.g., invoices, orders), which may feel distant or irrelevant to students. By leveraging a well-known cultural game and integrating it with core SQL learning objectives, this proposal seeks to foster active engagement and deeper understanding.

The main learning goals supported by this application include:

- Understanding the construction and execution of SQL queries.
- Applying SQL syntax in a game-based context.
- Practicing critical thinking using real-time database interaction.
- Exploring SQL queries in a non-traditional, motivating environment.

The project has been developed as a web application using two separate components to reflect the backend–frontend architecture commonly adopted in web development. This separation allows the backend to serve different types of interfaces (web, mobile, desktop) while maintaining a single shared database and consistent business logic. The backend, implemented as a RESTful API using Node.js, handles core functionalities such as user authentication and game session management. The frontend, built with ReactJS, provides a user-friendly interface that enables seamless interaction with the system.

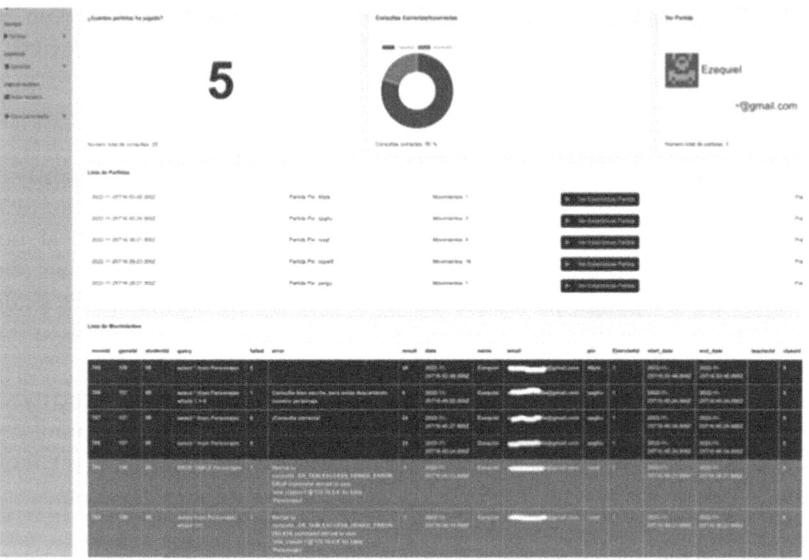

Fig. 1. Teacher dashboard showing class creation and statistics.

2.1 User Profiles

The application offers two distinct user profiles. The first is the **Teacher** profile, which allows instructors to create sessions across different game modes. This profile is designed to support the preparation of exercises aligned with curricular objectives and to facilitate student monitoring through access to detailed analytics on activity, errors, and performance (Fig. 1).

The available analytics include exercise-level reports, class-wide performance summaries, and individual student progress tracking. This data enables instructors to identify specific areas of student misunderstanding and to tailor their instruction accordingly, reinforcing the most challenging concepts for each particular group of students.

The second profile is the **Student** profile, which enables learners to join sessions, whether live games or assigned practice activities by means of a PIN provided by the instructor, without requiring prior registration. This lightweight access model allows students to participate during in-class activities or engage in independent practice at home, reviewing content at their own pace.

2.2 Game Modes

The application offers two distinct game modes tailored to different teaching scenarios. On the one hand, the **Live** game mode allows the instructor to launch a real-time session using a PIN code, which is shared with students so they can join without the need for registration. Once connected, students compete to identify the hidden character by writing SQL queries. The instructor can monitor a real-time leaderboard, which may be projected in the classroom to provide immediate feedback and enhance the competitive dynamic (Fig. 2).

Fig. 2. Live classroom leaderboard interface.

On the other hand, the **Practice** game mode is focused for asynchronous use. In this mode, instructors can create virtual class groups (Fig. 3). Students join using a class PIN and receive personalized exercises to complete at their own pace. This mode enables individual practice outside the classroom while still tracking progress.

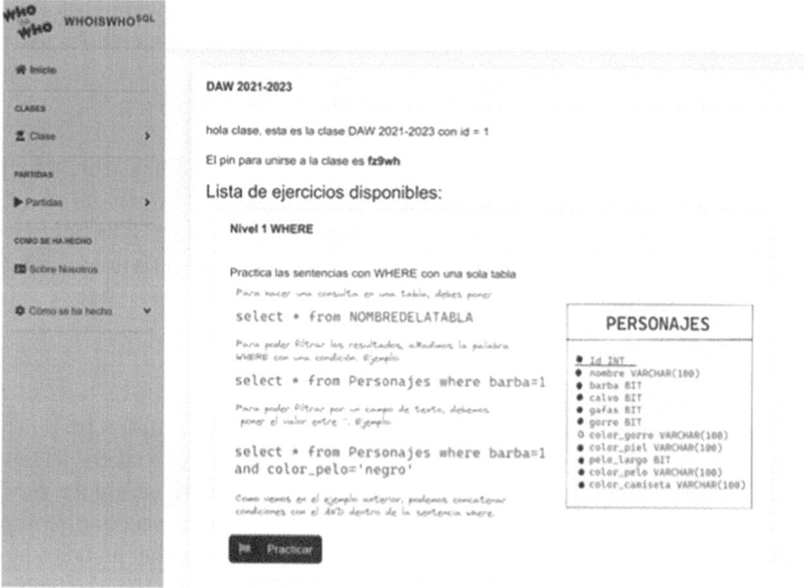

Fig. 3. Interface displaying assigned exercises for home practice.

2.3 Game Mechanics

The game interface displays 24 character avatars with various attributes (e.g., beard, glasses, shirt color) as shown in Fig. 4. A hidden character is selected at random, and the student must write SQL queries to narrow down the possibilities. For example:

```
SELECT * FROM Characters WHERE hat = 1;
```

The system provides immediate feedback based on the query result:

- **Green**: Correct syntax and relevant condition; non-matching characters are eliminated.
- **Yellow**: Correct syntax but incorrect condition; the hidden character would be excluded.
- **Red**: Syntax error; an error message is returned to guide correction.

Fig. 4. Game interface.

The game ends when the student successfully identifies the hidden character, with a celebratory message displayed on both the student and classroom screens.

The current version of the application includes two levels of gameplay aligned with different assessment criteria. The first, more basic level focuses on simple SQL statements, allowing students to practice queries on single tables. The more advanced level introduces exercises involving multiple tables and more complex query structures. This structure enables instructors to select the appropriate level of difficulty based on the learning objectives of their course. In this way, students can play independently at home while practicing SQL, the language used to interact with relational databases.

Overall, the application not only promotes learning through play but also serves as a scalable, secure, and pedagogically aligned platform. It empowers teachers with actionable data while offering students a highly interactive and meaningful way to practice SQL.

3 Evaluation

Below is an evaluation of the tool based on various questionnaires completed by both students and instructors involved in database-related courses. First, we aimed to assess whether students found the activities proposed by the tool to be engaging and effective for learning and practicing SQL queries. In this context, 97 computer science vocational training students participated in the evaluation. As shown in Fig. 5a, the vast majority of students surveyed believe that the tool can support the learning of SQL, while only four respondents disagreed or remained neutral on this point.

Figure 5b presents students' views on the competitive aspect of the tool. Although most students responded positively, a larger proportion expressed some

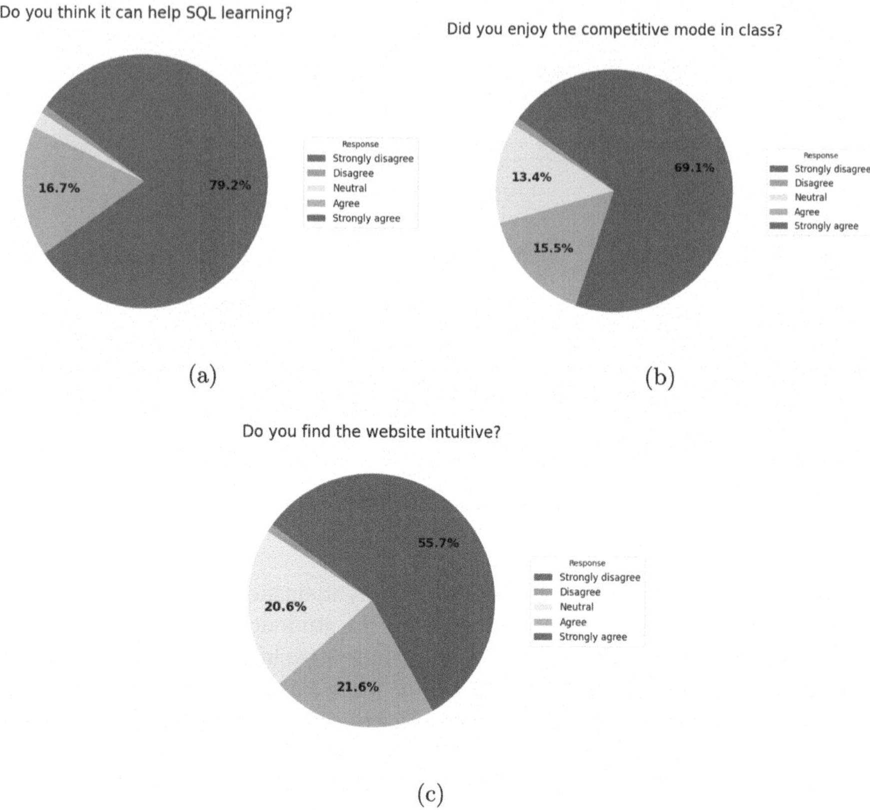

Fig. 5. Students' opinions on the tool: (a) Effectiveness for learning SQL; (b) Appeal of the competitive mode; (c) Intuitiveness of the interface.

level of disagreement compared to the previous question. This may be due to the pressure associated with competition, such as having to display answers publicly and compare performance with peers.

The aspect with the lowest rating among students was usability. Figure 5c shows responses to a question regarding this topic, where a smaller percentage of students considered the application to be very or fairly intuitive. We therefore identify usability as a key area for improvement in future iterations.

In addition to the students, a survey was conducted among 14 instructors teaching database-related subjects. Overall, as shown in Fig. 6a, instructors viewed the tool positively as a resource for helping students understand the *WHERE* clause syntax. Notably, none of the respondents disagreed with this statement. Furthermore, instructors also responded favorably when asked whether students would use the tool independently to practice at home (Fig. 6b).

As it can be observed, the results show that both students and instructors perceive the tool as an effective resource for learning SQL. While students

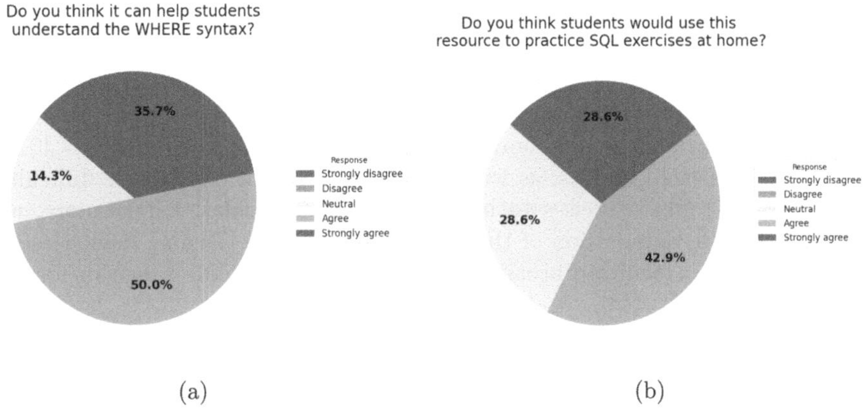

Fig. 6. Instructors' opinions on the tool: (a) Effectiveness in understanding the 'WHERE' syntax; (b) Likelihood that students would use the tool for self-directed practice.

responded particularly positively regarding its educational value and the gamified features, there is room for improvement in terms of usability and accessibility.

4 Conclusions

SQL is a fundamental component of database education, yet it often presents significant challenges for students due to its abstract and syntactic complexity. Recent pedagogical approaches have increasingly explored the use of gamification to enhance student motivation, engagement, and learning outcomes. By incorporating game-like elements into instructional design, educators aim to create more dynamic and participatory learning environments that address the motivational and cognitive barriers commonly associated with technical subjects like SQL.

In this context, we have presented *WhoSQLWho*, a web-based application that integrates gamified mechanics inspired by the classic game *Guess Who?* to support SQL learning. The tool invites students to write SQL queries in order to uncover a hidden character, thereby reinforcing core SQL concepts through iterative practice and immediate feedback. The application includes features tailored for both classroom use (via real-time competitive sessions) and individual study through asynchronous practice modes. Additionally, a teacher dashboard provides analytics for monitoring student progress and identifying common errors.

The evaluation of the tool, based on responses from both students and instructors, indicates that *WhoSQLWho* is perceived as an effective and engaging resource for practicing SQL. Students reported high levels of motivation and perceived learning benefits, particularly in the classroom-based game mode. Instructors highlighted the usefulness of the tool for reinforcing syntactic understanding and facilitating targeted instruction based on class performance data.

However, usability was identified as an area for improvement, suggesting the need for further refinement of the user interface and interaction design.

Future work will focus on addressing the usability issues raised during the evaluation, as well as expanding the pedagogical scope of the tool. Planned improvements include the incorporation of additional SQL topics such as `JOIN`, `GROUP BY`, and nested queries, as well as the development of adaptive difficulty mechanisms to tailor exercises to individual learner profiles. Furthermore, we aim to explore the integration of the tool within existing learning management systems (LMS) to facilitate broader adoption and streamline usage in formal educational settings.

Acknowledgements. This work is partially supported by Innovation and Educative Improvement project PIME-1951 from Universitat Politècnica de València, MINECO/FEDER RTI2018-095390-B-C31 project of the Spanish government, project TED2021-131295B-C32 from the State Research Agency, DIGITAL2022 CLOUDAI02/S8760000 from the European Commission, project PID2021-123673OB-C31 COSASS and project OPRES PID2021-124975OB-I00 partially funded by the Spanish Ministry of Science and Innovation and FEDER funds.

References

1. Butgereit, L.: Four NOSQLs in four fun fortnights: exploring NOSQLs in a corporate it environment. In: Proceedings of the Annual Conference of the South African Institute of Computer Scientists and Information Technologists, pp. 1–6 (2016)
2. González-Alonso, F.: La gamificación como estrategia de aprendizaje en educación secundaria: un estudio de caso. Revista de Educación a Distancia (RED) **54**, 1–25 (2017)
3. Kawash, J., Meston, L., Lane, H.C., Zvacek, S., Uhomoibhi, J.: Challenges with teaching and learning theoretical query languages. In: CSEDU (2), pp. 382–389 (2020)
4. Lai, P.P.Y.: Engaging students in SQL learning by challenging peer during the pandemic. In 2020 IEEE International Conference on Teaching, Assessment, and Learning for Engineering (TALE), pp. 205–212. IEEE (2020)
5. Morales-Trujillo, M.E., García-Mireles, G.A.: Gamification and SQL: an empirical study on student performance in a database course. ACM Trans. Comput. Educ. (TOCE) **21**(1), 1–29 (2020)
6. Obionwu, V., Broneske, D., Hawlitschek, A., Köppen, V., Saake, G.: Sqlvalidator-an online student playground to learn SQL. Datenbank-Spektrum **21**, 73–81 (2021)
7. Prabhu, S., Jaidka, S.: SQL and PL-SQL: analysing teaching methods. In: CITRENZ Conference (2019)
8. Rodríguez, J., Ginez, N., Martinez, R., Salazar, M., Cecchi, L.: Enfoque didáctico para la ensenanza de base de datos en la escuela secundaria. In: XIV Congreso Nacional de Tecnología en Educación y Educación en Tecnología (TE&ET 2019),(Universidad Nacional de San Luis, 1 y 2 de julio de 2019). (2019)
9. Sánchez-Anguix, V., Alberola, J.M., Julián, V.: Towards adaptive gamification in small online communities. In: 16th International Conference on Soft Computing Models in Industrial and Environmental Applications (SOCO 2021), pp. 48–57. Springer (2022)

10. Soflano, M., Connolly, T.M., Hainey, T.: An application of adaptive games-based learning based on learning style to teach SQL. Comput. Educ. **86**, 192–211 (2015)
11. Werbach, K., Hunter, D.: For the win, revised and updated edition: the power of gamification and game thinking in business, education, government, and social impact. University of Pennsylvania Press (2020)
12. Zichermann, G., Cunningham, C.: Gamification by design: implementing game mechanics in web and mobile apps. " O'Reilly Media, Inc." (2011)

Mixed Reality Training in University Courses

Alberto Martínez-Gutiérrez[1](✉), Pablo Alonso-Diez[2], Rubén Ferrero-Guillén[1], Iván Sánchez-Calleja[1], Jesús Pérez-González[1], Madalena Araújo[3], and Javier Díez-González[1]

[1] Department of Mechanical, Computer and Aerospace Engineering, Universidad de León, León 24071, Spain
amartg@unileon.es
[2] Department of General Didactics, Subject-Specific Didactics, and Theory of Education, Universidad de León, León 24071, Spain
[3] ALGORITMI, Universidade do Minho, Escola de Engenharia, Campus Azurem, Guimaraes 4800-058, Braga, Portugal

Abstract. Education plays a fundamental role in the development of individuals and societies by transmitting knowledge, skills, and values. With the rise of digital technologies, new opportunities have emerged to enhance how people learn and interact with content. However, many digital platforms still face challenges when it comes to creating meaningful, immersive learning experiences. In this context, mixed reality (MR) offers a promising alternative by combining physical and virtual elements in real time. In this work, a practical training activity supported by MR headsets was developed and applied with engineering students. The experience was structured in phases to evaluate performance indicators under controlled conditions. Additionally, students' perceptions were collected through surveys to assess motivation, usability, and potential limitations. The results show improvements in task execution with repeated exposure and high acceptance of the MR methodology. These findings support the idea that MR can be a useful tool for improving engagement and learning in higher education environments.

Keywords: Digital Training · Immersive Learning · Digital Twin · Mixed Reality · Education 5.0

1 Introduction

The transmission of knowledge across generations has always been central to human development and societal advancement. From oral traditions to institutionalized education, humanity has relied on pedagogical structures to disseminate values, skills, and competencies essential for survival and progress [12].

However, in recent years, the educational paradigm has undergone a profound transformation catalyzed by digital technologies. This shift has not only

E. Corchado et al. (Eds.): SOCO 2025, CCIS 2806, pp. 422–432, 2026.
https://doi.org/10.1007/978-3-032-19763-4_39

facilitated greater access to information but has also reshaped how knowledge is acquired and applied [9].

Within this broader context of digital transition, the concept of immersive learning environments has gained traction, propelled by innovations such as virtual reality (VR), augmented reality (AR), and more recently, mixed reality (MR). These technologies enable the creation of virtual spaces that blend physical and digital elements to enrich the learning process [5]. Compared to traditional 2D simulators and fully immersive VR systems, MR offers a more balanced approach by allowing users to remain aware of their physical environment while interacting with virtual elements in real time.

According to Martínez-Gutiérrez et al., the use of digital platforms and educational metaverses introduces a new dimension of interaction where both human agents and virtual elements coexist and co-construct knowledge in real time. This model fosters not only the transmission of factual information but also the cultivation of critical thinking, creativity, and collaboration—competencies aligned with the principles of Education 5.0 and the ambitions of Industry 5.0 [12].

While VR has already demonstrated significant potential in educational settings—particularly in enhancing engagement and conceptual understanding [8,16]—it is not without limitations. Fully virtual environments may isolate learners from real-world referents, leading to challenges in long-term retention and applicability [1].

Mixed reality, in contrast, aims to address these concerns by merging the tangible and intangible: it retains the immersive affordances of VR while anchoring experiences within physical contexts, thereby enhancing cognitive association and user agency [15].

Evidence suggests that MR fosters active learning, multisensory engagement, and higher-order thinking. For instance, Huang et al.'s meta-analysis reports a moderate effect size (0.56) in learning effectiveness when MR is integrated into instructional settings, with significant improvement in both motivation and knowledge retention [5]. These results corroborate earlier findings that underscore MR's value in supporting constructivist approaches and experiential learning [3].

Similarly, empirical data highlight that VR and MR environments substantially increase learner motivation and engagement. According to Díaz-García et al., 85% of students reported heightened interest in subject matter through VR use, while 92% described their experience as more immersive and participatory than traditional methods [4]. Nevertheless, current scholarship has primarily focused on isolated use cases or discipline-specific implementations, with few studies exploring MR's integration across diverse educational contexts or assessing its impact through longitudinal methodologies.

As highlighted by Martínez-Gutiérrez et al., the current state of MR in education calls for scalable, inclusive platforms that not only incorporate technological innovations but also respect pedagogical integrity and human-centered design principles [9]. Additionally, the integration of MR within Industry 5.0 frameworks emphasizes the need for systems that foster hybrid intelligence—where human creativity is amplified through machine precision [12].

In this light, the present study seeks to examine the educational applications of mixed reality from a cross-disciplinary and methodological standpoint. By synthesizing insights from existing literature and experimental studies, this paper aims to contribute to the ongoing dialogue on how MR can be leveraged to enhance learning outcomes, overcome pedagogical limitations, and meet the evolving demands of both learners and institutions.

Though the findings to date are promising, substantial work remains to fully realize the transformative potential of MR in education—particularly in terms of accessibility, scalability, and instructional design. This study emerges as a timely intervention that seeks to address these gaps and offer empirical guidance for the future integration of mixed reality in educational settings.

Therefore, the main educational and technological contributions of this work are:

- Design and implementation of a mixed reality-based educational platform to support practical training in engineering contexts.
- Experimental evaluation of performance metrics (task time and failures) across sequential training phases to assess learning progression.
- Analysis of student perceptions regarding motivation, usability, and physical interaction using MR headsets through structured survey data.

2 Case Study

This educational innovation has been implemented within the course Theory of Machines and Mechanisms, part of the third-year curriculum of the Bachelor's Degree in Mechanical Engineering at the University of León (Spain). The course focuses on the analysis and modeling of machine kinematics (i.e., mobility), with the aim of enabling students to understand mechanical systems' responses to known inputs.

Traditionally, after attending lecture-based theoretical sessions, students participate in practical classes involving computer-based simulations. However, these activities often suffer from perceptual limitations due to the constraints of two-dimensional interfaces such as computer screens.

To address this limitation and foster experiential learning, a new practical activity has been introduced in which students interact with a real mobile robot to explore and internalize its kinematic behavior. The underlying pedagogical hypothesis is that enhanced control over the physical system promotes deeper understanding of its mobility characteristics.

Nevertheless, due to the large number of enrolled students, the time required for individual hands-on training, and the high cost of the robotic platform (which limits the availability to a single unit), it becomes necessary to incorporate complementary instructional strategies. In this context, MR technologies offer an effective and scalable solution for skill development and concept reinforcement.

The objective of this educational intervention is to evaluate the effectiveness of mixed reality in the transfer of procedural knowledge and skills. To this end, students are presented with a challenge-based learning activity in which they must navigate the mobile robot through a predefined path with physical obstacles. The learning outcomes are assessed based on two key performance indicators: task completion time (i.e., agility) and number of errors committed (i.e., precision).

3 Technological Implementation

The successful execution of this case study requires the development of a MR platform specifically designed for educational purposes [6]. This platform must integrate a wide array of technological components—including Digital Twins (DTs), Graphical User Interfaces (GUIs), real-time rendering engines, mixed reality headsets, and physical robotic systems—as illustrated in Fig. 1. The convergence of these diverse technologies presents a significant technical challenge due to the heterogeneity of the underlying architectures and interaction protocols [14].

At the core of the system is a DT [10], which accurately replicates the behavior of the mobile robot in response to both environmental conditions and user-generated inputs. The DT is implemented within the Robot Operating System (ROS) framework, where the robot's kinematic behavior is mathematically modeled and simulated with high fidelity.

The output data generated by the DT is processed by a real-time graphics engine. This engine is responsible not only for rendering a high-fidelity 3D model of the robot but also for simulating physical interactions consistent with the model dynamics. Furthermore, the engine continuously updates the robot's virtual position based on the user's control actions, thereby enhancing realism and immersion and attaining an effective human-machine interaction.

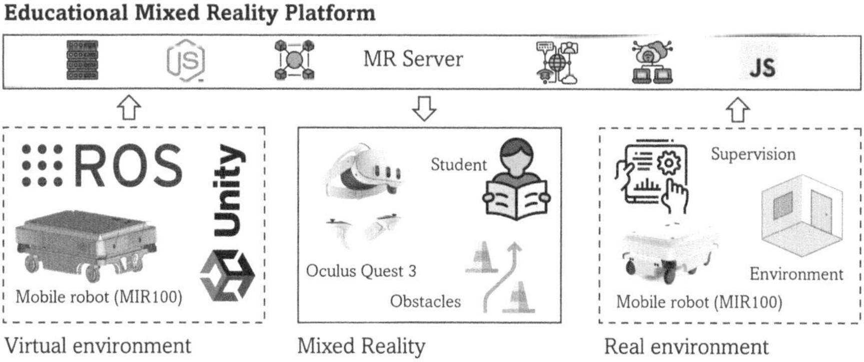

Fig. 1. Educational Mixed Reality platform integrating virtual, mixed, and real environments for robotics training

Mixed reality headsets (MR devices) play a crucial role in bridging the virtual and physical worlds. These devices convert digital outputs into multisensory stimuli—visual, auditory, and haptic—which are perceived by the learner in real time. This immersive experience allows students to interact with a virtual robot while simultaneously perceiving and navigating the physical environment surrounding them.

An additional feature of the MR-based training platform is the implementation of visualization interfaces for real-time performance monitoring. These interfaces provide key Performance Indicators (KPIs) that track student progress throughout the activity, supporting adaptive learning and enabling personalized instructional feedback. Instructors can also use these interfaces to view the virtual environment and deliver guidance or clarification during the exercise.

It is also critical to consider the integration with the physical mobile robot. This connection enables the assessment of skills transfer from the virtual simulation to any real-world application. Through the same graphical interface, all user interactions and performance data are logged and visualized, facilitating a comprehensive evaluation of the educational methodology.

4 Experiments

To rigorously assess the transfer of procedural skills and conceptual understanding related to machine kinematics through the use of MR technologies, a structured experimental protocol was implemented. The study was conducted with a sample population of 63 undergraduate students enrolled in the Theory of Machines and Mechanisms course of the Mechanical Engineering degree program at the University of León.

The experimental procedure, summarized in Fig. 2, was structured into three sequential phases designed to collect KPIs under controlled and replicable conditions.

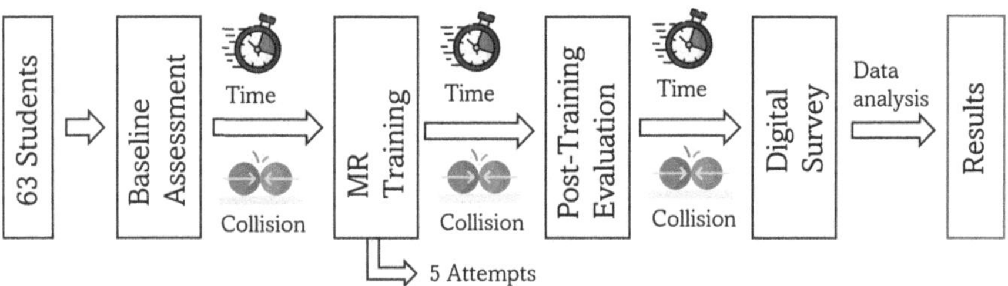

Fig. 2. Experimental protocol for assessing MR-based training through baseline, training, and post-training evaluations.

A) Baseline Assessment: Each participant performed a driving task with a physical mobile robot for the first time, without prior hands-on experience. This initial trial was designed to evaluate baseline motor and cognitive skills, as well as to account for interindividual variability in the sample [13]. Performance metrics—including task completion time and number of execution errors—were recorded to establish each student's initial proficiency level.

B) Training Phase with MR Simulation: Subsequently, all participants engaged in a structured MR-based training session replicating the same robot control conditions. The virtual environment maintained the kinematic and dynamic fidelity of the real system, allowing immersive interaction via mixed reality headsets. Each student completed five consecutive simulation attempts, with KPIs (execution time and errors) recorded for every trial to evaluate learning progression and adaptation within the virtual environment.

C) Post-training Evaluation: Upon completing the MR training, each participant performed a final trial with the physical robot. The objective of this stage was to measure the extent of skill transfer from the virtual to the real-world context, comparing final performance metrics against those obtained during the initial real-world attempt.

Following the practical sessions, participants were asked to complete an anonymous digital survey designed to capture qualitative insights regarding their

Fig. 3. Visual depiction of the mixed reality environment as perceived by students, showing the seamless integration of virtual objects within the physical space during the experimental activities.

perceived learning outcomes, usability of the MR system, and overall training experience. The survey included both closed-ended and open-ended items aimed at evaluating cognitive engagement, perceived realism, and motivational aspects of the MR-based learning process.

To better understand the students' perception, Fig. 3 presents a frame illustrating the superimposition of virtual objects onto the real environment during the experimental sessions. This visualization highlights one of the key advantages of mixed reality systems: the seamless integration of digital content within the learner's physical context. Such an approach enhances spatial awareness and interaction fidelity, allowing students to engage more intuitively and effectively with the learning materials. Moreover, by maintaining a direct view of the real environment, mixed reality minimizes disorientation and physical discomfort often associated with fully immersive virtual reality systems, thereby promoting longer engagement and improved knowledge retention.

5 Results

The KPIs collected in this study are categorized into two main domains: performance metrics (i.e., task completion time and number of errors) and learning experience indicators, the latter derived from the post-activity survey designed to assess students' subjective perceptions.

Performance Metrics and Relative Improvement. To ensure a fair evaluation of performance gains, it is essential to account for participants' initial skill levels. For this purpose, we introduce the metric Relative Improvement (RI), defined as the percentage change in performance between the first and final trials with the physical robot. This approach allows normalization across individuals with different prior abilities.

Figure 4 presents the evolution of performance metrics across the different stages of the training protocol, including the baseline assessment (T0), five mixed reality (MR) training attempts (T1–T5), and the final post-training evaluation (T6) with the physical system.

A clear downward trend was observed in task completion time throughout the MR training phase. The average time decreased from 33.92 s at baseline (T0) to 14.83 s by the fifth MR attempt (T5), representing a 56.3% reduction. This suggests progressive motor and cognitive adaptation to the task. The time increased slightly to 23.28 s in the post-training evaluation (T6), indicating a partial transfer of acquired skills from the virtual to the real environment. While not matching the final MR performance, this value was still significantly lower than the baseline.

The number of execution errors, measured as failures (e.g., collisions), followed a similar pattern. The mean number of failures decreased from 1.42 at T0 to 0.80 at T5, reaching 0.69 in the final physical trial (T6). This reduction indicates improved control and task understanding. The slight increase in errors

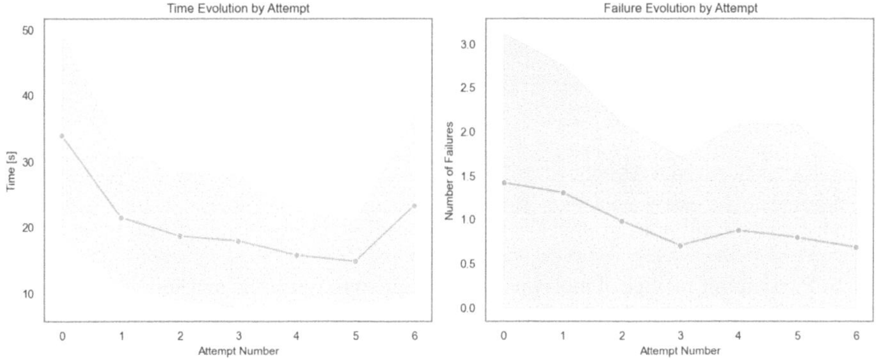

Fig. 4. Mean time (left) and failures (right) per training attempt with standard deviation bands. Both metrics decrease over attempts, indicating improved performance and accuracy.

from T3 to T4 may reflect a greater tendency to take more risks to reduce times due to a sense of control after overcoming the first attempts with MR.

Overall, the performance trends suggest effective learning during MR training and a measurable transfer of skills to the real-world context, albeit with a moderate decrease in efficiency and precision, which is consistent with the change in interaction medium.

Student Experience and Perception of MR Training. To complement the quantitative analysis, a digital survey was administered to gather qualitative insights into the learners' experience with the MR platform. The survey included the following Likert-scale questions:

- *Have you been motivated by the use of mixed reality glasses?*
- *Would you consider using this methodology in future practical sessions across other courses?*
- *Did you consciously slow down your body movements while wearing the mixed reality headset due to fear of colliding with real-world obstacles?*

The analysis of participant feedback reveals distinct trends in the perception and impact of the applied learning methodology utilizing MR headsets, as illustrated in Fig. 5.

Firstly, motivation levels toward the employed methodology were notably high. Specifically, 69.84% of respondents assigned the highest possible rating of 5 out of 5, while an additional 28.57% rated it 4 out of 5. This overwhelmingly positive reception underscores the methodology's effectiveness in capturing learners' attention and fostering intrinsic motivation—an essential driver for successful educational interventions and sustained engagement [2].

Secondly, regarding the willingness to reutilize this pedagogical approach across different academic courses, the response was similarly affirmative.

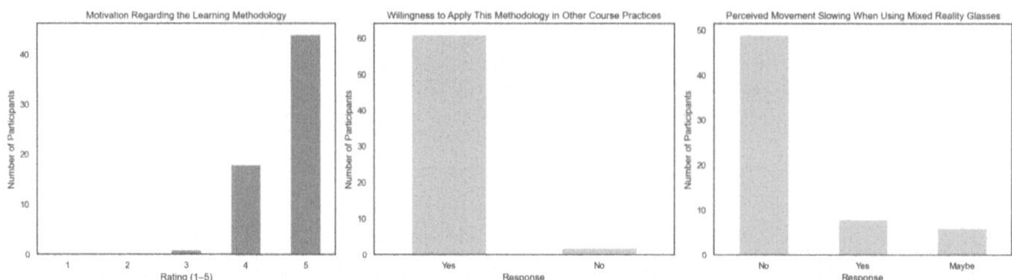

Fig. 5. Participant ratings of motivation, willingness to reuse the methodology, and self-reported movement slowing with mixed reality glasses. Data reflect unique respondents' answers.

A remarkable 96.83% of participants expressed openness to adopting this technology-enhanced instructional strategy in other educational contexts. This finding highlights the perceived value, scalability, and transferability of the methodology, emphasizing its potential for broad implementation across diverse curricula. Therefore, this research reinforces other studies highlighting the users' willingness for learning in MR environments [7].

Lastly, participants were queried about whether they experienced any deceleration in their physical movements due to apprehension about potential collisions while wearing the MR headsets. Responses were more heterogeneous in this regard: the majority (77.78%) reported no perceptible slowing of movement, whereas 22.22% acknowledged some degree of deceleration ("Yes" 12.70%) or uncertainty ("Maybe" 9.52%). These results suggest that, while the MR setup is generally comfortable and does not significantly hinder physical interaction for most users, a subset of participants may experience cautious behavior. This phenomenon likely stems from the MR system's preservation of real-world environmental cues, as opposed to Virtual Reality (VR) platforms that fully occlude physical surroundings, thus limiting spatial awareness. In fact, the results of this study objectively show an improvement in physical mobility with regard to one of our previous studies conducted in a fully virtual environment [11].

In sum, these findings contribute valuable insights into the integration of mixed reality technologies in higher education settings, demonstrating their capacity to enhance learner motivation, engagement, and adaptability while maintaining physical comfort and spatial orientation. The results advocate for further exploration of MR-based pedagogies within the framework of immersive and active learning paradigms.

6 Conclusions

This study contributes to the growing body of evidence supporting the integration of MR technologies into educational practice. By examining both performance metrics and user feedback across controlled experimental phases, the results demonstrate that MR environments can enhance learning effectiveness while maintaining physical comfort and user engagement. Notably, improvements in task performance were observed with repeated exposure, suggesting a learning curve that stabilizes over time and supports skill acquisition in an immersive setting.

Participants feedback further underscores the pedagogical viability of MR. High levels of motivation and willingness to reuse the methodology across other academic contexts reflect its perceived value and adaptability. While some users expressed caution regarding physical movement, the majority reported no hindrance, affirming the ergonomic and perceptual advantages of MR systems over fully immersive VR platforms. These insights are consistent with previous findings on the role of MR in promoting multisensory learning and fostering student agency.

Taken together, the results highlight MR's potential as a scalable and effective tool within higher education, particularly in the context of Education 5.0 and Industry 5.0 goals. However, the study also points to the need for continued research into long-term outcomes, cross-disciplinary applications, and inclusive design principles. Future work should focus on refining instructional strategies, ensuring accessibility, and expanding the evidence base through larger, more diverse samples and longitudinal designs.

Acknowledgements. This work has been funded by the project of the Spanish Ministry of Science and Innovation grant number PID2023-153047OBI00, by the Consejería de Educación de la Junta de Castilla y León and by the Universidad de León and the teaching innovation group MECACOM of the Universidad de León. The author Iván Sánchez-Calleja acknowledges funding for doctoral studies of the University of León. This work has been supported by FCT – Fundação para a Ciência e Tecnologia within the R&D Unit Project Scope UID/00319/Centro ALGORITMI (ALGORITMI/UM)

References

1. Campos, E., Hidrogo, I., Zavala, G.: Impact of virtual reality use on the teaching and learning of vectors. In: Frontiers in Education, vol. 7, p. 965640. Frontiers Media SA (2022)
2. Cook, D.A., Artino, A.R., Jr.: Motivation to learn: an overview of contemporary theories. Med. Educ. **50**(10), 997–1014 (2016)
3. Díez-González, J., Verde, P., Ferrero-Guillén, R., Álvarez, R., Juan-González, N., Martínez-Gutiérrez, A.: Evaluation of the skills' transfer through digital teaching methodologies. In: Computational Intelligence in Security for Information Systems Conference, pp. 340–349. Springer (2023). https://doi.org/10.1007/978-3-031-42519-6_32
4. García, V.D., Márquez, O.C., García-Chamizo, F.: Impacto del uso de gafas de realidad virtual en el aprendizaje de los alumnos: estudio empírico. Eur. Public Soc. Innov. Rev. **10**, 1–20 (2025)
5. Huang, T.C., Chen, C.H., Tseng, C.Y.: Exploring learning effectiveness of integrating mixed reality in educational settings: a systematic review and meta-analysis. Comput. Educ. 105327 (2025)

6. Kostov, G., Wolfartsberger, J.: Designing a framework for collaborative mixed reality training. Procedia Comput. Sci. **200**, 896–903 (2022)
7. Lee, L.K., et al.: A systematic review of the design of serious games for innovative learning: augmented reality, virtual reality, or mixed reality? Electronics **13**(5), 890 (2024)
8. Lin, X.P., Li, B.B., Yao, Z.N., Yang, Z., Zhang, M.: The impact of virtual reality on student engagement in the classroom-a critical review of the literature. Front. Psychol. **15**, 1360574 (2024)
9. Martínez-Gutiérrez, A., et al.: Educational metaverse. In: International Conference on EUropean Transnational Education, pp. 279–287. Springer (2024). https://doi.org/10.1007/978-3-031-75016-8_26
10. Martínez-Gutiérrez, A., Díez-González, J., Ferrero-Guillén, R., Verde, P., Álvarez, R., Perez, H.: Digital twin for automatic transportation in industry 4.0. Sensors **21**(10), 3344 (2021)
11. Martínez-Gutiérrez, A., Díez-González, J., Guillén, R.F., Sánchez-Calleja, I., Arampatzis, G., Araújo, M.: Impact of the metaverse on humans: physical exertion and perception shifts (2024)
12. Martínez-Gutiérrez, A., Díez-González, J., Perez, H., Araújo, M.: Towards industry 5.0 through metaverse. Robot. Comput. Integr. Manuf. **89**, 102764 (2024)
13. Martínez-Gutiérrez, A., Díez-González, J., Verde, P., Perez, H.: Convergence of virtual reality and digital twin technologies to enhance digital operators' training in industry 4.0. Int. J. Hum Comput Stud. **180**, 103136 (2023)
14. Martínez-Gutiérrez, A., Díez-González, J., Verde, P., Ferrero-Guillén, R., Perez, H.: Hyperconnectivity proposal for smart manufacturing. IEEE Access **11**, 70947–70959 (2023). https://doi.org/10.1109/ACCESS.2023.3294308
15. Pérez, P.G., Lema, J.M.M.: La realidad virtual para la enseñanza y aprendizaje de la perspectiva en el dibujo. Edutec, Revista Electrónica de Tecnología Educativa **83**, 188–207 (2023)
16. Zapatero Guillén, D.: La realidad virtual como recurso y herramienta útil para la docencia y la investigación. TE & ET (2011)

Special Session: Machine Learning and Computer Vision in Industry 4.0

VR Escape Room to Prevent Cognitive Decline in Older Adults

David Mulero-Pérez(✉), Lucila Vázquez-Soriano, Manuel Benavent-Lledo, David Ortiz-Perez, and Jose Garcia-Rodriguez

Department of Computer Technology, University of Alicante, Alicante, Spain
{dmulero,mbenavent,dortiz,jgarcia}@dtic.ua.es

Abstract. As cognitive decline becomes an increasingly prevalent concern in ageing populations, the exploration of innovative, engaging, and preventive interventions is of growing importance. This study presents a virtual reality (VR) escape room aimed at supporting cognitive health through immersive, gamified experiences. Building upon prior developments in advanced hand-tracking and realistic object manipulation in Unreal Engine 4, this work integrates refined interaction mechanisms to create a responsive and engaging virtual environment. The escape room incorporates puzzles and challenges specifically designed to stimulate key cognitive functions such as memory, attention, and problem-solving. Additionally, a performance tracking system was developed to collect user interaction data in real time, enabling post-session analysis by healthcare professionals. By merging cognitive training principles with immersive VR technologies, this research contributes to the development of accessible, non-pharmacological tools for early-stage cognitive deterioration prevention and monitoring.

Keywords: virtual reality · cognitive decline · ageing population · escape room · serious games · hand tracking

1 Introduction

In recent years, virtual reality (VR) has witnessed exponential growth in both adoption and technological advancement. The global sales of VR devices have steadily increased, and industries such as healthcare, education, and tourism have begun to integrate this technology beyond its traditional association with digital entertainment [2,21]. This widespread adoption has also driven research into more immersive and intuitive forms of interaction within virtual environments, focusing on eliminating physical controllers in favour of direct hand-based interaction, thereby enhancing the sense of presence and user engagement.

Within this context, this paper explores the application of VR to cognitive healthcare, specifically targeting the prevention and rehabilitation of cognitive decline in older adults. To achieve this, a set of interactive *escape rooms* has been developed, leveraging controller-free hand-tracking technologies based on the

E. Corchado et al. (Eds.): SOCO 2025, CCIS 2806, pp. 435–444, 2026.
https://doi.org/10.1007/978-3-032-19763-4_40

UnrealHandGrasp [11] system and UnrealGrasp [13]. These systems enable natural interaction with virtual environments by relying solely on the user's hands, thus fostering a fully immersive experience. Figure 1 provides an overview of six representative puzzles and themed rooms developed for this project, illustrating the diversity of cognitive challenges and contexts addressed.

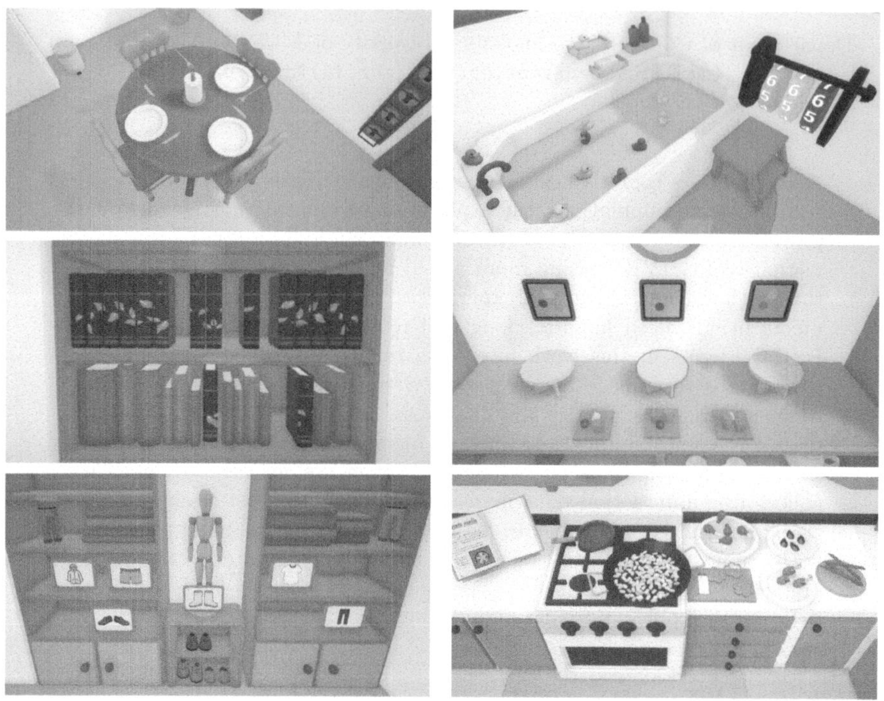

Fig. 1. Illustrative examples of six different puzzles and themed rooms implemented in the VR escape room application, covering a range of cognitive skills and interaction styles.

These *escape rooms* are designed as a therapeutic tool to stimulate key cognitive functions such as memory, attention, and logical reasoning. Recent research has highlighted the potential of VR environments in cognitive rehabilitation, owing to their capacity to simulate realistic, controllable, and repeatable scenarios that can be personalised to suit individual users' needs [17].

This work not only addresses the technical implementation of such environments but also adopts a user-centred design approach and ensures the scalability of the proposed solution. It introduces a modular puzzle system and emotionally evocative virtual settings tailored to the cognitive profile and preferences of elderly users.

The main contributions of this work are as follows:

1. **Development of an immersive VR environment without physical controllers**, utilising advanced hand-tracking technologies aimed at cognitive rehabilitation.
2. **Design and implementation of a modular and extensible system of therapeutic escape rooms**, with adaptable puzzles targeting specific cognitive functions.
3. **Integration of a user progress tracking system**, enabling continuous cognitive performance assessment and personalised adaptation of the experience.

The remainder of this paper is structured as follows. Section 2 reviews the current state of the art in virtual reality applications for cognitive rehabilitation and gesture-based interaction. Section 3 details the proposed system architecture, including the integration of hand-tracking technology, puzzle design methodology, and the modular development framework. The implementation details are explained in Sect. 4. Finally, Sect. 5 summarises the main findings and outlines directions for future work.

2 Related Work

This section provides an overview of current methodologies and relevant approaches. The study begins by examining the clinical and societal implications of cognitive impairment. The subsequent exploration delves into the domain of serious games, with a particular focus on escape rooms, which are utilised as therapeutic instruments that employ gamification to engage users in cognitively demanding tasks. Finally, the potential of VR as an immersive platform for delivering such interventions is reviewed, with a focus on recent advancements in interaction technologies such as hand tracking and their applications in VR-based cognitive training.

2.1 Cognitive Decline and Preventive Strategies

Cognitive impairment represents an escalating public health issue, particularly in ageing societies such as Spain, which ranks among the countries with the highest life expectancy. Age has been identified as a major risk factor for developing Mild Cognitive Impairment (MCI), with studies indicating that 15% of individuals over 65 with MCI progress to dementia within two years [15]. Global projections suggest dementia cases may nearly triple by 2050 [12]. These alarming trends underscore the pressing need for effective and scalable interventions to mitigate cognitive decline.

Recent studies have increasingly focused on the concept of cognitive reserve as a buffer against age-related cognitive decline [14]. Cognitive reserve, built through lifelong education and mentally stimulating activities, has been shown to delay the onset of dementia-related symptoms [8]. Activities such as reading, puzzles, and strategic games enhance various cognitive domains, such as memory, language, and attention, making them crucial in preventive interventions.

2.2 Serious Games in Elderly Care

Gamification can be defined as the integration of game-based components within non-entertainment settings with the objective of enhancing engagement and motivation. It has gained popularity in the fields of education, healthcare, and corporate training. When applied in elderly care, especially for cognitive rehabilitation, it offers several benefits: improving mental agility, emotional well-being, autonomy, and motivation among older adults [5,9].

One of the most promising applications of gamification in this domain is the development of *serious games*, such as thematic escape rooms. These game-like environments present users with logic puzzles and cognitive challenges that stimulate memory, reasoning, language, and problem-solving [16]. Designed as therapeutic tools, they offer a dynamic and engaging alternative to traditional repetitive cognitive exercises, which are often monotonous and quickly abandoned.

The concept of an escape room integrates multiple cognitive tasks into a coherent narrative experience. These environments can adapt to different levels of cognitive function and offer immediate feedback, a critical component in maintaining user motivation. By integrating game mechanics such as progression, challenge, and reward, serious games provide a powerful means of maintaining or even improving cognitive performance in ageing populations [1].

2.3 Virtual Reality for Cognitive Rehabilitation

Recent research supports the efficacy of virtual reality in cognitive rehabilitation, particularly for older adults with MCI. VR-based interventions have been shown to enhance memory, attention, and executive function, outperforming traditional non-immersive methods [10,24]. García-Betances et al. [20] further demonstrated that VR systems tailored for cognitive assessment and stimulation can improve not only cognitive performance but also emotional well-being and daily functioning. These findings are consistent with broader evidence highlighting the potential of immersive technologies in healthcare settings [6], reinforcing VR's role as a promising tool for accessible and personalised cognitive support.

VR escape rooms, which blend game mechanics with immersive environments, represent a particularly effective application of this technology. These experiences allow users to navigate virtual spaces, solve puzzles, and interact with the environment in a natural and intuitive manner. Whilst contemporary commercial offerings, such as *Escape Simulator* [19] provide thematic puzzle-solving experiences, they often lack support for advanced interaction techniques like hand tracking. Furthermore, this game was not developed for users with low expertise.

Hand-tracking technology enhances immersion by allowing users to interact with virtual objects using natural gestures, eliminating the need for traditional controllers. While promising, few applications currently support this feature in the context of escape rooms [18]. In summary, the integration of hand tracking and serious games into VR environments holds considerable potential for cognitive rehabilitation. By creating adaptive, engaging, and accessible experiences, this technological synergy addresses the motivational and functional needs of older adults in a scalable and impactful way.

3 Escape Room Design

The design of the VR escape room in this project is grounded in up-to-date research on cognitive health, gamification, and user-centred design for older adults. The primary objective of this study is to create a therapeutic experience that is both immersive and cognitively stimulating, with the aim of preventing or slowing cognitive decline. This section outlines how various cognitive domains were targeted through the design of specific tasks and puzzle mechanics, integrating principles from cognitive rehabilitation and serious games.

3.1 Cognitive Foundations and Design Criteria

The escape room was structured to target multiple cognitive domains implicated in Mild Cognitive Impairment, including working memory, executive function, attention, visuospatial skills, and language. As studies have demonstrated, engaging older adults in varied cognitive activities helps promote cognitive reserve and delay the progression of dementia-related symptoms [14, 24].

Puzzles has been designed following the guidelines proposed by Elumir [3] and Wiemker et al. [22], ensuring clarity, fairness, and a single correct solution. Each puzzle includes a clear goal, an obstacle to overcome, and a reward, contributing to a compelling loop of engagement and reinforcement. Moreover, tasks were tailored to avoid frustration and promote autonomy, allowing users to progress at their own pace.

3.2 Cognitive Domains Addressed

To ensure that the experience provides comprehensive cognitive stimulation, each puzzle was deliberately designed and mapped to target specific cognitive domains:

- **Memory and language**: word-chain puzzles and memory-based recipe tasks stimulate semantic and episodic memory.
- **Executive functions and reasoning**: sequencing tasks (e.g., planning a shower routine) and abstract reasoning puzzles foster logic and planning.
- **Attention and visuospatial skills**: visual matching, object placement, and shape reconstruction activities engage sustained attention and spatial perception.
- **Motor and perceptual skills**: using hand-tracking interaction to grasp and move virtual objects enhances fine motor skills and perceptual coordination.

3.3 Difficulty Scaling and Personalisation

To accommodate a range of cognitive abilities, each puzzle includes multiple difficulty modes (e.g., reduced options, visual hints). This approach allows professionals to tailor the experience based on user performance or stage of cognitive decline. A modular design also facilitates future expansion or integration of adaptive difficulty algorithms.

In addition to cognitive stimulation, the design incorporates elements of reminiscence therapy, which has shown benefits for emotional well-being and cognitive engagement [23]. Familiar objects such as rotary phones, traditional chocolate bars, and old radios were modelled and integrated into the game, evoking memories and increasing emotional resonance with the experience.

3.4 Structure and Level Progression

The environment mimics a familiar domestic setting, divided into themed rooms such as the kitchen, living room, bathroom, and bedroom. Each room contains three puzzles and a final challenge requiring all prior tasks to be completed. This structure aligns with best practices in escape room design and promotes a non-linear yet guided user journey. As illustrated in Fig. 2, the project adopts a hybrid progression model, primarily open, allowing players to attempt multiple puzzles independently, but incorporating sequential dependencies where specific tasks must be solved to unlock others. This approach balances player autonomy with structured advancement and is particularly suitable for single-user therapeutic settings, ensuring meaningful cognitive engagement while maintaining narrative and functional coherence.

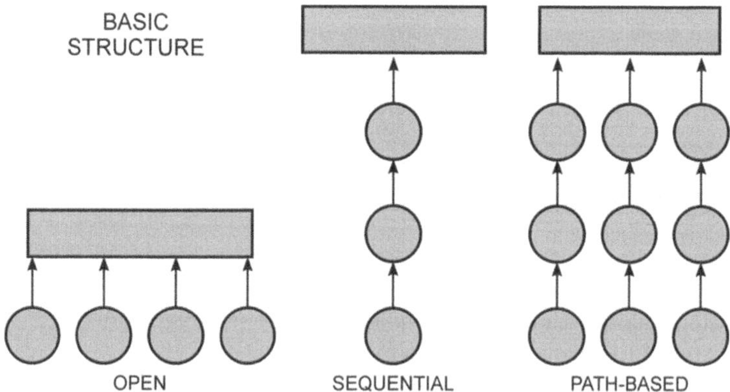

Fig. 2. Puzzle sequence models, adapted from Heikkinen and Shumeyko [7].

4 Implementation Details

The escape room was implemented using Unreal Engine 4 (UE4) in combination with Meta Quest 2 headsets and a custom hand-tracking interaction system. This section provides a high-level overview of the development process, highlighting the key technical components that enable the functionality, immersion, and data collection capabilities of the application.

4.1 Development Environment and Tooling

UE4 was selected for its robust support of VR development, compatibility with Meta Quest devices, and its hybrid visual and code-based scripting system (Blueprints and C++). The project leveraged previous work [11], which provided advanced hand-tracking and realistic object interaction frameworks. Custom 3D models and textures were created using Blender, allowing complete control over puzzle design and environmental assets. The models were optimised for performance and imported into UE4 with UV mapping and physics-based properties defined for realistic interactions.

4.2 Core Functionality and Game Architecture

The application is modular and driven by reusable blueprint classes:

- **PuzzleManager**: a global system instantiated at runtime to manage puzzle state, object associations, and resolution tracking.
- **BP_PuzzleObject and BP_Puzzle**: components assigned to interactable items and their corresponding puzzle logic.
- **BP_InteractiveObject and BP_ObjectPlacement**: allow users to pick up, move, and place objects using natural gestures tracked via the Oculus hand-tracking API.

A feedback system using dynamic materials, sound effects, and animated transitions provides immediate visual and auditory reinforcement to the user.

4.3 Tracking and Logging System

A key component of the system is the integrated logging mechanism, which captures and records user performance metrics in real time. The data collected includes the time required to complete each puzzle, interaction events such as object pickup, placement, and release, as well as the position and orientation of the VR headset, sampled at regular time intervals. These metrics are stored in structured JSON files using a hybrid Blueprint and C++ implementation. This logging infrastructure enables external tools and healthcare professionals to analyse session data for therapeutic assessment, progress tracking, and personalisation of future interventions. Moreover, the collected behavioural and interaction data can serve as input for deep learning models aimed at detecting early signs of cognitive decline [4].

4.4 Data Integrity and System Performance

To ensure consistent data collection and user progress tracking, the system includes a dedicated session management module that generates uniquely identified log files for each playthrough. These files are organised by timestamp and game level, enabling structured storage and facilitating post-session analysis. Each virtual room is encapsulated within its own data structure, allowing developers or clinicians to assess cognitive performance at a granular level. Additionally, the application incorporates an automatic saving mechanism that stores puzzle completion states, reducing the risk of data loss in the event of unexpected interruptions such as user dropouts or hardware issues. This functionality is especially important in clinical or home-based contexts where reliability and data continuity are essential.

From a technical standpoint, the system was developed in Unreal Engine version 4.26 to maintain compatibility with the original hand-tracking frameworks used in prior research. The application was executed on Meta Quest 2 headsets connected via Oculus Link, which allowed for stable hand-tracking performance while offloading computation to a desktop-class GPU. This setup ensures both visual fidelity and real-time responsiveness. In line with the project's focus on accessibility for older adults, all user interfaces were carefully designed with usability in mind. In-world widgets and visual indicators feature high-contrast designs and minimal clutter, promoting clarity and ease of interaction, even for users with limited technological experience or age-related visual impairments.

5 Conclusions

This paper has presented the design and implementation of a virtual reality escape room aimed at mitigating cognitive decline in older adults through immersive, gamified interaction. By integrating principles of cognitive rehabilitation with accessible hand-tracking technology and familiar virtual environments, the system offers a compelling therapeutic alternative to conventional cognitive training. The experience engages multiple cognitive domains, such as memory, attention, language, and reasoning; through a structured series of puzzles embedded in a modular and extensible game framework.

The results of this work highlight the broader potential of virtual reality as a platform for delivering accessible and personalised neurocognitive interventions. The system's modular design, natural interaction model, and integrated performance tracking establish a flexible foundation not only for clinical environments but also for decentralised, home-based care. By reducing reliance on traditional infrastructure and embracing controller-free interaction, the application supports more inclusive therapeutic models, particularly for users who may face technological or physical barriers. Moreover, the emotionally evocative and familiar virtual settings are deliberately designed to foster engagement among older adults, many of whom are underrepresented in digital health initiatives.

As ageing populations continue to expand, technologies such as the one presented here may play a transformative role in reshaping how cognitive health is supported, monitored, and preserved across diverse contexts.

Future development will focus on enriching the system with adaptive difficulty mechanisms, multisensory feedback, and expanded content tailored to a broader range of cognitive profiles and comorbidities. In parallel, longitudinal clinical validation involving diverse user groups will be essential to assess the therapeutic efficacy and real-world scalability of the platform. In sum, this project represents a promising step toward the development of engaging, evidence-based VR tools that promote cognitive well-being in ageing societies.

Acknowledgment. We would like to thank CIAICO/2022/132 Consolidated group project "AI4-Health" funded by the Valencian government. This work has also been supported by a national grant and two regional grants for PhD studies from the Spanish and Valencian governments, respectively, FPU21/00414, CIACIF/2021/430 and CIACIF/2022/175.

References

1. Abd-Alrazaq, A., Alhuwail, D., Aljafar, E.A., Househ, M.: The effectiveness and safety of serious games in improving cognitive abilities among elderly people with cognitive impairment: a systematic review and meta-analysis. JMIR Ser. Games **10**(1), e34592 (2022). https://doi.org/10.2196/34592, https://www.researchgate.net/publication/357440471
2. Bailenson, J.N., Yee, N., Blascovich, J., et al.: The use of immersive virtual reality in the learning sciences: digital transformations of teachers, students, and social context. J. Learn. Sci. **17**(1), 102–141 (2008). https://doi.org/10.1080/10508400701793141
3. Elumir, E.: 13 rules for escape room puzzle design (2018). https://thecodex.ca/13-rules-for-escape-room-puzzle-design/. Accessed on 07 May 2025
4. Fernández Montenegro, J.M., Villarini, B., et al.: A survey of alzheimer's disease early diagnosis methods for cognitive assessment. Sensors **20**(24) (2020). https://doi.org/10.3390/s20247292
5. Guzmán, D.E., Rengifo, C.F., García-Cena, C.E.: Serious games for cognitive rehabilitation in older adults: a conceptual framework. Multimodal Technol. Interact. **8**(8), 64 (2024). https://doi.org/10.3390/mti8080064, https://www.mdpi.com/2414-4088/8/8/64
6. Guzmán, D.E., Rengifo, C.F., Guzmán, J.D., García-Cena, C.E.: Virtual reality games for cognitive rehabilitation of older adults: a review of adaptive games, domains and techniques. Virtual Reality **28**, 92 (2024). https://doi.org/10.1007/s10055-024-00968-3, https://link.springer.com/article/10.1007/s10055-024-00968-3
7. Heikkinen, O.K., Shumeyko, J.: Designing an escape room with the experience pyramid model (2016). https://api.semanticscholar.org/CorpusID:113930399
8. Maragall, F.P.: ¿qué es y qué podemos hacer para aumentar la reserva cognitiva? https://blog.fpmaragall.org/reserva-cognitiva (2024)

9. Morán, J.F.O., Pagador, J.B., Preciado, V.G., et al.: A serious game for cognitive stimulation of older people with mild cognitive impairment: design and pilot usability study. JMIR Aging **7**, e41437 (2024). https://doi.org/10.2196/41437
10. Mulero-Pérez, D., Benavent-Lledo, M., Garcia-Rodriguez, J., Azorin-Lopez, J., Vizcaya-Moreno, F.: Holodemtect: a mixed reality framework for cognitive stimulation through interaction with objects. In: García Bringas, P., Pérez García, H., Martínez de Pisón, F.J., et al. (eds.) 18th International Conference on Soft Computing Models in Industrial and Environmental Applications (SOCO 2023), pp. 226–235. Springer, Cham (2023). https://doi.org/10.1007/978-3-031-42536-3_22
11. Mulero Pérez, D.: Interacción con objetos en realidad virtual utilizando tracking de manos (2021). http://hdl.handle.net/10045/115937
12. Nichols, E.: Estimation of the global prevalence of dementia in 2019 and forecasted prevalence in 2050: an analysis for the global burden of disease study 2019. Lancet Public Health **7**(2), e105–e125 (2022)
13. Oprea, S., Martinez-Gonzalez, P., Garcia-Garcia, A., Castro-Vargas, J.A., Orts-Escolano, S., Garcia-Rodriguez, J.: A visually realistic grasping system for object manipulation and interaction in virtual reality environments. Comput. Graph. **83**, 77–86 (2019)
14. Orueta, U.D., Buiza-Bueno, C., Yanguas-Lezaun, J.: Reserva cognitiva: evidencias, limitaciones y lineas de investigación futura. Revista Española de Geriatría y Gerontología (2010)
15. Petersen, R.C., Lopez, O., Armstrong, M.J., et al.: Practice guideline update summary: mild cognitive impairment: Report of the guideline development, dissemination, and implementation subcommittee of the American academy of neurology. NCBI (2018)
16. Rincón, S.X.J., and, A.T.M.: The learning behind the escape room. Med. Teach. **42**(4), 480–481 (2020). https://doi.org/10.1080/0142159X.2019.1654090, pMID: 31448982
17. Rizzo, A., Koenig, S.T., et al.: Is clinical virtual reality ready for primetime? Neuropsychology **31**(8), 877 (2017)
18. Studio, C.: Mindset. https://www.meta.com/es-es/experiences/mindset/5981642121902237/ (2023)
19. Studio, P.: Escape simulator. https://pinestudio.com/games/escape-simulator/ (2021)
20. Tortora, C., et al.: Virtual reality and cognitive rehabilitation for older adults with mild cognitive impairment: a systematic review. Ageing Res. Rev. **93**, 102146 (2024). https://doi.org/10.1016/j.arr.2023.102146, https://www.sciencedirect.com/science/article/pii/S1568163723003057
21. Tussyadiah, I.P., Wang, D., Jung, T.H., tom Dieck, M.: Virtual reality, presence, and attitude change: empirical evidence from tourism. Tourism Manage. **66**, 140–154 (2018). https://doi.org/10.1016/j.tourman.2017.12.003, https://www.sciencedirect.com/science/article/pii/S0261517717302662
22. Wiemker, M., Elumir, A.: Escape Room Games: "Can you transform an unpleasant situation into a pleasant one?" (2015)
23. Woods, B., O'Philbin, L., Farrell, E.M., Spector, A.E., Orrell, M.: Reminiscence therapy for dementia. Cochrane Database System. Rev. **2018**(3), CD001120 (2018)
24. Zhong, D., Chen, L., Feng, Y., Song, R., et al.: Effects of virtual reality cognitive training in individuals with mild cognitive impairment: a systematic review and meta-analysis. Int. J. Geriatr. Psychiatry **36**(12), 1829–1847 (2021). https://doi.org/10.1002/gps.5603

Multimodal Egocentric Action Recognition via Object-Aware Representations

Hugo Hernandez-Lopez, Manuel Benavent-Lledo(✉), David Mulero-Perez, David Ortiz-Perez, Javier Rodriguez-Juan, and Jose Garcia-Rodriguez

Department of Computer Technology, University of Alicante, Alicante, Spain
{mbenavent,dmulero,dortiz,jrodriguez,jgarcia}@dtic.ua.es

Abstract. Egocentric human action recognition is critical for applications across various domains. Unfortunately, existing models often overlook fine-grained human-object interactions that are essential in first-person views. In this work we introduce EgoObject, a multimodal action recognition architecture that fuses (1) spatio-temporal video features from TimeSformer, (2) visual object embeddings from ViT, and (3) textual object embeddings from BERT, contributing to a more robust egocentric representation. To avoid redundancy, we introduce a frame selection mechanism to select key frames that are input to GroundingDINO for object detection. EgoObject achieves 58.82% in Top-1 accuracy using the ADL dataset.

Keywords: action recognition · object detection · deep learning · activities of daily living

1 Introduction

Human Action Recognition (HAR) has become a key challenge in computer vision, with applications ranging from surveillance and robotics, to assistive technologies and autonomous systems. The task involves identifying and classifying human activities from sequences of visual data, typically RGB video.

Over the years, a variety of architectures have been proposed for HAR. While methods based on convolutional neural networks (CNNs) or recurrent neural networks (RNNs) have dominated for several years, Transformer-based methods have outperformed them in this task. This architecture, initially proposed for natural language processing [33], can produce a better spatio-temporal representation of the input video. Nonetheless, they often miss domain or view specific cues. For example, the egocentric perspective significantly benefits from hand and object detection to model their interactions [2,3,15].

In this work, we explore such benefits by introducing a multimodal architecture that combines spatio-temporal video features with object-centric cues extracted from the detected objects in the scene. We propose a multimodal fusion strategy that integrates both visual and textual representations of objects, aiming at improving action recognition performance. Additionally, we assess how

E. Corchado et al. (Eds.): SOCO 2025, CCIS 2806, pp. 445–454, 2026.
https://doi.org/10.1007/978-3-032-19763-4_41

different types of visual and textual features extracted from objects, as well as their fusion, influence overall performance. Our contributions can be summarized as follows:

- We propose EgoObject, a multimodal action recognition architecture capable of integrating object-level context alongside spatio-temporal video features.
- We conduct a detailed study on the impact of different object feature modalities (visual, textual, and combined) on action recognition accuracy.
- We analyze the effect of frame sampling strategies on performance, highlighting the trade-off between context granularity and computational cost.

The remainder of this paper is organized as follows. Section 2 reviews related work on action recognition and object-guided learning. Section 3 introduces the proposed architecture and its components. Section 4 presents the experimental setup and results. Finally, Sect. 5 summarizes the conclusions drawn from this work and discusses directions for future work.

2 Related Work

Recent advances in deep learning have significantly made progress across many areas of computer vision [4,12–14,24,30,31,34], including human action recognition and object detection. In this section, we review prior work in these key areas relevant to our study.

2.1 Action Recognition

Action recognition seeks to classify activities in videos. Early approaches adopted two-stream 2D CNNs that process visual modalities separately, and then aggregated them with segmental consensus as in TSN [35]. This allowed capturing coarse temporal structure without inferring significant computational costs. However, in order to model longer dynamics, other approaches have proposed aggregating CNN features through LSTM layers (e.g. Long-term Recurrent Convolutional Network [9]), leveraging recurrent memory, at the risk of suffering the gradient vanishing problem.

Alternatively, some works propose the use of 3D CNNs such as I3D [7] and SlowFast [11]. This type of CNNs, extend convolution into the temporal axis, jointly encoding motion at multiple frame rates, thus achieving a better performance than 2D CNNs at a higher compute cost. To reduce this cost, similar methods simulate 3D receptive fields within 2D budgets (e.g. TSM [20]).

Transformer-based methods currently represent the state-of-the-art in action recognition. TimeSformer [5] splits each frame into 2D patches and applies divided attention: temporal attention across the same spatial patch in different frames, followed by spatial attention within each frame ViViT [1] groups pixels into small 3D "tubelets" projects each tubelet to a token, and processes tokens with factorized stacks of spatial and temporal Transformer layers. VideoMAE [32] adopts ViViT [1] tubelet mechanism, and randomly masks 90% of tokens to train an encoder-decoder structure. The encoder provides the video representations and the decoder predicts the missing volumes. After pretraining, the resulting encoder yields robust representations even with limited data.

2.2 Object Detection

Similar to HAR, object detection has evolved from early handcrafted pipelines to data-driven deep learning models. The initial CNN-based approaches introduced two-stage detectors such as R-CNN [16], and Faster R-CNN [27], where a set of coarse candidate boxes is proposed and then refined and classified. For latency-critical setups, one-stage families (e.g. YOLO [26]) predict classes and boxes in a single pass, exchanging a small accuracy drop for real-time throughput. However, the anchor-based heuristics used for these last methods can reduce models generalization capacity. Recent anchor-free designs (e.g. CornerNet [18]) propose the use keypoints to simplify, and reduce the computational cost.

As in the previous domain, Transformers have outperformed existing approaches for object detection. One of the most widely used transformer-based frameworks for object detection is DETR [6]. This architecture brings bipartite matching that refines a set of boxes that are provided by a Transformer encoder-decoder. However, due to its high temporary consumption during training and its low performance with small objects, DINO [36] proposes the use of contrastive de-noising and mixed-query selection to speed up the training convergency and reduce the amount of false positives.

Open-set requirements motivate works enabling few-shot or zero-shot performance. To this end, vision-language models are employed as in GroundingDINO [21], which consists of a DETR-based architecture with a cross-modal attention mechanism to localize novel categories from textual prompts.

3 EgoObject

This section describes EgoObject, a multimodal architecture that predicts egocentric actions from RGB videos by using a divided-attention video encoder along with visual and textual cues from objects as represented in Fig. 1.

3.1 Video Module

From the input video, frames are downsampled to generate 16-frame clips. The clips are then processed by the video module. This module consists of a video encoder used to extract a spatio-temporal representation of the input sequence. Particularly, we use a frozen TimeSformer [5] model pretrained on Kinetics400 [17]. The divided attention mechanism in this architecture enables our proposal to extract more effective representations compared to other approaches as detailed by experimental results in Sect. 4. After computing the features using TimeSformer [5], our model uses the normalized `[cls]` token as the video representation.

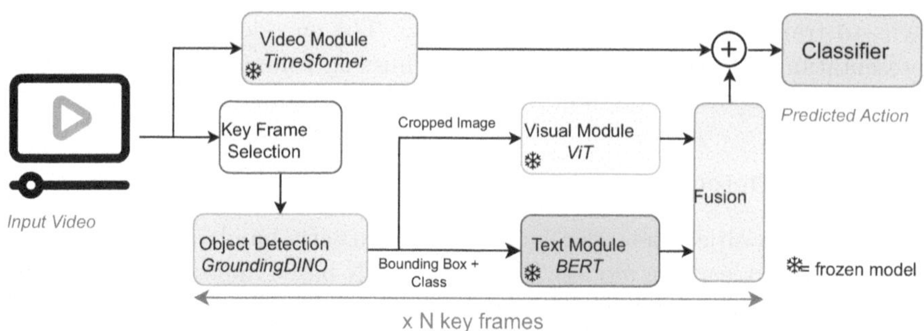

Fig. 1. Overview of EgoObject. Video frames processed by a video module to extract spatio-temporal features. Representative frames are selected from each video and processed by an object detector. Its outputs are processed by visual and textual modules, and the resulting features are fused using a Transformer encoder. Object and video features are concatenated to produce a final representation used for action recognition.

3.2 Key Frame Selection Mechanism

As previously detailed, spatio-temporal cues provide the most general information about the ongoing actions. However, we may find cases where such representation is not enough to disambiguate between two different but similar actions. In such cases, view-specific cues can contribute to a more robust representation of the action. In the egocentric perspective in particular, such cues are provided by the hands, objects and their interactions, as discussed in prior work [3]. To this end, we first select a set of key frames to extract object-aware information with an object detector.

For this task, we employ a stride that determines how many frames of the 16-frame sub-clip are skipped from the last one obtained. Detected object classes are aggregated and ranked by frequency, then the top four distinct classes are chosen. After this, in order to ensure rich contextual variability, one representative instance of each object class is obtained from each selected frame. In our experiments (see Sect. 4), the best performance is achieved when sampling every 4 frames. This selection strategy maximizes contextual variability while avoiding redundant detections of the objects.

3.3 Object Detector

In order to get a better understanding of the scene, we employ an object detector to get the objects present in the scene in which the action is being performed. During the development of this work, we have evaluated pretrained GroundingDINO models along with a pretrained GLIP [19] model. We have also tested the use of the dataset's own groundtruths to simulate how our architecture would work with a perfect detector. However, a GroudingDINO [21] model that was pre-trained on 1.8 million data samples achieved the highest mAP score on the ADL dataset.

3.4 Object Encoding

To represent the detected objects before fusing them with video embeddings we extract visual features from each cropped object, and a textual representation comprising bounding box location and label.

Visual Module: For every detected object we extract its visual embeddings to complement the global spatiotemporal features. Each bounding box provided by the object detector is cropped and processed by a frozen ViT-B/16 [10] model pretrained on the ImageNet-22k [28] dataset, due to superior performance compared to Swin-B [23]. As for the video module, the `[cls]` token is used as visual representation of the object.

Textual Module. In addition to the visual features, we encode every object label and position predicted by the object detector (e.g. "*90 120 110 140 book*") with a frozen BERT [8] model, which showed superior performance compared to RoBERTa [22] and DistilBERT [29]. Textual features provide additional high-level semantic knowledge that, when fused with motion and appearance cues, helps the classifier discriminate among visually similar egocentric actions.

Joint Object Representation. Visual and textual embeddings from the objects are fused with a 2-layer Transformer encoder to obtain the final object representation.

3.5 Video-Object Fusion

Video and object embeddings are concatenated into a single vector for classification. To this end, we employ a multi-label Multi-Layer Perceptron (MLP) classifier with Rectified Linear Unit (ReLU) activation and dropout to avoid overfitting. This MLP has an intermediate layer with 512 nodes and an output layer with 32. We use cross-entropy loss during training.

4 Experiments

This section describes the experimental setup on the ADL dataset, results of EgoObject, and ablation studies to determine the best configuration.

4.1 Experimental Setup

Dataset. We select the ADL dataset [25] for experimentation. Among egocentric datasets this dataset provides a reasonable trade-off between size and variability to verify the performance of our method, while avoiding overfitting. Moreover, it meets the requirements of providing action and object annotations. It comprises 20 videos recorded at 30 FPS of different people in home environments performing daily activities. Given the lack of splits we use 12 videos for training, 4 for validation and the remaining 4 for test, following a stratified approach on the classes. Videos are downsampled with a rate of 5 for efficiency.

Evaluation. We evaluate EgoObject on Top-k accuracy, particularly we report Top-1 and Top-5 following prior work on action recognition [1,5,7]. We use mean Average Precision (mAP) to evaluate performance of object detection models.

Training and Implementation Details. EgoObject is trained for 20 epochs using a learning rate of 0.0005, a batch size of 64, and AdamW optimizer with a weight decay of 0.1. Dropout is used to prevent overfitting and set to 0.1. To improve generalization, noise is introduced into video features during training. Implementation details may be found in our GitHub repository[1].

4.2 Results

The main results of EgoObject are shown in Table 1. It compares a unimodal approach consisting of a classification of video embeddings extracted with TimeSformer, with the proposed method. This method employs TimeSformer, ViT, and BERT as encoders. Regarding frame sampling, Table 3 shows that using the objects present in every 4 frames also gives us higher performance than the other tested ones. Therefore, our proposed method employs this strategy together with GroundingDINO [21] to detect the different objects present in the scene and obtain a better understanding of the scene.

Table 1. Top-1 and Top-5 scores results of testing an unimodal model and the EgoObject multimodal model. Best results have been highlighted in bold.

Model	Top-1	Top-5
Unimodal	**62.35**	83.53
Multimodal	58.82	**84.71**

Results on Tables 1 and 2 show that TimeSformer [5] yields a more accurate result as represented by Top-1 accuracy. We attribute the lower performance in EgoObject to 2 main factors. Firstly, the availability of bounding boxes affecting the key frame selection mechanism. Object annotations in the ADL [25] dataset are provided every 30 frames, *i.e.* every second, reducing the amount of examples available to adequately fine-tune an object detector, as demonstrated by the low mAP results in object detectors (Table 4). Secondly, the selected fusion strategy between video and object features, a concatenation, limits an effective fusion of embedding spaces. Additionally, it significantly increases the input size of the classifier.

In order to determine which are the best components for EgoObject, we first evaluated video-only configurations (Table 2). TimeSformer [5] emerged as the strongest, yielding 62.35% Top-1 and 83.53% Top-5 accuracy and outperforming ViViT and VideoMAE-based alternatives. Therefore, it forms the video encoder of our video module.

[1] https://github.com/3dperceptionlab/tfg_hhernandez/.

Table 2. Action recognition results on the ADL dataset. First part indicates RGB-only approaches. Second part indicates on Top-1 and Top-5 scores results of testing different models for our proposal using the every-2-frame sampling strategy.

Model	Top-1	Top-5
ViViT	43.04	79.75
VideoMAE	60.00	**84.71**
TimeSformer	**62.35**	83.53
TimeSformer+BERT	54.12	76.47
TimeSformer+RoBERTa	47.06	76.47
TimeSformer+DistilBERT	52.94	78.82
TimeSformer+ViT	52.94	<u>82.35</u>
TimeSformer+Swin	54.12	78.82
TimeSformer+BERT+Swin	55.29	78.82
TimeSformer+BERT+ViT	<u>56.47</u>	80.00

We also conduct different experiments to determine which are the best models for visual and textual feature extraction. For extracting textual features, we use natural language processing models, specifically BERT [8], DistilBERT [29], and RoBERTa [22]. For object features, we employ ViT [10] and Swin [23]. It is worth mentioning that these experiments have been conducted obtaining the 4 most present objects in every 2 frames. Results in Table 2 show that it is better to combine both features, since it allows to recover part of the loss when used in isolation. The table also shows the importance of aligning the different features. Although Swin [23] provides better results than ViT when used independently, ViT [10] improves Swin's performance when combined with BERT [8] features because their features are better aligned. Therefore, from now the different experiments are conducted using TimeSformer, BERT and ViT as the encoders.

Table 3 reports our key-frame sampling ablation. At one extreme, we analyzed detecting objects every second frame, supplying the richest context to the classification head. Reducing the rate to one detection every four frames retained temporal coverage while reducing the computational cost. This "moderate" setting proved to be the most effective. Pushing the stride to eight frames, or collapsing detection to a single snapshot taken either at the beginning or at the temporal center of the clip, lowered the cost further but missed objects that briefly appear or disappear, and performance dropped. From these results, we can conclude that a dense sampling strategy adds redundancy, thus making the classifier overly specialized. However, retrieving objects from frames too far apart causes too much information about their influence to be lost, reducing the classifier's ability to distinguish between actions.

Object detection results are represented in Table 4, this table compares results on GroundingDINO [21] with a zero-shot model and a fine-tuned model

Table 3. Top-1 and Top-5 scores results of testing different sampling strategies. Best results have been highlighted in bold.

Strategy	Top-1	Top-5
Every-2-Frame	56.47	80.00
Every-4-Frame	**58.82**	**84.71**
Every-8-Frame	55.29	76.47
First-Frame	55.29	81.12
Ninth-Frame	55.29	76.47

on different Intersection over Union (IoU) values. Results show that fine-tuning in this domain improves performance. As previously discussed, the performance of this model is limited. We attribute this to semantically and visually similar classes such as "tv" and "monitor".

Table 4. Comparison between Grounding DINO zero-shot and fine-tuned using mAP COCO metrics at different IoU thresholds and object sizes.

Metric	Zero-shot	Fine-tuned
$IoU = 0.5, area = A$	0.264	**0.442**
$IoU = 0.75, area = A$	0.203	**0.371**
$IoU = 0.5 : 0.95, area = A$	0.188	**0.330**
$IoU = 0.5 : 0.95, area = S$	0.047	**0.070**
$IoU = 0.5 : 0.95, area = M$	0.147	**0.263**
$IoU = 0.5 : 0.95, area = L$	0.259	**0.432**

A = All objects, S = Small objects, M = Medium objects, L = Large objects

5 Conclusions

In this work we have introduced EgoObject, a multimodal architecture designed to identify human-performed actions by integrating spatiotemporal features from TimeSformer with object-centered visual and textual context obtained through a ViT-B/16 and a BERT encoder. The results have shown that the simultaneous use of the two object-centered features provides a better result than using them independently. This work has also shown that extracting these features from the 4 most present objects in each 4 frames that make up the videos provides the best balance between diversity and compactness. Nevertheless, concatenating contextual and video features reduces performance due to the increased dimensionality of the input layer of the final classifier head. Therefore, future work will attempt to use more complex fusion strategies, such as the employment of a Transformer encoder to fuse video and contextual features. Furthermore, we believe that using a larger dataset with more general and diverse annotations can significantly reduce the noise introduced by poor object detections and increase the performance of the model.

References

1. Arnab, A., et al.: Vivit: a video vision transformer. arXiv arXiv:2103.15691 (2021)
2. Benavent-Lledo, M., Oprea, S., Castro-Vargas, J.A., et al.: Predicting human-object interactions in egocentric videos. In: IJCNN, pp. 1–7 (2022)
3. Benavent-Lledó, M., et al.: Interaction estimation in egocentric videos via simultaneous hand-object recognition. In: SOCO, pp. 439–448 (2022)
4. Benavent-Lledo, M., et al.: A comprehensive study on pain assessment from multimodal sensor data. Sensors **23**(24), 9675 (2023)
5. Bertasius, G., et al.: Is space-time attention all you need for video understanding? arXiv arXiv:2102.05095 (2021)
6. Carion, N., et al.: End-to-end object detection with transformers. arXiv **2005**, 12872 (2020)
7. Carreira, J., et al.: Quo vadis, action recognition? a new model and the kinetics dataset. arXiv arXiv:1705.07750 (2018)
8. Devlin, J., et al.: Bert: pre-training of deep bidirectional transformers for language understanding. arXiv arXiv:1810.04805 (2018)
9. Donahue, J., et al.: Long-term recurrent convolutional networks for visual recognition and description (2016)
10. Dosovitskiy, A., et al.: An image is worth 16x16 words: Transformers for image recognition at scale. In: ICLR (2021)
11. Feichtenhofer, C., et al.: Slowfast networks for video recognition. In: ICCV (2019)
12. Garcia-Garcia, A., et al.: Multi-sensor 3d object dataset for object recognition with full pose estimation. Neural Comput. Appl. **28**(5), 941–952 (2017)
13. Garcia-Garcia, A., et al.: A study of the effect of noise and occlusion on the accuracy of convolutional neural networks applied to 3d object recognition. Comput. Vis. Image Underst. **164**, 124–134 (2017)
14. Garcia-Garcia, A., et al.: Interactive 3d object recognition pipeline on mobile gpgpu computing platforms using low-cost rgb-d sensors. J. Real Time Image Proc. **14**(3), 585–604 (2018)
15. Girdhar, R., Grauman, K.: Anticipative video transformer. In: ICCV, pp. 13505–13515 (2021)
16. Girshick, R., et al.: Rich feature hierarchies for accurate object detection and semantic segmentation. arXiv arXiv:1311.2524 (2014)
17. Kay, W., et al.: The kinetics human action video dataset. arXiv arXiv:1705.06950 (2017)
18. Law, H., Deng, J.: Cornernet: detecting objects as paired keypoints. arXiv arXiv:1808.01244 (2019)
19. Li, L.H., et al.: Grounded language-image pre-training. In: CVPR (2022)
20. Lin, J., et al.: Tsm: temporal shift module for efficient and scalable video understanding on edge device. arXiv arXiv:2109.13227 (2021)
21. Liu, S., et al.: Grounding dino: marrying dino with grounded pre-training for open-set object detection. arXiv preprint arXiv:2303.05499 (2023)

22. Liu, Y., et al.: Roberta: a robustly optimized bert pretraining approach. arXiv arXiv:1907.11692 (2019)
23. Liu, Z., et al.: Swin transformer: hierarchical vision transformer using shifted windows. arXiv arXiv:2103.14030 (2021)
24. Orts, S., et al.: Gpgpu implementation of growing neural gas: application to 3d scene reconstruction. J. Parall. Distrib. Comput. **72**(10), 1361–1372 (2012)
25. Pirsiavash, H., et al.: Detecting activities of daily living in first-person camera views. In: CVPR, pp. 2847–2854 (2012)
26. Redmon, J., Divvala, S., Girshick, R., Farhadi, A.: You only look once: unified, real-time object detection. In: CVPR (2016)
27. Ren, S., et al.: Faster r-cnn: towards real-time object detection with region proposal networks. arXiv arXiv:1506.01497 (2016)
28. Ridnik, T., et al.: Imagenet-21k pretraining for the masses. arXiv arXiv:2104.10972 (2021)
29. Sanh, V., et al.: Distilbert, a distilled version of bert: smaller, faster, cheaper and lighter. arXiv arXiv:1910.01108 (2020)
30. Saval-Calvo, M., et al.: Three-dimensional planar model estimation using multi-constraint knowledge based on k-means and ransac. Appl. Soft Comput. **34**, 572–586 (2015)
31. Thalhammer, S., et al.: Challenges for monocular 6-d object pose estimation in robotics. IEEE Trans. Rob. **40**, 4065–4084 (2024)
32. Tong, Z., et al.: Videomae: masked autoencoders are data-efficient learners for self-supervised video pre-training. arXiv arXiv:2203.12602 (2022)
33. Vaswani, A., et al.: Attention is all you need. arXiv **1706**, 03762 (2023)
34. Viejo, D., et al.: Combining visual features and growing neural gas networks for robotic 3d slam. Inf. Sci. **276**, 174–185 (2014)
35. Wang, L., et al.: Temporal segment networks: towards good practices for deep action recognition. In: ECCV, pp. 20–36 (2016)
36. Zhang, H., et al.: Dino: detr with improved denoising anchor boxes for end-to-end object detection. arXiv arXiv:2203.03605 (2022)

Impact of Data Augmentation on Woven Fabrics Defect Detection by Means of Deep Learning

Beatriz Gil-Arroyo[1(✉)], Nuria Velasco-Pérez[1], Juan Marcos Sanz[2], Ángel Arroyo[1], Daniel Urda[1], and Álvaro Herrero[1]

[1] Grupo de Inteligencia Computacional Aplicada (GICAP), Departamento de Digitalización, Escuela Politécnica Superior, Universidad de Burgos, Av. Cantabria s/n, 09006 Burgos, Spain
{bgarroyo,nuriavp,aarroyop,durda,ahcosio}@ubu.es
[2] Textil Santanderina, Cabezón de la Sal, Spain
juanmarcos@tsanta.es
http://www.ubu.es , http://www.textilsantanderina.com

Abstract. In this study, the impact of various data augmentation techniques on defect detection in Batavia textiles was analyzed using a publicly available dataset consisting of high-resolution images from textile manufacturing processes. A baseline model trained without augmentation was compared with models trained using both individual and combined augmentation techniques, including horizontal and vertical flips, rotations, ZCA whitening, and brightness adjustments. Experiments were conducted using 5-fold stratified cross-validation, and the performance of the convolutional neural network DenseNet121 was evaluated using metrics suited for imbalanced binary classification tasks. The results show that not all augmentation strategies led to substantial improvements; while certain combinations of augmentation techniques resulted in the best performance-time balance, other transformations either had minimal impact or even degraded performance. Furthermore, the obtained results proof that AU-ROC remains stable across most simple transformations, suggesting that data augmentation, while useful, does not always yield a noticeable improvement in model performance.

Keywords: Defect detection · Data augmentation · Textile · Deep Learning · Image analysis

1 Introduction

The textile industry, within the framework of Industry 4.0, faces the ongoing challenge of maintaining high quality standards while minimizing production defects. In this context, the detection of anomalies in fabric surfaces is essential for ensuring product quality, maintaining competitiveness, and reducing economic losses [9]. Automated visual inspection systems based on deep learning

E. Corchado et al. (Eds.): SOCO 2025, CCIS 2806, pp. 455–465, 2026.
https://doi.org/10.1007/978-3-032-19763-4_42

techniques, particularly Convolutional Neural Networks (CNNs), have gained significant attention owing to their ability to identify visual defects efficiently and accurately. CNNs have demonstrated strong performance in image classification and anomaly detection tasks by learning hierarchical visual representations directly from raw image data [4,10]. These capabilities, coupled with the increasing availability of annotated datasets and advances in computational power, have enabled the deployment of non-assisted, high-performance inspection systems across a wide range of domains, including healthcare, manufacturing, and security.

DenseNet121, a deep convolutional neural network architecture characterized by dense connectivity between layers (as introduced by Huang et al. 2017 [5]), has proven to be a versatile and powerful architecture for various industrial applications, particularly in image classification tasks across different domains [6–8]. Its densely connected layers facilitate feature reuse and improve the gradient flow during training, which helps mitigate common issues such as the vanishing gradient problem [3]. These characteristics make DenseNet121 particularly suitable for binary classification tasks in texture-rich image domains, such as fabric inspection [13]. However, one of the primary limitations of CNNs in industrial settings is the limited availability of labeled training data, especially for defective samples, which often results in significant class imbalance [14]. This imbalance can severely impact the model's ability to generalize and maintain a high detection performance under real-world production conditions.

To overcome this challenge, data augmentation techniques [11] have been widely adopted to artificially increase the diversity of the training datasets [15]. By applying geometric and photometric transformations, such as rotations, horizontal and vertical flips, brightness variations, and statistical transformations like Zero-phase Component Analysis (ZCA), models can be trained to learn more invariant and robust features even when exposed to limited or unbalanced data. Despite their widespread use, the specific impact of different augmentation strategies on model performance in fabric defect detection tasks remains underexplored.

This study aims to evaluate the effectiveness of various data augmentation techniques in enhancing the performance of the DenseNet121 model for defect detection in Batavia fabric. All experiments were conducted using a publicly available dataset that was previously benchmarked [2] using DenseNet121, which yielded the highest AU-ROC score among the tested architectures [1]. Building on this prior work, the objective is to analyze the extent to which specific augmentation strategies influence the generalization capability and robustness of the model in the context of automated textile inspection.

The remainder of this paper is organized as follows: Sect. 2 presents the dataset used in this study. Section 3 describes the methods and experimental design. Section 4 reports the results obtained, and Sect. 5 summarizes the main conclusions and outlines the directions for future work.

2 Dataset

A public dataset for defect detection in textile manufacturing was used in this study. Collected in November 2022 at Textil Santanderina, the dataset contains high-resolution images of Batavia and Sarga fabrics. These images were processed (downscaled from 16-bit to 8-bit, cropped, and classified into cases and controls) [2].

The dataset used in this study focuses exclusively on the Batavia fabric and comprises 28,693 images of size 365 × 365 pixels. Among these, 19,911 images corresponded to controls, whereas 8,782 images represented cases. The dataset was notably imbalanced, with control images accounting for approximately 69% of the total and case images making up the remaining 31%.

In Fig. 1, two representative samples of Batavia fabric are shown. On the left is an image of a "case," which corresponds to a sample with defects in the fabric, while on the right is an image of a "control," which represents a sample without defects.

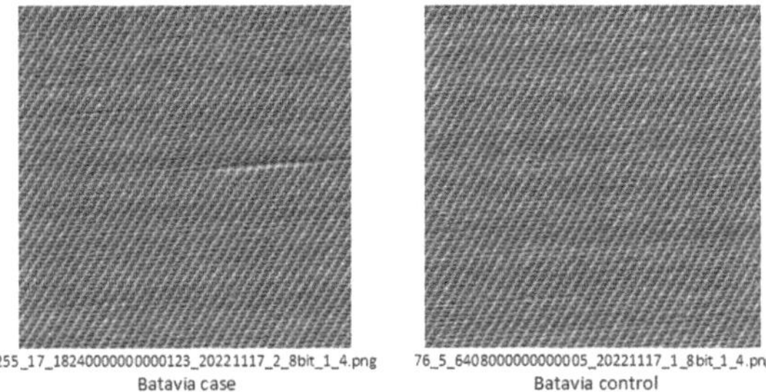

Fig. 1. Representative Batavia fabric samples. Left: case. Right: control.

3 Methods and Experimentation

This section describes the methods and experimental procedures employed to develop and validate a deep learning-based model for defect detection in Batavia fabrics. This approach includes the selection of the architecture, data augmentation techniques, stratified cross-validation, and evaluation metrics tailored to a binary classification problem with imbalanced classes.

3.1 Model Architecture

DenseNet121, a convolutional neural network introduced by Huang et al. [5], was adopted as the base architecture for transfer learning in this study. The model consisted of 121 layers. The key innovation lies in the implementation of dense connectivity, where each layer receives the feature maps of all the preceding

layers and transmits its own feature maps to all the subsequent layers. This design mitigates the vanishing gradient problem, enhances feature reuse and propagation, and significantly reduces the number of parameters compared with traditional deep networks of similar depth.
The input to each layer is the concatenation of all previous feature maps, as described by the following equation:

$$x_\ell = H_\ell([x_0, x_1, \ldots, x_{\ell-1}]) \quad (1)$$

where $[x_0, x_1, \ldots, x_{\ell-1}]$ represents the concatenation of the feature maps produced by all preceding layers, and H_ℓ is the composite function applied to this input, consisting of batch normalization (BN), rectified linear unit (ReLU) activation, and convolutional operations.

The implementation in this study was based on the DenseNet121 architecture, where the model's parameters were fine-tuned using transfer learning on a custom dataset. The architecture includes batch normalization layers, Rectified Linear Unit (ReLU) activation functions, and a global average pooling layer, which contribute to improved training stability and reduced overfitting. For this study, the classifier head was customized with two fully connected layers of 8192 units each, using ReLU activations. Dropout regularization was not applied, given the robustness of the network and the size of the training dataset.

Training was performed using adam model with an initial learning rate of 0.002, with a decay factor of 0.975 applied every 645 steps. The network was trained for a maximum of 100 epochs, with early stopping enabled when the AU-ROC metric failed to improve over 10 consecutive epochs. A batch size of 128 was used consistently across the training, validation, and test phases.

To illustrate the evolution of training and validation performance, Fig. 2 is included as an example. It shows the evolution of the AUROC metric for DA 1 + DA 2. These results correspond to one of the five folds used in cross-validation.

To improve model generalization and robustness, data augmentation was applied in real time (on the fly) to both the training and validation images. The transformations included random rotations, horizontal and vertical flipping, zooming, and spatial shifts. Classification thresholds were explored in the range [0.05, 0.95] with a step size of 0.025, allowing for a detailed performance analysis under varying decision boundaries.

3.2 Data Augmentation

In this study, several data augmentation techniques were applied to training images in real time (on the fly) to increase the diversity of the dataset and enhance the model's generalization capacity [12]. These transformations were chosen to simulate real world variations in fabric appearance, defect orientation, and lighting conditions while preserving the semantic structure of the images. The specific transformations applied were as follows:

- **Horizontal flipping**: This transformation reverses the image across its horizontal axis, introducing mirrored versions of the input. This enhances the

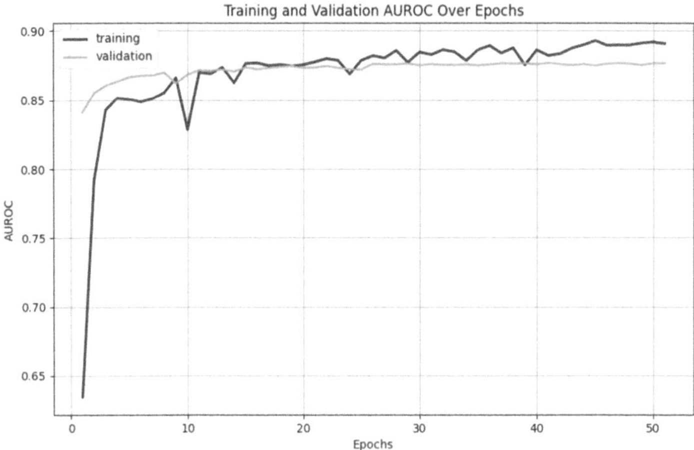

Fig. 2. Example of evolution of AUROC during training and validation using data augmentation strategy DA 1+2 in one of the five folds.

ability of the model to recognize defects regardless of their left-right orientation, which is especially advantageous in fabrics with repetitive or symmetrical patterns.

- **Vertical flipping**: By inverting the image across the vertical axis, this augmentation exposes the model to variations that may appear in top-down or bottom-up configurations. It is particularly effective for textile images with consistent structural patterns, improving the defect detection performance regardless of the orientation.
- **Random rotation (up to 90°)**: Introducing rotations simulates the arbitrary orientation of defects during acquisition. This enhances the rotational invariance of the model, which is especially important in cases where defects may not align with the primary axes of the fabric weave.
- **ZCA whitening**: Zero-phase Component Analysis (ZCA) whitening was used to reduce pixel-level correlations by linearly transforming the data while preserving spatial locality. This preprocessing method decorrelates features and can improve the convergence speed and performance of convolutional neural networks by making feature distributions more isotropic.
- **Brightness adjustment**: Random brightness variations were applied to simulate changes in illumination conditions, which can occur due to sensor variability or environmental factors during the fabric imaging. This helps the model become invariant to global lighting fluctuations.

Each technique was applied individually and in combination to the training set images during the training process on fly.

3.3 Evaluation Strategy

In this study, the DenseNet121 architecture, as detailed previously, was trained using a stratified 5-fold cross-validation approach. The dataset was divided into five non-overlapping subsets of equal size (5,738) to ensure that the class distribution remained consistent across all folds. Three boxes were used for training, one box for validation, and another for testing. The data were further split into 60% for training, 20% for validation, and 20% for testing. To reduce the bias introduced by a particular data partition and to enhance the robustness of the results, this process was repeated five times, rotating the folds used for training and testing in each round.

3.4 Performance Metrics

Given the significant class imbalance present in the dataset, this study evaluated a set of complementary performance metrics to provide a more comprehensive understanding of model effectiveness. In practical quality control applications, interest may vary depending on whether the objective is to improve the identification of defect-free samples (negative class), defects (positive class), or both. The selected metrics are based on the fundamental classification outcomes: True Positives (TP), True Negatives (TN), False Positives (FP), and False Negatives (FN).

- **Accuracy**: This metric measures the proportion of correct predictions among all evaluated samples, incorporating TP, TN, FP, and FN. However, in the context of imbalanced datasets, the accuracy can be misleading because a model that consistently predicts the majority class may still achieve a deceptively high score.
- **Precision**: Precision quantifies the ratio of true positives to all samples predicted as positive. This reflects the reliability of positive predictions and is especially relevant in domains where false positives carry high costs.
- **Recall**: Also referred to as sensitivity or true positive rate, recall assesses the proportion of actual positive cases correctly identified by the model. This is crucial in scenarios where missing a positive instance (false negative) is costly, such as anomaly detection.
- **F1-score**: The F1-score provides a single measure of a model's accuracy by computing the harmonic mean of precision and recall. This is particularly useful when a balance between precision and recall is desired.
- **Geometric Mean (G-mean)**: The G-mean evaluates the balance between sensitivity (recall for the positive class) and specificity (recall for the negative class). It is particularly informative for imbalanced datasets because it ensures that both classes are accurately represented.
- **Area Under the Precision-Recall Curve (AU-PR)**: AU-PR is a suitable metric for evaluating performance in imbalanced classification tasks, as it emphasizes the model's ability to correctly detect positive instances. It is derived from the area under the curve that plots the precision versus recall at

various threshold settings. A value closer to 1 indicates a favorable trade-off between precision and recall.
- **Area Under the ROC Curve (AU-ROC)**: AU-ROC is a threshold-independent metric that quantifies the model's ability to distinguish between positive and negative classes. It is based on the ROC curve, which graphs the true positive rate against the false positive rate across different thresholds. An AU-ROC of 1 signifies perfect separation, whereas 0.5 indicates performance equivalent to random guessing. This metric is often used to compare classification models across different applications.

4 Results

This section presents the results obtained, comparing the performance of the models trained with and without data augmentation techniques. The results are presented in the form of tables and graphs for easy interpretation.

The performance metrics of the models trained using different DA strategies are summarized in Table 1. These results provide a comprehensive comparison between models trained without data augmentation (without DA), those trained using individual DA techniques, and those trained using combinations of DA techniques. The DA strategies are described in detail in Sect. 3 and are referenced using the following labels, which are used in both Table 1 and Fig. 3:

- **Without DA**: No data augmentation techniques applied. Baseline model.
- **DA 1**: Horizontal flip.
- **DA 2**: Vertical flip.
- **DA 3**: Random rotation up to 90°.
- **DA 4**: ZCA Whitening.
- **DA 5**: Brightness adjustment.
- **DA ALL**: All data augmentation techniques applied (Horizontal flip, Vertical flip, Random rotation up to 90°, ZCA Whitening, and Brightness adjustment).
- **DA 1+2**: Combination of horizontal and vertical flips.

An additional experiment was conducted using the DA1+2 configuration, as the AU-ROC for both DA1 (Horizontal flip) and DA2 (Vertical flip) was slightly higher than the baseline model.

The comparison presented in Table 1 illustrates how each DA configuration impacts the model performance and training efficiency. In the table, the best-performing metrics are highlighted in bold, providing a clear indication of the best results for each metric across the different DA techniques evaluated.

Upon reviewing the results, certain data augmentation techniques demonstrated slight improvements over the baseline model. Techniques such as horizontal and vertical flipping (DA 1, DA 2, DA 1+2) and brightness adjustment (DA 5) show small but consistent positive effects on key performance metrics. For

Table 1. Performance metrics comparison for different types of data augmentation

DA Type	Threshold	Accuracy	Precision	Recall	F1-Score	G-mean	AU-PR	AU-ROC	Time (min)
Without DA	0.600	0.81832	0.69071	**0.73594**	0.71261	0.79308	0.81824	0.88200	23.60
DA 1	0.525	0.81989	0.70185	0.71544	0.70858	0.78711	0.81875	0.88203	36.99
DA 2	0.650	0.81689	0.70151	0.69927	0.70039	0.77942	0.81742	0.88202	33.29
DA 3	0.300	0.80382	0.66527	0.72261	0.69276	0.77893	0.79053	0.86016	116.58
DA 4	0.450	0.82226	0.70041	0.73263	0.71616	**0.79459**	0.81715	0.88124	**22.76**
DA 5	0.575	**0.82633**	0.71331	0.72330	**0.71827**	0.79408	0.81793	0.88148	34.80
DA ALL	0.325	0.77608	0.62563	0.66830	0.64626	0.74190	0.79961	0.86640	114.77
DA 1+2	0.675	0.82372	**0.71906**	0.69597	0.70733	0.78262	**0.81952**	**0.88348**	43.90

instance, DA 1+2, which combines both flips, results in a marginally higher AU-ROC (0.88348) and precision (0.71906), suggesting that these simple geometric transformations contribute to better generalization. This could be because flipping the images allows the model to learn invariance to horizontal and vertical orientations, which is particularly useful for textile images that may appear in various orientations. Similarly, DA 5, which adjusts brightness, achieves a slight increase in accuracy (0.82633) and F1-score (0.71827), which may be beneficial for simulating lighting variations typically encountered in real-world industrial settings. By adjusting the brightness, the model becomes more robust to different lighting conditions, making it better at recognizing features under varying illumination. DA 4, which applies ZCA Whitening, also shows a modest improvement by achieving the highest G-mean (0.79459) while maintaining relatively short training times (22.76 min). This suggests that whitening helps reduce correlations between pixel values, making the features more uniform and aiding in model convergence, without adding significant computational cost.

In contrast, more complex augmentation strategies, such as random rotation (DA 3) and the combination of all techniques (DA ALL), seem to reduce the model's performance. DA 3, which introduces random rotations, results in a slight decline in performance and the longest execution time (116.58 min). This could be due to rotations distorting the textile patterns in ways that do not represent natural variations, leading the model to learn irrelevant features or introduce noise. Similarly, DA ALL, which applies all augmentations simultaneously, increases the processing time and reduces the performance, particularly in terms of AU-ROC. This suggests that combining multiple augmentations may overcomplicate the data, making it more difficult for the model to discern meaningful patterns. Although combining transformations might be useful in some contexts, in this case, it appears to introduce unnecessary complexity that negatively impacts performance. Therefore, based on these results, it is evident that the best-performing augmentations are simpler ones, such as flipping and brightness adjustments, which enhance generalization without disrupting the natural structure of the textile images.

The execution time and performance are not always aligned. Some transformations that require more computational time (such as DA 3 and DA ALL) do not exhibit improvements in the performance metrics. This suggests that it is not always beneficial to invest more computational time in techniques that do not add predictive value.

Figure 3 provides a visual representation of the performance metrics for each DA configuration, allowing for a direct and clear comparison of the results. It can be observed that the DA ALL configuration has the lowest metrics compared to the other data augmentation techniques. These results indicate that, overall, the application of DA ALL does not improve the model's performance compared with other augmentation configurations and, in some cases, even reduces the model's ability to discriminate between classes. This suggests that applying multiple transformations simultaneously is not always beneficial, and further research is required to explore the specific combinations and appropriate intensities for each augmentation technique.

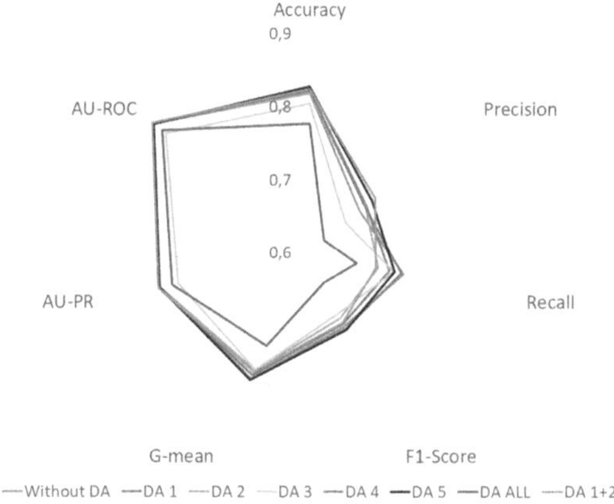

Fig. 3. Radar plot illustrating the performance of the models trained with and without data augmentation techniques on the Batavia weave dataset. The plot compares the models across key metrics.

AU-ROC remains a robust metric against most simple transformations. Except for DA 3 and DA ALL, AU-ROC remains nearly stable at approximately 0.882, indicating that the model continues to discriminate well between classes in most cases, even after applying data augmentation techniques.

5 Conclusions and Future Work

This study has examined the impact of various data augmentation (DA) techniques on the performance of a DenseNet121 convolutional neural network applied to a binary classification task involving Batavia textile images. The dataset consists of grayscale images of textile fragments labeled as defective or non-defective. The augmentation techniques evaluated included horizontal and vertical flipping, rotation, ZCA whitening, and brightness adjustment, both individually and in combination, to enhance model generalization and robustness.

The experimental results revealed that not all data augmentation strategies led to performance improvements. While some configurations, such as DA 1+2 (horizontal and vertical flipping), produced modest gains in AU-ROC and Precision (0.88348 and 0.71906, respectively), others, such as DA 3 (rotation) and DA ALL (all techniques combined), degraded performance and significantly increased execution time. Brightness adjustment (DA 5) was particularly notable for its relevance in simulating realistic lighting variations in grayscale textile images, offering promising improvements in some metrics without penalizing the runtime. Given that the data capture is in black-and-white (BN), and considering the considerable variations in brightness across different image fragments, DA 5 is an encouraging finding. It can simulate changes in tissue luminosity or color tone, which are typically detected as shifts in grayscale but are common in industrial applications. This feature of DA 5 provides valuable insights that can be leveraged to enrich the overall analysis.

These findings highlight that simplistic or indiscriminate application of data augmentation can be counterproductive. One of the key insights from this study is the need to employ domain-specific data augmentation techniques. Although traditional methods such as rotation, flipping, and scaling are commonly used, they may not fully capture the complex deformation characteristics of fabric defects. The results suggest that the Batavia dataset already contains a high degree of intrinsic variability, stemming from fabric textures, patterns, and lighting conditions, which provides sufficient generalization capability for the model. Consequently, augmentations like rotation, which disrupt the natural orientation of textile structures, may introduce noise rather than useful diversity. Therefore, careful selection and tuning of DA strategies are essential, particularly when working with domain-specific image data.

Future work will explore domain-specific data augmentation techniques, including sequential transformations and GAN-based synthetic data. While this study focuses on DenseNet121 to isolate the effect of data augmentation, future research could investigate alternative or lightweight models and assess generalizability across different datasets to broaden applicability.

References

1. Gil-Arroyo, B., Velasco-Pérez, N., Sanz, J., Casas, A., Arroyo, A., Urda, D.: Defect detection in batavia and sarga woven fabrics by means of convolutional neural networks. J. Appl. Logics (2025), Accepted for publication
2. Gil-Arroyo, B., Sanz, J.M., Arroyo, Á., Urda, D., Basurto, N., Herrero, Á.: Dataset for defect detection in textile manufacturing. Data Brief **59**, 111451 (2025). https://doi.org/10.1016/j.dib.2025.111451
3. Goodfellow, I., Bengio, Y., Courville, A.: Deep Learning. MIT Press, Cambridge, MA (2016). https://www.deeplearningbook.org/
4. Gu, J., et al.: Recent advances in convolutional neural networks. Pattern Recogn. **77**, 354–377 (2018)

5. Huang, G., Liu, Z., Van Der Maaten, L., Weinberger, K.Q.: Densely connected convolutional networks. In: 2017 IEEE Conference on Computer Vision and Pattern Recognition (CVPR), pp. 2261–2269 (2017). https://doi.org/10.1109/CVPR.2017.243
6. Eunice, J., Popescu, D.E., Chowdary, M.K., Hemanth, J.: Deep learning-based leaf disease detection in crops using images for agricultural applications. Agronomy **12**(10), 2395 (2022). https://doi.org/10.3390/agronomy12102395
7. Mao, Y., Kim, J., Podina, L., Kohandel, M.: Dilated se-densenet for brain tumor mri classification. Sci. Rep. **15**(1) (2025). https://doi.org/10.1038/s41598-025-86752-y
8. Mishra, S., Pandey, D., Yaduvanshi, R., Pandey, A.K., Verma, S., Rajpoot, P.: An integrated deep-learning model for smart waste classification. Environ. Monit. Assess. **196**(3) (2024). https://doi.org/10.1007/s10661-024-12410-x
9. Ongbali Samson, O., Afolalu Sunday, A., Salawu Enesi, Y.: Bottleneck problems arising in inter-industry production setting and vertical integration: a review. Technology **10**(5), 606–612 (2019)
10. Saberironaghi, A., Ren, J., El-Gindy, M.: Defect detection methods for industrial products using deep learning techniques: a review. Algorithms **16**(2) (2023)
11. Shanmugamani, R.: Deep Learning for Computer Vision. Packt Publishing, Birmingham, UK (2018)
12. Shorten, C., Khoshgoftaar, T.M.: A survey on image data augmentation for deep learning. J. Big Data **6**(1), 1–48 (2019). https://doi.org/10.1186/s40537-019-0197-0
13. Tan, L., Fu, Q., Li, J.: An improved neural network model based on densenet for fabric texture recognition. Sensors (Basel, Switzerland) **24**(23), 7758 (2024). https://doi.org/10.3390/s24237758
14. Tayeh, T., Myers, R., Aburakhia, S., Shami, A.: Distance-based anomaly detection for industrial surfaces using triplet networks. In: 2020 IEEE International Conference on Engineering and Emerging Technologies (IEMCON), pp. 372–377. IEEE (2020). https://doi.org/10.48550/arXiv.2011.04121
15. Um, T.T., et al.: Data augmentation of wearable sensor data for parkinson's disease monitoring using convolutional neural networks. In: Proceedings of the 19th ACM International Conference on Multimodal Interaction (ICMI), pp. 216–220. ACM (2017). https://doi.org/10.1145/3136755.3136817

Diffuse Radiation Forecasting Using a Hybrid Neural Network Architecture Integrating CNN, LSTM, and MLP Models

Gonzalo Surribas-Sayago[1,2], Jose David Fernández-Rodríguez[1,2](✉), and Enrique Domínguez[1,2]

[1] ITIS Software, University of Málaga, Málaga, Spain
{surribasg,enriqued}@uma.es

[2] Department of Computer Science, University of Málaga, 29071 Málaga, Spain
josedavid@uma.es

Abstract. Accurate forecasting of diffuse solar radiation is essential for improving photovoltaic (PV) energy predictions, particularly under variable atmospheric conditions. This capability is especially relevant for optimizing the scheduling of green hydrogen production, which depends on surplus renewable energy. In this study, we propose a deep learning framework based on data from the Cabauw station (Netherlands), employing a hybrid architecture that integrates Convolutional Neural Networks (CNN), Long Short-Term Memory (LSTM) units, and a Multilayer Perceptron (MLP). The CNN layers extract local temporal patterns from lagged input sequences, while the LSTM layers capture longer-term dependencies. Temporal features such as Fourier-based components are included to represent daily and sub-daily cycles. The model is evaluated using standard metrics, including Mean Absolute Error (MAE) and Root Mean Squared Error (RMSE), and is compared with traditional statistical and machine learning baselines. Results demonstrate that our approach significantly enhances the prediction of diffuse radiation, ultimately improving PV forecasting and enabling more efficient green hydrogen production.

Keywords: diffuse radiation · deep learning · CNN · LSTM · photovoltaic forecasting · green hydrogen

1 Introduction

The transition to renewable energy sources, such as photovoltaic (PV) power, is critical to achieving global sustainability goals and reducing greenhouse gas emissions. In this context, green hydrogen production through electrolysis has emerged as a promising solution to decarbonize sectors like transport and industry [18], relying heavily on the availability of surplus renewable electricity and the

E. Corchado et al. (Eds.): SOCO 2025, CCIS 2806, pp. 466–475, 2026.
https://doi.org/10.1007/978-3-032-19763-4_43

integration of renewable energy systems [9]. However, PV generation is strongly influenced by atmospheric variability, making it difficult to predict accurately.

Although total solar irradiance is typically used in PV forecasting models, diffuse radiation plays a key role under cloudy or partially shaded conditions. Prior studies have explored machine learning methods to forecast this component, including artificial neural networks and random forests [1,2,16]. Accurately modeling diffuse radiation can significantly improve PV power estimation and enable more efficient hydrogen scheduling [4].

However, direct measurements of diffuse radiation are not widely available due to the high cost and complexity of the required instruments. As a result, estimating diffuse radiation through modeling becomes essential for the design and planning of solar energy systems [11].

Traditional time series methods, such as persistence models and SARIMA [3], often fall short in capturing the nonlinear dynamics of solar data [13]. To address these limitations, we propose a hybrid deep learning framework trained on high-quality data from the Cabauw station (Netherlands). Our model combines Convolutional Neural Networks (CNN), Long Short-Term Memory (LSTM) units [15], and Multilayer Perceptrons (MLP), optimized for temporal modeling. Recent research also supports LSTM–BiGRU combinations in this context [14,17].

To enhance temporal awareness, we incorporate Fourier-based features encoding daily and sub-daily cycles [6,10]. The model's performance is evaluated using Mean Absolute Error (MAE) and Root Mean Squared Error (RMSE), and benchmarked against traditional statistical and machine learning baselines. The results highlight the potential of deep learning for improving PV forecasting and supporting green hydrogen optimization.

2 Materials and Methods

2.1 Dataset Description

The dataset used in this study was collected from the Cabauw Experimental Site for Atmospheric Research (CESAR), located in the Netherlands. This station is part of the Baseline Surface Radiation Network (BSRN) and is known for the high quality and frequency of its atmospheric measurements [8]. The dataset includes one-minute resolution observations of diffuse radiation (DIF [W/m^2]) along with key meteorological variables such as temperature, relative humidity, and atmospheric pressure. The analysis focuses on the winter period of December 2023, which presents high atmospheric variability and lower solar elevation, making it ideal for testing forecasting models under challenging conditions.

2.2 Feature Engineering

To model the temporal behavior of diffuse radiation, we constructed a set of features that included:

- **Lags:** Past values of DIF up to 48 h prior to each prediction point, enabling the model to learn short-term dependencies over a longer temporal horizon.

- **Exogenous variables:** Relative humidity, temperature, and pressure values at the current and lagged time steps.
- **Fourier components:** Sinusoidal and cosinusoidal features to capture daily and sub-daily periodicity. These were generated from the time index to preserve temporal cyclic patterns.

All features were scaled individually to the [0, 1] range using MinMaxScaler from scikit-learn. Each feature was normalized based on its own minimum and maximum values. This prevents any feature from dominating due to scale differences.

2.3 Baseline Models

To evaluate the effectiveness of our proposed model, we selected three widely used forecasting approaches from the literature as baselines:

- **SARIMAX:** A classical time series model that captures both seasonality and external influences through exogenous variables. It is commonly used for solar radiation forecasting due to its interpretability and solid statistical foundation [3].
- **XGBoost:** A state-of-the-art machine learning algorithm based on gradient boosting, known for its strong performance in regression tasks and its ability to model nonlinear relationships [5].
- **Stacked LSTM:** A deep learning approach that uses layers of Long Short-Term Memory (LSTM) units to capture complex temporal dependencies in time series data. It has shown success in renewable energy forecasting problems [7].

2.4 Forecasting Model Architecture

The proposed architecture is a hybrid deep learning model specifically designed to capture both short- and long-term temporal dependencies in diffuse radiation time series. This architecture integrates three main components: Convolutional Neural Networks (CNN), Long Short-Term Memory units (LSTM), and a Multilayer Perceptron (MLP), each contributing complementary mechanisms to model complex temporal patterns (Fig. 1).

Input Structure and Feature Design: the model receives a multivariate time series as input, composed of 72 historical time steps (lookback window) and 24 features. These include lagged values of the target variable (diffuse radiation), exogenous meteorological variables (temperature, relative humidity, and pressure), and temporal features derived from Fourier series [6,10]. Additionally, 15 auxiliary features are processed through a separate branch before feature fusion. All features were normalized using Min-Max scaling to ensure consistent ranges and training stability.

CNN Layers: the convolutional component is responsible for extracting local patterns from the temporal history. Specifically, one-dimensional convolutional layers are applied along the temporal axis to identify short-term dependencies

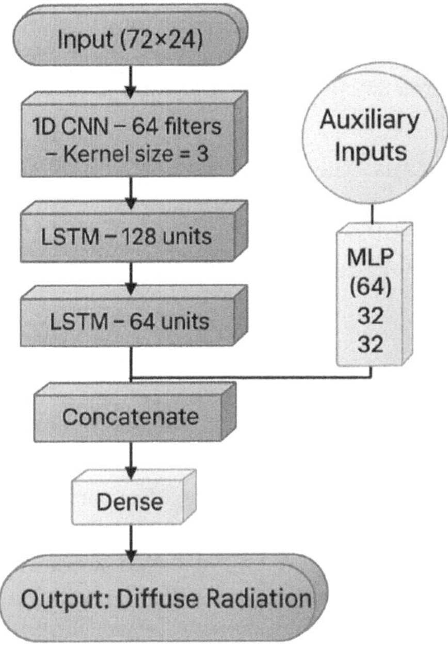

Fig. 1. Architecture of the hybrid CNN + LSTM + MLP model used for diffuse radiation forecasting.

and abrupt variations, such as those caused by cloud movements. Convolutional filters learn invariant representations, and max-pooling layers are used to reduce dimensionality while retaining relevant signals [12].

LSTM Layers: the CNN output is then passed to a stack of LSTM layers. These gated units can retain information over time and capture long-term dependencies while avoiding the vanishing gradient problem inherent in simple recurrent networks [15].

This work employs a two-layer LSTM structure with 128 and 64 units, respectively. Dropout is used between layers for regularization, and the first LSTM layer is configured with `return_sequences=True` to maintain the temporal dimension.

Auxiliary Branch and Feature Fusion: in parallel to the main branch, the 15 auxiliary features (including Fourier components) are processed through a small MLP composed of two dense layers with 64 and 32 neurons. This branch captures non-linear interactions among higher-level temporal signals and cyclic patterns that may not be easily captured in the sequential input.

The outputs of the LSTM stack and the auxiliary MLP are concatenated and passed through a final fully connected dense layer with a single neuron that generates the predicted diffuse radiation value for the next time step.

Training Strategy: the model was trained using the Adam optimizer with an initial learning rate of 0.005. The Mean Squared Error (MSE) loss function was selected for its ability to penalize large errors, which is desirable in solar radiation forecasting tasks.

To prevent overfitting and enhance generalization, early stopping was applied with a patience of 10 epochs, and batch normalization was used after each dense layer. The batch size was set to 32, and the model was trained for a maximum of 100 epochs.

Implementation Details: the model was fully implemented in TensorFlow using the Keras API. Training was performed on a workstation with an Intel Core i5 processor and 20 GB of RAM. Despite the lack of GPU acceleration, training times remained manageable due to the efficient model design and preprocessing pipeline.

3 Experiments and Results

3.1 Experimental Setup

The experiments were conducted to assess the performance of the proposed hybrid deep learning model, comparing it against traditional forecasting methods like SARIMAX, XGBoost, and LSTM-based models. The dataset consists of one-minute resolution observations of diffuse radiation (DIF [W/m^2]) and key meteorological variables (temperature, relative humidity, atmospheric pressure) collected from the Cabauw Experimental Site for Atmospheric Research (CESAR) in the Netherlands. The study focuses on the winter period of December 2023, which presents challenging atmospheric conditions for radiation forecasting due to high variability and lower solar elevation.

Training and Testing Split: The dataset was split into training and validation sets. The training data included the first 80% of the data, while the remaining 20% was used for testing.

Evaluation Metrics: The models were evaluated using two standard metrics: Mean Absolute Error (MAE) and Root Mean Squared Error (RMSE). These metrics provide insights into both the average error magnitude (MAE) and the penalty for large errors (RMSE), which are critical for assessing the performance in forecasting.

Model Training Specifications:

- **Hybrid CNN + LSTM + MLP:** Trained using 72 time steps as input, with 24 features (including lags and temporal variables), and 15 auxiliary features. The model was trained for 100 epochs with a batch size of 32, using the Adam optimizer with a learning rate of 0.005.
- **SARIMAX:** Tuned with seasonal parameters (0,1,1,24) and autoregressive components.
- **XGBoost:** Optimized using GridSearchCV with 24 lags and 3 exogenous variables.
- **LSTM Stacked:** Trained with 24 lags and all exogenous variables.

3.2 Model Comparison

To compare the performance of the proposed hybrid model against traditional and widely used forecasting approaches, each model was evaluated using its best configuration, selected through empirical tuning. The same training and validation sets were used across all experiments to ensure consistency.

SARIMAX: The SARIMAX model was configured with optimal seasonal and non-seasonal orders determined via grid search and AIC minimization. The final parameters were: $(p = 2, d = 1, q = 2)$ with seasonal order $(P = 1, D = 1, Q = 1, s = 24)$, and exogenous inputs including temperature, humidity, and pressure [3].

XGBoost: The XGBoost regressor achieved its best performance with the following hyperparameters: 300 estimators, a learning rate of 0.05, maximum tree depth of 5, and a subsample ratio of 0.8. It was trained on lagged features, exogenous variables, and Fourier-transformed temporal components, all of which were normalized [5].

Stacked LSTM: This deep learning model used two LSTM layers with 100 and 50 units respectively. ReLU activation functions and a dropout rate of 0.2 were applied, along with the Adam optimizer (learning rate = 0.001). Early stopping was used with a patience of 10 epochs [7].

Hybrid CNN + LSTM + MLP: The final hybrid model included a convolutional layer (64 filters, kernel size = 3), followed by max pooling and a 64-unit LSTM. An auxiliary MLP processed 15 engineered features through two dense layers (64 and 32 units). The two branches were merged and passed through a final dense layer. Training was conducted using the Adam optimizer (learning rate = 0.005, clipnorm = 1.0), batch size = 32, and early stopping based on validation loss.

Each model was evaluated using RMSE and MAE metrics on the validation dataset. Detailed results and visual comparisons are provided in the following section.

3.3 Results

Table 1 summarizes the performance of the different models. The hybrid CNN + LSTM + MLP with an aggressive learning rate (`lr = 0.005`) achieved the best results, outperforming all other models in both RMSE and MAE. The model captures both short- and long-term temporal patterns, while also leveraging auxiliary inputs such as relative humidity, temperature, pressure, and temporal Fourier components. Table 2 summarizes the best-performing model for each category.

The results show clear performance differences among the evaluated approaches for diffuse radiation forecasting. The hybrid CNN + LSTM + MLP model consistently outperformed traditional methods and other deep learning variants, indicating that combining multiple neural architectures is particularly effective in this domain.

Table 1. Comparison of forecasting models by type. Metrics computed on the validation set.

Model	Description	RMSE	MAE
SARIMAX			
SARIMAX	Optimized: order=(3,1,3), seasonal_order=(0,1,1,24)	7.4336	4.5099
XGBoost			
XGBoost Base	Without tuning, only lags + exogenous features	12.8603	5.8267
XGBoost + EarlyStopping	1000 trees, early stopping applied	12.8096	5.6610
XGBoost Optimized	With GridSearchCV tuning	12.7393	5.7150
LSTM and Stacked LSTM			
LSTM Base	50 units, no early stopping	14.4947	7.8444
LSTM + EarlyStopping	Same as base with early stopping	14.9092	7.9996
Stacked LSTM (100 + 50)	Two LSTM layers	14.0820	7.3625
LSTM (100 units + EarlyStopping)	Wider model	14.2978	7.2075
LSTM Stacked + All Exogenous	24 lags + all exogenous variables	13.6373	6.4148
Hybrid CNN + LSTM + MLP			
CNN + LSTM + MLP(base, lr = 0.0004)	Simple hybrid architecture	**5.2207**	**3.1886**
CNN + LSTM + MLP(adjusted, lr = 0.0006)	Intermediate tuned version	6.0124	3.6792
CNN + LSTM + MLP(lr = 0.005)	Final tuned hybrid model	5.2918	3.1974

Table 2. Best Models per Category.

Model	Category	RMSE	MAE
CNN + LSTM + MLP (lr = 0.0004)	Hybrid	**5.2207**	**3.1886**
SARIMAX	Statistical	7.4336	4.5099
XGBoost (GridSearchCV)	Ensemble	12.7393	5.7150
LSTM Stacked + All Exogenous	Deep Learning	13.6373	6.4148

The SARIMAX model, while grounded in solid statistical theory, exhibited limitations in handling abrupt changes and nonlinear dynamics, especially under variable winter weather conditions. This outcome is consistent with prior findings that ARIMA-based models perform best under regular seasonal trends and linear processes [3].

XGBoost, despite its strong performance in general regression tasks [5], lagged behind the hybrid model. Its tree-based structure cannot explicitly model sequential dependencies, and its performance depends heavily on manual feature engineering. In contrast, deep neural networks can learn representations from raw temporal inputs more effectively.

LSTM-based models performed moderately well and were superior to traditional statistical approaches. However, they still failed to match the accuracy of the hybrid model. This is likely due to the absence of convolutional filters, which are essential for detecting local and abrupt temporal patterns, and the lack of auxiliary nonlinear processing of cyclic features.

The hybrid CNN + LSTM + MLP model owes its superior accuracy to its architectural synergy. The CNN layers capture short-term fluctuations, LSTM units retain temporal memory across sequences [7], and the MLP branch incorporates engineered features, including Fourier components [6,10], to represent cyclic behavior and exogenous effects. This combination enables the model to generalize effectively across different meteorological regimes and time scales.

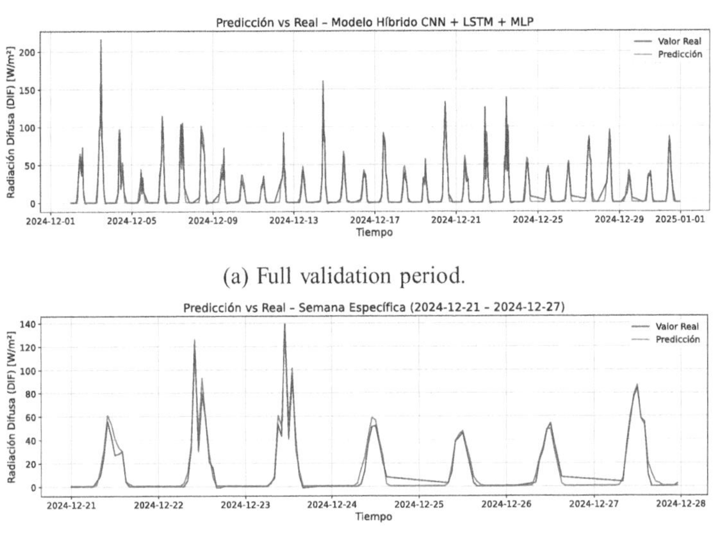

(a) Full validation period.

(b) Specific week (Dec 21–27).

Fig. 2. Predicted vs real diffuse radiation using the hybrid CNN + LSTM + MLP model.

These findings support the adoption of hybrid deep learning frameworks for complex time series forecasting, especially in energy-related applications where physical variability and external dependencies are significant.

These structural advantages, combined with effective training (using a high learning rate and regularization), allow the model to better generalize and adapt to the challenging meteorological conditions of the winter period in the Netherlands.

Figure 2 illustrates how the hybrid model closely matches the real data, confirming its ability to adapt to seasonal variability.

SARIMAX, while effective in modeling seasonality and trends, is limited by its linear assumptions and inability to leverage nonlinear exogenous interactions. XGBoost performed better than the LSTM variants but still lagged behind the hybrid model, highlighting the importance of sequential memory in time series forecasting.

Pure LSTM models, even with stacked configurations, failed to reach the same level of accuracy, underlining the added value of combining CNN and MLP structures, as well as incorporating temporal cyclic components.

4 Conclusions

This study introduces a novel hybrid deep learning architecture combining CNN, LSTM, and MLP components for accurate short-term forecasting of diffuse solar radiation. Using high-resolution atmospheric data from the Cabauw station in the Netherlands, the proposed model demonstrates superior performance over traditional statistical (SARIMAX), machine learning (XGBoost), and deep learning (Stacked LSTM) baselines in both RMSE and MAE metrics.

The hybrid model effectively captures complex temporal dependencies by integrating lagged diffuse radiation values, key meteorological variables, and periodic Fourier-based features. The CNN layers extract local temporal features, the LSTM units encode memory across time steps, and the MLP branches incorporate nonlinear interactions and cyclic temporal behavior. This architectural synergy enables better generalization, particularly under challenging winter conditions characterized by high variability in solar irradiance.

The results highlight the practical relevance of accurate diffuse radiation forecasting for the optimization of photovoltaic energy generation and green hydrogen scheduling. In scenarios where solar irradiance fluctuates due to cloud coverage or low solar elevation, improved predictions of the diffuse component enable more efficient use of renewable resources and enhance the reliability of grid-integrated energy systems.

Beyond its forecasting performance, the modularity and scalability of the proposed architecture offer a strong foundation for future extensions. These may include incorporating satellite-derived features, using attention mechanisms for long-range temporal reasoning, or adapting the framework to other forecasting domains within renewable energy and climate modeling.

In conclusion, this work reinforces the effectiveness of hybrid neural network models for complex time series forecasting tasks and contributes a robust and transferable framework for intelligent energy systems operating under dynamic meteorological conditions.

References

1. Almonacid, F., Pérez-Higueras, P., Fernández, E.F., Hontoria, L.: A methodology based on dynamic artificial neural network for short-term forecasting of the power output of a pv generator. Energy Convers. Manage. **85**, 389–398 (2014). https://doi.org/10.1016/j.enconman.2014.05.061
2. Benali, L., Notton, G., Fouilloy, A., Voyant, C., Dizene, R.: Solar radiation forecasting using artificial neural network and random forest methods: application to normal beam, horizontal diffuse and global components. Renew. Energy **132**, 871–884 (2019)
3. Box, G.E.P., Jenkins, G.M., Reinsel, G.C., Ljung, G.M.: Time Series Analysis: Forecasting and Control. John Wiley & Sons, 5th edn. (2015). https://doi.org/10.1002/9781118675021
4. Caramizaru, A., Uihlein, A.: Hydrogen from renewable resources: an overview of the potential in the eu. Renew. Sustain. Energy Rev. **120**, 109656 (2020). https://doi.org/10.1016/j.rser.2019.109656

5. Chen, T., Guestrin, C.: Xgboost: a scalable tree boosting system. In: Proceedings of the 22nd ACM SIGKDD International Conference on Knowledge Discovery and Data Mining, pp. 785–794 (2016). https://doi.org/10.1145/2939672.2939785
6. Cheng, X., Zhao, T., Ma, X., Tian, H., Li, K.: A fourier series-based deep learning approach for solar radiation forecasting. Energy Convers. Manage. **213**, 112837 (2020). https://doi.org/10.1016/j.enconman.2020.112837
7. Hochreiter, S., Schmidhuber, J.: Long short-term memory. Neural Comput. **9**(8), 1735–1780 (1997). https://doi.org/10.1162/neco.1997.9.8.1735
8. Knap, W.: Basic and other measurements of radiation at station cabauw (2024), https://www.pangaea.de/?q=BSRN+Cabauw+2024, Accessed July 2025
9. Liu, L., Zhai, R., Hu, Y.: Performance evaluation of wind-solar-hydrogen system for renewable energy generation and green hydrogen generation and storage: energy, exergy, economic, and enviroeconomic. Energy (Oxford) **276**, 127386– (2023). https://doi.org/10.1016/j.energy.2023.127386
10. Marks, R.J., Gravagne, I.A., Davis, J.M.: A generalized fourier transform and convolution on time scales. J. Math. Anal. Appl. **340**(2), 901–919 (2008). https://doi.org/10.1016/j.jmaa.2007.08.056
11. Mirzabe, A.H., Hajiahmad, A., Keyhani, A.: Assessment and categorization of empirical models for estimating monthly, daily, and hourly diffuse solar radiation: a case study of iran. Sustain. Energy Technol. Assess. **47**, 101330 (2021). https://doi.org/10.1016/j.seta.2021.101330
12. S., K.S., K., S., Satheesh, R.: Enhanced solar power forecasting with cnns using sky imagery and cloud detection. In: Proceedings of the 2024 International Conference on Advances in Computing and Communications (ICACC), IEEE (2024). https://doi.org/10.1109/ICACC63692.2024.10845330
13. Sivhugwana, K.S., Ranganai, E.: Intelligent techniques, harmonically coupled and sarima models in forecasting solar radiation data: a hybridisation approach. J. Energy Southern Africa **31**(3), 1–24 (2020). https://doi.org/10.17159/2413-3051/2020/v31i3a7754
14. Surribas-Sayago, G., Fernández-Rodríguez, J.D., Domínguez, E.: Advancing photovoltaic forecasting with neural networks: Integrating n-beats and sequential models with fourier analysis. In: Rodriguez, S., Paredes-Valverde, M., Ortega, A., Rodríguez-González, A., Rojek, I., Trawiński, B. (eds.) Iberamia 2024: Advances in Artificial Intelligence – 18th Ibero-American Conference. Lecture Notes in Computer Science, vol. 14460, pp. 312–325. Springer, Montevideo, Uruguay (2024). https://doi.org/10.1007/978-3-031-80366-6_24
15. Wang, Y., Chen, Y., Liu, H., Ma, X., Su, X., Liu, Q.: Day-ahead photovoltaic power forcasting using convolutional-lstm networks. In: 2021 3rd Asia Energy and Electrical Engineering Symposium (AEEES), pp. 917–921. IEEE (2021). https://doi.org/10.1109/AEEES51875.2021.9403023
16. Yona, A., Senjyu, T., Urasaki, N., Funabashi, T.: Application of neural network to 24-hour-ahead generating power forecasting for pv system. Sol. Energy **79**(3), 329–340 (2005). https://doi.org/10.1016/j.solener.2004.11.005
17. Zhang, Y., Hu, Q., Wang, J., Guo, C., Wang, F.: Hybrid deep learning model for photovoltaic power forecasting using lstm and bidirectional gru. Appl. Intell. (2024). https://doi.org/10.1007/s10489-024-05555-8
18. Zhao, H., Yuan, Z.Y.: Progress and perspectives for solar-driven water electrolysis to produce green hydrogen. Adv. Energy Mater. **13**(16), 2300254 (2023). https://doi.org/10.1002/aenm.202300254

Rotation-Invariant Deep Learning for 2D Detection: A Step Towards 3D LiDAR Object Recognition

Geovanny Satama Bermeo[1,2], Julian Estevez[1,3], Manuel Graña[1,4], Iñigo Aramendia[2], and Jose Manuel Lopez-Guede[1,2](✉)

[1] Department of System Engineering and Automation Control, Faculty of Engineering of Vitoria-Gasteiz, University of the Basque Country (UPV/EHU), Nieves Cano, 12, 01006 Vitoria-Gasteiz, Spain

[2] Faculty of Engineering of Vitoria-Gasteiz, University of the Basque Country (UPV/EHU), C/Nieves Cano 12, 01006 Vitoria-Gasteiz, Spain

jm.lopez@ehu.es

[3] Faculty of Engineering of Gipuzkoa, University of the Basque Country (UPV/EHU), Europa Plaza 1, 20018 San Sebastian, Spain

[4] Faculty of Computer Science, University of the Basque Country (UPV/EHU), Manuel Lardizabal Pasealekua 1, 20018 San Sebastian, Spain

Abstract. This work presents a methodology for 2D object detection and classification under extreme rotations, systematically evaluating pre-trained models with images rotated from 0° to 360° on X and Y axes. Lightweight architectures like YOLOX and YOLOv3 (Tiny) consistently outperform more complex models such as YOLOv8 and SSD-ResNet50 in detection. For classification, AlexNet and Inceptionv3 show superior rotational invariance over newer models like ResNet50, indicating that higher complexity does not guarantee geometric generalization. Rotational robustness varies notably between axes, with models showing distinct performance patterns and critical drops at specific angles. The pipeline stabilizes processing times quickly after initialization, supporting real-world deployment. These results emphasize the need for axis-specific evaluation in model selection and lay the groundwork for future 3D and LiDAR-based computer vision applications.

Keywords: Robust object detection · Rotation invariance · Deep learning

1 Introduction

Robust object detection under extreme variations in orientation and background remains a fundamental challenge in computer vision, particularly for advanced applications such as identifying urban furniture in three-dimensional environments using LiDAR point clouds, as illustrated in Fig. 1. In these scenarios, the orientation and arrangement of objects can vary unpredictably due to urban

E. Corchado et al. (Eds.): SOCO 2025, CCIS 2806, pp. 476–486, 2026.
https://doi.org/10.1007/978-3-032-19763-4_44

dynamics, sensor perspective, and partial occlusion, demanding perception systems capable of maintaining high levels of accuracy and reliability. Despite progress achieved with deep architectures such as YOLOv8 [3] and Faster R-CNN [11], several studies have reported that the performance of these models drops significantly when objects undergo rotations greater than 90°, limiting their applicability in real and safety-critical scenarios [4,12].

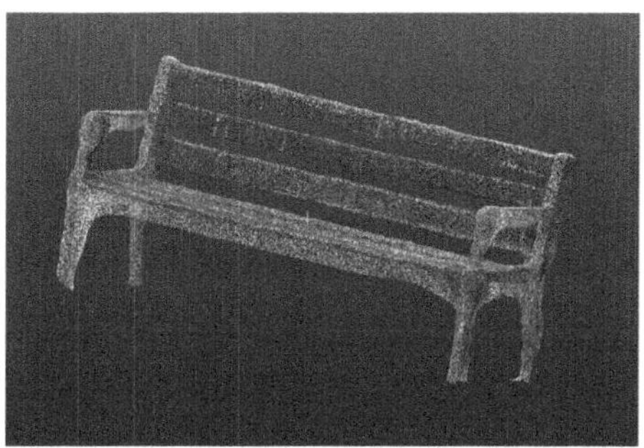

Fig. 1. LiDAR Point Cloud of an Urban Bench.

This work, which is part of an ongoing project, aims to establish the methodological foundations for robust 3D detection by systematically evaluating 2D models under controlled rotations. As a first step, everyday objects (e.g., mugs) are used to validate the methodology before scaling up to real cases of urban furniture in LiDAR, as depicted in Fig. 1. The significance of this approach lies in ensuring that future urban perception systems can maintain accuracy regardless of orientation, thereby contributing to safety and efficiency in applications such as traffic monitoring or agricultural automation [2,3].

Various strategies have been proposed to improve rotational invariance, including attention mechanisms and hybrid models that combine transformers and convolutional networks. The integration of saliency maps with Faster R-CNN has reduced false positives in complex scenarios [12], while the combination of transformers and CNNs has enhanced generalization in aerial images captured by UAVs [4]. However, challenges related to computational efficiency and scalability persist, especially on edge hardware or embedded systems [15].

Despite advancements, the scarcity of rotationally diverse datasets and the difficulty of transferring models trained in synthetic environments to real-world scenarios remain open challenges. Tools such as Simulink have facilitated the generation of synthetic datasets with controlled rotations [9], but domain adaptation still limits accuracy in real applications [8]. Furthermore, the potential of MATLAB for industrial integration and optimization remains underexplored in industrial integration and optimization contexts [15].

The remainder of this article is organized as follows: Sect. 2 reviews the related works in rotation-invariant 2D detection; Sect. 3 describes the methodology and experimental design; Sect. 4 presents and analyzes the results obtained;

Sect. 5 discusses their implications and compares with previous works; and Sect. 6 summarizes the main conclusions and suggests future research directions.

2 Related Work

Achieving robust 2D object detection under varying orientations and complex backgrounds remains a key challenge, despite advances in CNN architectures like YOLO and Faster R-CNN [11] that excel on standard datasets but struggle with angular variations [3,12]. Recent efforts focus on synthetic data augmentation, attention mechanisms, and CNN-transformer hybrids to enhance generalization and reduce false positives in demanding scenarios [4,9]. However, bridging the sim-to-real gap persists as a critical hurdle, particularly for industrial and aerial applications, as evidenced by emerging semi-supervised transfer learning approaches [8].

In parallel, the integration of attention modules, such as saliency-guided mechanisms, has shown notable improvements in reducing false positives and enhancing detection accuracy in environments saturated with visual information [12]. Additionally, the combination of transformers and CNNs has proven effective in aerial imagery, enabling more robust object recognition from unmanned aerial vehicles (UAVs) despite significant rotations and scale variations [4]. Nevertheless, these advances are often accompanied by increased computational requirements, which can limit their deployment on edge devices or real-time systems [15].

Recent studies have also explored the potential of domain transfer learning to mitigate the gap between synthetic and real data. For example, semi-supervised approaches based on YOLOv5 have demonstrated success in detecting SAR targets by leveraging limited annotated data and large-scale synthetic simulations [8]. Similarly, frameworks for face mask detection in UAVs using transfer learning have shown that pre-trained models can efficiently adapt to rotated objects in aerial images, provided they are fine-tuned with domain-specific augmentations [1]. These works underscore the importance of combining synthetic data generation with adaptive learning strategies to achieve robustness in real-world implementations.

The application of deep learning in specialized domains, such as agriculture and industrial automation, has underscored the need for rotation-invariant detection. For example, in detecting veins in clustered tobacco leaves, YOLOv8 with transfer learning achieved high accuracy under arbitrary orientations, highlighting the value of task-specific adaptations [3]. In traffic surveillance, combining radial basis function networks (RBF-FDLNN) with optical flow improved moving object detection in dynamic, variable backgrounds [2]. These advances suggest that hybrid approaches, merging geometric priors with data-driven learning, are essential for progress in rotation-sensitive object detection.

3 Methodology

The experimental evaluation of the proposed framework focused on exhaustively analyzing the performance of various detection and classification models under extreme rotation conditions in 2D images. Synthetic and real images were rotated from 0° to 360° along the X and Y axes, and multiple pre-trained architectures were evaluated in MATLAB. All tests used the same reference objects (mugs, chairs, etc.) with constant lighting and focal length to ensure fair comparison. Each model was assessed with default parameters, without rotation-specific adjustments, allowing evaluation of their intrinsic robustness to geometric transformations. Including architectures from different generations enabled analysis of the evolution of rotational invariance in computer vision and set a methodological precedent for future research.

The workflow was structured into four phases: model selection/analysis, data generation/augmentation, systematic evaluation, and comparative performance analysis. In the first phase, widely recognized pre-trained models were selected: YOLOv8, YOLOv9, YOLOX, Faster R-CNN, and SSD for detection; and VGG16, ResNet50, EfficientNet, AlexNet, and GoogLeNet for classification [3,11–14]. Each model was progressively evaluated, starting with images of everyday objects while varying angles and contexts, establishing baseline performance and identifying architectural strengths/limitations.

In the second phase, images were generated and augmented to enhance system robustness against geometric and contextual variability. Data augmentation included systematic rotations, reflections, scaling, and background perturbations, following recommendations from [6] to maximize dataset diversity and representativeness. This strategy facilitated consistency evaluation under adverse conditions.

The systematic evaluation phase analyzed the performance of pre-trained models without additional fine-tuning, assessing their out-of-the-box capacity to handle extreme rotations. Evaluation was conducted on both synthetic and real datasets using standard metrics (precision, recall, false positive rate) alongside angular robustness metrics measuring accuracy across rotation intervals [7,8]. This approach identified model behavior in edge-case scenarios and provided key insights for future improvements.

Performance evaluation quantified each architecture's rotational robustness by analyzing correctly detected/classified objects versus rotation angle. Processing time per iteration was also recorded for each model, revealing the framework's computational efficiency. The system's versatility was validated across application contexts, demonstrating its potential for detecting arbitrarily oriented objects in industrial and agricultural environments [3,5,6].

4 Results

The results section offers a rigorous comparative analysis of detection and classification models under extreme 2D rotation scenarios. All experiments were performed on a Windows 10 Enterprise LTSC system with Intel Xeon Gold 5120 @

2.20GHz processors (2 CPUs) and 96 GB RAM, without GPU acceleration. This setup highlights the framework's applicability in real-world, resource-constrained environments.

Figure 2 provides a comprehensive, continuous visualization of model performance across the full 0°–360° rotation range for both X and Y axes. The top panels reveal differences in angular robustness among detection architectures such as YOLOv3 (Tiny), YOLOX (small-coco), and SSD (ResNet50), while the central panels expose the binary accuracy patterns of classifiers like AlexNet and Inceptionv3, underscoring their sensitivity to rotation. The lower panel tracks processing time per rotation, which stabilizes near 2 s after an initial spike, confirming the computational efficiency of the system. This integrated layout enables immediate identification of critical failure zones, transition regions, and overall trends, establishing a robust foundation for the quantitative analysis that follows.

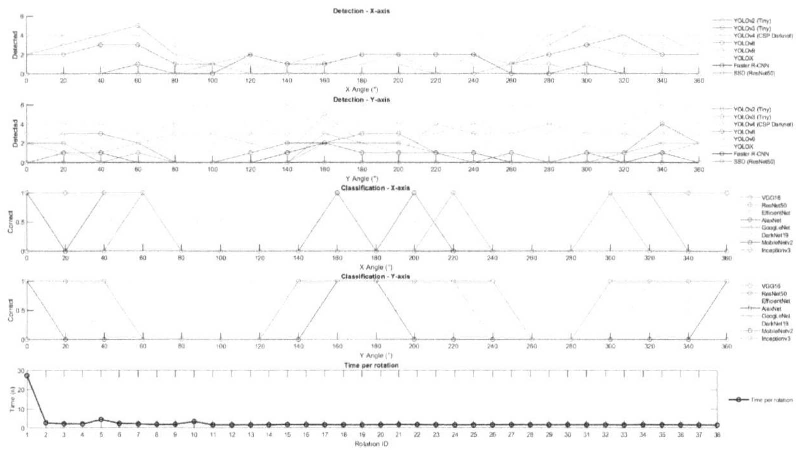

Fig. 2. Detection and Classification Comparison by Angle.

Building on the insights from Fig. 2, we present a focused quantitative evaluation to validate and expand upon the observed rotational patterns. The subsequent results tables report performance metrics at strategic 40-degree intervals—deliberately avoiding predictable fixed angles and blind spots—to facilitate direct, side-by-side architectural comparisons and precise assessment of rotational robustness along both axes. This dual approach, combining continuous visualization with targeted numerical data, ensures an exhaustive and transparent evaluation of each model's geometric invariance and highlights performance transitions that may not be immediately apparent from the graphical overview alone.

Table 1 summarizes detection performance under X-axis rotations sampled at 40-degree intervals. This interval selection captures both the stability and abrupt failures of each architecture across the full rotation cycle, avoiding redundancy at obvious or blind angles. YOLOX (small-coco) and YOLOv3 (Tiny) consistently outperform other models, but all architectures display pronounced drops between 80°–120° and near 240°, revealing persistent challenges in feature extraction under geometric stress. The elevated processing time at X = 0° (2.68 s) is due to model initialization, with subsequent iterations stabilizing at 2 s, confirming that computational efficiency is unaffected by rotation angle.

Table 1. Detection Model Performance under Rotations on the X-axis.

Rotation	YOLOv2 (Tiny)	YOLOv3 (Tiny)	YOLOv4 (CSP Darknet)	YOLOv8 (yolov8x)	YOLOv9 (Yolov9e)	YOLOX (small-coco)	Faster R-CNN (fasterRCNN)	SSD (ResNet50)	Time (s)
X = 0°	2	2	3	2	2	4	0	0	2.68
X = 40°	3	4	3	3	4	4	0	0	2.14
X = 80°	1	2	3	1	2	2	0	0	4.13
X = 120°	2	0	0	0	1	0	2	0	1.91
X = 160°	2	0	0	0	1	2	1	0	1.85
X = 200°	1	0	0	0	2	1	2	0	1.51
X = 240°	0	0	0	0	2	1	2	0	1.64
X = 280°	2	3	3	2	4	5	0	1	1.57
X = 320°	2	4	3	4	4	4	0	0	1.58

Table 2 presents detection results for Y-axis rotations, also sampled every 40°C to provide a balanced and representative view of angular robustness. YOLOX (small-coco) demonstrates exceptional stability, achieving peak detections at 140° and 340° and maintaining strong performance throughout the cycle. In contrast, SSD (ResNet50) and Faster R-CNN (fasterRCNN) show limited or no detection capability at most intervals, highlighting their vulnerability to off-axis rotations. Processing times remain consistent after initialization, underscoring the framework's suitability for real-time, resource-constrained environments.

Table 2. Detection Model Performance under Rotations on the Y-axis.

Rotation	YOLOv2 (Tiny)	YOLOv3 (Tiny)	YOLOv4 (CSP Darknet)	YOLOv8 (yolov8x)	YOLOv9 (Yolov9e)	YOLOX (small-coco)	Faster R-CNN (fasterRCNN)	SSD (ResNet50)	Time (s)
Y = 0°	2	2	3	2	2	4	0	0	1.66
Y = 40°	1	0	3	3	1	4	1	0	1.55
Y = 80°	0	0	0	0	3	4	0	0	1.54
Y = 120°	0	0	3	1	2	3	0	0	1.52
Y = 160°	2	3	3	2	5	3	2	0	1.53
Y = 200°	2	2	3	3	1	4	1	0	1.55
Y = 240°	0	0	3	0	3	4	1	0	1.64
Y = 280°	0	0	0	0	4	3	0	0	1.54
Y = 320°	1	1	3	1	3	4	0	0	1.55

Table 3 analyzes classification performance under X-axis rotations at 40-degree increments. This sampling strategy exposes the binary nature of model predictions and reveals the resilience or fragility of each architecture. Classical networks like AlexNet (AlexNet) and Inceptionv3 (Inceptionv3) sustain correct classifications across a wider range of angles, while deeper models such as ResNet50 (ResNet50) and EfficientNet (EfficientNet) are more susceptible to abrupt accuracy losses at specific intervals. These findings emphasize the importance of angular robustness over sheer model complexity in practical scenarios.

Table 3. Classification Model Performance under Rotations on the X-axis.

Rotation	VGG16 (VGG16)	ResNet50 (ResNet50)	EfficientNet (EfficientNet)	AlexNet (AlexNet)	GoogLeNet (GoogLeNet)	DarkNet19 (DarkNet19)	MobileNetv2 (MobileNetv2)	Inceptionv3 (Inceptionv3)	Time (s)
$X = 0°$	lipstick	lipstick	coffee mug	coffee mug	cup	coffee mug	coffee mug	cup	2.68
$X = 40°$	lipstick	lipstick	coffee mug	coffee mug	cup	coffee mug	coffee mug	cup	2.14
$X = 80°$	washbasin	washbasin	tray	iPod	tray	washbasin	washbasin	washbasin	4.13
$X = 120°$	lipstick	thimble	lighter	joystick	rubber eraser	thimble	lipstick	spotlight	1.91
$X = 160°$	lipstick	lipstick	lipstick	cup	lipstick	thimble	lipstick	lipstick	1.85
$X = 200°$	lipstick	lipstick	lipstick	cup	lipstick	lipstick	lipstick	thimble	1.51
$X = 240°$	lipstick	thimble	rubber eraser	pick	rubber eraser	thimble	lipstick	lampshade	1.64
$X = 280°$	washbasin	washbasin	tray	iPod	remote control	washbasin	washbasin	washbasin	1.57
$X = 320°$	lipstick	coffee mug	coffee mug	coffee mug	cup	coffee mug	coffee mug	cup	1.58

Table 4 displays classification outcomes for Y-axis rotations at 40-degree steps, reinforcing the patterns observed for the X-axis. AlexNet (AlexNet) and Inceptionv3 (Inceptionv3) again demonstrate superior invariance, maintaining correct predictions across diverse orientations. Newer models, including EfficientNet (EfficientNet) and ResNet50 (ResNet50), exhibit increased instability and critical failures at certain angles. This consistency across axes suggests that robust model selection must prioritize performance under realistic, non-ideal angular conditions.

Table 4. Classification Model Performance under Rotations on the Y-axis.

Rotation	VGG16 (VGG16)	ResNet50 (ResNet50)	EfficientNet (EfficientNet)	AlexNet (AlexNet)	GoogLeNet (GoogLeNet)	DarkNet19 (DarkNet19)	MobileNetv2 (MobileNetv2)	Inceptionv3 (Inceptionv3)	Time (s)
$Y = 0°$	lipstick	lipstick	coffee mug	coffee mug	cup	coffee mug	coffee mug	cup	1.66
$Y = 40°$	ballpoint	lipstick	lighter	coffee mug	coffee mug	soap dispenser	soap dispenser	cup	1.55
$Y = 80°$	lighter	washbasin	lighter	matchstick	lighter	pedestal	candle	tray	1.54
$Y = 120°$	ballpoint	washbasin	lighter	can opener	lighter	soap dispenser	pick	toilet tissue	1.52
$Y = 160°$	lipstick	lipstick	coffee mug	coffee mug	coffee mug	vase	measuring cup	coffee mug	1.53
$Y = 200°$	lipstick	lipstick	coffee mug	coffee mug	coffee mug	coffee mug	cocktail shaker	coffee mug	1.55
$Y = 240°$	cocktail shaker	loudspeaker	binder	mouse	lighter	ballpoint	iron	coffee mug	1.64
$Y = 280°$	refrigerator	washbasin	projectile	matchstick	ballpoint	letter opener	iron	soap dispenser	1.54
$Y = 320°$	lipstick	home theater	coffee mug	coffee mug	coffee mug	coffee mug	cocktail shaker	cup	1.55

The decision to run experiments on CPU validates the framework's viability in industrial and embedded environments, where access to specialized hardware is limited. To enhance readability and capture critical performance variations, results are presented at 40° intervals rather than every degree. This sampling approach strategically avoids both obvious fixed angles (0°, 90°, 180°) and complete blind spots, providing a balanced representation of rotational behavior that highlights the most relevant angular ranges where models transition between stable and unstable performance zones. The results confirm the robustness and efficiency of the 2D approach and pave the way for its extension to 3D detection and LiDAR point cloud analysis. Moving forward, we propose adapting this framework for Z-axis rotations and three-dimensional scenarios while maintaining the achieved computational efficiency, facilitating its application in industrial automation, autonomous vehicles, and inspection systems, where robustness to geometric variations is essential.

5 Discussion

The results of this work should be interpreted as a foundational step toward developing robust systems for detecting urban furniture in 3D LiDAR point clouds, where object orientation and geometric variability pose critical challenges for automated perception in real urban environments. The marked asymmetry in rotational robustness between the X and Y axes, along with the identification of critical performance drop zones (80°–120° on X-axis), reinforces the warning from [2,10] about the need to evaluate models beyond conventional benchmarks by incorporating multidimensional geometric transformations. This approach is especially relevant for smart city applications, industrial automation, and autonomous vehicles, where reliable detection of benches or traffic signs depends on generalization under arbitrary rotations and occlusions.

The comparative analysis reveals that lightweight models like YOLOX exhibit superior angular stability, particularly along the Y-axis, while complex architectures like SSD (ResNet50) show critical limitations, aligning with [16]. This suggests that future LiDAR applications should prioritize rotationally robust models, as sensor dynamics induce frequent vertical/lateral rotations. The identified critical angular ranges provide actionable insights for 3D data augmentation strategies following [9].

In classification tasks, classical models (AlexNet, Inceptionv3) outperform modern architectures under extreme rotations, consistent with [4,6,12]. This indicates that architectural simplicity and geometric feature preservation outweigh network depth for orientation-variable tasks. Computational efficiency remains stable (2 s/iteration post-initialization on CPU), validating deployment viability in embedded systems as proposed by [5,7,15].

Although our study maintained constant lighting, real-world deployment must address the impact of brightness variations on performance. Changes in illumination can degrade detection accuracy; low-light conditions reduce feature discriminability, while overexposure causes detail loss. Future extensions should incorporate brightness-variation datasets and augmentation techniques to enhance robustness across lighting conditions in practical applications.

Extending this work to 3D LiDAR point-cloud analysis represents a strategic contribution. The experimental findings enable the identification of critical angular ranges and robust architectures essential for the augmentation of 3D data, following [17]. This establishes methodological foundations for urban perception systems that operate reliably under extreme geometric variability in smart cities and industrial automation.

6 Conclusions

This study demonstrates that lightweight models such as YOLOX and YOLOv3 (Tiny), together with classical classifiers like AlexNet and Inceptionv3, show superior rotational robustness compared to more modern architectures, even under extreme rotations from 0° to 360°. Systematic evaluation in MATLAB and execution on an Intel Xeon Gold 5120 CPU (96 GB RAM) confirm that these models maintain stable and efficient performance (2 s per iteration after initialization), while more complex models like SSD-ResNet50 experience critical drops in specific angular ranges, especially between 80°–120° on the X-axis and 240°–280° on the Y-axis. Importantly, the observed increase in processing time during the first rotation is due solely to model loading, not to orientation complexity.

To ensure clarity and focus on the most informative angular transitions, all results are reported at 40° intervals. This sampling strategy was deliberately chosen to avoid obvious fixed angles (0°, 90°, 180°) and persistent blind spots, highlighting the most relevant zones where models transition between stable and unstable performance. These findings challenge traditional criteria for model selection in computer vision, emphasizing that angular invariance and computational efficiency should take priority over conventional benchmarks. The ability to operate without specialized hardware and the identification of critical performance zones make this approach viable for robotics, urban surveillance, and industrial inspection, where object orientation is unpredictable and resources may be limited.

As future work, we propose expanding the dataset with real-world images of urban furniture and dynamic scenarios to further validate model robustness. Additionally, the framework will be extended to include Z-axis rotations and LiDAR point cloud processing, with performance evaluated on low-power hardware and under the occlusion and geometric variability typical of real urban environments. This will support the development of robust and efficient computer vision systems for smart cities and industrial automation.

Acknowledgements. The authors were supported by the Vitoria-Gasteiz Mobility Lab Foundation, a governmental organization of the Provincial Council of Araba and the local council of Vitoria-Gasteiz under the following project grant: "Generación de Inventario Automatizado de Señalética mediante drones e Inteligencia Computacional".

References

1. Alinra, R.R., Utomo, S.B., Rahardi, G.A., Anam, K.: Detector face mask using uav-based cnn transfer learning of yolov5. In: 2022 IEEE International Conference on Cybernetics and Computational Intelligence (CyberneticsCom), pp. 329–334, June 2022. https://doi.org/10.1109/CyberneticsCom55287.2022.9865243

2. Chandrakar, R., Raja, R., Miri, R., Sinha, U., Kumar Singh Kushwaha, A., Raja, H.: Enhanced the moving object detection and object tracking for traffic surveillance using rbf-fdlnn and cbf algorithm. Expert Syst. Appl. **191**, 116306 (2022). https://doi.org/10.1016/j.eswa.2021.116306
3. Chen, Y.: Bundled tobacco leaves vein detection based on yolov8 and transfer learning. In: 2024 IEEE 7th International Conference on Automation, Electronics and Electrical Engineering (AUTEEE), pp. 1–4, December 2024. https://doi.org/10.1109/AUTEEE62881.2024.10869717
4. Hendria, W.F., Phan, Q.T., Adzaka, F., Jeong, C.: Combining transformer and cnn for object detection in uav imagery. ICT Express **9**(2), 258–263 (2023). https://doi.org/10.1016/j.icte.2021.12.006
5. Ismail, F.N., Yassin, I.M., Ahmad, A., Ali, M.S.A.M., Baharom, R.: Motorcycle detection using deep learning convolution neural network. In: 2020 IEEE 10th International Conference on System Engineering and Technology (ICSET), pp. 49–54, November 2020. https://doi.org/10.1109/ICSET51301.2020.9265361
6. Jose, R., Balasubramanian, K.: Leveraging deep learning for spice identification: a comparative study of alexnet, vggnet, and yolo. In: 2025 International Conference on Electronics and Renewable Systems (ICEARS), pp. 1311–1316, February 2025. https://doi.org/10.1109/ICEARS64219.2025.10940112
7. Kwaghe, O.P., Gital, A.Y., Madaki, A., Abdulrahman, M.L., Yakubu, I.Z., Shima, I.S.: A deep learning approach for detecting face mask using an improved yolo-v2 with squeezenet. In: 2022 IEEE 6th Conference on Information and Communication Technology (CICT), pp. 1–5, November 2022. https://doi.org/10.1109/CICT56698.2022.9997956
8. Liao, L.: Semi-supervised sar target detection with cross-domain transfer learning based on yolov5. In: IGARSS 2024 - 2024 IEEE International Geoscience and Remote Sensing Symposium, pp. 7401–7404, July 2024. https://doi.org/10.1109/IGARSS53475.2024.10641642
9. Mittal, P., Sharma, A., Singh, R.: A simulated dataset in aerial images using simulink for object detection and recognition. Int. J. Cogn. Comput. Eng. **3**, 144–151 (2022). https://doi.org/10.1016/j.ijcce.2022.07.001
10. Ray, K.S., Chakraborty, S.: Object detection by spatio-temporal analysis and tracking of the detected objects in a video with variable background. J. Vis. Commun. Image Represent. **58**, 662–674 (2019). https://doi.org/10.1016/j.jvcir.2018.12.002
11. Sawant, K., et al.: Fruit identification using yolo v8 and faster r-cnn –a commparative study. In: 2024 International Conference on Distributed Computing and Optimization Techniques (ICDCOT), pp. 1–6, March 2024. https://doi.org/10.1109/ICDCOT61034.2024.10515677
12. Sharma, V.K., Mir, R.N.: Saliency guided faster-rcnn (sgfr-rcnn) model for object detection and recognition. J. King Saud Univ. Comput. Inf. Sci. **34**(5), 1687–1699 (2022). https://doi.org/10.1016/j.jksuci.2019.09.012
13. Si, R., Phoophuangpairoj, R.: A hybrid yolo-vgg16 for fallen motorcycle detection. In: 2025 13th International Electrical Engineering Congress (iEECON), pp. 1–5, March 2025. https://doi.org/10.1109/iEECON64081.2025.10987817
14. Singh, G., Guleria, K., Sharma, S.: Advanced fruit sorting: pre-trained resnet50 model for rotten and fresh fruit classification. In: 2024 4th Asian Conference on Innovation in Technology (ASIANCON), pp. 1–5, August 2024. https://doi.org/10.1109/ASIANCON62057.2024.10837782
15. Terven, J.R., Córdova-Esparza, D.M.: Kinz an azure kinect toolkit for python and matlab. Sci. Comput. Program. **211**, 102702 (2021). https://doi.org/10.1016/j.scico.2021.102702

16. Xu, G., Wang, Z., Cheng, Y., Tian, Y., Zhang, C.: A man-made object detection algorithm based on contour complexity evaluation. Chin. J. Aeronaut. **30**(6), 1931–1940 (2017). https://doi.org/10.1016/j.cja.2017.09.001
17. Zhou, S., et al.: Improved yolo for long range detection of small drones. Sci. Rep. **15**(1), 12280 (2025). https://doi.org/10.1038/s41598-025-95580-z

Automated Detection of Fibromyalgia Using 3D Residual Networks on Structural Brain Imaging

Rebeca Piñol-Galera, David Ortiz-Perez(✉), Manuel Benavent-Lledo, David Mulero-Perez, Javier Rodriguez-Juan, and Jose Garcia-Rodriguez

Department of Computer Technology, University of Alicante, Alicante, Spain
{dortiz,mbenavent,dmulero,jrodriguez,jgarcia}@dtic.ua.es

Abstract. Fibromyalgia is a chronic pain condition with a complex clinical profile and no definitive biomarkers, often leading to delayed diagnosis and patient frustration. Recent neuroimaging studies have identified structural brain differences in individuals with fibromyalgia, offering a potential pathway to objective diagnosis. This study investigates the use of deep learning techniques for automatically classifying fibromyalgia using structural brain MRI scans. We propose a three-dimensional Convolutional Neural Network (3D CNN), specifically the 3DResNet-18 model, to extract and analyze relevant features from the images. Before the model training, the MRI images underwent a preprocessing step using SPM12 to align and normalize the images, addressing variability introduced during acquisition. Experiments have been conducted on the Emo-Fibro dataset, which comprises 33 T1-weighted MRI scans from fibromyalgia patients and 33 from healthy controls. To mitigate the impact of limited data, data augmentation techniques were applied. Furthermore, an ablation study has been performed to compare the proposed model with alternative deep learning architectures and assess relative performance. The proposed method yielded promising results, achieving an accuracy of up to 65% and an area under the curve (AUC) of 0.62 for fibromyalgia detection.

Keywords: Fibromyalgia · Chronic pain · Structural MRI · Deep learning · Data augmentation

1 Introduction

Fibromyalgia is a complex chronic condition affecting approximately 2.10% of the global population, with a significantly higher prevalence among women [10]. Despite its widespread occurrence, the condition remains poorly understood in medical and societal contexts. Fibromyalgia is characterized by widespread musculoskeletal pain, persistent fatigue, sleep disturbances, and cognitive impairments [18]. The lack of objective biomarkers further complicates its diagnosis,

E. Corchado et al. (Eds.): SOCO 2025, CCIS 2806, pp. 487–497, 2026.
https://doi.org/10.1007/978-3-032-19763-4_45

posing a substantial challenge for healthcare professionals in achieving accurate clinical identification [22].

The diagnostic process for fibromyalgia is often prolonged and emotionally taxing. A recent study involving 616 patients with fibromyalgia, 92.2% of whom were women, reported an average diagnostic delay of 3.45 years [24]. During this time, patients commonly consult multiple specialists and undergo numerous tests, which typically fail to identify any definitive abnormalities. The absence of objective findings frequently leads to disbelief and invalidation from both healthcare providers and the patient's social environment. These experiences significantly exacerbate the individual's suffering and have a profound negative impact on their quality of life.

Currently, the most widely accepted and diagnostically reliable criteria for fibromyalgia (FM) are those established by the American College of Rheumatology (ACR). These criteria are based on two primary assessment tools: the Widespread Pain Index (WPI) and the Symptom Severity Scale (SSS). The WPI evaluates the distribution of pain by asking patients to indicate whether they have experienced pain in 19 specific body regions over the past week, resulting in a score ranging from 0 to 19. In contrast, the SSS measures the severity of key symptoms commonly associated with fibromyalgia, such as fatigue, unrefreshing sleep, and cognitive disturbances, as well as additional somatic symptoms like headaches or abdominal pain, with a total score ranging from 0 to 12 [32,33].

However, despite these advancements, the diagnostic process for fibromyalgia remains inconsistent. This inconsistency contributes to overdiagnosis and underdiagnosis, partly due to the reliance on specific criteria versions on subjective self-reporting by patients. These limitations underscore the urgent need for more objective and standardized diagnostic tools to enhance the accuracy and reliability of fibromyalgia diagnosis [29]. Recent research has identified consistent structural brain differences between individuals with fibromyalgia and healthy controls, as detected through structural magnetic resonance imaging (MRI) and voxel-based morphometry (VBM) [26]. These findings present promising opportunities for developing imaging-based diagnostic approaches.

To mitigate the challenges posed by the lack of objective diagnostic tools for fibromyalgia, we propose the use of convolutional neural networks (CNNs) applied to T1-weighted brain MRI scans from the Emo-Fibro dataset to detect structural patterns that differentiate fibromyalgia patients from healthy controls. The MRI data were preprocessed using SPM12 to ensure spatial alignment and intensity normalization, thereby minimizing inter-subject variability caused by differences in image acquisition. Given the relatively small size of the dataset, data augmentation strategies were employed to improve model generalization and reduce overfitting. We implemented a 3D ResNet-18 architecture, which achieved a classification accuracy of up to 65% and an area under the curve (AUC) of 0.62. To further assess the reliability of our approach, we conducted an ablation study comparing the performance of this model with alternative deep learning architectures. By leveraging CNNs for automated analysis of structural neuroimaging data, this work contributes to the development of objective, reproducible tools that may assist in the diagnosis of fibromyalgia and ultimately promote more timely and validated clinical assessments for patients.

2 Related Works

The lack of objective biomarkers for fibromyalgia diagnosis has been a major contributor to delayed or inaccurate clinical assessments, prompting increasing interest in the use of artificial intelligence (AI) methods [2]. These techniques have shown strong potential to enhance diagnostic accuracy through automated analysis of large-scale medical data. In recent years, machine learning and deep learning approaches have been applied to a variety of data sources, including self-reported clinical measures, biosignals such as EEG [6], and neuroimaging modalities like MRI and fMRI [9,27].

Studies utilizing brain imaging techniques such as MRI and fMRI have proven particularly promising in the context of fibromyalgia diagnosis. For example, the study presented in [31] employs a combined structural and machine learning approach to differentiate between fibromyalgia patients and healthy controls, achieving an AUC of 95%, a significant improvement over traditional questionnaire-based methods. These findings underscore the capability of machine learning techniques to detect complex and subtle patterns in brain imaging data that are often imperceptible to human observers. This capability is crucial in the case of fibromyalgia, where chronic pain symptoms frequently arise in the absence of identifiable anatomical abnormalities.

Biomarker discovery studies using machine learning techniques in chronic pain conditions, particularly those leveraging MRI and fMRI data, are grounded in a broad body of neuroimaging research [19,30]. This literature has consistently identified both structural and functional abnormalities in key regions of the central nervous system involved in pain processing. These include areas associated with the sensory dimension of pain, such as the thalamus, primary and secondary somatosensory cortices, and posterior insula, the affective dimension, including the anterior cingulate cortex, amygdala, hypothalamus, and hippocampus, and the cognitive-evaluative dimension, notably the dorsolateral prefrontal cortex and the thalamus. In addition to regional activation patterns, studies have also reported changes in gray matter density and volume, as well as alterations in white matter integrity, which may reflect underlying neural mechanisms of chronic pain [16,23]. Given that this type of pain often manifests in the absence of observable tissue damage, there is increasing interest in using these central nervous system alterations as potential objective biomarkers—providing a means to classify chronic pain conditions more accurately, without relying solely on subjective patient reports [9].

In [9], machine learning techniques are applied to MRI and fMRI data to classify patients with fibromyalgia. Among the algorithms used, support vector machines (SVMs) [11,12] are the most common, although other methods such as logistic regression [15], naive Bayes [20], and k-nearest neighbors (KNN) [14,25] have also been explored. A key limitation of traditional machine learning models is their reliance on prior assumptions about the underlying data patterns. For example, if a condition reliably caused a specific reduction in the thickness of a brain region, a simple linear model might be sufficient to detect it. In practice, however, chronic pain disorders are highly complex and heterogeneous, with symptoms and neural patterns varying significantly between individuals. This complexity makes it particularly challenging to select an appropriate model [7, 28].

In this context, deep learning offers a powerful alternative. Unlike traditional machine learning methods, deep neural networks can automatically learn complex hierarchical representations directly from raw data, such as pixels in medical images, without the need for manual feature extraction. These models function as multi-level pattern detectors, progressively capturing more abstract and invariant features from the input. This capability has proven especially effective in image recognition tasks, contributing to the widespread adoption of deep learning in medical imaging applications [17].

Deep learning techniques have shown promising results with neuroimaging data for diagnosing various neurological disorders, including Alzheimer's disease [13] and brain tumors [8]. While their application to fibromyalgia remains limited, early studies suggest that these methods hold significant potential for identifying subtle brain alterations that could serve as objective diagnostic biomarkers.

3 Methodology

This section presents the methodology designed to detect fibromyalgia using T1-weighted structural MRI data. An overview of the pipeline is presented in Fig. 1, involving image preprocessing with SPM12, data augmentation to address dataset limitations, and classification using a 3D ResNet-18 model.

3.1 Image Preprocessing

The T1-weighted images extracted from the dataset were preprocessed to enhance image quality and ensure comparability across subjects. This process was performed using SPM12 within MATLAB R2023b[1], following a standardized pipeline that included the following steps:

1. **Segmentation:** Brain tissues were segmented into gray matter, white matter, and cerebrospinal fluid. Furthermore, a bias-corrected image was generated to enhance intensity homogeneity across the scan.
2. **Skull-stripping:** Non-brain structures were removed by multiplying the segmented tissue maps with the bias-corrected image, resulting in a clean representation of the brain.

[1] https://github.com/PhilKuhnke/fMRI_analysis/blob/main/1Preprocessing/SPM12/preprocessing_script.m.

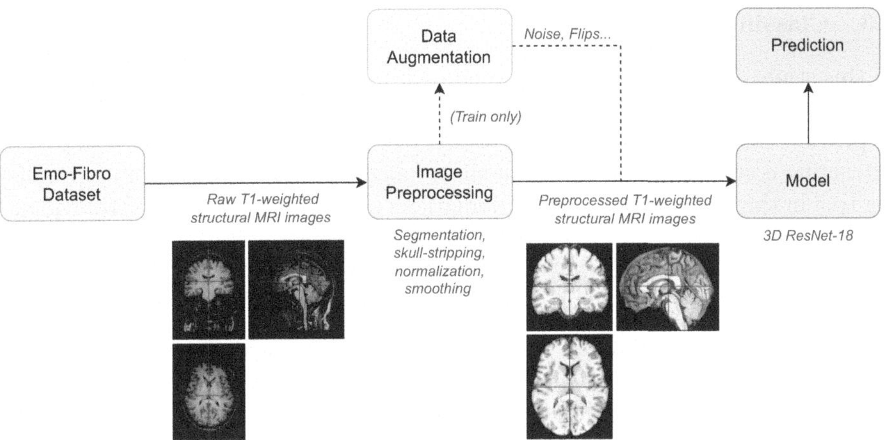

Fig. 1. Overview of the proposed deep learning pipeline. T1-weighted MRI scans from the Emo-Fibro dataset are preprocessed using SPM12 for alignment and normalization, followed by data augmentation in the train set. The processed images are input into a 3D ResNet-18 model for automatic feature extraction and classification of fibromyalgia patients versus healthy controls.

3. **Normalization:** The images were spatially normalized to the Montreal Neurological Institute (MNI) space to ensure anatomical consistency across participants.
4. **Smoothing:** A Gaussian smoothing filter with a kernel size of $6 \times 6 \times 6$ mm (approximately twice the original voxel size) was applied to reduce noise and enhance the signal-to-noise ratio.

3.2 Data Augmentation

To mitigate the limitations posed by the small dataset size, a data augmentation strategy was implemented to artificially increase the diversity of the training samples and improve the model's generalization capability. This process was achieved using TorchIO [21], a Python library specifically designed for preprocessing and augmenting 3D medical images. All transformations were applied exclusively to the training set and included the following:

- **Resampling:** Adjustment of spatial resolution to (1.0, 1.8, 1.8) mm to ensure consistency across samples.
- **RandomAffine:** Application of small random rotations (up to 10°) and scaling transformations with factors ranging from 0.97 to 1.03.
- **RandomNoise:** Addition of low-intensity Gaussian noise with a standard deviation of 0.02 to simulate acquisition variability.

3.3 Classification Models

In this study, we selected convolutional neural network (CNN) models designed to perform well even with limited data and that have demonstrated strong results in previous medical imaging applications. CNNs have proven highly effective in extracting complex features from medical images [8,13], making them a promising tool in the search for objective biomarkers in conditions such as fibromyalgia. Based on these criteria, three models were evaluated: a 3D ResNet-18, a simplified version, referred to as the Simplified 3D ResNet, and a transfer learning model adapted from a brain MRI-based Alzheimer's classification task.

3D ResNet-18. This architecture is widely used in medical image classification due to its balance between computational efficiency and robustness against vanishing gradients through residual connections. The 3D ResNet-18 replaces standard 2D convolutions with 3D convolutional blocks to accommodate the volumetric nature of MRI data. The model follows a classical ResNet structure: an initial convolutional layer, four residual blocks with increasing filter sizes, and a final fully connected layer for classification.

Simplified 3D ResNet. To investigate the impact of a lighter architecture and reduce the risk of overfitting, a simplified version of the 3D ResNet-18 was implemented. Key modifications included reducing the initial number of channels, decreasing the number of blocks per layer, and adding a Dropout3D layer before the final classification stage to enhance regularization.

Transfer Learning. A transfer learning approach was also explored, using a pre-trained model initially developed for Alzheimer's disease classification from brain MRI scans. Since transfer learning tends to perform best when source and target domains are similar [1], this model was adapted for fibromyalgia classification by freezing all but the final layers, replacing the fully connected layers, and fine-tuning the new components with differentiated learning rates.

4 Experiments

This section outlines the experimental setup used to evaluate the performance of the proposed architecture for fibromyalgia classification.

4.1 Experiments Setup

All models were trained for 30 epochs using a batch size of 8, a learning rate of 1×10^{-5}, and the Adam optimizer. To address the limited dataset size, stratified 5-fold cross-validation was applied, ensuring a balanced distribution of classes across all folds. In each fold, 56 samples were used for training and 13 for validation. The experiments were implemented in PyTorch and executed on an NVIDIA TITAN Xp GPU with 12 GB of memory, allowing for efficient processing of the 3D MRI volumes.

4.2 Emo-Fibro Dataset

The Emo-Fibro dataset [3,5] comprises data from 33 women diagnosed with fibromyalgia and 33 healthy controls, all drawn from a Mexican population. It includes T1- and T2-weighted magnetic resonance imaging (MRI) scans, resting-state and task-based functional MRI (fMRI) related to an emotional regulation task involving visual stimuli, clinical, behavioral, and demographic information.

The dataset is distributed across two platforms: OpenNeuro [3], which hosts the neuroimaging data (T1, T2, resting-state fMRI, and task-based fMRI), and Zenodo [5], which provides access to the associated non-imaging data.

For this study, only the T1-weighted structural MRI scans were used, with the aim of training models to detect fibromyalgia based solely on anatomical brain features.

4.3 Results

The performance of the model was evaluated using accuracy and AUC metrics. The results represent the mean and standard deviation calculated across a stratified 5-fold cross-validation over the Emo-Fibro dataset. Each model was evaluated through experimentation with and without the incorporation of data augmentation techniques. For the augmented condition, transformations such as resampling, random rotations, and Gaussian noise were applied using the TorchIO library. A summary of the results is presented in Table 1.

Table 1. Comparison of model performance using accuracy and AUC on the original (base) and augmented datasets. Results are reported as mean ± standard deviation over 5-fold cross-validation.

Model	Accuracy		AUC	
	Base	Aug.	Base	Aug.
3D ResNet-18	**0.65 ± 0.09**	0.62 ± 0.07	0.58 ± 0.09	**0.62 ± 0.07**
Simplified 3D ResNet	0.59 ± 0.08	0.61 ± 0.06	0.55 ± 0.09	0.59 ± 0.08
Transfer Learning	0.55 ± 0.08	0.57 ± 0.08	0.46 ± 0.09	0.52 ± 0.10

The results indicate that data augmentation does not consistently improve performance across all models. For the 3D ResNet-18, a slight decrease in accuracy is observed when using the augmented dataset (from 0.65 ± 0.09 to 0.62 ± 0.07), while the AUC shows a modest improvement (from 0.58 ± 0.09 to 0.62 ± 0.07). These results suggest that, although the model gains robustness in ranking predictions, it may lose some precision in final classification. The reduction in accuracy could be attributed to the variability introduced by the augmented samples, which may not fully reflect the distribution of the original data.

For the Simplified 3D ResNet, the use of the augmented dataset results in slight improvements in both accuracy (from 0.59 ± 0.08 to 0.61 ± 0.06) and AUC (from 0.55 ± 0.09 to 0.59 ± 0.08). These results suggest that a lighter model may benefit more from data augmentation when trained on a limited dataset. However, the observed improvements fall within the range of the standard deviations, indicating that the gains are not statistically significant and may vary across different cross-validation folds.

In contrast, the transfer learning model achieved the lowest performance among the three architectures. Although data augmentation led to slight improvements in both accuracy (from 0.55 ± 0.08 to 0.57 ± 0.08) and AUC (from 0.46 ± 0.09 to 0.52 ± 0.10), the overall results remain comparatively weak. These results may be attributed to the fact that the pretrained model was originally developed for a different task, Alzheimer's disease classification, and the learned features may not generalize effectively to fibromyalgia, despite fine-tuning of the final layers. Additionally, the relatively high standard deviations across folds indicate instability and suggest a strong dependence on specific data partitions.

These findings suggest that while data augmentation introduces variability that can support model generalization, particularly reflected in AUC improvements, it may also introduce noise that negatively impacts final classification accuracy.

Although the Emo-Fibro dataset has been used to examine functional connectivity and emotional processing in fibromyalgia, no prior studies have reported classification performance metrics, such as accuracy or AUC, using machine learning methods [4]. As a result, direct comparisons with previous work are not possible, further underscoring the novelty of our approach.

5 Conclusions

In this study, we proposed a neuroimaging-based approach for the detection of fibromyalgia using structural MRI. Experiments were conducted on the Emo-Fibro dataset, where T1-weighted images underwent preprocessing and data augmentation. Three classification models were evaluated: a 3D ResNet-18, a simplified variant, and a transfer learning model.

The results demonstrate the potential of deep learning applied to structural brain MRI for the automatic classification of fibromyalgia. Although overall performance remains moderate, with a maximum AUC of 0.62, the models were able to identify meaningful patterns that distinguish patients with fibromyalgia from healthy controls. Data augmentation yielded limited but encouraging improvements, particularly in enhancing model robustness. These findings support the feasibility of using neuroimaging and convolutional neural networks as a step toward more objective diagnostic tools. With access to larger datasets and further model refinement, this approach holds considerable promise for enabling earlier and more reliable diagnosis of fibromyalgia.

Acknowledgment. We would like to thank CIAICO/2022/132 Consolidated group project "AI4-Health" funded by the Valencian government. This work has also been supported by two national grants and two regional grants for PhD studies from the Spanish and Valencian governments, respectively, FPU21/00414, FPU23/00532, CIACIF/2021/430 and CIACIF/2022/175.

References

1. Alzubaidi, L., et al.: Towards a better understanding of transfer learning for medical imaging: A case study. Appl. Sci. **10**, 4523 (2020). https://doi.org/10.3390/app10134523
2. Atmakuru, A., et al.: Fibromyalgia detection and diagnosis: a systematic review of data-driven approaches and clinical implications (2013–2023). IEEE Access **13**, 25026–25044 (2025). https://doi.org/10.1109/ACCESS.2025.3539196
3. Balducci, T., et al.: A behavioral, clinical and brain imaging dataset with focus on emotion regulation of females with fibromyalgia (2022). https://doi.org/10.18112/openneuro.ds004144.v1.0.2
4. Balducci, T., et al.: A behavioral and brain imaging dataset with focus on emotion regulation of women with fibromyalgia. Sci. Data **9**(1), 581 (2022)
5. Balducci, T., et al.: Emo-fibro: dataset of woman with fibromyalgia (2022). https://doi.org/10.5281/zenodo.7032997
6. Barua, P.D., et al.: Innovative fibromyalgia detection approach based on quantum-inspired 3lbp feature extractor using ecg signal. IEEE Access **11**, 101359–101372 (2023). https://doi.org/10.1109/ACCESS.2023.3315149
7. Bauer, D., et al.: Challenges for monocular 6d object pose estimation in robotics. IEEE Trans. Robot. (2024)
8. Bhanumathi, V., Sangeetha, R.: Cnn based training and classification of mri brain images. In: 2019 5th ICACCS, pp. 129–133 (2019). https://doi.org/10.1109/ICACCS.2019.8728447
9. Boissoneault, J., et al.: Biomarkers for musculoskeletal pain conditions: Use of brain imaging and machine learning. Curr. Rheumatol. Rep. **19**(1), 5 (2017). https://doi.org/10.1007/s11926-017-0621-0
10. Cabo-Meseguer, A., et al.: Fibromialgia: prevalencia, perfiles epidemiológicos y costes económicos. Med. Clin. **149**(10), 441–448 (2017). https://doi.org/10.1016/j.medcli.2017.06.008
11. Camlica, Z., et al.: Medical image classification via svm using lbp features from saliency-based folded data. In: 2015 IEEE 14th ICMLA, pp. 128–132 (2015). https://doi.org/10.1109/ICMLA.2015.29
12. Chandra, M.A., Bedi, S.S.: Survey on svm and their application in image classification. Int. J. Inf. Technol. **13**(5), 1–11 (2021). https://doi.org/10.1007/s41870-021-00703-6
13. El-Assy, A.M., et al.: A novel cnn architecture for accurate early detection and classification of alzheimer's disease using mri data. Sci. Rep. **14**(1) (2024). https://doi.org/10.1038/s41598-024-53733-6
14. Leo, M.J.: Mri brain image segmentation and detection using k-nn classification. In: Journal of Physics: Conference Series, vol. 1362, p. 012073 (2019). https://doi.org/10.1088/1742-6596/1362/1/012073

15. Li, Q., et al.: Correlated logistic model with elastic net regularization for multilabel image classification. IEEE Trans. Image Process. **25**(8), 3801–3813 (2016). https://doi.org/10.1109/TIP.2016.2572687
16. Liu, D., et al.: Alterations of the resting-state brain network connectivity and gray matter volume in patients with fibromyalgia in comparison to ankylosing spondylitis. Sci. Rep. **14**(1), 29960 (2024). https://doi.org/10.1038/s41598-024-79246-w
17. Mamoshina, P., et al.: Applications of deep learning in biomedicine. Mol. Pharm. **13**(5), 1445–1454 (2016). https://doi.org/10.1021/acs.molpharmaceut.5b00982
18. Mayo Clinic: Fibromialgia - síntomas y causas. https://www.mayoclinic.org/es/diseases-conditions/fibromyalgia/symptoms-causes/syc-20354780 (2021)
19. Mosch, B., et al.: Brain morphometric changes in fibromyalgia and the impact of psychometric and clinical factors: a volumetric and diffusion-tensor imaging study. Arthritis Res. Therapy **25**(1), 1–12 (2023). https://doi.org/10.1186/s13075-023-03064-0
20. Nayak, M.M., Kengeri Anjanappa, S.D.: An efficient hybrid classifier for mri brain images classification using machine learning based naive bayes algorithm. SN Comput. Sci. **4**(3), 223 (2023). https://doi.org/10.1007/s42979-023-01956-z
21. Pérez-García, F., et al.: TorchIO: a Python library for efficient loading, preprocessing, augmentation and patch-based sampling of medical images in deep learning. Comput. Methods Program. Biomed. 106236 (2021). https://doi.org/10.1016/j.cmpb.2021.106236
22. Qureshi, A.G., et al.: Diagnostic challenges and management of fibromyalgia. Cureus **13**(10) (2021). https://doi.org/10.7759/cureus.18523
23. Robinson, M.E., et al.: Gray matter volumes of pain-related brain areas are decreased in fibromyalgia syndrome. J. Pain **12**(4), 436–443 (2011). https://doi.org/10.1016/j.jpain.2010.10.003
24. Salaffi, F., et al.: Delay in fibromyalgia diagnosis and its impact on the severity and outcome: a large cohort study. Clin. Exp. Rheumatol. **42**, 1198–1204 (2024)
25. Saval-Calvo, M., et al.: Three-dimensional planar model estimation using multi-constraint knowledge based on k-means and ransac. Appl. Soft Comput. **34**, 572–586 (2015)
26. Schmidt-Wilcke, T., et al.: Striatal grey matter increase in patients suffering from fibromyalgia–a voxel-based morphometry study. Pain **132**, S109–S116 (2007). https://doi.org/10.1016/j.pain.2007.05.010
27. Sevel, L., et al.: (337) mri based classification of chronic fatigue, fibromyalgia patients and healthy controls using machine learning algorithms: a comparison study. J. Pain **17**(4), S60 (2016). https://doi.org/10.1016/j.jpain.2016.01.225
28. Shehab, M., et al.: Machine learning in medical applications: a review of state-of-the-art methods. Comput. Biol. Med. **145**, 105458 (2022). https://doi.org/10.1016/j.compbiomed.2022.105458
29. Sierra, P.B., Rueda, J.Y.L.: Evolución de la epidemiología y diagnóstico de fibromialgia en los últimos diez años: Revisión bibliográfica. Pro Sciences: Revista de Producción, Ciencias e Investigación **6**(44 (esp)), 132–146 (2022)
30. Suñol, M., et al.: Brain structural changes during juvenile fibromyalgia: relationships with pain, fatigue, and functional disability. Arthritis Rheumatol. **74**(7), 1284–1294 (2022). https://doi.org/10.1002/art.42073
31. Thanh Nhu, N., et al.: Identification of resting-state network functional connectivity and brain structural signatures in fibromyalgia using a machine learning approach. Biomedicines **10**(12), 3002 (2022). https://doi.org/10.3390/biomedicines10123002

32. Velasco, M.: Dolor musculoesquelético: fibromialgia y dolor miofascial. Revista Médica Clínica Las Condes **30**(6), 414–427 (2019). https://doi.org/10.1016/j.rmclc.2019.10.002
33. Wolfe, F., et al.: The american college of rheumatology 1990 criteria for the classification of fibromyalgia. Arthritis Rheum. Official J. Am. Coll. Rheumatol. **33**(2), 160–172 (1990). https://doi.org/10.1002/art.1780330203

Characterization of Photovoltaic Power with SOM and Clustering: An Approach Based on Meteorological Data

Anibal Mantilla-Guerra[1](✉), Christian Mejia-Escobar[1], Jorge Azorin-Lopez[2], and Jose Garcia-Rodriguez[2]

[1] Central University of Ecuador, Quito, Ecuador
armantilla@uce.edu.ec
[2] Department of Computer Science and Technology, University of Alicante, Alicante, Spain

Abstract. A challenge for the widespread adoption of solar energy has always been its variability, largely due to changing weather conditions. Identifying weather patterns and their impact on energy is essential for improving the efficiency and performance of photovoltaic systems. However, traditional methods require human intervention and are not effective at capturing and visualizing the interaction between multiple highly complex and irregular variables. This work proposes the application of Self-Organizing Maps (SOM) for the discovery of weather and photovoltaic power patterns. As a result, well-defined behavior groups and their specific value ranges have been obtained. This allows for the establishment of scenarios or "typical days" based on weather conditions and their impact on photovoltaic generation, which helps define the particular conditions of each behavior pattern. This tool becomes part of smarter energy management and can contribute to informed decision-making.

Keywords: Artificial intelligence · Machine learning · SOM · Clustering · Solar power · Weather

1 Introduction

The global transition to sustainable energy is a crucial challenge for carbon neutrality and environmental impact [1]. Renewable solar energy systems have inherent intermittency due to weather fluctuations, creating uncertainty in their production [2,3]. Weather conditions are key to photovoltaic performance, influencing production. Knowing these factors allows for real-time management, improving network stability, optimizing facilities, predicting fluctuations, and promoting the adoption of renewables [4].

Traditionally, statistical models often have limitations when analyzing complex data such as photovoltaic power and climatic factors. The high dimensionality of the variables, their nonlinear interactions, irregularity, sensitivity to

E. Corchado et al. (Eds.): SOCO 2025, CCIS 2806, pp. 498–508, 2026.
https://doi.org/10.1007/978-3-032-19763-4_46

outliers or extreme values, difficulty of visualization, and human intervention due to their lack of self-organization are some of the challenges these methods face in extracting valuable and in-depth knowledge. These limitations require new solutions for solar intermittency.

Artificial Intelligence (AI) is emerging as a transformative alternative, with machine learning (ML) proving highly successful [5,6]. AI systems better understand and predict weather fluctuations [7]. The intersection of meteorological factors and AI is a promising frontier. ML is crucial for optimizing the efficiency of variable renewable energy systems, applying supervised and unsupervised methods in forecasting and monitoring [8].

The scientific literature generally focuses on the application of AI in solar forecasting. For example, Random Forests (RF) and Extra Trees (ET) predict hourly PV production, with ET outperforming RF and SVR in computational cost [9]. A hybrid k-means-SVR algorithm predicts daily solar radiation with high accuracy ($R^2 > 0.98$) [10]. Deep learning is powerful for PV systems, driven by IoT and applying CNN [11]. For PV power generation prediction, a hybrid deep learning framework uses Self-Organizing Maps (SOM) as preprocessing, clustering meteorological data to improve forecast robustness [12]. SOM has proven its value in optimizing PV systems, improving prediction accuracy by aggregating irradiance data [13], optimizes neural network training in electrical characterization [14], recommended for noise mitigation [15], is key in fault detection [16], improved accuracy in hourly forecasting [17], identifies "representative days" [18], and improved thermal profile clustering in fault detection [19].

This research aims to analyze the interaction of meteorological factors in photovoltaic production using unsupervised machine learning, specifically SOM. With this, AI will contribute to a more reliable and stable solar supply, alerting about conditions that affect production and maintaining efficiency. Data from the 67kWp Baltra photovoltaic plant (meteorological and operational) from January 2017 to July 2018 is used. The objectives are to apply and evaluate SOM to group data and identify patterns that impact solar energy production.

2 Methodology

By using AI techniques, specifically ML, we seek to understand the influence of climate variables on photovoltaic power. The objective is to identify the optimal weather conditions that maximize the electricity generation of a solar panel. The workflow of the developed model is illustrated in Fig. 1.

The workflow begins with the collection of data on climate variables and power output from the Baltra plant. This is followed by data exploration, where distributions are analyzed to understand their nature. Pre-processing is crucial for scaling the data, making it suitable for the algorithm. SOM is selected over other clustering or dimensionality reduction techniques (e.g., K-means, PCA, t-SNE) because of its spatial nature and its ability to preserve neighborhood relationships. These qualities are especially advantageous for visualizing continuity and transitions between different patterns of solar energy and meteorological behavior without prior labels. Training adjusts the model to the data, followed by an evaluation of the quality of the resulting clusters. The interpretation of these clusters explains how different combinations of meteorological factors impact PV production.

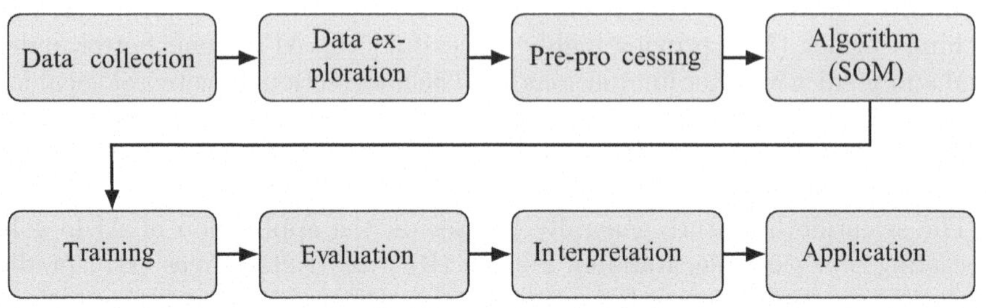

Fig. 1. Methodology workflow.

In the field of machine learning, data is the starting point. Acquiring data involves first defining the variables to be considered. Photovoltaic power is directly influenced by climatic factors. For this reason, standard meteorological data is used, which is essential for analysis and prediction tasks [20]. The Table 1 summarizes the variables considered in this study.

Table 1. Description of the variables considered.

Name	Definition	Unit	Type
Global radiation	Energy received that affects surface temperature and atmospheric energy	W/m^2	Meteorological
Temperature	Indicator of heat present in the atmosphere and its thermal patterns	°C	Meteorological
Relative humidity	Water vapor in the air relative to its maximum amount at a given temperature	%	Meteorological
Precipitation	Measurement of water falling from clouds to the Earth's surface	mm/h	Meteorological
Photovoltaic power	Energy generated by a system from sunlight	kWh	Electrical

The meteorological data was downloaded from the website NASA POWER Project Data Access [21] for the period from January 01-01-2017 to 31-07-2018 and hourly frequency. These data are numerical and correspond to five variables: temperature in degrees Celsius, relative humidity in %, wind in m/s, and precipitation in mm/h. The geographical location of the weather station has the following coordinates: latitude -0.4612 and longitude -90.2775, both in decimal degrees. The hourly average values of the power in kW supplied by the 67 kWp photovoltaic power plant on Baltra Island (Galapagos, Ecuador) are considered.

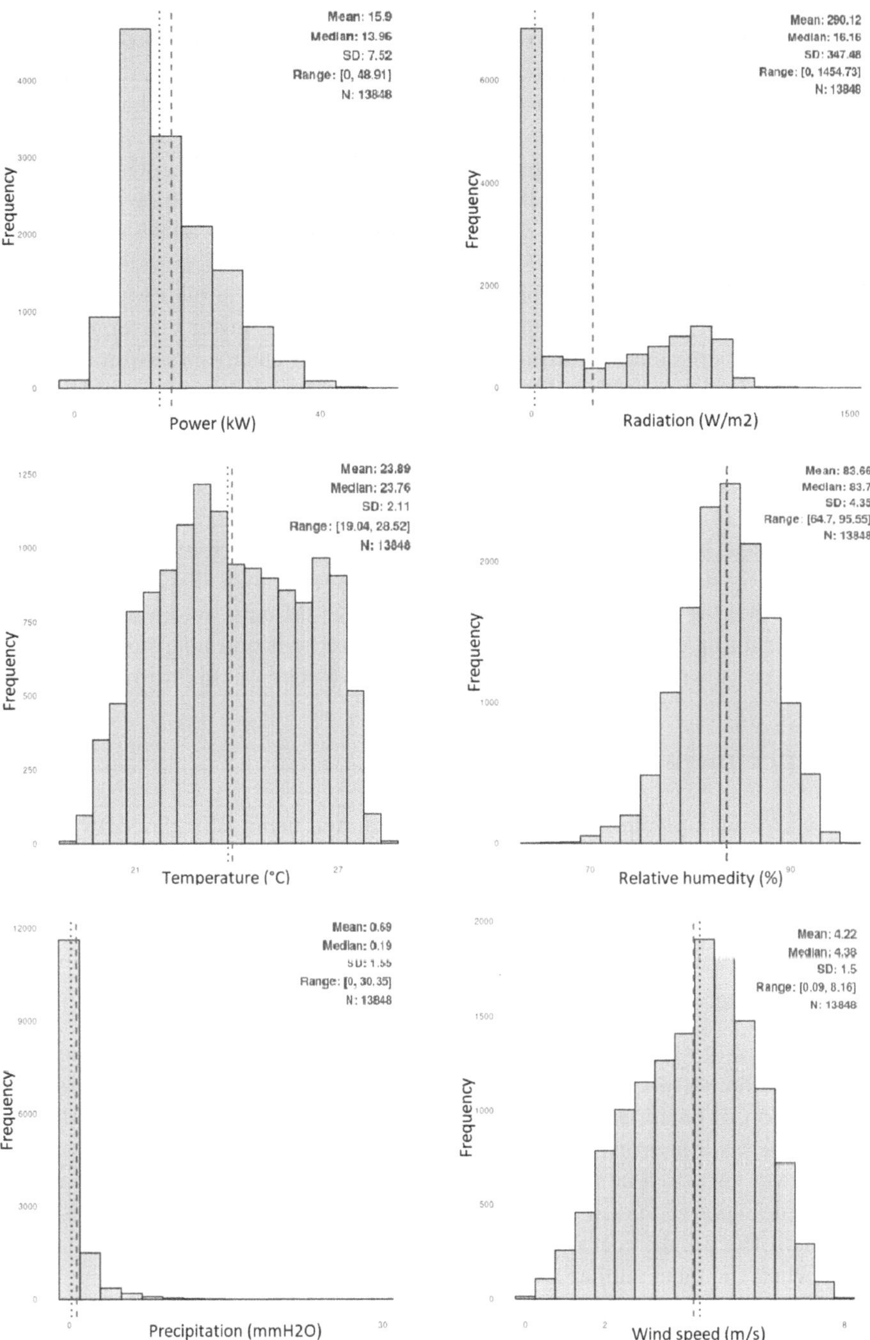

Fig. 2. Histograms of photovoltaic power, global radiation, temperature, relative humidity, precipitation, and wind speed (from top to bottom).

Through a basic statistical exploration of this data, its nature and behavior can be understood. Figure 2 presents the distributions of each variable, along with their main descriptive indicators.

Based on the analysis of the histograms, it was determined that no transformation of the variables is necessary. Power, temperature, relative humidity, and wind speed exhibit an approximately normal distribution, which is ideal and beneficial for the study. Although radiation and precipitation show some bias, maintaining their original distribution is essential. This allows crucial patterns to be identified, such as the detection of outliers or extreme values. Finally, standardizing the values is key for scaling the variables in a similar way, eliminating differences in units of measurement, and facilitating faster convergence and better model performance during training. This also ensures an equitable contribution from the variables. The min-max scaling technique was used to bring all variables to the same range between 0 and 1. Once the data is prepared, the analysis of the influence of climate on photovoltaic power can be addressed using unsupervised machine learning. *SOM* (Self-organizing map) algorithm projects each instance of a dataset onto a two-dimensional grid of nodes, called a *neural map*. The data is organized into zones, where the proximity of nodes indicates similarity. To establish the correlation between meteorological variables and PV power, the strategy consists of selecting the best SOM map by evaluating different sizes, training errors, and the distribution of records and neurons in clusters. This ensures optimal class assignment for the objective of this work.

3 Experimentation

The model was implemented, trained, and evaluated using RStudio software and the R programming language, with support from libraries such as *RSNNS* for data standardization, *kohonen* for the creation of the SOM and *dplyr* to calculate statistics. All of these tools are free and open source.

3.1 Training

The objective of the algorithm is to organize the data into a grid of neurons, placing similar data close to each other to form clusters. This facilitates the identification of patterns and relationships between them. After several tests, a map size of 5×5 neurons with a hexagonal topology and bubble neighborhood function was chosen. The input dataset consists of 13,848 records or instances of six variables (power and meteorological). Each record (vector of values) is compared with the weight vector of each neuron using a distance measure. Learning occurs when the most similar neuron and its neighbors adjust their weights (code vectors) to more closely resemble the processed record. Training is performed for 200 epochs (iterations of the dataset). The learning rate starts at 0.05 to allow for significant adjustments in the weights and decreases progressively to 0.01, which facilitates stabilization and refinement of the map. Therefore, the SOM self-organizes, and nearby neurons end up representing very similar data patterns, while distant ones indicate very different patterns.

The evolution of the training process is displayed in a bar chart 'changes' (Fig. 3). The training is considered successful, given that the mapping error metric, i.e., the average distance between the data vectors and their nearest neurons, progressively decreases toward zero in each iteration. This indicates an optimal fit and, therefore, an accurate representation of the data.

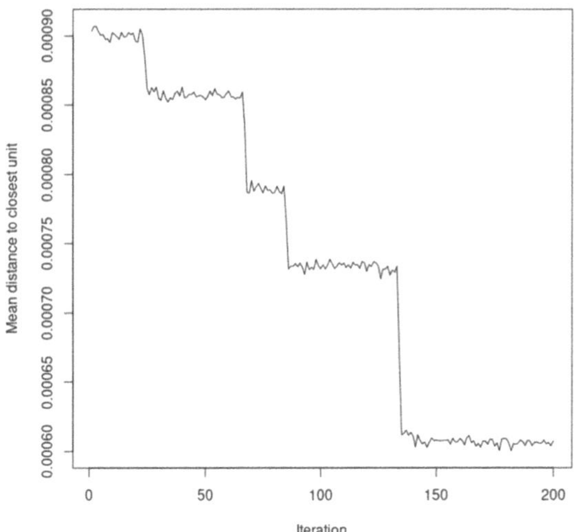

Fig. 3. Model training learning curve.

3.2 Evaluation

The quality of the SOM model is evaluated using visualization tools that allow us to understand two crucial criteria of its internal structure after training: the number of neurons and the distances between neighbors. Both criteria are illustrated in Fig. 4.

The 'counts' graph is useful for identifying the distribution of data on the map. The data count per neuron reveals a minimum of 122 and a maximum of 972, with an average of 554 and a median of 530. This indicates that there are no neurons with excessive load, very low load, or no data. The Self-Organizing Map (SOM) uses all of its neurons for a well-distributed representation of data. The graph of distances between neighboring neurons (U-matrix) visualizes these relationships: light colors indicate neurons with similar weight vectors (homogeneous data areas), while color transitions reveal significant differences between weight vectors, facilitating the identification of clusters or data groups.

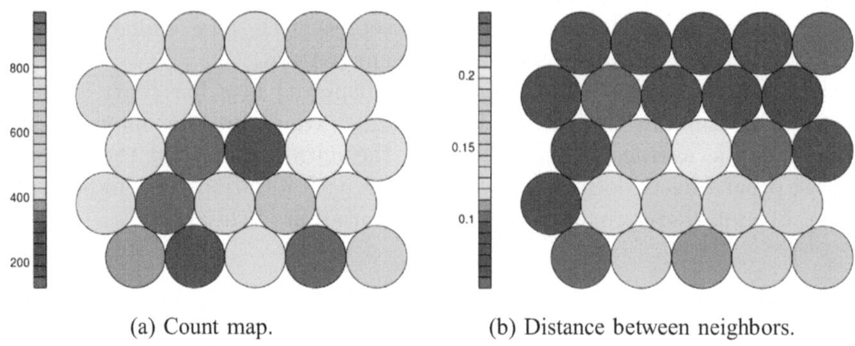

(a) Count map.

(b) Distance between neighbors.

Fig. 4. SOM evaluation maps.

Table 2. Identified clusters and their characteristics.

Cluster	Performance	Characterization
1	Best	Very high energy production, very high global radiation, moderately high temperature, high relative humidity, low precipitation, moderate wind speed.
2	High	Medium-high production, very low global radiation, temperature similar to cluster 1, higher humidity, higher precipitation range than cluster 1, moderate-low wind speed.
3	Medium	Very low production within a narrow range, extremely low global radiation, very stable and moderate temperatures, extremely high and stable relative humidity, low precipitation but within a very stable range, moderate wind speed.
4	Low	Low to medium production, very low global radiation, lower temperatures than other clusters, very high relative humidity, low precipitation, higher wind speed.
5	Very low	Low-medium production, very high global radiation, lower temperatures than cluster 1, high relative humidity, very low precipitation, higher wind speed.

3.3 Interpretation

The 'codes' graph is the most notable of the maps generated from the training. Each node includes a fan diagram, representing the instances of the dataset assigned to that neuron. Figure 5 presents this type of map, which is useful for visualizing the code vectors associated with each neuron and understanding which characteristics are dominant in different regions of the map.

The map presented shows certain patterns in the data and how the different variables covary across the map. To define several groups (clusters) in the SOM, the *hierarchical clustering* technique is applied by using the code vectors [22]. Table 2 details the internal characteristics of each SOM group, showing the behavior and interrelationship of its variables.

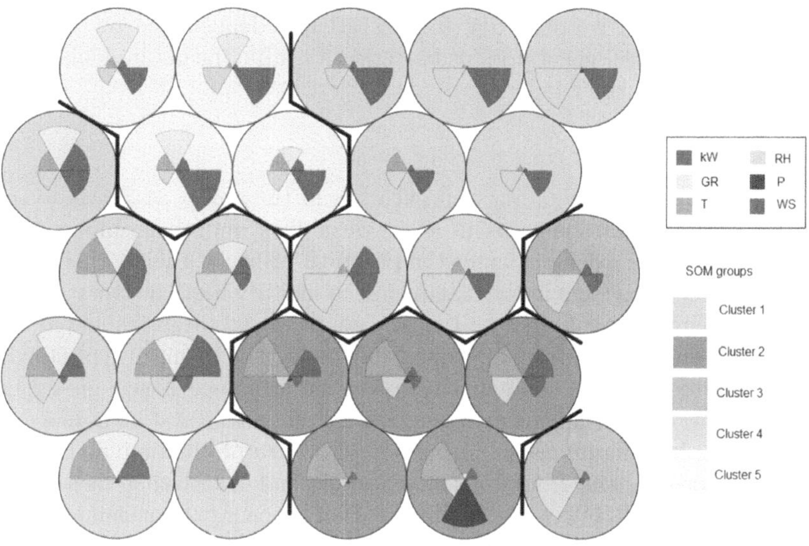

Fig. 5. Code vector and clustering map.

To facilitate the interpretation of the groupings formed by the SOM, the minimum and maximum values of the code vectors of each cluster were extracted to define the ranges and thresholds that characterize each group. Table 3 displays these values, which give physical meaning to the data groups and are essential for optimizing the management of the photovoltaic plant.

Table 3. Weather threshold values for each cluster.

Cluster	kW	GR	T	RH	P	WS
1	[17.86;34.28]	[550.02;836.58]	[23.51;27.11]	[75.35;86.33]	[0.17; 2.04]	[2.11;4.69]
2	[12.39;25.32]	[16.85;105.06]	[25.44;27.00]	[77.93;84.79]	[0.42;10.04]	[1.65;4.01]
3	[13.00;13.80]	[10.09;29.08]	[24.48;24.49]	[88.53;88.69]	[0.50;0.80]	[2.18;3.89]
4	[7.68;25.03]	[19.72;54.91]	[20.42;23.34]	[81.67;89.26]	[0.15;1.00]	[4.32;6.17]
5	[10.55;15.87]	[412.95;772.30]	[21.23;23.29]	[80.32;84.36]	[0.19;0.49]	[4.72;6.35]

It should be noted that some related work has improved sensor data using artificial vision and artificial intelligence techniques in many different fields, which affects the final accuracy of the model [23–30]. Lastly, the clusters identified by the SOM represent specific types of days on which the Baltra plant experiences different operating and performance conditions. It can be seen that the maximum power generated is 34.28 kWp under optimal environmental conditions, which represents 51% of its 67 kWp capacity. For low or medium energy production, installed overcapacity implies greater dependence on alternative sources (such as diesel generators or batteries) or adjustments in demand.

4 Conclusion

The methodological value of our approach lies in the ability of SOMs to reveal complex and nonlinear patterns in solar variability and their impact on PV generation. This work demonstrates the practical utility of this methodology on Baltra, a unique ecosystem in the Galapagos, to identify best and worst scenarios and improve energy management. Two key contributions are highlighted: i) the identification and classification of "typical day" scenarios to improve photovoltaic generation forecasting by anticipating plant performance based on expected weather conditions, optimizing grid scheduling and backup system management, and planning plant maintenance during periods of low production; and ii) meaningfully interpreting the resulting feature maps and translating those clusters and their range values into practical information for energy management. Global radiation drives solar energy, but climate optimizes it: moderate temperatures and light winds increase efficiency, while high humidity and cloudiness reduce it. Wind can cool the panels, but sometimes reduces radiation. Occasional rain is good because it cleans the panels and improves performance. Finally, this powerful tool can help the Baltra power plant understand and manage its solar energy production in the context of its specific weather conditions, transforming complex data into useful information for decision making.

References

1. Abdessadak, A., Ghennioui, H., Thirion-Moreau, N., Elbhiri, B., Abraim, M., Merzouk, S.: Digital twin technology and artificial intelligence in energy transition: a comprehensive systematic review of applications. Energy Rep. **13**, 5196–5218 (2025)
2. Abisoye, B.O., Sun, Y., Zenghui, W.: A survey of artificial intelligence methods for renewable energy forecasting: methodologies and insights. Renew. Energy Focus **48** (2024)
3. Ahmad, T., et al.: Artificial intelligence in sustainable energy industry: status quo, challenges and opportunities. J. Clean. Prod. **289** (2021)
4. Energy evolution expo: the role of AI in solar power forecasting. https://energyevolutionconference.com/the-role-of-ai-in-solar-power-forecasting/
5. Islam, M., Ammirrul, M., Noor, S.: AI-driven optimization for solar energy systems: theory and applications (2025)
6. Barhmi, K., Heynen, C., Golroodbari, S., van Sark, W.: A review of solar forecasting (2024)
7. EasySolar: how AI predicts extreme weather for solar systems
8. Gupta, M., Arya, A., Varshney, U., Mittal, J., Tomar, A.: A review of PV power forecasting using machine learning techniques. Prog. Eng. Sci. **2**(1) (2025)
9. Ahmad, M.W., Mourshed, M., Rezgui, Y.: Tree-based ensemble methods for predicting PV power generation and their comparison with support vector regression. Energy **164**, 465–474 (2018)

10. Ayodele, T.R., Ogunjuyigbe, A.S.O., Amedu, A., Munda, J.L.: Prediction of global solar irradiation using hybridized k-means and support vector regression algorithms. Renew. Energy Focus **29**, 78–93 (2019)
11. Chang, Z., Han, T.: Prognostics and health management of photovoltaic systems based on deep learning: a state-of-the-art review and future perspectives. Renew. Sustain. Energy Rev. **205** (2024)
12. Du, J., et al.: A theory-guided deep-learning method for predicting power generation of multi-region photovoltaic plants. Eng. Appl. Artif. Intell. **118** (2023)
13. Kazem, H.A., Yousif, J.H.: Comparison of prediction methods of photovoltaic power system production using a measured dataset. Energy Convers. Manage. **148**, 1070–1081 (2017)
14. Piliougine, M., Elizondo, D., Mora-López, L., Sidrach-de-Cardona, M.: Multilayer perceptron applied to the estimation of the influence of the solar spectral distribution on thin-film photovoltaic modules. Appl. Energy **112**, 610–617 (2013)
15. Gandomzadeh, M., et al.: Dust mitigation methods and multi-criteria decision-making cleaning strategies for photovoltaic systems: advances, challenges, and future directions. Energy Strategy Rev. **57** (2025)
16. Sepúlveda-Oviedo, E.H., Travé-Massuyès, L., Subias, A., Pavlov, M., Alonso, C.: Fault diagnosis of photovoltaic systems using artificial intelligence: a bibliometric approach. Heliyon **9**(11) (2023)
17. Sharadga, H., Hajimirza, S., Balog, R.S.: Time series forecasting of solar power generation for large-scale photovoltaic plants. Renew. Energy **150**, 797–807 (2020)
18. Ribeiro, R., Fanzeres, B.: Identifying representative days of solar irradiance and wind speed in Brazil using machine learning techniques. Energy AI **15** (2024)
19. Korkmaz, D., Acikgoz, H.: An efficient fault classification method in solar photovoltaic modules using transfer learning and multi-scale convolutional neural network. Eng. Appl. Artif. Intell. **113** (2022)
20. Neumann, O., Turowski, M., Mikut, R., et al.: Using weather data in energy time series forecasting: the benefit of input data transformations. Energy Inf. **6**, 44 (2023)
21. NASA langley research center (LaRC): prediction of worldwide energy resources (POWER)
22. Murtagh, F., Contreras, P.: Algorithms for hierarchical clustering: an overview. WIREs Data Min. Knowl. Discovery **2**, 86–97 (2012)
23. Saval-Calvo, M., Azorin-Lopez, J., Fuster-Guillo, A., Garcia-Rodriguez, J.: Three-dimensional planar model estimation using multi-constraint knowledge based on k-means and RANSAC. Appl. Soft Comput. **34**, 572–586 (2015)
24. Bauer, D., Hönig, P., Weibel, J.B., Garcia-Rodriguez, J., Vincze, M.: Challenges for monocular 6D object pose estimation in robotics. IEEE Trans. Robot. **26** (2024)
25. Viejo, D., Garcia-Rodriguez, J., Cazorla, M.: Combining visual features and growing neural gas networks for robotic 3D SLAM. Inf. Sci. **276**, 174–185 (2014)
26. Garcia-Garcia, A., Garcia-Rodriguez, J., Orts-Escolano, S., Oprea, S., et al.: A study of the effect of noise and occlusion on the accuracy of convolutional neural networks applied to 3D object recognition. Comput. Vis. Image Underst. **164**, 124–134 (2017)
27. Orts, S., Garcia-Rodriguez, J., Viejo, D., Cazorla, M., Morell, V.: GPGPU implementation of growing neural gas: application to 3D scene reconstruction. J. Parall. Distrib. Comput. **72**(10), 1361–1372 (2012)
28. Martinez-Gonzalez, P., Oprea, S., Castro-Vargas, J.A., Garcia-Garcia, A., et al.: Unrealrox+: an improved tool for acquiring synthetic data from virtual 3D envi-

ronments. In: International Joint Conference on Neural Networks (IJCNN), pp. 1–8 (2021)
29. Garcia-Garcia, A., Orts-Escolano, S., Garcia-Rodriguez, J., Cazorla, M.: Interactive 3D object recognition pipeline on mobile GPGPU computing platforms using low-cost RGB-D sensors. J. Real-Time Image Proc. **14**, 585–604 (2018)
30. Garcia-Garcia, A., Orts-Escolano, S., Oprea, S., Garcia-Rodriguez, J., et al.: Multisensor 3D object dataset for object recognition with full pose estimation. Neural Comput. Appl. **28**, 941–952 (2017)

Enhanced Skin Lesion Segmentation and Classification Using U-Net and Ensemble Deep Learning Models

Aboubakr Aakaou[3(✉)], Karl Thurnhofer-Hemsi[1,2,3], and Enrique Domínguez[1,2,3]

[1] ITIS Software, Universidad de Málaga, Málaga, Spain
{karlkhader,enriqued}@lcc.uma.es
[2] Instituto de Investigación Biomédica de Málaga y Plataforma en Nanomedicina–IBIMA Plataforma BIONAND, Málaga, Spain
[3] Department of Computer Languages and Computer Science, Universidad de Málaga, Málaga, Spain
aboubakr.aakaou@uma.es

Abstract. Accurate skin lesion detection and segmentation play a vital role in medical imaging, and specifically dermatology. However, challenges such as varying skin tones, lighting conditions, and multiple skin areas complicate this task. In this study, we propose a novel end-to-end pipeline that integrates enhanced preprocessing, multi-region segmentation, and an innovative classification approach using a convolutional neural network (CNN). The preprocessing stage employs CLAHE and sonar-like enhancement to improve contrast and lesion visibility. U-Net is then used for precise segmentation of lesion areas. We combine the image and segmentation mask into a six-channel input, enabling joint spatial-semantic learning. Unlike standard models that treat segmentation and classification as independent tasks or rely solely on RGB input, our integrated framework allows better focus on lesion regions while preserving contextual information. Experimental results on a dataset of 401,059 dermoscopic images from ISIC-2016 to ISIC-2020 demonstrate high performance compared with pre-trained models, with notable improvements in classification metrics. The proposed framework shows promising potential for applications in medical diagnosis, image analysis, and biometric systems, offering a reliable and scalable solution for real-world skin detection challenges.

Keywords: Skin detection · U-Net · Image segmentation · Skin lesion classification · Medical imaging · Melanoma · ISIC dataset

1 Introduction

Skin cancer, especially melanoma, is among the most aggressive cancers and requires early detection for improved therapeutic results [9]. Dermoscopic image

E. Corchado et al. (Eds.): SOCO 2025, CCIS 2806, pp. 509–519, 2026.
https://doi.org/10.1007/978-3-032-19763-4_47

analysis plays a vital role in non-invasive melanoma diagnosis, with deep learning (DL) and hybrid models advancing the field [14]. Traditional methods like active contour models and GrabCut assist in delineating lesion boundaries [5,21]. Convolutional neural networks, especially with transfer learning and attention mechanisms, have demonstrated strong performance in classifying pigmented lesions [22]. Hybrid architectures combining CNNs with recurrent layers (e.g., RCNN-LSTM) enhance feature extraction, while lightweight models such as MobileNet support real-time classification on edge devices [16]. Accurate segmentation is facilitated by the integration of U-Net with multiscale feature fusion. Challenges persist due to lesion variability, overlapping structures, lighting artifacts, and dataset inconsistencies [18,20]. These insights inform the foundation of our proposed approach. Building on these insights, this study proposes a comprehensive pipeline that integrates robust preprocessing, segmentation, and classification. Prior research has shown that combining segmentation with CNN-based classification enhances interpretability and diagnostic performance [6,8]. Despite progress in skin lesion analysis, challenges like low contrast, morphological variability, imaging artifacts (e.g., hair, shadows), and multi-lesion complexity persist. Existing methods often treat segmentation and classification separately, limiting performance. To address this, we propose an end-to-end pipeline with Contrast Limited Adaptive Histogram Equalization (CLAHE) and sonar-like enhancement for preprocessing, U-Net for multi-region segmentation, and a custom CNN that processes input by merging the original image and its segmentation mask. This fusion enables the model to learn both spatial and semantic features, improving lesion focus and classification accuracy. Our integrated method improves performance across key metrics and provides a scalable solution for real-world melanoma detection.

2 Related Works

Skin cancer, particularly melanoma, remains one of the most life-threatening cancers worldwide, making early detection through dermoscopic analysis vital. The advent of deep learning has dramatically improved skin lesion segmentation and classification in recent years. One of the most influential models in segmentation is U-Net, originally proposed by Ronneberger et al. Numerous enhancements to the original U-Net have been proposed, including Attention U-Netwhich, which improves attention and multiscale fusion. Further developments such as RA-Net, and dual-encoder architectures [2] have refined segmentation accuracy through advanced attention and multiscale mechanisms[13]. Active contour models [5] have also proven useful in capturing lesion boundaries. Hair removal techniques [1] and CLAHE-based preprocessing [7] are vital for reducing image artifacts and enhancing lesion contrast. On the classification side, CNNs such as MobileNet [19], AlexNet [10], and hybrid RCNN-LSTM models [4] are effective for melanoma detection, ensemble learning [12], and deep regularizers [3] have further enhanced model robustness. Several works [8] highlight that segmentation-aware classifiers outperform traditional models, particularly by focusing on lesion-specific regions. Overall, pipelines that integrate advanced preprocessing, accurate segmentation, and CNN-based classification form the foundation for high-performance skin lesion analysis systems.

3 Methodology

This study presents an automated system for melanoma detection using dermoscopic images, combining advanced preprocessing, U-Net-based segmentation, and a dual-path classification with pre-trained networks (Fig. 1).The pre- processing pipeline standardizes and enhances input images.

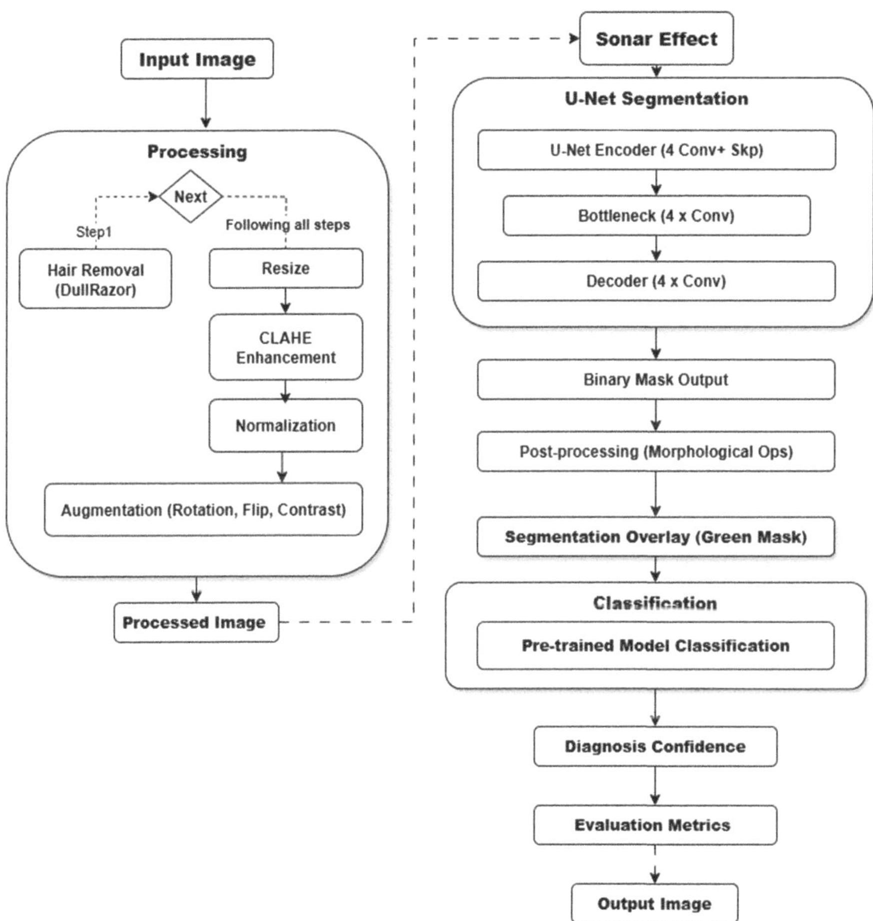

Fig. 1. Overview of the proposed melanoma detection system.

It begins with the DullRazor algorithm to remove hair artifacts using morphological operations and inpainting (Fig. 2).

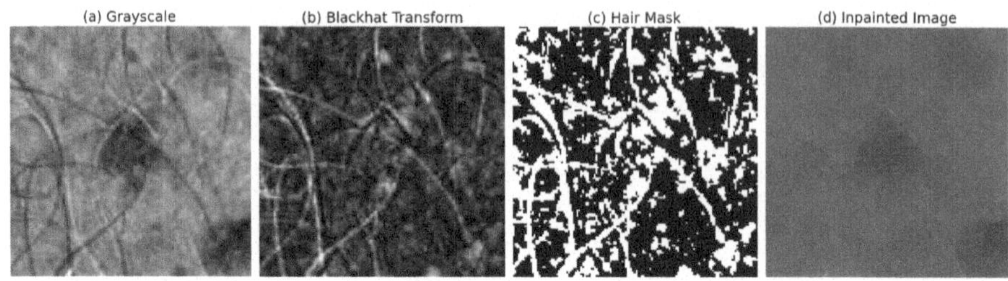

Fig. 2. Hair removal steps with the DullRazor algorithm.

Figure 3 illustrates the sequential preprocessing pipeline applied to dermoscopic images in this study. The process begins by resizing the original RGB image to a fixed resolution of 256×256 pixels using bilinear interpolation. To enhance local contrast and highlight subtle lesion features, CLAHE is applied specifically to the L channel in the LAB color space with a clip limit of 8.0 and a tile grid size of 8×8. After contrast enhancement, pixel intensities are normalized to the range $[0, 1]$. To enrich the dataset and improve model robustness, data augmentation includes random rotations between $-30°$ and $+30°$, brightness/contrast adjustments with $\alpha \in [0.8, 1.2]$ and $\beta \in [-20, 20]$, and random flips (horizontal, vertical, or both).

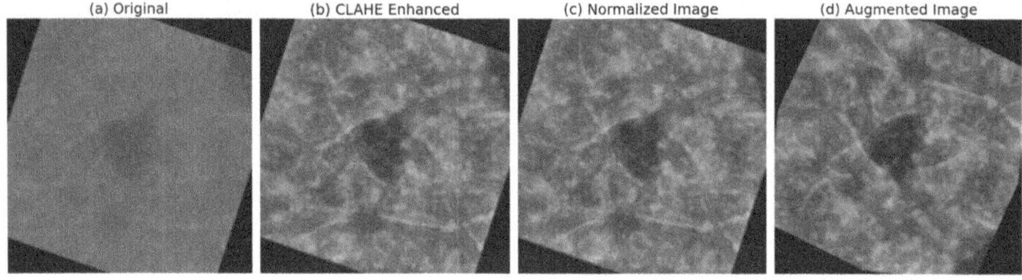

Fig. 3. Preprocessing steps after hair removal.

To illustrate the output of the proposed U-Net segmentation model, Fig. 4 shows an example of the segmentation results. The left image is the sonar-enhanced input, emphasizing lesion texture with a jet colormap. The center image presents the binary segmentation mask produced by the model. The right image overlays the mask on the original dermoscopic image using a blue tint to highlight the predicted lesion area. The U-Net uses a pretrained ResNet34 encoder and is trained with Dice loss, a batch size of 16, an initial learning rate of 1×10^{-4}, and the Adam optimizer, with early stopping applied after 10 epochs of no improvement.

Once the lesion has been segmented, the pipeline moves on to the classification stage. This step uses deep learning models to determine the kind

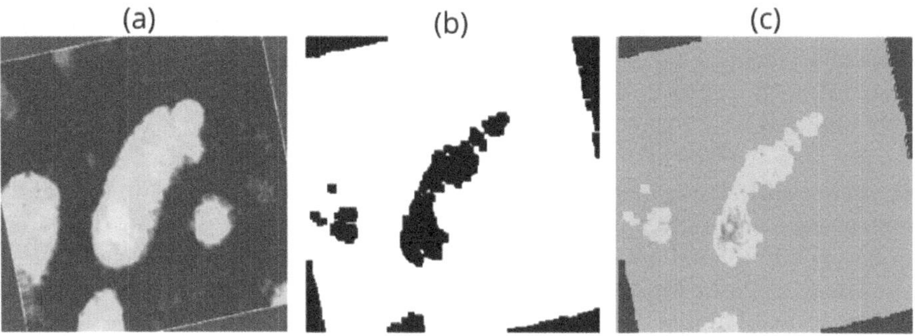

Fig. 4. (a): Sonar effect image; (b) Binary segmentation mask output from the U-Net; (c) Segmentation overlay on the original image using a blue tint to highlight the lesion area. (Color figure online)

of lesion. The segmented Region of Interest (ROI) is input into a fine-tuned deep Convolutional Neural Network (CNN) using transfer learning from models pre-trained on large datasets like ImageNet. This enables adaptation to lesion-specific patterns while leveraging general visual features. The tested models are ResNet (50, 101), VGG (16, 19), DenseNet (121, 169, 201), MobileNetV2, EfficientNet (B0, B1, B7), InceptionResNetV2, Xception, NASNet (Mobile, Large), CoatNet, HRNet, AlexNet, Inception-V3, and ConvNeXt. To enhance prediction reliability, ensemble methods like majority voting or weighted averaging are applied. The proposed framework was evaluated on segmentation and classification tasks using standard metrics. For segmentation, U-Net performance was measured using Intersection over Union (IoU), Dice Coefficient, Jaccard Index, Sensitivity (Recall), and Accuracy—each quantifying overlap, similarity, or detection correctness between predicted and actual lesion regions. For the classification of lesions into benign or melanoma, we used Accuracy, Precision, Recall, F1-score, and ROC-AUC. These metrics collectively assess prediction correctness, detection ability, balance between false positives and negatives, and overall class discrimination across thresholds. Evaluations were performed on a dataset of 401,059 dermoscopic images, providing a comprehensive assessment of the model's performance in both lesion localization and classification.

4 Experimental Results

Experiments were conducted using an NVIDIA RTX 3080 GPU (10 GB VRAM) with CUDA 12.4 and Python frameworks (TensorFlow, PyTorch) for efficient model training. Peak memory usage was 6964 MiB, and the system maintained stable thermal and power conditions.

Figure 5 presents a pie chart categorizing the IoU results into three performance levels. Notably, 92.4% of the segmented regions achieved an IoU score greater than 0.75 (Excellent), 7.6% fell within the 0.5–0.75 range (Good), and 0% were classified as Poor (IoU < 0.5), indicating a high-quality segmentation outcome. Our method demonstrates strong performance on the ISIC 2016–2020 dataset, achieving 86% segmentation accuracy, enhanced by CLAHE and the Sonar Effect for better lesion boundary visibility (Fig. 6).

In addition, Fig. 7 shows the comparison between normalized accuracy (blue) and the Dice coefficient (orange). Both metrics are skewed toward higher values, indicating strong performance overall. The Dice distribution is slightly more peaked and right-shifted, suggesting that it captures more precise overlap between predicted and ground truth lesion regions, which is crucial for segmentation, while in Fig. 8 here, the Jaccard Index (blue) and sensitivity (red) are almost overlapping, both centered around the 0.7 mark. This strong overlap implies consistent model behavior in capturing lesion areas (sensitivity) and measuring region agreement (Jaccard). The sharp peak around 0.7 reflects the stable segmentation capability across images.

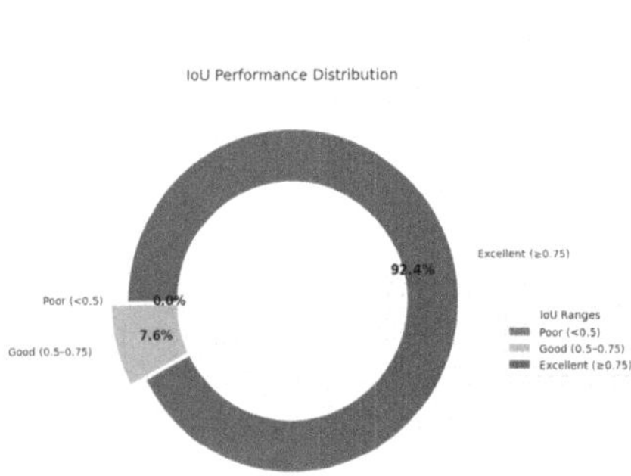

Fig. 5. IoU for U-Net segmentation.

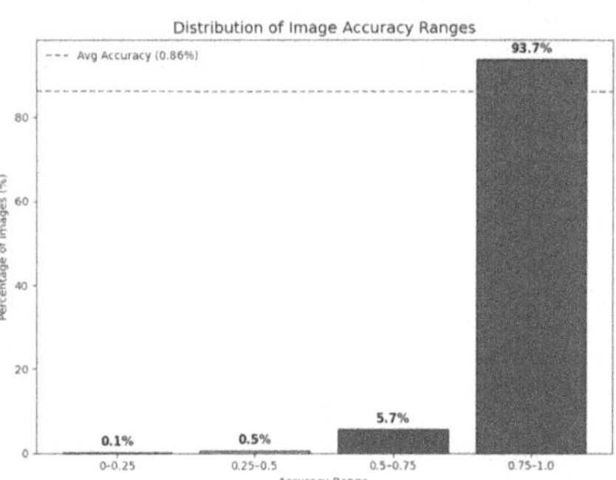

Fig. 6. Average accuracy.

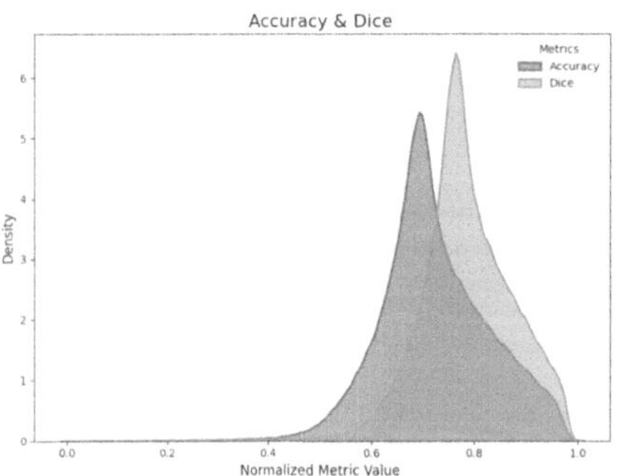

Fig. 7. Accuracy and Dice coefficient performance.

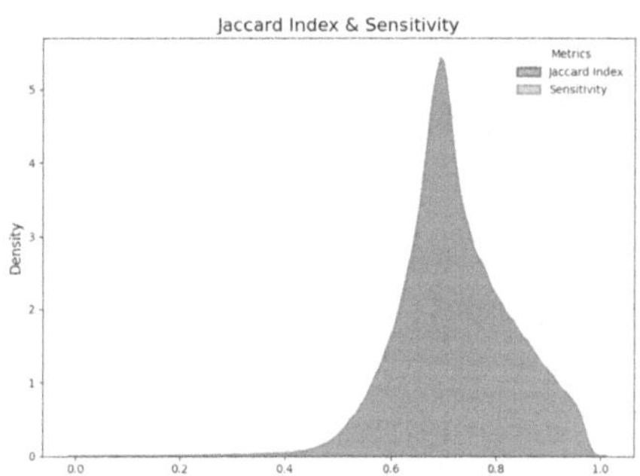

Fig. 8. Jaccard Index and Sensitivity performance.

Following segmentation, the extracted Regions of Interest (ROIs) were used for lesion classification. We evaluated several ranges of pre-trained deep convolutional neural networks fine-tuned via transfer learning. Among all pre-trained models tested, our proposed model achieved competitive performance, with an accuracy and precision of 99%. Figures 9–10 show the visualization techniques of our proposal, including the confusion matrix and ROC curve. Further, the ROC-AUC curve (Fig. 10) shows an AUC of 1.00, confirming excellent class separability by EfficientNetB7. This is driven by a large, diverse dataset (401,059 images), which minimizes overfitting and improves generalization. Together, these results validate the model's robustness, reliability, and consistent performance across lesion classes.

Additionally, Table 1 presents a comprehensive comparison of deep learning models for skin lesion classification. Our proposed method, based on a fine-tuned EfficientNetB7 architecture, achieves the highest overall performance, with an ROC AUC of 0.99 and strong classification metrics—Accuracy of 0.97 and Precision, Recall, and F1-score each at 0.96—surpassing all compared models. Among the other pre-trained architectures, EfficientNetB7 continues to stand out with an F1-score of 0.96, closely followed by EfficientNetB1, CoatNet, and HRNet, each scoring 0.95. Their ability to handle the complexity of dermoscopic images can be attributed to efficient scaling and hybrid feature representations. Models like ResNet101 and InceptionResNetV2 also perform well, with F1-scores of 0.94 and 0.93, respectively. In contrast, mid-tier models such as DenseNet121, DenseNet169, and MobileNetV2 show moderate effectiveness, with F1-scores between 0.83 and 0.88, indicating challenges in capturing subtle lesion features. Interestingly, earlier architectures like VGG16, NASNetMobile, AlexNet, and ConvNeXt maintain competitive F1-scores in the 0.89–0.91 range, suggesting that model success depends not only on depth or novelty but also on alignment with domain-specific characteristics.

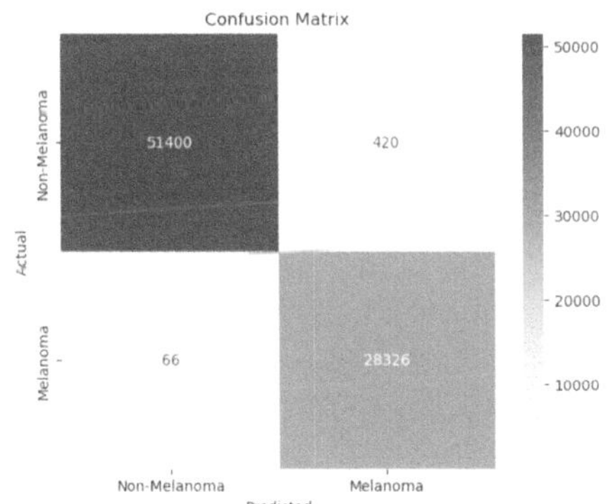

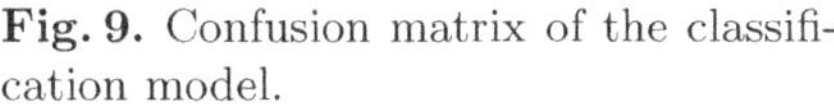
Fig. 9. Confusion matrix of the classification model.

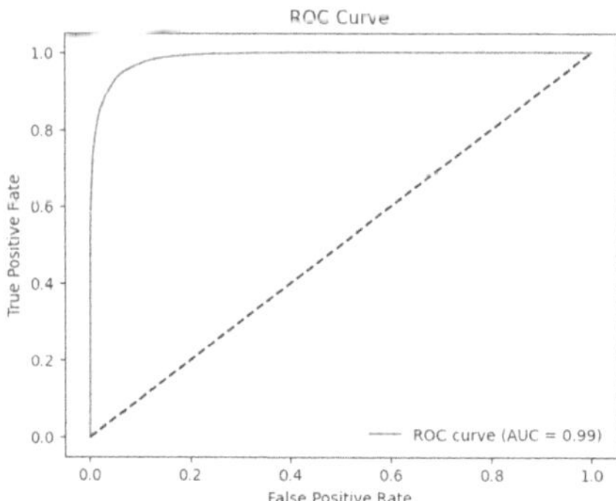

Fig. 10. ROC-AUC curve showing classification performance.

Table 1. Performance comparison of deep classification models.

Model	Accuracy	Precision	Recall	F1-score
ResNet50	0.92	0.91	0.90	0.92
ResNet101	0.94	0.94	0.93	0.94
VGG16	0.91	0.91	0.91	0.91
VGG19	0.90	0.90	0.90	0.90
DenseNet169	0.88	0.87	0.86	0.87
DenseNet121	0.89	0.88	0.87	0.88
DenseNet201	0.91	0.90	0.89	0.90
MobileNetV2	0.84	0.83	0.82	0.83
EfficientNetB0	0.95	0.94	0.94	0.94
EfficientNetB1	0.96	0.95	0.95	0.95
EfficientNetB7	**0.97**	**0.96**	**0.96**	**0.96**
InceptionV3	0.93	0.92	0.91	0.92
InceptionResNetV2	0.94	0.93	0.93	0.93
Xception	0.95	0.94	0.94	0.94
AlexNet	0.90	0.89	0.89	0.89
ConvNeXt	0.91	0.90	0.90	0.90
NASNetMobile	0.89	0.88	0.88	0.88
NASNetLarge	0.91	0.90	0.90	0.90
CoatNet	0.96	0.95	0.95	0.95
HRNet	0.95	0.94	0.94	0.94

For classification, our EfficientNetB7-based model achieves 97% accuracy with consistent predictions, fast inference, and high robustness. Achieving an AUC of 0.99, the model matches or exceeds other recent classification methodologies in accuracy, efficiency, and interpretability (Table 2), making it suitable for real-time clinical use, especially in resource-limited settings.

Table 2. Comparison of different state-of-the-art classification methodologies.

Model	Year	Dataset	Accuracy	Precision	Recall	F1 Score	AUC-ROC
Our Work	2025	ISIC 2016–2020	**97%**	**96%**	**96%**	**96%**	**0.99**
Albahar et al. [3]	2021	ISIC Archive	98.3%	–	–	–	0.983
Khan et al. [11]	2020	ISBI 2017	95.3%	95.5%	95.5%	–	0.98
Behara et al. [5]	2024	ISIC 2020 & HAM10000	98%	98.87%	98.82%	98%	0.973
Mahbod et al. [15]	2020	ISIC 2017	–	–	81.2%	–	0.856
Arjun et al. [4]	2024	ISIC 2018–2020	94.6%	95.13%	95.67%	95.13%	–
Shakya et al. [17]	2025	ISIC 2018	92.87%	93.1%	91.2%	92.1%	–
Hosny et al. [10]	2019	ISIC	96.86%	96.92%	96.90%	–	–
Sae-Lim et al. [16]	2019	HAM10000	83.23%	82.0%	85.0%	82.0%	–
Lee et al. [12]	2018	HAM10000	89.9%	–	–	–	–

5 Conclusions

This study presented an efficient approach for skin lesion analysis using image preprocessing and deep learning techniques. The method achieved 97% classification accuracy and 86% segmentation accuracy on the ISIC 2016–2020 dataset. Enhancements like CLAHE and the Sonar Effect significantly improved lesion boundary detection across diverse imaging conditions. Compared to state-of-the-art methods, our approach offers competitive performance and more reliable inference without relying on complex architectures or extensive data augmentation. Its simplicity, reproducibility, and explainability make it well-suited for clinical integration where trust and transparency are essential. Future work will focus on extending the framework to handle multi-class lesion classification and integrating real-time decision support systems.

Acknowledgments. This work is partially supported by the Ministry of Science and Innovation of Spain, grant number PID2022-136764OA-I00, project name Automated Detection of Non Lesional Focal Epilepsy by Probabilistic Diffusion Deep Neural Models. It includes funds from the European Regional Development Fund (ERDF). It is also partially supported by the Fundación Unicaja under project PUNI-003_2023, project name Intelligent System to Help the Clinical Diagnosis of Non-Obstructive Coronary Artery Disease in Coronary Angiography. The authors thankfully acknowledge the computer resources, technical expertise, and assistance provided by the SCBI (Supercomputing and Bioinformatics) center of the University of Málaga. They also gratefully acknowledge the support of NVIDIA Corporation with the donation of an RTX A6000 GPU with 48 Gb. The authors also thankfully acknowledge the grant of the Universidad de Málaga and the Instituto de Investigación Biomédica de Málaga y Plataforma en Nanomedicina-IBIMA Plataforma BIONAND.

References

1. Abbas, Q., Celebi, M.E., García, I.F.: Hair removal methods: a comparative study for dermoscopy images. Biomed. Signal Process. Control **6**(4), 395–404 (2011)
2. Ahmed, A., Sun, G., Bilal, A., Li, Y., Ebad, S.A.: Precision and efficiency in skin cancer segmentation through a dual encoder deep learning model. Sci. Rep. **15**(1), 4815 (2025)
3. Albahar, M.A.: Skin lesion classification using convolutional neural network with novel regularizer. IEEE Access **7**, 38306–38313 (2019)
4. Arjun, K., Kumar, K.S., Dhanaraj, R.K., Ravi, V., Kumar, T.G.: Optimizing time prediction and error classification in early melanoma detection using a hybrid RCNN-LSTM model. Microsc. Res. Tech. **87**(8), 1789–1809 (2024)
5. Behara, K., Bhero, E., Agee, J.T.: An improved skin lesion classification using a hybrid approach with active contour snake model and lightweight attention-guided capsule networks. Diagnostics **14**(6), 636 (2024)
6. Bi, L., Kim, J., Ahn, E., Kumar, A., Fulham, M., Feng, D.: Dermoscopic image segmentation via multistage fully convolutional networks. IEEE Trans. Biomed. Eng. **64**(9), 2065–2074 (2017)
7. Codella, N., Cai, J., Abedini, M., Garnavi, R., Halpern, A., Smith, J.R.: Deep learning, sparse coding, and SVM for melanoma recognition in dermoscopy images. In: International Workshop on Machine Learning in Medical Imaging, pp. 118–126. Springer, Heidelberg (2015)
8. Esteva, A., et al.: Dermatologist-level classification of skin cancer with deep neural networks. Nature **542**(7639), 115–118 (2017)
9. Gayatri, E., Aarthy, S.: Challenges and imperatives of deep learning approaches for detection of melanoma: a review. Int. J. Image Graph. **23**(03), 2240012 (2023)
10. Hosny, K.M., Kassem, M.A., Foaud, M.M.: Classification of skin lesions using transfer learning and augmentation with Alex-net. PLoS ONE **14**(5), e0217293 (2019)
11. Khan, M.A., Javed, M.Y., Sharif, M., Saba, T., Rehman, A.: Multi-model deep neural network based features extraction and optimal selection approach for skin lesion classification. In: 2019 International Conference on Computer and Information Sciences (ICCIS), pp. 1–7. IEEE (2019)
12. Lee, Y.C., Jung, S.H., Won, H.H.: Wonderm: skin lesion classification with fine-tuned neural networks. arXiv preprint arXiv:1808.03426 (2018)
13. Li, R., Zheng, S., Duan, C., Yang, Y., Wang, X.: Classification of hyperspectral image based on double-branch dual-attention mechanism network. Remote Sens. **12**(3), 582 (2020)
14. Liu, Z., Hu, J., Gong, X., Li, F.: Skin lesion segmentation with a multiscale input fusion U-Net incorporating Res2-SE and pyramid dilated convolution. Sci. Rep. **15**(1), 7975 (2025)
15. Mahbod, A., Schaefer, G., Ellinger, I., Ecker, R., Pitiot, A., Wang, C.: Fusing fine-tuned deep features for skin lesion classification. Comput. Med. Imaging Graph. **71**, 19–29 (2019)
16. Sae-Lim, W., Wettayaprasit, W., Aiyarak, P.: Convolutional neural networks using mobilenet for skin lesion classification. In: 2019 16th International Joint Conference on Computer Science and Software Engineering (JCSSE), pp. 242–247. IEEE (2019)
17. Shakya, M., Patel, R., Joshi, S.: A comprehensive analysis of deep learning and transfer learning techniques for skin cancer classification. Sci. Rep. **15**(1), 4633 (2025)

18. Shu, X., Li, Z., Tian, C., Chang, X., Yuan, D.: An active learning model based on image similarity for skin lesion segmentation. Neurocomputing **630**, 129690 (2025)
19. Srinivasu, P.N., SivaSai, J.G., Ijaz, M.F., Bhoi, A.K., Kim, W., Kang, J.J.: Classification of skin disease using deep learning neural networks with mobilenet v2 and lstm. Sensors **21**(8), 2852 (2021)
20. Tschandl, P., Rosendahl, C., Kittler, H.: The HAM10000 dataset, a large collection of multi-source dermatoscopic images of common pigmented skin lesions. Sci. Data **5**(1), 1–9 (2018)
21. Ünver, H.M., Ayan, E.: Skin lesion segmentation in dermoscopic images with combination of YOLO and grabcut algorithm. Diagnostics **9**(3), 72 (2019)
22. Zia Ur Rehman, M., Ahmed, F., Alsuhibany, S.A., Jamal, S.S., Zulfiqar Ali, M., Ahmad, J.: Classification of skin cancer lesions using explainable deep learning. Sensors **22**(18), 6915 (2022)

Towards Traffic Optimization Using Reinforcement Learning

Hicham Affou[1,2](✉), Daniel Caballero-Martin[1,2], Geovany Satama[1,2], Jose Antonio Ramos-Hernanz[2], Koldo Portal-Porras[2], and Jose Manuel Lopez-Guede[1,2]

[1] Department of System Engineering and Automation Control, Faculty of Engineering of Vitoria-Gasteiz, University of the Basque Country (UPV/EHU), Nieves Cano, 12, 01006 Vitoria-Gasteiz, Spain
haffou001@ikasle.ehu.eus , jm.lopez@ehu.es

[2] Faculty of Engineering of Vitoria-Gasteiz, University of the Basque Country (UPV/EHU), C/Nieves Cano, 12, 01006 Vitoria-Gasteiz, Spain

Abstract. Traffic congestion in urban environments causes significant delays, increased fuel consumption, and environmental pollution. While optimization algorithms have been explored, traditional traffic control strategies struggle with real-time dynamism and complexity. This paper addresses these limitations by proposing an adaptive traffic signal control system powered by reinforcement learning, specifically leveraging a Q-Network algorithm. The framework is implemented and validated using Eclipse SUMO (Simulation of Urban Mobility), a widely utilized microscopic traffic simulation software that integrates with Python to enable the implementation and customization of reinforcement learning techniques for specific environments. The efficacy of the proposed method is demonstrated through a simulation of one of Vitoria-Gasteiz's most complex and congested intersections: the América Latina roundabout, chosen for its high traffic volume and critical role as a connection point between main roads. The results demonstrate improved traffic management and environmental sustainability within intelligent transportation systems, offering a scalable solution for real-time optimization in evolving urban networks.

Keywords: Traffic optimization · Reinforcement learning · Traffic congestion · Intelligent transportation system · SUMO

1 Introduction

Urban areas are grappling with an unprecedented surge in vehicular traffic. In 2023, the EU alone reported approximately 265 million registered vehicles, marking a 6.5% increase compared to 2018 [2]. This explosive growth has significantly intensified congestion, contributing to longer travel times, increased fuel consumption, and elevated greenhouse gas emissions. A central component in

E. Corchado et al. (Eds.): SOCO 2025, CCIS 2806, pp. 520–529, 2026.
https://doi.org/10.1007/978-3-032-19763-4_48

managing urban mobility and mitigating these issues is the effective regulation of traffic signals, which governs flow across a wide range of intersection types, including roundabouts, T-junctions, and both signalized and unsignalized intersections.

Traffic Signal Control (TSC) research, initiated with fixed-time schedules [14] and later actuated systems, remains challenged by unpredictable traffic dynamics like congestion or accidents [10,16]. Advances in AI have enabled Adaptive Traffic Signal Control (ATSC) systems, categorized as model-based (using mathematical optimization with real-time and historical data [17]) or model-free (learning policies directly from environmental interactions [9]).

In traffic management, Reinforcement Learning (RL) particularly model-free implementations enables continuous adaptation to reduce delays, minimize stops, and enhance traffic flow in complex urban networks [6]. Despite these advantages, challenges persist in adapting to dynamic traffic patterns and heterogeneous environments. Q-learning, a specialized RL variant, offers compelling solutions by exploring dynamic conditions to identify optimal signal actions. Validated in optimizing individual intersections [18] and multi-agent traffic frameworks [13], Q-learning's autonomous adaptation capacity positions it as a promising approach for adaptive traffic signal control. This study specifically investigates its application to signalized intersections to address scalability and coordination gaps in existing systems.

We propose a Q-learning-based traffic light control system a clever method applied at the most complex roundabout in Vitoria-Gasteiz, Spain. This reinforcement learning approach autonomously optimizes signal timing using real-time traffic data to resolve severe congestion bottlenecks.

The manuscript unfolds through six integrated sections: Sect. 2 surveys extant scholarship addressing the research problem. Section 3 specifies materials comprising the experimental substrate. Section 4 elaborates the methodological paradigm, explicitly characterizing the proposed optimization framework. Empirical results are rigorously evaluated in Sect. 5, while Sect. 6 synthesizes conclusions.

2 Related Works

Traffic light control has evolved significantly. It started with traditional fixed-time (FT) systems that relied on historical data to set predetermined schedules [20]. Later, actuated methods emerged, dynamically adjusting signals using real-time sensors. While actuated control, including semi-actuated (minor-road detection) [5] and fully actuated (all-approach detection) [22] variants, improved responsiveness, its benefits were limited to individual intersections, preventing broader network optimization.

Modern approaches have led to Adaptive Traffic Signal Control (ATSC), which now enables continuous, network-wide optimization through advanced algorithmic models [19]. These frameworks generally divide into two main types.

Model-based systems use mathematical techniques like model predictive control [4] and dynamic traffic assignment [11], best suited for structured environments. In contrast, model-free reinforcement learning (RL) approaches utilize methods such as value-based techniques (like Q-learning and its DQN variants) or policy-based methods (like PPO and DPG) [15]. These RL techniques are particularly effective because they learn optimal policies directly from real-time traffic dynamics, performing exceptionally well in complex, unpredictable scenarios where older, traditional methods fail. For example, two RL methods (deep policy-gradient and value-function based agents) were suggested by Seyed Sajad Mousavi [12] to anticipate the optimal traffic signal for a traffic intersection. The value-function based agent first estimates values for all valid control signals, whereas the policy-gradient based agent translates its observations based on a snapshot of the present state of a graphical track simulator directly to the control signal. The agent could then base the choice on the value that is the highest.

Research applying deep reinforcement learning (DRL) to traffic management addresses distinct operational scales. M. Li et al. [7] specifically target the microscopic level of individual vehicle control, developing a driving strategy using Deep Deterministic Policy Gradient (DDPG) to enhance traffic safety and mitigate undesirable stop-and-go oscillations; their SUMO simulations demonstrate the strategy's efficacy in reducing accident probability compared to existing techniques like adaptive cruise control. Conversely, Z. Li, Xu, and Zhang [8], focus on the macroscopic challenge of traffic light network optimization, employing the Multi-Agent Deep Deterministic Policy Gradient (MADDPG) framework within a centralized learning and decentralized execution paradigm to coordinate signal control across intersections; their SUMO-based single-modal simulations confirm the approach's effectiveness in optimizing traffic flow regulation.

Additionally, the authors Xiaoqiang et al. [21], address scalability and cooperative learning challenges in applying Multi-Agent Reinforcement Learning (MARL) to large-scale traffic signal control (TSC). They propose Cooperative double Q-learning (Co-DQL), which enhances performance through independent double Q-learning with UCB to reduce over-estimation, mean field approximation for agent interactions, innovative reward, and state sharing mechanisms for stability. The algorithm's convergence is theoretically analyzed, and empirical validation against state-of-the-art decentralized MARL methods demonstrates Co-DQL's efficacy in minimizing vehicle waiting times in simulated traffic scenarios.

Finally, The study [23] tackled autonomous highway driving challenges by employing a Double Deep Q-Network (DDQN)-based DRL framework to enhance decision-making and vehicle interaction, addressing Q-learning's tendency to overestimate action values. The host vehicle agent learns through trial-and-error interactions within a SUMO-based simulation platform, which enables flexible testing of control algorithms. Results demonstrate that the trained agent successfully navigates highway scenarios, achieving near-maximum safe speeds without collisions, validating the model's ability to balance efficiency and safety.

The framework highlights the potential of DDQN in mitigating overestimation risks while enabling reliable, real-time autonomous driving decisions in dynamic environments.

3 Material

3.1 Simulator

SUMO (Simulation of Urban MObility) was selected among traffic simulators as AnyLogic, Vissim, and Aimsun for its open-source flexibility and microscopic multi-modal control enabling individual vehicle parameterization, despite limitations like static runtime behavior. Chosen for its alignment with Q-learning methodologies, it provides granular traffic dynamics, configurable demand and routing essential for creating diverse training scenarios. Key advantages include its discrete-action environment enabling precise phase control, the TraCI API facilitating real-time state retrieval and immediate reward computation after each action, and deterministic high-speed execution that supports the extensive exploration required for Q-learning convergence through comprehensive performance metrics.

3.2 Traffic Network

Vitoria-Gasteiz is a city in Spain, situated in the autonomous community of the Basque Country. As of 2024, it has a population of 257,968 inhabitants, according to the Spanish National Institute of Statistics [1]. The city is distinguished by its unique geographical setting. Unlike coastal cities, Vitoria-Gasteiz lies inland on a plateau, encircled by the Green Belt, a network of parks and natural areas that blend urban and rural landscapes. It spans an area of 276.8 km^2, resulting in a population density of approximately 929.9 inhabitants/km^2 [3]. The city is connected to major transportation networks, including the A-1 and AP-1 highways, which link it to Bilbao, San Sebastián, and Madrid, as well as the N-102 road that integrates it into regional transit systems.

4 Methodology

This study conducts an evaluation of the proposed system through the application of the Simulation of Urban MObility (SUMO) traffic simulator and RL techniques, with the city of Vitoria-Gasteiz, Spain, serving as an empirical case study. The subsequent section delineates the methodological framework, which comprises the structural design of the road network and the development of the optimization algorithm.

4.1 Road Network Design and Simulation Scope

Vitoria-Gasteiz, Spain, was selected for this traffic optimization study due to its significant congestion challenges, exemplified by the América Latina roundabout a critical hotspot near key neighborhoods (Lakua, Gazalbide, El Pilar), Arriaga Park, and the Bus Station. This roundabout forms part of the city's urban ring road, connecting major routes including Portal de Foronda (to Bilbao, Madrid, and Burgos), Honduras Street, Juan de Garay Avenue, and Bulevar de Euskal Herria. To focus on car traffic dynamics, the simulation simplified the complex real-world layout (originally featuring three lanes, a bus lane, and tram tracks) by excluding public transport infrastructure. The road network was reconstructed using OpenStreetMap data imported into SUMO via netconvert, with manual corrections in netedit applied to address OSM inaccuracies and ensure geometric fidelity.

4.2 Design of the Optimization Algorithm

The algorithm uses a decentralized Q-learning framework to optimize traffic light control in a SUMO simulation. Every traffic signal has its own independent operation, maintaining its own Q-table that maps three discrete phase actions (green, yellow, red indices [0, 1, 2]) to learned values representing their long-term utility. There are hyperparameters that guide the learning process: a learning rate ($\alpha = 0.1$) to moderate update sensitivity, a discount factor ($\gamma = 0.9$) to prioritize future rewards, and an epsilon-greedy policy ($\epsilon = 0.1$) to balance exploration (10% random actions) and exploitation (90% greedy selection of the highest Q-value action). At each simulation step, traffic lights select a phase using this policy, apply it via TraCI, and update their Q-tables using the Bellman equation. Here, the reward is derived from the negative sum of waiting times across all edges, incentivizing congestion reduction. The algorithm is stateless, relying solely on phase transitions rather than explicit traffic metrics (e.g., vehicle queues), which simplifies computation but limits context-aware adaptation.

4.3 Reward Mechanism and Data Collection

By penalizing accumulated waiting times, the reward function directly targets system-wide efficiency, ensuring traffic lights optimize holistically rather than locally. During training, data on waiting times, phase durations, and rewards are logged. Post-simulation, three layers of visualization are produced: (1) a waiting time trendline to assess policy efficacy, (2) phase distribution bar plots to identify dominant strategies for each traffic light, and (3) reward trajectories to track learning progress. While the decentralized design ensures scalability, it sacrifices coordination between traffic lights, potentially missing synergies that could further reduce congestion (Fig. 1).

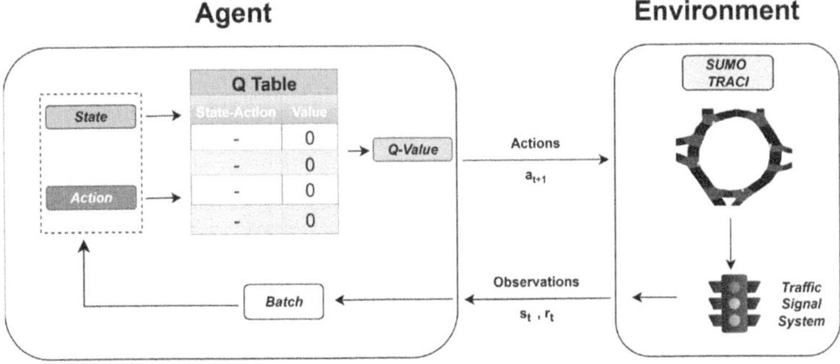

Fig. 1. Overview of the optimization process.

5 Interpretation of Results

Waiting Time Patterns: A decreasing trend in the total waiting time plot indicates the algorithm successfully learns to reduce congestion over iterations. The curve indicates convergence to an ideal policy if it stabilizes. However, persistent fluctuations might signal underfitting due to the stateless design or insufficient exploration (e.g., overly low ϵ) (Fig. 2).

Phase Distribution Analysis: The algorithm appears to favour optimizing throughput at those intersections, as seen by bar plots that display extended green phases for particular traffic lights. Conversely, frequent yellow and red phases might reflect adaptive responses to irregular traffic flows. Disproportion-

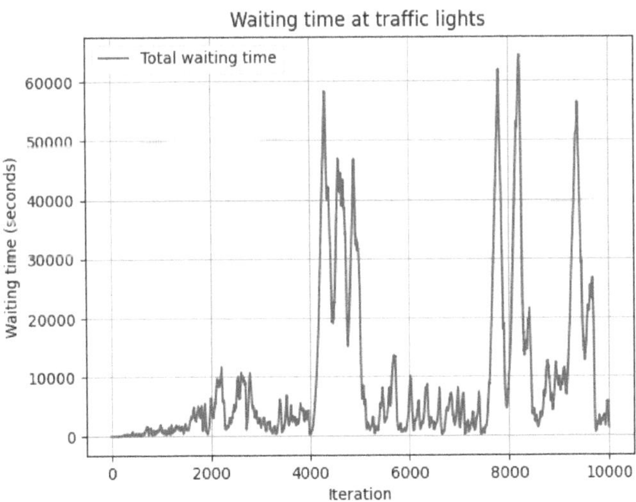

Fig. 2. Total vehicle waiting time.

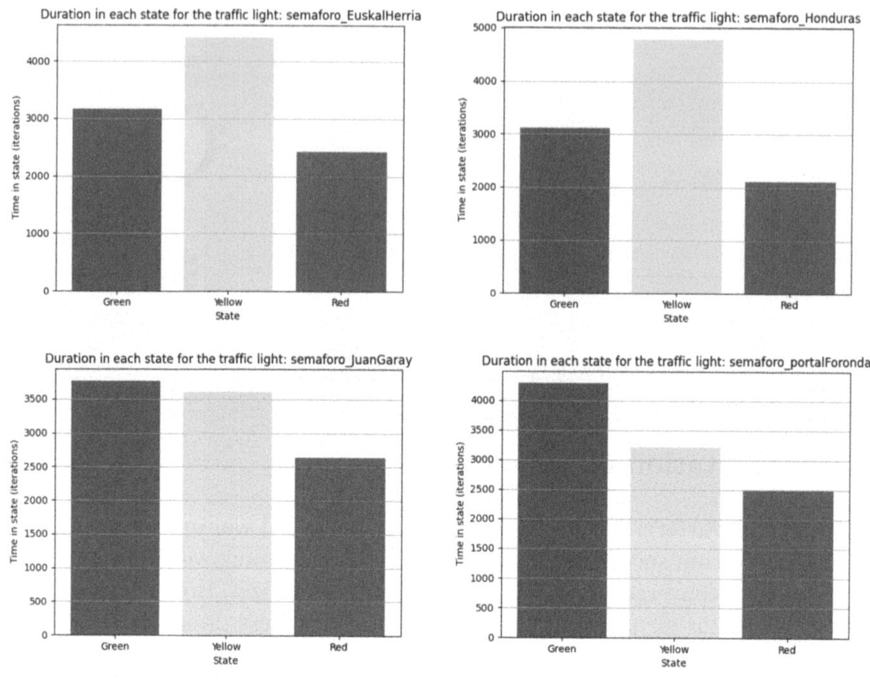

Fig. 3. Traffic Signal Phase Duration Distribution.

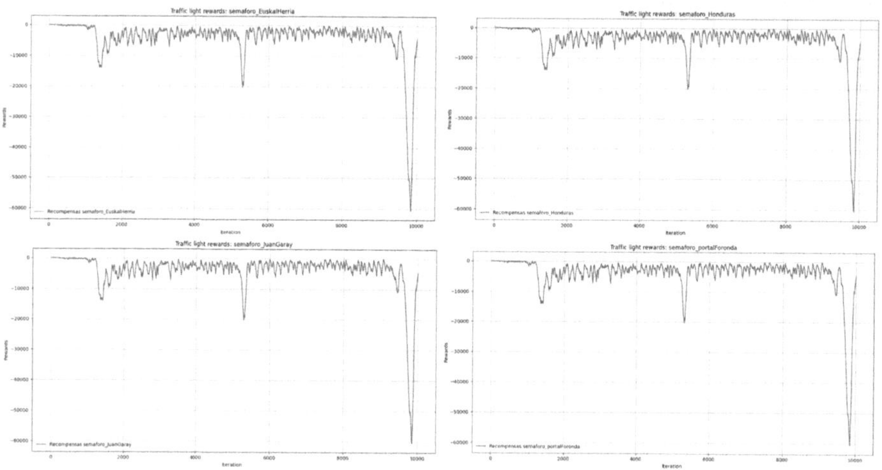

Fig. 4. Rewarding learning by reinforcement.

ate phase durations across traffic lights could highlight imbalances in traffic demand or suboptimal local policies (Fig. 3).

Reward Trajectories: Rewards that increase steadily (less negative values) Verify that learning is occurring effectively. Divergent reward trends between traffic lights (e.g., one agent improving while others stagnate) may indicate competition rather than cooperation, a limitation of the decentralized approach. Sharp drops in rewards could correlate with exploration steps, where random phase changes temporarily increase congestion (Fig. 4).

6 Conclusion

Our research has demonstrated the value of RL as a strategy for optimizing traffic light operation at urban intersections with complex traffic conditions. The implementation of this technique has shown that, through an appropriate approach, it is possible to achieve significant reductions in vehicle waiting times, as well as notable improvements in overall traffic flow.

The use of SUMO, combined with the integration of the TraCI library, has facilitated the simulation of various scenarios, allowing traffic light control parameters to be analysed and adjusted efficiently. Although multiple challenges were faced, especially related to the initial configuration and refinement of the algorithms, the results obtained validate the effectiveness of the approach adopted.

Furthermore, this article highlights the importance of further research at the intersection between artificial intelligence and urban transport. The inclusion of advanced techniques in traffic planning not only addresses current problems, but also offers scalable and sustainable solutions for the future.

Finally, opportunities for improvement were identified, such as the incorporation of other transport modes and the consideration of external factors, which could enrich the analyses and contribute to a more holistic representation of the urban mobility ecosystem.

Acknowledgments. This work was supported by the Vitoria-Gasteiz Mobility Lab Foundation, an organization of the Provincial Council of Araba and the City Council of Vitoria-Gasteiz, through the following grants for the project "Dynamic optimization of traffic lights using Computational Intelligence techniques".

References

1. Araba/Álava: Population by municipality and sex (2854). https://www.ine.es/jaxiT3/Datos.htm?t=2854. Accessed 02 May 2025
2. Passenger cars in the EU. https://ec.europa.eu/eurostat/statistics-explained/index.php?title=Passenger_cars_in_the_EU. Accessed 07 July 2025
3. Vitoria-Gasteiz (Araba, Basque Country, Spain) - Population Statistics, Charts, Map, Location, Weather and Web Information. https://www.citypopulation.de/en/spain/paisvasco/araba/01059__vitoria_gasteiz/. Accessed 02 May 2025

4. Berg, M.v.d., Hegyi, A., Schutter, B.d., Hellendoorn, H.: Integrated traffic control for mixed urban and freeway networks: a model predictive control approach. Eur. J. Transp. Infrastruct. Res. **7**(3) (2007). https://doi.org/10.18757/ejtir.2007.7.3.3390. https://journals.open.tudelft.nl/ejtir/article/view/3390
5. Dabiri, S., Kompany, K., Abbas, M.: Introducing a cost-effective approach for improving the arterial traffic performance operating under the semi-actuated coordinated signal control. Transport. Res. Rec. J. Transport. Res. Board **2672**(18), 71–80 (2018). https://doi.org/10.1177/0361198118772691. arXiv:1804.04960 [stat]
6. El-Tantawy, S., Abdulhai, B., Abdelgawad, H.: Multiagent reinforcement learning for integrated network of adaptive traffic signal controllers (MARLIN-ATSC): methodology and large-scale application on downtown toronto. IEEE Trans. Intell. Transport. Syst. **14**(3), 1140–1150 (2013). https://doi.org/10.1109/TITS.2013.2255286. https://ieeexplore.ieee.org/document/6502719
7. Li, M., Li, Z., Xu, C., Liu, T.: Deep reinforcement learning-based vehicle driving strategy to reduce crash risks in traffic oscillations. Transp. Res. Rec. **2674**(10), 42–54 (2020)
8. Li, Z., Xu, C., Zhang, G.: A deep reinforcement learning approach for traffic signal control optimization. arXiv preprint arXiv:2107.06115 (2021)
9. Michailidis, I.T., Manolis, D., Michailidis, P., Diakaki, C., Kosmatopoulos, E.B.: Autonomous self-regulating intersections in large-scale urban traffic networks: a chania city case study. In: 2018 5th International Conference on Control, Decision and Information Technologies (CoDIT), pp. 853–858 (2018). https://doi.org/10.1109/CoDIT.2018.8394910. https://ieeexplore.ieee.org/document/8394910. iSSN: 2576-3555
10. Michailidis, I.T., Michailidis, P., Rizos, A., Korkas, C., Kosmatopoulos, E.B.: Automatically fine-tuned speed control system for fuel and travel-time efficiency: a microscopic simulation case study. In: 2017 25th Mediterranean Conference on Control and Automation (MED), pp. 915–920 (2017). https://doi.org/10.1109/MED.2017.7984236. https://ieeexplore.ieee.org/document/7984236. iSSN: 2473-3504
11. Mitsakis, E., Salanova, J.M., Giannopoulos, G.: Combined dynamic traffic assignment and urban traffic control models. Procedia Social Behav. Sci. **20**, 427–436 (2011). https://doi.org/10.1016/j.sbspro.2011.08.049. https://linkinghub.elsevier.com/retrieve/pii/S1877042811014297
12. Mousavi, S.S., Schukat, M., Howley, E.: Traffic light control using deep policy-gradient and value-function-based reinforcement learning. IET Intel. Transp. Syst. **11**(7), 417–423 (2017)
13. Mushtaq, A., Haq, I.U., Sarwar, M.A., Khan, A., Khalil, W., Mughal, M.A.: Multi-agent reinforcement learning for traffic flow management of autonomous vehicles. Sensors **23**(5), 2373 (2023). https://doi.org/10.3390/s23052373. https://www.mdpi.com/1424-8220/23/5/2373
14. Noaeen, M., et al.: Reinforcement learning in urban network traffic signal control: a systematic literature review. Expert Syst. Appl. **199**, 116830 (2022). https://doi.org/10.1016/j.eswa.2022.116830. https://www.sciencedirect.com/science/article/pii/S0957417422002858
15. Papadopoulos, G., et al.: Deep reinforcement learning in service of air traffic controllers to resolve tactical conflicts. Expert Syst. Appl. **236**, 121234 (2024). https://doi.org/10.1016/j.eswa.2023.121234. https://linkinghub.elsevier.com/retrieve/pii/S0957417423017360
16. Pell, A., Meingast, A., Schauer, O.: Trends in real-time traffic simulation. Transport. Res. Procedia **25**, 1477–1484 (2017). https://doi.org/10.1016/j.trpro.2017.05.175. https://www.sciencedirect.com/science/article/pii/S2352146517304684

17. Qadri, S.S.S.M., Gökçe, M.A., Öner, E.: State-of-art review of traffic signal control methods: challenges and opportunities. Eur. Transp. Res. Rev. **12**(1), 55 (2020). https://doi.org/10.1186/s12544-020-00439-1
18. Rosyidi, M., Bismantoko, S., Widodo, T.: Reinforcement learning with the classical Q-learning algorithm for optimizing single intersection performance. In: 2020 International Conference on Data Analytics for Business and Industry: Way Towards a Sustainable Economy (ICDABI), pp. 1–5 (2020). https://doi.org/10.1109/ICDABI51230.2020.9325657. https://ieeexplore.ieee.org/document/9325657
19. Spall, J., Chin, D.: A model-free approach to optimal signal light timing for system-wide traffic control. In: Proceedings of 1994 33rd IEEE Conference on Decision and Control, vol. 2, pp. 1868–1875 (1994). https://doi.org/10.1109/CDC.1994.411110. https://ieeexplore.ieee.org/document/411110
20. Thunig, T., Scheffler, R., Strehler, M., Nagel, K.: Optimization and simulation of fixed-time traffic signal control in real-world applications. Procedia Comput. Sci. **151**, 826–833 (2019). https://doi.org/10.1016/j.procs.2019.04.113. https://linkinghub.elsevier.com/retrieve/pii/S1877050919305770
21. Wang, X., Ke, L., Qiao, Z., Chai, X.: Large-scale traffic signal control using a novel multiagent reinforcement learning. IEEE Trans. Cybern. **51**(1), 174–187 (2020)
22. Xu, H., Zhang, K., Zhang, D., Zheng, Q.: Traffic-responsive control technique for fully-actuated coordinated signals. IEEE Trans. Intell. Transport. Syst. **23**(6), 5460–5469 (2022). https://doi.org/10.1109/TITS.2021.3054054. https://ieeexplore.ieee.org/document/9349151/
23. Zhao, J., Qu, T., Xu, F.: A deep reinforcement learning approach for autonomous highway driving. IFAC-PapersOnLine **53**(5), 542–546 (2020)

Exploring Temporal Action Segmentation Techniques for Enhanced Bird Behavior Recognition

Bruno Sancho-Deltell, Javier Rodriguez-Juan(✉), Manuel Benavent-Lledo, David Mulero-Perez, David Ortiz-Perez, and Jose Garcia-Rodriguez

Department of Computer Technology, University of Alicante, Alicante, Spain
bruno.sancho@ua.es,
{jrodriguez,mbenavent,dmulero,dortiz,jgarcia}@dtic.ua.es

Abstract. Bird behavior recognition is an underexplored research topic within wildlife monitoring despite its significant implications for ecological conservation. This work employs Temporal Action Segmentation (TAS) methods based on deep learning to make progresses in the analysis of bird behaviors in the wild. We conduct a comprehensive research using two different strategies: a frame-level approach, which classifies behaviors based on single images, and segment-level pipelines, which leverage sequences of frames to explicitly model temporal context. Both approaches utilize state-of-the-art deep learning architectures, and their effectiveness is evaluated through extensive ablation studies on the Visual-WetlandBirds dataset. Furthermore, data augmentation methods, including background replacement, silhouette masking, and occlusion simulation, are developed to address dataset biases and class imbalance. This research offers practical insights into applying TAS techniques in ecological video analysis.

Keywords: bird behavior recognition · temporal action segmentation · ecological video surveillance

1 Introduction

Understanding animal behavior is essential for advancing ecological research and conservation efforts. In particular, birds serve as effective bioindicators due to their widespread presence and sensitivity to environmental disturbances [31]. Analyzing bird behavior can offer critical information about ecosystem health, biodiversity trends, and the consequences of climate change [32]. Despite its importance, the study of bird behavior in natural habitats remains challenging, mainly because traditional monitoring techniques depend on manual annotations by experts, which are often time-consuming and difficult to scale. One promising direction is Temporal Action Segmentation (TAS) [2], which involves classifying each frame in untrimmed videos with an action classes. While TAS has been widely explored in human activity recognition [1], its application to

E. Corchado et al. (Eds.): SOCO 2025, CCIS 2806, pp. 530–539, 2026.
https://doi.org/10.1007/978-3-032-19763-4_49

animal behavior, particularly in the wild, remains limited and presents unique challenges, including subtle inter-class variations, environmental complexity, and limited annotated data [17].

In this work, we explore the usage of TAS techniques to fulfill the bird behavior recognition task in videos. To this end, we used the Visual-Wetland Birds dataset [22], which captures diverse species and behavior types in natural wetland habitats. This dataset is appropriate for our goal as it provides fine-grained bird behavior annotations. Throughout this work, we evaluated two approaches: a frame-level pipeline that treats each frame independently, and a segment-level pipeline that models temporal structure to capture behavioral dynamics and motions. Additionally, we design and apply domain-specific data augmentation strategies to mitigate visual biases and enhance generalization, especially for underrepresented classes. Both pipelines rely on deep learning architectures, such as convolutional neural networks [6] and transformers [11].

The main contributions of this work are as follows:

- We evaluate the transferability of human-centric TAS methods to the birds behavior recognition task, taking into account domain-specific challenges and data scarcity.
- We introduce tailored data augmentation strategies designed for bird videos in the wild, improving robustness to species-specific appearance and background biases.
- We perform a comprehensive evaluation of both frame-level and segment-level TAS approaches for bird behavior recognition, including detailed ablation studies of key architectural choices.

These contributions provide a comprehensive evaluation framework for bird behavior recognition and offer practical insights for applying TAS methods in ecological video data. Code used for experimentation is available on GitHub[1].

2 Related Works

TAS is a key problem in video understanding, focused on partitioning untrimmed video streams into semantically meaningful action segments, and assigning a behavioral label to each frame [2]. Unlike action recognition [3], which classifies pre-trimmed clips, or action detection [4] and spatio-temporal localization [5], which require identifying boundaries or spatial regions, TAS demands continuous, fine-grained labeling throughout long videos.

Recent progress in TAS has been driven by deep learning [39–42]. Early methods combined convolutional neural networks for spatial feature extraction [7] with recurrent neural networks, especially long short-term memory networks [8], for temporal modeling. More recently, temporal convolutional networks [9] and multi-stage variants [10] have achieved state-of-the-art results by capturing multi-scale temporal context, while transformer-based models [11,12]

[1] https://github.com/3dperceptionlab/tfg_bsancho.

and hybrid architectures like Pyramid Dilated Attention Network (PDAN) [13] and ActionFormer [14] leverage self-attention for long-range temporal dependencies.

The field has benefited from large-scale human activity datasets such as 50Salads [15], and Breakfast [16], which offer frame-level annotations for structured scenarios. However, transferring these advances to animal behavior, and specifically to birds, presents several challenges. Ecological videos exhibit high scene diversity, frequent occlusions, rare or subtle behaviors, and significant class imbalance [18,19].

Public datasets for bird behavior recognition remain limited. Furthermore, most of them lack continuous, frame-level segmentation, and the combination of high intra-class variability with subtle visual differences makes this domain challenging. VB100 [20] focuses on fine-grained species classification with short video clips, while datasets like 3D-POP [21] addresses pose estimation rather than behavior. The recent Visual-WetlandBirds dataset [22], with frame-level annotations of behaviors and species in wetland habitats, represents a significant step for advancing TAS research in avian ecology.

Recent works have explored TAS in broader animal contexts, leveraging deep learning to automate behavior analysis. Common approaches rely on pose-estimation methods, such as DeepLabCut [33] and SLEAP [34], which provide skeletal representations for downstream behavior modeling. Recent transformer-based models further extend this direction, such as ADPT [35], which enhances cross-species pose estimation robustness, AnimalMotionCLIP [36], which integrates motion-aware embeddings, and PandaFormer [37], which employs joint time and spatial attention mechanisms. Nevertheless, dedicated TAS pipelines remain scarce in ecological settings, particularly for long, untrimmed sequences.

3 Methodology

Our experimental study is conducted on the Visual-WetlandBirds dataset [22], which contains 178 videos (almost one hour) of 13 bird species performing 7 distinct behaviors. Videos from the dataset were recorded in a natural park, located in Alicante, Spain. Each video is densely annotated at the frame level with species, behavior class, and spatial localization via bounding boxes, enabling supervised training and detailed evaluation of action segmentation models.

To address dataset-specific challenges such as class imbalance and background bias, we introduce targeted data augmentation strategies. These include: (1) replacing original backgrounds with uniform water to minimize correlation between species and aquatic scenes, (2) silhouette augmentation, which fills bird masks with black to suppress color biases, and (3) simulated occlusions using naturalistic shapes and brush strokes to enhance robustness against real-world obstructions. Frame samples applying this techniques are shown in Fig. 1. To generate the augmented data, we use Segment Anything Model (SAM) [29], which creates accurate bird masks for water and silhouette augmentations. In order to generate simulated occlusions, geometric and free-form shapes are added over the bird regions.

Fig. 1. Example of the three data augmentation filters applied to a frame. Top left shows the original frame, top right the uniform water background replacement, bottom left the silhouette, and bottom right the occlusions transformation.

As mentioned in Sect. 2, TAS involves per-frame classification of videos. To address this task, we propose two different processing pipelines: a frame-level pipeline, which classifies each frame independently, and a segment-level pipeline, which processes multiple frames simultaneously.

The frame-level pipeline is limited to analyzing a single frame at a time for classification. This constraint forces the model to rely on static visual cues, such as the bird's pose, to infer behavior. This pipeline consists of a pretrained feature extractor followed by an MLP classifier. We evaluate three MLP variants to explore the effect of classifier complexity on performance: a single linear layer, a two-layer MLP with one hidden activation, and a deeper three-layer MLP. In all experiments, feature extractors are kept frozen, and only classifiers are trained. Conversely, the segment-level pipeline utilizes multiple consecutive frames to classify behavior, allowing it to incorporate dynamic information such as the motion of birds across the sequence. This pipeline employs the PDAN [13], adapted for a multi-class setting, and is evaluated using various backbone architectures for spatio-temporal feature extraction on 16-frame video segments.

Both pipelines are trained using a 70/15/15 split for training, validation, and test sets, with careful split assignment to ensure balanced representation of bird behaviors and species. The main metrics for evaluation are mean Average Precision (mAP) and accuracy, which are systematically reported in all results. Due to the relatively small size of the dataset, all quantitative results are reported on the combined validation and test sets. This choice maximizes the amount of data used for model selection while providing a more reliable and statistically meaningful performance estimate, mitigating the risk of high variance that may arise from reporting on small test subsets alone.

To obtain the best performing setting, ablation studies on the main components of the architectures were conducted. Within the frame-level approach, we assessed the influence of data augmentation techniques, feature extractors, and classifier heads. For the segment-level pipeline, the studies explored the best architectural hyperparameters (*i.e.*, number of stages, layers, and embedding dimension) of the PDAN model.

4 Experiments and Results

This Section is organized as follows: we first present the results for the frame-level pipeline, including the impact of data augmentation; then analyze the segment-level pipeline; and finally compare both approaches.

4.1 Frame-Level Pipeline Results

The frame-level experiment evaluates the performance of image classification models for bird behavior recognition. In this approach we assess the models' capability to extract meaningful features for behavior classification from a single input image. We used five different pretrained backbones: DINO [23], ResNet-152 [7], ViT-Base [11], EfficientNet-B0 [24], and MobileNetV3-L [25]. This selection of extractors provides architectural diversity, covering conventional convolutional neural networks, transformer-based models, and self-supervised representations. Features extracted are then fed to a MLP classifier for behavior classification. As mention in Sect. 3, three MLP variants with different architectural complexities are tested to identify the best performing one.

Table 1. Combined mAP (%) and Accuracy (%) for each feature extractor and classifier configuration in the frame-level pipeline. Each extractor is evaluated with three MLP variants: shallow (1 hidden layer), medium (2 hidden layers), and deep (2 hidden layers with normalization). Bold indicates overall best. Underline indicates best augmentation setting for each extractor.

Extractor	Shallow		Medium		Deep	
	mAP	Acc.	mAP	Acc.	mAP	Acc.
DINO (ViT-B/16)	**49.2**	63.38	<u>48.03</u>	62.82	47.35	<u>62.79</u>
ResNet-152	47.94	63.66	47.39	63.60	<u>47.43</u>	62.57
ViTBase	46.62	61.56	46.50	61.93	46.27	60.54
EfficientNet-B0	48.12	63.49	46.13	62.86	46.15	61.89
MobileNetV3-L	47.37	**64.18**	46.08	<u>64.13</u>	47.31	62.76

Table 1 reports mAP and accuracy for each combination of feature extractor and classifier depth in the frame-level pipeline. The DINO (ViT-B/16) backbone achieves the highest mAP (49.2%), while MobileNetV3-L yields the best overall accuracy (64.18%), both using a shallow MLP as classification head. This difference likely reflects the optimization biases of each model: DINO, trained with self-supervised contrastive learning, is particularly effective at capturing semantically rich features that benefit mean average precision, an especially relevant metric in class-imbalanced settings. In contrast, MobileNetV3-L, a lightweight CNN optimized for classification, prioritizes discriminative performance at the instance level, which may explain its advantage in accuracy. Nonetheless, overall differences between extractors remain modest, likely due to the challenging nature of the dataset and the absence of backbone fine-tuning, which was not feasible given limited computational resources.

To further investigate the effect of data augmentation and class imbalance, Table 2 reports per-class AP for each extractor using the best classifier (shallow MLP), under baseline conditions and various augmentation strategies. The table also reports overall combined mAP and accuracy for each feature extractor under the augmentation setting that yielded the highest results for both metrics.

Table 2. Per-class AP (%) for each extractor using the shallow MLP, under baseline and best augmentation settings. Columns on the right show combined mAP and accuracy. Augmentation methods: W = water background replacement, S = silhouette masking, E = occlusion simulation. Bold indicates overall best. Underline indicates best extractor on the MLP.

Extractor	Aug.	Feeding	Swimming	Walking	Preening	Alert	Flying	Resting	mAP	Acc.
DINO (ViT-B/16)	Baseline	**90.42**	75.79	**88.28**	29.26	23.28	4.39	<u>24.78</u>	48.03	62.82
	W+S+E	89.00	**79.03**	74.12	<u>31.60</u>	<u>24.63</u>	<u>13.69</u>	24.53	**48.09**	<u>64.69</u>
ResNet-152	Baseline	86.16	75.98	77.67	34.64	24.93	9.24	23.09	47.39	63.60
	W+E	<u>87.71</u>	<u>77.73</u>	<u>81.79</u>	33.81	<u>25.40</u>	5.05	23.86	<u>47.91</u>	63.34
	S+E	86.22	75.73	69.87	<u>35.04</u>	24.52	**16.73**	<u>24.63</u>	47.54	<u>63.99</u>
ViTBase	Baseline	88.85	70.43	76.76	32.56	28.89	<u>5.59</u>	<u>22.39</u>	46.50	61.93
	Silhouette	<u>89.07</u>	<u>72.56</u>	<u>77.97</u>	31.93	**29.29**	5.04	22.18	<u>46.86</u>	61.67
	W+E	87.68	70.55	74.95	<u>32.65</u>	27.27	4.44	21.41	45.56	<u>62.11</u>
EfficientNet-B0	Baseline	<u>89.16</u>	75.83	68.88	32.92	24.60	8.72	<u>22.78</u>	46.13	62.86
	S+E	88.77	<u>78.27</u>	<u>68.97</u>	<u>33.53</u>	<u>24.72</u>	<u>13.67</u>	22.62	<u>47.22</u>	<u>64.19</u>
MobileNetV3-L	Baseline	87.05	70.06	74.65	37.04	<u>23.96</u>	6.63	23.17	46.08	64.13
	W+E	85.31	70.99	<u>74.83</u>	36.54	22.07	8.97	25.35	46.30	**65.23**
	W+S+E	<u>87.96</u>	<u>72.44</u>	73.95	**37.21**	20.88	<u>11.54</u>	**26.57**	<u>47.22</u>	64.21

Data augmentation proved particularly valuable for improving performance on underrepresented behavior classes, and in several cases, also led to modest gains in overall accuracy. For instance, with the DINO (ViT-B/16) extractor, *flying* class AP increased substantially from 4.39% (baseline) to 13.69% using W+S+E, while *swimming* improved from 75.79% to 79.03%. Similarly, MobileNetV3-L showed notable improvements: *flying* rose from 6.63% to 11.54%, and *resting* from 23.17% to 26.57%.

Overall, data augmentation resulted in the highest accuracy with MobileNetV3-L (65.23% using W+E) and the highest mAP with DINO (48.09% using W+S+E), demonstrating consistent improvements across all extractors. Compared to the results in Table 1, all extractors achieved their highest accuracy: DINO improved from 63.38% to 64.69%, ResNet-152 from 63.66% to 63.99%, and so on. Similarly, mAP scores increased for all extractors, with gains ranging from +0.06 to +1.14% points.

4.2 Segment-Level Pipeline Results

For the segment-level pipeline, we evaluate the PDAN, a TAS model that combines hierarchical dilated convolutions and attention mechanisms to capture both short and long-range dependencies in video sequences. Our ablation studies systematically vary three key architectural parameters: the number of stages, the number of layers per stage, and the feature map dimensionality, across three different video backbones: I3D [26], R(2+1)D [27], and MViT-B [28].

Table 3 shows that the best-performing configuration uses the I3D backbone with one stage, 7 layers, and a feature map size of 512 (experiment A4). This setup achieves 52.0% mAP and 64.53% accuracy, representing the highest values obtained in this set of experiments. For R(2+1)D, the optimal result is 39.1% mAP and 55.56% accuracy (A3), while for MViT-B, the best configuration (A1) yields 46.9% mAP and 62.0% accuracy.

Table 3. Segment-level ablation: Combined mAP (%) and Accuracy (%) for each architecture setting (A0A6) and backbone.

Exp.	Stages	Layers	F-maps	I3D		R(2+1)D		MViT-B	
				mAP	Acc.	mAP	Acc.	mAP	Acc.
A0	1	5	512	49.00	63.86	36.85	55.11	46.30	62.00
A1	2	5	512	47.61	64.30	38.03	53.50	46.93	63.06
A2	3	5	512	43.52	62.16	22.66	49.79	39.65	59.17
A3	1	3	512	48.53	63.83	39.12	55.56	45.39	61.34
A4	1	7	512	**52.01**	**64.53**	37.59	52.65	46.05	62.78
A5	1	5	256	48.57	63.60	35.59	54.18	43.38	60.50
A6	1	5	1024	49.80	63.64	35.53	53.09	44.34	61.41

4.3 Comparison Between Frame-Level and Segment-Level Pipelines

A direct comparison between the best frame-level and segment-level models reveals that the performance gap is narrower than might be expected. The top-performing frame-level pipeline (DINO with shallow MLP) achieves a combined mAP of 49.2%, while the best segment-level PDAN configuration with the I3D backbone reaches 52.0%. Although the segment-level approach leverages temporal information, the observed improvement is relatively modest. A key factor is the difference in models' pretrainings: frame-level feature extractors are initialized with ImageNet [38] weights, which include around 40 bird categories, providing strong visual knowledge about the birds domain. In contrast, segment-level video backbones are pretrained on Kinetics-400 [26], which does not contain any specific bird-related classes. As a result, frame-level models benefit from more specialized representations than their video-level counterparts.

Moreover, the mostly static or repetitive nature of behaviors in Visual-WetlandBirds, combined with short sequences and limited transitions, reduces the benefit of temporal modeling. Additionally, the dataset's size and class imbalance constrain the capacity of deep temporal models to extract and learn representative features from videos. Sophisticated segment-level models may show greater benefit on larger or more diverse datasets with a higher prevalence of subtle or rapidly changing behaviors.

5 Conclusions

This work systematically explores a range of TAS approaches for fine-grained bird behavior recognition in ecological videos, using both frame-level and segment-level deep learning pipelines on the challenging Visual-WetlandBirds dataset. Our systematic evaluation demonstrates that frame-level models, provide strong baselines for most simple bird behaviors. Segment-level architectures, although theoretically more powerful due to their ability to model temporal dependencies, yield only modest improvements in this context. We further observe that targeted data augmentation strategies such as background replacement, silhouette masking, and occlusion simulation, produce modest improvements in overall performance and are especially helpful for underrepresented behaviors.

Nonetheless, several remaining challenges suggest directions for future work, such as integrating optical flow to enhance motion understanding, and leveraging domain-specific pretraining, or self-supervised learning to address data scarcity, and improve performance in complex ecological scenarios. Future research directions include expanding the dataset with more annotated videos covering a broader range of species and behaviors, or integrating existing ecological datasets to construct a larger and more balanced resource. These steps would facilitate the application of domain-specific pretraining strategies for both image and video models, potentially enhancing model performance in complex ecological settings.

Acknowledgment. We would like to thank "A way of making Europe" European Regional Development Fund (ERDF) and MCIN/AEI/10.13039/501100011033 for supporting this work under the "CHAN-TWIN" project (grant TED2021-130890B-C21. HORIZON-MSCA-2021-SE-0 action number: 101086387, REMARKABLE, Rural Environmental Monitoring via ultra wide-ARea networKs And distriButed federated Learning. The work was also supported by two national grants and two regional grants for PhD studies from the Spanish and Valencian governments, respectively, FPU21/00414, FPU23/00532, CIACIF/2021/430 and CIACIF/2022/175. Additional thanks go to the CIAICO/2022/132 Consolidated group project "AI4-Health" funded by the Valencian government. This work is also part of the HELEADE project (TSI-100121-2024-24), funded by Spanish Ministry of Digital Processing, and by the European Union NextGeneration EU.

References

1. Yi, F., et al.: ASFormer: transformer for action segmentation. arXiv preprint arXiv:2110.08568 (2021)
2. Ding, G., Sener, F., Yao, A.: Temporal action segmentation: an analysis of modern techniques. arXiv preprint arXiv:2210.10352 (2023)
3. Krizhevsky, A., Sutskever, I., Hinton, G.E.: ImageNet classification with deep convolutional neural networks. NeurIPS **25**, 1097–1105 (2012)
4. Escorcia, V., et al.: DAPS: deep action proposals for action understanding. In: ECCV, pp. 768–784 (2016). https://doi.org/10.1007/978-3-319-46454-1_47
5. Girdhar, R., et al.: ActionVLAD: learning spatio-temporal aggregation for action classification. In: CVPR, 971–980 (2017). https://doi.org/10.1109/CVPR.2017.110
6. LeCun, Y., Bengio, Y., Hinton, G.: Deep learning. Nature **521**(7553), 436–444 (2015). https://doi.org/10.1038/nature14539
7. He, K., et al.: Deep residual learning for image recognition. In: CVPR, pp. 770–778 (2016). https://doi.org/10.1109/CVPR.2016.90
8. Hochreiter, S., Schmidhuber, J.: Long short-term memory. Neural Comput. **9**(8), 1735–1780 (1997). https://doi.org/10.1162/neco.1997.9.8.1735
9. Lea, C., et al.: Temporal convolutional networks: a unified approach to action segmentation. In: ECCV Workshops, pp. 47–54 (2016)
10. Farha, Y.A., Gall, J.: MS-TCN: multi-stage temporal convolutional network for action segmentation. In: CVPR, pp. 3575–3584 (2019)
11. Dosovitskiy, A., et al.: An image is worth 16×16 words: transformers for image recognition at scale. In: ICLR (2021)
12. Arnab, A., et al.: ViViT: a video vision transformer. In: ICCV, pp. 6836–6846 (2021)
13. Dai, R., et al.: PDAN: pyramid dilated attention network for action detection. In: WACV, pp. 2970–2979 (2021)
14. Zhang, C., Wu, J., Li, Y.: ActionFormer: localizing moments of actions with transformers. In: CVPR, pp. 8745–8754 (2022)
15. Stein, S., McKenna, S.J.: Combining embedded accelerometers with computer vision for recognizing food preparation activities. In: Proceedings of the 2013 ACM UbiComp, pp. 729–738 (2013)
16. Kuehne, H., et al.: Recognizing breakfast activities. In: ECCV, pp. 1–13. Springer, Cham (2014). https://doi.org/10.1007/978-3-319-10578-9_36
17. Fazzari, E., et al.: Animal behavior analysis methods using deep learning: a survey. arXiv preprint arXiv:2405.14002 (2024)
18. Barlow, C.R., et al.: Birds as bioindicators of river pollution and beyond: specific and general lessons from an apex predator. Ecol. Indicators **156**, 111366 (2023). https://doi.org/10.1016/j.ecolind.2023.111366
19. Biro, D., et al.: New technologies for automated animal behavior monitoring: a review. Curr. Biol. **32**(23), R1266–R1279 (2022)
20. Ge, Z., et al.: Exploiting temporal information for DCNN-based fine-grained object classification. arXiv preprint arXiv:1608.00486 (2016)
21. Naik, H., et al.: 3D-POP: an automated annotation approach to facilitate markerless 2D-3D tracking of freely moving birds with marker-based motion capture. In: CVPR, pp. 21274–21284 (2023)
22. Rodriguez-Juan, J., et al.: Visual WetlandBirds dataset: bird species identification and behaviour recognition in videos. Zenodo (2024)

23. Caron, M., et al.: Emerging properties in self-supervised vision transformers. In: ICCV, pp. 9650–9660 (2021)
24. Tan, M., Le, Q.: EfficientNet: rethinking model scaling for convolutional neural networks. In: ICML, pp. 6105–6114 (2019)
25. Howard, A., et al.: Searching for MobileNetV3. In: ICCV, pp. 1314–1324 (2019)
26. Carreira, J., Zisserman, A.: Quo vadis, action recognition? A new model and the kinetics dataset. In: CVPR, pp. 6299–6308 (2017)
27. Tran, D., et al.: A closer look at spatiotemporal convolutions for action recognition. In: CVPR, pp. 6450–6459 (2018)
28. Fan, H., et al.: Multiscale vision transformers. In: ICCV, pp. 6824–6835 (2021)
29. Kirillov, A., et al.: Segment Anything. arXiv preprint arXiv:2304.02643 (2023)
30. Nichols, J.D., Williams, B.K.: Monitoring for conservation. Trends Ecol. Evol. **21**(12), 668–673 (2006)
31. Morrison, M.L.:. Bird populations as indicators of environmental change. In: Current Ornithology, vol 3. Springer, Boston (1986)
32. Furness, R.W., Greenwood, J.J. (eds.): Birds as Monitors of Environmental Change. Springer, Heidelberg (2013)
33. Mathis, A., et al.: DeepLabCut: markerless pose estimation of user-defined body parts with deep learning. Nat. Neurosci. **21**, 1281–1289 (2018)
34. Pereira, T.D., et al.: SLEAP: a deep learning system for multi-animal pose tracking. Nat. Methods **19**, 486–495 (2022)
35. Tang, G., et al.: Anti-drift pose tracker (ADPT): a transformer-based network for robust animal pose estimation cross-species. eLife (2025)
36. Zhong, E., et al.: AnimalMotionCLIP: embedding motion in CLIP for animal behavior analysis. arXiv preprint arXiv:2505.00569 (2025)
37. Liu, J., et al.: A joint time and spatial attention-based transformer approach for recognizing the behaviors of wild giant pandas. Ecol. Inf. **83**, 102797 (2024). https://doi.org/10.1016/j.ecoinf.2024.102797
38. Deng, W.J., et al.: ImageNet: a large-scale hierarchical image database. In: CVPR, Miami, FL, USA, pp. 248–255 (2009). https://doi.org/10.1109/CVPR.2009.5206848
39. Saval-Calvo, M., et al.: Three-dimensional planar model estimation using multi-constraint knowledge based on k-means and RANSAC. Appl. Soft Comput. **34**, 572–586 (2015)
40. Bauer, D., et al. Challenges for monocular 6d object pose estimation in robotics. IEEE Trans. Rob. (2024)
41. Viejo, D., et al.: Combining visual features and growing neural gas networks for robotic 3D SLAM. Inf. Sci. **276**, 174–185 (2014)
42. Ortiz-Perez, D., et al.: A deep learning-based multimodal architecture to predict signs of dementia. Neurocomputing **548**, 126413 (2023)

Advanced Multistep Prediction and Dynamic Thresholding for Early Detection of Crowd Anomalies in Surveillance Systems

Rafael Rodrigo Guillén, Higinio Mora Mora, and Jorge Azorín-López(✉)

University of Alicante, Carretera San Vicente del Raspeig S/N, Alicante, Spain
jazorin@ua.es

Abstract. In the era of smart cities, the rapid increase in surveillance cameras has generated vast amounts of data, providing opportunities for advanced real-time analytics to improve public safety. However, effectively utilizing this data for proactive anomaly detection remains challenging. This study proposes a hybrid Conv1D-LSTM model integrating multistep forecasting and adaptive dynamic thresholding to detect crowd anomalies early, significantly before they escalate into critical events. The dynamic threshold adapts based on recent prediction errors, enhancing sensitivity and minimizing false alarms. Experimental validation using synthetic data demonstrates superior performance compared to traditional methods (ARIMA, N-BEATS, TFT), confirming the method's robustness and practical applicability. Future research directions include integrating additional contextual variables and optimizing computational efficiency for real-time applications.

1 Introduction

The motivation of this work is to design a system capable of issuing early warnings about anomalies before they escalate to their peak severity, thereby providing a "detection horizon" that allows adequate time for effective intervention. To achieve this, we use multistep forecasting based on Long Short-Term Memory (LSTM) networks, which are effective at capturing long-term dependencies in time series data. This forecasting approach is integrated with a dynamic threshold, continuously adjusted according to recent variations in the prediction error. The adaptive threshold minimizes false positives during stable periods and enhances sensitivity during critical situations.

In the existing literature, deep learning methods such as LSTM, Gated Recurrent Units (GRU), and autoencoders are widely employed for time series modeling and anomaly detection tasks. Additionally, Generative Adversarial Networks (GANs) are used to predict future video frames, anticipating unusual behaviors [6]. Other approaches integrate spatial-temporal features through Convolutional Neural Networks (CNNs) combined with attention mechanisms, enabling detection of both individual and collective anomalies [5, 13], or merge sensor data and time series correlations using autoencoders combined with hyperparameter optimization techniques [11, 12]. Furthermore, hybrid methods incorporate external variables (such as mobility data and policy information) to capture complex urban dynamics, as demonstrated by models like MobCovid

E. Corchado et al. (Eds.): SOCO 2025, CCIS 2806, pp. 540–549, 2026.
https://doi.org/10.1007/978-3-032-19763-4_50

[6]. In the field of public safety, the integration of real-time video surveillance and deep learning techniques is employed to identify high-risk areas and generate proactive alerts [11], predict risk levels in crowded scenarios [14], and extract anomalous patterns from video data using autoencoders [8, 9]. CNN-based methods are utilized to detect dangerous objects such as weapons and knives [10, 11], and models focusing on crowd dynamics and violent behavior recognition are developed to anticipate events like stampedes or riots [12, 13]. Recent literature reviews have consolidated deep learning approaches within smart surveillance systems to enhance urban safety through timely alerts [14, 15], and multi-view predictive models have been employed for crime forecasting in urban contexts [16]. The combination of an extended forecasting horizon, adaptive thresholding, and the incorporation of both real and synthetic anomalies positions our methodology beyond conventional "just-in-time" approaches, offering an advanced, customizable solution. The main objectives of this study are: (1) developing a multilayer LSTM network capable of multistep predictions accurate enough to detect minor deviations early; (2) defining an adaptive threshold based on a moving average and standard deviation of prediction errors; (3) validating the system on synthetic datasets with artificially introduced anomalies; and (4) quantitatively comparing our method against fixed-threshold methods and traditional detection approaches, demonstrating superior performance in terms of prediction lead time and detection accuracy.

Our proposed solution integrates predictive models (LSTM, light transformers, etc.) with anticipatory metrics designed to enhance the detection of gradual anomalies. The adaptive threshold, guided by historical variation and cyclical data patterns, minimizes false alarms while ensuring timely identification of anomalies in real-world scenarios. The key contribution of this study is the unified workflow combining multistep forecasting, synthetic anomaly generation, and dynamic threshold adaptation. Thanks to the adaptive thresholding, the system maintains low false positive rates during normal predictive behaviors and increases sensitivity as predictions deviate from expected patterns, thus guaranteeing early detection. Experimental results indicate that anomaly alerts are triggered several timesteps ahead of peak severity, highlighting the practical utility of our approach for proactive monitoring in real-life settings.

2 Method

The workflow implemented to detect anomalies before they reach their peak severity is described in detail. Our objective is to develop a pipeline capable of issuing early warnings based on the analysis of time series data containing both normal and anomalous conditions. This pipeline integrates a multistep forecasting model with a dynamically adaptive threshold, which adjusts according to the uncertainty observed at each moment. Figure 1 provides a graphical representation of the proposed methodology.

2.1 Network Architecture and Training

A hybrid Conv1D-LSTM model has been designed to effectively capture both local features and long-term dependencies inherent in time series data. The network architecture begins with a one-dimensional causal convolutional layer consisting of 64 filters,

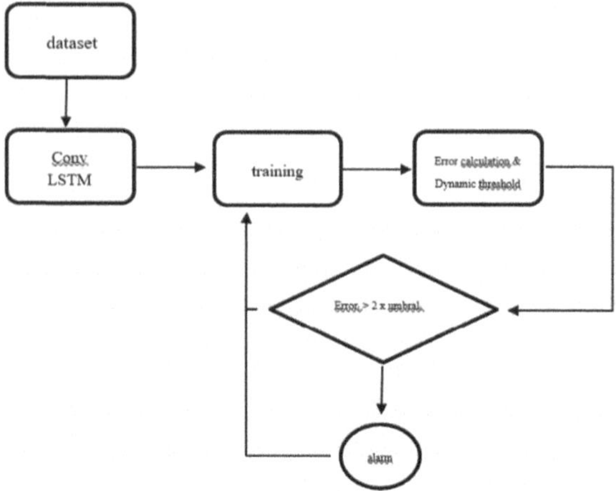

Fig. 1. Graphical representation of the model

each with a kernel size of 5. This layer extracts local patterns from sequences of five consecutive values. Being causal, it ensures each prediction uses exclusively historical data, preserving temporal coherence and preventing leakage of future information. Additionally, it acts as a smoothing filter that attenuates high-frequency noise present in the original signal. The output from this convolutional layer feeds into a block containing two LSTM layers: the first LSTM layer, comprising 256 units, generates a sequence of hidden states, enabling the model to learn long-term temporal patterns. To mitigate overfitting, a dropout layer with a dropout rate of 20% follows. The second LSTM layer, with 128 units, aggregates the entire previous sequence into a single state vector that encapsulates the most significant aspects of the signal's recent behavior. Another dropout layer of 20% is applied immediately afterward to further enhance regularization. The internal representation is then refined through two dense layers employing ReLU activation functions (first with 50 neurons and subsequently with 10) allowing flexible adjustment of the information extracted by the LSTM layers. Finally, a single neuron with linear activation provides the multistep prediction output, forecasting a fixed horizon of $h = 10$ steps ahead. The model training process is carried out by minimizing the Huber loss function, chosen for its robustness to outliers, using the Adam optimizer with an initial learning rate of 1×10^{-4}. To prevent overfitting, we implement early stopping based on the mean absolute error (MAE) computed on a validation set, which comprises 10% of the original dataset. Training halts if no improvement in MAE is observed for five consecutive epochs. In practice, this training strategy typically converges within 20 to 30 epochs.

2.2 Definition of Error and Dynamic Threshold

Having described the origin of the dataset utilized for training and testing our model, it is essential to clarify the definitions of prediction error and dynamic threshold, as well as their interrelation with the k-factor. These concepts are crucial for accurately calibrating and modelling the proposed anomaly detection system.

Definition of Error

At each time instant t (or each sample in the test set), the prediction error is defined as the difference between the actual value of the series ($\mathrm{real_t}$) and the value predicted by the model ($\widehat{\mathrm{y_t}}$). In this study, we use the absolute value of this difference.

In a forecasting scenario (e.g., when using an LSTM model to predict the next value in a time series, or a horizon of 10 future steps as in our example), every time a prediction $\widehat{\mathrm{y_t}}$ for a time t, it is compared with the actual value of the series at that same instant ($\mathrm{real_t}$). The error at each point t in the dataset is thus defined as $error_t = |\mathrm{real_t} - \widehat{\mathrm{y_t}}|$.

This measure indicates how closely the prediction matches reality. A small error implies that the model (in this case, the LSTM network) has produced an accurate prediction, while a large error reflects a significant deviation from the actual value.

Although other error metrics such as the squared error $(\mathrm{real_t} - \widehat{\mathrm{y_t}})^2$ can be used, the absolute error is most commonly employed for anomaly detection tasks due to its interpretability and simplicity.

Definition of dynamic threshold.

Unlike a fixed threshold, the dynamic threshold evolves over time in response to the recent behavior of the prediction error, as measured within a sliding window of the last N data points. In this way, the system "learns" the typical dispersion of recent errors and adjusts the anomaly detection threshold accordingly. An anomaly is only flagged if the current error exceeds this normal range (possibly plus a scaling factor).

Formally, at a given time i, we consider the most recent N errors: window_err $=$ $\{\mathrm{error}_{i-N}, \ldots, \mathrm{error}_{i-1}, \mathrm{error}_i\}$ If i $< N$, we use all available error values from the beginning of the series. For the current window, we then compute both the mean and standard deviation of the errors:

- Define $\overline{\mathrm{e}}$ as the average of those errors in the window,
- Define σ as the standard deviation of these errors.
- The base threshold at a given time *I* can thus be defined as: $\mathrm{threshold}_i = \overline{e} + k + \sigma$, where *k* is a tunable parameter that determines the strictness of the anomaly detection criterion.

It is important to emphasize the dynamic nature of this threshold: as i increases, the sliding window advances, and both e and σ are recalculated at each step. This allows the threshold to adapt continuously to the recent statistical properties of the prediction error. If there is a period during which the errors in the series increase, both the mean and variance will rise, thereby relaxing the detection threshold (as illustrated by the orange lines in Figs. 2, 3, and 4). This adjustment helps prevent the system from generating constant alarms during volatile phases. Conversely, if the series remains stable with consistently low errors, the threshold decreases, making the system more sensitive so that even a minor deviation may be sufficient to flag an anomaly. In summary, the threshold dynamically adapts to the recent variability observed in the last N samples.

When an error occurs that substantially exceeds the typical range, the system promptly triggers anomaly detection.

Relationship Between Error and Threshold
The crucial step in the detection process is to compare the error at the current instant with the corresponding threshold. In the basic approach, an anomaly is flagged if $\text{error}_t > \textit{threshold}_t$. To increase the strictness of anomaly detection and reduce the likelihood of false positives, this condition can be tightened to $\text{error}_t > 2 \times \textit{threshold}_t$.

This adjustment requires the actual error to be at least twice the dynamic threshold, making the detection criterion considerably more rigorous. When this threshold is surpassed, the system triggers an alarm at that point in time. By imposing the condition $\text{error}_\text{t} > 2 \times \textit{threshold}_\text{t}$ only errors that are significantly larger than those observed in the recent window will result in an anomaly being detected.

3 Experimentation

A framework for the early detection of anomalies in time series has been presented, combining multistep forecasting with a Conv1D-LSTM model and a dynamic threshold derived from local error statistics. In the following sections, we provide a detailed synthesis and analysis of the results obtained using a synthetic signal defined as $y(t) = \sin(\frac{t}{10})$. Subsequently, these results are compared with those obtained from other predictive models.

3.1 Synthetic Model Generation and Results

To isolate and better understand the core behavior of the anomaly detector, we begin with a synthetic signal defined by the function $y(t) = \sin(\frac{t}{10})$, sampled at regular intervals $\Delta t = 0.1$ over the range $t \in [0,4000)$. This results in a dataset of 40,000 data points representing the pure sine wave. Two subsets are defined:

Training set [0,2800): This subset consists exclusively of the undisturbed sine wave, allowing the model to learn normal behavior with complete traceability.

Test set [2800,4000): In this segment, we manually inject four anomalies of varying types. These anomalies are introduced at random positions within the test set (approximately at t = 3200, 3600, 4000 and 4400), encompassing a range of deviation patterns while ensuring the training set remains untouched. This annotated dataset serves as the basis for our subsequent performance evaluations.

By applying our methodology to the synthetic sine wave series with manually injected anomalies in the test phase, we observe that the model exhibits highly stable behavior in the anomaly-free regions and provides clear responses in the segments containing disturbances, as illustrated in Fig. 2.

The detector triggered an alarm, on average, 6.3 steps before the actual peak deviation (approximately 0.63 time units with a sampling interval of $\Delta\text{t} = 0, 1$), demonstrating the rapid responsiveness of the dynamic threshold as soon as the error deviates from its historical mean. For instance, as shown in Fig. 3, during the smooth decline of the sinusoidal signal at $\text{t} \approx 3600$, the error increases prior to the maximum deviation, and the

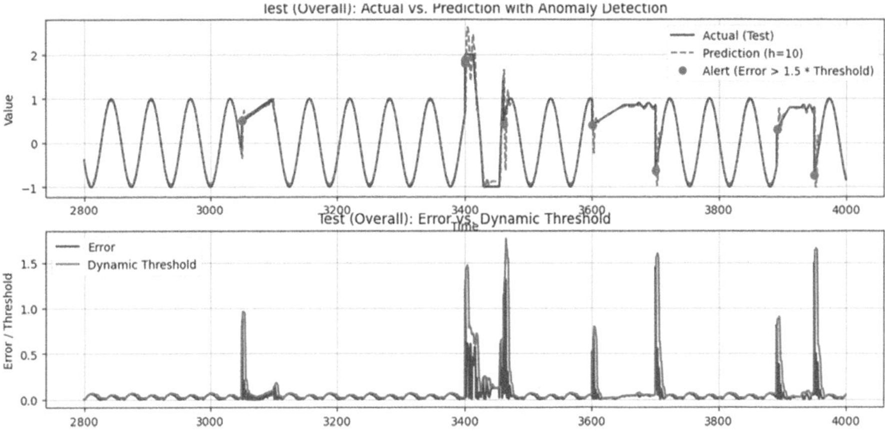

Fig. 2. Shows anomaly detection only in the test section

threshold τ_t (defined as mean $+2\sigma$) rises almost in parallel, thereby avoiding the need to wait for the peak. At the sharpest drop around t$\approx$3700, the growing variability causes a gradual increase in τ_t, and the final threshold ($2 \times \tau_t$) triggers the alarm early enough to capture the initial transition. After the first drop and throughout the subsequent plateau, the threshold readjusts after each event, responding first to the pulse and then preparing for the ramp, thereby maintaining sensitivity to further changes.

Regarding Fig. 4, which illustrates the other two anomalies, we observe that around t $\approx$ 3890 as soon as minor high-frequency fluctuations appear, the error rapidly surpasses the previously established noise level. The dynamic threshold increases after a few elevated error values but rises quickly enough to trigger the alarm precisely at the onset of these oscillations, thereby preventing missed detections at intermediate peaks.

At t $\approx$ 3950 there is a sharp decline in the actual signal that the LSTM model fails to anticipate (as shown by the red dashed prediction), resulting in a peak error that exceeds that of the previous oscillation. Although the threshold was already elevated due to the preceding fluctuation, it remains sufficiently sensitive to trigger the alert just before the maximum error is reached. After each alarm, the threshold is recalculated using the moving window, gradually returning to normal levels as the signal stabilizes. This demonstrates, as shown in the first figure, that the dynamic threshold remains responsive to both recurring anomalies and abrupt, high-intensity changes. Across the 1,200 test points, alarms were triggered in approximately 1.8% of cases (corresponding to 8 true positives and 50 false positives), predominantly occurring in regions of gradual transition where the error surpassed the dynamic threshold.

This low percentage indicates that the system effectively discriminates between normal behavior and significant deviations, although this analysis does not distinguish between true and false alarms. The sine wave, despite its simplicity, displays periodic behavior and an absence of noise, enabling us to isolate and evaluate the effectiveness of the dynamic threshold. Achieving consistent early detections more than half a second before the actual anomaly supports the hypothesis that a threshold based on the moving

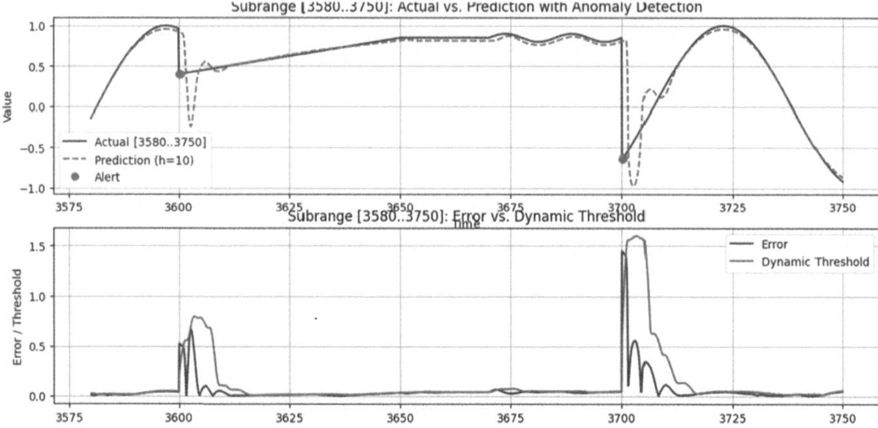

Fig. 3. Detection of two anomalies in the subrange [3580,3750]

average and standard deviation of errors is sufficient to generate timely warnings without additional manual adjustment.

3.2 Comparative Analysis

To assess the effectiveness of various forecasting approaches for anomaly detection in time series, we conducted a systematic comparison of four representative models: ARIMA, LSTM, N-BEATS (Neural Basis Expansion Analysis for Time Series), and TFT (Temporal Fusion Transformer). This selection includes both traditional statistical methods and state-of-the-art deep learning techniques, enabling a direct comparison of approaches with varying temporal modeling capabilities and levels of complexity.

All models were trained using the same univariate time series, free of anomalies in their training part, and subsequently evaluated on a test set containing manually injected anomalies at certain points in the series. The evaluation metrics employed in this study include True Positives (TP), Recall (%), False Positives (FP), and False Positive Rate (FPR %).

Table 1 presents the results obtained and provides a discussion of the significance of these metrics.

The evaluation of the four prediction models on the test set with artificially introduced anomalies reveals substantial differences in their anomaly detection capabilities. Table 1 reports the metrics obtained for each approach, including True Positives (TP), Recall, False Positives (FP), and False Positive Rate (FPR). Our LSTM model demonstrated the best overall performance, achieving a recall of 1.98%. While this value may be modest in absolute terms, it significantly outperformed the other models by detecting 8 out of 404 actual anomalies. However, this improved sensitivity came at the cost of a higher number of false positives (50), resulting in a false positive rate (FPR) of 0.42%. This outcome reflects the model's increased sensitivity, accompanied by a corresponding rise in false detections. ARIMA and N-BEATS exhibited nearly equivalent performance in anomaly detection, each achieving a recall of 0.99% with 4 true positives. However, their false

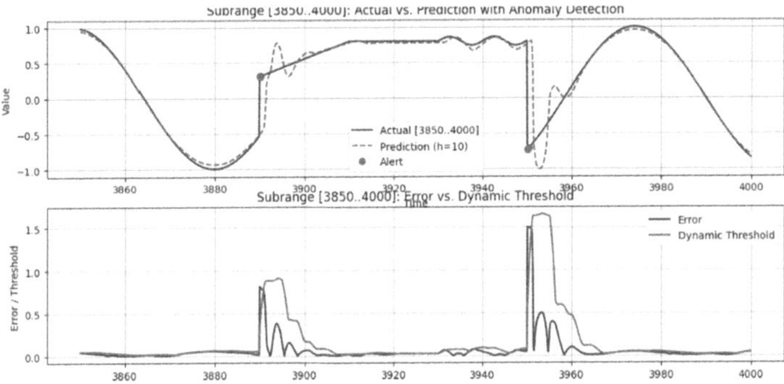

Fig. 4. Example of the detection of two anomalies during a transition characterized by a rapid fluctuation.

Table 1. Comparison of models

Model	True Positives	TP	Recall (%)	False Positives	TFP (%)
Conv1D-LSTM	8	404	1.98	50	0.42
ARIMA	4	404	0.99	12	0.10
N-BEATS	4	404	0.99	14	0.12
TFT	2	404	0.5	25	0.2

positive rates differed slightly: ARIMA produced 12 false positives (FPR of 0.10%), while N-BEATS resulted in 14 (FPR of 0.12%). These results indicate that, compared to the LSTM model, ARIMA and N-BEATS are more specific but have a reduced capacity to identify anomalous events. The TFT model, although the most recent among the four, exhibited the poorest performance in this experiment, detecting only 2 true positives (recall of 0.5%) and yielding 25 false positives (FPR of 0.2%). This outcome suggests that the model was unable to fully leverage its covariate-oriented architecture, likely due to the univariate nature of the dataset and the absence of supplementary temporal context. Overall, the results indicate that, under the conditions of this study (specifically, univariate time series with localized anomalies and no explicit covariates) our LSTM model offers the best trade-off between sensitivity and the cost of false positives. In contrast, ARIMA and N-BEATS demonstrate superior specificity. Conversely, the TFT model does not provide significant advantages, which may be attributable to the low complexity of the dataset relative to the model's architectural capabilities.

4 Conclusion and Future Lines of Research

The results obtained in both controlled settings and with real, high-noise data demonstrate that employing a dynamic threshold (derived from local error statistics) in combination with multistep prediction is an effective and practical strategy for the early detection of anomalies in time series. This approach proves robust across a range of conditions, highlighting its applicability in both synthetic and real-world scenarios. These findings also open numerous avenues for further refinement and extension of the proposed methodology. While our dynamically thresholded Conv1D-LSTM model has demonstrated effectiveness in the early detection of anomalies in both synthetic and real datasets, there remain opportunities for technical enhancements to further improve its accuracy, robustness, and practical applicability. Potential improvements include incorporating contextual variables (such as weather data, event types, and calendars), employing adaptive thresholds that account not only for previous observations but also for the rate of change and asymmetry in the error distribution to better capture subtle or irreversible anomalies, and evaluating computational efficiency and latency in real-time streaming environments.

References

1. Emad, M., Ishack, M., Ahmed, M., Osama, M., Salah, M., Khoriba, G.: Early-anomaly prediction in surveillance cameras for security applications. In: 2021 International Mobile, Intelligent, and Ubiquitous Computing Conference, MIUCC 2021, pp. 124–128 (2021)
2. Nauman, M.A., Shoaib, M.: Identification of anomalous behavioral patterns in crowd scenes. Comput. Mater. Continua **71**(1), 925–939 (2022)
3. Kendule, J.A., Karande, K.J.: Crowd anomaly detection and localization via deep convolutional model with improved spatio temporal textures. Multimed. Tools Appl. **83**(18), 55053–55074 (2024). https://doi.org/10.1007/s11042-023-17375-6
4. Saini, M., Sengupta, E., Singh, H.: Artificial intelligence inspired IoT-fog based framework for generating early alerts while train passengers traveling in dangerous states using surveillance videos. Multimed. Tools Appl. **83**, 13613–13635 (2024)
5. Syahira, S.D., Ihsan, A.F., Hasmawati: anomaly detection on natural gas pipeline operational data using GRU method. In: 2024 12th International Conference on Information and Communication Technology (ICoICT), pp. 539–544 (2024). https://doi.org/10.1109/icodsa62899.2024.10651865
6. Chen, J., et al.: MobCovid: confirmed cases dynamics driven time series prediction of crowd in urban hotspot. IEEE Trans. Neural Netw. Learn. Syst. **35**(10), 13397–13410 (2023). https://doi.org/10.1109/TNNLS.2023.3268291
7. Chang, J.H., Jung, H., Park, W.: Development of a risk space prediction model based on CCTV images using deep learning: crowd collapse. Int. J. Adv. Sci. Eng. Inf. Technol. **14**(1), 189–195(2024). https://doi.org/10.18517/ijaseit.14.1.19764
8. Khaire, P., Kumar, P.: A semi-supervised deep learning based video anomaly detection framework using RGB-D for surveillance of real-world critical environments. Forensic Sci. Int. Digit. Investig. **40** (2022). https://doi.org/10.1016/j.fsidi.2022.301346
9. Fang, J., Zhang, X., Yang, B., Chen, S., Li, B.: An attention-based U-net network for anomaly detection in crowded scenes. In: 2022 IEEE 14th International Conference on Computer Research and Development, ICCRD 2022, pp. 202–206 (2022). https://doi.org/10.1109/ICCRD54409.2022.9730481
10. Galab, M.K., Taha, A., Zayed, H.H.: Adaptive technique for brightness enhancement of automated knife detection in surveillance video with deep learning. Arab. J. Sci. Eng. **46**(4) (2021). https://doi.org/10.1007/s13369-021-05401-4
11. Fernandez-Carrobles, M.M., Deniz, O., Maroto, F.: Gun and knife detection based on faster R-CNN for video surveillance. In: Morales, A., Fierrez, J., Sánchez, J., Ribeiro, B. (eds.)

IbPRIA 2019. LNCS, vol. 11868, pp. 441–452. Springer, Cham (2019). https://doi.org/10.1007/978-3-030-31321-0_38

12. Alqaysi, H.H., Sasi, S.: Detection of abnormal behavior in dynamic crowded gatherings. In: Proceedings - Applied Imagery Pattern Recognition Workshop (2013). https://doi.org/10.1109/AIPR.2013.6749309
13. Yahuarcani, I.O., et al.: Recognition of violent actions on streets in urban spaces using Machine Learning in the context of the Covid-19 pandemic. In: International Conference on Electrical, Computer, and Energy Technologies, ICECET 2021 (2021). https://doi.org/10.1109/ICECET52533.2021.9698762
14. Yang, C.-L., Wu, T.-H., Lai, S.-H.: Moving-object-aware anomaly detection in surveillance videos. In: AVSS 2021 - 17th IEEE International Conference on Advanced Video and Signal-Based Surveillance (2021). https://doi.org/10.1109/AVSS52988.2021.9663742
15. Oladipo, I.D., AbdulRaheem, M., Awotunde, J.B., Bhoi, A.K., Adeniyi, E.A., Abiodun, M.K.: Machine learning and deep learning algorithms for smart cities: a start-of-the-art review. In: Nath Sur, S., Balas, V.E., Bhoi, A.K., Nayyar, A. (eds.) IoT and IoE Driven Smart Cities. EAI/SICC, pp. 143–162. Springer, Cham (2022). https://doi.org/10.1007/978-3-030-82715-1_7
16. Zheng, Z., Xia, Y., Chen, X., Yao, J.: Security alert: Generalized deep multi-view representation learning for crime forecasting. Comput. Intell. (2022). https://doi.org/10.1111/coin.12504

Thickness Estimation of Plastic Polymers in SWIR Imaging Using Deep Learning Methods

Dominik Stursa(✉) and Vitek Rais

University of Pardubice, nám. Čsl. legií 565, Pardubice, Czech Republic
{dominik.stursa,vitek.rais}@upce.cz

Abstract. This work demonstrates accurate non-contact measurement of polymer thickness from a single short-wave infrared (1350 nm) image. We captured 1 211 ABS samples of ten nominal thicknesses (0.5 to 5.2 mm) cropped to 224 × 224 images for supervised regression. Four model families were benchmarked: a five region if interest mean values as features to feed FFNN, a full-image FFNN, a compact CNN, and five YOLOv11 backbones modified with a linear head. After 5 to 10 independent trainings per model, the tiny YOLOv11-n achieved the best median of mean square errors of 0,0046 (0.08 mm RMSE), outperforming CNN (0.015) and both dense baselines (greater than 0.07). The results confirm that SWIR imaging combined with modern detection backbones enables fast polymer metrology suitable for thickness estimation.

Keywords: short-wave infrared (SWIR) imaging · polymer thickness · neural network regression

1 Introduction

Plastic components are critical elements in various industrial sectors, including automotive, electronics, packaging, and consumer goods. Ensuring their reliable performance and long service life requires thorough quality control, particularly in measuring physical parameters such as thickness and the number of material layers. These properties have a significant impact on mechanical strength and durability. Traditional inspection methods, typically based on contact measurements or analysis within the visible spectrum, often face limitations when dealing with plastics. These challenges highlight the need for more advanced approaches, among which short-wave infrared (SWIR) imaging combined with deep learning techniques has gained increasing attention.

Recent research demonstrates that SWIR imaging provides substantial advantages in material classification and analysis as a result of the distinct spectral responses of plastics in this wavelength range. For example, Song et al. (2024) [7] reported near-perfect material classification accuracy (∼99%) using SWIR imaging, in contrast to ∼77% when using only visible light. Similarly,

E. Corchado et al. (Eds.): SOCO 2025, CCIS 2806, pp. 550–560, 2026.
https://doi.org/10.1007/978-3-032-19763-4_51

Martinez-Hernandez et al. (2024) [1] showed that even low-cost multispectral systems, when coupled with convolutional neural networks (CNN), can yield competitive performance. In the context of plastic recycling, Singh et al. (2023) [3] and Serranti et al. (2024) [8] applied hyperspectral imaging in the 900 to 1700 nm range, reaching classification accuracies that are approaching 90% with both neural networks and traditional models like PLS-DA or SVM, depending on particle size and spectral resolution. Complementary unsupervised approaches, as demonstrated by Henriksen et al. (2022) [9], have also proven effective for in-line polymer classification without prior labeling.

Beyond classification, SWIR data has been leveraged for physical parameter estimation. Ozawa (2023) [2] applied the LambertBeer law to hyperspectral data to infer the number and thickness of overlapping plastic layers, while Zhong et al. (2025) [10] used Mask R-CNN segmentation to track coating thickness on pharmaceutical pellets in real-time production environments. Likewise, works by Ficzere et al. (2022) [11] and Jun et al. (2024) [12] highlight the effectiveness of deep learning in predicting coating thickness or nanostructure quality from optical or hyperspectral data. Furthermore, SWIR and deep learning have enabled robust defect detection in polymer applications, from injection molding (Zhang et al., 2024; Liu et al., 2021; Ha & Jeong, 2021) [13–15] to fuel cell membranes (Yan et al., 2023) [16] and composite materials (Azad & Kim, 2025) [18]. These techniques extend even to thermography (Bang et al., 2020) [22] and CT imaging modalities (Vorobev et al., 2023) [17] and are increasingly being used in edge or IoT contexts.

Given these findings, the combination of SWIR imaging and neural networks represents a promising pathway not only for identifying plastic types but also for quantitatively determining their physical properties, such as thickness.

In this paper, we focus on evaluating the ability to accurately measure the thickness of simple plastic samples under laboratory conditions. The goal is to validate the potential of SWIR imaging combined with deep neural networks for the thickness regression task, thereby opening the door to broader applications of these methods in quality control, defect detection, and damage assessment in industrial practice.

2 Problem Formulation

Accurately estimating the physical thickness h of circular polymer samples from a single short-wave infrared (SWIR) transmission image is cast here as an inverse optical measurement regression problem that must cope with a markedly non-uniform illumination field.

2.1 Optical Forward Model

The acquisition geometry (Fig. 1) consists of a SWIR camera mounted normal to a backlight composed of four 1350 nm LEDs. For a homogeneous polymer sample

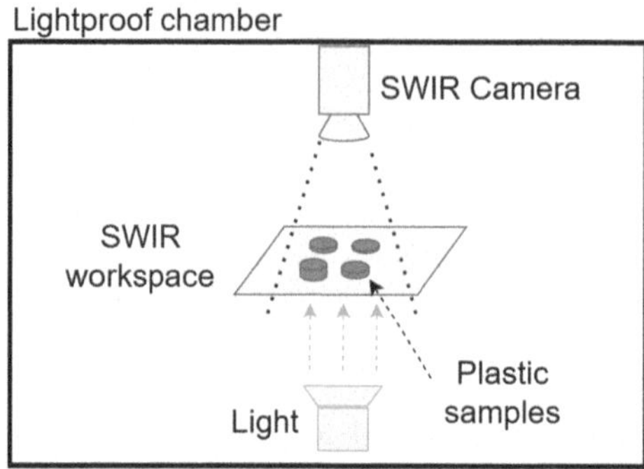

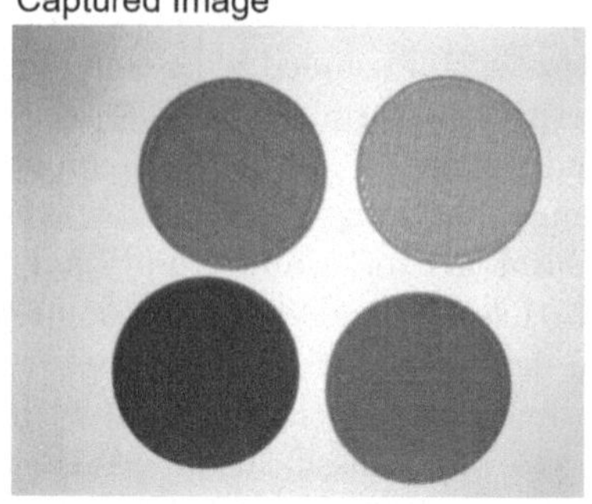

Fig. 1. Setup of the image acquisition system (left) and an example of a captured image from the dataset (right). The left part shows the hardware configuration used to collect the data, including the illumination source and camera placement. The right part presents an image containing four samples (tokens) of varying thicknesses, illustrating the arrangement and imaging conditions under which the dataset was acquired.

with a spectrally averaged absorption coefficient $\alpha(\lambda_0)$, the BeerLambert law [5] predicts

$$h(x, y) = -\frac{1}{\alpha(\lambda_0)} \ln\left(\frac{I(x, y)}{I_0(x, y)}\right), \tag{1}$$

where $I(x, y)$ is the converted irradiance and $I_0(x, y)$ is the (unknown) position-dependent reference field. Equation 1 is widely used for thin material thickness metrology but is based on two assumptions:

- **Uniform back-illumination** (I_0 constant across the field),
- **Known absorption coefficient** (temperature independent, minimal dispersion, etc.).

Manufacturing tolerances in the light guide, the emission cone of each LED, and losses in the camera lens cause gradual vignetting, which reduces the intensity in the corners of the image. This is combined with local dark spots caused by the non-homogeneity of the diffuser. As a result, two identical samples located in different places in the image may appear to have different optical densities, making direct pixel inversion unreliable.

2.2 Empirical Evidence of Position-Dependent Bias

A part of the original dataset (see Sect. 3.1) with 235 images, each containing up to four randomly placed samples and covering ten nominal thickness levels, was recorded in a light-proof chamber to eliminate ambient light. Crop-outs of the original images with 905 images of samples were used for analysis. The samples distribution is shown in Table 1. Exploratory analysis (Fig. 2) shows that the variance of the average pixel intensities within a class (one thickness) is as great as the difference between classes (different thicknesses). This makes a purely statistical fit to Eq. 1 insufficient.

Table 1. Distribution of samples by nominal thickness category.

Thickness [mm]	0.58	1.20	1.58	2.17	2.57	3.22	3.58	4.16	4.57	5.16
Samples	75	93	93	93	93	93	93	86	93	93

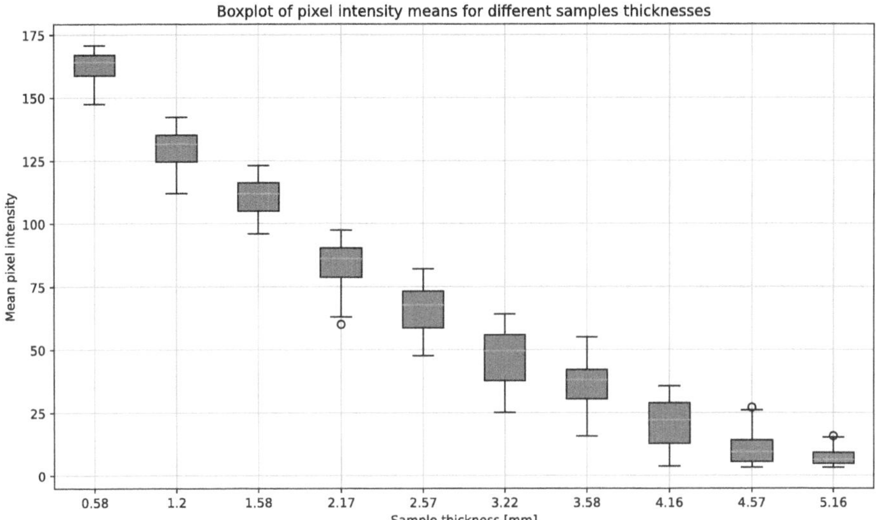

Fig. 2. Box-plots of the mean grayscale intensities computed from segmented samples regions.

2.3 Learning Based Re-formulation

To compensate for the unknown shading field $I_0(x, y)$ and for residual deviations from the Beer-Lambert linearity, the problem is reformulated as a supervised regression task:

$$f_\theta : I(x, y) \longmapsto \hat{h} \in R, \tag{2}$$

where f is a trainable neural network with parameters θ that produces a single continuous thickness value. This article evaluates four progressively more expressive models, which are described in detail in the following chapters. Summary of the problem itself as follows:

- The backlight field $I_0(x, y)$ is unknown and position-dependent, making direct inversion according to BeerLambert's law impossible.
- Empirical statistics show strong class overlap and high variability, which rule out threshold-based approaches.
- Formulating the problem as a regression task allows the model to implicitly learn the shading structure from the data itself.

3 Methods and Experiments

This section outlines the complete processing pipeline that converts raw trans-illumination images of polymer samples into quantitative thickness estimates. After introducing the custom dataset, we describe three custom and one well-known strategies of increasing representational capacity: a lightweight feed forward network (FFNN) driven by five regions of interest (ROI) statistics; a FFNN trained directly on flattened pixel intensities; a compact convolutional neural network (CNN); and models based on the You Only Look Once (YOLO) family. The procedure with the methods used is shown in Fig. 3.

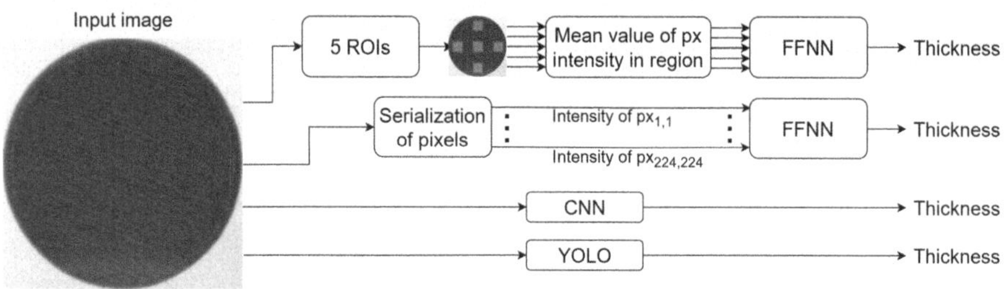

Fig. 3. Thickness estimation pipeline for the four evaluated methods. Each pipeline row illustrates the data processing steps from input image to predicted thickness value.

3.1 Dataset Creation

For the purposes of this study, we prepared a new dataset. The objects of interest were circular samples made of ABS polymer, each with a radius of 30 mm and manufactured in ten uniformly distributed thicknesses between 0.5 mm and 5.2 mm. To eliminate the potential influence of ambient lighting, all measurements were performed within a light-proof chamber (see Fig. 1).

The NIT WiDy SWIR 640 camera with the appropriate lens was placed in the upper part of the chamber perpendicular to the paper surface. The light Smart Vision Light R80 of wavelength 1350 nm was placed in the lower part of the chamber, positioned directly beneath, also perpendicular to the surface, which was illuminated by this light from below. As a workspace diffuser we chose paper because of its opacity but also because of its excellent light transmission of the wavelengths involved. The output format was set to raw data, in the form of 8-bit grayscale images, with a resolution of 640×512 pixels. Each image contains 1 to 4 non-overlapping objects of interest. A total of 362 images containing 1211 objects of interest were produced.

For the purposes of thickness regression, the images were then divided into individual objects of interest, each with a resolution of 224×224 pixels. We used a dataset organization similar to the classification task; thus, the images were divided into folders that have a name identical to the thickness of the object of interest. Then these were randomly split into a training set with 905 images and a test set with 306 images.

3.2 ROI Statistics with FFNN

The first learning strategy couples lightweight, physically motivated features with a small FFNN regressor. For every image, we compute the mean grayscale intensity inside five local regions of interest (ROI): one at the image center; and four in fixed offsets (up, down, left, right). These five averages jointly describe the light transmitted through the various regions of the sample. A modest fully connected network then maps the five-element vector to a single thickness value. These five numbers are passed to a small feed forward neural network that contains two hidden layers with 64 and 32 rectified linear units, respectively, followed by a single linear output neuron that returns the predicted thickness. The network is trained for 50 epochs with a mini-batch size of 32 under the MSE loss.

This design purposefully keeps the parameter count low and the feature space interpretable while still allowing the model to learn some of the non-linear backlight dependencies.

3.3 Feed Forward Neural Network

The second method eliminates the selected features altogether. Each 224×224 image is flattened into a one-dimensional array of raw pixel intensities, which serves as input to a deeper FFNN regressor. Because every pixel is treated as an independent scalar, this model has no built-in spatial bias; instead, it relies on the dense weight matrix to discover any relevant patterns. Two fully connected layers, the first with 512 and the second with 128 rectified units, gradually compress the signal before a linear output neuron produces the thickness estimate.

Although this architecture contains far more trainable parameters than the previous model, it provides an unbiased reference for how well a purely dense network can exploit the spatial statistics embedded in raw intensity patterns.

3.4 Convolutional Neural Network

The third technique introduces spatial inductive bias through convolutional filters. It begins with two convolutionpooling stages: the first applies 32 filters of size 3×3 with rectified activations and retains spatial resolution through same padding, after which 2×2 max-pooling halves the spatial dimensions; the second stage repeats the pattern with 64 filters. A final convolutional layer with 128 filters further refines the representation, and global average pooling then collapses each feature map to a single scalar, thereby aggregating global context without incurring the parameter overhead of large dense layers. A linear output neuron maps the resulting 128-dimensional vector to the predicted thickness.

Compared with the pixel-wise perceptron, this approach trades some capacity for greater data efficiency and robustness to estimation and local noise.

3.5 Custom Models Explanation and Experimental Setup

The choice of models was guided by the nature of the task, namely predicting the scalar value of the thickness from image data, where lightweight architectures are sufficient and desirable. We deliberately used compact networks to avoid overfitting and ensure practical usability, reflecting the simplicity of the regression problem. Furthermore, the selected approaches represent two distinct modeling paradigms: ROI-based models that use interpretable regional statistics and pixel-based models that rely on learned features across the entire image. This contrast allows us to investigate how prior structure and spatial inductive bias influence predictive performance in a low-complexity setting.

Each architecture is trained ten times with independent weight initializations. Every run comprises 50 epochs of optimization with identical mini-batch schedules, after which the MSE on the test part of the dataset is recorded. None of the prevention of overfitting and the structural complexity of each topology was considered here due to the straightforwardness of the problem.

3.6 You Only Look Once Architecture

The YOLO was used as the fourth method. In recent years, many different architectures of neural networks have been developed for a variety of vision tasks. One of the most popular architectures in recent times has been YOLO [4] and its later versions, mainly due to the combination of very good precision and inference speed.

For this comparison, we selected all models of YOLOv11 [6], which is the latest version at the time of this study. Regression task is not directly supported. Therefore, we used models for classification, which were only slightly modified for this purpose. Specifically, we have changed the model head where the last layer of the neural network does not use a softmax function but a linear transformation. Consequently, the output of the model is no longer a class, but rather a continuous value that indicates the thickness of the object of interest.

As a loss function, we used Mean Square Error (MSE), which is a standard metric in regression problems. As a stopping criterion, we chose the total number of epochs, which was set to 1000. During the training process, the model with the best value of the loss function was always stored to prevent overtraining and to ensure that the most accurate model was obtained. In order to reduce the stochastic nature of training, training was performed five times for each variant. Due to the relatively small size of the dataset, we used data augmentation techniques to enrich the dataset. However, due to the nature of the task, it was necessary to omit all methods that affected the color.

4 Results

The experimental evaluation confirms that the regression head attached to the YOLOv11 backbone delivers by far the lowest prediction error on the test set.

Across all independent training runs, its mean square error (MSE) clusters tightly around 5.7×10^{-3}, resulting in a deviation of the root mean square thickness of only 0.08 mm in the full range of 0.5 to 5.2 mm. Table 2 sums the mean, median, and spread of the mean square error (MSE) of the test set obtained from repeated trainings of each method, while Fig. 4 visualizes the full distributions. The inference time and parameter count for all evaluated models are summarized in Table 3. All measurements were performed on an NVIDIA Quadro P5000 GPU with 16,384 MB of memory.

Table 2. Test set MSE statistics across independent initializations.

Model	Runs	Mean MSE	Median	Std. σ	RMSE [mm]
YOLOv11 all	25	5.69×10^{-3}	5.40×10^{-3}	1.37×10^{-3}	**0.075**
CNN	10	1.46×10^{-2}	1.40×10^{-2}	4.15×10^{-3}	0.121
5F+FFNN	10	7.06×10^{-2}	7.05×10^{-2}	2.00×10^{-3}	0.266
FFNN (pixels)	10	8.60×10^{-2}	4.52×10^{-2}	9.57×10^{-2}	0.293

Table 3. Comparison of average inference time and parameter count for all evaluated models.

Model	Parameters	Inference Time [ms]
FFNN (ROI)	2.5K	1.58
FFNN (pixels)	25.7M	9.98
CNN	92.8K	1.29
YOLOv11n	1.5M	0.85
YOLOv11s	5.5M	0.90
YOLOv11m	10.4M	1.64
YOLOv11l	12.8M	2.18
YOLOv11x	28.4M	3.32

The compact CCN was second best. By introducing locality through convolutional filters, it reduces the average MSE to 1.46×10^{-2} (approx. 0.12 mm), an order of magnitude better than the purely dense pixelwise perceptron yet still roughly twice as large as the YOLOv11 error.

Replacement of convolution with a small set of ROI features yields a further drop in performance. The intensity signature of five ROIs combined with a two-layer feedforward network consistently converges, but its mean error remains around 7.1×10^{-2} (approx. 0.27 mm).

Finally, the FFNN trained directly on flattened pixel intensities exhibits the largest variance. Although several runs reach MSE values below 5×10^{-2}, two

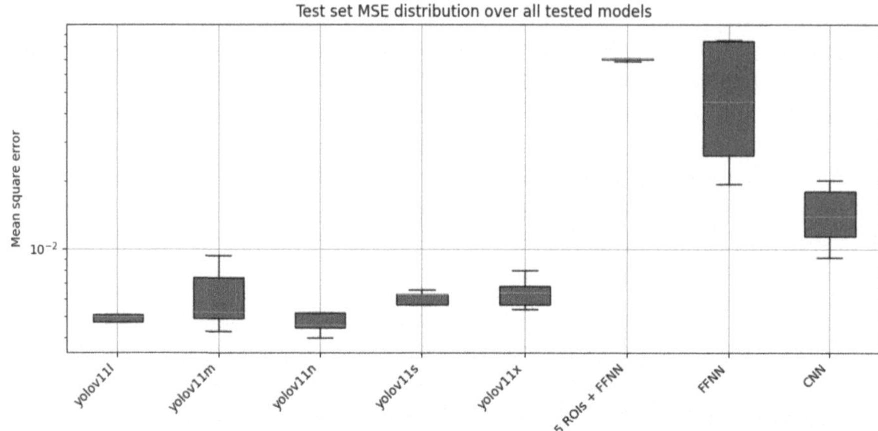

Fig. 4. Boxplot of test set MSE for all tested methods with YOLO models separated.

instances diverge markedly, pushing the mean to 8.6×10^{-2} (approx. 0.29 mm). These fluctuations reflect the difficulty of optimizing a high-capacity dense network on a limited dataset when no spatial inductive bias is available to guide learning.

Taken together, the results support the learning-based reformulation of Eq. 2. A detection-grade convolutional backbone, even when repurposed for single-value regression, can achieve thickness estimates that meet typical requirements without any explicit backlight field correction or contact measurement.

In addition to accuracy, Table 3 highlights the practical trade-offs between model complexity and runtime efficiency. While the FFNN with ROI features is by far the lightest model (2.5K parameters, 1.58 ms), its accuracy is limited. The pixel-wise FFNN, despite having over 25 million parameters, performs reasonably well (9.9 ms inference), though still less robust. The CNN offers a balanced compromise with under 100K parameters and 1.29 ms inference time. Among the YOLOv11 models, smaller variants such as YOLOv11n and YOLOv11s achieve the best accuracy (0.85 ms and 0.90 ms respectively), while keeping parameter count and latency manageable. In contrast, the larger YOLOv11m/l/x variants show diminishing returns–requiring significantly more resources without improving prediction quality, and in fact showing increased error.

5 Conclusions

This study demonstrated that accurate polymer thickness measurements can be achieved from a single transmission image at a wavelength of 1350 nm by reformulating the problem as a supervised regression and combining it with modern neural network-based detection methods trained on a proprietary dataset.

Of the four methods evaluated, the lightweight YOLOv11-n model with a linear head achieved the lowest mean error of 0.0046 MSE, corresponding to

0.08 mm RMSE, significantly outperforming compact CNNs and FFNN based models. These results confirm that SWIR imaging combined with deep neural networks can meet typical industrial tolerances without explicit backlight correction, enabling the use of inexpensive wavelength-specific lighting panels in industrial applications and thus minimizing costs.

In future work, we want to focus on observing defects in plastic parts (weakened layer leads to intensity growth), but also on physical properties in 3D space, namely determining the volume of a part using the same configuration when evaluating transparency (3rd axis data). Future work should also extend the validation to a larger multispectral dataset, along with testing different types of polymers and their suitability for this type of image-based determination in the SWIR spectrum.

Acknowledgment. The work was supported from ERDF/ESF "Cooperation in Applied Research between the University of Pardubice and companies, in the Field of Positioning, Detection and Simulation Technology for Transport Systems (PosiTrans)" (No. CZ.02.1.01/0.0/0.0/17_049/0008394); and SGS project realized at the University of Pardubice, Faculty of Electrical Engineering and Informatics.

References

1. Martinez-Hernandez, U., West, G., Assaf, T.: Low-cost recognition of plastic waste using deep learning and a multi-spectral near-infrared sensor. Sensors **24**(9) (2024). https://doi.org/10.3390/s24092821
2. Ozawa, K.: Identification of overlapping plastic sheets using short-wavelength infrared hyperspectral imaging. Opt. Express **31**(8), 12328–12338 (2023). https://doi.org/10.1364/OE.485039
3. Singh, M.K., Hait, S., Thakur, A.: Hyperspectral imaging-based classification of plastics for plastics recycling using artificial neural network. Process Saf. Environ. Prot. **179**, 593–602 (2023). https://doi.org/10.1016/j.psep.2023.09.052
4. Redmon, J., Divvala, S., Girshick, R., Farhadi, A.: You only look once: unified, real-time object detection. In: Proceedings of the IEEE Conference on Computer Vision and Pattern Recognition, pp. 779–788 (2016). https://doi.org/10.1109/CVPR.2016.91
5. Mayerhöfer, T.G., Pahlow, S., Popp, J.: The Bouguer–Beer–Lambert law: shining light on the obscure. ChemPhysChem **21**(18), 2029–2046 (2020). https://doi.org/10.1002/cphc.202000464
6. Jocher, G., Qiu, J.: Ultralytics YOLO11 (2024)
7. Song, H., et al: Short-wave infrared (SWIR) imaging for robust material classification: overcoming limitations of visible spectrum data. Appl. Sci. (Switzerland) **14**(23) (2024). https://doi.org/10.3390/app142311049
8. Serranti, S., Capobianco, G., Cucuzza, P., Bonifazi, G.: Efficient microplastic identification by hyperspectral imaging: a comparative study of spatial resolutions, spectral ranges and classification models to define an optimal analytical protocol. Sci. Total Environment **954**, 176630 (2024). https://doi.org/10.1016/j.scitotenv.2024.176630

9. Henriksen, M.L., Kolås, T., He, L. Plastic classification via in-line hyperspectral camera analysis and unsupervised machine learning. Vib. Spectrosc. **18**, 103329 (2022). https://doi.org/10.1016/j.vibspec.2021.103329
10. Zhong, L., Cheng, H., Gao, L., Li, L., et al.: Real-time film thickness monitoring in complex environments using deep learning-based visual imaging. Powder Technol. 120795 (2025) https://doi.org/10.1016/j.powtec.2025.120795
11. Ficzere, M., Mészáros, L.A., Kállai, N., Kovács, A., et al.: Real-time coating thickness measurement and defect recognition of film-coated tablets with machine vision and deep learning. Int. J. Pharm. **627**, 121957 (2022). https://doi.org/10.1016/j.ijpharm.2022.121957
12. Jun, S., et al.: Detection of defective chips from nanostructures with a high-aspect ratio using hyperspectral imaging and deep learning. J. Micro/Nanopattern. Mater. Metrol. **23**(4), 044002 (2024). https://doi.org/10.1117/1.JMM.23.4.044002
13. Zhang, R., Li, Y., Guan, Z.: Research on injection molded parts defect detection algorithm based on multiplicative feature fusion and improved attention mechanism. Sci. Rep. **14**, 30864 (2024). https://doi.org/10.1038/s41598-024-81430-x
14. Liu, J., Guo, F., Gao, H., Li, M., Zhang, Y., Zhou, H.: Defect detection of injection molding products on small datasets using transfer learning. J. Manuf. Process. **70**(2), 400–413 (2021). https://doi.org/10.1016/j.jmapro.2021.08.034
15. Ha, H., Jeong, J.: CNN-based defect inspection for injection molding using edge computing and industrial IoT systems. Appl. Sci. **11**(14), 6378 (2021). https://doi.org/10.3390/app11146378
16. Yan, A., Lazzari, M., Park, J., et al.: Towards deep computer vision for in-line defect detection in polymer electrolyte membrane fuel cell materials. Int. J. Hydrogen Energy **48**(50), 18978–18995 (2023). https://doi.org/10.1016/j.ijhydene.2023.01.257
17. Vorobev, R., Vasilev, I., Kremnev, I.: Segmentation of structural defects in polymer composite computed tomography images with deep learning models. Tomogr. Mater. Struct. **3**(1), 100014 (2023). https://doi.org/10.1016/j.tmater.2023.100014
18. Azad, M.M., Kim, H.S.: An explainable artificial intelligence-based approach for reliable damage detection in polymer composite structures using deep learning. Polym. Compos. **46**(2), 1536–1551 (2025). https://doi.org/10.1002/pc.29055
19. Moroni, M., Balsi, M., Bouchelaghem, S.: Plastics detection and sorting using hyperspectral sensing and machine learning algorithms. Waste Manag. **203**, 114854 (2025). https://doi.org/10.1016/j.wasman.2025.114854
20. Ji, T., Fang, H., Zhang, R., Yang, J., Wang, Z., Wang, X.: Plastic waste identification based on multimodal feature selection and cross-modal swin transformer. Waste Manag. **192**, 58–68 (2025). https://doi.org/10.1016/j.wasman.2024.11.027
21. Leger, P.A., Ramesh, A., Ulloa, T., Wu, Y., et al.: Machine learning enabled fast optical identification and characterization of 2D materials. Sci. Rep. **14**, 27808 (2024). https://doi.org/10.1038/s41598-024-79386-z
22. Bang, H.T., Park, S., Jeon, H.: Defect identification in composite materials via thermography and deep learning techniques. Compos. Struct. **246**, 112405 (2020). https://doi.org/10.1016/j.compstruct.2020.112405

Content-Based Image Retrieval in Echocardiography Using Textures and Convolutional Neural Networks

Jesús Benito-Picazo[1,2(✉)], David Elizondo[4], Juan-José Torres-Palomo[3], and Miguel A. Molina-Cabello[1,2]

[1] ITIS Software, University of Málaga, Málaga, Spain
{jpicazo,mamc}@uma.es
[2] Instituto de Investigación Biomédica de Málaga – IBIMA, Málaga, Spain
[3] Department of Computer Languages and Computer Science, University of Málaga, Málaga, Spain
juanjosetp02@uma.es
[4] De Montfort University, Leicester, UK
elizondo@dmu.ac.uk

Abstract. Cardiology is one of the most intricate specialities in the area of medicine. In order to detect potential diseases, echocardiography is widely used as a non-invasive imaging technique in medical practice. This medical test obtains complete images that cardiologists can evaluate for detecting possible anomalies. In this work, a deep learning-based Content-Based Image Retrieval (CBIR) system is developed to assist specialists in heart disease diagnosis. Several experiments have been carried out to determine the optimal parameters for the CBIR system. Our findings suggest that the consistency and balance of the information input into the CBIR system are more important than the amount of data used.

Keywords: deep learning · neural network · content based image retrieval · image classification · echocardiograms

1 Introduction

In 2019, it is estimated that 19 million people died due to heart or circulatory diseases and this number is expected to continue to rise [3]. This is why the focus of research on improving the ability to diagnose heart and circulatory diseases in a early stages takes on great relevance. Apart from the advances that could be developed through discoveries of new types of tests or treatments, it is crucial to find ways to exploit all the information we already have with existing tests. Here is where deep learning and computer vision take a really important part [12]. In fact, in the last decades, the use of different deep learning techniques applied to the field of medicine has been increasing [4].

Among the various neural network models available, Convolutional Neural Networks (CNNs) are specifically designed to handle image data, making them

E. Corchado et al. (Eds.): SOCO 2025, CCIS 2806, pp. 561–571, 2026.
https://doi.org/10.1007/978-3-032-19763-4_52

convenient for different applications such as object detection or classification, which aligns with the objectives of this study [9].

The Content-Based Image Retrieval (CBIR) system presented in this work leverages pre-trained CNNs, in particular AlexNet [6], to extract both semantic information from the pictures in a similar way a human vision system would do.

The primary contribution of this study lies in analyzing the performance of the CBIR system, which was developed using a fine-tuned AlexNet CNN, that shows to deliver accurate results. This analysis is conducted by applying the system with various feature configurations and two distinct datasets, focusing on the critical and complex field of medicine. The study generates both qualitative and quantitative comparative reports, discussing the outcomes in detail and providing insights into the system's effectiveness.

2 Related Work

In recent years, significant progress has been made in applying CNNs to various image processing tasks, including medical image classification and retrieval based on content [5,9]. CBIR systems have generated increasing attention for their potential to enhance medical diagnostics. However, challenges remain, particularly in the construction of effective semantic search mechanisms within CBIR systems for complex domains such as echocardiography. The study [7] serves as a successful example of applying CNNs to enhance a CBIR system for medical diagnostics.

CBIR techniques have been widely applied in medicine for indexing and searching images. Methods such as the Gray Level Co-occurrence Matrix (GLCM) have proven effective for feature extraction [14], and studies like [11] have examined the role of feature selection in enhancing retrieval performance.

Regarding the CBIR's clinical utility, it is further supported by applications in diverse areas such as breast cancer diagnosis [2], radiology-assisted image analysis [1], and cardiac MRI pathology detection [10].

By synthesizing insights from the studies presented in this section, we identify AlexNet as a robust foundation for our CBIR system, leveraging its proven success in bridging the gap between low-level and high-level features.

3 Methodology

In this work, a Content-Based Image Retrieval (CBIR) system using an ensemble of CNN is proposed. First of all, let D represent the number of object classes, and let $\mathcal{S}$ be the set of all the training image samples:

$$\mathcal{S} = \{\mathbf{X}_i \mid i \in \{1, ..., N\}\} \quad (1)$$

where $\mathbf{X}_i$ is the i-th image.

Regarding the neural network architecture, it contains an output layer that includes one neuron for each object class in the training image database. So that, the output layer has D neurons.

The CNN must be trained by using the images of the training database, where the desired output for each image is a unit vector of size D. This vector has a "1" at the position corresponding to the class of the object shown in the training image, and zeros in all other positions. The CNN is designed to estimate the probabilities that the input object belongs to any of the D classes being considered:

$$f(\mathbf{X}) = (P(C_1|\mathbf{X}), ..., P(C_D|\mathbf{X})) \in [0,1]^D \tag{2}$$

where $\mathbf{X}$ is the input image, and C_i is the i-th class, with $i \in \{1, ..., D\}$.

Now, let us assume that an image texture processing operator G_j, where $j \in 1, \ldots, M$, affects the images while the ground-truths remain unchanged, where M represents the number of different textures considered, such as mean, std, contrast, dissimilarity or ASM, for example. Thus, the j-th operator takes the image $\mathbf{X}_i$ as input and the processed image $\tilde{\mathbf{X}}_{i,j}$ is obtained.

$$\tilde{\mathbf{X}}_{i,j} = G_j(\mathbf{X}_i) \tag{3}$$

Then, the model $\mathcal{D}_j$ is generated after a training process by using those images obtained by the operator G_j.

Once the CNN has been trained in this manner, the system is prepared to handle user queries. Let $\mathbf{X}Query$ be the query image and $\mathbf{X}j$ be the training database images, where $j \in 1, \ldots, N$ and N represents the total number of images in the database. Then, the database images are ranked based on the Euclidean distances between the probability vector $f(\mathbf{X}Query)$ of the query image and the probability vectors $f(\mathbf{X}j)$ of the database images. This way, the most similar image in the database can be identified as follows:

$$s = \arg \min_{j \in \{1,\ldots,N\}} \|f(\mathbf{X}_{Query}) - f(\mathbf{X}_j)\| \tag{4}$$

Therefore, the k most similar images in the database can then be retrieved, ranked by the Euclidean distances $\|f(\mathbf{X}Query) - f(\mathbf{X}j)\|$.

Figure 1 offers a graphical abstract of the architecture followed in this work. First, the query image in raw format is classified by the AlexNet model previously finetuned with our training partition of the CAMUS dataset. Additionally, nine images that are different textures of the original image are generated in parallel. In the next step, the standard deviation for each texture image is calculated and stored together with the probabilities generated by the classifier model in an array. Then, we calculate the Euclidean distances of that information with respect to each of the images we have in the database. Finally, we retrieve the K images with the smallest Euclidean distances, which means that they are the K most similar images.

4 Experimental Results

This study is focused in analyzing the performance of a CBIR system built upon Convolutional Neural Networks and statistical parameters searches, applied to a

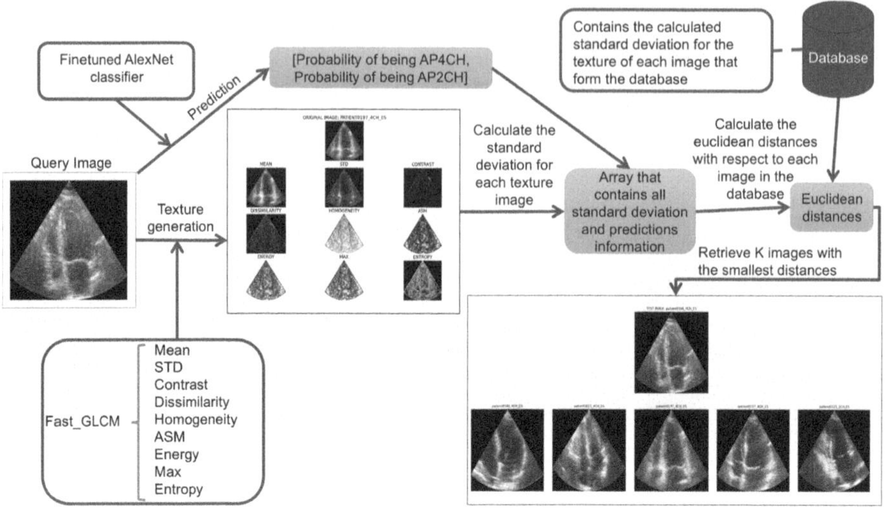

Fig. 1. Scheme of the proposed CBIR system.

specific subject such as cardiology and more specifically echocardiograph images. During the course of this work, several experiments have been carried out with different parameters selection, even the combination of two datasets has been explored, and the resulting performance is detailed in this section. The methods considered are described in Sect. 4.1, followed by parameter values selection for the CNN model in Sect. 4.2. Datasets are described in Sect. 4.3, while results are presented in Sections 4.4 and 4.5. Finally, the conclusions are depicted in Sect. 5.

4.1 Methods

The CBIR system is built around AlexNet, a pre-trained CNN known for its efficiency optimizations such as overlapping pooling and GPU parallelization. AlexNet achieved a 15.3% top-5 error in the 2012 ImageNet Challenge [6]. In our main experiment, it was fine-tuned using 60% of the CAMUS dataset for training, 20% for validation, and 20% for testing. A cross-validation strategy was applied to mitigate bias due to the dataset's limited size [13].

In addition to the CNN, the system includes texture analysis via a fast Gray-Level Co-Occurrence Matrix (GLCM) library, which generates multiple texture images derived from the original input. These texture images capture various gray-level intensity patterns, contributing additional features to the CBIR system. We have used the following github fast GLCM library implemented using numpy[1]. Together, the CNN and the texture features extracted by the GLCM library form the core of the knowledge database.

[1] https://github.com/tzm030329/GLCM.

Table 1 shows the original image and its corresponding GLCM-based textures, which reveal structural patterns useful for image retrieval in the CBIR system.

Table 1. Original echocardiography image from CAMUS dataset patient 1 and generated textures by GLCM.

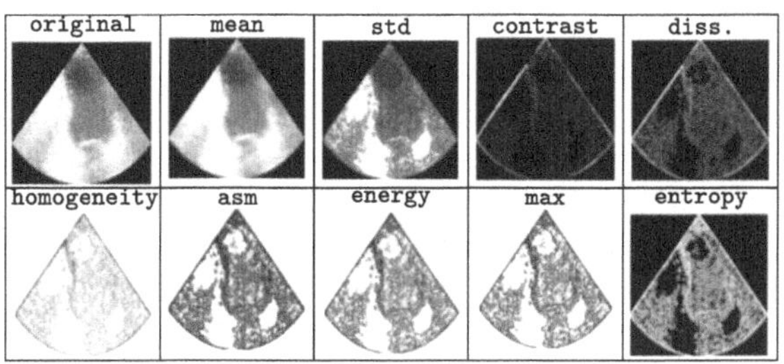

4.2 Classifier Model Parameters Selection

To build the CBIR system, it was necessary to identify the optimal configuration for fine-tuning the pre-trained AlexNet model. This involved experimenting with several hyperparameter combinations. The final parameters selected to fit the pre-trained AlexNet model in the training process are detailed in Table 2.

Table 2. Param. values

Parameter name	Value
Learning rate	0.001
# epochs	20
Momentum	0.9
Batch Size	32

The selected parameters were chosen based on systematic testing and evaluated using standard metrics: confusion matrix, AUC curve, F-score, Spatial Accuracy, Specificity, Accuracy, Recall, and Precision. Training performance was also tracked through precision/loss vs. epoch plots. Results confirmed that the fine-tuned AlexNet model offers robust performance as the core of the CBIR system.

4.3 Datasets

In this work, two datasets were utilized: CAMUS, serving as the primary dataset, and EchoNet-Dynamic, used to extend and complement the experimental analysis developing additional experiments.

Table 3. Results from different experiments.

Attributes	Relevants retrieved		Precision	Recall
	AP4CH	AP2CH		
mean	1390	1418	0.936	0.0039
mean, std, median	1123	1092	0.738	0.0031
mean, std	1230	1221	0.817	0.0034
std	1407	1417	0.9413	0.00392

CAMUS Dataset. The CAMUS dataset contains heart images extracted from clinical examinations of 500 anonymized patients from the University Hospital of St. Etienne, France. No specific prerequisites were set for the acquisition of the images, ensuring a more realistic dataset with diverse imaging conditions, resulting in varying image quality and anatomical visibility. The raw dataset consists of 500 directories, each representing a patient, containing images for different views (AP4CH and AP2CH), and sequence types (ED, ES, and half_sequence). This gives a total of 3000 images, taking into account all the variations in view and sequence [8].

EchoNet-Dynamic Dataset. This dataset consists of 10.300 .avi video files from patients undergoing routine clinical care at Stanford Hospital between 2016 and 2018. These videos capture electrocardiograph sequences from dynamic cardiac exams. All the videos in the dataset correspond to the apical four-chamber (AP4CH) view [15]. For our research, we extracted from the EchoNet-Dynamic videos those exact frames that correspond to ED and ES sequences, obtaining a total of 20.048 images of this dataset that aligns with the same type of images we have in CAMUS.

4.4 Quantitative Results

To objectively assess the performance of the CBIR system and compare different configurations, we used well-established metrics: precision and recall, which are widely applied in information retrieval and machine learning systems. These metrics are defined by the following functions:

$$\text{Precision} = \frac{\#\,\text{relevant images retrieved}}{\#\,\text{retrieved images}}; \text{Recall} = \frac{\#\,\text{relevant images retrieved}}{\#\,\text{of relevant images}}$$

Both metrics yield values between 0 and 1, providing a measure of retrieval effectiveness. The system is better as metrics are as nearest as possible to 1.

In order to determine the optimal configuration for our proposal, initial experiments were carried out with a fixed number of images to retrieve, denoted by the parameter k. For these experiments, we have set $k = 5$ which means that the system will retrieves the 5 images with less distance to the query image, so that they must be the most similar ones in the database. While doing this initial experiment we determined which is the approach that returned better results using precision and recall as evaluation metrics. The approaches used were: Experiment using only the **mean**; experiment using **mean, standard deviation and median**; experiment using **mean and standard deviation** and experiment using **standard deviation**. The results of these initial experiments are summarized in Table 3.

Precision and recall reflect distinct aspects of system performance. Precision measures how many retrieved images are relevant, while recall evaluates how many relevant images are retrieved from the dataset. The significant difference observed in these metrics (high precision, but dramatically low recall) is due to the large number of relevant images for each class in the dataset (1200 relevant images for AP4CH class and 1200 relevant images for AP2CH class) and the small number of retrieved images $k=5$. In this configuration, the high precision indicates that almost all retrieved images are relevant, showcasing the system's accuracy. However, the recall is naturally low because the portion of retrieved images relative to the total number of relevant images is very small.

To further evaluate recall performance, we ran experiments with larger k values. As shown in Fig. 2, recall increases linearly with the k value. This confirms that the system consistently retrieves relevant images as k grows, demonstrating that recall is working correctly despite its low values for small k. Higher k values were not explored, as retrieving too many images becomes impractical and less useful in real-world diagnostic settings.

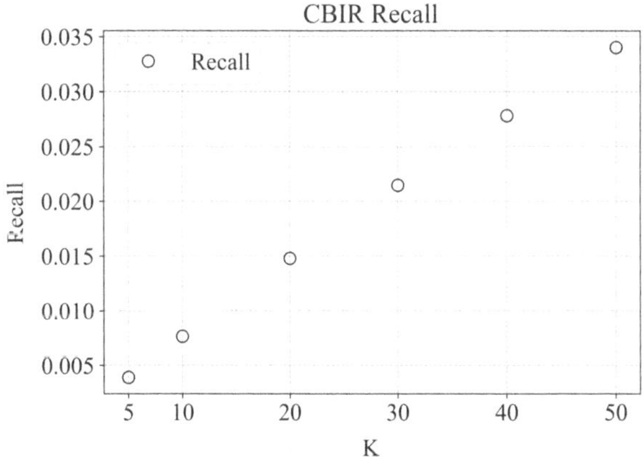

Fig. 2. Recall results for different k values.

Given the promising performance of our CBIR system on the CAMUS dataset, we extended the experiments to a combined dataset with 3,000 CAMUS and 20,048 EchoNet-Dynamic images. While fine-tuning showed robust overall performance, the confusion matrix revealed difficulty classifying AP2CH views, likely due to class imbalance. Despite this, the F-measure, suitable for imbalanced data, reached 0.7362, indicating reasonable performance given the challenges of the dataset.

4.5 Qualitative Results

In the context of medical imaging, qualitative analyses are equally critical, as they offer practical insights that complement numerical evaluations. Notably, in CBIR systems applied to medical imaging, retrieving images from a different class than the query image does not necessarily indicate a failure as the system might be identifying structural or feature-based similarities that hold clinical significance. This section presents representative cases to assess the CBIR system's behaviour using the best-performing configuration. Visual results include query images and their top-5 retrieved matches. Each case includes at least three retrieved images from a different class for a better understanding of the system's reasoning process.

Evaluation of CBIR System with only CAMUS Dataset. In Table 4, a relevant case, focusing first on the results obtained from the CBIR system trained on the single CAMUS dataset. It can be observed how two images of the same patient (patient 129) have been recovered at different times, the second and third images from left to right. This indicates that the system has found similarities with the organ of this particular patient. On the other hand, with regard to the recovered images of a different class (AP4CH) to the class of query image (AP2CH), it can be observed that, in spite of the difference in class, there are some prominent white vertical lines in the left area of the images and other high-intensity regions in the lower areas, which correspond to some walls of the cardiac structures and are in the same way relevant in the query image, which is possibly why they have been recovered.

Table 4. Example of images retrieved by the CBIR system trained only with the CAMUS dataset. The leftmost image corresponds to the query image and the rest of them correspond to the images retrieved by the CBIR system.

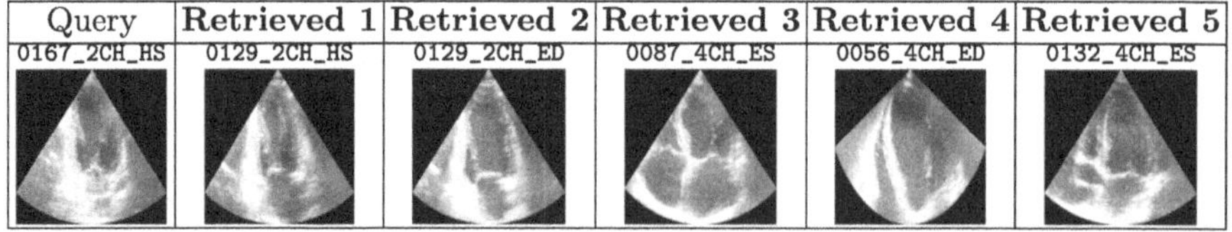

Query	**Retrieved 1**	**Retrieved 2**	**Retrieved 3**	**Retrieved 4**	**Retrieved 5**
0167_2CH_HS	0129_2CH_HS	0129_2CH_ED	0087_4CH_ES	0056_4CH_ED	0132_4CH_ES

Evaluation of CBIR System with both CAMUS and EchoNet-Dynamic Datasets. Building on the initial qualitative evaluation of the CBIR system trained only on CAMUS dataset, we now turn our attention to the results obtained when utilizing a model trained on a combined dataset that integrates both CAMUS and EchoNet-Dynamic images. This extended dataset provides the opportunity to evaluate the system's performance in a diverse data environment, challenging its ability to retrieve meaningful and clinically relevant results across varying contexts. In Table 5, two representative cases are shown to evaluate and discuss over the results obtained by the CBIR system in this combined dataset scenario.

Table 5. Examples of images retrieved by the CBIR system trained with both CAMUS and EchoNet-Dynamic datasets.

Test	Query	Retrieved 1	Retrieved 2	Retrieved 3	Retrieved 4	Retrieved 5
1	0212_4CH_HS	0477_4CH_ED	0301_4CH_ES	0428_4CH_ED	0281_4CH_ES	0292_4CH_ED
2	DD98BD6CD12	CEDF5353B79D	CEDF5353B79D	127FC335FF07	127FC335FF07	097BE78FA04F

The CBIR system showed clear dataset separation: as seen in the test images, queries only retrieved results from the same dataset, with no cross-retrieval between CAMUS and EchoNet-Dynamic, even in large-scale tests with K=5.

5 Conclusions

This paper presents a study on the implementation, integration, and performance of a CBIR system designed to aid in heart disease diagnosis using echocardiographic images. The proposed methodology integrates a CNN-based model with statistical parameters, specifically the standard deviation of textures derived from GLCM calculations, as the core features of the system. A series of experiments and analyses were conducted to justify the exclusive use of the standard deviation feature, leading to an optimized and efficient model. The results of these experiments were discussed from both quantitative and qualitative perspectives.

Qualitative analysis revealed that the retrieved images, despite belonging to a different class from the query, exhibited considerable visual similarity highlighting the robustness of the system in identifying visually similar images.

When combining the CAMUS and EchoNet-Dynamic datasets, the system consistently retrieved images from the same dataset as the query, rather than across both datasets. This limitation is likely due to the imbalance in the datasets, with the EchoNet-Dynamic dataset being significantly larger. The base

CNN model's training might have been skewed by this imbalance, preventing cross-dataset retrievals.

This study provides valuable insights for the design of CBIR systems in medical imaging, particularly for echocardiographic applications. As a baseline, this work opens up opportunities for further research, such as the inclusion of additional datasets and images to enable broader testing and validation, or exploring techniques to address dataset imbalances potentially improving the ability to retrieve relevant information.

Acknowledgements. This work is partially supported by projects UMA20-FEDERJA-108, PID2022-136764OA-I00, PUNI-003_2023 and ATECH-25-02. The authors thankfully acknowledge the computer resources, technical expertise and assistance provided by the SCBI (Supercomputing and Bioinformatics) center of the University of Malaga. They also gratefully acknowledge the support of NVIDIA Corporation with the donation of a RTX A6000 GPU with 48Gb. The authors also thankfully acknowledge the grant of the Universidad de Málaga and the Instituto de Investigación Biomédica de Málaga y Plataforma en Nanomedicina-IBIMA Plataforma BIONAND.

References

1. Akgül, C.B., Rubin, D.L., Napel, S., Beaulieu, C.F., Greenspan, H., Acar, B.: Content-based image retrieval in radiology: current status and future directions. J. Digit. Imaging **24**(2), 208–222 (2011). https://doi.org/10.1007/s10278-010-9290-9
2. Carvalho, E.D., et al.: Breast cancer diagnosis from histopathological images using textural features and CBIR. Artif. Intell. Med. **105**, 101845 (2020). https://doi.org/10.1016/j.artmed.2020.101845, https://www.sciencedirect.com/science/article/pii/S0933365719306621
3. Foundation, B.H.: Global heart & circulatory diseases factsheet. https://www.bhf.org.uk/-/media/files/for-professionals/research/heart-statistics/bhf-cvd-statistics-global-factsheet.pdf. Accessed 20 Feb 2024
4. Jiang, X., Hu, Z., Wang, S., Zhang, Y.: Deep learning for medical image-based cancer diagnosis. Cancers **15**(14) (2023). https://doi.org/10.3390/cancers15143608, https://www.mdpi.com/2072-6694/15/14/3608
5. Koziarski, M., Cyganek, B.: Image recognition with deep neural networks in presence of noise - dealing with and taking advantage of distortions. Integr. Comput.-Aid. Eng. **24**, 1–13 (2017). https://doi.org/10.3233/ICA-170551
6. Krizhevsky, A., Sutskever, I., Hinton, G.E.: ImageNet classification with deep convolutional neural networks. In: Pereira, F., Burges, C., Bottou, L., Weinberger, K. (eds.) Advances in Neural Information Processing Systems, vol. 25. Curran Associates, Inc. (2012). https://proceedings.neurips.cc/paper_files/paper/2012/file/c399862d3b9d6b76c8436e924a68c45b-Paper.pdf
7. Kruthika, K., Rajeswari, Maheshappa, H.: Cbir system using capsule networks and 3D CNN for Alzheimer's disease diagnosis. Inform. Med. Unlocked **14**, 59–68 (2019). https://doi.org/10.1016/j.imu.2018.12.001, https://www.sciencedirect.com/science/article/pii/S235291481830176X
8. Leclerc, S., Smistad, E., Pedrosa, J., et al., A.O.: Deep learning for segmentation using an open large-scale dataset in 2D echocardiography. IEEE Trans. Med. Imaging **37**, 2198–2210 (2019). https://doi.org/10.1109/TMI.2019.2900516, https://www.creatis.insa-lyon.fr/Challenge/camus/databases.html

9. Litjens, G., et al.: A survey on deep learning in medical image analysis. Med. Image Anal. **42**, 60–88 (2017). https://doi.org/10.1016/j.media.2017.07.005, https://www.sciencedirect.com/science/article/pii/S1361841517301135
10. Michno, T., Jelonek, M.: A report on work: cardiac MRI CBIR for pathologies detetion. In: Proceedings of the 25th International Conference on Enterprise Information Systems - Volume 1: ICEIS, pp. 667–674. INSTICC, SciTePress (2023). https://doi.org/10.5220/0012006800003467
11. Popova, A.A., Neshov, N.N.: Combining features evaluation approach in content-based image search for medical applications. In: Kountchev, R., Iantovics, B. (eds) Advances in Intelligent Analysis of Medical Data and Decision Support Systems. Stud. Comput. Intell., pp. 113–123. Springer International Publishing, Heidelberg (2013). https://doi.org/10.1007/978-3-319-00029-9_10
12. Razzak, M.I., Naz, S., Zaib, A.: Deep learning for medical image processing: overview, challenges and the future. In: Dey, N., Ashour, A., Borra, S. (eds) Classification in BioApps. Lecture Notes in Computational Vision and Biomechanics, pp. 323–350. Springer International Publishing, Cham (2018). https://doi.org/10.1007/978-3-319-65981-7_12
13. Refaeilzadeh, P., Tang, L., Liu, H.: Cross-validation. In: LIU, L., ÖZSU, M.T. (eds) Encyclopedia of Database Systems, pp. 532–538. Springer US, Boston, MA (2009). https://doi.org/10.1007/978-0-387-39940-9_565
14. Sudhish, D.K., Nair, L.R., S, S.: Content-based image retrieval for medical diagnosis using fuzzy clustering and deep learning. Biomed. Signal Process. Control **88**, 105620 (2024). https://doi.org/10.1016/j.bspc.2023.105620, https://www.sciencedirect.com/science/article/pii/S1746809423010534
15. University, S.: A large new cardiac motion video data resource for medical machine learning. https://echonet.github.io/dynamic/index.html. Accessed 20 Jun 2024

Photovoltaic Power Estimation Based on Meteorological Data Using LSTM and Attention

Anibal Mantilla-Guerra[1(✉)], Christian Mejia-Escobar[1], Jorge Azorin-Lopez[2], and Jose Garcia-Rodriguez[2]

[1] Central University of Ecuador, Quito, Ecuador
armantilla@uce.edu.ec
[2] Department of Computer Science and Technology, University of Alicante, Alicante, Spain

Abstract. Given the increasing adoption of solar energy and the need for reliable prediction to optimize energy production and the stability of electric service, this study presents an artificial intelligence model to estimate the generation of photovoltaic power. Historical data of generated power (kW) and meteorological variables such as temperature, relative humidity, precipitation, and wind speed with time windows of 24, 168, 720, 1440, and 2160 h are used. The results indicate that LSTM networks outperform traditional methods such as moving average. Despite incorporating an attention mechanism, the standard LSTM model performed better in extended time windows with a mean MSE of 0.0023 on the test set. The larger number of parameters due to a longer time window suggests an overfitting of the model with the attention layer. The tool obtained is strategic and replicable in energy planning, reducing the use of fossil fuels and extending the lifetime of photovoltaic systems.

Keywords: LSTM · Attention mechanism · Time series · Power · Weather

1 Introduction

The world is moving towards renewable energies, away from polluting sources [1]. For solar power plants, predicting their behavior is essential to ensure sufficient energy supply and meet future demand. This enables proper planning, forecasting, and prevention [2]. However, the problem is the uncertainty of photovoltaic production and demand due to the high variability of weather, socioeconomic, and consumption patterns. The complexity of the interactions between these factors makes it challenging to accurately predict the dynamics of solar panel power generation [3].

Traditionally, energy estimation has been approached with classical statistical methods such as moving average, exponential smoothing, and regression

E. Corchado et al. (Eds.): SOCO 2025, CCIS 2806, pp. 572–581, 2026.
https://doi.org/10.1007/978-3-032-19763-4_53

analysis [4]. Although useful, these methods have important limitations in complex time series [5]. Their main purpose is to identify trends; they are sensitive to outliers and tend to assume linear relationships, which limits their effectiveness in scenarios of high variability and abrupt changes, such as power generation.

The scientific literature reveals that artificial intelligence techniques based on classical Machine Learning (ML), Deep Learning (DL) and hybrid approaches outperform traditional models in predicting solar generation using environmental factors [6] [7] [8].

Among the regression algorithms Gradient Boosting (GB), XGBoost, k-Nearest Neighbors (kNN), LightGBM and CatBoost, the latter was the most remarkable [9]. Random Forest outperformed Linear Regression (LR), Decision Tree and XGBoost, confirming that the accuracy tends to decrease as the prediction time horizon extends [10]. Support vector machine regression (SVR) proved to be more efficient than multivariate regression (MR) and Gaussian regression (GR) [11]. Hybrid models integrating SVR, RF and gradient improve accuracy over individual models but increase the complexity of the algorithm [12]. Comparison of RL, kNN, SVR, RF and GBR together with Multi-Level Perceptron (MLP) and Conv2D-LSTM shows the best performance of deep learning in predicting solar energy [13]. The use of DL with LSTM and Bi-LSTM showed improved performance over tree-based (LSBoost, XGBoost, Bagging and RF) and traditional (GRNN, SVM, MLANN and RBANN) techniques [14]. A hybrid approach of clustering and LSTM improves day-ahead prediction accuracy [15]. By combining a recurrent neural network (RNN), a Transformer model and LSTM, it is possible to capture complex dependencies on various time scales and make reliable predictions [16].

This research implements a standard LSTM network and one combined with an auto-attention mechanism to predict hourly PV power from meteorological data. As a case study, a photovoltaic plant without hybrid back-up or batteries is used, which allows an accurate validation of the behavior of solar generation. The purpose is to contrast the results obtained and analyze the multivariate temporal correlations, improving performance with respect to previous methods [17]. It is validated with real data and allows for improved estimation in the short and medium term. The proposed model aims to support efficient energy management and be replicable in other remote communities integrating renewable energies [18]. The created dataset and the generated code are available on GitHub (github.com/cimejia/powerlstm).

2 Model

The estimation of solar plant power output using historical weather data is framed in a multivariate time series forecasting task. It focuses on an LSTM network, which can capture the irregular oscillations of the considered variables, complemented by an attention mechanism. The model is shown in Fig. 1.

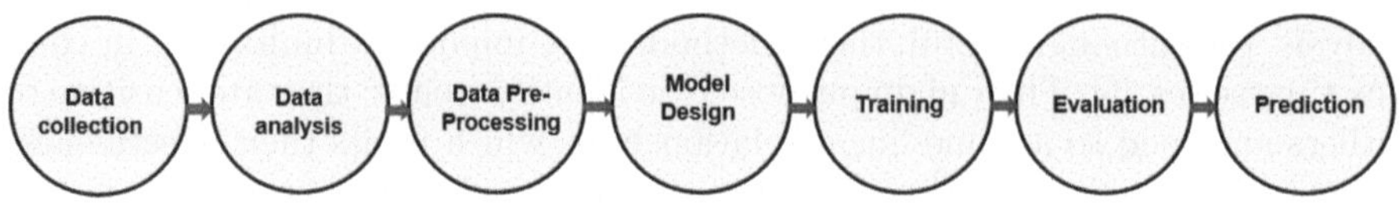

Fig. 1. Workflow followed for the development of this useful tool for the prediction task.

2.1 Data Collection

Data is the main resource of machine learning. The average hourly values of the power in kW supplied by the 67 kWp photovoltaic generation plant at Baltra Island (Galapagos, Ecuador), are considered. Meteorological data were downloaded from the NASA POWER Project Data Access [19] site for the indicated period and hourly frequency. These data are numerical and correspond to 4 variables: temperature in degrees centigrade, relative humidity in %, wind in m/s, and precipitation in mm/h. The geographical location of the weather station has as coordinates: latitude −0.4612 and longitude −90.2775, both in decimal degrees.

2.2 Data Analysis

Conventional statistical techniques allow to know the behavior of the variables of the study. Table 1 and Fig. 2 show the most important graphs and indicators, respectively. The site of interest is medium to high temperature, high humidity, light wind and little or no rainfall. The time series of electrical power has a direct relationship with temperature and an inverse relationship with wind speed.

Table 1. Descriptive statistics of the variables.

measure	kW	T	RH	P	WS
count	13848.00	13848.00	13848.00	13848.00	13848.00
mean	15.89	23.89	83.66	0.69	4.22
std	7.52	2.11	4.35	1.55	1.50
min	0.00	19.04	64.70	0.00	0.09
max	48.91	28.52	95.55	30.35	8.16

2.3 Data Preprocessing

The data needs to be transformed into a format suitable for LSTM training. To simulate a real prediction scenario, the data splitting must maintain chronological order [20]. The training set comprises 60% of the data, i.e., 8308 in quantity, and ranges from 2017-01-01 00:00:00 to 2017-12-13 03:00:00. The validation set is 20% of the data, i.e., 2769, and ranges from 2017-12-13 04:00:00 to 2018-04-07 12:00:00. The remaining 20% belongs to the test set, 2771 in total, and they range from 2018-04-07 13:00:00 to 2018-07-31 23:00:00. Data scaling helps the model converge faster and perform better [21]. Min-Max normalization, a technique that converts all values to a range of [0, 1], was used. Sequence creation is a distinctive and key operation with sequential data [22]. The time window is the basis for structuring the data into input-output pairs for training the LSTM model.

2.4 Model Architecture

The solution explores two approaches with LSTM networks: a standard LSTM model and an LSTM combined with an attention mechanism (see Fig. 3).

The input sequence has the shape (timesteps, 5), i.e., the time window and the five climatic variables considered. For processing, two alternatives are used based on [23]: a) Standard LSTM: Uses 64 neurons and predicts the next value based on the last output. b) LSTM with attention: Returns the complete sequence of outputs to assign weights and determine the relevance of each step in the final prediction.

3 Experimentation

The model was implemented, trained, and evaluated on *Google Colaboratory*, which offers an Intel Xeon processor, 12.7 GB of RAM, and an NVIDIA Tesla T4 GPU with 15 GB of RAM. The programming language is Python 3.11.12, and the support libraries are Pandas, NumPy, Matplotlib, Seaborn, Scikit-learn, Keras, and TensorFlow.

3.1 Training

A total of ten trainings were performed, five for each model: standard LSTM and LSTM with attention. Each training corresponds to a time window: 24, 168, 720, 1440 and 2160 h, i.e., 1 day, 7 d, 1 month, 2 and 3 months, respectively. These different time horizons are useful for evaluating short-, medium- and long term performance. Figure 4 indicates that LSTM models learn well, with training and validation losses decreasing. However, LSTM with attention shows an irregular validation curve, suggesting overfitting, possibly due to the increase in attention weights with the time window.

3.2 Evaluation

The test data are used, the model is run with each time window set, and the mean square error (MSE) is calculated as a performance metric. The LSTM with attention has an average MSE of 0.0057, while the standard LSTM is 0.0023. The standard LSTM error decreases as the length of the time window increases. To contrast the efficacy of these methods, the *Moving Average* was also used, which yielded an MSE of 31.29.

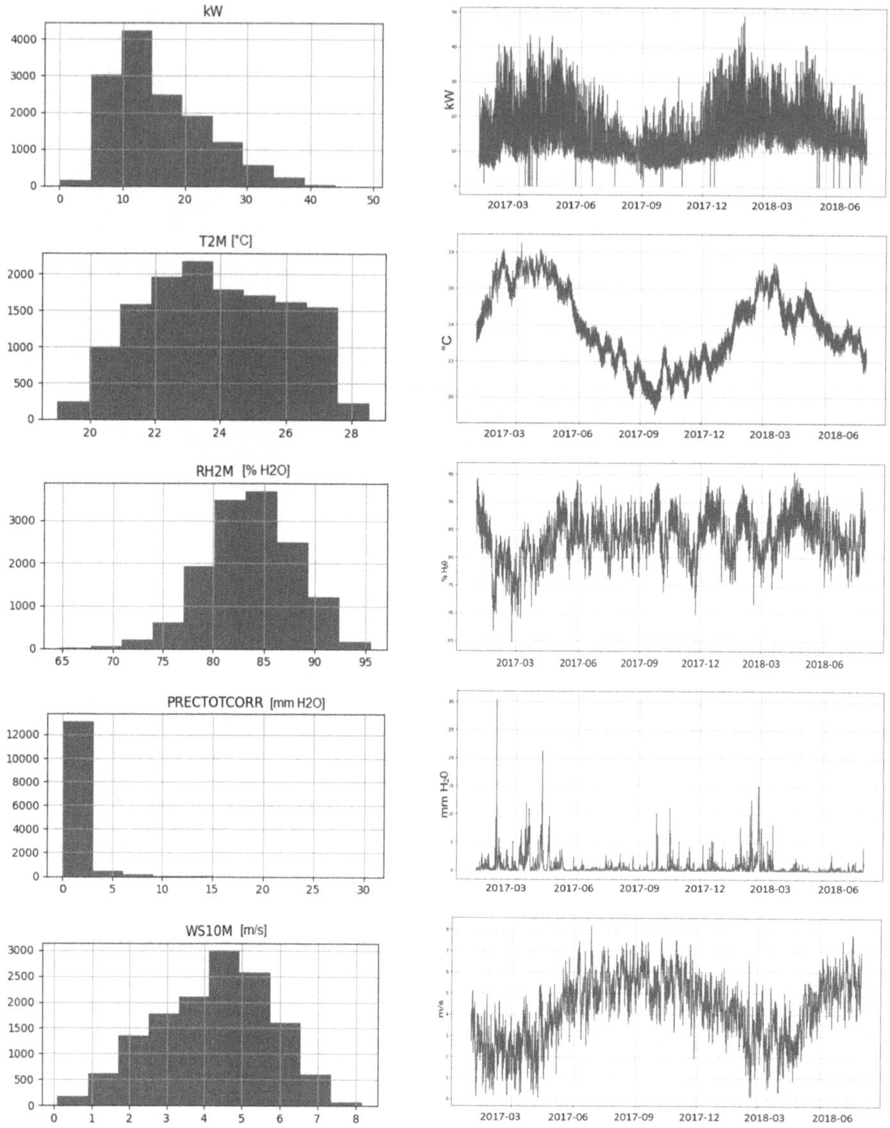

Fig. 2. Descriptive exploration of the variables considered: electrical power, temperature, relative humidity, precipitation, and wind speed. The histograms are shown in the left column, while their variation over time is shown in the right column.

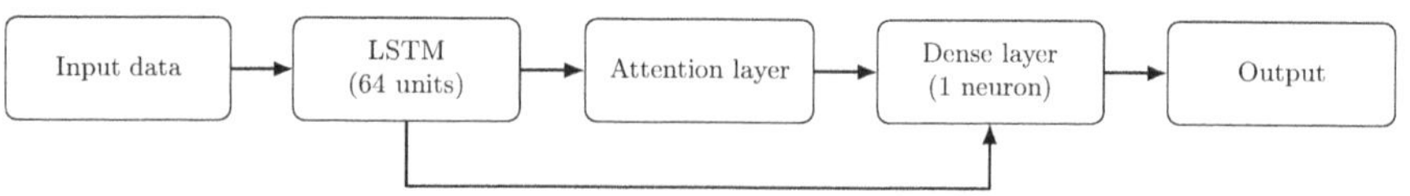

Fig. 3. Solution architecture: standard LSTM and LSTM+attention.

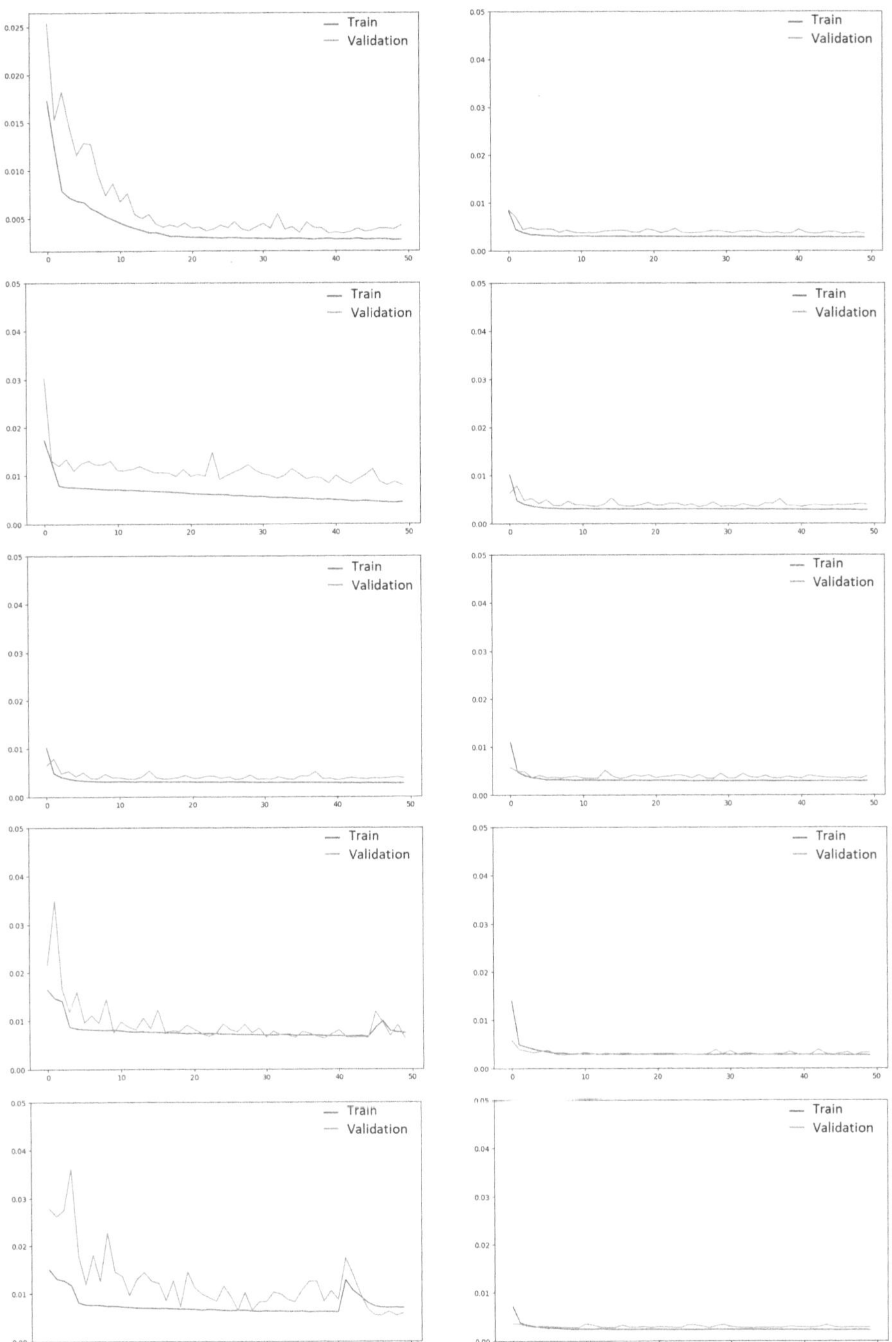

Fig. 4. Model training learning curves: LSTM+attention (left) and standard LSTM (right) for each time window: 24 h, 168 h, 720 h, 1440 h and 2160 h, from top to bottom, respectively. In each graph, the vertical axis corresponds to the loss and the horizontal axis to the periods.

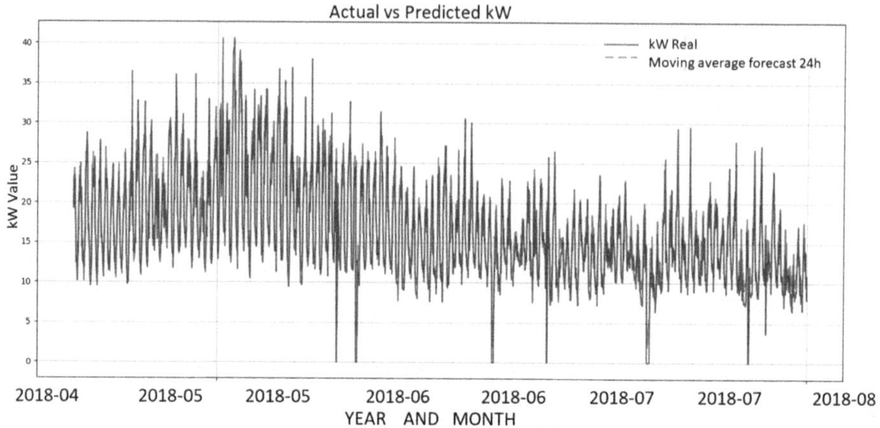

(a) LSTM+attention

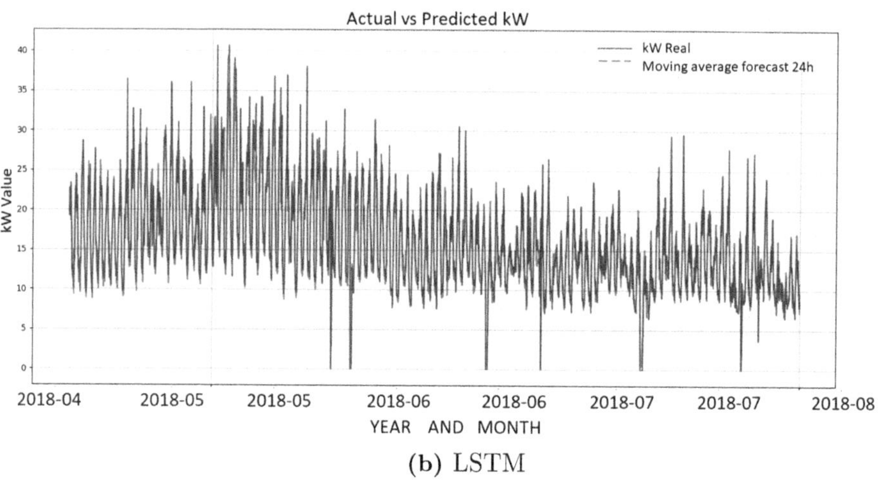

(b) LSTM

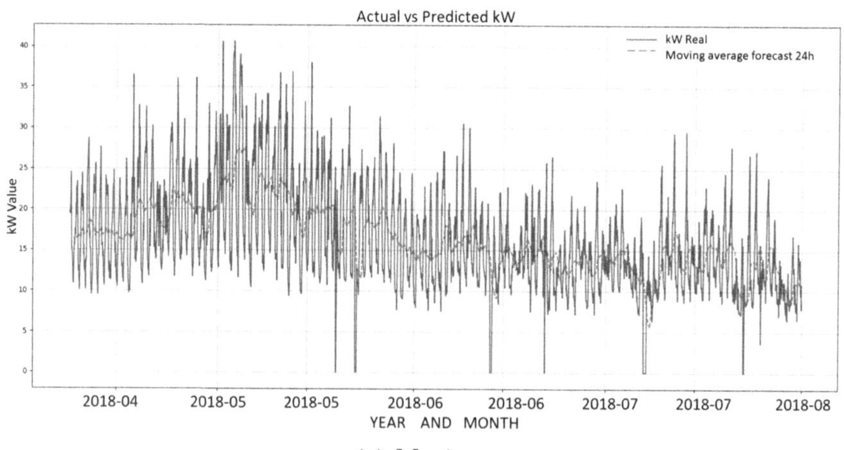

(c) Moving average

Fig. 5. Model prediction on test data.

3.3 Prediction

This is the inference phase, where the model obtained is used to make real predictions about the future [24]. Figure 5 presents the predictions of each model on the test set using a 168 h (1 week) time window, which could be considered as an ideal forecast horizon for electrical planning. LSTM models predict future photovoltaic power with high accuracy, outperforming the moving average. LSTM with attention stands out for generating robust predictions that avoid extremes, unlike standard LSTM, which tends to underestimate minimum power. On the other hand, it should also be noted that some related work has improved sensor data using artificial vision and artificial intelligence techniques in many different fields, which affects the final accuracy of the model [25–32].

4 Conclusion

This work implements LSTM networks, both standard and combined with attention, for a critical multivariate time series forecasting task. Our primary contribution lies in addressing the unique complexities and yielding novel insights specific to reliable PV power prediction, which is inherently challenging due to the high climatic variability and the interplay of several meteorological factors. By using a comprehensive range of prediction horizons (time windows of 24, 168, 720, 1440, and 2160 h), our study demonstrates that LSTM networks consistently outperform traditional methods such as moving average, affirming their suitability for this complex domain. However, a key finding emerges from our comparative analysis of the LSTM models: the attention layer provided higher accuracy only for shorter prediction periods and decreased in accuracy for longer periods. This suggests that the increase in attention layer parameters due to longer time windows increases the complexity of the model, which may lead to overfitting. This specific observation challenges the assumption that attention mechanisms universally guarantee an improvement in accuracy, which provides a crucial insight for researchers and practitioners, particularly in contexts characterized by high climatic variability. The proposed model is useful and replicable for a more sustainable energy management in remote communities. Future lines of research could focus on the development of DL-based architectures that seek more interpretable models for these types of tasks, improving the efficiency, reliability, and sustainability of energy systems.

References

1. Hassan, Q., et al.: The renewable energy role in the global energy transformations. Renewable Energy Focus **48** (2024)
2. Ukoba, K., Olatunji, K., Adeoye, E., Jen, T., Madyira, D.: Optimizing renewable energy systems through artificial intelligence: review and future prospects. Energy Environ. **35**, 3833–3879 (2024). https://doi.org/10.1177/0958305X241256293
3. Eze, V., Richard, K., Ukagwu, K., Okafor, W.: Factors influencing the efficiency of solar energy systems. J. Eng. Technol. Appl. Sci. (JETAS) **6**, 119–131 (2024)

4. Mystakidis, A., Koukaras, P., Tsalikidis, N., Ioannidis, D., Tjortjis, C.: Energy forecasting: a comprehensive review of techniques and technologies. Energies **17**, 1662 (2024)
5. Abdelwali, H., Abdelati, M., et al.: Enhancing time series forecasting with the advanced cumulative weighted moving average technique. Int. J. Adv. Sci. Innov. **6** (2024)
6. Gupta, M., Arya, A., Varshney, U., Mittal, J., Tomar, A.: A review of PV power forecasting using machine learning techniques. Progress Eng. Sci. **2**, 100058 (2025)
7. Miller, J., et al.: A survey of deep learning and foundation models for time series forecasting. arXiv preprint arXiv:2401.13912 (2024)
8. Ukoba, K., Onisuru, O., Jen, T.: Harnessing machine learning for sustainable futures: advancements in renewable energy and climate change mitigation. Bull. Natl. Res. Centre **48**, 99 (2024)
9. Nguyen, H., Tran, Q., Ngo, C., Nguyen, D., Van Quan, T.: Solar energy prediction through machine learning models: a comparative analysis of regressor algorithms. PLOS ONE **20**(1) (2025)
10. Lari, A., Sanfilippo, A., Bachour, D., Perez-Astudillo, D.: Using machine learning algorithms to forecast solar energy power output. Electronics **14** (2025)
11. Tripathi, A., et al.: Advancing solar PV panel power prediction: a comparative machine learning approach in fluctuating environmental conditions. Case Stud. Thermal Eng. **59**, 104459 (2024)
12. Ahmed, S., Bhuiyan, K., Rahman, I., Salehfar, H., Selvaraj, D.: Reliability of regression-based hybrid machine learning models for the prediction of solar photovoltaics power generation. Energy Rep. **12**, 5009–5023 (2024)
13. Shah, A., Viswanath, V., Gandhi, K., Ranka, P., Dedhia, U., Patil, N.: Predicting solar energy generation with machine learning based on AQI and weather features. Res. Square (2023). https://doi.org/10.21203/rs.3.rs-3178713/v1
14. Demir, V.: Evaluation of solar radiation prediction models using AI: a performance comparison in the high-potential region of Konya, Türkiye. Atmosphere **16** (2025)
15. Yang, F., Fu, X., Zhang, Y.: LSTM-based day-ahead photovoltaic power prediction. In: Statistical Relational Artificial Intelligence in Photovoltaic Power Uncertainty Analysis, pp. 35–70 (2025)
16. Song, D., et al.: Accurate solar power prediction with advanced hybrid deep learning approach. Eng. Appl. Artif. Intell. **148**, 110367 (2025)
17. Hong, D., Li, F., Ma, J., Man, K., Wen, H., Wong, P.: Temporal environment-informed photovoltaic performance prediction framework with multi-spatial attention LSTM. Sol. Energy **296**, 113550 (2025)
18. Hu, Z., Gao, Y., Ji, S., Mae, M., Imaizumi, T.: Improved multistep-ahead photovoltaic power prediction model based on LSTM and self-attention with weather forecast data. Appl. Energy **359**, 122709 (2024)
19. NASA Langley Research Center (LaRC): prediction of worldwide energy resources (POWER) project data access. https://power.larc.nasa.gov. Accessed 15 Jan 2025
20. Yunpeng, L., Di, H., Junpeng, B., Yong, Q.: Multi-step ahead time series forecasting for different data patterns based on LSTM recurrent neural network. In: 14th Web Information Systems and Applications Conference (WISA), pp. 305–310 (2017)
21. Sharma, V.: A study on data scaling methods for machine learning. Int. J. Global Acad. Sci. Res. **1**, 31–42 (2022)
22. Zargar, S.: Introduction to sequence learning models: RNN, LSTM. North Carolina State University, GRU. Department of Mechanical and Aerospace Engineering (2021)

23. Baqi, A.: How do I make an LSTM model with multiple inputs? Data Science Dojo (2024). https://datasciencedojo.com/blog/how-do-i-make-an-lstm-model-with-multiple-inputs/
24. Lindas, E., Goude, Y., Ciais, P.: Towards accurate forecasting of renewable energy: building datasets and benchmarking machine learning models for solar and wind power in France (2025)
25. Saval-Calvo, M., Azorin-Lopez, J., Fuster-Guillo, A., Garcia-Rodriguez, J.: Three-dimensional planar model estimation using multi-constraint knowledge based on k-means and RANSAC. Appl. Soft Comput. **34**, 572–586 (2015)
26. Bauer, D., Hönig, P., Weibel, J.B., Garcia-Rodriguez, J., Vincze, M.: Challenges for monocular 6D object pose estimation in robotics. IEEE Trans. Rob. **26** (2024)
27. Viejo, D., Garcia-Rodriguez, J., Cazorla, M.: Combining visual features and growing neural gas networks for robotic 3D SLAM. Inf. Sci. **276**, 174–185 (2014)
28. Garcia-Garcia, A., Garcia-Rodriguez, J., Orts-Escolano, S., Oprea, S., et al.: A study of the effect of noise and occlusion on the accuracy of convolutional neural networks applied to 3D object recognition. Comput. Vis. Image Underst. **164**, 124–134 (2017)
29. Orts, S., Garcia-Rodriguez, J., Viejo, D., Cazorla, M., Morell, V.: GPGPU implementation of growing neural gas: application to 3D scene reconstruction. J. Parall. Distrib. Comput. **72**(10), 1361–1372 (2012)
30. Martinez-Gonzalez, P., Oprea, S., Castro-Vargas, J.A., Garcia-Garcia, A., et al.: Unrealrox+: an improved tool for acquiring synthetic data from virtual 3D environments. In: International Joint Conference on Neural Networks (IJCNN), pp. 1–8 (2021)
31. Garcia-Garcia, A., Orts-Escolano, S., Garcia-Rodriguez, J., Cazorla, M.: Interactive 3D object recognition pipeline on mobile GPGPU computing platforms using low-cost RGB-D sensors. J. Real-Time Image Proc. **14**, 585–604 (2018)
32. Garcia-Garcia, A., Orts-Escolano, S., Oprea, S., Garcia-Rodriguez, J., et al.: Multi-sensor 3D object dataset for object recognition with full pose estimation. Neural Comput. Appl. **28**, 941–952 (2017)

Ensemble Learning for Polymer Injection Molding Quality Prediction

Iván Sánchez-Calleja[1](✉), Jesús Pérez-González[1], Andrea Fernández-Gorgojo[2], Rubén Ferrero-Guillén[1], Alberto Martínez-Gutiérrez[1], Javier Díez-González[1], and Hilde Perez[1]

[1] Department of Mechanical, Computer and Aerospace Engineering, Universidad de León, León 24071, Spain
{isanc,javier.diez}@unileon.es

[2] Department of Mechanical Engineering, Universidad Politécnica de Madrid, Madrid 28006, Spain

Abstract. The adjustment of input parameters in plastic injection molding remains one of the primary challenges within this manufacturing process. To determine the optimal adjustment parameters previous to the injection, it is essential to predict the critical quality indicators, such as the weight of the injected parts. Thus, in this paper, we propose the use of two different Machine Learning (ML) ensembling strategies for a robust prediction of its actual value. To build the necessary dataset, 197 parts were injected while systematically recording both the injection parameters and the resulting part weights, capturing the relationship between process parameters and part quality for the subsequent metamodel development. In this sense, this paper evaluates the bagging and stacking ensemble strategies to accurately predict part quality in the complex and environmentally sensitive injection molding process. The achieved results present a superior performance of the stacking ensemble method, offering a 99.96% of accuracy for predicting quality features, contributing to the definition of a comprehensive final quality index.

Keywords: Injection Molding · Machine Learning · Quality Prediction · Ensembling Models

1 Introduction

Injection molding is one of the most widely used techniques in the plastics industry. Currently, more than 30% of plastic products are manufactured using this industrial process [12]. One of its main advantages lies in its ability to produce components with complex geometries at high speeds and relatively low cost. However, a major challenge associated with injection molding is the tuning of a vast number of parameters that significantly impact the final quality of the molded parts [5]. These parameters are influenced by factors such as part geometry, mold design, material characteristics, and environmental conditions [2].

E. Corchado et al. (Eds.): SOCO 2025, CCIS 2806, pp. 582–592, 2026.
https://doi.org/10.1007/978-3-032-19763-4_54

The plastic injection molding process involves a wide array of parameters, including mold and injection temperature, injection rate, injection and holding pressure, and the durations of both the holding and cooling stages. Among these parameters, injection pressure and injection temperature are regarded as the most critical variables due to their significant influence on the quality and dimensional consistency of the final product [8]. As a result, each parameter contributes to a wide range of potential outcomes, substantially increasing the complexity of achieving optimal process control. This inherent variability highlights the pressing need for advanced monitoring and optimization strategies to ensure consistency, precision, and efficiency in injection molding operations.

Traditionally, injection parameters optimization has relied on manual adjustment from the experience of one or more operators, and typically carried out through a trial-and-error approach [8]. This method is inherently inefficient, time-consuming and costly, wasting energy, raw material, and overall productivity. In this context, the integration of emerging technologies into industrial environments plays a pivotal role in mitigating these challenges.

In the context of injection molding, a substantial volume of data is continuously generated throughout the process. This data can be effectively captured using sensors strategically placed across various components of the injection molding machine, including within the mold cavity itself [12]. Within this new paradigm of Smart Manufacturing [11], advanced tools such as artificial intelligence (AI), the Internet of Things (IoT), Machine Learning (ML) and intelligent process automation are being introduced to reduce the costs associated with parameter adjustment [10,11,14].

The quality establishment of the produced parts is a key factor in defining the appropriate input parameters in the plastic injection molding production process. Because of this, the evaluation of part quality in injection molding can be carried out by examining three main categories of quality indicators: (1) dimensional and weight stability of the injected parts, (2) surface characteristics, including roughness, weld lines, and sink marks, and (3) physical properties, such as mechanical, optical, and electrical performance [12]. Current works consider the weight of injected parts to be one of the most reliable and widely employed indicators of product quality, due to its correlation to the piece quality and the accessibility to a robust measurement without the need of costly equipment [3].

For this reason, in this paper, we propose the development of a metamodel from real experimentation to accurately predict the weight of a part before the injection molding process. This metamodel is based on the elaboration of multiple experiments, varying the input parameters of the injection molding machine, such as injection and holding pressure, injection temperature, and holding time.

To this end, ensemble learning strategies are employed. These strategies combine multiple models to improve predictive performance and generalization, thereby enhancing the algorithm's robustness across varying process conditions.

This approach provides a novel solution from actual data for accurately predicting the weight of injected pieces, thus representing a critical cornerstone for injection molding process parameter optimization.

2 Related Works

The weight of injection molding parts is a widely researched topic in the literature due to the inherent complexity of the process. Different authors have approached this challenge from three principal perspectives to adjust the injection parameters: (1) analyzing the effect of injection parameters on the final part weight, (2) developing predictive models to forecast the feasibility of the injected parts given a predefined weight, and (3) developing predictive models to forecast the final part weight.

Regarding influential parameters, Hassan et al. [6] demonstrated the impact of various injection parameters on the final part weight. Their results showed that increasing the packing pressure, packing time or injection pressure, or decreasing injection time or injection temperature leads to a higher final part weight. Moreover, López et al. [9] analyzed the effect of injection parameters on part weight considering the part geometrical complexity. In this sense, the study demonstrates the variability of the parameters' influence on the final part weight depending on the complexity of the geometry.

In the case of predictive models, Párizs et al. [12] applied four base ML classification algorithms to predict part quality based on 19 injection parameter features, using part weight as the quality indicator. This model categorizes parts as underweight, acceptable, or overweight according to the input parameters. Similarly, Zhao et al. [15] employed a Support Vector Classifier (SVC) to predict the deviation type relative to the standard part weight. In their approach, the kernel function derived from the SVC is considered representative of the standard weight, classifying parts as underweight or overweight based on their deviation from this function. However, the classification approach provides a limited insight into the magnitude of deviation from the target weight, which could be critical for process optimization and quality control.

To address the limitations of classification approaches, other authors have proposed direct part weight prediction through regression models. Chen et al. [4] developed a prediction part weight model using a combination of Design of Experiments (DOE) and Taguchi Design to identify and optimize the most influential parameters, and Fuzzy Logic to predict the part weight considering the uncertainty of the process. Their model relies on variables such as part volume, injection temperature, packing pressure, and cooling time, and fuzzy rules that associate these parameters with part weight variations. Ke et al. [7] proposed a hybrid machine learning model for predicting multiple quality indicators (i.e., final dimensions, part weight, and residual stress). However, both studies address this challenge using base models, which can be susceptible to inefficiencies in other injection conditions and compromise the prediction model.

In this sense, ensemble strategies in ML provide a robust prediction upon complex systems (e.g., injection molding process), but there is a lack in the literature regarding applying ensemble learning in this field. Considering these aspects, it becomes necessary to design a robust prediction model for enabling an accurate quantification of this critical quality attribute, establishing a solid foundation for the subsequent development of a robust global quality index.

Therefore, this paper proposes the development of a ML model for predicting the actual weight of injection molded parts through ensemble learning strategies. Various ensemble learning models are considered to address the challenge and evaluated to enhance predictive accuracy and robustness of the approach.

3 Methodology

In this section, a systematic experimental methodology is proposed to develop a robust model for quality part prediction based on part weight. Figure 1 illustrates the steps of the methodology used to generate the metamodel based on the weight index quality. First, the injection process parameters are defined in order to manufacture the various parts. The resulting plastic components are weighed using a precision balance, and a database is created with all the collected information. Once the database is established, machine learning algorithms are applied to build an ensemble-based metamodel. This metamodel is capable of predicting the weight of non-injected parts with high accuracy.

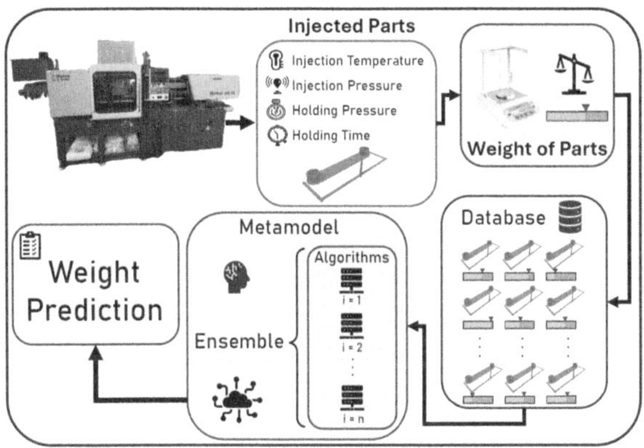

Fig. 1. Methodology procedure for part weight prediction.

To conduct the study, the Metcor 165/55 injection molding machine from Mateu & Solé [1] is used. This equipment allows for the injection of 197 samples while systematically varying the following input parameters: (1) injection temperature, (2) injection pressure, (3) holding pressure, and (4) holding time. These parameters are widely considered the most relevant in the plastic injection molding process [2]. Each parameter is varied within a predefined range, as shown in Table 1, resulting in injected parts from multiple configurations, from which the dataset is constructed. This strategy ensures a diverse dataset that captures a wide variety of process conditions, ultimately enabling the creation of a robust and representative database for subsequent modeling.

The geometry of a connecting rod is selected for the injected part. This geometry presents significant challenges during the mold filling stage of the injection molding process. Since the material is injected through two gates located at the ends of the runner, there is a risk that the molten plastic does not properly flow towards the central region of the part. This may result in the formation of cold spots, leading to weld line defects, visible joint marks, or, in the worst case, incomplete filling if the injection parameters are not adequately controlled.

Such defects can be easily detected by monitoring variations in the weight of the produced parts, from which conclusions can be drawn regarding the quality of each individual piece based on its measured weight.

Each part will be weighed, including the sprue, runners, gates, and any flash produced during the injection process, in order to obtain the most accurate measurement possible. Weighing will be performed using an RS PRO ES-HA 300 precision balance, offering a resolution of 0.01 g, thus providing sufficient sensitivity to detect injection-induced defects.

In this study, a real dataset is constructed based on experimental trials conducted on the injection molding machine. The dataset comprises 197 distinct instances, each corresponding to the injected parts produced under varying processing conditions. For each instance, the injection process parameters are systematically varied within predefined ranges. Table 1 presents the specific range and step size used for each parameter during the experimentation.

Table 1. Injection Parameters range and interval step employed to create the database.

Injection Parameter	Range of Values	Step
Injection Temperature ($T_{injection}$)	210 °C - 230 °C	20 °C
Injection Pressure ($P_{injection}$)	40 MPa - 70 MPa	10 MPa
Holding Pressure ($P_{holding}$)	20% - 80%	30%
Holding Time ($t_{holding}$)	2 seg - 6 seg	4 seg

*Holding Pressure is expressed as a percentage of Injection Pressure.

Following the injection of each part, its weight was measured using a precision balance with an accuracy of 0.01 g, thereby serving as the target variable and primary quality indicator. This procedure ensures that the dataset reflects real-world variability in the process and provides a reliable foundation for the subsequent training and evaluation of machine learning models.

In this regard, the database distribution is evaluated using F-test and Mutual Information (MI) to analyze the data sparsity behavior, facilitating the subsequent selection of models based on their characteristics and suitability to the observed data distribution. To properly train the ML models and avoid inefficiencies, it is crucial to consider a preprocessing of the data effectively depending on the model (e.g., normalization, standardization, minmax). These models are trained using cross-validation and GridSearchCV to determine the best hyperparameters. To improve robustness and address the limitations of individual models, two ensemble strategies are employed: bagging, which aggregates predictions using the mean and median of the base models, and stacking, which combines three base models with the remaining model serving as the meta-learner.

The proposed methodology aims to enable the prediction of the injection parts' weight using ensemble learning. The different ensembling strategies may address the problem with different rankings, which requires the use of evaluation metrics (e.g., R^2, MeAE, RMSE, and MAPE). In the next section, the obtained results are discussed to evaluate the performance and draw insightful conclusions.

4 Results

After implementing the proposed methodology, the F-test and MI of $P_{holding}$ (19.11 and 1.36, respectively) and $t_{holding}$ (1.77 and 0.71, respectively) are significantly low, which indicates that linear models (e.g., Lasso, Support Vector Machines) cannot handle this problem efficiently. Based on these analyses, the selected models are k-Nearest Neighbors (kNN), Decision Tree (DT), Random Forest (RF), and Extreme Gradient Boost (XGBoost), due to their good performance at predicting non-linear relationships. However, different preprocessing methods are considered to achieve an adequate fitting of the models to the available database.

In this sense, the models based on trees (e.g., DT, RF, and XGBoost) are poorly sensitive to raw information, but kNN models can be susceptible to bad predictions in large attribute domains. Furthermore, the stacking ensemble strategy needs the same data scale for the training when considering the original attributes. Therefore, all data is normalized to reduce potential inefficiencies.

Once preprocessing is complete, the base models are trained using cross-validation combined with GridSearchCV for hyperparameter optimization, followed by the implementation of ensemble strategies. These models are trained using 5-fold cross-validation and GridSearchCV to determine the best hyperparameters objectively [13]. This GridSearchCV is conducted in all models, tuning their specific hyperparameters, and introducing another 5-fold cross-validation to the training set for objectively optimizing their values.

For kNN, the number of neighbors is varied among {3, 5, 7}, with 7 yielding the best performance. Additionally, two weighting schemes (e.g., uniform and distance) are considered, with distance performing best.

For DT, a maximum tree depth of {5, 10, unlimited} is considered, where a depth of 5 gave the best results. The minimum number of samples searched to split an internal node is evaluated over {2, 4}, with 4 being optimal, and the minimum number of samples searched to be at a leaf node is tested over {1, 2, 3}, resulting in 3 being found to be best.

RF shares the same tree-specific hyperparameters tested for DT, being optimal with the same values, but also the number of estimators, with {100, 120} evaluated. 100 estimators led to the best performance.

For the XGBoost model, the same tree structure parameters as RF are considered, and additionally, learning rates in the set {0.2, 0.3}. The optimal hyperparameters coincide with RF, and a learning rate of 0.3.

Table 2 provides a summary of the performance metrics obtained for each prediction model.

Table 2. Evaluation metrics of the considered prediction models.

Prediction Model	R^2	MeAE	RMSE	MAPE
kNN	0.983	0.201 g	0.496 g	1.996%
DT	0.983	0.249 g	0.496 g	1.996%
RF	0.985	0.199 g	0.467 g	1.857%
XGBoost	0.983	0.201 g	0.496 g	1.996%
Bagging_median	0.983	0.173 g	0.496 g	1.996%
Bagging_weight_mean	0.985	0.169 g	0.467 g	1.857%
Stacking_KNN	0.969	0.276 g	0.661 g	2.939%
Stacking_DT	0.982	0.009 g	0.510 g	2.102%
Stacking_RF	0.994	0.119 g	0.292 g	1.300%
Stacking_XGBoost	0.998	0.002 g	0.167 g	0.431%

By comparing the distinct metrics between proposals, Table 2 seemingly indicates that the best predictive model is Stacking_XGBoost. Although Stacking_RF and Stacking_XGBoost exhibit comparable performance metrics, Stacking_XGB presents a lower RMSE, MeAE and MAPE, indicating a smaller prediction error and a more accurate estimation of the actual part weight. Furthermore, the feature weights in all trained models, which represent the influence of features on class prediction, agree that $P_{injection}$ is the most important parameter, followed by the $T_{injection}$, and of lesser importance in $P_{holding}$ and $t_{holding}$.

Additionally, a standard deviation plot and a series of heat maps are generated to support the identification of optimal parameter configurations for producing high-quality injected parts. Prior to the analysis, the process experts have considered 21 g as the target weight for the part. Figure 2 shows an example of good and wrongly injected parts according to experts' criteria. This target serves as a reference for assessing the quality of the parts and evaluating the prediction model's performance.

To this end, the standard deviation plot represents the variability in part weights across different parameter configurations of this problem. In this sense, a low standard deviation indicates process stability, which is generally correlated with good part quality, as defects typically introduce significant weight variation. Figure 3 illustrates the standard deviation of each injection parameter configuration used for the injection of the 197 parts included in the database. Notably, the lowest variability is observed in configurations whose average part

Fig. 2. Injected parts of (a) parameter set 24, and (b) parameter set 36, which present examples of unfeasible and well-injected parts, according to experts' criteria. The injected piece is 180 mm long, 32 mm wide, 19 mm high, and 2 mm of target thickness.

weights are close to the 21-gram target (highlighted with a blue frame), which corroborates the proposed hypothesis.

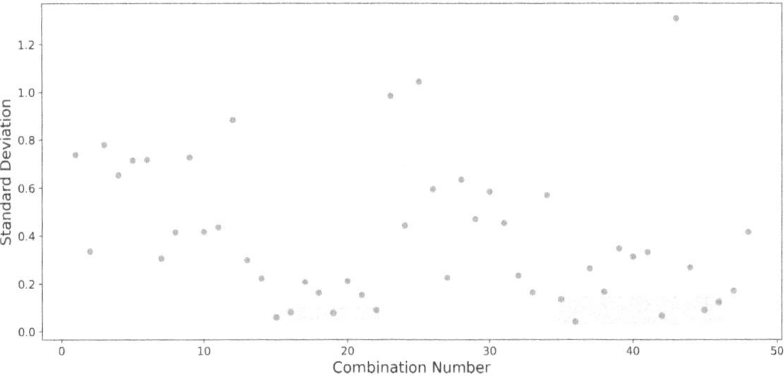

Fig. 3. Standard deviation by combination number.

In addition, the heat maps from Fig. 4 provide a visualization of the predicted deviation from the target weight over the design space, using combinations of two selected injection parameters as axes. These maps offer an intuitive representation of the most adequate parameter ranges for achieving the target weight. Figure 4 exposes the different heat maps, where each adequate parameter range can be derived.

As can be observed, the heat maps present clear *target weight* zones for specific configurations, which coincide with the low-variance configurations of Fig. 3. In this regard, based on the above conclusions in the standard deviation plot, and the relationship among the heat maps through a precise predictive model, this approach promotes an accessible interpretation of the process and facilitates the real implementation in the industry.

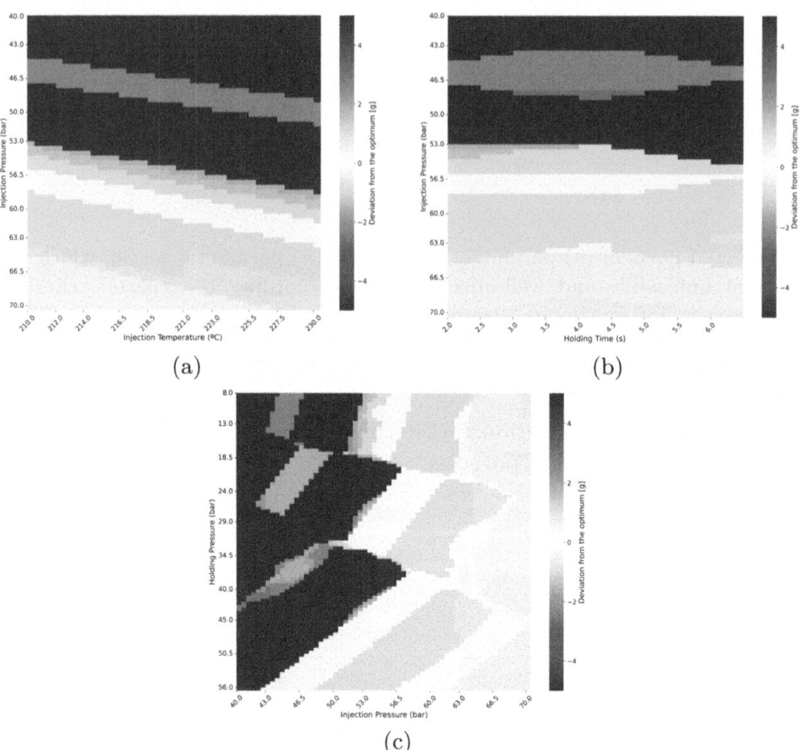

Fig. 4. Heat maps of (a) $T_{injection}$ - $P_{injection}$, (b) $t_{holding}$ - $P_{injection}$, and (c) $P_{injection}$ - $P_{holding}$ based on Stacking_XGBoost prediction model, where red zones implicates *overweighting*, white zones *target weight*, and blue zones *underweighting*.

5 Conclusions

This work proposes an ensemble learning-based metamodel to predict the final part quality in plastic injection molding processes, addressing the challenge posed by the nonlinear behavior and parameter sensitivity of the process. In this sense, the part weight is used as the quality indicator due to its ability to reflect other defects. Thus, a dataset is constituted by 197 injected parts through systematic experimentation under 47 different injection parameter configurations.

For this, four ML base algorithms (kNN, DT, RF, and XGBoost) and two ensembling strategies (bagging and stacking) are considered to address these inefficiencies optimally through the transfer of real injection information. After tuning and evaluating each model, Stacking_XGBoost demonstrates an accuracy of 99.96% for handling the injection molding behavior. Furthermore, the generation of standard deviation plots and target deviation heat maps provides an intuitive interpretation of the process and facilitates the parameter adjustment.

However, since the prediction of injected part weight is essential for defining and ensuring product quality, this work not only provides an effective ensemble learning-based approach to enhance predictive accuracy, but also demonstrates the potential of such models to process complex manufacturing processes, contributing to the advancement of redefining quality assessment of the plastic injection molding process.

Acknowledgments. This work has been funded by the project of the Spanish Ministry of Science and Innovation grant number PID2023-153047OBI00, by the Consejería de Educación de la Junta de Castilla y León and by the Universidad de León. The author Iván Sánchez-Calleja acknowledges funding for doctoral studies of the University of León.

References

1. "Mateu Solé Meteor 165/55 datasheet". https://mateusole.com/wp-content/uploads/2018/12/FICHAS_METEOR_150.pdf
2. Chaves, M.L., Sánchez-González, L., Díez, E., Pérez, H., Vizán, A.: Experimental assessment of quality in injection parts using a fuzzy system with adaptive membership functions. Neurocomputing **391**, 334–344 (2020)
3. Chen, J.Y., Liu, C.Y., Huang, M.S.: Enhancement of injection molding consistency by adjusting velocity/pressure switching time based on clamping force. Int. Polym. Proc. **34**(5), 564–572 (2019)
4. Chen, J., Krisshnasamy, R.: Development of eDART-based weight prediction system in injection molding via taguchi design and fuzzy logic. In: Journal of Physics: Conference Series, vol. 2631, p. 012013. IOP Publishing (2023)
5. Dang, X.P.: General frameworks for optimization of plastic injection molding process parameters. Simul. Model. Pract. Theory **41**, 15–27 (2014)
6. Hassan, H.: An experimental work on the effect of injection molding parameters on the cavity pressure and product weight. Int. J. Adv. Manuf. Technol. **67**, 675–686 (2013)
7. Ke, K.C., Wu, P.W., Huang, M.S.: Multi-quality prediction of injection molding parts using a hybrid machine learning model. Int. J. Adv. Manuf. Technol. **131**(11), 5511–5525 (2024)
8. Li, D., Zhou, H., Zhao, P., Li, Y.: A real-time process optimization system for injection molding. Polym. Eng. Sci. **49**(10), 2031–2040 (2009)
9. López, A., Aisa, J., Martinez, A., Mercado, D.: Injection moulding parameters influence on weight quality of complex parts by means of doe application: case study. Measurement **90**, 349–356 (2016)
10. Martínez-Gutiérrez, A., Díez-González, J., Ferrero-Guillén, R., Verde, P., Álvarez, R., Perez, H.: Digital twin for automatic transportation in industry 4.0. Sensors **21**(10), 3344 (2021)
11. Martínez-Gutiérrez, A., Díez-González, J., Verde, P., Ferrero-Guillén, R., Perez, H.: Hyperconnectivity proposal for smart manufacturing. IEEE Access **11**, 70947–70959 (2023)
12. Párizs, R.D., Török, D., Ageyeva, T., Kovács, J.G.: Machine learning in injection molding: an industry 4.0 method of quality prediction. Sensors **22**(7), 2704 (2022)
13. Ranjan, G., Verma, A.K., Radhika, S.: K-nearest neighbors and grid search cv based real time fault monitoring system for industries. In: 2019 IEEE 5th International Conference for Convergence in Technology (I2CT), pp. 1–5. IEEE (2019)

14. Sánchez-Calleja, I., Martínez-Gutiérrez, A., Ferrero-Guillén, R., Díez-González, J., Perez, H.: Contact system method for the precise interaction between cobots and mobile robots in smart manufacturing. Int. J. Precis. Eng. Manuf. **25**(2), 303–318 (2024)
15. Zhao, P., et al.: Optimization of injection-molding process parameters for weight control: converting optimization problem to classification problem. Adv. Polym. Technol. **2020**(1), 7654249 (2020)

Personalised Visualisation of Home Monitoring and Treatment Adherence with AI2EPD

David Díaz-Jiménez[1](✉), José L. López-Ruiz[1], Jesús González-Lama[2], Juan F. Gaitán-Guerrero[1], and Macarena Espinilla[1]

[1] Department of Computer Science, University of Jaén, Jaén, Spain
{ddjimene,llopez,jgaitan,mestevez}@ujaen.es

[2] Maimonides Biomedical Research Institute of Cordoba, Córdoba, Córdoba - Andalusia, Spain
jegonla@telefonica.net

Abstract. This paper presents AI2EPD, a platform oriented to healthcare professionals for monitoring therapeutic adherence in assisted living environments. The system combines home sensors with inference models to estimate adherence to activities defined in an individualised therapeutic contract. The focus is on the design and evaluation of different forms of visual representation of adherence levels, with the aim of facilitating clinical interpretation and decision making. Through multiple interfaces developed, different visualisation strategies are explored and tested in a study with healthcare professionals. The results suggest that the way data is presented has a significant impact on the perceived usefulness of the system, highlighting the need for representations adapted to different clinical profiles. This approach reinforces the role of soft technologies and interactive systems in supporting patient-centred clinical decision making.

Keywords: therapeutic adherence · home sensors · data visualisation · clinical interfaces · fuzzy logic · therapeutic contract · assisted living

1 Introduction

Adherence to treatment is a key determinant in the clinical outcome of patients with chronic diseases, neurodegenerative disorders or conditions requiring long-term follow-up. The literature has consistently shown that poor adherence is associated with a higher rate of hospital readmissions, poorer symptom control and a considerable increase in healthcare costs. Despite its importance, its assessment in clinical practice still presents significant challenges, especially in home-based settings where direct supervision is limited [16].

In recent years, advances in the field of environmental monitoring and assisted living systems have led to the development of technologies capable of continuously recording certain indicators of everyday behaviour. Among them, activity

E. Corchado et al. (Eds.): SOCO 2025, CCIS 2806, pp. 593–601, 2026.
https://doi.org/10.1007/978-3-032-19763-4_55

recognition (AR) has gained prominence as a viable strategy to infer indirect compliance with therapeutic guidelines. Detecting when a patient has cooked, rested adequately or maintained their hygiene routine allows us to estimate, with a certain degree of certainty, their level of commitment to the established treatment [6,13,14].

At the same time, there has been a growing interest in incorporating assisted decision-making models to transform raw data into clinically relevant information. Key to this is not only the collection of data through sensors, but also its appropriate representation. The information must be adapted to the different professional profiles - doctors, nurses, therapists - and provide understandable visualisations that favour both timely analysis and patient follow-up [8,11].

This paper presents `AI2EPD`, a modular platform oriented to healthcare professionals, which allows a personalised monitoring of therapeutic adherence at home. The platform relies on a set of non-intrusive sensors deployed in the patient's home (detection of movement, doors, steps, etc.), and on a previously defined therapeutic contract that specifies the expected activities. From these elements, representations of daily adherence are constructed using fuzzy logic-based inference models [15]. In contrast to other approaches that focus exclusively on the accuracy of event detection, `AI2EPD` is oriented towards the clinical visualisation of information. Different interactive interfaces have been designed to explore adherence levels from different approaches: temporal, categorical and global. These interfaces have been evaluated by healthcare professionals in order to identify the most effective form of representation to support clinical decision-making in non-face-to-face care contexts [8].

The paper is structured as follows. Section 1 introduces the motivation and context of this work, highlighting the challenges of monitoring treatment adherence at home. Section 2 reviews previous studies in activity recognition and information visualisation in healthcare. Section 3 describes the architecture of the proposed *AI2EPD* platform, including the sensors used and data processing infrastructure. Section 4 details the design principles and visual components of the user interfaces. Section 5 presents the evaluation procedure and the results obtained from healthcare professionals. Section 6 discusses the implications of the results and preferences observed. Finally, Sect. 7 summarises the main findings and outlines future work directions.

2 Related Work

Adherence has been widely recognised as an essential component in the treatment of chronic diseases and in the success of long-term interventions. Several studies have shown that poor adherence is directly related to poorer clinical outcomes, increased hospitalisations and higher healthcare costs [6,9,10,12].

However, its objective measurement in home settings remains a challenge, especially in older or cognitively impaired patients [4].

In this sense, activity recognition approaches have emerged as a viable alternative to infer patient behaviour from their interaction with the environment.

These systems can identify patterns of daily life - such as sleeping, eating or leaving the house - through the use of environmental sensors, wearable devices and signal processing techniques [5, 18].

Hybrid architectures combining PIR, aperture and humidity sensors with devices such as smart wristbands or smartphones have been proposed to improve the granularity of detections, thus facilitating a more accurate assessment of therapeutic behaviour [1–3].

In addition, some work has addressed the use of fuzzy inference or machine learning models to translate detected activities into indicators of adherence. Although these approaches show promising results, in many cases the representation of the information is limited to aggregate metrics or specific alerts, without sufficient attention to their clinical utility [6, 7].

Furthermore, the design of visual interfaces for healthcare professionals has received relatively little attention in the field. The way in which data is presented can significantly influence the perception of the system, confidence in the results and the integration of the tool into the clinical workflow [13, 17]. Although there are proposals for dashboards and interfaces in hospital settings, the visual representation of adherence in assisted living settings remains an under explored area.

This proposal sits at the intersection of these approaches, integrating home-based sensors, contextual inference and visual interfaces adapted to diverse clinical profiles. In particular, the focus is on how information representation can facilitate decision-making in non-face-to-face care settings.

3 System Architecture

The *AI2EPD* platform is based on a distributed architecture that uses environmental sensors and location devices to monitor the patient's daily behaviour in their home environment. This infrastructure is designed to be non-intrusive, low-cost and easily scalable.

3.1 Environmental Sensors

- **Aqara Door and Window Sensor:** Wireless magnetic sensor that detects opening and closing events of doors or windows. It uses the Zigbee protocol and offers a battery life of more than two years with a button-type battery.
- **Aqara Motion Sensor P1:** Passive infrared (PIR) motion sensor, with a detection angle of up to 170° and a range of 7 m. Sensitivity and time delay between detections can be configured. It also incorporates a luminosity sensor and communicates via Zigbee 3.0.
- **Aqara Temperature and Humidity Sensor:** Environmental measurement sensor that records temperature and relative humidity. It is accurate to within $\pm0.3^{\circ}$C for temperature and $\pm3\%$ for humidity, and integrates easily with the home's Zigbee network.

3.2 Network Infrastructure and Sensor Management

Communication with the Zigbee sensors takes place via a USB ConBee II coordinator connected to the central Raspberry Pi, which acts as the control node. This unit runs the Home Assistant environment, from which the sensor network is managed, events are received and the collected data is structured.

All events generated by the sensors are published in real time on different MQTT topics, which allows decoupling data acquisition from post-processing. This architecture facilitates the integration of new devices and the extension of functionalities.

3.3 Location and Presence of the Patient

- **Raspberry Pi 4 Model B:** Several units are deployed in different rooms, each acting as an anchor node to estimate patient location. These units are equipped with low-cost Bluetooth USB adapters and wireless connectivity.
- **Xiaomi Mi Band 3:** Smart bracelet worn by the patient, which emits Bluetooth signals detected by Raspberry Pi. The RSSI (Received Signal Strength Indicator) is monitored to infer the patient's stay.
- **Local processing and hybrid inference:** Before being sent to the cloud, the data collected by the anchors are processed locally using hybrid models. This step allows reducing signal noise and obtaining the location of the patient in the monitored environment.

4 User Interface

The adherence visualisation module of the *AI2EPD* platform has been developed using the **Django** framework, with the aim of providing a clear, intuitive and useful experience for healthcare professionals. While the system includes complete functionalities for the management of patients, therapeutic contracts, sensors, dwellings and authorised personnel, this section specifically describes the design of the interfaces to represent the level of compliance with the therapeutic contract.

The design has been guided by the principles of user-centred design and established standards in the literature for effective information representation in clinical environments. Different visualisation strategies, both categorical and temporal, have been applied to facilitate decision-making in non-face-to-face monitoring contexts.

Each interface includes key components aimed at facilitating clinical interpretation: patient identification and the evaluation period, including the start date in the system and summary data; an activity panel displaying, for each activity, the expected goal, the current recorded value, and a completion percentage, with the aid of color codes and status labels (e.g., Compliant, Partial) to enable rapid assessment; horizontal bar charts that allow immediate comparison of adherence levels across categories such as physical activity, rest, hygiene, nutrition, and medication; detailed tables containing precise quantitative values (goal, daily average, current value, completion percentage) along with status indicators; and a textual summary that translates numerical adherence data into qualitative assessments, using expressions that are easily interpretable by healthcare professionals (e.g., "very good adherence", "irregular hygiene", "excellent treatment engagement"). A key feature of the system is the ability to generate adherence summaries on different time scales, including daily, weekly and monthly reports. In addition, the user can set a customised date range to suit specific situations, such as one-off assessments, clinical reviews or specific phases of the therapeutic intervention. This flexibility facilitates longitudinal monitoring and comparative analysis of the patient's evolution.

All these representations have been designed to ensure fast, standardised visualisation that is compatible with the different levels of technological expertise of healthcare staff. Attention to visual design, the use of icons and colour schemes help to reduce the cognitive load during data review and support smooth integration into clinical assessment routines.

5 Evaluation of Display Interfaces

In order to determine which of the developed visualisation options is most useful and effective for clinical decision-making, a questionnaire was designed and applied to healthcare professionals. The questionnaire focuses exclusively on the representation of information related to adherence, as this is considered the most critical aspect in the process of remote patient follow-up.

The questionnaire, entitled 'Assessment of patient data display interfaces', was implemented digitally and consists of a series of structured items that allow both quantitative assessments and open-ended responses to be collected. Each participant is presented with several interfaces (named Interface 1, Interface 2, etc.) corresponding to different visual strategies applied on the same set of real data: summary cards with percentages, horizontal bar charts, coloured tables with status indicators, and generated narrative reports.

For each interface, ratings were requested on a 5-point Likert scale on the following aspects:

- Clarity in the presentation of information.
- Ease of understanding the fulfilment of each activity.
- Perceived visual attractiveness.
- Degree of completeness of the information presented.
- Preference for maintaining that format or opting for a different one.

In addition, two open questions were included for each interface to identify the most valued elements and possible areas for improvement:

- Which part of the interface did you like the most?
- What would you improve or change?

Finally, at the end of the questionnaire, a general question on the preferred type of visualisation was incorporated, allowing the participant to select from the available options: individual cards, bar charts, colour-coded status tables or narrative report. A field for additional comments was also provided.

6 Discussion of the Results

In order to identify the most suitable interface for representing adherence in clinical settings, a comparative evaluation of four different visual approaches was carried out: a table with colours and status (Interface 1), Fig. 1, a narrative report (Interface 2), Fig. 2, a combination of progress bars with textual summary (Interface 3), Fig. 3, and a system of individual cards with percentages (Interface 4), Fig. 4.

Contract id: 1

Activity	Objective	Average per day	Progress	Percentage
Physical activity	60 minutes	225 minutes		97%
Rest	300 minutes	344 minutes		80%
Self-care: brushing teeth	2 times	1.7 times		78%
Self-care: shower	1 time	0.6 times		56%
Meals	3 times	2.8 times		94%
Medication	2 times	3.6 times		97%

Fig. 1. Interface 1.

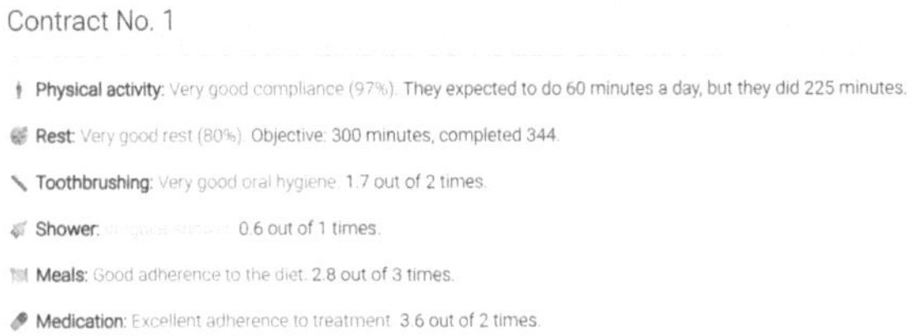

Contract No. 1

Physical activity: Very good compliance (97%). They expected to do 60 minutes a day, but they did 225 minutes.

Rest: Very good rest (80%). Objective: 300 minutes, completed 344.

Toothbrushing: Very good oral hygiene. 1.7 out of 2 times.

Shower: 0.6 out of 1 times.

Meals: Good adherence to the diet. 2.8 out of 3 times.

Medication: Excellent adherence to treatment. 3.6 out of 2 times.

Fig. 2. Interface 2.

Each participating professional rated the four interfaces independently, answering the same questions for each. The dimensions assessed included clarity, ease of interpretation, visual appeal, completeness of information and preference of use, all measured by Likert-type scales from 1 to 5 points.

The results show that the table-based interface was the highest rated in terms of clarity (4.75/5), visual appeal (5.00/5) and completeness (4.75/5), and also received the fewest suggestions for improvement. The narrative report (Interface 2) was also highly rated, being notable for its ease of interpretation (4.25/5) and

Fig. 3. Interface 3.

Fig. 4. Interface 4.

aesthetics (4.50/5), although some participants indicated that it was slightly overloaded with information.

The mixed interface (progress bars and explanatory text) showed intermediate scores in all categories, with a mean of 4.0 for clarity and ease, and 3.75 for visual appeal. The cards interface, while perceived as complete (4.50/5) and relatively clear (4.00/5), scored lower on visual appeal (3.75/5) and received more suggestions for aesthetic improvement.

Finally, participants were asked to select their preferred type of visualisation. The most chosen options were the narrative report and bar charts, followed by cards and the coloured table. This result reinforces the need to offer multiple forms of representation, adapted to different professional profiles and clinical scenarios.

Overall, the results show that clarity of presentation, together with the ability to synthesise relevant information for decision-making, are key aspects when designing interfaces oriented towards healthcare professionals in adherence monitoring platforms such as *AI2EPD*.

7 Conclusion and Future Work

This work has presented *AI2EPD*, a platform aimed at monitoring therapeutic adherence using home sensors, activity-based inference and visualisations for healthcare professionals. Through the deployment of non-intrusive sensors, the use of hybrid inference models and the design of interfaces, the aim is to offer a tool that facilitates clinical decision-making in non-face-to-face care contexts.

The results obtained in the evaluation of different visualisation interfaces have shown that, although there are general patterns of preference, such as the clarity of the tables or the interpretability of the narrative report, there is no single ideal representation for all users. The ratings show subtle but relevant differences, and some interfaces, such as cards or mixed views, were well received by certain professional profiles. In this context, it is concluded that a clinical decision support system such as *AI2EPD* should not impose a single form of data representation. On the contrary, it should allow the user to choose between different visual alternatives, or even configure their view according to their needs, experience or time of the process. EThis flexibility is key to encourage the real adoption of this type of technology, avoiding the need for professionals to adapt to a single interface that does not always align with their cognitive or functional preferences.

As future work, we plan to incorporate automatic personalisation mechanisms, as well as to expand the tests with a larger number of participants and clinical contexts, in order to validate the usefulness of the system in other real home care scenarios.

Acknowledgments. This result has been partially supported by grant PID2021-127275OB-I00 funded by MICIU/AEI/10.13039/501100011033 and by "ERDF A way of making Europe" and grant PDC2023-145863-I00 funded by MICIU/AEI/10.13039/501100011033 and by "European Union NextGenerationEU/PRTR".

Disclosure of Interests. The authors have no competing interests to declare that are relevant to the content of this article.

References

1. Abidine, M., Fergani, B.: Activity recognition from smartphone data using weighted learning methods. Intelligenza Artificiale (2021)
2. Alhammad, N., Al-Dossari, H.: Recognizing physical activities for spinal cord injury rehabilitation using wearable sensors. Sensors (2021)

3. Arigbabu, O.: Entropy decision fusion for smartphone sensor based human activity recognition. arXiv (2020)
4. Bitencourt, A., Reis, S.M.: Adherence to treatment in patients with chronic kidney disease undergoing hemodialysis. HSJ (2024)
5. Chen, L., et al.: A method of human activity recognition in transitional period. Information (2020)
6. Cherian, J., et al.: An activity recognition system for taking medicine using in-the-wild data to promote medication adherence. IUI 2021 (2021)
7. Cleves-Valencia, J., et al.: Beyond therapeutic adherence: alternative pathways for understanding medical treatment in type 1 diabetes mellitus. IJERPH (2024)
8. Díaz-Jiménez, D., López, J.L., González-Lama, J., Espinilla, M.: A new horizon in healthcare: an innovative methodology for sensor-based adherence platforms. IEEE Internet Things J. **11**, 33064–33076 (2024)
9. Ferrara, F., et al.: The slow path to therapeutic adherence. Hospital Pharmacy (2022)
10. Gaviria-Mendoza, A., et al.: eHealth and mHealth: adherence to treatment in chronic diseases (2021)
11. Hussain, M., Afzal, M., Khan, W.A., Lee, S.: Clinical decision support service for elderly people in smart home environment. In: 2012 12th International Conference on Control Automation Robotics and Vision (ICARCV), pp. 678–683 (2012)
12. Religioni, U., et al.: Enhancing therapy adherence: impact on clinical outcomes, healthcare costs, and patient quality of life. Medicina (2025)
13. Rodríguez, M.D., Beltrán, J., Valenzuela-Beltrán, M., Cruz-Sandoval, D., Favela, J.: Assisting older adults with medication reminders through an audio-based activity recognition system. Pers. Ubiquit. Comput., pp. 1–15 (2020)
14. Roy, P.C., Abidi, S.S.R.: Monitoring activities related to medication adherence in ambient assisted living environments. In: Studies in Health Technology and Informatics, vol. 235, pp. 28–32 (2017)
15. Roy, P.C., Abidi, S.S.R.: Possibilistic activity recognition with uncertain observations to support medication adherence in an assisted ambient living setting. Knowl.-Based Syst. **133**, 156–173 (2017)
16. Sabati, N., Snyder, M., Edin-Stibbe, C., Lindgren, B., Finkelstein, S.: Facilitators and barriers to adherence with home monitoring using electronic spirometry. AACN Clin. Issues **12**(2), 178–185 (2001)
17. Triboan, D., et al.: A semantics-based approach to sensor data segmentation in real-time activity recognition. Future Gener. Comput. Syst. (2019)
18. Wijekoon, A., et al.: A knowledge-light approach to personalised and open-ended human activity recognition. Knowl.-Based Syst. (2020)

Comparative Evaluation of Feature Graphical Descriptors and Classifiers for Pavement Condition Assessment

María Inmaculada Rodríguez-García[1(✉)], M. G. Carrasco-García[2], Adriana Pabón Noguera[3], Pedro J. S. Cardoso[4], and I. J. Turias[1]

[1] Department of Computer Science Engineering, Higher School of Engineering (ETSI), University of Cádiz, Cádiz, Spain
inma.rodriguezgarcia@gm.uca.es, ignacio.turias@uca.es

[2] Department of Industrial and Civil Engineering, Algeciras School of Engineering and Tech-Nology (ASET), University of Cádiz, Algeciras, Spain
maria.carrasco@uca.es

[3] Department of Civil Engineering, Faculty of Engineering, University of Magdalena (Unimagdalena), Santa Marta 470003, Colombia
adriana.pabon@gm.uca.es

[4] NOVA Laboratory for Computer Science and Informatics (NOVA LINCS). Instituto Superior de Engenharia, University of Algarve, Campus da Penha, 8005-139 Faro, Portugal
pcardoso@ualg.pt

Abstract. The objective of this study is to evaluate and compare the effectiveness of different image feature descriptors—Histogram of Oriented Gradients (HOG), Local Binary Patterns (LBP), and Gray-Level Co-occurrence Matrix (GLCM)—combined with various classification algorithms for the automated assessment of pavement surface conditions. A custom dataset was built from 80 high-resolution images of concrete slabs, evenly split between cracked and intact surfaces. Each image was subdivided into smaller sections to increase the data volume, resulting in 180 samples per class. The study investigates the classification performance of Support Vector Machines (SVM), k-Nearest Neighbors (k-NN), Naïve Bayes, Decision Trees, and Artificial Neural Networks (ANNs) using these features. Through repeated 5-fold cross-validation, the models were evaluated based on accuracy, precision, recall, specificity, and F1-score. The goal is to identify the most effective combinations of descriptors and classifiers for reliable pavement defect detection, especially in resource-constrained environments or when dealing with texture-rich surfaces. The results highlight that SVM performs best with HOG, while ANNs yield superior results with LBP and GLCM features. This work contributes to the development of scalable, data-driven approaches for urban infrastructure monitoring and supports future improvements through hybrid descriptors and advanced neural architectures.

Keywords: Automated Graphical Inspection · Computer Vision · HOG · GCLM · LBP · Pattern Recognition

E. Corchado et al. (Eds.): SOCO 2025, CCIS 2806, pp. 602–612, 2026.
https://doi.org/10.1007/978-3-032-19763-4_56

1 Introduction

In recent years, smart cities have emerged as a new model for urban development, integrating ICT, sensor networks, and data-driven systems to enhance infrastructure efficiency and citizen quality of life [1, 2]. A key aspect is the automated monitoring of infrastructure using IoT, AI, and big data to optimize services like transportation and environmental control. High-resolution sensor data, such as air quality metrics, supports proactive urban management [3]. This shift from manual inspections to intelligent, real-time systems enables predictive maintenance and reduces disruptions. Urban pavement infrastructure—roads and sidewalks—is essential for safe mobility. Over time, wear and environmental stress cause defects such as cracks and potholes, impacting safety and accessibility. Accurate defect detection is critical for effective maintenance [4, 5]. Computer vision and machine learning technologies enable automated, large-scale assessments [6], using imagery from vehicles, drones, or smartphones [7–9]. Feature extraction remains a core challenge due to variability in surface textures and lighting. Early approaches to automated pavement defect detection extensively utilized classical machine learning classifiers and shallow neural networks for their ability to learn patterns from extracted features. Common examples include Support Vector Machines (SVMs), which are effective for classification tasks by finding an optimal hyperplane that separates data points into different classes [10]. Random Forests (RF) [11], an ensemble learning method, combine multiple decision trees to improve prediction accuracy and control overfitting, proving useful for robust defect classification (Ho & Huang, 2022). Additionally, Artificial Neural Networks (ANNs) [12–14], particularly those with a few hidden layers, have been employed to model complex non-linear relationships between image features and defect categories. These methods, while less computationally intensive than deep learning, rely heavily on the quality and discriminative power of the hand-crafted features derived from the pavement images. This study explores three proven image feature extraction techniques—HOG, LBP, and GLCM—combined with machine learning models for classifying pavement defects [15–17]. HOG captures local edge gradients, LBP encodes local textures robustly to lighting changes [11, 16], and GLCM quantifies pixel relationships to reflect surface texture [17]. These features are used to train classifiers to distinguish defective from intact pavement. This approach offers a scalable alternative to manual inspection, aiming to improve reliability in urban infrastructure monitoring.

2 Database

The dataset was constructed from 80 high-resolution images of pavement surfaces, comprising 40 images of slabs with visible cracks or fractures and 40 of intact slabs. To expand the dataset, each main image—already labeled as defective or intact—was subdivided into 9 smaller sections due to the high image resolution, significantly enlarging the database. A representative sample of these sub-images is illustrated in Fig. 1. This process significantly increased the amount of data available for analysis obtaining a database of 180 images from each class. For model validation, a 5-fold cross-validation method and the grid search method with 20 repetitions was employed, using 64 sub-images of defective slabs and 64 of intact slabs in each iteration. All images were captured under uniform natural lighting, with precautions taken to avoid direct sunlight and shadows, ensuring consistency during acquisition. Different defect morphologies were incorporated (linear cracks, branched fissures, isolated fractures, partial disintegration) and were visually documented to ensure separability between classes. Variability in surface conditions was also considered: presence of dirt, traffic-induced wear, and differences in material roughness. The images were subdivided into 9 equal parts, selecting the most representative ones.

The images are privately held and not accessible through any public repository. Variability among the slabs was intentionally considered during acquisition to ensure representation of different fracture types, which were carefully documented.

3 Methodology

This study follows a supervised machine learning approach for texture, edges, surface analysis and classification using three different feature extraction techniques and multiple classifiers. The methodological pipeline is composed of the following main stages: feature extraction, classification, and performance evaluation.

3.1 Feature Extraction

Three different texture descriptors were employed to capture complementary structural and statistical information from grayscale images:

HOG – Histograms of Oriented Gradients

The Histogram of Oriented Gradients (HOG) is a feature descriptor widely used in computer vision and image processing for object detection and texture analysis. It captures the edge or gradient structure of objects in an image. Proposed by Navneet Dalal and Bill Triggs in 2005 [18]. HOG works by counting the occurrences of gradient orientation in localized portions of an image. The following methodology outlines the steps for computing HOG descriptors from grayscale images:

- *Gradient Computation.* The first step is to compute the gradient magnitude and orientation at each pixel. Typically, gradients are computed using a simple Sobel operator:

$$G_x = I(x+1, y) - I(x-1, y) \tag{1}$$

$$G_y = I(x, y+1) - I(x, y-1) \tag{2}$$

$$G = \sqrt{G_x^2 + G_y^2}, \theta(x, y) = \arctan\left(\frac{G_y}{G_x}\right) \tag{3}$$

Where G_x is the gradient in the x-direction, G_y is the gradient in the y-direction, G is the gradient magnitude and $\theta(x, y)$ is the gradient orientation.

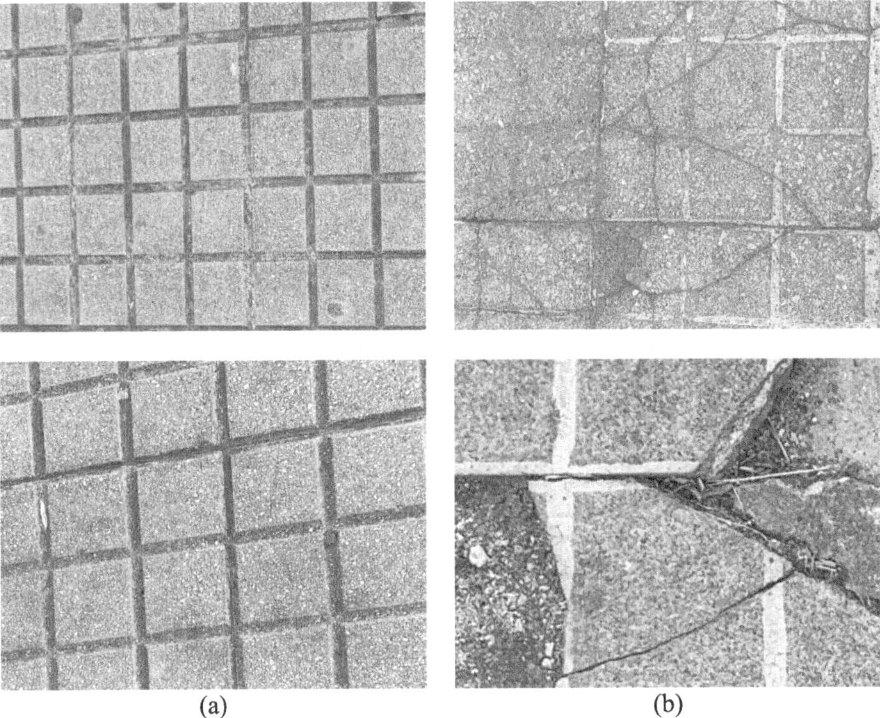

(a) (b)

Fig. 1. Pavement surface samples. (a) Uncracked surface example. (b) Cracked surface example.

- *Cell Division.* The image is divided into small cells, for each pixel in a cell, a histogram of gradient directions is created.
- *Orientation Binning.* The orientation range (usually 0°–180° or 0°–360°) is divided into bins (typically 9 bins of 20° each). Each pixel contributes to the histogram bin proportional to its gradient magnitude. Contribution can be interpolated between bins (bilinear voting).
- *Block Normalization.* To reduce the effect of lighting and contrast adjacent cells are grouped into larger blocks, the histograms are normalized in each block.

LBP – Local Binary Patterns

Local Binary Patterns, proposed by [19] in 1996, are used to describe the local texture of an image using binary patterns. For each pixel eight neighborhoods are taken, each neighboring pixel is compared to the central pixel value. If the neighbor's value ≥ central value, assign 1; otherwise, 0. A binary number is formed from the results and converted to decimal (LBP code). A histogram of the LBP codes is constructed as a feature vector (see Eq. 4).

$$LBP(x_c, y_c) = \sum_{p=0}^{P-1} s\left(i_p - i_c\right) \cdot 2^p \quad (4)$$

Where:

i_c is the intensity of the central pixel.
i_p is the intensity of the neighbor p.
$s(x) = 1$ if $x \geq 0$, and $s(x) = 0$ if $x < 0$.

GLCM/Grey – Level Co-occurrence Matrix
The Grey-Level Co-occurrence Matrix (GLCM) is an $N \times N$ matrix, where N represents the number of gray levels in the image. It captures the spatial relationship between pairs of pixels by calculating how often a pixel with gray-level value i appears in a specific spatial relationship to a pixel with gray-level value j. By default, the graycomatrix function computes the GLCM based on horizontal adjacency, using the offset [0 1], which refers to neighboring pixels in the same row. Given a grayscale image I, the command graycomatrix(I) generates a GLCM by counting the number of times a pixel with value i is located next to a pixel with value j in the specified direction. Each element (i, j) in the GLCM represents the frequency of occurrence of pixel value i relative to pixel value j, based on a chosen offset and direction (e.g., 0°, 45°, 90°, or 135°). This process is expressed in Table 1. The objective is to analyze the global texture through the spatial relationship of gray levels. This statistical representation provides valuable information about the texture of the image.

Statistical measures calculated from the matrix include contrast, energy, correlation and homogeneity [20]:

- *Contrast.* Measures the local intensity variation (Eq. 5).

$$\sum_{i,j} (i-j)^2 \cdot P(i,j) \tag{5}$$

Table 1. Table lists the offset values that specify common angles, given the pixel distance D.

Angle	Offset
0°	[0 D]
45°	[-D D]
90°	[-D 0]
135°	[-D -D]

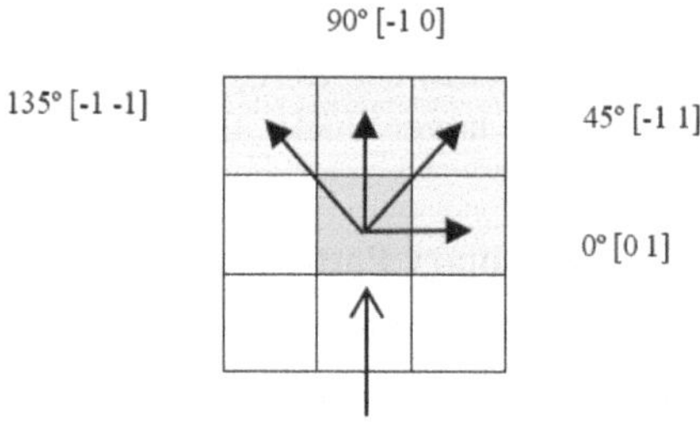

Where $P(i, j)$ is the pixel of interest, is the normalized value of the GLCM at position (i, j). N is the number of gray levels,

- *Energy (ASM).* Indicates homogeneity or uniformity (Eq. 6).

$$\sum_{i,j} P(i,\ j)^2 \tag{6}$$

- *Correlation.* Measures the dependency between pixels (Eq. 7).

$$\sum_{i,j} \frac{(i-\mu_i)(j-\mu_j)P(i,j)}{\sigma_i\sigma_j} \tag{7}$$

- *Homogeneity (IDM).* Reflects the closeness to the GLCM diagonal. This metric gives greater weight to co-occurrences near the diagonal, meaning it emphasizes local similarities in the image. $|i-j|$ represents the distance from the diagonal (the closer to 0, the more homogeneous) (Eq. 8).

$$Homogeneity = \sum_{i=0}^{N-1}\sum_{j=0}^{N-1} \frac{P(i,j)}{1+|i-j|} \tag{8}$$

3.2 Classification Algorithms

The extracted features were used to train and evaluate five classification models:

- **Support Vector Machines (SVM):** SVM is a supervised learning algorithm used for classification tasks. In this study, it was implemented with a Radial Basis Function (RBF) kernel to handle non-linear decision boundaries. SVM aims to find the optimal hyperplane that maximally separates the data classes [21].
- **k-Nearest Neighbors (k-NN):** k-NN is a non-parametric method that classifies data points based on the majority class among their nearest neighbours. The Euclidean distance was used to measure similarity, and the optimal number of neighbours was determined through internal validation. to enhance accuracy [22].
- **Naïve Bayes:** This is a probabilistic classifier based on Bayes' theorem, assuming independence between features. It is computationally efficient and often performs well on high-dimensional data [23].
- **Decision Trees:** Rule-based classifier built through recursive partitioning. Decision Trees classify instances by recursively partitioning the feature space into regions based on learned decision rules. They are intuitive, interpretable, and perform well with data containing non-linear relationships [24].
- **Artificial Neural Networks (ANNs):** ANNs use a multilayer perceptron (MLP) architecture with at least one hidden layer. In this case, the network was trained with ReLU activation and backpropagation. ANNs are capable of modelling complex, non-linear patterns, offering superior performance in image classification tasks [25].

In this study, four ANN topologies were tested to evaluate their effectiveness in modelling different feature descriptors (HOG, LBP, and GLCM) for pavement defect classification. The different ANN topologies were selected with the aim of evaluating how model performance varies according to the complexity of the architecture. Networks with one, two, and three hidden layers and different layer sizes were tested to observe the balance between generalization capacity and risk of overfitting. The process started with simpler configurations (ANN [100 10]) and progressively increased the number of layers and neurons (ANN [100 20 10], [100 15 8 4]) to identify the structure that provided the best results in terms of standard metrics: Accuracy, Precision, Sensitivity/Recall, Specificity, and F1-score. For each ANN topology, different feature descriptors—HOG, LBP, and GLCM—were evaluated separately. This approach aims to systematically analyse how both the network architecture and the type of feature extraction influence model performance. By applying each descriptor independently to every ANN configuration, it becomes possible to observe which combination of topology and descriptor provides the best balance between accuracy, generalization capacity, and the best model.

3.3 Evaluation Metrics

A 5-fold cross-validation procedure with 20 repetitions was employed to ensure reliable and statistically robust performance estimation. The dataset was partitioned into training and testing subsets, rotating folds across repetitions. The classifiers' performance was assessed using the standard metrics: Accuracy, Precision, Sensitivity/Recall, Specificity, and F1-score. These metrics were averaged across folds and repetitions to compute final results.

4 Results and Discussion

The classification results of the experimental procedure are collected in this section (see Tables 2, 3 and 4). The HOG method shows very good results (Table 2). The classification results based on HOG features demonstrate strong performance across both traditional machine learning classifiers and artificial neural networks (ANNs).

Table 2. HOG-based classifiers.

Classifier	Average Precision	Average Recall	Average Specificity	Average F1 Score
SVM	**0.9526**	**0.9692**	**0.9346**	**0.9560**
KNN	0.9142	0.9538	0.8731	0.9183
Naïve Bayes	0.8902	0.9538	0.8244	0.8994
Trees	0.8677	0.8615	0.8756	0.8669
ANN [100 10]	0.9209	0.9526	0.8885	0.9245
ANN [100 20 10]	0.9366	0.9846	0.8885	0.9417
ANN [100 50]	0.9372	0.9846	0.8897	0.9401
ANN [100 15 8 4]	0.9369	0.9538	0.9205	0.9384

SVM achieved the best overall performance with the highest values of precision (0.9526), recall (0.9692), specificity (0.9346), and F1-score (0.9560), demonstrating its effectiveness with HOG descriptors. KNN and Naïve Bayes delivered acceptable results, but with lower specificity. Decision Trees showed the weakest performance. Among neural networks, the three hidden layers with 100-20-10 units topology, stood out in recall (0.9846) and F1-score (0.9417), while the two hidden layers with 100-50 units topology offered better precision and the four hidden layers with [100 15 8 4] units topology offered the highest specificity.

Table 3. LBP-based classifiers.

Classifier	Average Precision	Average Recall	Average Specificity	Average F1Score
SVM	0.7151	0.7385	0.7077	0.6559
KNN	0.8674	0.9372	0.7974	0.8777
Naïve Bayes	0.8751	0.9064	0.8436	0.8795
Tree	0.8591	0.8590	0.8590	0.8590
ANN [100 10]	0.9222	0.9538	0.8923	0.9253
ANN [100 20 10]	0.9298	0.9692	0.8923	0.9331
ANN [100 50]	**0.9298**	**0.9692**	**0.8923**	**0.9331**
ANN [100 15 8 4]	0.9222	0.9538	0.8923	0.9253

The results from classification using LBP features show a contrasting performance pattern compared to HOG-based features (see Table 3). Surprisingly, SVM performs the worst among classical classifiers on LBP features, with the lowest precision (0.7151), recall (0.7385), specificity (0.7077), and F1-score (0.6559), likely due to LBP's less linearly separable feature space. In contrast, Naïve Bayes and KNN perform well, with Naïve Bayes slightly leading in precision and F1-score. Decision Trees show moderate but better performance than SVM, suggesting some capacity to capture LBP's nonlinearities. All ANN architectures outperform classical classifiers by a wide margin, with Topologies 2 and 3 achieving precision around 0.93, recall around 0.97, and F1-scores above 0.93. Specificity remains consistent (~0.89), highlighting their robustness. ANNs effectively model LBP's complex patterns, and performance stability across topologies suggests that even simpler networks are sufficient.

The classification performance using GLCM features shows a distinct pattern compared to HOG and LBP descriptors (see Table 4). SVM shows the lowest recall (0.4372) despite high specificity (0.9205) and moderate precision (0.6791), indicating an overly conservative model prone to false negatives (F1-score 0.5740). In contrast, the Decision Trees achieved the best performance among classical classifiers with high recall (0.8910), precision (0.8588), and F1-score (0.8637), effectively modeling the nonlinear patterns of GLCM features. KNN and Naïve Bayes perform reasonably well but fall short of Decision Trees. ANNs significantly outperform all classical models. Topology 3 [100 50] yields the best overall performance (precision 0.9382, recall 0.9385, F1-score

Table 4. GLCM-based classifiers.

Classifier	Average Precision	Average Recall	Average Specificity	Average F1Score
SVM	0.6791	0.4372	0.9205	0.5740
KNN	0.8369	0.8436	0.8295	0.8382
Naïve Bayes	0.8126	0.7513	0.8744	0.7969
Tree	0.8588	0.8910	0.8295	0.8637
ANN [100 10]	0.9298	0.9385	0.9231	0.9303
ANN [100 20 10]	0.9065	0.9231	0.8910	0.9076
ANN [100 50]	**0.9382**	**0.9385**	**0.9385**	**0.9377**
ANN [100 15 8 4]	0.9225	0.9077	0.9385	0.9212

0.9377), while other topologies show similarly strong results, each excelling in specific metrics. The results highlight GLCM's suitability for texture analysis, especially when paired with Decision Trees or ANNs. Poor SVM performance suggests GLCM features are not linearly separable. Overall, GLCM features combined with ANNs offer robust and reliable classification across architectures.

5 Conclusions

This study compares three feature extraction methods—HOG, LBP, and GLCM—combined with classical machine learning classifiers and Artificial Neural Networks (ANNs) for pavement image classification. HOG performed best with classical models, especially SVM and gradient descriptors. LBP worked better with Naïve Bayes and KNN, while ANNs achieved the best overall performance with LBP, effectively capturing its complex textures. GLCM yielded the weakest results with SVM, but Decision Trees handled its nonlinearity better. ANNs consistently outperformed classical classifiers across all feature types, with a topology with two hidden layers of 100 and 50 units, respectively, delivering the best results. The findings highlight the importance of matching feature types with suitable classifiers. SVM is optimal with HOG under low-resource conditions, while ANNs paired with LBP or GLCM are ideal for complex, texture-rich images. The study confirms the ongoing value of features, especially when data is scarce. Future work may include hybrid feature sets and deeper neural architectures to further improve performance, although more examples would be needed.

References

1. Caragliu, A., Del Bo, C., Nijkamp, P.: Smart cities in Europe. J. Urban Technol. **18**(2), 65–82 (2011). https://doi.org/10.1080/10630732.2011.601117
2. Albino, V., Berardi, U., Dangelico, R.M.: Smart cities: definitions, dimensions, performance, and initiatives. J. Urban Technol. **22**(1), 3–21 (2015). https://doi.org/10.1080/10630732.2014.942092

3. Hashem, I.A.T., et al.: The role of big data in smart city. Int. J. Inf. Manag. **36**(5), 748–758 (2016)
4. Gopalakrishnan, K.: Deep learning in data-driven pavement image analysis and automated distress detection: a review. Data **28** (2018)
5. Pan, Y., Zhang, X., Cervone, G., Yang, L.: Detection of asphalt pavement potholes and cracks based on the unmanned aerial vehicle multispectral imagery (2018)
6. Marasinghe, R., Yigitcanlar, T., Mayere, S., Washington, T., Limb, M.: Computer vision applications for urban planning: a systematic review of opportunities and constraints. Sustain. Cities Soc. **100**, 105047 (2024)
7. Shang, J., Zhang, A., Zishuo Dong, Zhang, H., Anzheng He, A.: Automated pavement detection and artificial intelligence pavement image data processing technology. Autom. Constr. **168**, Part 1, 105797 (2024)
8. Riid, A., Lõuk, R., Pihlak, R., Tepljakov, A., Vassiljeva, K.: Pavement distress detection with deep learning using the orthoframes acquired by a mobile mapping system. Appl. Sci. **9**(22), 4829 (2019). https://doi.org/10.3390/app9224829
9. Abbas, I.H., Ismael, M.Q.: Automated pavement distress detection using image processing techniques. Eng. Technol. Appl. Sci. Res. **11**(5), 7702–7708 (2021)
10. Gao, H., Chen, W., Lihua Dou, L.: Image classification based on support vector machine and the fusion of complementary features. Comput. Sci. Comput. Vis. Pattern Recognit. arXiv: 1511.01706 (2015)
11. Ojala, T., Pietikainen, M., Maenpaa, T.: Multiresolution gray-scale and rotation invariant texture classification with local binary patterns. IEEE Trans. Pattern Anal. Mach. Intell. **24**(7), 971–987 (2002)
12. Zhang, H., Guan, J., Sun, G.C.: Artificial neural network-based image patter recognition. In: ACM 30th Annual Southeast Conference (1992)
13. Bishop, C.M.: Pattern Recognition and Machine Learning. Springer, New York (2006)
14. Alhindi, T.J., Kalra, S., Ng, K.H., Afrin, A., Tizhoosh, H.R.: Comparing LBP, HOG and deep features for classification of histopathology images. In: Proceedings of the IEEE World Congress on Computational Intelligence (IEEE WCCI), Rio de Janeiro, Brazil, 8–3 July 2018 (2018)
15. Mohammed, M., et al.: Artificial intelligence approaches in predicting the mechanical properties of natural fiber-reinforced concrete: a comprehensive review (2025)
16. Picard, R.W., Kabir, M.: Finding similar patterns in large image databases. In: Proceedings of the 1993 IEEE International Conference on Acoustics, Speech, and Signal Processing, vol. 5, pp. 161–164. IEEE (1993)
17. Ojala, T., Valkealahti, K., Oja, E., Pietikäinen, M.: Texture discrimination with multidimensional distributions of signed gray level differences. Pattern Recognit. (2000) (in Press)
18. Dalal, N., Triggs, B.: Histograms of oriented gradients for human detection. In: 2005 IEEE Computer Society Conference on Computer Vision and Pattern Recognition (CVPR 2005) (2005). 0-7695-2372-2
19. Ojala, T., Pietikäinen, M., Harwood, D.: A comparative study of texture measures with classification based on feature distributions. Pattern Recognit. **29**, 51–59 (1996)
20. Haralick, R.M., Shanmugam, K., Dinstein, I.: Textural features for image classification. IEEE Trans. Syst. Man Cybern. **SMC-3**(6), 610–621 (1973)
21. Cortes, C., Vapnik, V.: Support-vector networks. Mach. Learn. **20**, 273–297 (1995)
22. Cover, T., Hart, P.: Nearest neighbor pattern classification. IEEE Trans. Inf. Theory **13**(1), 21–27 (1967)
23. Zhang, H.: The optimality of Naive Bayes. In: Proceedings of the International FLAIRS Conference (2004). ISBN 978-1-57735-201-3

24. Breiman, L., Friedman, J.H., Olshen, R.A., Stone, C.J.: Classification and Regression Trees. Monterey, CA: Wadsworth & Brooks/Cole Advanced Books & Software (1984). ISBN 978-0-412-04841-8
25. Rumelhart, D.E., Hinton, G.E., Williams, R.J.: Learning representations by back-propagating errors. Nature **319**, 321–324 (1986)

Automated Multi-UAV Scenarios Generation for Cooperative Aerial Missions

Daniel Caballero-Martin[1,2], Julian Estevez[1,3], Manuel Graña[1,4], Daniel Teso-Fz-Betoño[2], and Jose Manuel Lopez-Guede[1,2](✉)

[1] Computational Intelligence Group, University of the Basque Country, (UPV/EHU), Vitoria-Gasteiz, Spain
jm.lopez@ehu.eus

[2] Faculty of Engineering of Vitoria-Gasteiz, University of the Basque Country, (UPV/EHU), C/Nieves Cano 12, 01006 Vitoria-Gasteiz, Spain

[3] Faculty of Engineering of Gipuzkoa, University of the Basque Country, (UPV/EHU), Europa Plaza 1, 20018 San Sebastian, Spain

[4] Faculty of Computer Science, University of the Basque Country, (UPV/EHU), Manuel Lardizabal Pasealekua 1, 20018 Sebastian, Spain

Abstract. Simulation is a key pillar in the development of robotic systems, especially in aerial robotics, where multi-agent coordination and autonomous decision making present significant challenges. This work introduces a system for the automatic generation of simulation environments for multi-UAV formations using ROS 2 and Gazebo. The tool enables the creation of geometric or customized configurations, generating simulation packages ready for direct deployment in ROS 2 without manual setup. This automation accelerates experimentation, reduces errors, and improves reproducibility. The generator supports user defined parameters or internal position calculations, adapting to different mission profiles and cooperative strategies. The results demonstrate the system's capability to generate high fidelity scenarios useful for evaluating formation control and swarm behavior under realistic constraints. Aligned with the Sim2Real paradigm, this framework facilitates transfer to the real world. Future work envisions the integration of Artificial Intelligence (AI) techniques for adaptive scenarios.

Keywords: Multi-UAV Simulation · ROS 2 · Gazebo · Automated Scenario Generation · Swarm Robotics

1 Introduction

Simulation has become established as an essential tool in the field of robotic development, enabling researchers and developers to test and validate algorithms, control architectures, and mechanical designs in virtual environments before implementing them in real world scenarios, in order to reduce costs, minimize risks, and accelerate the development of complex robotic systems [32].

E. Corchado et al. (Eds.): SOCO 2025, CCIS 2806, pp. 613–622, 2026.
https://doi.org/10.1007/978-3-032-19763-4_57

The advancement in the field of simulation is closely linked to progress in the fields of computing and (AI) [1]. Early real time control systems, such as the Real Time Control System (RCS), lay the groundwork for the integration of sensors and actuators in robotic systems, enabling more dynamic interaction with their environment [2]. Initially developed in the 1970s by James S. Albus at the National Bureau of Standards (NBS), now known as the National Institute of Standards and Technology (NIST), the RCS is conceived as a reference architecture for the design and analysis of complex intelligent control systems [5]. The system's hierarchical and modular architecture allows for the decomposition of complex tasks into more manageable subtasks, which facilitates the real time control of robotic systems [2].

The RCS has undergone continuous evolution, incorporating advanced functionalities that have extended its applicability in intelligent control systems, such as environment modeling, task planning, and sensory processing [6]. These improvements significantly expand the system's scope of application, from factory automation [20] to autonomous vehicles [4] and space robotic systems [22]. In the context of simulation, the RCS provides a solid foundation for the development of simulated environments that replicate the behavior of robotic systems in the real world, allowing control and perception algorithms to be tested and validated in virtual settings. Subsequently, approaches such as evolutionary robotics emerge, based on the use of Evolutionary Algorithms (EA) for the design and optimization of both the morphology and behavior of robots within the simulation [21].

Over time, simulation in the field of robotics has evolved toward environments of greater realism and complexity. This has been made possible by the integration of three dimensional simulators with advanced physics engines, which allow for the modeling of complex physical dynamics and offer a wide range of sensors and actuators. The proliferation of open source platforms has significantly transformed the simulation landscape by facilitating access to advanced tools and fostering a collaborative community. These initiatives enable the exchange of knowledge and technological advancements, accelerating innovation and lowering the barriers to entry in the field of robotics. As a result of this collaborative paradigm, open source platforms play a fundamental role in the development and validation of complex robotic systems in simulated environments [23].

The remainder of the paper is as follows. Section 2 is dedicated to the explanation of the state of the art. Section 3 presents the technical specifications that the implemented code generator must meet, while Sect. 4 explores the results obtained using the implemented tool with various input configurations. Finally, the article ends in Sect. 5 with conclusions and future work in this field.

2 Related Works

Robot Operating System 2 (ROS 2) [8] and the Gazebo simulator [24] are key standards in robotics. ROS 2 offers a modular, scalable framework for robotic software development, improving component interaction. Gazebo provides a realistic 3D simulation environment with complex physical dynamics.

Beyond the widely adopted ROS 2 and Gazebo platforms, robotic simulation has diversified with tools tailored to various needs. Webots [16] offers open source 3D simulation with extensive model and sensor support. CoppeliaSim [25] features distributed control and multi physics engine integration. MuJoCo [34] and PyBullet [11] deliver fast, accurate physics, ideal for Reinforcement Learning (RL) and control tasks. NVIDIA Isaac Sim [36], built on Omniverse, provides photorealistic environments and precise dynamics, enhancing AI training. Other platforms like Unity and Unreal Engine also expand options for robotics applications.

Regarding middleware alternatives to ROS 2, various platforms offer specialized features. Yet Another Robot Platform (YARP) [29] prioritizes modularity and interoperability, especially in cognitive and humanoid robotics. Open Robot Control Software (OROCOS) [9] is suited for real time industrial applications, enabling precise and reliable control. Middleware for Robotic Applications (MIRA) [15] provides a decentralized architecture for efficient communication in distributed systems. Zenoh [10] is a recent middleware designed for distributed, real time systems, focusing on efficiency and flexibility. These alternatives allow for selecting middleware tailored to specific requirements such as scalability, integration, and latency.

Simulation applied to Unmanned Aerial Vehicle (UAV) systems emerges as an indispensable tool for development, validation, and training in aerial robotics. One of its main benefits lies in the ability to conduct tests in safe environments, avoiding risks associated with the use of real hardware and facilitating the iterative development of navigation, perception, and control algorithms. Various studies explore simulations based on ROS 2 and Gazebo, integrating modules for navigation, mapping, trajectory planning, and vision based guidance in environments. These simulations demonstrate realistic flight behaviors, comparable to those obtained in physical tests, reinforcing their value as a testing environment prior to real world implementation [35].

The use of open and customizable simulation platforms is crucial for addressing the challenges of interoperability and scalability in the development of UAV systems. Solutions based on open source hardware and software enable the reuse and cooperative validation of developments, overcoming the limitations imposed by proprietary controllers [14].

On the other hand, modular tools such as ROS2GazeboDrone are emerging, designed to facilitate the integration and evaluation of planning and perception algorithms in drones, enabling the optimization of the workflow from simulation to implementation [18]. At the level of systematic review, dozens of UAV simulators have been identified, classified according to their customization capabilities, fidelity, ease of integration, and intended use. This diversity highlights the need for clear criteria to select the appropriate platform based on project requirements [12].

An innovative area in UAV simulation is swarm robotics, where Software In The Loop (SITL) simulations validate coordination, collision avoidance, and fault tolerance in distributed networks. Communication constraints and localization errors impact swarm stability and efficiency [26]. Simulated environments also prove valuable in education, as structured exposure to tasks improves pilot performance. Studies report notable gains in accuracy and efficiency after simulator-based training [33].

The incursion of AI into UAV simulation has significantly transformed the development and validation of autonomous systems, enabling the training of navigation, perception, and control algorithms in safe and reproducible virtual environments [27]. As an innovative aspect, the use of Deep Reinforcement Learning (DRL) is introduced to train drone swarms in ground surveillance tasks, allowing for efficient and robust coordination in dynamic environments [7].

Simulation platforms like AirSim [30], developed by Microsoft, provide realistic environments for testing DL, computer vision, and RL in autonomous vehicles, enabling safe experimentation. AirSim also generates high quality synthetic data for training AI models in object detection, semantic segmentation, and path planning capabilities that are key to improving autonomy in complex environments. The availability of tools that allow the generation and simulation of scenarios for autonomous car based driving [17], inspires the development of a tool aimed at the control and testing of drone swarms. A versatile, flexible tool is presented, capable of integrating a wide variety of elements into the scene, enabling more complete and realistic simulations.

3 Autonomous Generation of Scenarios

The integration of ROS 2 and Gazebo has been established as a strategic option in the field of development and validation of advanced robotic systems, supported by widespread adoption within the scientific and technological communities. This combination provides a robust and flexible environment that facilitates the accurate simulation of robots in virtual settings, enabling the validation of control and perception strategies prior to their implementation in the real world. Various studies explore the effectiveness of the integration between ROS 2 and Gazebo in the development and validation of advanced robotic systems [13].

Open source tools have been specifically developed for the simulation of robots with elastic actuators, demonstrating the capability of this integration for validating control strategies in soft robots within high precision virtual environments [28]. Recent research has also evaluated the usability of ROS 2 together with Gazebo in the control of mobile robots such as the TurtleBot3, highlighting its applicability in educational, training, and research environments [3]. The evolution of ROS toward its second version has introduced substantial improvements in terms of security, real time communication, and native support for distributed architectures, significantly expanding its adoption in industrial applications and advanced research projects that require robustness, scalability, and reliability, supporting a smooth transition between simulation and real hardware implementation, and fostering the approach known as Sim2Real [19]. This approach is crucial for the development of control and perception algorithms that are robust and transferable to real world environments.

The choice as the development and simulation platform in this research is justified by its ability to provide a realistic and flexible simulation environment, its wide adoption within the scientific community, and its compatibility with advanced control and machine learning tools. These features allow for rigorous validation of the developed robotic systems, ensuring their effectiveness and reliability before implementation in real world environments. A quad-core processor at 2 GHz, 8 GB of RAM, and a GPU with at least 1 GB of VRAM provide a basic environment to run ROS 2 and Gazebo. The required computational capacity will depend on the scenarios complexity and the physical calculations involved.

In the context of this research, the development of a modular and extensible tool is proposed, aimed at automating multi-UAV simulation scenarios in ROS 2 and Gazebo environments. In its current version, it allows the generation of UAV spatial configurations in two modes: predefined geometric formations (dual, triangular, quadrangular), and fully customized configurations, in which the user can explicitly define the position of each aircraft within the simulated space. The tool automatically generates the scenario description files (in .sdf format), structuring the necessary packages for immediate execution without requiring additional manual adjustments (Fig. 1).

This code generator is designed to be parameterizable, allowing input data to be set from external files, graphical interfaces, or internally through generator functions, enabling easy scaling of the number of units, dynamic modification of the layout, or reuse of previous configurations. In advanced stages of the framework's development, the extension of the generator is planned to include additional elements in the environment (static obstacles, loading/unloading zones, beacons, etc.) and the implementation of capabilities to simulate cooperative load transport missions, in which multiple UAVs interact through physical or virtual coupling mechanisms. This evolution aims at a system capable of producing realistic environments oriented toward the validation of cooperative navigation algorithms and aerial logistics.

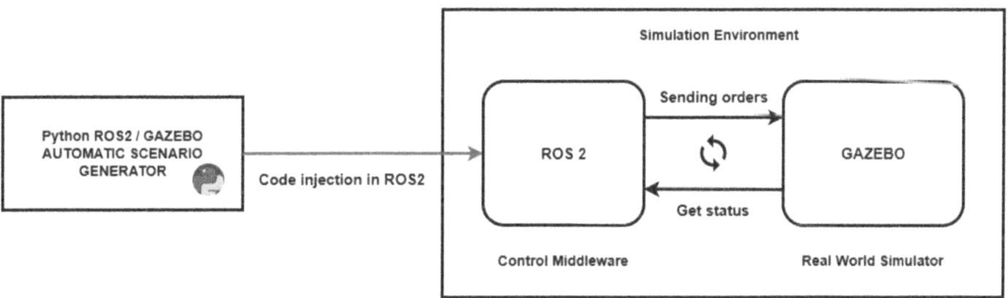

Fig. 1. Development and implementation workflow in a simulated environment.

The X3 UAV model provided by Gazebo is selected as the basis for the simulations. This model includes a detailed representation of the UAV's physical and morphological characteristics, allowing for realistic and accurate simulations in the Gazebo environment. This model is available on the Gazebo Fuel platform, where its specifications and related files can be found [31]. The choice of this model is based on its ease of integration with ROS2, ensuring that the simulations reflect appropriate physical behavior in the simulated environment. The system allows the introduction of different types of drones and other elements within the same scenario, as well as the configuration of custom behaviors for each one. The results can be presented individually or collectively, based on the data obtained from the simulator.

4 Results

This section presents a series of configurations generated using the system developed for the automatic creation of multi-UAV simulation environments in ROS 2 and Gazebo. The objective of this phase is to illustrate the generator's capability to construct complex arrangements of UAV, adapted to formation flights with precision and flexibility.

The implemented system has successfully achieved the initial objective of generating simulation scenarios with both geometric and customized drone formations. Through the developed tool, it is possible to automatically construct multi-UAV environments in ROS 2 and Gazebo, allowing for flexible configuration and immediate execution. The results presented in this section demonstrate the system's effectiveness in producing structured, reusable, and scalable simulation configurations, in line with the design specifications. As part of the ongoing work, the aim is to generate much more complete scenarios that include additional environmental elements, as well as to develop systems capable of collaboratively transporting loads using ropes, in order to validate navigation and control strategies in cooperative missions.

Multiple formations are presented, both standardized geometric ones (Fig. 2) as well as customized ones (Fig. 3), which allow the analysis of collective drone behavior under controlled conditions. All of them have been automatically generated by the system based on user defined parameters or through internal calculation rules (Fig. 4). Each resulting environment includes not only the spatial arrangement of the UAVs but also the complete structure required for direct simulation in ROS 2, thus facilitating an efficient transition from design to validation. This automation has proven effective in accelerating experimentation cycles, reducing configuration errors, and improving the analysis of collective behavior under different formation and flight space conditions.

The graphical representation of the configurations allows for a clear visualization of the expected behavior, as well as the detection of spatial inconsistencies before dynamic execution. This immediate visual validation is especially useful in the early stages of iterative design.

The set of examples presented in this section demonstrates the versatility of the developed system, confirming its applicability to various simulation contexts without the need for additional manual intervention. All of this reinforces its

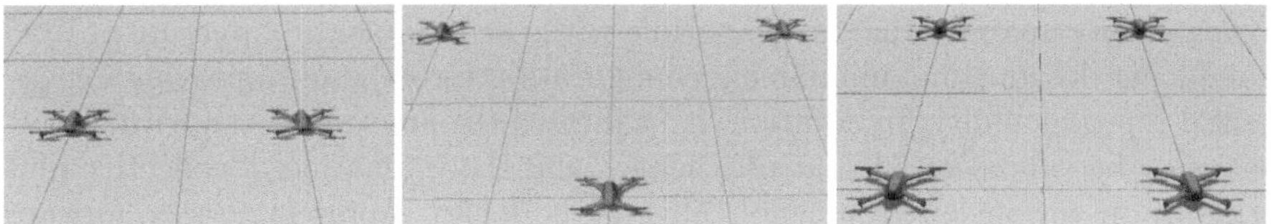

Fig. 2. Geometric formations.

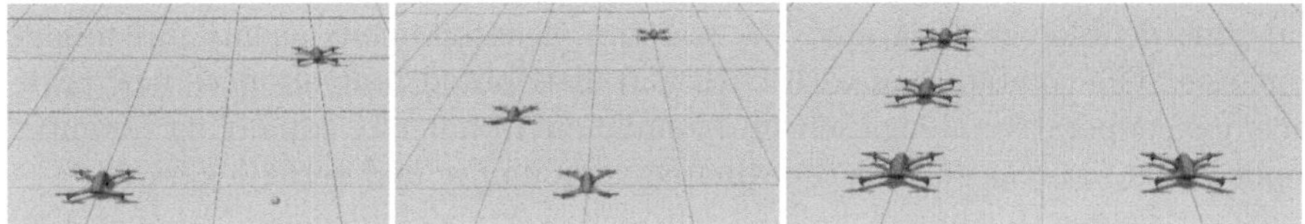

Fig. 3. Customized formations.

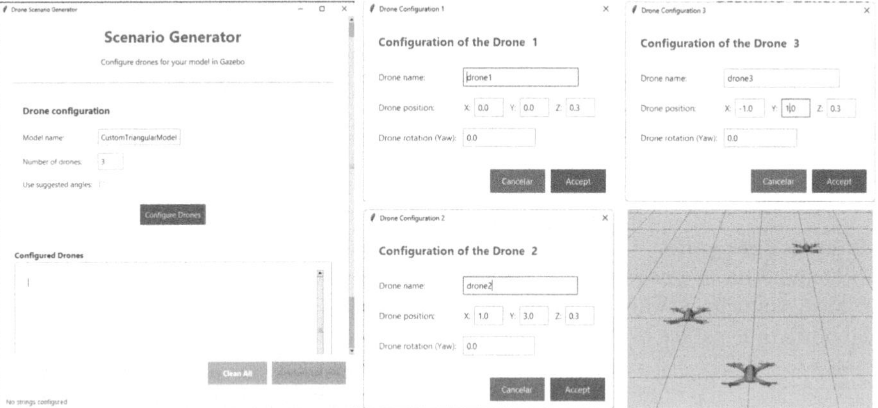

Fig. 4. Scenario generation for triangular model.

usefulness as a support tool for experimental design, rapid prototyping of multi-UAV strategies, and the preparation of standardized tests within the ROS 2 and Gazebo ecosystem.

5 Conclusion

Simulation is established as a key tool for the development of multi-UAV systems, as it enables the safe and controlled study of collaborative dynamics. The integration of ROS 2 and Gazebo provides a robust and flexible technological foundation, capable of recreating realistic physical conditions and allowing the validation of algorithms prior to their implementation on real platforms.

The automated design of scenarios represents an efficient solution for formation flight simulation, allowing configurations of geometric or customized UAV arrangements in a way that is compatible with the modular architecture of ROS 2, reducing design time, minimizing configuration errors, and improving experimental reproducibility. In addition, it facilitates the analysis of cooperative formations, the validation of control schemes, and the evaluation of collective performance in various environments. This methodology represents a step forward toward more standardized tools within the ROS 2-based simulation ecosystem.

The future outlook for this technology is highly promising, with a growing integration of AI techniques that will enable trajectory optimization, improved real time decision making, and the training of autonomous agents in complex scenarios. The combination with DRL and distributed control opens new pathways for cooperative navigation in dynamic environments, expanding potential applications to areas such as coordinated surveillance and aerial logistics.

Nevertheless, this line of work also presents certain challenges. Although simulation fidelity is high, it does not always accurately capture all the variables of the real physical environment, such as disturbances, latencies, or sensor errors. In addition, computational scalability can limit its use in scenarios with multiple agents or complex architectures. Overcoming these barriers will be key to consolidating simulated environments as robust and transferable tools for the development of cooperative robotic systems in the real world.

Future work will focus on the complete simulation of complex multi-UAV scenarios involving coordinated flight formations for collaborative suspended load transport.

Acknowledgements. The authors were supported by the Vitoria-Gasteiz Mobility Lab Foundation, a governmental organization of the Provincial Council of Araba and the local council of Vitoria-Gasteiz under the following project grant: "Transporte colaborativo de cargas asistido y controlado por algoritmos de Inteligencia Artificial".

References

1. Dissecting Robotics - historical overview and future perspectives (2025)
2. RCS: The NBS real-time control system-All Databases (2025). https://www.webofscience.com/wos/alldb/full-record/INSPEC:2509711
3. Alapati, S.D., Prakash, P.O.: Autonomous navigation of TurtleBot3 using DDPG reinforcement learning in ROS2 framework. In: 2024 International Conference on Sustainable Communication Networks and Application (ICSCNA), pp. 1356–1361. IEEE (2024)
4. Albus, J.S.: 4D/RCS - a reference model architecture for intelligent unmanned ground vehicles. In: Gerhart, G.R., Shoemaker, C.M., Gage, D.W. (eds.) Unmanned Ground Vehicle Technology IV, vol. 4715, pp. 303–310. SPIE-International Society Optical Engineering (2002)
5. Albus, J.S.: The NIST real-time control system (RCS): a reference model architecture for computational intelligence. In: Computational Intelligence and Its Impact on Future High-Performance Engineering Systems (1996)
6. Albus, J.: Outline for a theory of intelligence **21**(3), 473–509 (1991)

7. Arranz, R., Carraminana, D., de Miguel, G., Besada, J.A., Bernardos, A.M.: Application of deep reinforcement learning to UAV swarming for ground surveillance. Sensors **23**(21), 8766 (2023)
8. Bonci, A., Gaudeni, F., Giannini, M.C., Longhi, S.: Robot operating system 2 (ROS2)-based frameworks for increasing robot autonomy: a survey. Appl. Sci.-Basel **13**(23), 12796 (2023)
9. Bruyninckx, H.: Open robot control software: the OROCOS project. In: 2001 IEEE International Conference On Robotics And Automation. VOLS I-IV, Proceedings, pp. 2523–2528. IEEE, New York (2001)
10. Corsaro, A., et al.: Zenoh: unifying communication, storage and computation from the cloud to the microcontroller. In: Niar, S., Ouarnoughi, H., Skavhaug, A. (eds.) 2023 26TH Euromicro Conference on Digital System Design, DSD 2023, pp. 422–428. IEEE Computer Society, Los Alamitos (2023)
11. Coumans, E., Bai, Y.: PyBullet quickstart guide. PyBullet Quickstart Guide (2021). https://docs.google.com/document/u/1/d
12. Dimmig, C.A., et al.: Survey of simulators for aerial robots: an overview and in-depth systematic comparisons. IEEE Rob. Autom. Mag. (2024)
13. Djizi, H., Lakehal, A., Zahzouh, Z.: A quadrotor controlled in real-time using hand gestures and ROS2 multi-node communication within Gazebo 3D environment **18**(2) (2024)
14. Ebeid, E., Skriver, M., Terkildsen, K.H., Jensen, K., Schultz, U.P.: A survey of open-source UAV flight controllers and flight simulators. Microprocess. Microsyst. **61**, 11–20 (2018)
15. Einhorn, E., Langner, T., Stricker, R., Martin, C., Gross, H.M.: MIRA - middleware for robotic applications. In: 2012 IEEE/RSJ International Conference on Intelligent Robots and Systems (IROS), pp. 2591–2598. IEEE, New York (2012)
16. Franchi, M., et al.: Webots.HPC: a parallel simulation pipeline for autonomous vehicles. In: Practice and Experience In Advanced Research Computing 2022 (2022)
17. Goyal, S., Griggio, A., Tonetta, S.: System-level simulation-based verification of autonomous driving systems with the VIVAS framework and CARLA simulator. Sci. Comput. Programm. **242** (2025). https://doi.org/10.1016/j.scico.2024.103253
18. Haridevan, A.D., Kang, J., Yuan, M., Shan, J.: ROS2-gazebo simulator for drone applications. In: 2024 International Conference on Unmanned Aircraft Systems, ICUAS, pp. 1232–1238. IEEE, New York (2024)
19. Höfer, S., et al.: Sim2Real in robotics and automation: applications and challenges **18**(2), 398–400 (2021)
20. Jin, S., Yu, F., Wang, B., Zhang, M., Wang, Y.: Research on a real-time control system for discrete factories based on digital twin technology **14**(10), 4076 (2024)
21. Juricek, M., Parak, R., Kudela, J.: Evolutionary computation techniques for path planning problems in industrial robotics: a state-of-the-art review **11**(12), 245 (2023)
22. Kalaycioglu, S., De Ruiter, A.: Passivity based nonlinear model predictive control (PNMPC) of multi-robot systems for space applications **10**, 1181128 (2023)
23. Katara, P., Khanna, M., Nagar, H., Panaiyappan, A.: Open source simulator for unmanned underwater vehicles using ROS and Unity3D. In: 2019 IEEE Underwater Technology (UT), p. 7 (2019)
24. Koenig, N., Howard, A.: Design and use paradigms for Gazebo, an open-source multi-robot simulator. In: 2004 IEEE/RSJ International Conference on Intelligent Robots and Systems (IROS) (IEEE Cat. No.04CH37566), vol. 3, pp. 2149–2154 (2004)

25. Liu, C.: Quadrotor trajectory tracking and simultaneous localization and mapping implementation in CoppeliaSim. J. Inst. Eng. (India) Ser. C (Mech. Prod. Aerosp. Mar. Eng.) **106**, 529–540 (2025)
26. Marek, D., et al.: Swarm of drones in a simulation environment-efficiency and adaptation. Appl. Sci. **14**(9), 3703 (2024)
27. McEnroe, P., Wang, S., Liyanage, M.: A survey on the convergence of edge computing and ai for UAVs: opportunities and challenges. IEEE Internet Things J. **9**(17), 15435–15459 (2022)
28. Mengacci, R., Zambella, G., Grioli, G., Caporale, D., Catalano, M.G., Bicchi, A.: An open-source ROS-gazebo toolbox for simulating robots with compliant actuators **8** (2021)
29. Metta, G., Fitzpatrick, P., Natale, L.: YARP: yet another robot platform. Int. J. Adv. Rob. Syst. **3**(1), 8 (2006)
30. Microsoft: AirSim: A Simulator for Autonomous Vehicles (2025). https://microsoft.github.io/AirSim/
31. OpenRobotics: X3 UAV (2023). https://fuel.gazebosim.org/1.0/OpenRobotics/models/X3UAV
32. Singh, J., Patel, P.: Robotics in arthroplasty: historical progression, contemporary applications, and future horizons with artificial intelligence (AI) integration **16**(8), e67611 (2024)
33. Somerville, A., Lynar, T., Joiner, K., Wild, G.: Use of simulation for pre-training of drone pilots. Drones **8**(11), 640 (2024)
34. Todorov, E., Erez, T., Tassa, Y.: MuJoCo: a physics engine for model-based control. In: 2012 IEEE/RSJ International Conference on Intelligent Robots and Systems (IROS), pp. 5026–5033. IEEE, New York (2012)
35. Zhang, M., et al.: A high fidelity simulator for a quadrotor UAV using ROS and gazebo. In: IECON 2015 - 41st Annual COnference of the IEEE Industrial Electronics Society, pp. 2846–2851. IEEE, New York (2015)
36. Zhou, Z., et al.: Towards building AI-CPS with NVIDIA Isaac Sim: an industrial benchmark and case study for robotics manipulation. In: 2024 ACM/IEEE 44TH International Conference on Software Engineering: Software Engineering In Practice, ICSE-SEIP 2024, pp. 263–274. Association Computing Machinery, New York (2024)

X-RAPT: A Virtual Reality Platform for Adaptive Training in Robotic Programming

David Mulero-Pérez[1(✉)], Beatriz Zambrano-Serrano[2], Enrique Ruiz Zúñiga[3], Michael Fernández-Vega[1,4], and Jose Garcia-Rodriguez[1]

[1] Department of Computer Technology, University of Alicante, Alicante, Spain
{dmulero,jgarcia}@dtic.ua.es
[2] MetaMedicsVR, Malaga, Spain
beatriz@metamedicsvr.com
[3] University of Skövde, Skövde, Sweden
enrique.ruiz.zuniga@his.se
[4] University of Costa Rica, San Jose, Costa Rica
michael.fernandezvega@ucr.ac.cr

Abstract. This paper presents the design and development of X-RAPT (eXtended Reality Adaptive Programming Training), a simulation framework aimed at enhancing robotics education through immersive extended reality (XR) technologies. X-RAPT is a multi-user platform that enables students to learn and practice industrial robotic programming in safe, virtual environments. The system combines real-time VR interaction, hand-tracking, voice recognition, and a web-based task editor to provide an adaptive and scalable learning experience. Using realistic 3D environments and virtual replicas of industrial robots such as the UR3 and ABB IRB120, X-RAPT allows learners to engage in training scenarios that simulate tasks like object assembly, labeling, and packaging. The platform supports collaborative sessions with role-based functionality and cross-device synchronization, enabling intuitive interaction and immediate feedback. Preliminary results indicate that X-RAPT is an effective tool for lowering the entry barrier to robotics programming and fostering hands-on learning in virtual training environments.

Keywords: virtual reality · simulation-based training · immersive learning · human-robot interaction

1 Introduction

The demand for skilled professionals in industrial robotics and automation continues to grow, driven by increasing digitization and the need for flexible, adaptive production environments. However, traditional training approaches often struggle to keep pace with this evolution [3,19,24]. Robotic programming, in particular, presents steep learning curves, costly training setups, and limited

E. Corchado et al. (Eds.): SOCO 2025, CCIS 2806, pp. 623–632, 2026.
https://doi.org/10.1007/978-3-032-19763-4_58

access to real systems—challenges that hinder workforce readiness and industrial efficiency.

To address these limitations, the X-RAPT (eXtended Reality Adaptive Programming Training) project introduces a novel solution: a fully immersive and collaborative extended reality (XR) environment for teaching robotic programming skills. X-RAPT leverages state-of-the-art virtual reality technologies to simulate real-world robotic environments, tasks, and behaviors. The platform enables instructors to design customized training scenarios via a web-based editor, which can then be experienced by students within a multi-user VR environment that supports hand-tracking, voice commands, and role-specific functionality.

X-RAPT is grounded in the real needs of industrial partners like Volvo Group, ensuring that the simulations reflect authentic assembly and packaging tasks [10, 11]. By allowing trainees to interact with virtual versions of actual robots, such as the UR3 and ABB IRB120, the system bridges the gap between theoretical instruction and hands-on experience. Beyond technical fidelity, X-RAPT also offers pedagogical value, providing feedback mechanisms, performance tracking, and adaptive interfaces tailored to both instructors and students.

(a) Virtual Environment Scenes (b) VR Perspectives

Fig. 1. Illustrations of the X-RAPT training platform. (a) shows multiple scenes from the virtual industrial environment used in robotic programming tasks. (b) displays the immersive VR experience from both first-person and third-person viewpoints.

The paper details the architecture and development of the X-RAPT platform, including its use of Unity, VIROO, multiplayer networking, and backend infrastructure. Figure 1 provides a visual overview of the virtual training environment, including examples of simulated industrial scenes and user interaction within the VR system, as seen from both first-person and third-person perspectives.

The remainder of this paper is structured as follows. Section 2 reviews the current state of the art in technologies and tools for robotic environment simulation. Section 3 describes the development of the X-RAPT simulation framework, focusing on its architecture and interaction design. Section 4 discusses the capabilities and limitations of the proposed system. Finally, Sect. 5 summarizes the main findings and outlines directions for future work.

2 Related Work

The increasing demand for skilled professionals in industrial robotics has driven the development of new educational technologies and training methodologies. Extended reality (XR), including both virtual and augmented reality, has emerged as a promising tool for delivering immersive and interactive learning experiences. This section reviews prior work in VR-based robotics training, collaborative virtual environments, and simulation frameworks for technical education, situating the X-RAPT platform within this broader context.

2.1 Robotics Simulations

Simulation frameworks play a pivotal role in robotics education, enabling safe experimentation with programming and control algorithms. Early platforms like Gazebo provided foundational physics engines, while modern approaches integrate digital twins and mixed reality. Orsolits et al. [16] developed ARNO, combining desktop robots with AR interfaces to teach kinematics and safety protocols. Nava-Téllez et al. [14] demonstrated how VR-based digital twins improve error detection and process optimization before physical implementation.

2.2 VR Environments for Training

Recent research confirms VR's superiority over traditional methods for technical skill acquisition. Benotsmane et al. [4] highlighted VR's transformative impact on manufacturing training through digital twin creation and production line optimization. Collaborative systems like VR Co-Lab enable multi-user interaction for industrial workflows, supporting ergonomic collaboration and skill transfer [12]. Wang et al. [25] developed a mixed-reality platform for remote training using shared 3D CAD models and gesture-based interaction, demonstrating improvement in task accuracy compared to conventional methods [4,25].

Educational applications show particular promise, Andone [2] documented international VR projects improving digital literacy and cross-cultural collaboration, Berki [5] demonstrated VR's effectiveness in creating multi-user environments for practical engineering courses and lastly skill mastery rates is reported in vocational training using VR simulations for endotracheal intubation [17]. These implementations highlight VR's capacity for creating safe, repeatable training scenarios across disciplines [2,4,5,17].

Advanced VR systems enable detailed behavior tracking and feedback mechanisms. SurgTwinVR [9] automates tutorial creation and ergonomic assessment using digital human models for orthopedic surgery. VR Co-Lab [12] incorporates real-time analytics for formative assessment, such systems provide objective metrics for skill evaluation while reducing instructor workload.

2.3 Broader Trends and Challenges

Systematic reviews reveal rapid XR adoption across educational levels, though significant barriers persist. Van Mechelen et al. [23] identified curriculum integration challenges and ethical concerns in K-12 XR implementations. Bicalho et al. [6] noted infrastructure and hardware limitations in higher education. Within Key challenges are included implementation costs [15,21], technical complexity requiring instructor training [18], ethical and medical considerations in data collection [13] and general perspectives requirements [6,7].

Emerging solutions are increasingly combining XR with physical interfaces [20] and IoT technologies [1] to enable adaptive learning systems. Tisza et al. [22] demonstrated a digital twin laboratory for hydrogen production education using MQTT protocols, while intelligent tutoring systems are being adopted in Industry 5.0 workforce training [8]. However, long-term efficacy studies remain scarce—especially in the context of collaborative XR platforms [8,15].

Current research mainly establishes potential for enhancing engagement [5, 25], collaboration [2,12], medical skill transfer uses [9,17,18], education applicability at different levels including challenges [5,21–23] and industrial applications [14,25].

However, existing systems often lack integration of critical features such as combination of hand-tracking with haptic feedback [13], scenario authoring tools for multi-user settings [26] and lastly support for cross-device synchronization and usage. In summary, prior work analysis has demonstrated the potential of VR technologies to improve engagement and skill acquisition in robotics and technical education. However, few systems combine immersive multi-user experiences, real-time feedback, hand-tracking, and scenario authoring within a unified, flexible platform. X-RAPT aims to fill this gap by offering a robust framework for immersive, collaborative, and adaptive robotics training.

3 Method

This section presents the design principles and technical architecture of the X-RAPT platform, an immersive training framework for industrial robotics programming. The system is structured to support multi-user VR interaction, adaptive learning scenarios, and real-time performance monitoring. Key implementation choices, including interaction modalities, synchronization logic, and simulation strategies; are discussed in terms of their relevance to XR-based educational environments.

3.1 System Architecture and Interaction Design

The core system integrates several technologies: Unity serves as the primary development engine, VIROO provides the underlying XR infrastructure, and enables multi-user synchronization and interaction. The platform operates through a client-server model, ensuring consistent state management across multiple VR users in real time.

Interaction within X-RAPT is facilitated by hand-tracking, voice and chat commands, and controller-based inputs, allowing users to engage naturally with virtual robots and interfaces. The system supports both instructor and student roles, each with tailored permissions and interaction capabilities. For instance, instructors have the capacity to initiate scenarios and modify the variables they wish to select (levels of difficulty, number of pieces, task to complete), assign different types of users (main user, observers), and monitor student metrics while students interact directly with virtual robot models and virtual control panels.

To enhance realism, X-RAPT incorporates animation based interactions to accurately simulate robot behaviors and environmental responses. All user actions and state changes are logged by a backend system, which also supports real-time dashboards for performance monitoring and scenario adaptation. This architecture enables flexible deployment across various hardware configurations, from individual VR headsets to large-scale training labs with multiple simultaneous users.

Fig. 2. Main components of the X-RAPT simulation environment, including a programming tablet for task configuration and interaction, a conveyor belt for simulating assembly workflows, and two industrial robotic arms (UR3 and ABB IRB120) used for executing robotic operations within the virtual scene.

3.2 Simulation Tasks and Learning Workflow

Training scenarios within X-RAPT are structured around task-based simulations that replicate real-world industrial processes, such as assembly, packaging, and pick-and-place operations. Using a virtual programming tablet (see Fig. 2), users can configure and control various elements of the simulation, including robotic arms equipped with task-specific end-effectors, the conveyor belt system, and peripheral devices such as labeling machines. These scenarios are authored via a web-based interface that allows instructors to define objectives, configure robot workcells, and set success criteria, without requiring programming expertise. Once deployed, trainees can access the scenarios via VR devices or a PC, and the training unfolds in either guided or exploratory modes depending on the learner's progress and the instructor's configuration.

Each learning session is divided into four discrete stages: introduction, task briefing, task execution, and debriefing. The introduction stage is conducted once per group and provides a general overview of the training environment and learning goals. During the task briefing, each user receives specific instructions for the task they are about to perform, including its objectives and the available tools. In the execution stage, trainees carry out the task with contextual guidance provided through visual cues, instructional overlays, and voice prompts. The system also supports collaborative learning modes, allowing multiple users to work simultaneously or assume complementary roles, such as operator and supervisor. The final debriefing stage includes both automatic performance feedback and self-assessment. Users are prompted to complete a structured questionnaire reflecting on the task performed and their experience, while instructors have access to detailed session analytics. These include task completion times and error frequencies.

The workflow is designed to be iterative and adaptive: scenarios can be dynamically adjusted based on learner performance, and repeated practice is encouraged to reinforce skill acquisition. This structure aims to balance autonomy with guidance, supporting both self-paced exploration and instructor-led learning.

4 Discussion

The discussion reflects on the capabilities, implications, and limitations of the X-RAPT framework from a scientific and pedagogical perspective. We evaluate the system's potential to enhance robotics education through immersive, data-driven training, and examine current constraints in terms of usability, generalization, and scalability. Key design trade-offs and future development directions are also considered.

4.1 Educational Impact and Design Advantages

The X-RAPT platform is designed not only as a technological innovation but also as a pedagogical tool aimed at transforming how industrial robotics is taught and learned. Its immersive, interactive nature enhances learner engagement, promotes active learning, and provides access to training opportunities that are often constrained by cost, safety, or availability in traditional settings. One of the key educational impacts of X-RAPT lies in its support for experiential learning. By interacting directly with realistic simulations of robotic systems, learners can develop procedural knowledge and muscle memory in a safe, repeatable environment. The system enables both guided instruction and exploratory learning, catering to different pedagogical approaches and learner preferences. Scenarios can be scaffolded to increase in complexity, allowing progressive skill acquisition aligned with competency-based education models.

The design of X-RAPT also emphasizes collaboration, a critical skill in modern industrial environments. Multi-user functionality enables team-based tasks, role differentiation, and peer learning, reflecting real-world workflows and promoting soft skills such as communication, coordination, and problem-solving. Moreover, X-RAPT's integration of real-time feedback and post-session analytics empowers instructors to track performance, diagnose misconceptions, and tailor instruction to individual learner needs. This adaptability contributes to more efficient learning curves and helps bridge the gap between theoretical knowledge and practical competence.

From a design perspective, the selection of the UR3 and ABB IRB120 robotic arms was intentional and pedagogically motivated. These two models are widely used in industrial and academic settings, making them highly representative of real-world applications. The UR3, as a collaborative robot, supports safe humanâĂŞrobot interaction and reflects modern trends in flexible automation. In contrast, the ABB IRB120 is a traditional industrial robot that requires safety barriers and precise programming, exposing learners to more rigid automation workflows. By simulating both types, the platform enables trainees to gain experience across a spectrum of robotic systems. This choice enhances the realism and applicability of training scenarios, ensuring that acquired skills are transferable. Furthermore, the modular system architecture and scenario authoring tools allow educators to create or adapt simulations involving these robots without programming expertise, broadening access and reducing adoption barriers in educational environments.

4.2 Limitations and Future Improvements

Despite its promising capabilities, the X-RAPT platform still faces several limitations that affect its accessibility, fidelity, and scalability. One of the primary challenges is the high hardware requirement for immersive VR experiences. Running the simulation smoothly demands dedicated GPUs and mid- to high-end PCs, which may not be readily available in all educational institutions. Although the system supports non-VR access via desktop mode, the overall performance still relies on machines with dedicated graphics cards, limiting deployment in lower-end or shared hardware environments.

Network infrastructure also presents a significant constraint, particularly in multi-user settings. For optimal synchronization and real-time interaction, the system depends on stable, high-speed internet—ideally via wired Ethernet connections. Wi-Fi setups often introduce latency and inconsistencies in object synchronization, which can disrupt the learning experience in collaborative training sessions.

To improve accessibility, future versions of X-RAPT may incorporate lighter, hybrid alternatives such as augmented reality (AR) support or enhanced desktop-based interfaces. These would enable broader adoption in resource-limited contexts and reduce technical barriers for institutions. At the same time, the current fidelity of robotic behavior in the simulation, while adequate for basic training, does not yet capture the full complexity of physical systems. Key aspects such as haptics, force interactions, and high-precision manipulation remain difficult to emulate convincingly in VR, suggesting the need for integration with more advanced simulation techniques.

From a usability standpoint, while the platform includes features such as hand tracking and voice command support, interaction design still requires refinement. Some users report difficulties with precision tasks or navigating interface menus. Future work will focus on improving gesture recognition, simplifying the UI, and developing onboarding procedures to reduce cognitive load for novice users.

Pedagogically, X-RAPT allows for scenario customization and basic performance tracking, but lacks adaptive learning features that can tailor experiences to individual needs. Enhancements such as intelligent tutoring systems, real-time feedback, and competency-based progression would make the platform more effective in supporting long-term skill acquisition. In this direction, future development cycles will also include formal evaluation studies with student populations to assess learning outcomes, usability, and transferability to real-world robotic tasks.

5 Conclusions

This paper presented the design and development of X-RAPT, a virtual reality training platform for industrial robotic programming. The system combines immersive simulation, multi-user collaboration, and scenario authoring tools to support hands-on learning in virtual environments. By integrating realistic robotic components, task-based workflows, and role-specific interfaces, X-RAPT enables students to engage with complex industrial processes in a safe and accessible way. The platform has been implemented using Unity and VIROO, with support for VR and desktop access modes, hand tracking, and performance logging.

Future work will focus on the empirical validation of the platform through structured user studies. Additionally, planned improvements include enhanced performance analytics, adaptive learning features, and integration with learning management systems (LMS). Technical efforts will also aim to improve accessibility and scalability, exploring lightweight client options and cloud-based rendering solutions to reduce hardware dependencies and expand adoption across institutions with limited infrastructure.

Acknowledgment. This work has been conducted as part of the X-RAPT (Immersive XR-Based Adaptive Training for Robotics Programming in Assembly and Packaging) project, funded by the European Union under the Horizon Europe programme [Project ID: 101093079]. This work has also been supported by the Valencian regional government CIAICO/2022/132 Consolidated group project AI4Health, and International Center for Aging Research ICAR funded project "IASISTEM". It has also been funded by a regional grants for PhD studies from the Valencian government, CIACIF/2021/430.

References

1. Adel, A.: The convergence of intelligent tutoring, robotics, and IoT in smart education for the transition from industry 4.0 to 5.0. Smart Cities **7**(1), 325 (2024). https://doi.org/10.3390/smartcities7010014
2. Andone, D., Frydenberg, M.: Creating virtual reality in a business and technology educational context. In: tom Dieck, M., Jung, T. (eds.) Augmented Reality and Virtual Reality. Progress in IS. Springer, Cham (2019). https://doi.org/10.1007/978-3-030-06246-0_11
3. Bauer, D., Hönig, P., Weibel, J.B., García-Rodríguez, J., Vincze, M., et al.: Challenges for monocular 6D object pose estimation in robotics. IEEE Trans. Rob. (2024)
4. Benotsmane, R., Trohak, A., Bartók, R., Mélypataki, G.: Transformative learning: nurturing novices to experts through 3D simulation and virtual reality in education. In: 2024 25th International Carpathian Control Conference (ICCC), pp. 1–6. IEEE, Krynica Zdrój, Poland (2024). https://doi.org/10.1109/ICCC62069.2024.10569390
5. Berki, B., Sudár, A.: Virtual and augmented reality in education based on CoginfoCom and cVR conference insights. In: 2024 IEEE 15th International Conference on Cognitive Infocommunications (CogInfoCom), pp. 000115–000120. IEEE, Hachioji, Tokyo, Japan (2024). https://doi.org/10.1109/CogInfoCom63007.2024.10894726
6. Bicalho, D.R., Piedade, J.M.N., de Lacerda Matos, J.F.: The use of immersive virtual reality in educational practices in higher education: a systematic review. In: 2023 International Symposium on Computers in Education (SIIE), pp. 1–5. IEEE, Setúbal, Portugal (2023). https://doi.org/10.1109/SIIE59826.2023.10423711
7. Chimere, N.C., Isaac, O.A., Nwankwo, W.: Integration of emerging technologies into computing education: benefits and challenges. IUP J. Inf. Technol. **21**(1), 21–48 (2025). https://doi.org/10.71329/IUPJIT/2025.21.1.21-48
8. Flowers, B.A., Rebensky, S.: Are you seeing what I'm seeing?: Perceptual issues with digital twins in virtual reality. In: 2022 IEEE Conference on Virtual Reality and 3D User Interfaces Abstracts and Workshops (VRW), pp. 126–130. IEEE, Christchurch, New Zealand (2022). https://doi.org/10.1109/VRW55335.2022.00037
9. Hein, J., et al.: Virtual reality for immersive education in orthopedic surgery digital twins. In: 2024 IEEE International Symposium on Mixed and Augmented Reality Adjunct (ISMAR-Adjunct), pp. 628–629. IEEE, Bellevue, WA, USA (2024). https://doi.org/10.1109/ISMAR-Adjunct64951.2024.00183
10. Jain, T., Desai, A., Kurna, S.: Methodology development for E-Axle NVH performance evaluation through virtual simulation. Technical report, SAE Technical Paper (2024)
11. Lind, A.: Multi-objective optimisation of a logistics area in the context of factory layout planning. Prod. Manuf. Res. **12**(1), 2323484 (2024)
12. Maddipatla, Y., Tian, S., Liang, X., Zheng, M., Li, B.: VR co-lab: a virtual reality platform for human–robot disassembly training and synthetic data generation. Machines **13**(3), 239 (2025). https://doi.org/10.3390/machines13030239
13. van der Meijden, O.A., Schijven, M.P.: The value of haptic feedback in conventional and robot-assisted minimal invasive surgery and virtual reality training: a current review. Surg. Endosc. **23**(6), 1180–1190 (2009). https://doi.org/10.1007/s00464-008-0298-x
14. Nava-Téllez, I.A., Elias-Espinosa, M.C., et al.: Digital twins and virtual reality as means for teaching industrial robotics: a case study. In: 2023 11th International

Conference on Information and Education Technology (ICIET), pp. 29–33. IEEE, Fujisawa, Japan (2023). https://doi.org/10.1109/ICIET56899.2023.10111361
15. Nikolaos, P., et al.: Physics-based tool usage simulations in VR. Multimodal Technol. Interact. **9**(4), 29 (2025). https://doi.org/10.3390/mti9040029
16. Orsolits, H., Rauh, S.F., Estrada, J.G.: Using mixed reality based digital twins for robotics education. In: 2022 IEEE International Symposium on Mixed and Augmented Reality Adjunct (ISMAR-Adjunct), pp. 56–59. IEEE, Singapore, Singapore (2022). https://doi.org/10.1109/ISMAR-Adjunct57072.2022.00021
17. Rajeswaran, P., Varghese, J., Kumar, P., Vozenilek, J., Kesavadas, T.: AirwayVR: virtual reality trainer for endotracheal intubation. In: 2019 IEEE Conference on Virtual Reality and 3D User Interfaces (VR), pp. 1345–1346. IEEE, Osaka, Japan (2019). https://doi.org/10.1109/VR.2019.8797998
18. Ricci, S., et al.: RiNeo MR: a mixed-reality tool for newborn life support training. In: 2021 43rd Annual International Conference of the IEEE Engineering in Medicine & Biology Society (EMBC), pp. 5043–5046. IEEE, Mexico (2021). https://doi.org/10.1109/EMBC46164.2021.9629612
19. Saval-Calvo, M., Azorin-Lopez, J., Fuster-Guillo, A., Garcia-Rodriguez, J.: Three-dimensional planar model estimation using multi-constraint knowledge based on K-means and RANSAC. Appl. Soft Comput. **34**, 572–586 (2015)
20. Serebryakov, M.Y., Moiseev, I.S.: Current trends in the development of cyber-physical interfaces linking virtual reality and physical system. In: 2022 Conference of Russian Young Researchers in Electrical and Electronic Engineering (ElConRus), pp. 419–424. IEEE, Saint Petersburg, Russian Federation (2022). https://doi.org/10.1109/ElConRus54750.2022.9755545
21. Talbi, K., Zergout, I., Souad, A., Aithaddouchane, Z.: Analytical study of virtual reality in Moroccan higher education. In: 2023 7th IEEE Congress on Information Science and Technology (CiSt), pp. 503–507. IEEE, Agadir - Essaouira, Morocco (2023). https://doi.org/10.1109/CiSt56084.2023.10409978
22. Tisza, J., Ortega, D.: Methodological proposal for the use of digital twins in remote engineering education. In: 2024 IEEE 4th International Conference on Advanced Learning Technologies on Education & Research (ICALTER), pp. 1–4. IEEE, Tarma, Peru (2024). https://doi.org/10.1109/ICALTER65499.2024.10819216
23. Van Mechelen, M., et al.: Emerging technologies in K–12 education: a future HCI research agenda. ACM Trans. Comput.-Hum. Interact. **30**(3), 40 (2023). Article 47. https://doi.org/10.1145/3569897
24. Viejo, D., Garcia-Rodriguez, J., Cazorla, M.: Combining visual features and growing neural gas networks for robotic 3D slam. Inf. Sci. **276**, 174–185 (2014)
25. Wang, P., et al.: An MR remote collaborative platform based on 3D cad models for training in industry. In: 2019 IEEE International Symposium on Mixed and Augmented Reality Adjunct (ISMAR-Adjunct), pp. 91–92. IEEE, Beijing, China (2019). https://doi.org/10.1109/ISMAR-Adjunct.2019.00038
26. Wegner, K., Seele, S., et al.: Comparison of two inventory design concepts in a collaborative virtual reality serious game. In: CHI PLAY 2017 Extended Abstracts, pp. 323–329. Association for Computing Machinery, New York, NY, USA (2017). https://doi.org/10.1145/3130859.3131300

Sentinel-1 Potential to Identify Landslides in Mountainous Regions

Tannia M. Mayorga-Torres[1,2], Jose García-Rodríguez[2(✉)], and Juan M. Lopez-Sanchez[3]

[1] Facultad de Ingeniería en Geología, Minas, Petróleos y Ambiental, Universidad Central del Ecuador, Gilberto Gatto Sobral y Jerónimo Leiton, Ciudadela Universitaria, 170521 Quito, Ecuador
tmmayorga@uce.edu.ec

[2] Department of Computer Science and Technology, University of Alicante, Carr. de San Vicente del Raspeig, 03690 San Vicente del Raspeig, Spain
tmt10@alu.ua.es, jgarcia@dtic.ua.es

[3] Signals, Systems and Telecommunications Group, Institute for Computer Research (IUII), University of Alicante, 03080 Alicante, Spain
juanma.lopez@ua.es

Abstract. Landslides are a ubiquitous natural hazard that can cause severe economic losses and human casualties, particularly when they occur near populated areas and critical infrastructure. The impacts can be devastating, with the potential to bury entire towns, rupture pipelines, block waterways, and render roads unusable. In this study, we propose a comprehensive, multi-pronged approach to landslide detection and prediction. This approach integrates the analysis of Interferometric Synthetic Aperture Radar (InSAR) data from the Sentinel-1 satellite with an improved landslide inventory, a detailed examination of failure mechanisms, and the identification of key causal factors. We further enhance the methodology by employing deep learning techniques to enable the early detection of landslides. Specifically, we leverage a Long Short-Term Memory (LSTM) model to predict the movement of unstable slopes, as LSTM models have demonstrated significant promise in this field.

Keywords: automatic detection · landslide · Sentinel-1 · InSAR

1 Introduction

Landslides have been shown to affect all regions, regardless of their geographic location. They cause high annual death tolls and significant costs due to damage to roads and infrastructure. A landslide can be identified by a susceptibility, hazard, and vulnerability assessment [30]. UNOSAT [26] asserts the crucial role of Earth observation data in disaster risk management. Synthetic aperture radar (SAR) sensors [8] emit microwave beams to the ground surface, thereby acquiring the physical and chemical properties of the elements on the ground, providing

E. Corchado et al. (Eds.): SOCO 2025, CCIS 2806, pp. 633–643, 2026.
https://doi.org/10.1007/978-3-032-19763-4_59

data during day and night. Their capacity to pass through clouds makes them especially suitable for cloudy regions, such as the tropics. The Sentinel-1 (S-1) satellites operate at C-band and are freely accessible. This work proposes using InSAR data and deep learning methods to identify landslide hazards.

InSAR involves the processing of at least two observations of the same target, made at different times during the satellite's flight path, to obtain an interferometric pair. InSAR is useful to generate Digital Elevation Models (DEM) [16] and allows for the measurement of ground subsidence (vertical displacements) or horizontal displacements over a time span, hence identifying areas with geological deformation events, such as post-seismic movements and volcanic eruptions [17]. Remote sensing facilitates the examination of past landslides. Westen et al. [30] point out the importance of building an up-to-date landslide inventory as soon as a seismic event occurs. This serves as an input for research that aims to detect early precursors of landslides in advance [10]. Landslides are triggered by seismic and rainfall events, which constitute failure mechanisms. Tomás et al. [24] emphasize the advantage of exploiting InSAR for the identification, monitoring, and early warning of landslides to comprehend the triggering factors involved in the slope stability processes. Environmental factors are expected to have an impact on landslide occurrence and thus constitute causal factors in their prediction. The data used for landslide identification and prediction are divided into two categories: dynamic datasets, such as land cover that change over time; and static datasets, such as the geology [30]. The identification strategy is based on the selection of specific environmental factors associated with the landslide type [7] and failure mechanisms in each particular environment.

It is necessary to establish a historical database of landslide and seismic events in Ecuador that encompasses the geographical location and occurrence date of such events, facilitating the identification and extent of the affected area. This will in turn support the actions of decision-makers. The determination of the rainfall threshold is a contributing factor in the prediction of landslide hazards [18,30]. The objective of this study is, by employing S-1 images and deep learning methods, to facilitate the autonomous delineation of landslide occurrence in mountainous regions. This contribution is essential in developing countries, where despite the presence of national and local risk institutions dedicated to prevent hazards, these occurrences take authorities unaware, forcing them to respond only in the aftermath of the event.

This paper is structured as follows. Section 2 presents a comprehensive literature review, summarizing the relationships between key factors influencing landslide occurrence and the current state-of-the-art automatic detection models. Section 3 outlines the proposed framework, detailing the methodological approach. Section 4 shows the discussion. Finally, Sect. 5 concludes the potential application of SAR data for enhanced landslide assessment, with a particular focus on mountainous regions.

2 Related Work

SAR facilitates landslide detection using amplitude differences [9,14], getting valuable evidence of changes in land cover. Landslide detection involved the use of the Log-Ratio index [20] by measuring pixel-based changes between successive pairs of S-1 intensity images, which were compared to previously measured values, before and after the landslide. The extraction of landslide occurrence from amplitude radar images is accomplished through the implementation of segmentation and classification procedures [9], such as the segmentation module included in the open source GRASS GIS software [19]. Tomás et al. [24] have demonstrated the InSAR capability to study landslides using different approaches in diverse case studies. A semi-automated identification of clusters of active persistent scatterers [25] was carried out, through their association with types of deformational processes in extensive regions. In the context of rugged mountains and valleys, small sample areas utilize Small Baseline Subset InSAR (SBAS-InSAR) and high-resolution optical imagery to identify landslides [34]. Significant factors influencing landslides are the terrain slope, rainfall, distance to roads, and geological rock formation [25,34]. In mountainous areas, landslide occurrences are closely associated with particular elevation conditions [25]. The construction of roads on slopes alters the support structure at the base of the slope, leading to the formation and expansion of cracks [34].

Four approaches employ the landslide hazard assessment [23], including the landslide inventory-based probabilistic, heuristic, statistical, and deterministic methods. The heuristic approach does not necessitate geotechnical datasets [3]. The classification of landslide risk assessment [15] is divided into three categories: qualitative, semi-quantitative, and quantitative. The qualitative approach (heuristic) is based on expert opinion, and the risk areas are classified as follows: very high, high, moderate, low, and very low [3]. The resulting map shows that landslides are more prevalent on mid-altitude slopes than on high- and low-altitude slopes [1]. The quantitative or data-driven models [27] have been utilised in regional landslide susceptibility assessments, e.g. Weight of Evidence (WofE) [13]; Frequency Ratio (FR) [32]; Logistic Regression (LR) [1,4,32]; Artificial Neural Network (ANN) [4,32]; Support Vector Machine (SVM) [31]; and integrated learning algorithm [28]. To evaluate landslide sensitivity, the following models [34] are used: machine-learning, Random Forest (RF), Gradient Boosting Decision Tree (GBDT), Categorical Boosting (CatBoost), LR, and stacking ensemble strategies. Combining empirical modelling with InSAR time series provides an assessment of the landslide volume and area [33].

Deep learning algorithms, characterized by more levels of nonlinear operations [27], can also be applied in this context, e.g. ANN and SVM to map landslide susceptibility based on statistical learning theory that entails a training dataset comprising associated input and target output values [31]. Slope, altitude, and rainfall are the primary factors in the formation of landslides [27]. A Convolutional Neural Network (CNN) algorithm demonstrated a 4% higher performance in comparison with the Deep Neural Network (DNN). CNN and DNN models have been utilised for the assessment of natural disaster susceptibility [21,29], combined with InSAR that identifies hidden danger points from landslides [27]. It is necessary to work with suitable scale layers. For instance, if lithology maps were available on a large scale, all scars would be located in the same lithologic unit, which would not improve training in this case. Zuo et al. [35] propose a Feature-Selecting Long Short-Term Memory (FS-LSTM) to refine landslide susceptibility mapping accuracy by integrating feature selection techniques with sequence-based modeling, identifying the influential factors: soil type, elevation, and average annual cumulated rainfall. CNN combined with PhLSTM (Peephole long short-term memory) [11] considers spatial and time series to predict the spatiotemporal evolution trend of surface deformation, where faults and lithology determine the deformation mechanism. The critical factors are the number of network layers and hidden layer units. The hyperparameters contain the learning rate and training times. Then, the length of the historical sequence is 5, the number of PhLSTM network layers is 3, the number of hidden layer cells are 256 and 128. In addition, [5] compared the LSTM neural network, RNN (Recurrent Neural Network) and MLP (Multi Layer Perceptron) models to predict land subsidence. Using the LSTM model, the number of neurons is 32, and the number of hiddel layers is 2.

3 Methodology

The methodology delineates landslide hazards through the utilisation of InSAR data and deep learning methods. This approach involves the application of congruent thresholds within the scale factor to differentiate ground surface characteristics from S-1 InSAR data. Table 1 shows categories and conditioning factors (classes) that contribute to the formation of landslides, which function as input layers during the learning process. It is imperative to comprehend the behaviour of the conditioning factors, distinguishing whether they are dynamic or static, and dependent or independent variables, to find correlations to be adjusted to different scenarios in the landslide hazard assessment.

The procedure is divided into six stages, as illustrated in Fig. 1. Stage 1 defines the centre point of the geographic coordinates used to process the InSAR time series images. Stage 2 determines the scale of input data, which influences the output predictions. Stage 3 involves the collecting and processing datasets listed in Table 1. Stage 4 prepares the landslide inventory, whose scale will determine the predicted output. Stage 5 applies the LSTM neural network [2] by setting up and training the model using time series InSAR data. The model is then tested for prediction. Stage 6 performs the validation process.

3.1 Stage 1 - Case Study

Ecuador is distinguished by its biodiversity, characterised by the presence of four distinct climatic regions and altitudes: the Coastal Region, the Sierra Region, the Amazon Region, and the volcanic archipelago located 1,000 km from continental Ecuador in the Pacific Ocean. Ecuador's diverse climatic zones offer

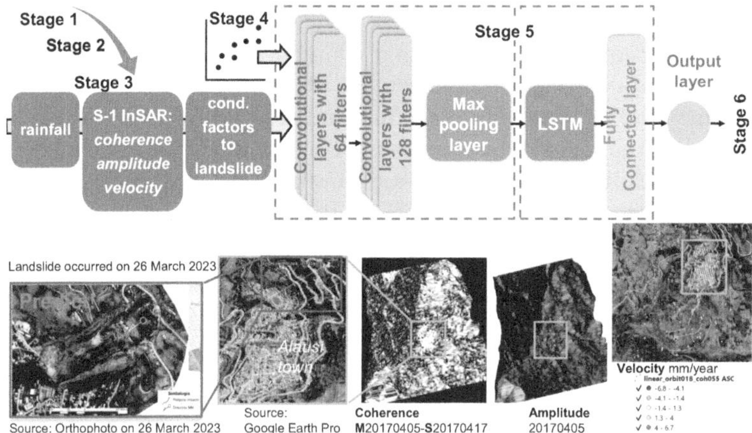

Fig. 1. General framework of the proposed algorithm with the LSTM model.

distinct advantages for agricultural and livestock farming activities. As with the geographical diversity of Ecuador, the social difficulties arise from the proximity of human settlements to recurrent occurrences of geological threats, including landslides, floods, earthquakes, volcanic eruptions, and droughts, which have inflicted substantial damage and losses in the past.

The city of Alausí (Lat. 2° 11′ 39.36″S, Long. 78° 50′ 36.04″W, Alt. 2,480 m.a.s.l.) is the capital of Alausí canton in the Chimborazo province, Andean region. Its population is 37,275, within an area of 1,707 Km^2. The Casual-Nuevo Alausí landslide occurred on 26 March 2023 at 21:30, resulting in the covering of houses and the burial of entire families. The National Secretary of Risk Management (SNGR) [22] reported a total of 65 fatalities, and 10 individuals were reported as missing. Several road sections were destroyed: 0.427 km of the railway line, 1.12 km of a national highway E35, and 1.2 km of a second-order pathway. Basic services collapsed, and the stadium and the coliseum were destroyed.

3.2 Stage 2 - Input Scale

The scale is upon the availability of data within the case study. The geographical coordinates (X, Y, Z) of the centre point facilitate the S-1 satellite data search. The proposal's objective is to facilitate the automatic detection of landslide components, which are the following: crown, crown cracks, scarp, and main body. Subsequently, the input data will be aligned with the specified output scale.

3.3 Stage 3 - Datasets

The following datasets are collected and processed: rainfall, S-1 InSAR satellite imagery, and conditioning factors for the landslide. Regarding rainfall, experts consider that a period of six years constitutes reliable information on which to base research, for which the availability of a sufficient number of landslide occurrence dates is essential [30]. The second component is the InSAR time series. The processed data correspond to coherence (0–1), amplitude (sigma0_dB), and velocity (mm/year) data, in both ascending and descending orbits. Table 1, based on [6,35], shows the third dataset with the conditioning factors, which are represented by the following categorisation of variables: dynamic or static, and dependent or independent. The objective of this categorisation is to ascertain the existence of a correlation among them.

Table 1. Analysis of input variables.

Category	Cond. factors	Explanation	Dynamic	Static	Dep.	Indep.
Topography	Slope angle	Slope steepness classes		x		x
	Slope aspect	Slope direction classes		x		x
Geology	Lithology			x		x
	Geol. rock formation			x		x
	Laboratory test			x		x
Hydrology	Drainage density	Distance from rivers		x	x	
	Watersheds			x		x
Geomorphology	Unit	Geological units		x		x
DEM	Altitude	Elevation classes		x		x
	Profile curvature slope	Curvature		x		x
Tectonic	Seismic event		x		x	
	Active fault		x		x	
Land use	Soil map	Soil type	x		x	
Land cover	Road	Distance from roads	x		x	
	Infrastruct.	Distance from infrastruct.	x		x	
Moisture content	TWI index		x		x	
Pluviosity	Rainfall	Daily precipitation	x		x	
Landslide	Scarp	Distance from ridges	x		x	

Datasets of Dynamic and Dependent Variables. *Road*: Impact of road construction may loosen material or alter the slope structure. *TWI*: Influence of topography on moisture accumulation, showing areas of spatial distribution of surface water. *Rainfall*: Higher rainfall increases both surface and subsurface moisture, reducing soil stability. *Scar*: The distance from ridges.

Datasets of Static and Dependent Variables. *Drainage density*: Distance map from rivers influences the erosional effects on slope stability.

Datasets of Static and Independent Variables. *Slope angle*: It is extracted from DEM. There is a clear relationship between slope steepness and landslide occurrence. *Slope aspect*: It is extracted from DEM. To use slope direction from the DEM, the relation is less clear. *Aspect* represents the direction of slope orientation and determines the extent of solar radiation received, related to moisture distribution and vegetation cover, which impact landslide risk. *Lithology*: It is extracted from the geology map and determines the resistance of rock and soil to weathering and erosion. With loose rock layers, the soil is likely to become unstable under external forces. *Geological units*: The relationship between landslides and the geological units is rather complex. To combine the geological map with slope classes to get the resulting contrast factors and avoid illogical combinations. The positive contrast factor for this geological unit for slopes between 0–10° is a good indication of the poor quality of the DEM, and the general low level of quality of this important data layer. *Altitude*: To analyze landslide distribution related to higher elevations, greater topographic relief, and erosion processes. There is a good correlation between the altitude and rainfall. *Profile curvature slope*: Curvature describes the concavity of the land surface, which affects water flow direction and catchment areas.

3.4 Stage 4 - Landslide Inventory

The known landslides inventory is the input for the training and testing procedures. The proposal involves the use of deep learning to generate a model trained with an up-to-date landslide inventory [35]. Unfortunately, there is no official landslide inventory produced by public institutions. The landslide inventory is based on point data, representing their geographic coordinates.

3.5 Stage 5 - LSTM

A Long Short-Term Memory (LSTM) is an improved version of the RNN [12] capable of learning long-term dependencies, especially in sequence prediction problems. Traditional RNN use a single hidden state passed through time. However, it faces difficulties in learning long-term dependencies from distant time steps, because the gradients (which help the model learn) can shrink passing through many steps. Another difficulty is that sometimes gradients can grow too large, causing instability. LSTM are used to learn over long-term dependencies in classified sequential data, making it useful for time series forecasting. It is suitable for this work because there are time series of deformation derived from S-1. To detect the difference in velocity, as an indicator of a landslide, the model will have as input the interferometric pairs, moving forward to the next date.

LSTM introduces a memory cell that has a long memory. Considering the cell as the centre point, the input data, such as the interferograms, the rainfall and the distinguished conditioning factors, are retained by the cells, and three gates do the memory manipulations. The memory cell is controlled by these gates: the input gate (add information to the cell), the forget gate (remove information from the cell), and the output gate. This allows LSTM networks to particularly retain or discard information in the long-term dependencies learning process. The hidden state corresponds to a kind of short-term memory, which is updated using the current input, the previous hidden state, and the current state of the memory cell. Amplitude, coherence, and velocity information from S-1 will be processed separately in the model. We start to analyze amplitude since it contains the backscatter signal of ground surface characteristics.

3.6 Stage 6: Validation

The output will contain old and recent fissures, escarpments, and body landslides, where the location of landslides constitutes a contribution to be corroborated in the field. The characterization is made on the field with geologists, with interpretation through aerial photographs, and by drones. Besides, information related to the landslide type and activity, e.g. active, dormant, stabilized, and undetermined, is collected in the field.

4 Discussion

LSTM neural network prediction is based on known time series subsidence, as the InSAR surface deformation, which are the basis of the predictive processes where the ground surface characteristics are distinguished. The current study propose several layers (Table 1) which will be tested in further research.

5 Conclusions

The increasing occurrence of geological hazards, driven by climate change and unusual weather patterns, has heightened the risk of landslides, even in urban areas. To mitigate casualties and infrastructure damage, it is essential to implement automatic detection systems as a proactive tool for hazard prevention. This study proposes a method tailored for national and local public institutions engaged in risk management. It combines a range of conditioning factors (as detailed in Table 1), InSAR data—particularly valuable in tropical regions—and deep learning techniques to deliver accurate and sensitive assessments of potential landslides. These assessments can then be validated through targeted field inspections. The precision of the results depends significantly on the scale factor of the input layers. Therefore, it is crucial for public institutions to provide up-to-date and high-quality data to get predictive models.

Acknowledgements. We thank the Universidad Central del Ecuador, under the project DOCT-DI-2023-10 "Inventario en tiempo real de la ocurrencia de movimientos en masa, limitados por clusters", and the University of Alicante in Spain for the opportunity to carry out a research stay.

References

1. Ayalew, L., Yamagishi, H.: The application of GIS-based logistic regression for landslide susceptibility mapping in the Kakuda-Yahiko mountains. Central Japan. Geomorphol. **65**(1), 15–31 (2005). https://doi.org/10.1016/j.geomorph.2004.06.010
2. Calzone, O.: An intuitive explanation of LSTM. Technical report, Medium (2022). medium.com/@ottaviocalzone/an-intuitive-explanation-of-lstm-a035eb6ab42c
3. Castellanos, E., Westen, C.J.V.: Qualitative landslide susceptibility assessment by multicriteria analysis: a case study from San Antonio del Sur, Guantánamo. Cuba. Geomorphol. **94**(3), 453–466 (2008). https://doi.org/10.1016/j.geomorph.2006.10.038
4. Chen, W., Pourghasemi, H., et al.: A GIS-based comparative study of Dempster-Shafer, logistic regression, and artificial neural network models for landslide susceptibility mapping. Geocarto Int. **32**, 367–385 (2016). https://doi.org/10.1080/10106049.2016.1140824
5. Chen, Y., He, Y., Zhang, L., et al.: Prediction of InSAR deformation time-series using a long short-term memory neural network. Int. J. Remote Sens. **42**(18), 6919–6942 (2021). https://doi.org/10.1080/01431161.2021.1947540
6. Van Westen, C.J.: National scale landslide susceptibility assessment for Dominica (2016). https://doi.org/10.13140/RG.2.1.4313.2400
7. Cruden, D.M., Varnes, D.J.: Landslide types and processes. Spec. Rep. Nat. Res. Counc. Transp. Res. Board **247**, 36–57 (1996)
8. Earthdata: NASA SAR Handbook. Synthetic Aperture Radar (SAR), NASA (2025). www.earthdata.nasa.gov/learn/earth-observation-data-basics/sar
9. Esposito, G., Mondini, A.C., Marchesini, I., Reichenbach, P., Salvati, P., Rossi, M.: An example of SAR-derived image segmentation for landslides detection. PeerJ Prepr. **6**, e27212 (2018). https://doi.org/10.7287/peerj.preprints.27212v2
10. Guerrero, B., Salvador, J., Garcia, J., Mejia, C.: Improving landslides prediction: meteorological data preprocessing based on supervised and unsupervised learning. Cybern. Syst. **55**(6), 1332–1356 (2024). https://doi.org/10.1080/01969722.2023.2240647
11. He, Y., Yan, H., Yang, W., Yao, S., et al.: Time-series analysis and prediction of surface deformation in the Jinchuan Mining area, Gansu Province, by using InSAR and CNN–PhLSTM network. IEEE J. Sel. Top. Appl. Earth Observ. Remote Sens. **15**, 6732–6751 (2022). https://doi.org/10.1109/JSTARS.2022.3198728
12. Hochreiter, S., Schmidhuber, J.: Long short-term memory. Neural Comput. **9**(8), 1735–1780 (1997). https://doi.org/10.1162/neco.1997.9.8.1735
13. Ilia, I., Tsangaratos, P.: Applying weight of evidence method and sensitivity analysis to produce a landslide susceptibility map. Landslides **13**(2), 379–397 (2016). https://doi.org/10.1007/s10346-015-0576-3
14. Konishi, T., Suga, Y.: Landslide detection using COSMO-SkyMed images: a case study of a landslide event on Kii Peninsula. Jpn. Eur. J. Remote Sens. **51**, 205–221 (2018). https://doi.org/10.1080/22797254.2017.1418185
15. Lee, E.M., Jones, D.K.C: Landslide risk assessment. Q. J. Eng. Geol. Hydrogeol. **39** (2004). https://doi.org/10.1144/1470-9236/05-102
16. Lu, Z., Jung, H.S., Zhang, L., et al.: Digital elevation model generation from satellite interferometric synthetic aperture radar: Chapter 5. In: Advances in Mapping from Aerospace Imagery, pp. 119–144 (2012)

17. Lu, Z., Kwoun, O., Rykhus, R.: Interferometric synthetic aperture radar (InSAR): its past, present and future. Photogramm. Eng. Remote Sens. **73**(3), 217–221 (2007). https://pubs.usgs.gov/publication/70157055
18. Mirus, B., Morphew, M., Smith, J.: Developing hydro-meteorological thresholds for shallow landslide initiation and early warning. Water **10**(9), 1274 (2018). https://doi.org/10.3390/w10091274
19. Momsen, E., Metz, M.: i.segment - identifies segments (objects) from imagery data (2017). grass.osgeo.org/grass-stable/manuals/i.segment.html
20. Mondini, A.C.: Measures of spatial autocorrelation changes in multitemporal SAR images for event landslides detection. Remote Sens. **9**, 554 (2017). https://doi.org/10.3390/rs9060554
21. Panahi, M., Jaafari, A., Shirzadi, A., et al.: Deep learning neural networks for spatially explicit prediction of flash flood probability. Geosci. Front. **12**(3), 101076 (2021). https://doi.org/10.1016/j.gsf.2020.09.007
22. SNGR: Informe de situac. nacional SitRep N. 97 - Deslizam. Casual-Alausí. Situación Nacional, Secretaría Nacional de Gestión de Riesgos, Alausí (2023). www.gestionderiesgos.gob.ec/informes-de-situacion-actual-por-eventos-adversos-ecuador/
23. Soeters, R., van Westen, C.J.: Slope instability recognition, analysis, and zonation. Landslides investigat. Mitigat. Transport. Res. Board Spec. Rep. **247**, 129–177 (1996). onlinepubs.trb.org/Onlinepubs/sr/sr247/sr247-008.pdf
24. Tomás, R., Zeng, Q., Lopez-Sanchez, J., et al.: Advances on the investigation of landslides by space-borne synthetic aperture radar interferometry. Geo-Spat. Inf. Sci. **27**(3), 602–623 (2024). https://doi.org/10.1080/10095020.2023.2266224
25. Tomás, R., et al.: Semi-automatic identification and pre-screening of geological-geotechnical deformational processes using persistent scatterer interferometry datasets. Remote Sens. **11**(14), 1675 (2019). https://doi.org/10.3390/rs11141675
26. UNOSAT: UNOSAT's Empowerment in Disaster Risk Management: Navigating Geospatial Technologies for Resilience in Lao PDR (2024). https://goo.su/U7gEHek
27. Wang, S., Zhuang, J., Mu, J., et al.: Evaluation of landslide susceptibility of the Ya'an–Linzhi section of the Sichuan-Tibet railway based on deep learning. Environ. Earth Sci. **81**(9), 250 (2022). https://doi.org/10.1007/s12665-022-10375-z
28. Wang, S., Zhuang, J., Zheng, J., et al.: Application of Bayesian hyperparameter optimized random forest and XGBoost model for landslide susceptibility mapping. Front. Earth Sci. **9** (2021). https://doi.org/10.3389/feart.2021.712240
29. Wang, Y., Fang, Z., Wang, M., et al.: Comparative study of landslide susceptibility mapping with different recurrent neural networks. Comput. Geosci. **138**, 104445 (2020). https://doi.org/10.1016/j.cageo.2020.104445
30. Westen, C.J.V., Castellanos, E., Kuriakose, S.L.: Spatial data for landslide susceptibility, hazard, and vulnerability assessment: an overview. Eng. Geol. **102**(3), 112–131 (2008). https://doi.org/10.1016/j.enggeo.2008.03.010
31. Yao, X., Tham, L.G., Dai, F.C.: Landslide susceptibility mapping based on support vector machine: a case study on natural slopes of Hong Kong. China. Geomorphol. **101**(4), 572–582 (2008). https://doi.org/10.1016/j.geomorph.2008.02.011
32. Yilmaz, I.: Landslide susceptibility mapping using frequency ratio, logistic regression, artificial neural networks and their comparison: a case study from Kat landslides (Tokat-Turkey). Comput. Geosci. **35**(6), 1125–1138 (2009). https://doi.org/10.1016/j.cageo.2008.08.007

33. Zhang, Y., Meng, X., Dijkstra, T., et al.: Forecasting the magnitude of potential landslides based on InSAR techniques. Remote Sens. Env. **241**, 111738 (2020). https://doi.org/10.1016/j.rse.2020.111738
34. Zhao, Z., Li, Z., Lv, P., et al.: The study on landslide hazards based on multi-source data and GMLCM approach. Rem. Sens. **17**(9) (2025). https://doi.org/10.3390/rs17091634
35. Zuo, P., Zhao, W., Yan, W., et al.: Landslide susceptibility mapping using an LSTM model with feature-selecting for the Yangtze river basin in China. Water **17**(2) (2025). https://doi.org/10.3390/w17020167

A Hybrid CNN-LSTM Model for Hyperspectral Image-Based Detection of Hydrocarbon and Clean Water

María Gema Carrasco-García[1(✉)], María Inmaculada Rodríguez-García[2], Javier González-Enrique[2], Juan Jesús Ruiz-Aguilar[1], Paloma Rocío Cubillas-Fernández[3], and Ignacio José Turias-Domínguez[2]

[1] Department of Industrial and Civil Engineering, School of Engineering, University of Cádiz, Ramón Puyol Av., 11202 Algeciras, Spain
maria.carrasco@uca.es

[2] Department of Computer Science Engineering, School of Engineering, University of Cádiz, Ramón Puyol Av., 11202 Algeciras, Spain

[3] Department of Thermal Machines and Motors, School of Engineering, University of Cádiz, Ramón Puyol Av., 11202 Algeciras, Spain

Abstract. This work presents a comparative analysis of deep learning architectures for hyperspectral image (hyperimage) classification, focusing on the discrimination of water-hydrocarbon mixtures. Two main architectures were evaluated: 3D Convolutional Neural Networks (3DCNNs) and hybrid 2DCNN–LSTM models, using different convolutional kernel sizes and dataset volumes. The study aimed to assess the performance, robustness, and computational efficiency of these models under varying data availability. Experimental results reveal that both architectures achieve high classification accuracy with large datasets, often surpassing 0.99 in overall accuracy (OA). However, hybrid models demonstrated superior robustness in data-scarce scenarios, outperforming 3DCNNs in stability and class-wise recall, particularly when classifying complex hydrocarbon mixtures. The hybrid models also proved computationally more efficient: they achieved comparable or better accuracy than 3DCNNs with significantly reduced training times, up to seven times faster under certain configurations. These findings highlight the potential of lightweight 2DCNN–LSTM hybrids and scalable solutions for real-time hyperspectral classification tasks, particularly when training data is limited or rapid retraining is required. The results further suggest that careful tuning of kernel size and recurrent unit complexity is critical to balancing performance and computational cost in practical applications.

Keywords: Hyperspectral image classification (HSI) · deep learning · 3DCNNs · 2DCNN–LSTM hybrid models

1 Introduction

The detection of hydrocarbon pollution in aquatic environments is a critical task in environmental monitoring and protection. Accidental spills, industrial discharges, and maritime activities contribute to hydrocarbon contamination, resulting in severe ecological

E. Corchado et al. (Eds.): SOCO 2025, CCIS 2806, pp. 644–654, 2026.
https://doi.org/10.1007/978-3-032-19763-4_60

damage and health risks to humans and wildlife [1]. With its ability to capture hundreds of contiguous spectral bands for each image pixel, HSI provides high-resolution spectral information that can distinguish between different water surface compositions, making it a powerful tool for pollution detection [2].

The high dimensionality and complexity of HSI data have positioned deep learning models as the most suitable approach for processing this type of data [3]. These models are not only capable of handling large volumes of information and extracting relevant spatial and spectral features but also excel at capturing intricate spatial-spectral relationships within the data [4]. This ability is particularly relevant when working with image data, where neighboring pixels often represent the same material or object, and preserving spatial coherence is essential for accurate classification. By learning both spectral signatures and their spatial context, deep learning models can provide robust and consistent predictions across complex and heterogeneous scenes.

In this context, 3D-CNNs [5] have been widely adopted for HSI analysis, particularly in applications such as environmental monitoring, mineral mapping, and agricultural assessment [6, 7]. The application of convolutional kernels simultaneously across all three dimensions of the hyperspectral data cube– two spatial and one spectral– allows the joint extraction of spatial and spectral features while preserving intrinsic spatio-spectral relationships. The combination of efficient data dimensionality reduction and hierarchical extraction of complex patterns within the hyperspectral cube enables a more complete understanding of the scene, improving classification accuracy and leading to greater robustness. This advantage has also been applied to solve the problem of identifying polluted water surfaces from hyperspectral remote sensing. For instance, Yang et al. [8] employed a CNN architecture for oil detection using multi-scale feature extraction, achieving classification accuracies above 85%. However, despite these promising results, the application of 3DCNNs to hyperspectral data presents specific challenges. On the one hand, the high dimensions associated with hyperspectral data often require the use of large kernel sizes along the spectral dimension to maintain controllable model complexity. This, in turn, can lead to the loss of fine spectral detail, information that can be critical for accurately detecting subtle or low-concentration hydrocarbon signatures. On the one hand, these deep learning models are computationally intensive, requiring considerable processing power and memory, and often resulting in long training and inference times. This characteristic represents a disadvantage in scenarios such as real oil spills, where the monitoring systems must often be retrained with new data, as each spill can present different spectral features depending on the type of hydrocarbon, environmental conditions, and lighting. This constant need for model updating imposes additional constraints on computational efficiency and adaptability, especially in incremental learning or real-time processing contexts.

Leveraging the strengths of convolutional models, recent trends explored the integration of CNNs with recurrent neural networks (RNNs) to further improve performance in hyperspectral image analysis [9]. This hybrid approach exploits the ability of CNNs to extract local spatial features while incorporating the ability of RNNs to model sequential dependencies along the spectral dimension. In hyperspectral data, where spectral signatures often exhibit high internal correlation and significant patterns may span multiple bands, the recurrent component facilitates learning spectral continuity and long-range relationships. Among the different types of RNNS, LSTM networks [10] have gained particular attention in this context. LSTMs are specifically designed to also learn short-long-term dependencies in sequential data. Their ability to retain and selectively update information over long sequences makes them especially suitable for modeling the spectral dimension of hyperspectral data. When integrating into hybrid CNN-LSTM architectures, LSTMs contribute to a more accurate representation of spectral continuity across spectral bands, especially in complex or noisy scenes. Several works have demonstrated the effectiveness of these hybrid CNN-LSTM architectures. For example, Hu et al. [11] proposed a spatial-spectral deep learning model that achieved remarkable improvements in classification accuracy for remote sensing applications, reporting overall accuracies (OA) greater than 95%. Using the same dataset, Tulapurkar et al. [12] further increased the OA to over 99% by incorporating a ConvLSTM architecture. Similarly, Farooque et al. [13] introduced a residual network based on ConvLSTM-CNN 3D spectral-spatial-CNN (SSCR), also achieving OA values above 99% on the same benchmark datasets. These results highlight the effectiveness of hybrid architecture in exploiting spatial and spectral dependencies to improve hyperspectral image classification.

Despite recent advances, the use of hybrid architectures has hardly been explored in the specific context of hydrocarbon contamination detection on aquatic surfaces. This study addresses this lack by proposing the application of a hybrid 2DCNN-LSTM architecture for the classification of hyperspectral images of water surfaces, distinguishing between uncontaminated water and water polluted with three common hydrocarbon types associated with maritime and bunkering activities: gasoil, C10, and fuel oil. The proposed architecture aims to develop an accurate and computationally efficient model for oil spills. Such characteristics are essential for HSI monitoring systems that must adapt to evolving scenarios through periodic retraining with newly acquired data.

In this work, the impact of key design parameters on model performance was systematically evaluated, including the number of neurons in the LSTM layer, which operates along the spectral dimension, the kernel size in the CNN layers for spatial feature extraction, and the size of the training dataset. Classification performance was assed using standard accuracy metrics, and results were compared to those obtained using analogous 3D-CNN architectures. In addition, computational time was also analysed with the aim of highlighting the suitability of hybrid 2DCNN-LSTM models over 3D-CNN for scenarios involving incremental learning or limited computational resources, such as continuous oil spill detection systems. This study contributes to the advancement of efficient and scalable solutions for environmental monitoring based on hyperspectral technology. Section 2 describes the methodology, including the models and the experimental procedure. Section 3 presents the main results and their discussion, while the conclusions of the research are summarised in Sect. 4.

2 Methodology

This section describes the deep learning architectures used in the classification task and the experimental procedure followed to evaluate their performance. Deep learning models have shown remarkable performance in processing hyperspectral data due to their ability to automatically extract meaningful spatial and spectral patterns. This capability greatly facilitates classification tasks. Among these models, CNNs and LSTM networks have emerged as particularly effective components [3].

Convolutional Neural Networks (CNNs). CNNs have demonstrated remarkable efficacy in a panoply of computer vision tasks, including image classification, attributable to their inherent ability to autonomously learn hierarchical spatial features from raw pixel data [4]. Within the domain of HSI, both 2D and 3D CNN architectures have been employed to exploit the spatial and spectral attributes of the data [6]. Two-dimensional CNNs typically treat each spectral band as a discrete channel, extracting spatial features independently across bands. In contrast, three-dimensional CNNs process the hyperspectral cube holistically, treating the spectral dimension as an additional spatial axis. This allows them to learn joint spatial-spectral features and capture more complex interdependencies within the data [9]. In both approaches, convolutional layers act as a feature extractor, generating feature maps. When stacked in multiple layers, CNNs can hierarchically learn increasingly abstract spatial features. In the case of 3DCNNs, the kernel spans not only the spatial dimensions but also the spectral dimension, allowing for joint spatial-spectral feature extraction. This increases model capacity but also significantly raises the computational cost.

Long Short-Term Memory (LSTM). LSTMs, a specialized variant of RNNs, are particularly adept at processing sequential data by maintaining a temporal memory of prior inputs [10]. In HSI, the spectral signature of each pixel or patch can be interpreted as a sequence, where each band represents a time step. LSTMs were specifically designed to overcome the vanishing and exploding gradient problems that affect the training of traditional RNNs through backpropagation through time. The mathematical formulation of LSTM units can be found in Hochreiter et al. [10]. LSTMs possess the capacity of modeling long-range dependencies in the spectral domain, allowing them to detect subtle variations that might indicate differences in water quality/contamination levels.

Convolutional Neural Network - Long Short-Term Memory (CNN-LSTM). The integration of CNNs and LSTMs has emerged as a promising paradigm for spectral analysis in HSI classification. While 2DCNNs are effective at extracting spatial features, they do not perform any reductions or modeling of the spectral dimension. This limitation is addressed by 3DCNNs, which can jointly capture spatial and spectral information. However, due to the high dimensionality of HSI data, large kernel sizes along the spectral axis are required, especially in the early layers, to maintain manageable computational performance. This can lead to a loss of fine spectral details, potentially affecting classification performance. In contrast, LSTM networks are well suited for modelling spectral dependencies, but if no other operation is performed beforehand and has to be applied directly to the raw data, it can be computationally expensive. The synergistic combination of 2DCNNs and LSTMs provides an efficient and powerful alternative. The CNN layers first extract spatial features, and then the resulting feature maps are flattened into sequences to feed the LSTM layers. The LSTM network then models spectral relationships by treating the data as a sequence of spectral observations.

Model Configurations. The study involves two different deep learning architectures for their implementation and evaluation: a hybrid 2DCNN-LSTM model, which constitutes the core of this study, and a conventional 3DCNN model, used for comparative purposes.

Both approaches aim to classify hyperspectral images into four classes: clean water and polluted water with one of three hydrocarbons: gasoil, C10, and fuel oil.

The hybrid model consists of two main components: a convolutional block for spatial feature extraction and a recurrent block for spectral sequence modelling. The convolutional block is composed of three stacked 2D convolutional layers, each followed by a ReLU activation function to introduce non-linearity and enhance the learning capacity. After the third convolutional layer, a max pooling layer with a kernel size of 3 is applied to reduce spatial dimensions and improve computational efficiency while retaining the most salient spatial features. All convolutional layers use the same kernel size in both dimensions, but differ in the number of filters, which increase progressively across layers: 8, 16, and 32, respectively. The resulting spatial feature maps are then reshaped into a sequence by a flatten layer to prepare the data for the recurrent block. This sequence is then passed to an LSTM layer, which models the spectral dependencies by capturing relationships across contiguous spectral bands. Finally, a fully connected layer computes the class probabilities through a softmax activation function.

The conventional 3DCNN model follows a similar layer-wise structure to the hybrid model but replaces the LSTM layer with additional spectral feature extraction using 3D convolutions. It consists of three 3D convolutional layers, each followed by a ReLu activation function. A 3D max pooling layer with a kernel size of 3 along all dimensions is also applied after the third convolutional layer. Regarding the spatial kernels, the strategy used in the hybrid is maintained. In contrast, the spectral kernel sizes are fixed and set to 130, 20, and 15 for the first, second, and third convolutional layers, respectively. This design allows the network to progressively abstract spectral information in a hierarchical manner. As in the hybrid model, the output is passed to a fully connected layer with softmax activation, producing the class probabilities.

Experimental Procedure. To evaluate the proposed architectures, a hyperspectral image dataset was specifically created for this study (Fig. 1). Four types of samples were prepared under controlled conditions: clean water, water mixed with gasoil, water mixed with C10, and water mixed with fuel oil. These hydrocarbons were selected because of their presence in maritime traffic and bunkering operations. The hyperimages were acquired using the Cubert Ultris X20 Plus hyperspectral camera, a compact and portable device suitable for laboratory and field applications. A total of 80 hyperspectral images were acquired from these samples (20 per class). Each hyperimage was segmented into non-overlapping patches of size 25×25 pixels, resulting in a balanced dataset of 6480 patches, evenly distributed across the four classes. Instead of performing standard percentage-based train-test splits over the full dataset, we defined three independent and balanced subsets of 512, 1024, and 2048 samples, created by random sampling from the complete dataset. These subsets were designed to simulate different data availability scenarios and to evaluate model robustness under limited data conditions. To ensure robust and unbiased evaluation, each dataset was randomly shuffled into a double cross-validation procedure. In each repetition, the selected subset was split into training and test sets according to the cross-validation protocol, ensuring class balance and reproducibility across runs. For each dataset size, Table 1 lists the hyperparameter configurations evaluated during the experiments. A total of three configurations were tested for the 3DCNN approach, and twelve for the hybrid 2DCNN-LSTM model.

This design enabled us to investigate how varying levels of recurrent complexity impact spectral modeling performance. Each configuration was trained twenty times to evaluate model stability and reduce random effects. Performance was assess using several metrics: (i) average accuracy (AA), defined as the mean accuracy across the twenty runs, (ii) overall accuracy (OA), defined as the highest accuracy achieve in the twenty runs, (iii) the mean recall per class, to assess classification performance at the class level, and (iv) average training time, to analyse computational efficiency.

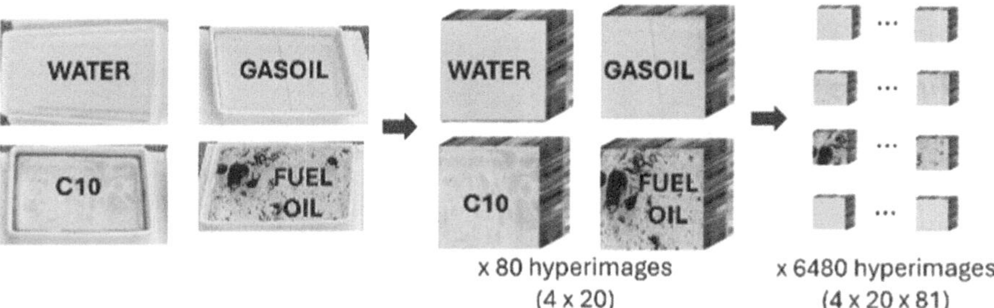

Fig. 1. Workflow of the dataset generation. The original hyperimage was divided into patches of 25 × 25 pixels, labelled according to its corresponding class: water, gasoil, C10, fuel oil.

Table 1. Hyperparameter configurations evaluated for each data size, including kernel sizes and LSTM units (only for the hybrid model).

Architecture	Kernel Size	LSTM unit
2DCNN-LSTM	[2,2] - [2,2] - [2,2] [3,3] - [3,3] - [3,3] [5,5] - [5,5] - [5,5]	32; 64; 128; 256
3DCNN	[2,2, 130] - [2,2, 20] - [2,2, 15] [3,3, 130] - [3,3, 20] [3,3, 15] [5,5, 130] - [5,5, 20] - [5,5, 15]	-

3 Results and Discussion

The objective was not only to assess the overall classification performance but also to analyse computational efficiency and the suitability of these models for real-time or incremental learning scenarios.

Figure 2 summarizes average and overall accuracy results across kernel sizes (2, 3, and 5), showing consistently high classification performance for both architectures. Most models reached OA values above 0.99, even in hybrid configurations with smaller LSTM units. With 2048 training samples, all setups achieved near-perfect performance with minimal variance (std < 0.05), indicating robust behavior. However, the trade-off between convolutional kernel size and LSTM capacity was evident: smaller kernels required fewer LSTM units (e.g., 2DCNN[2,2]-LSTM(32)), while larger kernels needed more (e.g., 2DCNN[5,5]-LSTM(256)). With only 512 samples, performance differences increased—3DCNN models performed poorly with larger kernels due to limited generalization, while hybrid models maintained stability and high AO values. For medium-sized datasets (1024 samples), all models achieved AO above 0.90, but 3DCNN[5,5] showed greater variability, unlike more stable 2DCNN-LSTM setups.

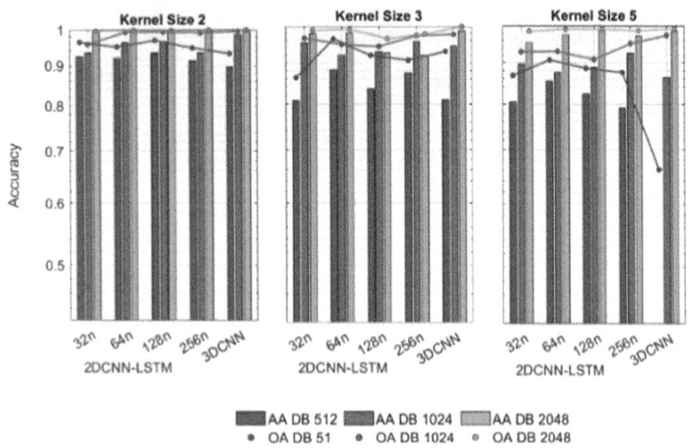

Fig. 2. OA (highest accuracy) and AA (average accuracy) for all configurations using different kernel sizes (2, 3, and 5) and dataset sizes (512, 1024, 2048).

Figures 3, 4 and 5 show recall by class across kernels and models. Detection of clean water (class 1) was near perfect in most cases, except for 3DCNN with kernel size 5 and only 512 samples. Hybrid models, including lightweight ones like 2DCNN[5,5]-LSTM(32), reached 0.99 recall. Greater recall variability appeared in classes 3 and 4 (hydrocarbon mixtures), which were harder to classify. With sufficient training data, all models performed well across classes, even with simplified hybrid configurations (e.g., 64 LSTM units). However, with only 512 samples, 3DCNN required small kernels to maintain a recall near 1.0 and still struggled with complex mixtures. Surprisingly, increasing LSTM size in low-data conditions did not guarantee improvements, highlighting the need for a careful balance between convolutional abstraction and recurrent complexity.

On the other hand, Fig. 6 compares training time (log-scale), highlighting the computational advantage of hybrid 2DCNN-LSTM models. These models required significantly less time than 3DCNNs while maintaining similar or superior accuracy. For instance, 3DCNN[2,2,2] trained on 1024 samples took 794 s to reach OA = 0.99, while 2DCNN[2,2]-LSTM(128) achieved the same OA in just 121 s—about seven times faster. This efficiency makes hybrid models more suitable for practical applications requiring fast or iterative training, such as incremental learning scenarios.

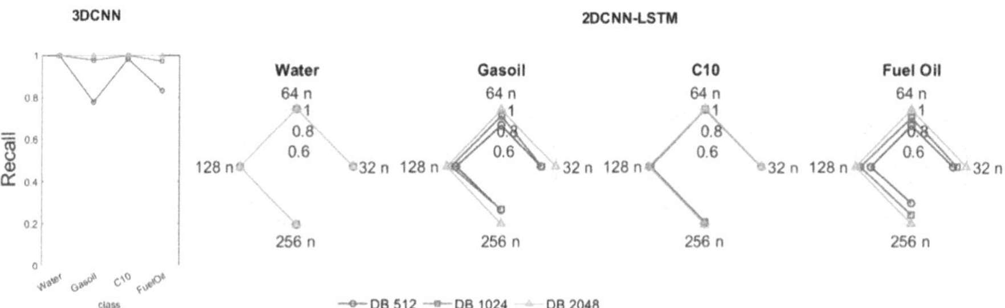

Fig. 3. Recall per class for all model configurations using a convolutional kernel size of 2.

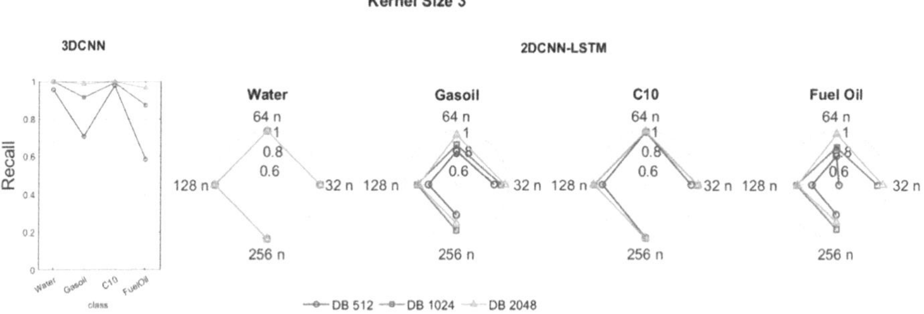

Fig. 4. Recall per class for all model configurations using a convolutional kernel size of 3.

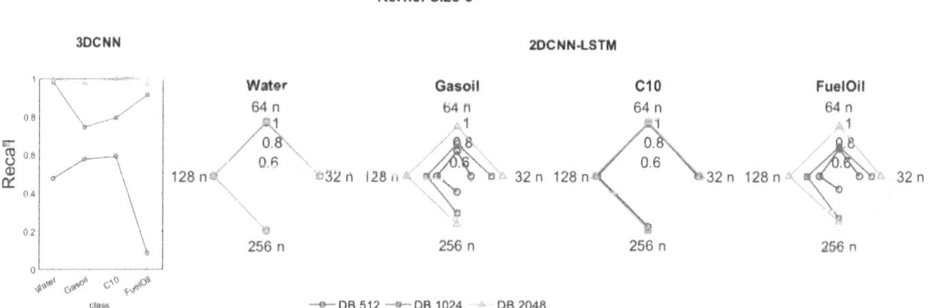

Fig. 5. Recall per class for all model configurations using a convolutional kernel size of 5.

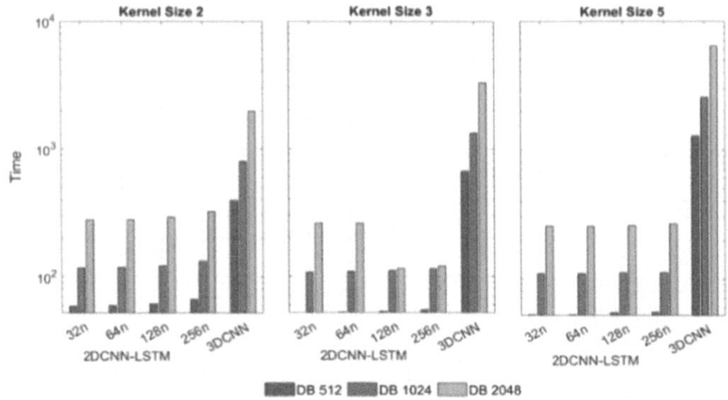

Fig. 6. Average training time for all model configurations (log-scales)

4 Conclusions

This study demonstrates the effectiveness of a hybrid 2DCNN-LSTM architecture for classifying clean and hydrocarbon-contaminated water surfaces from HSI. The integration of CNNs for spatial analysis and LSTMs for spectral sequence learning exhibits several advantages over the conventional 3DCNN architecture, as outlined:

- Superior computational efficiency: Although both 3DCNN and 2DCNN-LSTM approaches could achieve near-perfect classification performance under favorable data conditions, the training time of the hybrid model is significantly lower. For the fastest 3DCNN configuration that reached OA of 1, the training time was seven times slower than the 2DCNN-LSTM with the same OA. This computational efficiency becomes particularly relevant in real-time or incremental learning applications, such as scenarios requiring frequent model training.
- Greater robustness in data-scarce scenarios: The 3DCNN architecture showed significant sensitivity to limited training data, particularly when using larger kernel sizes. Only configurations with small spatial kernels were able to partially compensate for the data shortage. This suggests that 3DCNNs require large datasets to generalize effectively and are prone to performance degradation otherwise. In contrast, the 2DCNN-LSTM model consistently outperformed the 3DCNNs under these conditions, maintaining high accuracy and stable recall across all classes.
- Balance trade-off between complexity and performance: Among the hybrid configurations, models with a moderate number of LSTM units (e.g. 64 or 128) achieved an optimal balance between generalisation, stability, and computational cost.

Future work will address a key research challenge: extending the experimental validation to real-world conditions is essential. This involves deploying the hyperspectral system on unmanned aerial vehicles (UAVs) to capture images in uncontrolled environments, under varying lighting and weather conditions. To this end, we plan to integrate the imaging device with a UAV platform equipped with stabilization and georeferencing systems, allowing for accurate image acquisition during flight. Additionally, we will explore on-board preprocessing techniques to reduce noise caused by atmospheric effects and illumination variability. These steps will be complemented by domain adaptation or transfer learning strategies to retrain the models using field-acquired data. This extended framework aims to provide a more realistic assessment of model robustness and generalization in operational scenarios such as port areas, oil spill response, or industrial discharges.

Acknowledgements. This work is part of the technological contract OT2024/028 "Use of Hyperspetral imaging and UAV in smart management of urban waste landfills" supported by ARCGISA. It is also funded by the Plan de ayudas para el Fomento de la Investigación y Transferencia de Conocimiento en el Campus Bahía de Algeciras, promoted by the Algeciras Technological Campus Foundation, University of Cádiz.

References

1. Gbadamosi, F., Aldstadt, J.: The interplay of oil exploitation, environmental degradation and health in the Niger Delta: a scoping review. Trop. Med. Int. Health (2025). https://doi.org/10.1111/tmi.14108
2. Kang, X., Wang, Z., Duan, P., Wei, X.: The potential of hyperspectral image classification for oil spill mapping. IEEE Trans. Geosci. Remote Sens. **60**(5538415) (2022). https://doi.org/10.1109/TGRS.2022.3205966
3. Paoletti, M.E., Haut, J.M., Plaza, J., Plaza, A.: Deep learning classifiers for hyperspectral imaging: a review. ISPRS J. Photogramm. Remote Sens. **158**, 279–317 (2019). https://doi.org/10.1016/j.isprsjprs.2019.09.006
4. LeCun, Y., Bengio, Y., Hinton, G.: Deep learning. Nature **521**(7553), 436–444 (2015). https://doi.org/10.1038/nature14539
5. Ji, S., Xu, W., Yang, M., Yu, K.: 3D convolutional neural networks for human action recognition. IEEE Trans. Pattern Anal. Mach. Intell. **35**(1), 221–231 (2013). https://doi.org/10.1109/TPAMI.2012.59
6. Quan, Y., et al.: Spectral-spatial feature extraction based CNN for hyperspectral image classification. In: IGARSS 2020 - 2020 IEEE International Geoscience and Remote Sensing Symposium, Waikoloa, HI, USA, 2020, pp. 485–488 (2020). https://doi.org/10.1109/IGARSS39084.2020.9323629
7. Mu, Q., Kang, Z., Guo, Y., Chen, L., Wang, S., Zhao, Y.: Hyperspectral image classification of wolfberry with different geographical origins based on three-dimensional convolutional neural network. Int. J. Food Prop. **24**(1), 1705–1721 (2021). https://doi.org/10.1080/10942912.2021.1987457
8. Yang, J.F., Wan, J.H., Ma, Y., Zhang, J., Hu, Y.B., Jiang, Z.C.: Oil spill hyperspectral remote sensing detection based on DCNN with multi-scale features. J. Coastal Res. **90**, 332–339 (2019). https://doi.org/10.2112/SI90-042.1
9. Mou, L., Ghamisi, P., Zhu, X.X.: Deep recurrent neural networks for hyperspectral image classification. IEEE Trans. Geosci. Remote Sens. **55**(7), 3639–3655 (2017). https://doi.org/10.1109/TGRS.2016.2636241
10. Hochreiter, S., Schmidhuber, J.: Long short-term memory. Neural Comput. **9**(8), 1735–1780 (1997). https://doi.org/10.1162/neco.1997.9.8.1735
11. Hu, W.S., Li, H.C., Pan, L., Li, W., Tao, R., Du, Q.: Spatial-spectral feature extraction via deep ConvLSTM neural networks for hyperspectral image classification. IEEE Trans. Geosci. Remote Sens. **58**(6), 4237–4250 (2020). https://doi.org/10.1109/TGRS.2019.2961947

12. Tulapurkar, H., Banerjee, B., Krishna Mohan, B.: Effective and efficient dimensionality reduction of hyperspectral image using CNN and LSTM network. In: 2020 IEEE India Geoscience and Remote Sensing Symposium, InGARSS 2020 - Proceedings, pp. 213–216 (2020). https://doi.org/10.1109/InGARSS48198.2020.9358957
13. Farooque, G., Xiao, L., Yang, J., Sargano, A.B.: Hyperspectral image classification via a novel spectral-spatial 3D ConvLSTM-CNN. Remote Sens. **13**(21), 4348 (2021). https://doi.org/10.3390/rs13214348

Author Index

E. Corchado et al. (Eds.): SOCO 2025, CCIS 2806, pp. 655–657, 2026.
https://doi.org/10.1007/978-3-032-19763-4

The manufacturer's authorised representative in the EU is Springer Nature Customer Service Centre GmbH, Europaplatz 3, 69115 Heidelberg, Germany. If you have any concerns regarding our products, please contact ProductSafety@springernature.com

Printed and bound by CPI Group (UK) Ltd, Croydon, CR0 4YY
07/07/2026
02160917-0020